T0201734

X-Rays and Extreme Ultraviolet Radiation

With this fully updated second edition, you will gain a detailed understanding of the physics and applications of modern x-ray and extreme ultraviolet (EUV) radiation sources. Taking into account the most recent improvements in capabilities, coverage is expanded to include new chapters on free electron lasers (FELs), laser high harmonic generation (HHG), x-ray and EUV optics, and nanoscale imaging, a completely revised chapter on spatial and temporal coherence, and extensive discussion of the generation and applications of femtosecond and attosecond techniques. You will be guided step by step through the mathematics of each topic, with over 300 figures, 50 reference tables, and 600 equations enabling easy understanding of key concepts. Homework problems, a solutions manual for instructors, and links to YouTube lectures accompany the book online.

This is the "go-to" guide for graduate students, researchers, and industry practitioners looking to grasp the fundamentals of x-ray and EUV interaction with matter, or expand their knowledge of this rapidly growing field.

David Attwood is Professor Emeritus at the University of California, Berkeley. He is a co-founder of the Applied Science and Technology PhD program at Berkeley, and a Fellow of the American Physical Society and the Optical Society of America. He has published over 100 scientific papers and co-edited several books.

Anne Sakdinawat is a scientist at the SLAC National Accelerator Laboratory, where she leads a scientifically motivated imaging and nanofabrication group co-located at Stanford University. She is the recipient of the international Meyer-Ilse Award for advances in x-ray microscopy, and a US Department of Energy Early Career Award.

X-Rays and Extreme Ultraviolet Radiation

Principles and Applications

DAVID ATTWOOD

University of California, Berkeley

ANNE SAKDINAWAT

SLAC National Accelerator Laboratory

Illustrations by LINDA GENIESSE

CAMBRIDGE
UNIVERSITY PRESS

CAMBRIDGE
UNIVERSITY PRESS

University Printing House, Cambridge CB2 8BS, United Kingdom

One Liberty Plaza, 20th Floor, New York, NY 10006, USA

477 Williamstown Road, Port Melbourne, VIC 3207, Australia

4843/24, 2nd Floor, Ansari Road, Daryaganj, Delhi - 110002, India

79 Anson Road, #06-04/06, Singapore 079906

Cambridge University Press is part of the University of Cambridge.

It furthers the University's mission by disseminating knowledge in the pursuit of
education, learning and research at the highest international levels of excellence.

www.cambridge.org
Information on this title: www.cambridge.org/9781107062894

© Cambridge University Press 2016

First published 2016
Reprinted 2021

Printed in the United Kingdom by TJ Books Limited, Padstow Cornwall

A catalog record for this publication is available from the British Library

Library of Congress Cataloging in Publication data
Names: Attwood, David T., author. | Sakdinawat, Anne, author. | Geniesse, Linda, illustrator.
Title: X-rays and extreme ultraviolet radiation : principles and applications / David Attwood
(University of California, Berkeley), Anne Sakdinawat (SLAC National Accelerator Laboratory) ;
illustrations by Linda Geniesse.
Other titles: Soft x-rays and extreme ultraviolet radiation
Description: Second edition. | Cambridge, United Kingdom ; New York, NY : Cambridge
University Press, 2016. | Earlier edition published under title: Soft x-rays and extreme
ultraviolet radiation. | Includes bibliographical references and index.
Identifiers: LCCN 2016035981 | ISBN 9781107062894 (hardback ; alk. paper) |
ISBN 1107062896 (hardback ; alk. paper)
Subjects: LCSH: Grenz rays. | Ultraviolet radiation.
Classification: LCC QC482.G68 A88 2016 | DDC 539.7/222 – dc23
LC record available at https://lccn.loc.gov/2016035981

ISBN 978-1-107-06289-4 Hardback

Additional resources for this publication at www.cambridge.org/xrayeuv

To Professors Stanley Goldstein
and Nathan Marcuvitz, and to Ellie

Contents

Preface to the Second Edition

There has been a remarkable improvement in capabilities for probing matter with x-rays and extreme ultraviolet (EUV) radiation since the previous edition of this book appeared in 2000. The spectral brightness and coherence of available research facilities has increased by many orders of magnitude across the EUV and x-ray spectral regions, extending from photon energies of 30 eV (40 nm wavelength) to 50 keV (0.25 Å). The ability to probe electron dynamics in atoms, molecules, clusters and solids has been extended from picoseconds to femtoseconds and attoseconds. X-ray optics have improved dramatically, with reflective and diffractive optics now able to focus, or resolve images, to 10 nm across much of this spectrum. New techniques have emerged for attosecond temporal measurements, nanoscale tomographic imaging of individual cells, and ever more sophisticated coherent diffraction and imaging techniques. Commercial capabilities have also improved with brighter laboratory sources and widely available x-ray microscopes for nanoscale imaging in the research and commercial sectors. And it appears that high-volume manufacturing of computer chips with 13.5 nm EUV radiation will soon become a reality, likely reaching world markets at the 7 nm node in 2017.

Acknowledgments for the Second Edition

It is a great pleasure to acknowledge the significant assistance of many colleagues worldwide, from graduate students to world leaders, each of whom took significant amounts of time from their own activities to proofread, advise on how best to explain the math and physics, and to offer material to improve this edition. Among those who contributed to a better understanding of the new material and how it is presented in this edition are Christopher Behrens (DESY), Zhirong Huang and Andrew Aquila (SLAC), Makina Yabashi and Hitoshi Tanaka (SACLA), Claudio Pellegrini (SLAC and UCLA), Ivan Vartanyants and Andrej Singer (DESY), James Murphy (DOE/OS); Christian Ott (Max Planck, Heidelberg), Thomas Pfeifer (Max Planck, Heidelberg), Anne L'Huillier (Lund), Martin Schultze and Ferenc Krausz (Max Planck, Garching), Michael Zürch (Jena); Richard Walker (Diamond), Mau-Tsu Tang (NSRRC); Karolis Parfeniukas and Hans Hertz (KTH), Kazuto Yamauchi (U. Osaka), Donald Bilderback (Cornell), Erik Gullikson (LBL), Regina Soufli (LLNL), Marine Cotte and Irina Snigereva (ESRF); Gerd Schneider (HZB), Günter Schmahl (Göttingen), Stefano Marchesini (LBL), Vadim Banine (ASML), Winfried Kaiser (Zeiss), and Hiroo Kinoshita (U. Hyogo). Those who offered use of their own material are acknowledged in the accompanying figure captions. Those who offered assistance in the first edition are also acknowledged as their efforts endure, especially Kwang-Je Kim and James Underwood who tutored in their respective areas of expertise. We are indebted to Linda Geniesse who has advised and prepared all the artwork and figures in a manner to enhance understanding.

Preface to the First Edition

This book is intended to provide an introduction to the physics and applications of soft x-rays and extreme ultraviolet (EUV) radiation. These short wavelengths are located within the electromagnetic spectrum between the ultraviolet, which we commonly associate with sunburn, and harder x-rays, which we often associate with medical and dental imaging. The soft x-ray/EUV region of the spectrum has been slow to develop because of the myriad atomic resonances and concomitant short absorption lengths in all materials, typically of order one micrometer or less. This spectral region, however, offers great opportunities for both science and technology. Here the wavelengths are considerably shorter than visible or ultraviolet radiation, thus permitting one to see smaller features in microscopy, and to write finer patterns in lithography. Furthermore, optical techniques such as high spatial resolution lenses and high reflectivity mirrors have been developed that enable these applications to a degree not possible at still shorter wavelengths. Photon energies in the soft x-ray/EUV spectral region are well matched to primary resonances of essentially all elements. While this leads to very short absorption lengths, typically one micrometer or less, it provides a very accurate means for elemental and chemical speciation, which is essential, for instance, in the surface and environmental sciences. Interestingly, water is relatively transparent in the spectral region below the oxygen absorption edge, providing a natural contrast mechanism for imaging carbon-containing material in the spectral window extending from 284 to 543 eV. This provides interesting new opportunities for both the life and the environmental sciences.

Exploitation of this region of the spectrum is relatively recent. Indeed the names and spectral limits of soft x-rays and extreme ultraviolet radiation are not yet uniformly accepted. We have chosen here to follow the lead of astronomers, the lithography community, and much of the synchrotron and plasma physics communities in taking extreme ultraviolet as extending from photon energies of about 30 eV to 250 eV (wavelengths from about 40 nm to 5 nm) and soft x-rays as extending from about 250 eV (just below the carbon K edge) to several thousand eV (wavelengths from 5 nm to about 0.3 nm). The overlaps with ultraviolet radiation on the low photon energy side and with x-rays on the high photon energy side of the spectrum are not well defined. For comparison, green light has a photon energy in the vicinity of 2.3 eV and a wavelength of 530 nm. Recent developments involve advances in both science and technology, moving forward in a symbiotic relationship. Of particular importance is the development of nanofabrication techniques by the electronics industry. These provide well-defined structures with feature sizes similar to the wavelengths of interest here. The development of thin film multilayer coating capabilities by the materials science community has also been of great importance.

This book is intended for use by graduate students and researchers from physics, chemistry, engineering, and the life sciences. It is an outgrowth of classes I have taught during the past 14 years at the University of California at Berkeley. Typically the students in these classes were from the Ph.D. programs in Applied Science and Technology, Electrical Engineering and Computer Science, Physics, Chemistry, Materials Science, Nuclear Engineering, and Bioengineering. In some cases there were undergraduate students. This diversity of academic backgrounds has led to a text well suited for interdisciplinary pursuits. The text is intended to be comprehensive, covering basic knowledge of electromagnetic theory, sources, optics, and applications. It is designed to bring readers from these backgrounds to a common understanding with reviews of relevant atomic physics and electromagnetic theory in the first chapters. The remaining chapters develop understanding of multilayer coated optics with applications to materials science and EUV astronomy; synchrotron and undulator radiation; laser-produced plasmas; EUV and soft-x-ray lasers; coherence at short wavelengths; zone plate lenses and other diffractive structures with applications to biomicroscopy, materials microscopy and inspection of nanostructure patterns; and, finally, a chapter on the application of EUV and soft x-ray lithography to future high-volume production of sub-100 nm feature size electronic devices.

While the book is comprehensive in nature, it is meant to be accessible to the widest possible audience. Each chapter begins with a short summary of the important points in the material, illustrations that capture the main subject matter, and a few selected equations to whet the academic appetite. Most chapters have introductory sections designed for readers new to the field that include heuristic arguments and illustrations meant to clarify basic concepts. Each chapter also contains a mathematical development of equations for graduate students and specialists with particular interest in the chapter subject matter. To follow these mathematical developments, an undergraduate training in vector calculus and Fourier transforms is required. Descriptions of current applications in the physical and life sciences are incorporated. While there is a rigorous mathematical development, it is possible to absorb important concepts in the introductory material and then skip directly to the applications. Homework problems, which may be found at the website http://www.coe.berkeley.edu/AST/sxreuv, are designed to strengthen understanding of the material, to familiarize the reader with units and magnitudes, and to illustrate application of various formulas to current applications.

Over 600 references are provided to serve as an entry point to current research and applications. To facilitate use as a reference work many of the more important equations are boxed. In some cases the equations are repeated in numerical form, with common units, for more convenient use in a handbook fashion. Reference appendicies include tables of electron binding energies, characteristic emission lines, tables and graphs of real and imaginary scattering factors for many elements, graphs of calculated photo-absorption cross-sections, updated physical constants, and a convenient list of vector and mathematical relations. The International System of Units (SI) is also summarized, with lists of derived units and conversion factors commonly used in this field.

Berkeley, California
June 1999

Acknowledgments for the First Edition

It is my pleasure to acknowledge the sustained efforts, over several years, of Rudolf (Bob) Barton and Linda Geniesse. Bob typed and edited several versions of the text, carefully setting all the equations and showing great patience as I constantly revised the text and references. Linda, my wife, was responsible for all of the figures and created all of the original artwork, which I believe will benefit readers. She too showed patience far beyond reasonable expectations as we fine-tuned the artwork many times over for maximum clarity.

This book is a direct descendant of notes used at UC Berkeley in classes taught in thirteen of the past fourteen years. As such its content, method of presentation, and level of detail have been greatly influenced by Cal students. Their probing questions, discussions in class, occasional puzzled looks, contributions to homeworks, critical advice, and suggestions at semesters end have affected every paragraph of this book. I greatly appreciate their contributions. In particular I wish to acknowledge specific contributions by Kostas Adam, Junwei Bao, H. Raul Beguiristain, Kevin Bowers, Matt Brukman, Chang Chang, Gregory Denbeaux (Duke University), Eric DeVries, Daniel Finkenthal, Andrea Franke, Qian Fu, Ernie Glover, Kenneth Goldberg, Susanna Gordon, Joseph Heanue, Ronald Haff (UC Davis), John Heck, W.R. (Tony) Huff, Nasif Iskander, Ishtak Karim, Chih-wei Lai, Luke Lee, Sang Hun Lee, Yanwei Liu, Martin Magnuson (Uppsala University), Edward Moler, Vladimir Nikitin, Khanh Nguyen, Tai Nguyen, Tom Pistor, Nen-Wen Pu, Richard Schenker, Robert Socha, Regina Soufli, Alan Sullivan, Edita Tejnil, Akira Villar, Max Wei, Yan Wu, and Andrew Zenk.

The book has also benefited substantially from colleagues near and far. In preparing lectures I have sought advice and clarification from members of the Center for X-Ray Optics at Lawrence Berkeley National Laboratory. James Underwood provided original material and helpful insights on many occasions, Eric Gullikson modified many tables and graphs for use in the text, and Kwang-Je Kim, now at Argonne National Laboratory and the University of Chicago, patiently tutored me on the subject of synchrotron radiation. Werner Meyer-Ilse, Stanley Mrowka, Erik Anderson, Jeffrey Bokor (also of UC Berkeley), Patrick Naulleau, and Kenneth Goldberg each made contributions in their areas of expertise. Several of them also read particular chapters of the text and provided critical feedback. Michael Lieberman of UC Berkeley also read several early chapters and provided feedback. Portions of Chapters 2 and 6 follow lectures by Nathan Marcuvitz, then at New York University.

From a greater distance many other colleagues helped to improve the text by reading specific chapters and suggesting a wide range of improvements, corrections, and additions. For this I am grateful to Ingolf Lindau (Stanford and Lund Universities), Bernd Crasemann (University of Oregon), Joseph Nordgren (Uppsala University), David Windt (Lucent Technologies), Claude Montcalm (LLNL), Eric Ziegler (ESRF), Alexandre Vinogradov (Lebedev Physical Institute), Albert Hofmann (CERN), R. Paul Drake (University of Michigan), R. Kauffman (LLNL), Andrei Shikanov (Lebedev Physical Institute), Luiz DaSilva (LLNL), Syzmon Suckewer (Princeton University), Jorge Rocca (Colorado State University), Emil Wolf (University of Rochester), Günter Schmahl (Göttingen University), Janos Kirz (SUNY, Stony Brook), Alexei Popov (Instit. Terr. Magn. Iono. Rad. Prop., Troitsk), Franco Cerrina (University of Wisconsin), Donald Sweeney (LLNL), Richard Stulen (Sandia), Hiroo Kinoshita (Himeji University), Victor Pol (Motorola), David Williamson (SVGL), and Frits Zernike.

Finally I am grateful to those who contributed to the atmosphere of support for research and teaching in Berkeley. These include Louis Ianiello, Iran Thomas, William Oosterhuis, and Jerry Smith at the Department of Energy's Office of Basic Energy Sciences; Howard Schlossberg at the Air Force Office of Scientific Research, who supported student research activities in our group over many years; David Patterson, who heads DARPA's Advanced Lithography Program; and the Intel, Motorola, and Advanced Micro Devices Corporations. A special note of gratitude goes to John Carruthers of Intel for his continual support of these activities and of this book in particular. Daniel Chemla is warmly acknowledged for his support and encouragement, without which it would not have been possible to maintain a vibrant research group while teaching, advising, and writing a lengthy text.

Berkeley, California
June 1999

1 Introduction

1.1 The X-Ray and Extreme Ultraviolet Regions of the Electromagnetic Spectrum

One of the last regions of the electromagnetic spectrum to be developed is that extending from the extreme ultraviolet to hard x-rays, generally shown as a dark region in charts of the spectrum. It is a region where there are a large number of atomic resonances, leading to absorption of radiation in very short distances, typically measured in nanometers (nm) or micrometers (microns, μm), in all materials. This has historically inhibited the pursuit and exploration of the region. On the other hand, these same resonances provide mechanisms for both elemental (C, N, O, etc.) and chemical (Si, SiO_2, $TiSi_2$) identification, creating opportunities for advances in both science and technology. Furthermore, because the wavelengths are relatively short, it becomes possible to study nanoscale structures using the techniques of absorption, scattering and microscopy. To exploit these opportunities requires advances in relevant technologies, for instance in nanofabrication. These in turn lead to new scientific understandings, in subjects such as materials science, surface science, chemistry, biology and physics, providing feedback to the enabling technologies. Development of the extreme ultraviolet, soft and hard x-ray spectral regions is presently in a period of rapid growth and interchange among science and technology.

Figure 1.1 shows that portion of the electromagnetic spectrum extending from the infrared to the x-ray region, with wavelengths across the top and photon energies along the bottom. Major spectral regions shown are the infrared (IR), which we associate with molecular resonances and heat; the visible region from red to violet, which we associate with color and vision; the ultraviolet (UV), which we associate with sunburn and ionizing radiation; the regions of extreme ultraviolet (EUV), soft x-rays (SXR), and finally hard x-rays, which we associate with medical and dental x-rays and with the scientific analysis of crystals, materials, and biological samples through the use of diffraction, scattering, and other techniques. In this book we address techniques and opportunities based on the generation and use of radiation extending from the EUV through x-ray regions of the spectrum.

The extreme ultraviolet is taken here as extending from photon energies of about 30 eV to about 250 eV, with corresponding wavelengths in vacuum extending from

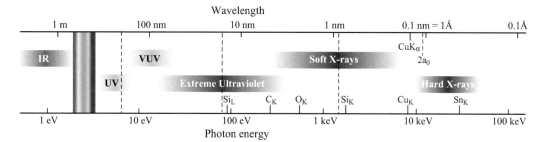

- See smaller features
- Write smaller patterns
- Elemental and chemical sensitivity
- Penetrate visibly opaque objects

Figure 1.1 The electromagnetic spectrum as it extends from the infrared (IR) to the x-ray regions. Visible light is shown with red (650 nm), green (530 nm), and blue (470 nm) wavelengths. At shorter wavelengths are ultraviolet (UV) radiation, extreme ultraviolet radiation (EUV), soft x-rays (SXR), and hard x-rays. Shown for reference are the silicon L-absorption edge at 99.2 eV (12.5 nm wavelength), the carbon K-absorption edge at 284 eV (4.37 nm), the oxygen K-absorption edge at 543 eV (2.28 nm), the silicon K-absorption edge at 1.84 keV (0.674 nm), the copper K-absorption edge at 8.98 keV (1.38 Å), the tin K-absorption edge at 29.2 keV (0.423 Å), the copper K_α-emission line at 1.54 Å (8.05 keV), and twice the Bohr radius at $2a_0 = 1.06$ Å, the diameter of the $n = 1$ orbit in Bohr's model of the hydrogen atom, but more generally a dimension within which resides most of the charge for all atoms. Vertical dashed lines correspond to the transmission limits of common window materials used to isolate vacuum. Shown are approximate transmission limits for common thicknesses of fused silica (pure SiO_2) at 200 nm, a thin film of silicon nitride ($\sim$100 nm thick Si_3N_4) at 15 nm, and an 8 μm thick beryllium foil at a wavelength of about 1 nm.

about 5 nm to 40 nm.* The soft x-ray region is taken as extending from about 250 eV (just below the carbon K-edge) to several keV, and the x-ray region of interest here extends from photon energies just below 10 keV to energies approaching 100 keV, all as shown in Figure 1.1. These spectral regions are characterized by the presence of the primary atomic resonances and absorption edges of most elements, from low to high Z, where Z is the atomic number (the number of protons in the nucleus). The primary atomic absorption edges† for selected elements are given in Table 1.1, along with $1/e$ absorption lengths at photon energies of 100 eV, 1 keV, and 10 keV. The K- and L-absorption edges, associated with the removal of a core electron by photoabsorption from the most tightly bound atomic states (orbitals of principal quantum numbers $n = 1$ and $n = 2$, respectively), are described later in this chapter. The K-absorption edges of carbon (C_K), oxygen, silicon, copper and tin are shown in Figure 1.1, as is the L-absorption edge of silicon (Si_L), just below 100 eV.

---

* It is common to express photon energies in this spectral region in electron volts (eV) or thousands of electron volts (keV), where the photon energy is $\hbar\omega$, $\hbar$ is Planck's constant divided by 2π, and $\omega = 2\pi f$ is the radian frequency. Wavelengths (λ) are commonly expressed in nanometers (1 nm $= 10^{-9}$ m) and ångströms (1 Å $= 10^{-10}$ m). See Appendix A for the values of physical constants and conversion factors.

† Standard reference data for this spectral region are given in Refs. 1–4.

Table 1.1 K- and L_3-absorption edges for selected elements. Also given are $1/e$ absorption depths at photon energies of 100 eV, 1 keV, and 10 keV. Energies are given to the nearest electron volt. They are measured from the vacuum level for gases (N_2, O_2), relative to the Fermi level for metals, and relative to the top of the valence band for semiconductors. Wavelengths are given to three significant figures. These K- and L-edge values can vary somewhat with the chemical environment of the atom. Values here are taken from Williams.[1] Absorption lengths are obtained from Henke, Gullikson, and Davis.[3]

| | | | | | | l_{abs} | | |
Element	Z	K_{abs}-edge (eV)	L_{abs}-edge (eV)	$\lambda_{K\text{-abs}}$ (nm)	$\lambda_{L\text{-abs}}$ (nm)	100 eV (nm)	1 keV (μm)	10 keV (μm)
Be	4	112	–	11.1	–	730	9.0	9600
C	6	284	–	4.36	–	190	2.1	2100
N	7	410	–	3.02	–	–	–	–
O	8	543	–	2.28	–	–	–	–
H_2O						160	2.3	2000
Al	13	1560	73	0.795	17.1	34	3.1	160
Si	14	1839	99	0.674	12.5	63	2.7	130
S	16	2472	163	0.502	7.63	330	1.9	100
Ca	20	4039	346	0.307	3.58	290	1.3	69
Ti	22	4966	454	0.250	2.73	65	0.38	20
V	23	5465	512	0.227	2.42	46	0.26	14
Cr	24	5989	574	0.207	2.16	31	0.19	10
Fe	26	7112	707	0.174	1.75	22	0.14	7.4
Ni	28	8333	853	0.149	1.45	16	0.11	5.4
Cu	29	8979	933	0.138	1.33	18	0.10	5.1
Se	34	12 658	1434	0.0979	0.865	63	0.96	52
Mo	42	20 000	2520	0.0620	0.492	200	0.19	12
Sn	50	29 200	3929	0.0425	0.316	17	0.17	11
Xe	54	34 561	4782	0.0359	0.259	–	–	–
W	74	69 525	10 207	0.0178	0.121	28	0.13	5.6
Au	79	80 725	11 919	0.0154	0.104	28	0.10	4.5

We see in Table 1.1 that many of these absorption edges lie in the combined x-ray and extreme ultraviolet spectral region. What differentiates these regions from neighboring spectral regions is the high degree of absorption in all materials. At lower photon energies, in the visible and ultraviolet, and at higher photon energies, well into the hard x-ray region, many materials become transparent and it is not necessary to utilize vacuum isolation techniques. For example, Figure 1.1 shows dashed vertical lines at the locations of common window materials that can hold vacuum over square centimeter areas while still transmitting radiation in the indicated regions. In the UV, fused silica, a form of pure SiO_2, is transmissive to wavelengths as short as 200 nm, in millimeter thicknesses. For shorter wavelengths one quickly enters the vacuum ultraviolet (VUV), where air and all materials are absorbing. Shown just below 1 nm wavelength is the transmission limit of a thin ($\cong 8\,\mu$m) beryllium foil that transmits photons of energy greater than about 1.5 keV. For many years these two materials defined the limits of available window materials. More recently thin films (~ 100 nm) such as silicon nitride (stoichiometrically Si_3N_4) have extended transmissive windows to photon energies just under 100 eV, as shown in Figure 1.1.

While this plenitude of atomic resonances and efficient photoabsorption has made the EUV and soft x-ray regions more difficult to access, it also provides a very sensitive tool for elemental and chemical identification, thus creating many scientific and technological opportunities. These opportunities are enhanced in this spectral region in that the wavelengths are short, but not so short as to preclude the development of high-resolution optical techniques, thus permitting direct image formation and spatially resolved spectroscopies, to spatial resolutions of order 10 nm. The relative transparency of water and its natural contrast with other elements further add to these opportunities, for instance for spectroscopy and imaging in the life and environmental sciences.

In the paragraphs that follow we will briefly review the basic processes of absorption, scattering, and photoemission; atomic energy levels and allowed transitions; and associated absorption edges and characteristic emission lines. We note two interesting features associated with wavelengths in the EUV/x-ray spectral region. In the EUV and soft x-ray regions the wavelengths are large compared to the Bohr radius, $\lambda \gg a_0$, where a_0 is the radius of the first ($n = 1$) stationary electron orbit in the Bohr model of hydrogen.[‡] More significantly here, the diameter $2a_0 = 1.06$ Å typically encompasses most of the electronic charge in multi-electron atoms,[¶] so that to a large degree the treatment of scattering in the EUV/SXR region simplifies as the various atomic electrons experience a rather uniform phase variation, an assumption that does not hold at shorter x-ray wavelengths. For the shorter wavelength, however, these same simplifications are valid for x-rays scattering into the forward direction, which contributes to refractive index. Furthermore, because the wavelengths of interest in this text are longer than the Compton wavelength[4, 5] ($\lambda \gg \lambda_C = h/mc = 0.0243$ Å), momentum transfer from the photon can be ignored during the scattering process, again simplifying analysis in the EUV/x-ray spectral region.

Finally, we close this section with some numerical relationships in units[7, 8] convenient for work in this spectral region. Based on the dispersion relation in vacuum, $f\lambda = c$ or $\omega = kc$, where c is the velocity of light[§] in vacuum and $\omega = 2\pi f$, the product of photon energy $\hbar\omega$ and wavelength λ is given by (see Appendix A for values of physical constants)

$$\boxed{\hbar\omega \cdot \lambda = hc = 1239.842 \text{ eV nm}} \tag{1.1}$$

[‡] Numerically $a_0 = 4\pi\epsilon_0\hbar^2/me^2 = 0.529$ Å, where m is the electron rest mass, e the electron charge, ϵ_0 the permittivity of free space, and $\hbar$ Planck's constant divided by 2π. See Eisberg and Resnick, Ref. 5, for a discussion of Bohr's model of the hydrogen atom (Chapter 4) through a discussion of wave mechanics for the multi-electron atom (Chapter 10). Also see Tipler and Llewelyn, Ref. 6, for a somewhat more introductory presentation.

[¶] In multi-electron atoms the inner shells typically have very small radii, of order a_0/Z, as they experience nearly the full Coulomb attraction of the higher-Z nucleus, with little shielding by the outer electrons. A few outer electrons typically orbit with a radius na_0. See Eisberg and Resnick, Ref. 5.

[§] The phase velocity of EUV and soft x-ray radiation is derived from Maxwell's equations in Chapters 2 and 3, for propagation in vacuum and materials.

The number of photons required for one joule of energy, with wavelength given in nanometers (nm), is

$$\boxed{1 \text{ joule} \Rightarrow 5.034 \times 10^{15} \lambda[\text{nm}] \text{ photons}}$$ (1.2a)

or in terms of powers

$$\boxed{1 \text{ watt} \Rightarrow 5.034 \times 10^{15} \lambda[\text{nm}] \frac{\text{photons}}{\text{s}}}$$ (1.2b)

where 1 nm = 10 Å. Thus for a wavelength $\lambda = 1$ nm, a power of one watt corresponds to a photon flux of 5.034×10^{15} photons/s, each photon having an energy $E \simeq 1240$ eV.

1.2   Basic Absorption and Emission Processes

In this section we briefly review the basic processes through which radiation interacts with matter. In Figure 1.2 we show simplified models of the atom, with point electrons in orbit around a nucleus of positive charge $+Ze$. In x-ray notation the electron orbits are labeled K, L, and M, corresponding to principal quantum numbers $n = 1, 2$, and 3, respectively. A more accurate model of the atom is discussed in the next section, but that shown in Figure 1.2 suffices for these introductory comments.

Shown in Figure 1.2(a) is a primary electron incident on a multi-electron atom, with sufficient energy to remove a core electron in a close encounter. Common nomenclature refers to the incident electron as a *primary* electron, shown as *scattered* (redirected) off at some new angle, and in this case with reduced energy E'_p, where the lost energy is used to overcome the binding energy needed to remove the core electron, now free and referred to as a *secondary* electron, and to supply kinetic energy to the electron (E_s). The core vacancy can then be filled by an electron from a higher-lying orbit, pulled by the strong nuclear potential, with the emission of a photon of characteristic energy equal to the difference between the two shells. In Figure 1.2(b), a related process, photo-ionization, is shown in which a photon of sufficient energy is absorbed by the atom, transferring the energy to an emitted *photoelectron* with a kinetic energy equal to that of the incident photon, minus the binding energy of an electron in the particular shell. As an L-shell electron is bound to the atom with less energy than a K-shell electron, it will emerge with greater kinetic energy. Electron binding energies for hydrogen through uranium are given in Appendix B, Table B.1.

In both of these ionization processes [(a) and (b)] the atom is left with a core vacancy. The atom can rearrange itself for minimal total energy by the transition of a higher-lying electron, pulled by the strong nuclear potential, to the vacancy by one of two competing processes.

In (c) the atom is shown rearranging in a process of fluorescence, in which the electron transition is accompanied by the emission of a photon of characteristic energy equal

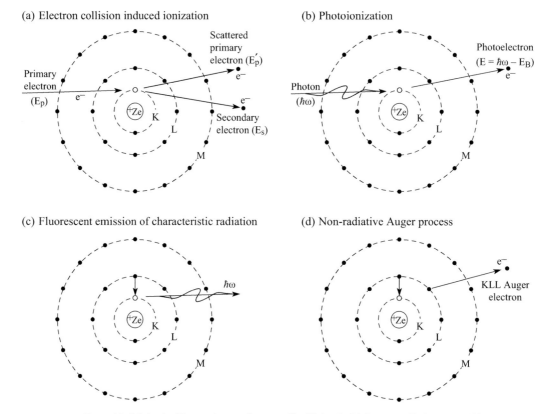

(a) Electron collision induced ionization

Primary electron (E_p)

Scattered primary electron (E'_p)

Secondary electron (E_s)

(b) Photoionization

Photon $(\hbar\omega)$

Photoelectron $(E = \hbar\omega - E_B)$

(c) Fluorescent emission of characteristic radiation

$\hbar\omega$

(d) Non-radiative Auger process

KLL Auger electron

Figure 1.2 (a) An incident *primary electron* of sufficiently high energy E_p is scattered by an atom as it knocks free a core electron from the K-shell. The primary electron now travels in a new direction, with a reduced energy E'_p. The lost energy is used to overcome the binding energy of the previously bound electron, and to impart kinetic energy to what is now referred to as a *secondary electron*. The core vacancy (K-shell in this case) can then be filled by a higher-lying L- or M-shell electron. (b) An incident photon of sufficient energy $\hbar\omega$ is absorbed by the atom with the emission of a *photoelectron* of kinetic energy equal to the photon energy minus the binding energy. Again a vacancy is created, eventually to be filled by an outer electron. (c) An atom with a core vacancy readjusts as a higher-lying electron makes a transition to the vacancy, with the emission of a photon of characteristic energy (fluorescent radiation). (d) The atom adjusts to the core vacancy through the non-radiative Auger process in which one electron makes a transition to the core vacancy, while a second electron of characteristic energy is emitted. The second electron is not necessarily emitted from the same shell.

to the difference between that of the initial and final atomic states. Characteristic emission energies are given in Appendix B, Table B.2. In a competing effect (d) the atom rearranges through the emission of a second *Auger* (pronounced ō −'zhā), electron, again of characteristic energy. The emitted Auger electron is labeled with three capital letters, the first representing the shell of the original vacancy, the second representing the shell from which the vacancy is filled, and the third representing the shell from which the Auger electron is ejected. In the competition between fluorescent emission and the Auger process, the probability tends to favor fluorescence for high Z atoms, as shown

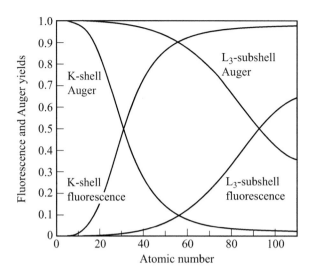

Figure 1.3 Fluorescence and Auger yields for the K-shell and the L_3-subshell as a function of atomic number Z. The Auger yields include all non-radiative contributions. Following M. Krause,[9] Oak Ridge National Laboratory.

in Figure 1.3, and the Auger process for low Z atoms.[9] Auger electron energies[9] for lithium through uranium are given in Appendix B, Table B.3. As the Auger electrons have a fixed characteristic energy, they are used extensively for elemental characterization in surface and interface analysis.

The study of atoms, molecules, and surfaces by the measurement of photoelectron kinetic energies, as a function of incident photon energy, is known as photoemission spectroscopy.[10] This process is widely used for the elemental identification and analysis of chemical bonding for atoms at or near surfaces. As generally employed, photons of fixed energy penetrate a surface or thin film, providing the energy required to lift bound electrons into the continuum, as shown in Figure 1.4. With well-known electron binding energies (Appendix B, Table B.1) the observed kinetic energies can be used to identify the elements present. As the binding energies of core electrons are affected by the orbital parameters of the outer electrons (chemical bonding in molecules, valence and conduction bands in solids), photoemission also provides a powerful tool for the study of chemical states.[11–17] As L-shell energies are more sensitive to the bonding of outer electrons than are the energies of the more tightly bound and shielded K-shell electrons, the L-shell electrons are more commonly used in photoemission studies.

If the emitted photoelectron travels any distance in a material, it is likely to lose energy quickly through interactions with other electrons (individual collisions or collective motion). Figure 1.5 shows typical electron range data, as a function of electron energy, in aluminum and gold,[18] as well as a *universal curve* for many materials.[19, 20] With incident photon energies characteristic of the EUV/soft-x-ray spectral region, it is clear that photoelectron ranges will be extremely short, of order 1 nm, so that these techniques are clearly limited to surface science.

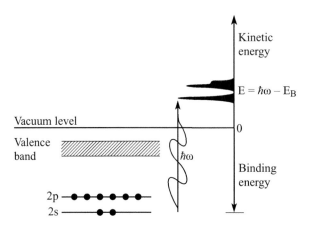

Figure 1.4 The process of photoemission in which an absorbed photon transfers its energy to a bound electron near the material–vacuum interface, resulting in a transition to a free electron state in the continuum with kinetic energy (E) equal to that of the incident photon ($\hbar\omega$) minus the binding energy (E_B).

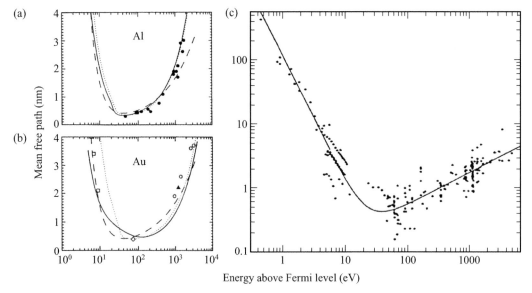

Figure 1.5 Electron mean free path, as a function of electron energy, for (a) aluminum, (b) gold, and (c) a combination of many materials. The data in (a) and (b) are from Penn,[18] while the data in (c) are from Seah and Dench.[19] The various curves reflect efforts to develop a universal model that describes inelastic scattering of electrons in a solid.

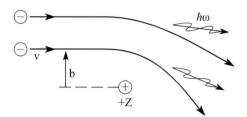

Figure 1.6 Bremsstrahlung radiation occurs predominantly when an incident electron is accelerated as it passes a nucleus, causing it to radiate. A broad continuum of radiation results when a large number of electrons interact randomly with nuclei at various distances of closest approach, b, resulting in wide variations in experienced acceleration and collision time.

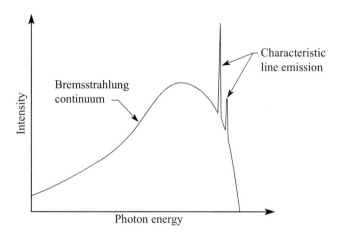

Figure 1.7 Continuum radiation and narrow line emission from a solid target as might be observed as cathode electrons strike the anode in an electrical discharge tube.

When observing the emission spectrum from a solid material bombarded by electrons it is typical to observe both characteristic line emission and continuum emission. This latter process is called *bremsstrahlung*, from the German word for "braking radiation." Figure 1.6 shows a simple diagram of the process, in which electrons of a given velocity v, or energy E, approach an electron or nucleus at various distances of closest approach, b (the *impact parameter*), experiencing a wide range of accelerations (depending on the closeness of the interaction) and thus emitting photons across a wide range of energies. With a large number of incident electrons and a wide variety of impact parameters, a rather broad continuum of radiation is produced. Where photoemission occurs due to direct impact with bound electrons, as described earlier in Figure 1.2(a), characteristic line emission is also observed. Both phenomena are illustrated, as they might typically be observed,[21] in Figure 1.7. The nature and nomenclature of the characteristic line emissions are discussed in the following section.

Historically, the process of photoabsorption [Figure 1.2(b)] has been observed macroscopically by passing radiation through thin foils and observing the resultant decrease in intensity as a function of thickness.[22] As shown in Figure 1.8, one observes that with

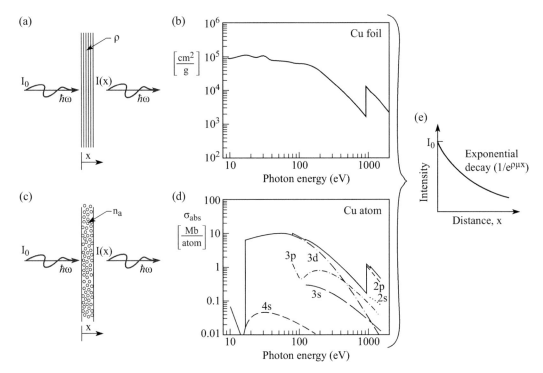

Figure 1.8 (a) Photoabsorption as observed with thin foils of increasing thickness x at fixed photon energy, with (b) an example of the linear absorption coefficient μ for copper (from Henke, Gullikson, and Davis[3]). The same process is described on an atomic level in (c), with the photoabsorption cross-section (photoionization) for a copper atom in (d) (from Yeh and Lindau[4]). Exponential attenuation of the radiation is shown in (e). Differences observed in comparing (b) and (d) are due to solid state effects in metallic copper foils, most noticeably for copper in the absence of the atomic 3d edge just above 10 eV photon energy.

incremental increases in thickness, Δx, there is an an incremental decrease in transmitted intensity I, relative to the incident intensity I_0, such that

$$\frac{\Delta I}{I_0} = -\rho\mu \, \Delta x$$

where ρ is the mass density and μ is an energy- and material-dependent absorption coefficient. Writing this in the differential limit ($\Delta x \to dx$, $\Delta I \to dI$), the equation integrates to a logarithmic dependence $\ln(I/I_0) = -\rho\mu x$, or in exponential form

$$\boxed{\frac{I}{I_0} = e^{-\rho\mu x}} \tag{1.3a}$$

where $\mu = \mu(E, Z)$, $E = \hbar\omega$ is the photon energy, Z represents the elemental dependence, and μ is the *linear absorption coefficient*. Standard values of μ are given in Appendix C for representative materials. This same expression can be written in terms of an atomic density n_a and a cross-section for photoabsorption, σ_{abs}, as

$$\frac{I}{I_0} = e^{-n_a \sigma_{abs} x}$$ (1.3b)

where σ_{abs} depends on both element (Z) and photon energy. Curves of σ_{abs}, also referred to as the photoionization cross-section, are given in Appendix C for representative elements, and more completely in Ref. 4.

The development of Eqs. (1.3a) and (1.3b), which represent macroscopic and microscopic descriptions of the same process, is given in Chapter 3, Section 3.2, where it is shown that $\mu = \sigma_{abs}/Am_u$, where A is the number of atomic mass units and m_u is an atomic mass unit (approximately the mass of a proton or neutron), as given in Appendix A. There are some differences in the two sets of data, as the thin foil absorption coefficients μ are experimentally derived and thus involve atoms in a particular solid material or molecular form. The cross-sections σ_{abs} are calculated for single isolated atoms. The latter have the benefit that they include separately identifiable contributions of the various atomic subshells,[4] as seen here in Figure 1.8(d).

While Figure 1.8(b) and (d) are macroscopic and microscopic manifestations of the same physical processes, some differences are notable. At low photon energies solid state effects in the metallic copper foil [Figure 1.8(b)] are important, and as a result the sharp 3d edge of the isolated copper atom [Figure 1.8(d)] just above 10 eV is not observed. Such data are of great interest to atomic and solid state researchers. Examples of the measured and calculated curves are given in Appendix C. A variety of spectroscopic techniques are employed to study atomic positions and binding energies within solids and on surfaces, based on details of the absorption and emission processes in the presence of near-neighbor atoms. Examples of the literature are given in Refs. 11–17.

1.3   Atomic Energy Levels and Allowed Transitions

The modern understanding of atomic energy levels, and allowed transitions between these levels, began with the Bohr–Rutherford model of the atom[5, 6] consisting of a small positive nucleus of charge $+Ze$, surrounded by electrons of charge $-e$ orbiting at relatively large radii, of order 1 Å. Based on Rutherford's experiments (1911) with the scattering of α-particles, which demonstrated the existence of a very small nucleus of positive charge, Planck's concept (1900) of radiation from quantized oscillators, and extensive spectroscopic data showing that atoms emit characteristic narrow lines with frequencies (or wavelengths) in specific numerical sequences, Bohr (1913) proposed the first partially successful quantum model of the atom. By equating the Coulomb force due to the positive nucleus, $Ze^2/4\pi\epsilon_0 r^2$, to the centripetal force mv^2/r for quantized circular orbits of angular momentum $mvr = n\hbar$ ($n = 1, 2, 3, \ldots$), Bohr found stationary electron orbits, for the single electron atom, of energy E_n and radius r_n, where

$$E_n = -\frac{mZ^2 e^4}{32\pi^2 \epsilon_0^2 \hbar^2} \frac{1}{n^2} \tag{1.4}$$

and

$$r_n = \frac{4\pi \epsilon_0 \hbar^2}{mZe^2} \cdot n^2 \tag{1.5}$$

where e and m are the electron charge and mass, respectively, Ze is the nuclear charge, ϵ_0 is the permittivity of free space, and $\hbar$ is Planck's constant divided by 2π.

The Bohr model, despite continuous acceleration of the electron, permits radiation only when the electron makes a transition from one stationary state (n_i) to another (n_f), with characteristic energies

$$\hbar\omega = E_i - E_f = \frac{mZ^2 e^4}{32\pi^2 \epsilon_0^2 \hbar^2} \left(\frac{1}{n_f^2} - \frac{1}{n_i^2} \right) \tag{1.6}$$

– characteristic in that there is a Z^2 dependence specific to the particular element radiating, and because of the numerical sequence involving the possible combinations of n_i and n_f. The constant $me^4/32\pi^2\epsilon_0^2\hbar^2 = hcR_\infty = 13.606$ eV, known historically as the Rydberg constant from earlier studies of hydrogen spectra, gives the ionization potential[||] of the ground state ($n_i = 1$, $n_f = \infty$) of the hydrogen atom ($Z = 1$). The value of the first Bohr radius of the hydrogen atom, $r_1 \equiv a_0$, is a common scale of atomic radii; from Eq. (1.5)

$$a_0 = \frac{4\pi \epsilon_0 \hbar^2}{me^2} = 0.5290 \text{ Å} \tag{1.7}$$

In terms of the Rydberg constant and first Bohr radius, the characteristic emission lines of a single electron atom of nuclear charge Z are

$$\hbar\omega = (13.606 \text{ eV})Z^2 \left(\frac{1}{n_f^2} - \frac{1}{n_i^2} \right) \tag{1.8}$$

and the radii are

$$r_n = \frac{a_0 n^2}{Z} \tag{1.9}$$

A great success of the Bohr model was its ability to accurately match well-known optical spectra of hydrogen, known as the Balmer series (1885), corresponding to $n_f = 2$ and $n_i = 3, 4, 5, \ldots$, and also to give an accurate theoretical formula for the experimentally known (1890) Rydberg constant. This was soon extended to the then unknown Lyman series, largely in the ultraviolet, with $n_f = 1$, $n_i = 2, 3, 4, \ldots$; the Paschen series

---

[||] The energy required to remove an electron from an atom.

with $n_f = 3$, $n_i = 4, 5, 6, \ldots$; the Brackett and Pfund series with $n_f = 4$ and $n = 5$, respectively, both in the infrared. Sommerfeld (1916) extended the success of the Bohr atom by introducing elliptical orbits and a second, azimuthal quantum number characterizing the ellipticity of the orbits. Additionally, taking account of the relativistic nature of the electron motion ($v/c \sim 10^{-2}$ in the hydrogen atom), Sommerfeld showed that quantized elliptical orbits introduce energetic fine structure in the spectra, as was observed experimentally.

These successes, however, raised questions about the model and the very nature of the physics. Among the specifics, not all predicted emission lines were observed, suggesting that among the possible quantum states, only some transitions were permitted. Indeed, the model said nothing regarding transition rates or line intensities. More generally, the model was perplexing in that it was based on continuous electron acceleration within the permitted orbits, but without radiation and thus loss of energy – clearly in conflict with classical radiation physics. Collectively, concepts contributing to the above model are now known as "the old quantum theory." Following a decade of intense creativity,[**] in the period from 1925 to 1930, Schrödinger, Heisenberg, Dirac and others developed a new quantum theory based on wave mechanics, in which the particles are described in terms of a probabilistic wave function $\Psi(\mathbf{r}, t)$. In combination with the introduction of electron spin, the new quantum mechanics provides a procedure for accurately predicting and matching experimental observations regarding properties of the atoms.

The quantum mechanical description[5, 6, 23–26] of a particle's motion is in terms of a wave function $\Psi(\mathbf{r}, t)$, which obeys Schrödinger's wave equation

$$-\frac{\hbar^2}{2m}\nabla^2\Psi(\mathbf{r}, t) + V(\mathbf{r}, t)\Psi(\mathbf{r}, t) = i\hbar\frac{\partial\Psi(\mathbf{r}, t)}{\partial t} \tag{1.10}$$

where m is the particle mass, $V(\mathbf{r})$ is the potential energy, and ∇ is the vector gradient. Particle energy and momentum are associated with the operators

$$E \rightarrow i\hbar\frac{\partial}{\partial t} \tag{1.11}$$

and

$$\mathbf{p} \rightarrow -i\hbar\nabla \tag{1.12}$$

respectively. In wave mechanics, the probability of finding a particle within coordinates $d\mathbf{r}$ is

$$P(\mathbf{r}, t)\,d\mathbf{r} = \Psi^*(\mathbf{r}, t)\Psi(\mathbf{r}, t)\,d\mathbf{r} \tag{1.13}$$

where Ψ^* is the complex conjugate of Ψ, and $d\mathbf{r}$ is shorthand notation for the scalar volume around the position $\mathbf{r}$, for instance $d\mathbf{r} = dx\,dy\,dz$ in rectangular coordinates. The function $\Psi(\mathbf{r}, t)$ is normalized to unity, so that

$$\underset{\text{all space}}{\iiint} |\Psi(\mathbf{r}, t)|^2\,d\mathbf{r} = 1 \tag{1.14}$$

** For a review see the texts by Tipler and Llewellyn (Ref. 6) and by Eisberg and Resnick (Ref. 5), for example.

Furthermore, expectation values for quantities such as the position vector, energy, and momentum are given by integrals of the following form: For the expectation value of vector position,

$$\bar{\mathbf{r}} = \iiint \mathbf{r} P(\mathbf{r}, t) \, d\mathbf{r} = \iiint \Psi^*(\mathbf{r}, t) \mathbf{r} \Psi(\mathbf{r}, t) \, d\mathbf{r} \tag{1.15}$$

which is a probabilistic description of where, on average, the particle can be expected to be found at time t. For the expectation value of energy,

$$\bar{E} = \iiint \Psi^*(\mathbf{r}, t) E \Psi(\mathbf{r}, t) \, d\mathbf{r} = i\hbar \iiint \Psi^*(\mathbf{r}, t) \frac{\partial \Psi(\mathbf{r}, t)}{\partial t} \, d\mathbf{r} \tag{1.16}$$

and for the expectation value of momentum,

$$\bar{\mathbf{p}} = \iiint \Psi^*(\mathbf{r}, t) \mathbf{p}(\mathbf{r}, t) \Psi(\mathbf{r}, t) \, d\mathbf{r} = -i\hbar \iiint \Psi^*(\mathbf{r}, t) \nabla \Psi(\mathbf{r}, t) \, d\mathbf{r} \tag{1.17}$$

where Eqs. (1.11) and (1.12) have been used.

Solution of Schrödinger's equation for the one-electron atom assumes a time dependence

$$\Psi(\mathbf{r}, t) = \Psi(\mathbf{r}) e^{-iEt/\hbar} \tag{1.18}$$

in a Coulomb potential

$$V(\mathbf{r}) = \frac{Ze^2}{4\pi \epsilon_0 r} \tag{1.19}$$

with separable functions in spherical coordinates

$$\Psi(\mathbf{r}) = \Psi(r, \theta, \phi) = R(r)\Theta(\theta)\Phi(\phi) \tag{1.20}$$

where θ is measured from the z-axis. Requiring that these functions be finite, continuous, single-valued and normalizable introduces three quantum numbers, n, l, and m_l, one for each coordinate. Negative energies correspond to bound electrons in orbits of discrete, quantized energy. For positive energies the states are continuous and the electron is free. Here n is the principal quantum number, associated with the radial coordinate, and having allowed integer values $n = 1, 2, 3, \ldots$ The orbital quantum number l, associated with the θ-coordinate, is related to the angular momentum by $L = \sqrt{l(l+1)}\hbar$, and is constrained to the integer values $l = 0, 1, 2, \ldots, n - 1$. The magnetic quantum number m_l, associated with continuity of the wave function in angle ϕ, is related to the z-component of angular momentum by $L_z = m_l \hbar$, and is constrained to the integer values $m_l = -l, -l + 1, \ldots, 0, 1, \ldots, l$.

The quantum mechanical description is completed with the introduction of a fourth quantum number, m_s, associated with the intrinsic electron angular momentum or spin, s. With s having a value of $\frac{1}{2}$, the quantum number m_s can have values of $\pm\frac{1}{2}$. This admits the Pauli Exclusion Principle, that no two electrons can have an identical set of quantum numbers. Electron spin additionally allows a spin–orbit coupling that energetically matches the fine structure observed in emission lines.

The enumerated constraints on allowable quantum numbers n, l, m_l, and m_s, along with the exclusion principle, dictate limits on the number of electrons in each shell. For instance, the first shell, with $n = 1$, can hold only two electrons, with quantum numbers $l = 0$, $m_l = 0$, $m_s = \pm\frac{1}{2}$. The second shell, with $n = 2$, can hold eight electrons, two in the $l = 0$ subshell as above; and six in the $l = 1$ subshell, with $m_l = 0, \pm1$ and $m_s = \pm\frac{1}{2}$. The third shell, with $n = 3$, can hold 18, with quantum number combinations $l = 0$, $m_l = 0$, $m_s = \pm\frac{1}{2}$; $l = 1$, $m_l = 0, \pm1$, and $m_s = \pm\frac{1}{2}$; and $l = 2$, $m_l = 0, \pm1, \pm2$, and $m_s = \pm\frac{1}{2}$; etc. In spectroscopic notation the electron configuration according to n and l, for an atom such as argon $(Z = 18)$, would be written as $1s^2\ 2s^2\ 2p^6\ 3s^2\ 3p^6$, where s refers to an $l = 0$ subshell and p refers to $l = 1$ (d refers to $l = 2$, f refers to $l = 3$, etc.), and where the historical use of s predates the later use of s for spin.

With the constraints on the quantum numbers n, l, m_l, the Schrödinger equation provides a set of wavefunctions, Ψ_{n,l,m_l}, with which to describe the atom. For instance, the probability of finding an electron within coordinates $d\mathbf{r}$ at a vector position $\mathbf{r}$, in a state described by Ψ_{n,l,m_l}, is given by Eq. (1.13) to be $|\Psi_{n,l,m_l}|^2\, d\mathbf{r}$. The expectation value of vector coordinates for this particular state ("orbit") of the atom is given by Eq. (1.15) to be

$$\bar{\mathbf{r}}_{n,l,m_l} = \iiint \Psi^*_{n,l,m_l}\, \mathbf{r}\Psi_{n,l,m_l}\, d\mathbf{r}$$

That is, the coordinates are only known probabilistically, in contrast with the Bohr model, where there were well-defined orbital coordinates. Interestingly, for the hydrogen atom, the expectation values of energy are equal to those of the Bohr atom [Eq. (1.4)], with a correction due to spin–orbit coupling. The explicit coordinate dependence of the hydrogen atom wavefunctions, and their energies including spin–orbit fine structure, are described in the literature.[5, 27, 28]

The probability of a transition between two stationary states of the atom, which we abbreviate here as Ψ_i and Ψ_f for initial and final states, is proportional to the square of the quantum mechanical dipole matrix element[29–31]

$$-e\bar{\mathbf{r}}_{if} = -e \int \Psi^*_i \mathbf{r}\Psi_f\, d\mathbf{r} \tag{1.21}$$

During a transition from the higher energy stationary state Ψ_i to the lower energy stationary state Ψ_f, the average position of the electron oscillates between the two states at a frequency equal to the difference in energies $\omega_{if} = (E_i - E_f)/\hbar$, as shown in Figure 1.9. Quantum mechanically the atom is in a mixed state in which the probability of finding the atom in the upper state gradually diminishes from unity to zero, while the probability of finding it in the lower state increases during this same transition period, or *lifetime*, from zero to unity. During the transition period the electron typically executes millions of oscillations. This provides a quantum mechanical description of the spontaneous emission of radiation that occurs after an atom is excited to a higher energy level by photoabsorption (the inverse process) or collision with an electron. The line width of the resultant emission depends on the time duration (lifetime) of the transition, as the latter affects the number of oscillations corresponding to the emitted photon or

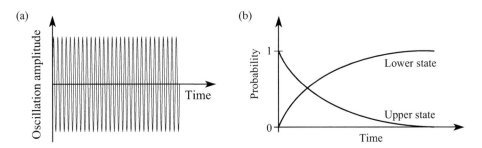

Figure 1.9 (a) Radiative decay from an upper state Ψ_i to a lower state Ψ_f involves a mixed atomic state in which the atom oscillates between the two at a frequency $\omega_{if} = (E_i - E_f)/\hbar$, (b) with the probability of finding the atom in the upper state slowly decaying to zero in a transition lifetime equal to many millions of cycles. Following Liboff.[31]

wavetrain. The longer the wavetrain, the better defined the wavelength and hence the narrower the line width.

The transition probability depends on the integral matrix element given by Eq. (1.21). The integral has a classical counterpart[30] in the current density $\mathbf{J} = -en\mathbf{v}$, whose time derivative is used to calculate radiation in Maxwell's equations, a subject we return to in Chapter 2. Here the particle *density* is given by $\Psi^*\Psi$, the charge density by $-e\Psi^*\Psi$, and the velocity $\mathbf{v}$ by $d\mathbf{r}/dt \rightarrow i\omega_{if}\mathbf{r}$, so that a time derivative of the classical current is analogous to that of the quantum mechanical *dipole moment* as given in Eq. (1.21). Transitions from an initial state Ψ_i to a final state Ψ_f occur quantum mechanically when the two wavefunctions yield a finite oscillation amplitude $\bar{\mathbf{r}}(t)$ as given by the matrix element in Eq. (1.21). If the wavefunctions Ψ_i and Ψ_f are such that the integral is zero, there is no oscillation leading to a transition, and the transition is said to be *not allowed*. Examining the integral, one notes that $\mathbf{r}$ is an odd function of the coordinates (replacing r by $-r$ changes the sign of the integrand), requiring that the initial and final wavefunctions be of opposite parity (one even, one odd in the coordinates of integration) for a non-zero integral. The parity of the wavefunctions is found to alternate with increasing quantum number l, leading to *selection rules* for allowed transitions in the hydrogen or single electron atom[5, 6, 23–31]:

$$\Delta l = \pm 1 \qquad (1.22)$$

Furthermore, the total angular momentum quantum number, j, determined by the vector sum of orbital and spin angular momentum, must satisfy

$$\Delta j = 0, \pm 1 \qquad (1.23)$$

where j can take the values $l + s$, $l - s$, or s when $l = 0$. The special case of a transition between $j = 0$ states is not allowed. Note that in order to conserve angular momentum in the allowed transitions, the emitted photon must carry away a quantum ($\hbar$) of angular momentum.[5, 29] These allowed transitions lead to the strong characteristic spectral

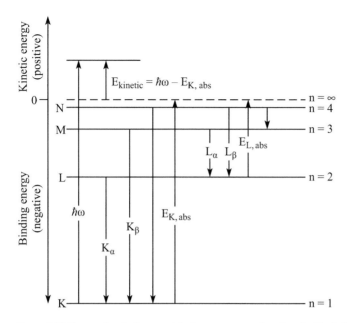

Figure 1.10 Energy levels for a multi-electron atom, showing the K-shell ($n = 1$), L-shell ($n = 2$), M-shell ($n = 3$), and N-shell ($n = 4$), with transitions that provide characteristic narrow line emission at well-defined photon energies. Examples shown include K_α ($n = 2$ to $n = 1$), K_β ($n = 3$ to $n = 1$), L_α ($n = 3$ to $n = 2$), etc. Also shown are the absorbtion edge energies, such as $E_{K,abs}$, the energy required to take an electron from the K-shell ($n = 1$) to the continuum limit ($n = \infty$). For an incident photon of energy $\hbar\omega > E_{K,abs}$, a K-shell electron can be lifted beyond the continuum limit to a state of positive kinetic energy, as shown. The binding energies are not drawn to scale. Subscript labeling of the emission lines is less systematic than implied here, as discussed in the text. Following Compton and Allison.[22]

emission lines observed experimentally. Furthermore, the atomic transition probabilities[29, 30] between any two states can be computed on the basis of matrix elements of the form given in Eq. (1.21).

Selection rules for transitions involving multi-electron atoms follow similar rules when the quantum numbers are assigned to a core level vacancy, as occurs in the photoemission process described earlier in Figure 1.2. Figures 1.10 and 1.11 illustrate the energy levels and several prominent transitions for the multi-electron atom. Figure 1.10 introduces the x-ray nomenclature[22] wherein the $n = 1$ state is referred to as the K-shell, $n = 2$ as the L-shell, $n = 3$ as the M-shell, etc. Emission lines terminating in the ground state ($n = 1$) are referred to as K-shell emissions, shown here as K_α, K_β, etc. The energy required to lift a K-shell electron to a free state of zero binding energy is referred to as the K-absorption edge $E_{K,abs}$. Excess energy beyond this value goes to kinetic energy of the liberated electron. Similar notation is shown for the L-shell emissions.

Figure 1.10 is useful for an introduction. It is simplified, however, in that it does not show shell substructure, and that it implies a systematic labeling of sequential emission lines within a given series by ordered Greek subscripts α, β, etc. In fact the lines

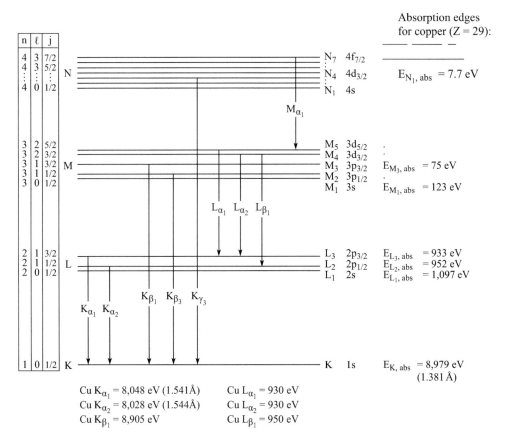

n	ℓ	j
4	3	7/2
4	3	5/2
⋮	⋮	⋮
4	0	1/2

Absorption edges
for copper (Z = 29):

N_7 $4f_{7/2}$
N_4 $4d_{3/2}$
N_1 $4s$

$E_{N_1, abs}$ = 7.7 eV

M_{α_1}

3	2	5/2
3	2	3/2
3	1	3/2
3	1	1/2
3	0	1/2

M_5 $3d_{5/2}$
M_4 $3d_{3/2}$
M_3 $3p_{3/2}$ $E_{M_3, abs}$ = 75 eV
M_2 $3p_{1/2}$
M_1 $3s$ $E_{M_1, abs}$ = 123 eV

L_{α_1} L_{α_2} L_{β_1}

2	1	3/2
2	1	1/2
2	0	1/2

L_3 $2p_{3/2}$ $E_{L_3, abs}$ = 933 eV
L_2 $2p_{1/2}$ $E_{L_2, abs}$ = 952 eV
L_1 $2s$ $E_{L_1, abs}$ = 1,097 eV

K_{β_1} K_{β_3} K_{γ_3}

K_{α_1} K_{α_2}

| 1 | 0 | 1/2 |

K $1s$ $E_{K, abs}$ = 8,979 eV
(1.381 Å)

Cu K_{α_1} = 8,048 eV (1.541Å)   Cu L_{α_1} = 930 eV
Cu K_{α_2} = 8,028 eV (1.544Å)   Cu L_{α_2} = 930 eV
Cu K_{β_1} = 8,905 eV   Cu L_{β_1} = 950 eV

Figure 1.11 Energy level diagram for copper ($Z = 29$) showing transitions allowed by the selection rules $\Delta l = \pm 1$ and $\Delta j = 0, \pm 1$, where n is the principal quantum number, l is the quantum number for orbital angular momentum, and j is the quantum number for total angular momentum (orbital plus spin). In x-ray notation the K-shell corresponds to $n = 1$, the L-shell to $n = 2$, etc. Absorption edge nomenclature is shown to the right. Following spectroscopic notation, angular momentum quantum numbers $l = 0, 1, 2, 3$ are represented by the letters s, p, d, and f, respectively. Sample energies (and wavelengths) are shown for various absorption edges and allowed transitions. Note that the energy levels are not to scale, but are approximately logarithmic. Following A. Sandström, Uppsala University.

have historical designations of limited value today. Figure 1.11 shows a more accurate version[28] of the energy levels and some well-known transitions (on a logarithmic scale) for the copper atom ($Z = 29$). Values of the quantum numbers n, l, and total angular momentum j are given for each subshell, along with the subshell designations and spectroscopic notation. Specific values of the various energy levels and a few well-known transition energies are given for the specific example of a copper atom. Note the substantial energy fine structure due to spin–orbit coupling in the various angular momentum states.[5, 27, 28] Note also that the 29 copper electrons are written in spectroscopic notation as $1s^2$ $2s^2$ $2p^6$ $3s^2$ $3p^6$ $3d^{10}$ $4s^1$, thus consisting of closed K-, L-, and M-shells,

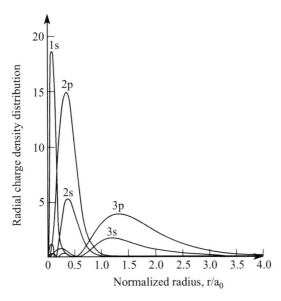

Figure 1.12 The probabilistic radial distribution of charge density in the argon atom, in units of electrons per unit radius $(e/\text{Å})$, plotted as a function of normalized radius r/a_0, where a_0 is the Bohr radius as given in Eq. (1.7). This is essentially the probability of finding an electron in the region between two concentric spheres of radius r and $r + \Delta r$. Probabilities are shown for the 1s K-shell ($n = 1, l = 0$), the 2s and 2p L-shells ($n = 2, l = 0$ and $n = 2, l = 1$), and the 3s and 3p M-shells. Note that much of the electronic charge is pulled close to the high Z nucleus. From *Quantum Physics of Atoms, Molecules, Solids, Nuclei, and Particles*, Second Edition, by R. Eisberg and R. Resnick[5]; reprinted by permission of John Wiley & Sons, Inc.

plus a single valance or conduction electron. Tabulated values of binding energies[1] and prominent emission lines[2] for elements through uranium ($Z = 92$) are given in Appendix B.

Of great interest to us in later chapters is the spatial distribution of charge in multi-electron atoms, as we will be calculating the scattering of electromagnetic radiation and are interested in where it is appropriate to assume that the wavelength λ is large compared to atomic radii for extreme ultraviolet and x-ray radiation. Toward that end, the probabilistic radial charge distribution density for filled quantum states of the argon atom ($1s^2\ 2s^2\ 2p^6\ 3s^2\ 3p^6$) is shown in Figure 1.12. Owing to the strong nuclear attraction ($Z = 18$), the K-shell electrons (1s; $n = 1, l = 0$) are pulled into a region of small radius, with highest probable radial coordinate much smaller than the hydrogenic Bohr radius a_0. The 2s and 2p L-shells ($n = 2, l = 0$ and $n = 2, l = 1$, respectively) have their charge distribution largely within a radius less than a_0, while only the M-shell 3s and 3p have significant probability of being located in the radial interval from a_0 to $3a_0$. Thus in EUV and soft-x-ray scattering calculations, with wavelengths of order 1 nm or longer (about $20a_0$), a reasonable approximation is that all electrons see approximately the same phase and scatter collectively (in phase) in all directions. This approximation cannot be made for shorter x-ray wavelengths where $\lambda \sim a_0$, at least not for the outer

Table 1.2 Key to the periodic table of the elements.[33, 34] See back inside cover of book for full periodic table.

References: International Tables for X-ray Crystallography (Reidel, London, 1983) (Ref. 33) and J.R. De Laeter and K.G. Heumann (Ref. 34, 1991); NIST Sept. 2014).

valence levels. Note that the valence electrons at radii beyond a_0 are those responsible for chemical bonding, and to first order set the apparent size of an atom, as in a molecule, crystal, or other solid.[32] This compact binding of the inner electrons explains the relatively small variation of volume occupied by atoms of widely different atomic number (Z) in solids and molecules: in the higher Z elements the additional electrons are largely confined to tight orbits nearer to the highly charged nucleus, with outer *valence* electrons having radial charge distributions, or equivalent mean radii, not too different from their lower Z cousins. For instance, in diamond the carbon atoms are separated by only 1.5 Å. For comparison, the hydrogen $n = 1$ mean diameter ($2a_0$) is 1.1 Å. For the face-centered cubic silicon crystal ($Z = 14$), adjacent silicon nuclei are separated by about 2.4 Å, and for common salt (NaCl) the separation distance between closest Na and Cl nuclei is about 2.8 Å.[27, 32] Indeed, for gold ($Z = 79$) with an atomic mass of 197 and a mass density of 19.3 g/cm^3, the atom to atom separation is only 2.9 Å. There is strikingly little difference in the separation distances between gold, or other high Z atoms, and low Z atoms in their natural states.

A periodic table of the elements, including atomic number, atomic mass, common mass density, atomic density, atomic separation distance, and spectroscopic notation of electron structure, can be found on the inside the back cover. The key to the periodic table of the elements is given here in Table 1.2. The atomic density is obtained from the relation $n_a = \rho N_A / A$, where ρ is the mass density, N_A is Avogadro's number (see Appendix A), and A is the atomic weight of the atom, as given in the periodic chart, expressed in atomic mass units (amu).

1.4   Scattering, Diffraction, and Refraction of Electromagnetic Radiation

This text assumes a familiarity with Maxwell's equations, from which one can describe the propagation of electromagnetic radiation. A wide range of literature is available that discusses the development of these equations.[35–42] In Chapters 2, 3, and 4 we will

(a) Isotropic scattering from a point object

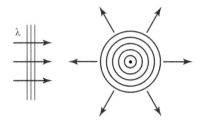

(b) Non-isotropic scattering from a partially ordered system

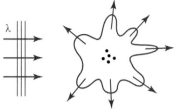

(c) Diffraction by an ordered array of atoms, as in a crystal

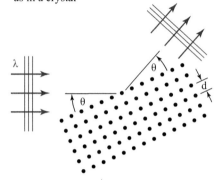

(d) Diffraction from a well-defined geometric structure, such as a pinhole

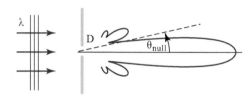

(e) Refraction at an interface

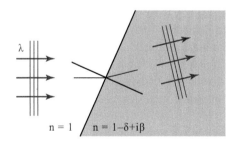

(f) Total external reflection

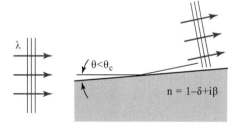

Figure 1.13 Scattering, diffraction, and refraction of electromagnetic radiation.

consider various aspects of wave propagation, scattering, diffraction, and refraction of radiation, with particular emphasis on application to the x-ray and extreme ultraviolet regions of the spectrum. In general we will use these words in the following senses.

Scattering is a process by which incident radiation[††] is redirected over a very wide angular pattern, perhaps even 4π sr, generally by disordered systems or rough surfaces, as shown in Figure 1.13(a) and (b). The angular pattern of scattering is related to the

[††] In this section we use "incident radiation" to mean nearly monochromatic radiation propagating in a well-defined direction.

spatial periodicities of the scattering object through the Fourier transform of their charge density correlation function. A point particle, for example, scatters radiation equally in all directions. The subject of scattering from free and bound electrons is discussed in Chapter 2. The term "scattering" is used in the same sense when discussing particles, as when high-energy electrons are scattered by individual nuclei.

Diffraction is generally used to describe the process whereby incident radiation is redirected into relatively well-defined directions by ordered arrays of scatterers. The diffraction patterns result from positive interference in certain directions. Examples include the diffraction of x-rays by a crystal in which the position of the atoms, in a periodic array, provide well-defined planes from which the radiation appears to reflect at well-defined angles, as described by Bragg's law (1913)[22, 27, 28, 32, 43–48]

$$m\lambda = 2d \sin\theta \qquad (1.24)$$

where d is the spatial periodicity, λ is the wavelength, m is an integer, and θ is measured from the reflecting planes, as shown in Figure 1.13(c) and discussed further in Chapter 3, Section 3.9. In fact, the atomic positions in a crystal generally describe many such planes, so that diffraction (positive interference) occurs in several directions. Diffraction of x-rays by crystals is discussed further in Chapter 3. Diffraction from one-dimensional ordered systems, known in the EUV and x-ray communities as multilayer mirrors, is also described in Chapter 3. These structures may be amorphous or partially ordered within the individual layers but well ordered in the stack direction, leading to strong positive interference in directions described by the Bragg condition, Eq. (1.24). This can lead to very high reflectance for appropriate choices of material, wavelength (photon energy), and angle. There is a rich literature available on the important subject of x-ray diffraction by crystals.[27, 28, 32, 43–45] Of particular interest for a modern description of x-ray scattering and diffraction is the text of Als-Nielsen and McMorrow.[47] The book by MacDonald contains many worked examples regarding x-ray sources, optics, and applications.[49]

"Diffraction" is also used to describe the situation in which incident radiation is redirected in some well-defined angular pattern, as when visible light or shorter wavelength radiation is diffracted by a small circular aperture (a pinhole), causing a divergence of the radiation with bright and dark angular interference rings, as shown in Figure 1.13(d). This is known as an Airy pattern (1835); the first dark null in the diffraction pattern occurs at a half angle (see Chapter 4, Section 4.7)

$$\theta_{\text{null}} = \frac{1.22\lambda}{d} \qquad (1.25)$$

where θ is measured from the axis of symmetry through the center of the pinhole, and d is the pinhole diameter. Similar characteristic angular patterns occur when radiation is diffracted from other well-defined objects, such as a knife edge or a sphere.

Finally, *refraction* is the turning of radiation at an interface of materials of dissimilar refractive index *n*, generally written in the EUV/x-ray region as[‡‡]

$$n = 1 - \delta + i\beta$$ (1.26)

where in this spectral region both δ and β are generally very small compared to unity, as is discussed in Chapter 3. Refractive turning at an interface occurs as the incident wave excites radiation among atoms at the surface of the second material. This launches a new wave through interference of radiation from the various atoms, in a manner that ensures continuity of the field quantities at the boundary. The process is sketched in Figure 1.13(e). Finally, Figure 1.13(f) shows an example of reflection unique to this spectral region, or nearly so. With a refractive index less than unity it is possible to have *total external reflection* in which most of the incident energy is redirected by the surface. There is little absorption even in the case of an otherwise absorptive material ($\beta \leq \delta$). Total external reflection occurs as long as the glancing angle of incidence θ is less than a critical angle $\theta_c \simeq \sqrt{2\delta}$, as described first by Compton and Allison[22] in 1922. This subject is described further in Chapter 3.

The development of electromagnetic theory with special emphasis on the application to EUV and x-ray wavelengths, based on Maxwell's equations, begins in Chapter 2. The wave equation in vacuum is obtained, Poynting's theorem and expressions for radiated power are developed, and scattering cross-sections are introduced. The cross-sections for scattering of radiation by free and bound electrons are obtained. These results are extended to scattering by multi-electron atoms and to atomic scattering factors, within certain approximations generally applicable in the EUV and x-ray regions of the spectrum.

In Chapter 3 Maxwell's equations are applied to the propagation of short wavelength radiation in a medium of uniform density, developing expressions for the complex refractive index, phase variation, and attenuation. Also discussed are reflection and refraction at arbitrary angle of incidence, total external reflectance and normal incidence reflection each as special cases, Brewster's angle, and topics such as the Kramers–Kronig relations among the real and imaginary parts of the atomic scattering factors. The transition from scattering to diffraction and reflection from periodic structures such as crystals and multilayer coatings is also discussed.

[‡‡] The choice of $\pm i\beta$ in Eq. (1.26) depends on the mathematical form by which waves are represented, and must be consistent with wave decay in the presence of absorption. Early x-ray workers such as Compton and Allison[22] employed a plane wave representation equivalent to $\exp[i(\omega t - kr)]$, where ω is the radian frequency, $k = 2\pi/\lambda$, and the refractive index *n* is defined by the dispersion relation $\omega = kc/n$, or equivalently $f\lambda = c/n$. Written in this form, *n* has a *negative* imaginary component in the presence of absorption. This form continues to dominate common usage in the x-ray community. However, in the broader community of modern electrodynamics and optical sciences (Sommerfeld,[41] Born and Wolf,[39] Fowles,[42] Jackson,[38] and Spiller[48]) plane wave representations are more commonly written in the form $\exp[-i(\omega t - kr)]$, where the imaginary component of refractive index is *positive* in the presence of absorption, as in Eq. (1.26). For this text the more modern approach has been adopted; however, its adaptation to the older one should cause the reader little inconvenience. Algebraic demonstration of the above is given in Chapter 3, Section 3.2.

In Chapter 4 the subject of spatial and temporal coherence is presented. Criteria for determining full spatial coherence and coherence length are developed. The van Cittert–Zernike theorem for characterizing partially coherent radiation is derived. Spatial filtering and applications of partially coherent radiation are discussed. Applications such as scanning transmission x-ray microscopy and x-ray holography are described.

In Chapters 5–9 the physics of intense, bright, and ultrafast sources of EUV and x-ray radiation are described. These include synchrotron radiation, free electron lasers (FELs), laser high harmonic generation (HHG), hot-dense plasmas, and atomic EUV and soft x-ray lasers. In Chapter 10 optics suitable for use in this spectral region are described, such as focusing optics based on total external reflection, zone plate lenses, multilayer mirrors and crystals. In Chapter 11 techniques for nanoscale imaging are described and illustrated with applications to the physical and life sciences.

References

1. A compilation by G.P. Williams of Brookhaven National Laboratory, "Electron Binding Energies," in *X-Ray Data Booklet* (Lawrence Berkeley National Laboratory Pub-490 Rev.2, 1999), based largely on values given by J.A. Bearden and A.F. Barr, "Reevaluation of X-Ray Atomic Energy Levels," *Rev. Mod. Phys.* **39**, 125 (1967); M. Cardona and L. Ley, Editors, *Photoemission in Solids I: General Principles* (Springer, Berlin, 1978); and J.C. Fuggle and N. Mårtensson, "Core-Level Binding Energies in Metals," *J. Electron. Spectrosc. Relat. Phenom.* 21, 275 (1980).
2. A compilation by J.B. Kortright, "Characteristic X-Ray Energies," in *X-Ray Data Booklet* (Lawrence Berkeley National Laboratory Pub. 490 Rev. 2, 1999), based on values given by J.A. Bearden, "X-Ray Wavelengths," *Rev. Mod. Phys.* **39**, 78 (1967).
3. B.L. Henke, E.M. Gullikson, and J.C. Davis, "X-Ray Interactions: Photoabsorption, Scattering, Transmission, and Reflection at E = 50–30,000 eV, Z = 1–92," *Atomic Data and Nucl. Data Tables* **54**, 181 (1993). Updated values are maintained by E.M. Gullikson at www.cxro .LBL.gov/optical_constants .
4. J.-J. Yeh and I. Lindau, "Atomic Subshell Photoionization Cross Sections and Asymmetry Parameters: $1 \leq Z \leq 103$," *Atomic Data and Nucl. Data Tables* **32**, 1–155 (1985); J.-J. Yeh, *Atomic Calculation of Photoionization Cross-Sections and Asymmetry Parameters* (Gordon and Breach, Langhorne, PA, 1993); I. Lindau, "Photoemission Cross Sections," Chapter 1, p. 3, in *Synchrotron Radiation Research: Advances in Surface and Interface Science*, Vol. 2 (Plenum, New York, 1992), R.Z. Bachrach, Editor.
5. R. Eisberg and R. Resnick, *Quantum Physics of Atoms, Molecules, Solids, Nuclei, and Particles* (Wiley, New York, 1985), Second Edition.
6. P.A. Tipler and R.A. Llewelyn, *Modern Physics* (Freeman, New York, 2012), Sixth Edition.
7. Numerical values of fundamental physical constants are obtained from P.J. Mohr, B.N. Taylor and D.B. Newell, "The Fundamental Physical Constants," *Phys. Today*, p. 52 (July 2007); http://physics.nist.gov/cuu/Constants/index.html
8. Use of metric units (SI) follows R.A. Nelson, "Guide for Metric Practice," *Phys. Today*, p. BG15 (August 1995).
9. M.O. Krause, "Atomic Radiative and Radiationless Yields for K and L Shells," *J. Phys. Chem. Ref. Data* **8**, 307 (1979); M.O. Krause and J.H. Oliver, "Natural Widths of Atomic K and L Levels, Kα X-Ray Lines and Several KLL Auger Lines," *J. Phys. Chem. Ref. Data*

8, 329 (1979); M.O. Krause, "Average L-Shell Fluorescence, Auger, and Electron Yields," *Phys. Rev. A* 22, 1958 (1980).

10. J.M. Hollas, *Modern Spectroscopy* (Wiley, 2004), Fourth Edition; J.M. Hollas, *High Resolution Spectroscopy* (Wiley, 1998), Second Edition.

11. F.J. Himpsel, "Photon-in Photon-out Soft X-ray Spectroscopy for Materials Science", *Phys. Status Solidi B* **248**(2), 292 (2011).

12. C.S. Fadley, "Basic Concepts of X-Ray Photoelectron Spectroscopy," pp. 1–156 in *Electron Spectroscopy, Theory, Techniques and Applications* (Pergamon Press, Oxford, 1978), C.R. Brundle and A.D. Baker, Editors; C.S. Fadley *et al.*, "Surface, Interface and Nanostructure Characterization with Photoelectron Diffraction and Photoelectron and X-ray Holography," *J. Surface Anal.* 3, 334 (1997); C.S. Fadley and P.M. Len, "Holography with X-rays," *Nature* 380, 27 (1996); G. Faigel and M. Tegze, "X-ray Holography," *Rep. Progr. Phys.* **62**, 355 (1999).

13. P.Y. Yu and M. Cardona, *Fundamentals of Semiconductors: Physics and Materials Properties* (Springer, Berlin, 1996), Chapter 8, "Photoelectron Spectroscopy."

14. S. Hüfner, *Photoelectron Spectoscopy: Principles and Applications* (Springer, Berlin, 1996); E. Rotenberg and A. Bostwick, "MicroARPES and nanoARPES at Diffraction-Limited Light Sources: Opportunities and Performance Gains," *J. Synchr. Rad.* 21, 1048 (September 2014).

15. J. Stöhr, *NEXAFS Spectroscopy*, Second Edition (Springer, Heidelberg, 2003); J. Stöhr and H.C. Siegmann, *Magnetism: From Fundamentals to Nanoscale Dynamics* (Springer, Heidelberg, 2006).

16. W. Eberhardt, Editor, *Applications of Synchrotron Radiation: High Resolution Studies of Molecular Adsorbates on Surfaces* (Springer-Verlag, Berlin, 1995).

17. E.J. Nordgren, S.M. Butorin, L.-C. Duda and J.H. Guo, "Soft X-ray Emission and Resonant Inelastic X-Ray Scattering Spectroscopy", pp. 595–659 in *Handbook of Applied Solid State Spectroscopy* (Springer, 2007), Edited by D.R. Vij; J. Guo, "Soft X-ray Absorption and Emission Spectroscopy in the Studies of Nanomaterials," in *X-Rays in Nanoscience: Spectroscopy, Spectromicroscopy and Scattering Techniques* (Wiley-VCH Verlag, Weinheim, 2010), Edited by J. Guo; T. Schmitt, F.M.F. de Groot and J.-E. Rubensson, "Prospects of High-Resolution Resonant X-Ray Inelastic Scattering Studies on Solid materials, Liquids, and Gases at Diffraction-Limited Storage Rings," *J. Synchr. Rad.* 21, 1065 (September 2014).

18. D.R. Penn, "Electron Mean-Free-Path Calculations Using a Model Dielectric Function," *Phys. Rev. B* **35**, 482 (1987).

19. M.P. Seah and W.A. Dench, "Quantitative Electron Spectroscopy of Surfaces: A Standard Data Base for Electron Inelastic Mean Free Paths in Solids," *Surface and Interface Anal.* **1**, 2 (1979).

20. C.J. Powell, "Attenuation Lengths of Low-Energy Electrons in Solids," *Surface Sci.* **44**, 29 (1974); C.J. Powell and M.P. Seah, "Precision, Accuracy, and Uncertainty in Quantitative Surface Analysis by Auger-Electron Spectroscopy and X-Ray Photoelectron Spectroscopy," *J. Vac. Sci. Technol. A* **8**, 735 (March/April 1990).

21. For example, see data for Mo ($Z = 42$) in the *Handbuch der Physik: Röntgenstrahlen*, Encyclopedia of Physics: X-Rays, Vol. XXX, (Springer-Verlag, Berlin, 1957), S. Flügge, Editor, pp. 10, 338.

22. A.H. Compton and S.K. Allison, *X-Rays in Theory and Experiment* (Van Nostrand, New York, 1935), Second Edition.

23. L.I. Schiff, *Quantum Mechanics* (McGraw-Hill, New York, 1968), Third Edition.

24. A. Messiah, *Quantum Mechanics* (Wiley, New York, 1961).

25. J.L. Powell and B. Crasemann, *Quantum Mechanics* (Addison-Wesley, Reading, MA, 1961).

26. D.I. Blokhintsev, *Principles of Quantum Mechanics* (Allyn and Bacon, Boston, 1964).

27. R.B. Leighton, *Principles of Modern Physics* (McGraw-Hill, New York, 1959).

28. F.K. Richtmyer, E.H. Kennard, and T. Lauritsen, *Introduction to Modern Physics* (McGraw-Hill, New York, 1955).

29. A. Corney, *Atomic and Laser Spectroscopy* (Clarendon, Oxford, 1977).

30. R. Louden, *The Quantum Theory of Light* (Clarendon Press, Oxford, 1983), Second Edition.

31. R.L. Liboff, *Introductory Quantum Mechanics* (Addison-Wesley, Reading, MA, 2002), Fourth Edition.

32. C. Kittel, *Introduction to Solid State Physics* (Wiley, New York, 1976), Fifth Edition.

33. *International Tables for X-Ray Crystallography, Vol. III*, International Union of Crystallography (Reidel, London, 1983), C.H. MacGillavry, G.D. Rieck, and K. Lonsdale, Editors.

34. J.R. DeLaeter and K.G. Heumann, "Atomic Weights of the Elements 1989," *J. Phys. Chem. Ref. Data 20*, 1313 (1991).

35. G.S. Smith, *Classical Electromagnetic Radiation* (Cambridge University Press, 1997).

36. J.A. Stratton, *Electromagnetic Theory* (McGraw-Hill, New York, 1941).

37. S. Ramo, J. Whinnery, and T. Van Duzer, *Fields and Waves in Communications Electronics* (Wiley, New York, 1984).

38. J.D. Jackson, *Classical Electrodynamics* (Wiley, New York, 1998), Third Edition.

39. M. Born and E. Wolf, *Principles of Optics* (Cambridge University Press, New York, 1999), Seventh Edition.

40. E. Hecht, *Optics* (Addison-Wesley, Reading, MA, 1998), Third Edition.

41. A. Sommerfeld, *Optics* (Academic Press, New York, 1964); A. Sommerfeld, *Electrodynamics* (Academic Press, New York, 1964).

42. G.R. Fowles, *Introduction to Modern Optics* (Dover, New York, 1975), Second Edition.

43. R.W. James, *The Optical Principles of the Diffraction of X-Rays* (Ox Bow Press, Woodbridge, CT, 1982).

44. J. Helliwell, *Macromolecular Crystallography with Synchrotron Radiation* (Cambridge University Press, New York, 1992); S.W. Lovesey and S.P. Collins, *X-ray Scattering and Absorption by Magnetic Materials* (Oxford University Press, 1996); J.R. Helliwell and P.M. Rentzepis, Editors, *Time-Resolved Diffraction* (Oxford University Press, 1997).

45. B.D. Cullity and S.R. Stock, *Elements of X-Ray Diffraction* (Prentice-Hall, NJ, 2001), Third Edition.

46. B.E. Warren, *X-Ray Diffraction* (Dover, New York, 2014).

47. J. Als-Nielsen and D. McMorrow, *Elements of Modern X-Ray Physics* (Wiley, 2011), Second Edition.

48. E. Spiller, *Soft X-Ray Optics* (SPIE, Bellingham, WA, 1994).

49. C. MacDonald, *Introduction to X-ray Physics, Optics and Applications* (Princeton University Press, 2016).

Homework Problems

Homework problems for each chapter will be found at the website:
www.cambridge.org/xrayeuv

2 Radiation and Scattering at EUV and X-Ray Wavelengths

$$r_e = \frac{e^2}{4\pi \epsilon_0 m c^2} \qquad (2.44)$$

$$\sigma_e = \frac{8\pi}{3} r_e^2 \qquad (2.45)$$

$$\sigma = \frac{8\pi}{3} r_e^2 \frac{\omega^4}{\left(\omega^2 - \omega_s^2\right)^2 + (\gamma\omega)^2} \qquad (2.51)$$

$$\frac{dP}{d\Omega} = \frac{e^2 |\mathbf{a}|^2 \sin^2 \Theta}{16\pi^2 \epsilon_0 c^3} \qquad (2.34)$$

$$f(\Delta\mathbf{k}, \omega) = \sum_{s=1}^{Z} \frac{\omega^2 e^{-i\Delta\mathbf{k}\cdot\Delta\mathbf{r}_s}}{\omega^2 - \omega_s^2 + i\gamma\omega} \qquad (2.66)$$

$$\sigma(\omega) = \frac{8\pi}{3} |f|^2 r_e^2 \qquad (2.69)$$

$$f^0(\omega) = \sum_{s=1}^{Z} \frac{\omega^2}{\omega^2 - \omega_s^2 + i\gamma\omega} \qquad (2.72)$$

In this chapter basics of electromagnetic theory are reviewed. Beginning with Maxwell's equations, the wave equation is developed and used to solve several problems of interest at short wavelengths. Poynting's theorem regarding the flow of electromagnetic energy is used to solve the power radiated by an accelerated electron. The concept of a scattering cross-section is introduced and applied to the scattering of radiation by free and bound electrons. A semi-classical model is used in the latter case. Scattering by a multi-electron atom is described in terms of a complex atomic scattering factor. Tabulated scattering factors, which are available in the literature for use in special circumstances, are described.

2.1 Maxwell's Equations and the Wave Equation

In this chapter we will consider radiation and scattering by accelerated charges. We will use these results to study scattering cross-sections and interesting phenomena at visible,

EUV, and soft x-ray wavelengths. In later chapters we will examine their relation to the refractive index (propagation effects) and the properties of undulator radiation.

Our study begins with *Maxwell's equations,*[1–5] written in a form appropriate to the use of MKS units:*

$$\nabla \times \mathbf{H} = \frac{\partial \mathbf{D}}{\partial t} + \mathbf{J} \quad \text{(Ampere's law)} \tag{2.1}$$

$$\nabla \times \mathbf{E} = -\frac{\partial \mathbf{B}}{\partial t} \quad \text{(Faraday's law)} \tag{2.2}$$

$$\nabla \cdot \mathbf{B} = 0 \tag{2.3}$$

$$\nabla \cdot \mathbf{D} = \rho \quad \text{(Coulomb's law)} \tag{2.4}$$

where $\mathbf{E}$ is the electric field vector, $\mathbf{H}$ is the magnetic field vector, $\mathbf{D}$ is the electric displacement, $\mathbf{B}$ is the magnetic flux density or magnetic induction, $\mathbf{J}$ is the current density, ρ is the charge density, ϵ_0 is the permittivity (dielectric constant) of free space, and μ_0 is the magnetic permeability. If the above are considered to describe fields in free space (vacuum), the constitutive relations take the form

$$\mathbf{D} = \epsilon_0 \mathbf{E} \tag{2.5}$$

$$\mathbf{B} = \mu_0 \mathbf{H} \tag{2.6}$$

where now the charge density ρ and current density $\mathbf{J}$ must be described in a self-consistent manner, i.e., where the fields affect the particles and the particles contribute to the fields. Note that, as is common in the literature, we have used ρ for both charge and mass density (Chapter 1). The reader will recognize the difference by the context.

As described by James Maxwell in 1865, these equations can be combined to form a vector wave equation describing the propagation of electromagnetic waves, as later demonstrated by Heinrich Hertz in 1888. The mathematical description covers electromagnetic phenomena extending from very long wavelengths, to radiowaves, microwaves, infrared, visible, ultraviolet, and x-rays and beyond. The *vector wave equation* can be obtained from Maxwell's equations by taking $\nabla \times$ [Eq. (2.2)] and using the vector identity[†] $\nabla \times \nabla \times \mathbf{A} = \nabla (\nabla \cdot \mathbf{A}) - \nabla^2 \mathbf{A}$ to obtain

$$\nabla \times (\nabla \times \mathbf{E}) = \nabla \times \left(-\frac{\partial \mathbf{B}}{\partial t} \right)$$

$$\nabla(\nabla \cdot \mathbf{E}) - \nabla^2 \mathbf{E} = -\mu_0 \frac{\partial}{\partial t}(\nabla \times \mathbf{H})$$

$$\nabla \left(\frac{\rho}{\epsilon_0} \right) - \nabla^2 \mathbf{E} = -\mu_0 \frac{\partial}{\partial t} \left(\frac{\partial D}{\partial t} + \mathbf{J} \right)$$

$$\nabla \left(\frac{\rho}{\epsilon_0} \right) - \nabla^2 \mathbf{E} = -\epsilon_0 \mu_0 \frac{\partial}{\partial t} \left(\frac{\partial \mathbf{E}}{\partial t} + \frac{\mathbf{J}}{\epsilon_0} \right)$$

---

* See Appendix A, Units and Physical Constants.
[†] See Appendix D, Mathematical and Vector Relationships.

Rearranging terms, we obtain

$$\epsilon_0\mu_0\frac{\partial^2 \mathbf{E}}{\partial t^2} - \nabla^2\mathbf{E} = -\mu_0\frac{\partial \mathbf{J}}{\partial t} + \frac{1}{\epsilon_0}\nabla\rho$$

which when properly grouped is recognized as the vector wave equation:

$$\left(\frac{\partial^2}{\partial t^2} - c^2\nabla^2\right)\mathbf{E}(\mathbf{r}, t) = -\frac{1}{\epsilon_0}\left[\frac{\partial \mathbf{J}(\mathbf{r}, t)}{\partial t} + c^2\nabla\rho(\mathbf{r}, t)\right] \tag{2.7}$$

where

$$c \equiv \frac{1}{\sqrt{\epsilon_0\mu_0}} \tag{2.8}$$

is identified as the phase velocity of an electromagnetic wave in vacuum, often referred to as the "speed of light in vacuum." Note that this is an inhomogeneous partial differential equation. The driving terms on the right-hand side of the equation can be linear or non-linear, leading to a wealth of interesting phenomena.

The wave equation (2.7) will serve as our point of departure in considering radiation, scattering, and refractive index in situations including free and bound electrons, single atoms, and various distributions of atoms. In this chapter we will treat the scattering of x-rays by individual electrons and atoms through appropriate representations of the induced source terms on the right side of the wave equation. We will obtain several interesting results, including expressions for the well-known Thomson and Rayleigh scattering cross-sections for free and bound electrons, as well as atomic scattering cross-sections for multi-electron atoms. We will employ very simple models of the atom, but will observe that the basic results are identical in form to those obtained with more sophisticated quantum mechanical models. In Chapter 3 we will treat wave propagation phenomena in relatively uniform media containing many atoms. There we will find it convenient to bring the uniformly distributed source terms to the left side of the wave equation, where the combined terms will lead to a modified phase velocity and thus the introduction of a refractive index. This will lead to several practical results, including equations governing the total external reflection of x-rays. We will also discuss how refractive indices and complex scattering cross-sections are determined and tabulated in practice.

Before proceeding to these topics, we note that an expression for conservation of charge, the so-called equation of charge continuity, is easily developed and will be useful later. This follows by taking $\nabla \cdot$ [Eq. (2.1)] and noting the vector identity $\nabla \cdot \nabla \times \mathbf{A} \equiv 0$. We have

$$\nabla \cdot \nabla \times \mathbf{H} = \frac{\partial}{\partial t}\nabla \cdot \mathbf{D} + \nabla \cdot \mathbf{J}$$

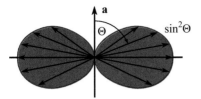

Figure 2.1 As observed at a great distance, fields radiated by an accelerated charge propagate over a broad angular range, but not in the direction of acceleration.

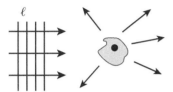

Figure 2.2 Irregularly shaped objects and isolated charges, including free electrons and electrons bound to isolated atoms, scatter radiation in many directions.

where the left-hand side is identically 0, and on the right-hand side $\nabla \cdot \mathbf{D} = \rho$. This gives the equation for conservation of charge:

$$\nabla \cdot \mathbf{J} + \frac{\partial \rho}{\partial t} = 0 \tag{2.9}$$

What we would like to accomplish in this chapter is a solution to Eq. (2.7) for $\mathbf{E}(\mathbf{r}, t)$ in the presence of source terms – for instance, radiated fields due to accelerated free and bound electrons (Figure 2.1), or the *scattered* fields caused by such oscillating charges in the presence of incident (x-ray) radiation. The latter is a process whereby well-directed energy, propagating in a given direction, is widely redirected by non-uniformly distributed charges or by irregularly shaped objects (Figure 2.2).

One can write the current density $\mathbf{J}(r, t)$ as

$$\mathbf{J}(\mathbf{r}, t) = qn(\mathbf{r}, t)\mathbf{v}(\mathbf{r}, t) \tag{2.10}$$

a product of charge density and velocity, both as functions of space and time. Note that the product $n\mathbf{v}$ raises the possibility of non-linear terms entering via particle motions into the otherwise linear electromagnetic fields. This interesting phenomenon, which we touch on in later chapters, occurs in media (plasmas, fluids, solids) in which both n and $\mathbf{v}$ vary, giving rise to beat frequencies, sum and difference frequencies, and such phenomena as harmonic generation. In this chapter we concern ourselves with isolated charges, both free and bound, for which Eq. (2.10) can be written in a somewhat different form convenient to this simple case. We will consider this shortly.

2.2   Calculating Scattered Fields

First we outline the approach to be taken in solving the wave equation (2.7) for $\mathbf{E}(\mathbf{r}, t)$ in the presence of a source term. Then we calculate the radiated power and its angular dependence. Treating the bracketed quantity on the left side of Eq. (2.7) as an operator, we can imagine solving for $\mathbf{E}(\mathbf{r}, t)$ in terms of arbitrary sources, with a formal solution of the form

$$\mathbf{E}(\mathbf{r}, t) = \int_{\text{volume}} [\text{Green's function}][\text{source terms}] \, d\mathbf{r} \qquad (2.11)$$

where the field $\mathbf{E}$ is described mathematically in terms of the response to a distribution of sources – embodied in the so-called *Green's function* – integrated over the full volume.

This is not convenient to do with Eq. (2.7) in the form of a vector partial differential equation. But it can be accomplished[6, 7] without too much difficulty by introducing space-time transformations of $\mathbf{E}(r, t)$ that algebraize the partial differential operators, ∇ and $\partial / \partial t$. Working then with transform amplitudes $\mathbf{E}(\mathbf{k}, \omega)$, abbreviated $\mathbf{E}_{k\omega}$, the differential operators become algebraic multipliers, the inversions required in Eq. (2.11) become simple, and the integration proceeds using standard techniques. Note that the vector field transform amplitudes $\mathbf{E}_{k\omega}$ are a generalization of the Fourier coefficients utilized in one-dimensional (scalar) problems. Using this technique to solve for $\mathbf{E}(\mathbf{r}, t)$, the remaining fields, including $\mathbf{H}(\mathbf{r}, t)$, etc., can be determined from transforms of Eqs. (2.1)–(2.6).

In order to simplify both the space (∇) and time ($\partial / \partial t$) operators, we introduce the *Fourier–Laplace transform*

$$\mathbf{E}(\mathbf{r}, t) = \int_{\mathbf{k}} \int_{\omega} \mathbf{E}_{k\omega} e^{-i(\omega t - \mathbf{k} \cdot \mathbf{r})} \frac{d\omega \, d\mathbf{k}}{(2\pi)^4} \qquad (2.12a)$$

and its inverse

$$\mathbf{E}_{k\omega} = \int_{\mathbf{r}} \int_{t} \mathbf{E}(\mathbf{r}, t) e^{i(\omega t - \mathbf{k} \cdot \mathbf{r})} \, d\mathbf{r} \, dt \qquad (2.12b)$$

where the symbols $d\mathbf{k}$ and $d\mathbf{r}$ are shorthand notation for scalar volume elements. These abbreviations correspond in rectangular coordinates to $d\mathbf{r} = dx \, dy \, dz$ and $d\mathbf{k} = dk_x \, dk_y \, dk_z$. We also understand that the Fourier–Laplace amplitudes $\mathbf{E}_{k\omega}$ are shorthand notation for $\mathbf{E}(\mathbf{k}, \omega)$, as they are vector field amplitudes and are functions of the wave vector $\mathbf{k}$ and frequency ω. We assume that the frequency has a small imaginary component such that the amplitude $\mathbf{E}_{k\omega}$, defined by the integral in Eq. (2.12b), is finite for real fields $\mathbf{E}(\mathbf{r}, t)$. Thus with $\omega = \omega_r + i\omega_i$, where ω_i is small but positive, the integrand in Eq. (2.12b) goes to zero as $t \to \infty$.

Representing all field quantities in a similar manner, i.e.,

$$\mathbf{J}(\mathbf{r}, t) = \int_{\mathbf{k}} \int_{\omega} \mathbf{J}_{k\omega} e^{-i(\omega t - \mathbf{k} \cdot \mathbf{r})} \frac{d\omega \, d\mathbf{k}}{(2\pi)^4} \qquad (2.13a)$$

$$\mathbf{J}_{k\omega} = \int_{\mathbf{r}} \int_{t} \mathbf{J}(\mathbf{r}, t) e^{i(\omega t - \mathbf{k} \cdot \mathbf{r})} d\mathbf{r} \, dt \qquad (2.13b)$$

$$\rho(\mathbf{r}, t) = \int_{\mathbf{k}} \int_{\omega} \rho_{k\omega} e^{-i(\omega t - \mathbf{k} \cdot \mathbf{r})} \frac{d\omega \, d\mathbf{k}}{(2\pi)^4} \qquad (2.14a)$$

$$\rho_{k\omega} = \int_{\mathbf{r}} \int_{t} \rho(\mathbf{r}, t) e^{i(\omega t - \mathbf{k} \cdot \mathbf{r})} d\mathbf{r} \, dt \qquad (2.14b)$$

etc., we can appreciate the algebraized nature of the operators. For instance, if we consider the time derivative, $\partial / \partial t$, acting on $\rho(\mathbf{r}, t)$, we have

$$\frac{\partial \rho(\mathbf{r}, t)}{\partial t} = \frac{\partial}{\partial t} \int_{\mathbf{k}} \int_{\omega} \left[\rho_{k\omega} e^{-i(\omega t - \mathbf{k} \cdot \mathbf{r})} \right] \frac{d\omega \, d\mathbf{k}}{(2\pi)^4}$$

Note that $\partial / \partial t$ passes through the integrals (to first order, $\mathbf{k}$ and ω are not functions of time). Since $\rho_{k\omega}$ is also not a function of time (to first order), the time derivative acts only on the exponent, giving

$$\frac{\partial \rho(\mathbf{r}, t)}{\partial t} = \int_{\mathbf{k}} \int_{\omega} (-i\omega) \rho_{k\omega} e^{-i(\omega t - \mathbf{k} \cdot \mathbf{r})} \frac{d\omega \, d\mathbf{k}}{(2\pi)^4}$$

Thus the time differential operator $\partial / \partial t$ results in an algebraic multiplier $(-i\omega)$ when acting on $\rho(\mathbf{r}, t)$, i.e., when operating in $\mathbf{k}$, ω-space. Similarily the gradient operator, taken component by component in some coordinate space, becomes

$$\nabla \rho(\mathbf{r}, t) = \nabla \int_{\mathbf{k}} \int_{\omega} \left[\rho_{k\omega} e^{-i(\omega t - \mathbf{k} \cdot \mathbf{r})} \right] \frac{d\omega \, d\mathbf{k}}{(2\pi)^4}$$

Although the process of algabraizing may be evident to the reader at this point, it is easily illustrated by introducing rectangular coordinates, such that the $\mathbf{k} \cdot \mathbf{r}$ term in the exponent becomes $k_x x + k_y y + k_z z$, the gradient becomes $\nabla = \mathbf{x}_0 \, \partial / \partial x + \mathbf{y}_0 \, \partial / \partial y + \mathbf{z}_0 \, \partial / \partial z$, and $\mathbf{k} = \mathbf{x}_0 k_x + \mathbf{y}_0 k_y + \mathbf{z}_0 k_z$, where $\mathbf{x}_0$, $\mathbf{y}_0$, $\mathbf{z}_0$ are unit vectors. Since the components $\partial / \partial x$, $\partial / \partial y$, $\partial / \partial z$ act only on the exponent, the expression for $\nabla \rho(\mathbf{r}, t)$ becomes

$$\nabla \rho(\mathbf{r}, t) = \int_{\mathbf{k}} \int_{\omega} (i\mathbf{x}_0 k_x + i\mathbf{y}_0 k_y + i\mathbf{z}_0 k_z) \left[\rho_{k\omega} e^{-i(\omega t - \mathbf{k} \cdot \mathbf{r})} \right] \frac{d\omega \, d\mathbf{k}}{(2\pi)^4}$$

$$\nabla \rho(\mathbf{r}, t) = \int_{\mathbf{k}} \int_{\omega} i\mathbf{k} \rho_{k\omega} e^{-i(\omega t - \mathbf{k} \cdot \mathbf{r})} \frac{d\omega \, d\mathbf{k}}{(2\pi)^4}$$

so that the ∇ operator is replaced by a multiplicative factor $i\mathbf{k}$ in $\mathbf{k}$, ω-space. Finally, if we consider the operator $\nabla^2 = \nabla \cdot \nabla = \partial^2 / \partial x^2 + \partial^2 / \partial y^2 + \partial^2 / \partial z^2$, we can readily demonstrate that

$$\nabla^2 \mathbf{E}(\mathbf{r}, t) = \int_{\mathbf{k}} \int_{\omega} (i\mathbf{k})^2 \mathbf{E}_{k\omega} e^{-i(\omega t - \mathbf{k} \cdot \mathbf{r})} \frac{d\omega \, d\mathbf{k}}{(2\pi)^4}$$

where now it can be seen that $\nabla^2 = \nabla \cdot \nabla$ algebraizes to $i\mathbf{k} \cdot i\mathbf{k} = -k^2$ in $\mathbf{k}$, ω-space.

The vector wave equation (2.7) can now be algebraized itself into a very convenient form in terms of the Fourier–Laplace amplitudes:

$$\int_{\mathbf{k}} \int_{\omega} (-i\omega)^2 \mathbf{E}_{k\omega} e^{-i(\omega t - \mathbf{k} \cdot \mathbf{r})} \frac{d\omega \, d\mathbf{k}}{(2\pi)^4} - c^2 \int_{\mathbf{k}} \int_{\omega} (-k)^2 \mathbf{E}_{k\omega} e^{-i(\omega t - \mathbf{k} \cdot \mathbf{r})} \frac{d\omega \, d\mathbf{k}}{(2\pi)^4}$$

$$= -\frac{1}{\epsilon_0} \left[\int_{\mathbf{k}} \int_{\omega} (-i\omega) \mathbf{J}_{k\omega} e^{-i(\omega t - \mathbf{k} \cdot \mathbf{r})} \frac{d\omega \, d\mathbf{k}}{(2\pi)^4} + c^2 \int_{\mathbf{k}} \int_{\omega} i\mathbf{k} \rho_{k\omega} e^{-i(\omega t - \mathbf{k} \cdot \mathbf{r})} \frac{d\omega \, d\mathbf{k}}{(2\pi)^4} \right]$$

where we note that every term includes the same $\mathbf{k}$, ω-integration, which can therefore be removed, and further, that each term contains an exponential factor $-i(\omega t - \mathbf{k} \cdot \mathbf{r})$, which can also be removed, leaving

$$(-i\omega)^2 \mathbf{E}_{k\omega} - c^2(-k^2)\mathbf{E}_{k\omega} = -\frac{1}{\epsilon_0}[(-i\omega)\mathbf{J}_{k\omega} + c^2(i\mathbf{k})\rho_{k\omega}]$$

In operator form, the wave equation in $\mathbf{k}$, ω-space is

$$(\omega^2 - k^2 c^2)\mathbf{E}_{k\omega} = \frac{1}{\epsilon_0}[(-i\omega)\mathbf{J}_{k\omega} + ic^2 \mathbf{k}\rho_{k\omega}] \qquad (2.15)$$

which, if inverted, can be solved for $\mathbf{E}_{k\omega}$:

$$\mathbf{E}_{k\omega} = \frac{(-i\omega)\mathbf{J}_{k\omega} + ic^2 \mathbf{k}\rho_{k\omega}}{\epsilon_0(\omega^2 - k^2 c^2)} \qquad (2.16)$$

This shows the source terms in the numerator, and poles at $\omega = \pm kc$ representing the system response in terms of incoming and outgoing waves. Our task now is to set models for the sources in a given problem, $\mathbf{J}(\mathbf{r}, t)$ and $\rho(\mathbf{r}, t)$, obtain their transforms $\mathbf{J}_{k\omega}$ and $\rho_{k\omega}$, determine a solution for $\mathbf{E}_{k\omega}$ from Eq. (2.16), and then return to the inverse transform [Eq. (2.12a)] to calculate the radiated field $\mathbf{E}(r, t)$ through the required $d\omega \, d\mathbf{k}$ integrations [Eq. (2.12a)].

Making the same substitutions as above in the equation for charge conservation [Eq.(2.9)],

$$\nabla \cdot \mathbf{J} + \frac{\partial \rho}{\partial t} = 0$$

permits similar simplification of it. Upon use of similar transforms and operations this becomes

$$i\mathbf{k} \cdot \mathbf{J}_{k\omega} - i\omega\rho_{k\omega} = 0$$

so that

$$\rho_{k\omega} = \frac{\mathbf{k} \cdot \mathbf{J}_{k\omega}}{\omega}$$

Thus the expression for the radiated fields $\mathbf{E}_{k\omega}$ [Eq. (2.16)] can be written as

$$\mathbf{E}_{k\omega} = -\frac{i\omega}{\epsilon_0}\left[\frac{\mathbf{J}_{k\omega} - (c^2/\omega^2)\mathbf{k}(\mathbf{k} \cdot \mathbf{J}_{k\omega})}{\omega^2 - k^2 c^2}\right]$$

or

$$\mathbf{E}_{k\omega} = -\frac{i\omega}{\epsilon_0}\left[\frac{\mathbf{J}_{k\omega} - \mathbf{k}_0(\mathbf{k}_0 \cdot \mathbf{J}_{k\omega})}{\omega^2 - k^2 c^2}\right] \qquad (2.17)$$

Here we have written the wave propagaton vector as $\mathbf{k} = k_0 k$. The unit vector $\mathbf{k}_0$ is in the propagation direction, and $k = 2\pi/\lambda$ is the scalar wavenumber. The equation $\omega = kc$ (equivalent to $f\lambda = c$) is most readily appreciated from Eq. (2.17) as the required condition for finite field amplitude $\mathbf{E}_{k,\omega}$ even in the absence of sources. In other words, $\omega = kc$ satisfies the homogeneous (vacuum) case where the right-hand side of Eq. (2.15)

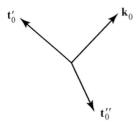

Figure 2.3 A vector coordinate system useful for decomposing vector fields into components parallel and transverse to the wave propagation direction. Polar coordinates in this k-space are k, θ, ϕ, where θ is the polar angle measured from $\mathbf{k}_0$ and ϕ lies in the transverse plane ($\mathbf{t}'_0$, $\mathbf{t}''_0$). The unit vectors are related by $\mathbf{k}_0 \times \mathbf{t}'_0 = \mathbf{t}''_0$.

is zero. These are the so-called resonances or natural modes of the system. The condition $\omega = kc$ is often referred to as a dispersion relation; we will discuss it further in Chapter 3.

We can simplify Eq. (2.17) for $\mathbf{E}_{k\omega}$ by introducing a coordinate system oriented around the wave propagation direction $\mathbf{k}_0$, as shown in Figure 2.3. For instance, we can decompose the vector $\mathbf{J}_{k\omega}$ into components along and transverse to the propagation direction $\mathbf{k}_0$:

$$\mathbf{J}_{k\omega} = \mathbf{J}_{T_{k\omega}} + \mathbf{J}_{L_{k\omega}}\mathbf{k}_0$$

If we do so, it is evident that with $\mathbf{k} = k\mathbf{k}_0$ the numerator in the bracketed factor in Eq. (2.17) becomes

$$\mathbf{J}_{k\omega} - \underbrace{\mathbf{k}_0(\mathbf{k}_0 \cdot \mathbf{J}_{k\omega})}_{\mathbf{k}_0 \text{ portion}} = \mathbf{J}_{T_{k\omega}}$$

where $\mathbf{J}_{T_{k\omega}}$ is the component of the vector $\mathbf{J}_{k\omega}$ transverse to $\mathbf{k}_0$. Note that $\mathbf{k}_0 \cdot \mathbf{J}_{k\omega}$ is the scalar component of $\mathbf{J}_{k\omega}$ in the $\mathbf{k}_0$-direction, and thus $\mathbf{k}_0(\mathbf{k}_0 \cdot \mathbf{J}_{k\omega})$ is the vector component of $\mathbf{J}_{k\omega}$ in the $\mathbf{k}_0$-direction, i.e., the longitudinal vector component. Thus, when it is subtracted from $\mathbf{J}_{k\omega}$, only the transverse vector portion remains. The solution for $\mathbf{E}_{k\omega}$, Eq. (2.17), then becomes

$$\mathbf{E}_{k\omega} = \frac{-i\omega}{\epsilon_0} \cdot \frac{\mathbf{J}_{T_{k\omega}}}{\omega^2 - k^2c^2} \tag{2.18}$$

Having a formal solution (2.18) to $\mathbf{E}_{k\omega}$ in terms of the source, we can now return to the Fourier–Laplace transform relations, specifically Eq. (2.12a), to find the space and time dependent field $\mathbf{E}(\mathbf{r}, t)$:

$$\mathbf{E}(\mathbf{r}, t) = \int_{\mathbf{k}} \int_{\omega} \mathbf{E}_{k\omega} e^{-i(\omega t - \mathbf{k}\cdot\mathbf{r})} \frac{d\omega\, d\mathbf{k}}{(2\pi)^4}$$

With Eq. (2.18) we can express the radiated electric field $\mathbf{E}(\mathbf{r}, t)$ in terms of the transverse component of current density as

$$\mathbf{E}(\mathbf{r}, t) = \int_{\mathbf{k}} \int_{\omega} \left(-\frac{i\omega}{\epsilon_0}\right) \frac{\mathbf{J}_{T_{k\omega}} e^{-i(\omega t - \mathbf{k}\cdot\mathbf{r})}}{(\omega^2 - k^2c^2)} \frac{d\omega\, d\mathbf{k}}{(2\pi)^4} \tag{2.19}$$

which is the form of a Green's function integration, albeit in k, ω-space. The question now is: what is $\mathbf{J}_{T_{k\omega}}$? This will depend on the specific problem of interest.

Let us start by considering a *point radiator* that is small compared to the radiating wavelength, and could be an oscillating free or bound electron. If this radiating particle is sufficiently small, we can represent its density $n(\mathbf{r})$ as a Dirac delta function[‡] such that the current density, $J(\mathbf{r}, t)$ [Eq. (2.10)], given by

$$\mathbf{J}(\mathbf{r}, t) = qn(\mathbf{r}, t)\mathbf{v}(\mathbf{r}, t)$$

can be written for an electron as

$$\mathbf{J}(\mathbf{r},\ t) = -e\delta(\mathbf{r})\mathbf{v}(t) \tag{2.20}$$

In rectangular coordinates $\delta(\mathbf{r})$ is shorthand notation for the product $\delta(x)\delta(y)\delta(z)$, where the delta function has the properties

$$\delta(x) = \begin{cases} 0 & \text{for } x \neq 0 \\ \infty & \text{for } x = 0 \end{cases}$$

The normalization condition of the delta function is

$$\int_{-\infty}^{\infty} \delta(x)dx = 1$$

The delta function has the additional property that

$$\delta(x - a) = \begin{cases} 0 & \text{for } x \neq a \\ \infty & \text{for } x = a \end{cases}$$

with the normalization condition

$$\int_{-\infty}^{\infty} \delta(x - a)dx = 1$$

This leads to the so-called *sifting property*

$$\int_{-\infty}^{\infty} f(x)\delta(x - a)dx = f(a)$$

The transformed current density $\mathbf{J}_{k\omega}$ for a point source radiator can now be determined for use in Eq. (2.19) by utilizing Eq. (2.13b):

$$\mathbf{J}_{k\omega} = \iint \mathbf{J}(\mathbf{r}, t)e^{i(\omega t - \mathbf{k}\cdot\mathbf{r})}d\mathbf{r}\,dt$$

Using Eq. (2.20) for the current density of a point source, one has

$$\mathbf{J}_{k\omega} = \int_{\mathbf{r}}\int_t [-e\delta(\mathbf{r})\mathbf{v}(t)]e^{i(\omega t - \mathbf{k}\cdot\mathbf{r})}d\mathbf{r}dt$$

$$\mathbf{J}_{k\omega} = -e\underbrace{\int_t \mathbf{v}(t)e^{i\omega t}\,dt}_{\mathbf{v}(\omega)}$$

$$\mathbf{J}_{k\omega} = -e\mathbf{v}(\omega)$$

[‡] See Appendix D.7.

This has a transverse component

$$\mathbf{J}_{T_{k\omega}} = -e\mathbf{v}_T(\omega) \qquad (2.21)$$

where $\mathbf{v}_T$ is the velocity component transverse to the propagation direction $\mathbf{k}_0$.

To determine the electric field, as given in Eq. (2.19), one must then perform the integration

$$\mathbf{E}(\mathbf{r}, t) = \frac{ie}{\epsilon_0} \int_{\mathbf{k}} \int_{\omega} \frac{\omega \mathbf{v}_T(\omega) e^{-i(\omega t - \mathbf{k} \cdot \mathbf{r})}}{\omega^2 - k^2 c^2} \frac{d\omega \, d\mathbf{k}}{(2\pi)^4} \qquad (2.22)$$

The $\mathbf{k}$-space integration is accomplished by introducing spherical coordinates oriented around the propagation vector $\mathbf{k}_0$, with differential volume element (a scalar quantity), for instance, as in Figure 2.3,

$$d\mathbf{k} = k^2 \underbrace{\sin\theta \, d\theta \, d\phi}_{d\Omega} dk \qquad (2.23)$$

where θ is measured from $\mathbf{k}_0$, and where $0 \le k \le \infty$, $0 \le \theta \le \pi$, and $0 \le \phi \le 2\pi$. In these coordinates, for a vector $\mathbf{r}$ at a polar angle θ to $\mathbf{k}$, the phase term that occurs in the exponent of Eq. (2.22) becomes

$$\mathbf{k} \cdot \mathbf{r} = kr \cos\theta$$

The angular integrations of Eq. (2.22) in polar coordinates are straightforward. This leaves the k- and ω-integrations. The k-integration can be performed in the complex k-plane using the Cauchy integral formula.[8] The integrand is seen to have two poles, at $k = \pm\omega/c$, representing incoming and outgoing waves. Assuming that the poles each have a small lossy component representing wave decay, the integration path can be closed in the upper half plane with a semicircle of infinite radius, which itself makes no contribution to the integral. The closed path then encloses a single pole, slightly displaced from the real axis, and the k-integral is readily evaluated. Details of the integration are given in Appendix E. The result of the k-integration is

$$\mathbf{E}(\mathbf{r}, t) = \frac{e}{4\pi \epsilon_0 c^2 r} \int_{-\infty}^{\infty} (-i\omega)\mathbf{v}_T(\omega) e^{-i\omega(t - r/c)} \frac{d\omega}{2\pi} \qquad (2.24)$$

which leaves only the ω-integration to be completed. The quantity $-i\omega$ in the integrand is recognized as equivalent to the differential operator d/dt, which acts only on the exponential term and can then be taken outside the ω-integral, leaving the transform of $\mathbf{v}_T(\omega)$. The result for the radiated electric field due to an oscillating point electron is then

$$\mathbf{E}(\mathbf{r}, t) = \frac{e}{4\pi \epsilon_0 c^2 r} \frac{d\mathbf{v}_T(t - r/c)}{dt}$$

Recognizing this as the transverse component of acceleration, the electric field associated with the radiated wave can be written as

$$\boxed{\mathbf{E}(\mathbf{r}, t) = \frac{e\mathbf{a}_T(t - r/c)}{4\pi \epsilon_0 c^2 r}} \qquad (2.25)$$

That is, the radiated electric field $\mathbf{E}(\mathbf{r}, t)$ is due to the component of electron acceleration transverse to the propagation direction, observed at a *retarded time* $t - r/c$, that is, after traveling a distance r to the observer at the speed of light, c. The dependence on only the transverse component of acceleration introduces angular effects into radiation and scattering problems, to be addressed in a following section.

2.3 Radiated Power and Poynting's Theorem

Radiated power, or more specifically power per unit area, is described in electromagnetic theory by the so-called *Poynting vector*[1-5]

$$\mathbf{S} = \mathbf{E}(\mathbf{r}, t) \times \mathbf{H}(\mathbf{r}, t) \qquad (2.26)$$

which gives the magnitude and direction of energy flow or power per unit area. This can be obtained directly from Maxwell's equations. We will derive it, and then calculate the power radiated by an accelerated electron, using the radiated electric field $\mathbf{E}(\mathbf{r}, t)$ given by Eq. (2.25).

To obtain Poynting's theorem for $\mathbf{S}$, we begin with Maxwell's equations (2.1)–(2.6), and form the difference of $\mathbf{H} \cdot$ [Eq. (2.2)] $- \mathbf{E} \cdot$ [Eq. (2.1)] to obtain

$$\mathbf{H} \cdot (\nabla \times \mathbf{E}) - \mathbf{E} \cdot (\nabla \times \mathbf{H}) = -\mathbf{H} \cdot \frac{\partial \mathbf{B}}{\partial t} - \mathbf{E} \cdot \frac{\partial \mathbf{D}}{\partial t} - \mathbf{E} \cdot \mathbf{J}$$

Recalling the vector identity $\nabla \cdot (\mathbf{A} \times \mathbf{B}) = \mathbf{B} \cdot (\nabla \times \mathbf{A}) - \mathbf{A} \cdot (\nabla \times \mathbf{B})$, and using Eqs. (2.5) and (2.6) for $\mathbf{B}$ and $\mathbf{D}$, this becomes

$$\nabla \cdot (\mathbf{E} \times \mathbf{H}) = -\mu_0 \mathbf{H} \cdot \frac{\partial \mathbf{H}}{\partial t} - \epsilon_0 \mathbf{E} \cdot \frac{\partial \mathbf{E}}{\partial t} - \mathbf{E} \cdot \mathbf{J}$$

or

$$\nabla \cdot \underbrace{(\mathbf{E} \times \mathbf{H})}_{\mathbf{S}} = -\frac{\partial}{\partial t} \left(\frac{\mu_0 H^2}{2} \right) - \frac{\partial}{\partial t} \left(\frac{\epsilon_0 E^2}{2} \right) - \mathbf{E} \cdot \mathbf{J} \qquad (2.27)$$

Equation (2.27) is the differential form of Poynting's theorem. The time derivative terms on the right side of the equation represent the rate of change of energy per unit volume (energy density) stored in the magnetic and electric fields, respectively. The rightmost term represents the rate of energy dissipation per unit volume associated with the current density $\mathbf{J}$. Identification of $\mathbf{S} = \mathbf{E} \times \mathbf{H}$ on the left is thus suggestive, as this would represent the net flow of energy into or out of a controlled volume. Certain ambiguities[3,4] in this interpretation can be removed by integrating (2.27) over a closed volume, e.g.,

$$\iiint\limits_{\text{vol.}} \nabla \cdot (\mathbf{E} \times \mathbf{H}) \, dV = -\frac{\partial}{\partial t} \iiint\limits_{\text{vol.}} \left(\frac{\mu_0 H^2}{2} + \frac{\epsilon_0 E^2}{2} \right) dV - \iiint\limits_{\text{vol.}} (\mathbf{E} \cdot \mathbf{J}) \, dV$$

which by Gauss's divergence theorem[¶] for a vector quantity $\mathbf{B}$,

$$\iiint_{\text{vol.}} (\nabla \cdot \mathbf{B}) \, dV = \iint_{\text{surface}} \mathbf{B} \cdot d\mathbf{A}$$

becomes the *integral form of Poynting's theorem:*

$$\iint_{\text{surface}} \underbrace{(\mathbf{E} \times \mathbf{H})}_{\mathbf{S}} \cdot d\mathbf{A} = -\frac{\partial}{\partial t} \iiint_{\text{vol.}} \underbrace{\left(\frac{\mu_0 H^2}{2} + \frac{\epsilon_0 E^2}{2} \right)}_{\text{stored energy density}} dV$$
$$- \iiint_{\text{vol.}} \underbrace{(\mathbf{E} \cdot \mathbf{J})}_{\text{energy dissipation}} dV \tag{2.28}$$

where $\mathbf{S}(\mathbf{r}, t) = \mathbf{E}(\mathbf{r}, t) \times \mathbf{H}(\mathbf{r}, t)$, the Poynting vector, represents the vector flow of energy per unit area in the direction orthogonal to $\mathbf{E}$ and $\mathbf{H}$, which we shall see shortly is the propagation direction $\mathbf{k}_0$ for plane waves. The units of $\mathbf{S}$ are those of energy per unit time and per unit area. The magnitude of $\mathbf{S}$ is often referred to as the *intensity*, I, typically with units of watts per square centimeter.

To complete the calculation of radiated power we must form the vector product $\mathbf{E} \times \mathbf{H}$, wherein to this point we have only calculated $\mathbf{E}$ [Eq. (2.25)]. To obtain $\mathbf{H}(\mathbf{r}, t)$ knowing $\mathbf{E}(\mathbf{r}, t)$ we return to Eq. (2.2):

$$\nabla \times \mathbf{E}(\mathbf{r}, t) = -\frac{\partial \mathbf{B}(\mathbf{r}, t)}{\partial t}$$

By Eq. (2.6),

$$\mathbf{B} = \mu_0 \mathbf{H}$$

which gives

$$\nabla \times \mathbf{E} = -\mu_0 \frac{\partial \mathbf{H}}{\partial t}$$

Recalling the algebraic equivalents of ∇ and $\partial / \partial t$ for fields transformed to a $\mathbf{k}$, ω-plane wave presentation, this takes the form

$$i\mathbf{k} \times \mathbf{E}_{k\omega} = +i\omega\mu_0 \mathbf{H}_{k\omega}$$

Thus in $\mathbf{k}$, ω-space

$$\mathbf{H}_{k\omega} = \sqrt{\frac{\epsilon_0}{\mu_0}} \mathbf{k}_0 \times \mathbf{E}_{k\omega}$$

where we have used the fact that for waves propagating in free space $\omega = kc$ and $c = 1/\sqrt{\epsilon_0 \mu_0}$. Now it is possible to replace both $\mathbf{E}_{k\omega}$ and $\mathbf{H}_{k\omega}$ by their inverse transforms, as in Eq. (2.12b). Noting that the rotation operator $\mathbf{k}_0 \times$ passes through the $d\mathbf{r} \, dt$ integrals on both sides of the resulting equation, one obtains an equation of identical form in $\mathbf{r}$,

---

[¶] See Appendix D.1.

t-space. That is, for plane waves propagating in free space the electric and magnetic fields are related by

$$\mathbf{H}(\mathbf{r}, t) = \sqrt{\frac{\epsilon_0}{\mu_0}} \mathbf{k}_0 \times \mathbf{E}(\mathbf{r}, t) \qquad (2.29)$$

For example, the magnetic field associated with radiation from an accelerated charge, with $\mathbf{E}(\mathbf{r}, t)$ given by Eq. (2.25), is

$$\mathbf{H}(\mathbf{r}, t) = \frac{e}{4\pi c r} \mathbf{k}_0 \times \mathbf{a}_T \left(t - \frac{r}{c} \right) \qquad (2.30)$$

which we note is both transverse to the propagation direction $\mathbf{k}_0$ and, by Eq. (2.29), transverse to $\mathbf{E}$.

The radiated power per unit area can be determined from Eqs. (2.26) and (2.29), and consideration of Poynting's vector, where for a plane wave in vacuum

$$\mathbf{S} = \mathbf{E} \times \mathbf{H} = \mathbf{E}(\mathbf{r}, t) \times \left[\sqrt{\frac{\epsilon_0}{\mu_0}} \mathbf{k}_0 \times \mathbf{E}(\mathbf{r}, t) \right]$$

Noting the vector identity $\mathbf{A} \times (\mathbf{B} \times \mathbf{C}) = (\mathbf{A} \cdot \mathbf{C})\mathbf{B} - (\mathbf{A} \cdot \mathbf{B})\mathbf{C}$, and that for a transverse wave $\mathbf{k}_0 \cdot \mathbf{E} = 0$, we have

$$\mathbf{S}(\mathbf{r}, t) = \sqrt{\frac{\epsilon_0}{\mu_0}} |\mathbf{E}|^2 \mathbf{k}_0 \qquad (2.31)$$

The quantity $\sqrt{\mu_0/\epsilon_0}$ is often referred to as the "impedance of free space," Z_0.

For the accelerated point charge with radiated electric field given by Eq. (2.25), Eq. (2.31) gives *the instantaneous power per unit area*, radiated in the direction $\mathbf{k}_0$:

$$\mathbf{S}(\mathbf{r}, t) = \frac{e^2 |\mathbf{a}_T|^2}{16\pi^2 \epsilon_0 c^3 r^2} \mathbf{k}_0 \qquad (2.32)$$

which decreases as the distance squared and is proportional to the square of the vector acceleration in a direction orthogonal to $\mathbf{k}_0$, i.e., to $|\mathbf{a}_T|^2$. Referring to the vector coordinates for propagation described in Figure 2.4, the acceleration $\mathbf{a}$ at angle Θ to the propagation (observation) direction $\mathbf{k}_0$ has a transverse component of magnitude

$$|\mathbf{a}_T| = |\mathbf{a}| \sin \Theta$$

Thus from Eq. (2.32) the *instantaneous power per unit area radiated by an accelerated electron* becomes

$$\mathbf{S}(\mathbf{r}, t) = \frac{e^2 |\mathbf{a}|^2 \sin^2 \Theta}{16\pi^2 \epsilon_0 c^3 r^2} \mathbf{k}_0 \qquad (2.33)$$

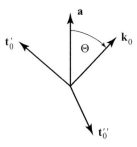

Figure 2.4 Vector coordinates for acceleration a and propagation (observation) direction $\mathbf{k}_0$, separated by an angle Θ measured from the acceleration direction **a**. Note that the vector components of the acceleration are $\mathbf{a}_L = \mathbf{k}_0 \cdot \mathbf{a} = a \cos \Theta$, and $\mathbf{a}_T = - \mathbf{k}_0 \times (\mathbf{k}_0 \times \mathbf{a})$, where $|\mathbf{a}_T| = a \sin \Theta$.

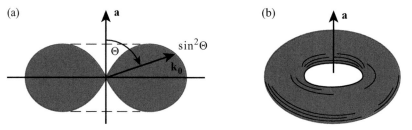

Figure 2.5. (a) The $\sin^2 \Theta$ radiation pattern of an accelerated charge, and (b) its three-dimensional toroidal appearance.

again showing the r^2 decrease with distance, the dependence on a^2, and the well-known $\sin^2 \Theta$ angular pattern of *dipole radiation*[1–5, 9], as observed in the far field when the oscillation amplitude is small compared to the wavelength, e.g., in the point source approximation. The resultant radiation pattern, sketched in Figure 2.5, displays a $\sin^2 \Theta$ toroidal pattern, with maximum radiation intensity orthogonal to the acceleration direction, and zero radiation in the direction of acceleration. The power per unit solid angle is obtained by noting that $\mathbf{S} = (dP/dA)\mathbf{k}_0$, and that the differential elements of area and solid angle are related by $dA = r^2 \, d\Omega$, so that it follows from Eq. (2.33) that in an outgoing $\mathbf{k}_0$-direction

$$\frac{dP}{d\Omega} = \frac{e^2 |\mathbf{a}|^2 \sin^2 \Theta}{16\pi^2 \epsilon_0 c^3} \tag{2.34}$$

The total power radiated, P, is determined by integrating $\mathbf{S}$ over the area of a distant sphere:

$$P = \iint_{\text{area}} \mathbf{S} \cdot d\mathbf{A} = \iint_{\text{solid angle}} \mathbf{S} \cdot (r^2 d\Omega \, \mathbf{k}_0) \tag{2.35}$$

where for $0 \leq \Theta \leq \pi$ and $0 \leq \phi \leq 2\pi$ we have $d\Omega = \sin \Theta \, d\Theta \, d\phi$; thus

$$P = \iint \left[\frac{e^2 |\mathbf{a}|^2 \sin^2 \Theta}{16\pi^2 \epsilon_0 c^3 r^2} \mathbf{k}_0 \right] \cdot r^2 \sin \Theta \, d\Theta \, d\phi \, \mathbf{k}_0$$

$$P = \frac{e^2 |\mathbf{a}|^2}{16\pi^2 \epsilon_0 c^3} \int_0^{2\pi} \underbrace{\int_0^{\pi} \sin^3 \Theta \, d\Theta}_{\int_0^{\pi} (1 - \cos^2 \Theta) \sin \Theta \, d\Theta = \frac{4}{3}} d\phi$$

Thus the *total power radiated* by an oscillating electron of acceleration $\mathbf{a}$ is

$$\boxed{P = \frac{8\pi}{3} \left(\frac{e^2 |\mathbf{a}|^2}{16\pi^2 \epsilon_0 c^3} \right)} \tag{2.36}$$

For sinusoidal fields we are often interested in the time-averaged power $\bar{P}$ and the time-averaged power per unit area, $\bar{\mathbf{S}}$. To form appropriate expressions it is necessary to take the product of real field quantities. For sinusoidal fields we have found it convenient to write the various vectors $\mathbf{E}$, $\mathbf{H}$, $\mathbf{a}$, etc., in the form

$$\mathbf{E}(\mathbf{r}, t) = \mathbf{E}_0 e^{-i(\omega t - \mathbf{k} \cdot \mathbf{r})}$$

where the amplitude $\mathbf{E}_0$ itself may include a complex phase factor $e^{i\phi}$. For such fields one can show by algebraic substitution[3, 4] that the time-averaged real power per unit area flowing in a particular direction is

$$\boxed{\bar{\mathbf{S}} = \tfrac{1}{2} \mathrm{Re}(\mathbf{E} \times \mathbf{H}^*)} \tag{2.37}$$

where Re refers to the real part, the asterisk refers to complex conjugation, and the factor $\frac{1}{2}$ arises from averaging a $\sin^2 \omega t$ or $\cos^2 \omega t$ product term over a full cycle.[3] In calculations of average power and intensity we will thus encounter quantities such as $\mathbf{E}_0 \cdot \mathbf{E}_0^*$, and $\mathbf{a} \cdot \mathbf{a}^*$, which we will abbreviate as $|\mathbf{E}_0|^2$ and $|\mathbf{a}|^2$, etc. Following these procedures, the average power radiated by an oscillating electron, $\bar{P}$, will be half that given by Eq. (2.36), and the average power per unit solid angle will be half that given by Eq. (2.34), when the field amplitudes $\mathbf{a}$ and $\mathbf{E}$ are expressed in terms of their peak values. Since the acceleration $\mathbf{a}$ depends on the local electric field, $\mathbf{E}$, which is described explicitly in the next section, radiated power is proportional to $|\mathbf{E}|^2$.

2.4 Scattering Cross-Sections

We are now in a position to calculate the power radiated by free and bound electrons experiencing an acceleration $\mathbf{a}(\mathbf{r}, t)$. An interesting problem is that of an oscillating electron, free or bound, accelerated by an incident electromagnetic field. As this process redirects radiation to a wide range of angles (Figure 2.6), it is generally referred to as *scattering*. A measure of the scattering power of an electron is given by its equivalent

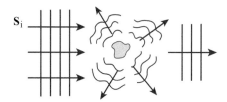

Figure 2.6 Scattering of incident radiation into many directions, leaving a less intense wave in the forward direction.

scattering cross-section σ, that is, its effective area for redirecting incident radiation. This cross-section is defined as the average power radiated divided by the average incident power per unit area, $|\bar{\mathbf{S}}_i|$, that is,

$$\boxed{\sigma \equiv \frac{\bar{P}_{\text{scatt.}}}{|\bar{\mathbf{S}}_i|}} \tag{2.38}$$

where $\bar{P}_{\text{scatt.}}$ is the average power scattered to all directions when an incident wave of electric field $\mathbf{E}_i(\mathbf{r}, t)$ excites an electron to acceleration $\mathbf{a}(\mathbf{r}, t)$. By Eqs. (2.37) and (2.31) the average power per unit area carried by the incident electromagnetic wave is given by

$$\bar{\mathbf{S}}_i = \frac{1}{2}\sqrt{\frac{\epsilon_0}{\mu_0}}|\mathbf{E}_i|^2 \mathbf{k}_0 \tag{2.39}$$

2.5   Scattering by a Free Electron

For a free electron the incident field causes an oscillatory motion described by Newton's second equation of motion, $\mathbf{F} = m\mathbf{a}$, where $\mathbf{F}$ is the Lorentz force on the electron; thus

$$m\mathbf{a} = -e\left[\mathbf{E}_i + \mathbf{v} \times \mathbf{B}_i\right] \tag{2.40}$$

From Eq. (2.29), $\mathbf{H} = \sqrt{\epsilon_0/\mu_0}\,\mathbf{k}_0 \times \mathbf{E}(\mathbf{r}, t)$ for an incident plane electromagnetic wave in vacuum, so that with $\mathbf{B} = \mu_0\mathbf{H}$,

$$\mathbf{B}_i(\mathbf{r}, t) = \frac{\mathbf{k}_0 \times \mathbf{E}_i(\mathbf{r}, t)}{c} \tag{2.41}$$

We see from the above that the term $\mathbf{v} \times \mathbf{B}_i(\mathbf{r}, t)$ in Eq. (2.40) is of order v/c compared to $\mathbf{E}_i(\mathbf{r}, t)$, and therefore negligible for non-relativistic oscillation velocities. The instantaneous acceleration of a free electron driven by a passing (incident) electromagnetic wave is then

$$\mathbf{a}(\mathbf{r}, t) = -\frac{e}{m}\mathbf{E}_i(\mathbf{r}, t) \tag{2.42}$$

The scattered electric field, given by Eq. (2.25), depends only on the transverse component of acceleration, which as seen in Figure 2.4 has scalar amplitude

$$a_T = a \sin \Theta = -\frac{e}{m} E_i \sin \Theta$$

From Eq. (2.25) the scalar electric field, scattered to an angle Θ with respect to the polarization direction of the incident electric field, can thus be expressed as

$$\mathbf{E}(\mathbf{r}, t) = -\frac{e^2 E_i \sin \Theta}{4\pi \epsilon_0 mc^2 r} e^{-i\omega(t-r/c)}$$

Introducing the classical electron radius, r_e, the electric field scattered by a free electron can be written more compactly as

$$\mathbf{E}(\mathbf{r}, t) = -\frac{r_e E_i \sin \Theta}{r} e^{-i\omega(t-r/c)} \qquad (2.43)$$

where

$$r_e = \frac{e^2}{4\pi \epsilon_0 mc^2} \qquad (2.44)$$

is defined[1] by equating the electrostatic energy of a uniform sphere of radius r and charge e, $e^2/4\pi \epsilon_0 r$, to its rest energy mc^2. In a later section we will calculate the scattered field due to a many-electron atom and will compare it with that for a single free electron, Eq. (2.43). In this way we will introduce an *atomic scattering factor*, a multiplying factor that compares the electric field scattered by a multi-electron atom with that scattered by a single free electron.

The average power scattered by an oscillating electron is obtained by combining Eqs. (2.36) and (2.42):

$$\bar{P}_{\text{scatt.}} = \frac{1}{2} \frac{8\pi}{3} \frac{e^2 \left(\frac{e^2}{m^2} |\mathbf{E}_i|^2 \right)}{16\pi^2 \epsilon_0 c^3}$$

where the factor $\frac{1}{2}$ appears due to time averaging the squared sinusoidal fields. The scattering cross-section given by Eq. (2.38) is then

$$\sigma = \frac{\bar{P}_{\text{scatt}}}{|\bar{\mathbf{S}}|} = \frac{\frac{4\pi}{3} \left(\frac{e^4 |\mathbf{E}_i|^2}{16\pi^2 \epsilon_0 m^2 c^3} \right)}{\frac{1}{2} \sqrt{\frac{\epsilon_0}{\mu_0}} |\mathbf{E}_i|^2}$$

where we have used Eq. (2.39) for the time averaged Poynting vector of the incident wave. Again using the classical electron radius r_e, the scattering cross-section for a free

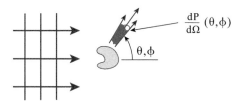

Figure 2.7 Angle-dependent scattering can be described in terms of a differential scattering cross-section that takes account of variations in power per unit solid angle.

electron can be expressed as

$$\sigma_e = \frac{8\pi}{3} r_e^2$$
(2.45)

where the subscript e denotes the fact that this is the scattering cross-section for a single electron. This result was first obtained by J.J. Thomson,[§] and is referred to as the *Thomson cross-section*[1] for scattering of electromagnetic waves by a free electron. Note that for a free (unbound) electron, Eq. (2.45) has no frequency (wavelength) dependence, thus indicating that the scattering cross-section is the same across the electromagnetic spectrum, from microwaves to visible light to x-rays. Limitations for very short wavelength x-rays and gamma rays, where the momentum of the incident photon is sufficient to cause recoil, are discussed in the literature.[10] Note that numerically

$$r_e = 2.818 \times 10^{-13} \text{ cm}$$
(2.46a)

$$\sigma_e = 6.652 \times 10^{-25} \text{ cm}^2$$
(2.46b)

A differential scattering cross-section per unit solid angle (Figure 2.7) can be obtained by the same procedures. Using Eq. (2.34), with a factor[2] for average power in terms of peak acceleration, and normalizing to $|\bar{S}_i| = \frac{1}{2}\sqrt{\epsilon_0/\mu_0}E_i^2$, the differential scattering cross-section is defined as

$$\frac{d\sigma_e}{d\Omega} \equiv \frac{1}{|S_i|}\frac{d\bar{P}}{d\Omega} = \frac{\frac{e^4 a^2 \sin^2\Theta}{32\pi^2\epsilon_0 c^3}}{\frac{1}{2}\sqrt{\frac{\epsilon_0}{\mu_0}}|\mathbf{E}_i^2|}$$

Again using Eq. (2.42), we now obtain the *differential Thomson scattering cross-section for a free electron*:

$$\frac{d\sigma_e}{d\Omega} = r_e^2 \sin^2\Theta$$
(2.47)

[§] See J.J. Thomson, *Conduction of Electricity Through Gases* (Cambridge University Press, 1906), Second Edition, p. 325. The "corpuscles" (see his p. 197 as well) are what we now call electrons. In the Third Edition (1933, with G.P. Thomson) the classical theory of scattering is presented in the manner followed by modern texts.

In addition to Thomson's early efforts to identify the nature of the electron, knowledge of the free-electron cross-section is widely used for other purposes. In modern studies of plasma physics, as in fusion research, Thomson scattering of laser light is widely used as a diagnostic of free electron density and of temperature (both electron and ion), and to determine the presence of various plasma waves.[11]

2.6   Scattering by Bound Electrons

A number of interesting phenomena can be explained on the basis of scattering by bound electrons. Topics of interest here include the scattering of x-rays by multi-electron atoms, the refractive index at x-ray wavelengths, and phenomena such as total external reflection of x-rays at glancing incidence to a material surface, as well as the more common scattering of visible sunlight in the atmosphere, which leads to the appearance of a blue sky and a red sunset. While the scattering is best described by quantum mechanical techniques,[12–16] much can be learned from a simple semi-classical model in which the atom is represented by a massive positively charged $(+Ze)$ nucleus surrounded by several (Z) electrons held at discrete binding energies. In this model the relatively massive nucleus does not respond dynamically to the high-frequency incident fields, but the electrons are caused to oscillate at the frequency ω imposed by the electric field $\mathbf{E}_i$ of a passing electromagnetic wave. The various electrons, being bound by differing restoring forces, respond differently to the impressed fields. The response depends on the resonant frequencies ω_s of the bound electrons and, more specifically, on the closeness of the driving (incident wave) frequency to the resonances, that is, on $\omega - \omega_s$. There is much literature regarding this semi-classical model and its relation to the more rigorous quantum mechanical model.[12–17]

To proceed we require an equation of motion for each of the bound electrons so that we may determine its acceleration $\mathbf{a}$ in the presence of an incident field – and from that determine the reradiated power in much the same manner as we did previously for free electrons. Thus we must determine an appropriate formulation of Newton's second law of motion ($\mathbf{F} = m\mathbf{a}$) for each of the bound electrons. In the semi-classical model we treat the multi-electron atom as a collection of harmonic oscillators, each with its own set of resonances, $\hbar\omega_s$, which we can associate with known transitions between stationary states of the atom. We note, before proceeding to the semi-classical equation of motion, that in a proper quantum mechanical model, the presence of a time-dependent external electric field perturbs the atomic system so that there is a time-dependent probability of finding the atom in various stationary states ψ_n – perhaps upper and lower states – oscillating continuously between the two at the impressed frequency and thus giving the sense of a time-dependent oscillation of charge distribution within the atom.

In the semi-classical model each bound electron is forced to execute simple harmonic motion by the impressed electric field while in the presence of the restoring central force field of the massive, positively charged nucleus. The equation of motion can then be written as follows:[||]

$$m\frac{d^2\mathbf{x}}{dt^2} + m\gamma\frac{d\mathbf{x}}{dt} + m\omega_s^2\mathbf{x} = -e(\mathbf{E}_i + \underbrace{\mathbf{v} \times \mathbf{B}_i}_{\simeq 0}) \qquad (2.48)$$

where the first term is the acceleration ($m\mathbf{a}$), the second term is a dissipative force term that accounts for energy loss (we assume $\gamma/\omega \ll 1$), and the third term is due to the restoring force for an oscillator of resonant frequency ω_s; where $-e(\mathbf{E}_i + \mathbf{v} \times \mathbf{B}_i)$ is the Lorentz force exerted by the incident fields; and where, as before, the $\mathbf{v} \times \mathbf{B}_i$ term is small for non-relativistic oscillation velocities $\mathbf{v}$. With oscillations impressed by an incident electric field of the form

$$\mathbf{E} = \mathbf{E}_i e^{-i\omega t}$$

we anticipate that the displacement $\mathbf{x}$, velocity, and acceleration will all have the same $e^{-i\omega t}$ time dependence. The time derivative in all terms can then be replaced by $-i\omega$, so that the equation of motion (2.48) becomes

$$m(-i\omega)^2\mathbf{x} + m\gamma(-i\omega)\mathbf{x} + m\omega_s^2\mathbf{x} = -e\mathbf{E}_i$$

where we have suppressed the explicit $e^{-i\omega t}$ factor that appears in each term. Combining factors, we see that the harmonic displacement is given by

$$\mathbf{x} = \frac{1}{\omega^2 - \omega_s^2 + i\gamma\omega}\frac{e\mathbf{E}_i}{m} \qquad (2.49)$$

and thus the acceleration is

$$\mathbf{a} = \frac{-\omega^2}{\omega^2 - \omega_s^2 + i\gamma\omega}\frac{e\mathbf{E}_i}{m} \qquad (2.50)$$

Following procedures used earlier for the free electron [Eqs. (2.38), (2.39), and (2.44)], we obtain the semi-classical *scattering cross-section for a bound electron* of resonant frequency ω_s:

$$\sigma = \frac{8\pi}{3}r_e^2\frac{\omega^4}{\left(\omega^2 - \omega_s^2\right)^2 + (\gamma\omega)^2} \qquad (2.51)$$

where we note that the scattering cross-section for bound electrons displays a strong frequency dependence, especially near the resonance.

This is substantially different from the free electron cross-section, which is frequency independent. The form of the bound electron cross-section is illustrated in Figure 2.8. This semi-classical result shows a strong resonance at $\omega \cong \omega_s$, with a peak cross-section very large compared to the free electron result. Near resonance the line shape approximates that of a Lorentzian of half width at half maximum $\gamma/2$. For very large

[||] J.D. Jackson, Ref. 1, p. 309.

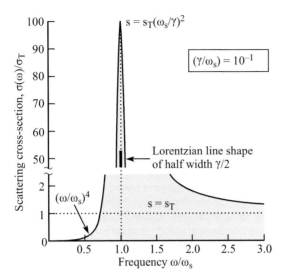

Figure 2.8 The semi-classical scattering cross-section, Eq. (2.51), for a bound electron of resonant energy $\hbar\omega_s$ and an assumed damping factor $\gamma/\omega_s = 0.1$.

frequencies the cross-section approaches Thomson's result [Eq. (2.45)] for the free electron. In this very high frequency limit, where $\omega^2 \gg \omega_s^2$, the bound electrons scatter as though they were free. In this case the oscillations forced by the incident radiation are too rapid to be affected by the natural response of the resonant system. For incident frequencies well below the resonant frequency, such that $\omega^2 \ll \omega_s^2$ and $\gamma \ll \omega_s$, the cross-section takes on a form first described by Lord Rayleigh:[18]

$$\sigma_R = \frac{8\pi}{3}r_e^2\left(\frac{\omega}{\omega_s}\right)^4 = \frac{8\pi}{3}r_e^2\left(\frac{\lambda_s}{\lambda}\right)^4 \tag{2.52}$$

which has a very strong (λ^{-4}) wavelength dependence.

Rayleigh first used this result in 1899 to explain the blue color of the sky. The photon energies ($\hbar\omega$) of visible light extend from about 1.8 eV (7000 Å) for red to about 2.3 eV (5300 Å) for green and about 3.3 eV (3800 Å) for blue light. The bound electrons of atmospheric oxygen and nitrogen, with UV resonances at 8.6 eV and 8.2 eV (1520 Å), respectively, cause very strong scattering at the shorter visible light wavelengths. Indeed, the λ^{-4} wavelength dependence of the scattering cross-section gives a factor of about 2^4, or 16 times more scattering for blue light than for red. This explains both the blue appearance of the sky when looking overhead, and the residual red appearance of the setting sun – the latter being observed in direct viewing after the light has propagated over a long path in which the blue, green, etc., have been preferentially scattered to the other directions, as illustrated in Figure 2.9. For a quantum mechanical description of resonant scattering, including lifetime and Doppler (motion) effects, refer to Loudon, Ref. 12, pp. 314–318 and 70–78.

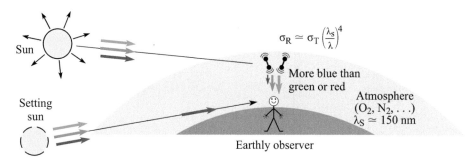

Figure 2.9 Looking upward, the earthly observer at night sees only blackness or the occasional star. However, during daylight, indirect light is scattered toward the observer when looking overhead. This scattering is caused by non-uniformities in the atmospheric density of O_2 and N_2, and appears blue because of the strong wavelength dependence of bound electron scattering. Upon direct viewing of the sun, particularly at sunset, the light path is long and passes through the most dense portion of the atmosphere. As much of the residual light reaching the observer at sunset is greatly depleted of blue and green, the sun appears red, as do clouds off which this reddish light reflects. Very fine atmospheric dust, such as volcanic ash, causes a similar effect.

2.7   Scattering by a Multi-Electron Atom

In this section we turn to the subject of scattering by an atom that contains many electrons. We again use a semi-classical model of point electrons, each with its own resonant frequency, excited by a continuous electromagnetic wave. Because we do not wish to make the assumption that the wavelength is long compared to atomic dimensions – which is often not true for x-rays – we permit each of the electrons to have separate coordinates. Although this is a very simple atomic model, it gives valuable insights into the angular and wavelength limitations of various scattering formulae.

In this semi-classical model we can write the electron distribution function within the atom as

$$n_e(\mathbf{r}, t) = \sum_{s=1}^{Z} \delta[\mathbf{r} - \Delta \mathbf{r}_s(t)] \qquad (2.53)$$

where $\mathbf{r}$ is the coordinate of the nucleus, $\Delta \mathbf{r}$ is the vector displacement from the nucleus, and Z is the total number of electrons held by the atom. The total current density can be written as

$$\mathbf{J}(\mathbf{r}, t) = -e \sum_{s=1}^{Z} \delta[\mathbf{r} - \Delta \mathbf{r}_s(t)]\mathbf{v}_s(t) \qquad (2.54)$$

where for the purposes of scattering calculations $\mathbf{v}_s(t)$ will be dominated by the incident field. The assumption that $\mathbf{v}_s$ is dominated by the incident field, ignoring the effect of waves scattered by neighboring electrons, is referred to as the *Born approximation*.

To make use of our radiated power and scattering cross-section formulae we follow the procedures developed earlier, first calculating the $\mathbf{k}$, ω transform of $\mathbf{J}(\mathbf{r}, t)$ [see Eq. (2.13b)]:

$$\mathbf{J}_{k\omega} = \int_{\mathbf{r}} \int_t \mathbf{J}(\mathbf{r}, t) e^{i(\omega t - \mathbf{k} \cdot \mathbf{r})} \, d\mathbf{r} \, dt$$

which with Eq. (2.54) becomes

$$\mathbf{J}_{k\omega} = -e \sum_{s=1}^{Z} \iint \delta(\mathbf{r} - \Delta\mathbf{r}_s) \mathbf{v}_s(t) e^{i(\omega t - \mathbf{k} \cdot \mathbf{r})} \, d\mathbf{r} \, dt$$

$$\mathbf{J}_{k\omega} = -e \sum_{s=1}^{Z} e^{-i\mathbf{k} \cdot \Delta\mathbf{r}_s} \underbrace{\int \mathbf{v}_s(t) e^{i\omega t} \, dt}_{\mathbf{v}_s(\omega)}$$

where we have assumed that the time dependence of $\Delta\mathbf{r}_s(t)$ due to positional variation of the electrons is on a slower time scale than ω^{-1} and thus separable to first order from the integration. Recognizing the time integral as the Laplace transform $\mathbf{v}_s(\omega)$, we have

$$\mathbf{J}_{k\omega} = -e \sum_{s=1}^{Z} e^{-i\mathbf{k} \cdot \Delta\mathbf{r}_s} \mathbf{v}_s(\omega) \tag{2.55}$$

Continuing to follow our earlier procedures, the electric field scattered by the Z electrons of the atom is given by Eq. (2.19):

$$\mathbf{E}(\mathbf{r}, t) = \int_{\mathbf{k}} \int_{\omega} \frac{-i\omega}{\epsilon_0} \frac{\mathbf{J}_{T_{k\omega}} e^{-i(\omega t - \mathbf{k} \cdot \mathbf{r})}}{\omega^2 - k^2 c^2} \frac{d\mathbf{k} \, d\omega}{(2\pi)^4}$$

$$\mathbf{E}(\mathbf{r}, t) = -\frac{e}{\epsilon_0} \sum_{s=1}^{Z} \int_{\mathbf{k}} \int_{\omega} \frac{(-i\omega) e^{-i\mathbf{k} \cdot \Delta\mathbf{r}_s} \mathbf{v}_{T,s}(\omega) e^{-i(\omega t - \mathbf{k} \cdot \mathbf{r})}}{\omega^2 - k^2 c^2} \frac{d\mathbf{k} \, d\omega}{(2\pi)^4}$$

$$\mathbf{E}(\mathbf{r}, t) = -\frac{e}{\epsilon_0} \sum_{s=1}^{Z} \int_{\mathbf{k}} \int_{\omega} \frac{(-i\omega) e^{i\mathbf{k} \cdot (\mathbf{r} - \Delta\mathbf{r}_s)} \mathbf{v}_{T,s}(\omega) e^{-i\omega t}}{(\omega - kc)(\omega + kc)} \frac{d\mathbf{k} \, d\omega}{(2\pi)^4}$$

where $\mathbf{v}_{T,s}(\omega)$ is the component of $\mathbf{v}_s(\omega)$ transverse to the direction of propagation (observation) $\mathbf{k}$. The quantity $\mathbf{r} - \Delta\mathbf{r}_s$ in the exponent is identified as the vector distance from the particular s-electron ($\Delta\mathbf{r}_s$) to the observation point ($\mathbf{r}$), as illustrated in Figure 2.10.

Shown in Figure 2.10 are the various point electrons at their positions $\Delta\mathbf{r}_s$ (measured from the nucleus at $\mathbf{r} = 0$), an incident wave vector $\mathbf{k}_i$, and the scattered wave vector $\mathbf{k}$ propagating toward the observer at $\mathbf{r}$. Note that the scattered wave vector $\mathbf{k}_0$ is at angle 2θ (lowercase) to the incident wave vector $\mathbf{k}_i$, and that this *scattering angle* is different from Θ (uppercase), which is the angle between E_i (not shown) and $\mathbf{k}$. Note also that the diagram defines the vector position of each electron as seen by the observer,

$$\mathbf{r}_s \equiv \mathbf{r} - \Delta\mathbf{r}_s \tag{2.56}$$

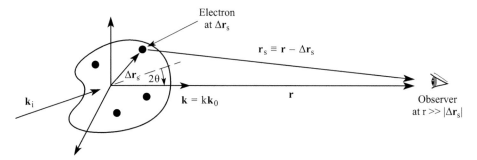

Figure 2.10 Scattering diagram for radiation incident on a many-electron atom, semi-classically described as a collection of point electrons surrounding a nucleus of charge $+Ze$ at $\mathbf{r} = 0$.

With this definition the expression for $\mathbf{E}(\mathbf{r}, t)$ becomes

$$\mathbf{E}(\mathbf{r}, t) = -\frac{e}{\epsilon_0} \sum_{s=1}^{Z} \int_{\mathbf{k}} \int_{\omega} \frac{(-i\omega)e^{i\mathbf{k}\cdot\mathbf{r}_s}\mathbf{v}_{T,s}(\omega)e^{-i\omega t}}{(\omega - kc)(\omega + kc)} \frac{d\mathbf{k}\, d\omega}{(2\pi)^4}$$

where now $\mathbf{r}_s$ is the vector distance to the point at which radiated fields are detected. Proceeding as before with the $d\mathbf{k}\, d\omega$ integrals, which led to Eq. (2.25), we find for the multi-electron atom that

$$\mathbf{E}(\mathbf{r}, t) = \frac{e}{4\pi\epsilon_0 c^2} \sum_{s=1}^{Z} \frac{1}{|\mathbf{r}_s|} \frac{d}{dt} \mathbf{v}_{T,s} \underbrace{\left(t - \frac{|\mathbf{r}_s|}{c}\right)}_{t'_s}$$

where there now appears a retarded time of observation $t'_s = t - |\mathbf{r}_s|/c$ appropriate to each electron. Defining the scalar magnitude of observer distance to the sth electron as

$$r_s \equiv |\mathbf{r}_s|$$

the expression for the electric field scattered from a multi-electron atom becomes

$$\mathbf{E}(\mathbf{r}, t) = \frac{e}{4\pi\epsilon_0 c^2} \sum_{s=1}^{Z} \frac{\mathbf{a}_{T,s}\,(t - r_s/c)}{r_s} \tag{2.57}$$

which is an evident extension of the earlier result for a single electron [Eq. (2.25)].

Using expressions for the transverse acceleration of each bound electron, $\mathbf{a}_{T,s}$, in terms of the incident field $\mathbf{E}_i$ which excites it to oscillation, we can calculate the scattered power and cross-section. We can write the equation of motion for each of these electrons as

$$m\frac{d^2\mathbf{x}_s}{dt^2} + m\gamma\frac{d\mathbf{x}_s}{dt} + m\omega_s^2\mathbf{x}_s = -e(\mathbf{E}_i + \underbrace{\mathbf{v}_s \times \mathbf{B}}_{\simeq 0}) \tag{2.58}$$

where in this case we must be careful to keep the spatial dependence of the incoming wave in order to account for the differing phase seen by each electron. To do so we rewrite the electric field as

$$\mathbf{E}_i(\mathbf{r}, t) \rightarrow \mathbf{E}_i e^{-i(\omega t - \mathbf{k}_i \cdot \Delta \mathbf{r}_s)} \tag{2.59}$$

where we explicitly label the incoming (incident) wave vector as $\mathbf{k}_i$. Combining Eqs. (2.58) and (2.59), we proceed as before to a solution of the equation of motion (at position $\Delta \mathbf{r}_s$) for the oscillatory motion of a bound electron in the presence of an incident electromagnetic field:

$$\mathbf{x}_s(t) = \frac{1}{\omega^2 - \omega_s^2 + i\gamma\omega} \frac{e}{m} \mathbf{E}_i e^{-i(\omega t - \mathbf{k}_i \cdot \Delta \mathbf{r}_s)} \tag{2.60}$$

and

$$\mathbf{a}_s(t) = \frac{-\omega^2}{\omega^2 - \omega_s^2 + i\gamma\omega} \frac{e}{m} \mathbf{E}_i e^{-i(\omega t - \mathbf{k}_i \cdot \Delta \mathbf{r}_s)} \tag{2.61}$$

The electric field scattered by a multi-electron atom to a distant observer can now be obtained by combining Eqs. (2.57) and (2.61), defining the transverse component of acceleration as was done previously for the single electron case, and introducing the classical electron radius [Eq. (2.44)], so that in terms of field amplitudes

$$\mathbf{E}(\mathbf{r}, t) = \frac{-e^2}{4\pi\epsilon_0 mc^2} \sum_{s=1}^{Z} \frac{-\omega^2 \mathbf{E}_i \sin \Theta}{\omega^2 - \omega_s^2 + i\gamma\omega} \frac{1}{r_s} e^{-i[\omega(t - r_s/c) - \mathbf{k}_i \cdot \Delta \mathbf{r}_s]}$$

where $\mathbf{r}_s \equiv \mathbf{r} - \Delta \mathbf{r}_s$ and $r_s = |\mathbf{r}_s|$. For $r \gg \Delta r_s$ we can write to good approximation (see the boxed note below)

$$r_s \simeq r - \mathbf{k}_0 \cdot \Delta \mathbf{r}_s \tag{2.62}$$

Note on the relative phase terms *for a multi-electron atom:* If $\mathbf{r}_s = \mathbf{r} - \Delta \mathbf{r}_s$ then

$$\mathbf{r}_s \cdot \mathbf{r}_s = (\mathbf{r} - \Delta \mathbf{r}_s) \cdot (\mathbf{r} - \Delta \mathbf{r}_s) = \mathbf{r} \cdot \mathbf{r} + \Delta \mathbf{r}_s \cdot \Delta \mathbf{r}_s - 2\mathbf{r} \cdot \Delta \mathbf{r}_s$$

$$r_s^2 = r^2 + \Delta r_s^2 - 2\mathbf{r} \cdot \Delta \mathbf{r}_s$$

For $r \gg \Delta r_s$,

$$r_s^2 \simeq r^2 - 2\mathbf{r} \cdot \Delta \mathbf{r}_s = r^2 \left(1 - \frac{2\mathbf{r} \cdot \Delta \mathbf{r}_s}{r^2}\right)$$

$$r_s \simeq r \left(1 - \frac{\mathbf{r} \cdot \Delta \mathbf{r}_s}{r^2}\right)$$

and for $\mathbf{r}$ in the $\mathbf{k}_0$-direction (see Figure 2.10)

$$\frac{\mathbf{r}}{r} \equiv \mathbf{r}_0 = \mathbf{k}_0$$

Thus for $r \gg \Delta r_s$

$$r_s \simeq r - \mathbf{k}_0 \cdot \Delta \mathbf{r}_s$$

as given in Eq. (2.62).

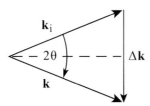

Figure 2.11 A vector scattering diagram for incident wave vector $\mathbf{k}_i$ and scattered wave vector $\mathbf{k}$. If both waves are at the same frequency (stationary scatterer), the vector diagram is isosceles.

We can simplify the expression for $\mathbf{E}(\mathbf{r}, t)$ by approximating r_s by r in the slowly varying amplitude term, while retaining it in the rapidly varying phase term. The electric field amplitude is then

$$E(r, t) = -r_e \sum_{s=1}^{Z} \frac{-\omega^2 E_i \sin\Theta}{\omega^2 - \omega_s^2 + i\gamma\omega} \frac{1}{r} \exp\left\{-i\left[\omega\left(t - \frac{r}{c}\right) + \omega\left(\frac{\mathbf{k}_0 \cdot \Delta\mathbf{r}_s}{c}\right) - \mathbf{k}_i \cdot \Delta\mathbf{r}_s\right]\right\}$$

Noting that $\omega/c = k$ and that $k\mathbf{k}_0 = \mathbf{k}$, the phase term can be written more compactly as

$$E(\mathbf{r}, t) = -r_e \sum_{s=1}^{Z} \frac{\omega^2 E_i \sin\Theta}{\omega^2 - \omega_s^2 + i\gamma\omega r} \frac{1}{r} \exp\left\{-i\left[\omega\left(t - \frac{r}{c}\right) + \underbrace{(\mathbf{k} - \mathbf{k}_i)}_{\Delta\mathbf{k}} \cdot \Delta\mathbf{r}_s\right]\right\}$$

where we have introduced the quantity $\Delta\mathbf{k}$ defined by

$$\Delta\mathbf{k} = \mathbf{k} - \mathbf{k}_i \tag{2.63}$$

where $\Delta\mathbf{k}$ is the vector periodicity associated with the inhomogeneity of the medium that results in a wave of propagation vector $\mathbf{k}_i$ being scattered into a direction characterized by the scattered wave vector $\mathbf{k}$. This *density fluctuation wave vector*, denoted here by $\Delta\mathbf{k}$, is a quantity one encounters generally in the study of scattering processes, including the scattering of light or x-rays from crystals, plasma waves, and a host of other non-uniform density distributions.

Since both $\mathbf{k}$ and $\mathbf{k}_i$ propagate in vacuum, the magnitudes of the wave vectors satisfy $|\mathbf{k}| = |\mathbf{k}_i| = \omega/c$, so that the scattering diagram, illustrated in Figure 2.11, is isosceles with

$$\boxed{|\Delta\mathbf{k}| = 2k_i \sin\theta} \tag{2.64}$$

With simple identifications this will be recognized as the Bragg equation, $\lambda = 2d\sin\theta$, where $k_i = 2\pi/\lambda$ and for a crystal the periodicity is $\Delta k = 2\pi/d$. A powerful insight into this simple equation (2.64) is that electron density distributions of periodicity d scatter radiation of wavelength λ through an angle 2θ, and can thus provide a Fourier analysis technique useful in many fields of study. See Section 3.9 for further discussion.

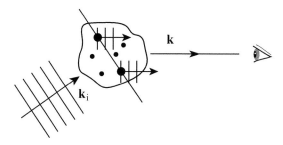

Figure 2.12 Scattering diagram for a multi-electron atom. Note that the two electrons shown as large dots see the same incident phase, but are of different phase to the observer.

Using the definition $\Delta\mathbf{k}$ in the expression for $E(\mathbf{r}, t)$, we have

$$E(\mathbf{r}, t) = -\frac{r_e}{r} \underbrace{\left[\sum_{s=1}^{Z} \frac{\omega^2 e^{-i\Delta\mathbf{k}\cdot\Delta\mathbf{r}_s}}{\omega^2 - \omega_s^2 + i\gamma\omega} \right]}_{f(\Delta\mathbf{k},\, \omega)} E_i \sin\Theta e^{-i\omega(t-r/c)} \qquad (2.65)$$

where the quantity $f(\Delta\mathbf{k}, \omega)$ is the *complex atomic scattering factor*

$$f(\Delta\mathbf{k}, \omega) = \sum_{s=1}^{Z} \frac{\omega^2 e^{-i\Delta\mathbf{k}\cdot\Delta\mathbf{r}_s}}{\omega^2 - \omega_s^2 + i\gamma\omega} \qquad (2.66)$$

a function of the incident wave frequency ω, the various resonance frequencies ω_s of the bound electrons, and the phase terms due to their various positions within the atom, $\Delta\mathbf{k} \cdot \Delta\mathbf{r}_s$. The atomic scattering factor describes the electric field amplitude of the scattered wave relative to that scattered by a free electron:

$$E(\mathbf{r}, t) = -\frac{r_e f(\Delta\mathbf{k}, \omega) E_i \sin\Theta}{r} e^{-i\omega(t-r/c)} \qquad (2.67)$$

where for the referenced (single) free electron, Eq. (2.43), ω_s, γ, and $\Delta\mathbf{r}_s$ are zero, so that $f(\Delta\mathbf{k}, \omega)$ is unity. Note that the expression $\Delta\mathbf{k} \cdot \Delta\mathbf{r}_s$ gives the phase variation of the scattered fields, due to differing electron positions, as seen by the observer. That product contains both the different incident phases seen by the various electrons and the phase variations due to the different path lengths to the observer (see Figure 2.12).

Having determined $\mathbf{E}(\mathbf{r}, t)$ for the semi-classical multi-electron atom, we can now calculate the *differential and total scattering cross-sections* following the procedures used earlier in this chapter for the free electron, obtaining

$$\frac{d\sigma(\omega)}{d\Omega} = r_e^2 |f|^2 \sin^2\Theta \qquad (2.68)$$

$$\sigma(\omega) = \frac{8\pi}{3} |f|^2 r_e^2 \qquad (2.69)$$

where the complex atomic scattering factor (2.66) is

$$f(\Delta\mathbf{k}, \omega) = \sum_{s=1}^{Z} \frac{\omega^2 e^{-i\Delta\mathbf{k}\cdot\Delta\mathbf{r}_s}}{(\omega^2 - \omega_s^2 + i\gamma\omega)}$$

and where the cross-sections (2.68) and (2.69) now display not only the various resonances and a damping term, but also explicit phase terms $e^{-i\phi_s}$ accounting for the discrete positions of the various electrons in the semi-classical atom – factors that are significant when λ is less than or similar to the atomic radius. In general the $\Delta\mathbf{k} \cdot \Delta\mathbf{r}_s$ phase terms do not simplify, and treatment of the complex atomic scattering factor is problematical. However, in two special situations $f(\Delta\mathbf{k}, \omega)$ does simplify. To understand these two special cases we reconsider the vector scattering diagram of Figure 2.12. We recall Eq. (2.64):

$$|\Delta\mathbf{k}| = 2k_i \sin\theta$$

where $k_i = 2\pi / \lambda$, so that

$$\Delta k = \frac{4\pi}{\lambda} \sin\theta$$

With the charge distribution within the atom largely contained within dimensions of the order of the Bohr radius,** traditionally written as a_0, the phase term in Eq. (2.66) for the complex atomic scattering function is bounded by

$$|\Delta\mathbf{k} \cdot \Delta\mathbf{r}_s| \leq \frac{4\pi a_0}{\lambda} \sin\theta \tag{2.70}$$

where the inequality results from the nature of the vector dot product. The phase expression in (2.70) clearly simplifies in two special cases:

$$|\Delta\mathbf{k} \cdot \Delta\mathbf{r}_s| \to 0 \quad \text{for } a_0/\lambda \ll 1 \quad \text{(long wavelength limit)} \tag{2.71a}$$

$$|\Delta\mathbf{k} \cdot \Delta\mathbf{r}_s| \to 0 \quad \text{for } \theta \ll 1 \quad \text{(forward scattering)} \tag{2.71b}$$

In each of these two cases the atomic scattering factor $f(\Delta\mathbf{k}, \omega)$ reduces to

$$f^0(\omega) = \sum_{s=1}^{Z} \frac{\omega^2}{\omega^2 - \omega_s^2 + i\gamma\omega} \tag{2.72}$$

where we denote these special cases by the superscript zero. According to Eq. (2.71a) scattering of radiation at the relatively long soft x-ray and EUV wavelengths is isotropic and the scattering factor $f(\Delta\mathbf{k}, \omega)$ can be accurately approximated by $f^0(\omega)$, and with less accuracy to wavelengths approaching 2 Å. For forward scattering, including the very important description of refractive index, the approximation is valid even for the shortest x-ray wavelengths of interest in this text [Eq. (2.71b)].

** For example see Chapter 1, Figure 1.12. Note that the Bohr radius for the ground state of the hydrogen atom ($n = 1$) is $a_0 = 4\pi\epsilon_0\hbar^2/me^2 = 0.5292$ Å.

It is convenient at this point to introduce the concept of *oscillator strengths*, g_s, which in the simple semi-classical model are integers that indicate the number of electrons associated with a given resonance frequency ω_s. In such a model one could associate two electrons with a K-shell resonance, six with an L-shell resonance, etc. Spectroscopists have long taken this model a step further, introducing fractional oscillator strengths to accommodate known probabilities for transition to various higher-lying energy states for each atom. Thus we take the sum of oscillator strengths as equal to the total number of electrons:

$$\sum_s g_s = Z \qquad (2.73)$$

A shortcoming of the semi-classical model, among others, is that while it gives the proper form of scattering cross-sections and refractive index, it does not provide a basis for calculating oscillator strengths. In the quantum mechanical description[12–15] these oscillator strengths[††] arise naturally as non-integer *transition probabilities, g_{kn}*, between stationary states ψ_k and ψ_n of the atom, leading to an expression similar to Eq. (2.73) when summed over final states n from an initial state k:

$$\sum_n g_{kn} = Z \qquad (2.74)$$

Equation (2.74) is known as the Thomas–Reiche–Kuhn sum rule[14, 15, 19]. It was deduced before wave mechanics provided the modern understanding of quantum mechanics.

Introducing the oscillator strengths, we can rewrite the atomic scattering cross-sections (2.72) for the special cases of long wavelength ($\lambda \gg a_0$) or small angles ($\theta \ll \lambda/a_0$) as

$$\frac{d\sigma(\omega)}{d\Omega} = r_e^2 |f^0(\omega)|^2 \sin^2 \Theta \qquad (2.75)$$

and

$$\sigma(\omega) = \frac{8\pi}{3} r_e^2 |f^0(\omega)|^2 \qquad (2.76)$$

where now

$$f^0(\omega) = \sum_s \frac{g_s \omega^2}{\omega^2 - \omega_s^2 + i\gamma\omega} \qquad (2.77)$$

[††] In the literature the symbol f is commonly used to represent both scattering factor and oscillator strength. To avoid confusion we use g to represent oscillator strength in this chapter and in Chapter 3. In the EUV/SXR laser Chapter 9, where the scattering factor does not occur, we use f for oscillator strength.

where again $\sum_s g_s = Z$. Although based on a very simple semi-classical model, Eqs. (2.75)–(2.77) give a solution for the scattering of radiation by a multi-electon atom, which, except for the different interpretations of oscillator strengths, is identical in functional form to that derived by modern quantum mechanical techniques. These limitations require that either the wavelength be large compared to atomic dimensions [Eq. (2.71a)] or that the scattering be in the forward direction [Eq. (2.71b)], and also that the photon energy be not too close to an atomic resonance, as that case requires an understanding of lifetimes (damping rates γ), which is not addressed by the semi-classical model.[20]

An interesting limit of scattering involves low-Z atoms and relatively long wavelength soft x-rays and extreme ultraviolet radiation, for which λ/a_0 is much greater than unity. Such scattering played a role in the development of early atomic theory[19] in that it gave evidence of the number of electrons bound to the atom. In this very special case one can simultaneously satisfy the conditions that $\omega^2 \gg \omega_s^2$ and $\lambda/a_0 \gg 1$, so that the atomic scattering factor $f(\Delta \mathbf{k}, \omega)$ reduces to[‡‡]

$$f(\Delta \mathbf{k}, \omega) \rightarrow f^0(\omega) \rightarrow \sum_s g_s = Z \tag{2.78a}$$

and thus

$$\frac{d\sigma(\omega)}{d\Omega} \simeq Z^2 r_e^2 \sin^2 \Theta = Z^2 \frac{d\sigma_T}{d\Omega} \tag{2.78b}$$

and

$$\sigma(\omega) \simeq \frac{8\pi}{3} r_e^2 Z^2 = Z^2 \sigma_e \tag{2.78c}$$

Thus, for example, a carbon atom ($Z = 6$) scatters 4 Å wavelength radiation about 36 times more than that of a single free electron. In this case we say that the six electrons are scattering *coherently* in all directions, that is, the scattered electric fields from all atomic electrons add in phase at all distant points of observation. In this case of carbon at 4 Å wavelength, the photon energy is $\hbar\omega \cong 3$ keV, well above the binding energy of the most tightly held electrons, the K-shell electrons, for which the binding energy is about 284 eV. This is indeed much less than the photon energy (3 keV), and λ is much greater than the Bohr radius, $a_0 \cong 0.5$ Å.

For both scattering and refractive index (the latter to be considered in the next chapter) we will want to determine the real and imaginary parts of $f^0(\omega)$, which we will write as[23][¶¶]

$$f^0(\omega) = f_1^0(\omega) - i f_2^0(\omega) \tag{2.79}$$

[‡‡] For high-Z atoms some orbits bring electrons very close to the nucleus, at highly relativistic velocities, thus increasing their mass and decreasing their scattering strength from that of a free electron. Corrections of order $(Z\alpha)^2$, where α is the fine structure constant, are discussed in the literature.[21]

[¶¶] Note that at high frequencies, such that $\omega^2 \gg \omega_s^2$, Eq. (2.77) approaches the limit given by Eq. (2.78a). Consequently some authors[14] write $f_1^0(\omega)$ as a decrement from the total electronic charge, $Z + \Delta f_1^0(\omega)$.

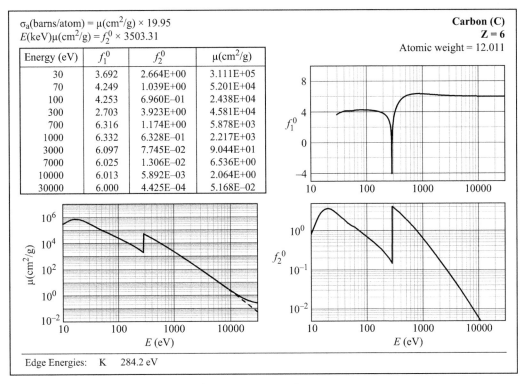

Energy (eV)	f_1^0	f_2^0	μ(cm^2/g)
30	3.692	2.664E+00	3.111E+05
70	4.249	1.039E+00	5.201E+04
100	4.253	6.960E–01	2.438E+04
300	2.703	3.923E+00	4.581E+04
700	6.316	1.174E+00	5.878E+03
1000	6.332	6.328E–01	2.217E+03
3000	6.097	7.745E–02	9.044E+01
7000	6.025	1.306E–02	6.536E+00
10000	6.013	5.892E–03	2.064E+00
30000	6.000	4.425E–04	5.168E–02

σ_a(barns/atom) = μ(cm^2/g) × 19.95
E(keV)μ(cm^2/g) = f_2^0 × 3503.31

Carbon (C)
Z = 6
Atomic weight = 12.011

Edge Energies: K   284.2 eV

Figure 2.13 Tabulated real (f_1^0) and imaginary (f_2^0) parts of the atomic scattering factor for the carbon atom in the limit (superscript zero) of long wavelength or small scattering angle. Note that the sign of f_2^0 is consistent with the mathematical representation of forward wave propagation as exp[$-i(\omega t - \mathbf{k} \cdot \mathbf{r})$]. Data are from Henke, Gullikson, and Davis[22]; with updated tabulations for all elements maintained online by Gullikson.[22, 23]

where the sign of the imaginary portion is chosen to be consistent with our use of $e^{-i\omega t}$, time dependence. In general these are not calculable for the many electron atom. We will see in Chapter 3 that there is a very close relation between forward scattering ($\theta = 0$), where we can use the f^0 approximation, and refractive index. Indeed, we will determine that the refractive index $n(\omega)$ can be written as[19]

$$n(\omega) = 1 - \delta + i\beta = 1 - \frac{n_a r_e \lambda^2}{2\pi}\left(f_1^0 - i f_2^0\right) \qquad (2.80)$$

relating the complex atomic scattering factor for forward scattering ($\theta = 0$) to both phase velocity variation, through f_1^0, and wave amplitude decay due to absorption, through the imaginary component f_2^0. We will address the experimental determination of f_1^0 and f_2^0 in the next chapter. At this point we simply note that these quantities are tabulated,[22, 23] and attach as an example the data for carbon as Figure 2.13. Note that for the special case of low-Z atoms (such as carbon), the approximation $\lambda \gg a_0$ combined with $\hbar\omega \gg \hbar\omega_s$ does work well. In the case of carbon cited above, the tabulated data give $f_1^0 \cong 6.1$, which is very close to the value of Z. The tabulated value $f_2^0 \cong 0.071$

shows that absorption is relatively weak well above the binding energy. Note, however, that in general f_1^0 and f_2^0 are very strong functions of photon energy, particularly near absorption edges. In the limit of very high photon energy the binding energies become relatively unimportant and all electrons scatter as though they were free. In this limit f_1^0 goes to Z, and f_2^0 goes to zero. Discussions of the oscillator strength sum rules for intermediate energies are given by Soufli[24] and Wooten.[17] Note too that the tabulations are for $f^0(\omega)$, not $f(\Delta \mathbf{k}, \omega)$, and thus do not address angular effects at short wavelengths. Nonetheless, the data are very useful for refractive index and long-wavelength (soft x-ray) scattering where these specialized approximations are valid.

References

1. J.D. Jackson, *Classical Electrodynamics* (Wiley, New York, 1998), Third Edition.
2. S. Ramo, J. Whinnery, and T. Van Duzer, *Fields and Waves in Communication Electronics* (Wiley, New York, 1984), Chapter 3.
3. J.A. Stratton, *Electromagnetic Theory* (McGraw Hill, New York, 1941), pp. 135–137.
4. M. Born and E. Wolf, *Principles of Optics* (Cambridge Univ. Press, 1999), Seventh Edition.
5. G.S. Smith, *Classical Electromagnetic Radiation* (Cambridge University Press, 1997).
6. Following N. Marcuvitz, *Notes on Plasma Turbulence*, unpublished (New York University, 1969).
7. P.C. Clemmow, *The Plane Wave Spectrum Representation of Electromagnetic Fields* (Pergamon, Oxford, 1966).
8. E. Kreyszig, *Advanced Engineering Mathematics* (Wiley, New York, 1993), Seventh Edition, p. 770.
9. R Leighton, *Principles of Modern Physics* (McGraw-Hill, New York, 1959), p. 410.
10. W. Heitler, *The Quantum Theory of Radiation* (Clarendon Press, Oxford, 1954).
11. D. Evans and J. Katzenstein, *Rpts. Progress Phys. (London)* **32**, 207 (April 1969).
12. R. Loudon, *The Quantum Theory of Light* (Oxford Univ. Press, London, 1983), Second Edition.
13. D.I. Blokhintsev, *Quantum Mechanics* (Gordon and Breach, New York, 1964).
14. R.W. James, *The Optical Principles of the Diffraction of X-Rays* (Bell, London, 1962), Chapter IV.
15. R.L. Liboff, *Introductory Quantum Mechanics* (Addison-Wesley, 2002), Fourth Edition, Section 13.9; also see J.C. Slater, *The Quantum Theory of Matter* (McGraw Hill, New York, 1968), Second Edition, Chapter 14.
16. R. Eisberg and R. Resnick, *Quantum Physics of Atoms, Molecules, Solids, Nuclei and Particles* (Wiley, New York, 1985).
17. F. Wooten, *Optical Properties of Solids* (Academic Press, 1972), Chapter 3.
18. Lord Rayleigh, *Phil. Mag. (London)* **XLVII**, 375–384 (1899); *Scientific Papers, Vol. IV* (Dover, New York, 1964), p. 397.
19. A.H. Compton and S.K. Allison, *X-Rays in Theory and Practice* (Van Nostrand, New York, 1935), Second Edition.
20. See J.C. Slater, Ref. 15, pp. 276–280, comparing the semi-classical and quantum mechanical interpretations. References 12–14 have similar discussions.

21. L. Kissel and R.H. Pratt, "Rayleigh Scattering: Elastic Photon Scattering by Bound Electrons," p. 465 in *Atomic Inner Shell Physics* (Plenum, New York, 1985), B. Crasemann, Editor.

22. B.L. Henke, E.M. Gullikson, and J.C. Davis, "X-Ray Interactions: Photoabsorption, Scattering, Transmission and Reflection," *Atomic and Nuclear Data* **54**, 181–342 (1993); E.M. Gullikson maintains an updated website of x-ray optical constants and other data at www.cxro.LBL.gov/optical_constants.

23. E.M. Gullikson, "Optical Properties of Materials," Chapter 13, pp. 257–270, in *Vacuum Ultraviolet Spectroscopy I* (Academic Press, New York, 1998), J.A.R. Samson and D.L. Ederer, Editors.

24. R. Soufli, "Optical Constants of Materials in the EUV/Soft X-Ray Region for Multilayer Mirror Applications," PhD thesis, Department of Electrical Engineering and Computer Science, University of California at Berkeley (1997).

Homework Problems

Homework problems for each chapter will be found at the website:
www.cambridge.org/xrayeuv

3 Wave Propagation and Refractive Index at X-Ray and EUV Wavelengths

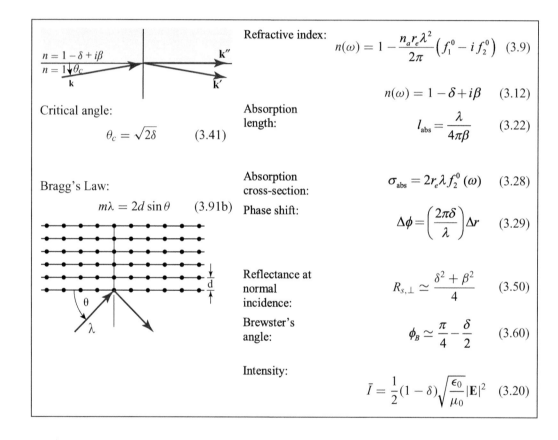

Refractive index:

$$n(\omega) = 1 - \frac{n_a r_e \lambda^2}{2\pi}\left(f_1^0 - i f_2^0\right) \quad (3.9)$$

$$n(\omega) = 1 - \delta + i\beta \quad (3.12)$$

Critical angle:

$$\theta_c = \sqrt{2\delta} \quad (3.41)$$

Absorption length:

$$l_{abs} = \frac{\lambda}{4\pi\beta} \quad (3.22)$$

Bragg's Law:

$$m\lambda = 2d\sin\theta \quad (3.91b)$$

Absorption cross-section:

$$\sigma_{abs} = 2r_e\lambda f_2^0(\omega) \quad (3.28)$$

Phase shift:

$$\Delta\phi = \left(\frac{2\pi\delta}{\lambda}\right)\Delta r \quad (3.29)$$

Reflectance at normal incidence:

$$R_{s,\perp} \simeq \frac{\delta^2 + \beta^2}{4} \quad (3.50)$$

Brewster's angle:

$$\phi_B \simeq \frac{\pi}{4} - \frac{\delta}{2} \quad (3.60)$$

Intensity:

$$\bar{I} = \frac{1}{2}(1-\delta)\sqrt{\frac{\epsilon_0}{\mu_0}}|\mathbf{E}|^2 \quad (3.20)$$

In this chapter wave propagation in a medium of uniform atomic density is considered. Expressions for the induced motion of bound atomic electrons are used in combination with the wave equation to obtain the complex refractive index for EUV, soft and hard x-ray propagation. This is then expressed in terms of the atomic scattering factors of Chapter 2. Phase velocity, absorption, reflection, and refraction are then considered. Results such as the total external reflection of x-rays at glancing incidence from the surface of a lossy material, the weak normal incidence reflection of x-rays, Brewster's angle, and Kramers–Kronig relations are obtained. For materials with periodic density distributions in one direction, such as the atomic planes of a natural crystal, or the layered coatings of a multilayer mirror, individual atomic scatterings evolve collectively by constructive interference to diffraction in a well-defined direction described by the Bragg equation.

3.1 The Wave Equation and Refractive Index

Here we consider the subjects of electromagnetic wave propagation, reflection, and refraction,[1-5] with particular emphasis on application to the EUV and x-ray regions of the spectrum. Here the wavelengths are very short, reaching atomic dimensions, and the photon energies are comparable to the binding energies of atomic electrons. Whereas in the previous chapter we considered scattering from a single atom, in this chapter we consider scattering from many atoms, each containing many electrons. We can imagine that in general this could be a very complicated problem; however, if we restrict ourselves to propagation in the forward direction, the problem simplifies significantly, leading to relatively simple expressions for the refractive index in terms of atomic scattering factors obtained in the previous chapter. Indeed, it is the sum of forward-scattered radiation from all atoms that interferes with the incident wave to produce a modified propagating wave, different from that in vacuum. As the scattering process involves both elastic (lossless) and inelastic (dissipative) processes, the resultant refractive index is in general a complex quantity, describing not only a modified phase velocity, compared to that in vacuum (c), but also a wave amplitude that decays as it propagates.

Our point of departure for the study of refractive index is the vector wave equation [Chapter 2, Eq. (2.7)]

$$\left(\frac{\partial^2}{\partial t^2} - c^2 \nabla^2\right) \mathbf{E}(\mathbf{r}, t) = \frac{-1}{\epsilon_0}\left[\frac{\partial \mathbf{J}(\mathbf{r}, t)}{\partial t} + c^2 \nabla \rho(\mathbf{r}, t)\right]$$

which, as seen in Chapter 2, follows directly from Maxwell's equations with the identification that [Eq. (2.8)]

$$c \equiv \frac{1}{\sqrt{\mu_0 \epsilon_0}}$$

is the phase velocity for propagation in vacuum. For the propagation of transverse waves ($\mathbf{E}$ perpendicular to $\mathbf{k}$) the $\nabla \rho$ term does not contribute, nor does the longitudinal component of $\mathbf{J}$, i.e., the component of $\mathbf{J}$ in the direction of propagation [see the arguments in Chapter 2 leading to Eqs. (2.16) and (2.18)]. Thus for transverse electromagnetic waves of the form $\exp[-i(\omega t - \mathbf{k} \cdot \mathbf{r})]$, propagating in the vector $\mathbf{k}$-direction, we need consider only field components transverse to $\mathbf{k}$,

$$\left(\frac{\partial^2}{\partial t^2} - c^2 \nabla^2\right) \mathbf{E}_{\mathrm{T}}(\mathbf{r}, t) = -\frac{1}{\epsilon_0}\frac{\partial \mathbf{J}_{\mathrm{T}}(\mathbf{r}, t)}{\partial t} \tag{3.1}$$

where the subscript T denotes a direction transverse to $\mathbf{k}$, as illustrated previously in the vector coordinate system shown in Figure 2.3. The two possible transverse coordinates correspond to the two possible states of polarization. Equation (3.1) is recognized as the inverse Fourier–Laplace transform of Eq. (2.18), the transverse wave equation in $\mathbf{k}$, ω-space.

To determine $\mathbf{J}_T$ for the many-atom situation we must sum the contributions of all electrons. We recall from Chapter 2 that a passing electromagnetic wave of frequency ω induces an oscillatory electron motion of the same frequency, with an amplitude oscillation given by [Eq. (2.49)]

$$\mathbf{x}(\mathbf{r}, t) = \frac{e}{m} \frac{1}{\left(\omega^2 - \omega_s^2\right) + i\gamma\omega} \mathbf{E}(\mathbf{r}, t)$$

where in this semi-classical model of the atom ω_s is the electron's natural frequency of oscillation, γ is a dissipative factor, and $\mathbf{E}(\mathbf{r}, t)$ is the electric field of the passing wave. For small amplitude oscillations, the oscillation velocity is thus

$$\mathbf{v}(\mathbf{r}, t) = \frac{e}{m} \frac{1}{\left(\omega^2 - \omega_s^2\right) + i\gamma\omega} \frac{\partial \mathbf{E}(\mathbf{r}, t)}{\partial t} \tag{3.2}$$

The total current density $\mathbf{J}(\mathbf{r}, t)$ must sum the contributions of all such bound electrons within an atom, and sum over all atoms. Were we interested in the scattering to all angles within this many atom system, we would have the formidable problem of describing not only the positions of all electrons within the atom, as was considered in Chapter 2, but also the relative positions of all atoms. However, in this chapter we restrict our interests to propagation only in the forward direction ($\theta = 0$), which as we saw in the previous chapter leads to a significant simplification. Indeed, we found that in the forward direction the positions of the individual electrons are irrelevant [Eq. (2.67)], as the forward-scattered radiation has the same phase, with respect to the incident radiation, for all electrons of like resonant frequency, independent of their positions. It is the interaction of these forward-scattered waves with the incident wave that contributes to modified propagation characteristics that we refer to as the refractive index – both the modified phase velocity and the amplitude decay.

As the electron positions do not affect forward propagation, we can simplify the current density expression for these purposes by introducing an expression with subscript zero, $\mathbf{J}_0(\mathbf{r}, t)$, referring to the special case of forward scattering ($\theta = 0$) where now all similar atoms contribute identically, and the summation is only over like resonances, that is,

$$\mathbf{J}_0(\mathbf{r}, t) = -en_a \sum_s g_s \mathbf{v}_s(\mathbf{r}, t) \tag{3.3}$$

where n_a is the average density of atoms, and where the oscillator strengths for the various resonances sum to the total number of electrons per atom,[*] i.e., [Eq. (2.73)]

$$\sum_s g_s = Z$$

---

[*] As in Chapter 2, we used g_s for the oscillator strength to avoid confusion with the use of f for the scattering factor. In a quantum mechanical, multi-electron model the normalization condition, summed over final states n, is $\sum_n g_{nk} = Z$, known as the Thomas–Reiche–Kuhn rule, Eq. (2.74). In Chapter 9, Extreme Ultraviolet and Soft X-Ray Lasers, the oscillator strength is represented by the more traditional f_{lu}.

where Z is the number of electrons per atom. Combining Eqs. (3.2) and (3.3), the total current density $\mathbf{J}_0$ contributing to propagation in the forward direction is

$$\mathbf{J}_0(\mathbf{r}, t) = -\frac{e^2 n_a}{m} \sum_s \frac{g_s}{(\omega^2 - \omega_s^2) + i\gamma\omega} \frac{\partial \mathbf{E}(\mathbf{r}, t)}{\partial t} \tag{3.4}$$

Substituting this into the transverse wave equation (3.1), one has

$$\left(\frac{\partial^2}{\partial t^2} - c^2 \nabla^2\right) \mathbf{E}_T(\mathbf{r}, t) = \frac{e^2 n_a}{\epsilon_0 m} \sum_s \frac{g_s}{(\omega^2 - \omega_s^2) + i\gamma\omega} \frac{\partial^2 \mathbf{E}_T(\mathbf{r}, t)}{\partial t^2}$$

Combining terms with similar differential operators, one has

$$\left[\left(1 - \frac{e^2 n_a}{\epsilon_0 m} \sum_s \frac{g_s}{(\omega^2 - \omega_s^2) + i\gamma\omega}\right) \frac{\partial^2}{\partial t^2} - c^2 \nabla^2\right] \mathbf{E}_T(\mathbf{r}, t) = 0 \tag{3.5}$$

which can be rewritten in the standard form of the wave equation

$$\left[\frac{\partial^2}{\partial t^2} - \frac{c^2}{n^2(\omega)} \nabla^2\right] \mathbf{E}_T(\mathbf{r}, t) = 0 \tag{3.6}$$

where the frequency dependent refractive index $n(\omega)$ is identified as

$$n(\omega) \equiv \left[1 - \frac{e^2 n_a}{\epsilon_0 m} \sum_s \frac{g_s}{(\omega^2 - \omega_s^2) + i\gamma\omega}\right]^{1/2} \tag{3.7}$$

Note that we have used n for both the refractive index $n(\omega)$ and the number density n_a (atoms or electrons per unit volume), as is also common. The reader will have to be alert to these differences, generally differentiated by subscripts or indicated independent variables.

The refractive index $n(\omega)$ has a strong frequency dependence, particularly near the resonant frequencies ω_s, and is thus said to be *dispersive*. That is, waves of different frequencies propagate at different phase velocities and thus tend to separate (disperse). It is a simple matter to show that for EUV and x-ray radiation ω^2 is very large compared to $e^2 n_a / \epsilon_0 m$, so that to a high degree of accuracy the index of refraction can be written as

$$n(\omega) = 1 - \frac{1}{2} \frac{e^2 n_a}{\epsilon_0 m} \sum_s \frac{g_s}{(\omega^2 - \omega_s^2) + i\gamma\omega} \tag{3.8}$$

This equation predicts both positive and negative dispersion, depending on whether the frequency ω is less or greater than ω_s. This sign convention follows experience with visible light, where the resonances, ω_s are generally in the ultraviolet region for common glass lenses and prisms. Among early researchers this led to what became known as *normal dispersion*. Radiation for which $\omega > \omega_s$ was considered "anomalous." Figure 3.1 illustrates a generic refractive index across the electromagnetic spectrum with resonances in the infrared (IR), in the ultraviolet (UV), and the x-ray region.

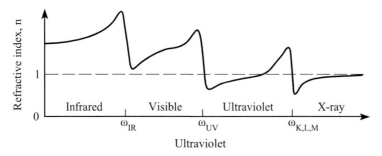

Figure 3.1 A sketch of refractive index showing the strong variations near IR, UV, and x-ray resonances (ω_s), and the general tendency toward unity for very short wavelengths where the frequencies are higher than all atomic resonances. Only the real part of the refractive index is shown here.

Note that Eq. (3.8) can be rewritten in somewhat simpler form in terms of the classical electron radius [Eq. (2.44)]

$$r_e = \frac{e^2}{4\pi\epsilon_0 mc^2}$$

and the complex atomic scattering factor ($\theta = 0$, superscript zero) [Eq. (2.77)]

$$f^0(\omega) = \sum_s \frac{g_s\omega^2}{\omega^2 - \omega_s^2 + i\gamma\omega}$$

which can be written in terms of its complex components [Eq. (2.79)]

$$f^0(\omega) = f_1^0(\omega) - if_2^0(\omega)$$

Making these substitutions, the refractive index [Eq. (3.8)] can be rewritten as

$$n(\omega) = 1 - \frac{n_a r_e \lambda^2}{2\pi}\left[f_1^0(\omega) - if_2^0(\omega)\right] \tag{3.9}$$

where λ is the wavelength in vacuum. Note that this relationship shows explicitly the link between forward scattering and refractive index. In a later section we will discuss the experimental and computational determinations of $f_1^0(\omega)$ and $f_2^0(\omega)$ and their utilization in various x-ray experiments.

Some further comments can be made on the role of the refractive index as embodied in the wave equation (3.6), which can be factored into the product of two operators, viz.,

$$\left(\frac{\partial}{\partial t} - \frac{c}{n(\omega)}\nabla\right)\left(\frac{\partial}{\partial t} + \frac{c}{n(\omega)}\nabla\right)\mathbf{E}_{\mathrm{T}}(\mathbf{r}, t) = 0 \tag{3.10}$$

The equation can be satisfied for non-zero electric field when either factor is zero. These two conditions correspond to left- and right-propagating waves, as illustrated in Figure 3.2.

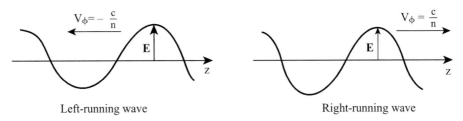

$V_\phi = -\dfrac{c}{n}$ E   z

$V_\phi = \dfrac{c}{n}$ E   z

Left-running wave   Right-running wave

Figure 3.2 Left- and right-propagating waves.

We observe that Eq. (3.10) for a wave propagating in a medium of uniform atomic density the phase velocity (the speed with which crests of fixed phase move) is not equal to c as in vacuum, but rather is modified to a value

$$v_\phi = \frac{c}{n(\omega)} \tag{3.11}$$

That this is so can be seen by examining Eq. (3.10) for the case of fixed amplitude, for instance traveling with fixed phase at the peak amplitude of the wave. This requires that one of the brackets, which correspond to left- and right-propagating waves, contributes as a zero multiplier. For the right-running one-dimensional wave illustrated in Figure 3.2, with space-time dependence $E = E_0 \exp[-i\,(\omega t - kz)]$, setting the operator to zero gives the condition

$$-i\left(\omega - \frac{ck}{n}\right) = 0$$

or

$$\frac{\omega}{k} = \frac{c}{n}$$

which we recognize as a *phase velocity* v_ϕ, different from the vacuum value c, in a medium of refractive index n. Thus for visible light, with $\omega < \omega_s$, the refractive index is greater than unity, typically 1.5 or so for common glass, which corresponds to a relatively low phase velocity, less than c. Typical phenomena affected by low phase velocity propagation of visible wavelengths include reflection and refractive turning at tilted interfaces, focusing by lenses, dispersive separation of wavelengths by prisms, and total internal reflection, as in a prism or a fish tank. For EUV and x-ray radiation, where ω is greater than ω_s for many of the atomic electrons, the refractive index is less than unity, but only slightly so, indicating that x-rays propagate in materials at phase velocities somewhat greater than in vacuum (c). This gives rise to the interesting and important phenomenon of *total external reflection* of x-rays, whereby reflection occurs with little absorption at glancing incidence from material interfaces. Note that although the phase velocity can be greater than c for x-rays, the group velocity, which represents energy flow, is less than c. For further discussion of phase and group velocities, see Refs. 1–3.

As the refractive index for EUV and x-ray radiation differs only a small amount from unity, it is common[6] to write it in the following form:

$$n(\omega) = 1 - \delta + i\beta \tag{3.12}$$

where the choice of a positive sign for the β term is consistent with the form of exponential wave description used throughout these notes,[†] $\exp[-i(\omega t - \mathbf{k} \cdot \mathbf{r})]$. As we shall see shortly, this choice of sign for the imaginary term leads to an appropriate decay of wave intensity in a lossy medium.

Comparing Eqs. (3.9) and (3.12), we observe that

$$\delta = \frac{n_a r_e \lambda^2}{2\pi} f_1^0(\omega) \tag{3.13a}$$

$$\beta = \frac{n_a r_e \lambda^2}{2\pi} f_2^0(\omega) \tag{3.13b}$$

That these are small quantities can be verified by a simple example. We consider carbon with mass density $\rho = 2.26$ g/cm^3, and thus atomic density 1.13×10^{23} atoms/cm^3 (see the periodic table, Chapter 1, Table 1.2, and the inside rear cover). For carbon at a wavelength of 4 Å, $f_1^0 = 6.09$ and $f_2^0 = 0.071$ (see Appendix C). Thus with $r_e = 2.82 \times 10^{-13}$ cm [Eqs. (2.44) and (2.46)], we have $\delta = 4.90 \times 10^{-5}$ and $\beta = 5.71 \times 10^{-7}$, indeed much less than unity. For longer wavelengths or higher Z (atomic number on the periodic chart) materials, the values of δ and β will be larger, but still much less than unity. Generally, the λ^2 dependence dominates, while f_2^0 scales slowly with Z, somewhat less than linearly. Thus even for a high Z element such as gold, δ and β will have values of order 10^{-2} or less for soft x-ray wavelengths, still far less than unity, and of order 10^{-4} to 10^{-8} for hard x-rays. Values of $f_1^0(\omega)$ and $f_2^0(\omega)$ have been tabulated by Henke, Gullikson, and their colleagues[7–9] for all elements from hydrogen to uranium ($Z = 92$), and for photon energies extending from 30 eV to 30 keV (from 10 eV for f_2^0), by techniques we will return to at the end of this chapter. See Appendix C for representative values for several common elements. Currently updated data is maintained at the CXRO website by Gullikson.[8]

3.2   Phase Variation and Absorption of Propagating Waves

Having convenient relations for the refractive index [Eqs. (3.12) and (3.13a)] one can now describe phase variation and absorption for EUV and x-ray wave propagation. In Chapter 1 we considered absorption and transmission by thin foils in terms of a

---

[†] See the footnote following Eq. (1.26) in Chapter 1. The consistency of this choice is clarified in the algebra leading to Eq. (3.17) in Section 3.2 below.

so-called[‡] mass-dependent absorption coefficient, $\mu(\lambda)$. We now inquire as to how the absorption coefficient μ is related to the complex refractive index. We can answer this by considering a plane wave of the form

$$\mathbf{E}(\mathbf{r}, t) = \mathbf{E}_0 e^{-i(\omega t - \mathbf{k} \cdot \mathbf{r})} \tag{3.14}$$

propagating in some material with an initial amplitude $\mathbf{E}_0$ and having a complex dispersion relation given by

$$\boxed{\frac{\omega}{k} = \frac{c}{n} = \frac{c}{1 - \delta + i\beta}} \tag{3.15}$$

Solving for k, we obtain

$$k = \frac{\omega}{c}(1 - \delta + i\beta) \tag{3.16}$$

Substituting this into Eq. (3.14), in the propagation direction defined by $\mathbf{k} \cdot \mathbf{r} = kr$ one has

$$\mathbf{E}(\mathbf{r}, t) = \mathbf{E}_0 e^{-i[\omega t - (\omega/c)(1 - \delta + i\beta)r]}$$

or

$$\mathbf{E}(\mathbf{r}, t) = \underbrace{\mathbf{E}_0 e^{-i\omega(t - r/c)}}_{\text{vacuum propagation}} \underbrace{e^{-i(2\pi\delta/\lambda)r}}_{\phi\text{-shift}} \underbrace{e^{-(2\pi\beta/\lambda)r}}_{\text{decay}} \tag{3.17}$$

where the first exponential factor represents the phase advance had the wave been propagating in vacuum, the second factor (containing $2\pi\delta r/\lambda$) represents the modified phase shift due to the medium, and the factor containing $2\pi\beta r/\lambda$ represents decay of the wave amplitude.

To compute the intensity of the wave whose electric field is given by Eq. (3.17), we must first determine the associated magnetic field and take the cross product of $\mathbf{E}$ and $\mathbf{H}$ to obtain the Poynting vector $\mathbf{S}$. We follow the same general procedure as was used in Chapter 2 for a plane electromagnetic wave in vacuum, but now use the refractive index appropriate to propagation in a uniform, isotropic material. For a plane wave in any medium, we found in Chapter 2 [above Eq. (2.29)] that the field components are given by

$$i\mathbf{k} \times \mathbf{E}_{k\omega} = i\omega\mu_0 \mathbf{H}_{k\omega}$$

while for a material of refractive index n, according to Eq. (3.15),

$$\frac{\omega}{k} = \frac{c}{n}$$

[‡] The choice of name is not the best, as it is photons that are absorbed. The name is meant to differentiate this absorption coefficient from that defined in terms of an atomic density, that is, in terms of n_a rather than ρ.

Thus with $\mathbf{k} = k\mathbf{k}_0$, where $\mathbf{k}_0$ is a unit vector in the direction of propagation, the field components are related by

$$ik\mathbf{k}_0 \times \mathbf{E}_{k\omega} = i(kc/n)\mu_0\mathbf{H}_{k\omega}$$

or

$$\mathbf{H}_{k\omega} = \frac{n}{c\mu_0}\mathbf{k}_0 \times \mathbf{E}_{k\omega}$$

With $c \equiv 1/\sqrt{\epsilon_0\mu_0}$ and refractive index n varying slowly with frequency, the transformed fields in real space are given by

$$\mathbf{H}(\mathbf{r}, t) = n\sqrt{\frac{\epsilon_0}{\mu_0}}\mathbf{k}_0 \times \mathbf{E}(\mathbf{r}, t) \tag{3.18}$$

which is similar in form to Eq. (2.29), but more general in that Eq. (3.18) includes the effect of the refractive index n.

The scalar average intensity of a plane electromagnetic wave, $\bar{I}$, in units of power per unit area, is given for sinusoidal fields by the magnitude of the Poynting vector averaged over one period (as denoted by the bar). Following the discussion in Chapter 2 [Eq. (2.37)], the average intensity[1, 3] is

$$\bar{I} = |\bar{\mathbf{S}}| = \tfrac{1}{2}|\mathrm{Re}(\mathbf{E} \times \mathbf{H}^*)| \tag{3.19}$$

which for a plane wave in a medium of complex refractive index n, with use of Eq. (3.18), is given by

$$\bar{I} = \frac{1}{2}\mathrm{Re}(n)\sqrt{\frac{\epsilon_0}{\mu_0}}|\mathbf{E}_0|^2 \tag{3.20a}$$

where in Eqs. (3.19) and (3.20a) the field values are those at the peak of the cycle. Thus for the plane wave described by the electric field of Eq. (3.17), and with $\mathrm{Re}(n) = (1 - \delta)$, the average intensity is given by

$$\bar{I} = \frac{1}{2}(1 - \delta)\sqrt{\frac{\epsilon_0}{\mu_0}}|\mathbf{E}_0|^2 e^{-2(2\pi\beta/\lambda)r} \tag{3.20b}$$

which can be written completely in terms of the intensity $\bar{I}_0$ at some reference plane in the material (for instance just on the material side of an interface with vacuum), as

$$\bar{I} = \bar{I}_0 e^{-(4\pi\beta/\lambda)r} \tag{3.21}$$

that is, the wave decays with distance r into the material, with an exponential decay length

$$\boxed{l_{\mathrm{abs}} = \frac{\lambda}{4\pi\beta}} \tag{3.22}$$

where we recall that β is the absorptive portion of the complex refractive index, as seen in Eq. (3.12). Referring back to Eq. (3.13b), we can write the absorption length in terms of the imaginary portion of the complex atomic scattering coefficient, $f_2^0(\omega)$, as

$$l_{abs} = \frac{1}{2n_a r_e \lambda f_2^0(\omega)} \tag{3.23}$$

In Chapter 1 we considered experimentally observed absorption in thin foils, writing

$$\frac{\bar{I}}{I_0} = e^{-\rho \mu r} \tag{3.24}$$

where ρ is the mass density,[¶] μ is the linear absorption coefficient, and r is the foil thickness. Comparing Eqs. (3.21) to (3.24) shows that macroscopic (μ) and atomic (f_2^0) absorption factors are related by

$$\rho \mu = 2n_a r_e \lambda f_2^0(\omega)$$

Since the mass density ρ is related to the atomic density n_a by

$$\rho = m_a n_a = A m_u n_a \tag{3.25}$$

where m_a is the atomic mass, m_u is the atomic mass unit,[§] and A is the number of atomic mass units (as given in the periodic chart, Table 1.2), the macroscopic-to-atomic relationship can be written as

$$\mu = \frac{2r_e \lambda}{A m_u} f_2^0(\omega) \tag{3.26}$$

Thus we have a relationship between the macroscopically observed absorption of x-rays by thin foils, $\mu(\omega)$, and the absorptive portion of the atomic scattering factor for a single atom, f_2^0.

For some applications the absorption of radiation by thin films is expressed in terms of an atomic cross-section for absorption, σ_{abs}, through a relation similar to Eq. (3.24), but in terms of the atomic density n_a rather than the mass density ρ [see Eq. (1.3)]. In this case one expresses the absorption as

$$\frac{\bar{I}}{I_0} = e^{-n_a \sigma_{abs} r} \tag{3.27}$$

[¶] The choice of name is not the best, as it is photons that are absorbed. The name is meant to differentiate this absorption coefficient from that defined in terms of an atomic density, that is, in terms of n_a rather than ρ.

[§] The atomic mass unit is given in Appendix A.2 as $m_u = 1.66054 \times 10^{-24}$ g. The mass of one mole expressed in grams is $m = A m_u N_A$, where A is the atomic weight and N_A is Avogadro's number [Appendix A.2].

so that a comparison with Eqs. (3.21)–(3.23) gives an expression between the atomic absorption cross-section and β or f_2^0, viz.,

$$\sigma_{\text{abs}} = 2r_e\lambda f_2^0(\omega) \tag{3.28a}$$

or equivalently

$$\sigma_{\text{abs}} = Am_u\mu(\omega) \tag{3.28b}$$

An example showing the photon energy dependence of μ and σ_{abs} for copper atoms is given in Chapter 1, Figure 1.8.

Returning to Eq. (3.17), we note that we have written the electric field in terms of an initial value $\mathbf{E}_0$ multiplied by factors involving δ and β that take account of phase shifting and absorption in the medium. We see from Eq. (3.17) that the phase of the wave is shifted because of its propagation through the medium of atoms (n_a), and that the relative phase shift $\Delta\phi$, compared to propagation in vacuum, is given by

$$\Delta\phi = \left(\frac{2\pi\delta}{\lambda}\right)\Delta r \tag{3.29}$$

where Δr is the thickness or propagation distance. We will see in later chapters that this phase shift plays an important role in diffractive optics, multilayer mirrors, interferometry, and many other subjects.

The measurement of phase shift in a material is most directly accomplished with the use of an interferometer, an instrument in which a wavefront is dissected into parts, one of which is propagated through the object and one of which is propagated through a reference path, typically vacuum or air. The two waves are then recombined to form an interference pattern where differences in optical path ($\Delta\phi$) are manifested as localized shifts of the fringe pattern.[||] Interferometry is a widely used technique[1, 10] at visible and longer wavelengths, with many variations based on available optics, parameters of interest, and coherence properties of the radiation.

To introduce the general concept, Figure 3.3 illustrates an interferometer introduced in the 1880s by Mach and Zehnder to study airflow patterns with incoherent visible light. Knowing the material thickness Δr and measuring the phase shift $\Delta\phi$, one can deduce δ, the real part of the refractive index, through use of Eq. (3.29), and thus for our purposes find f_1^0 as a function of probe wavelength or photon energy. Note that the beamsplitters (BS) and mirrors (M) must be optically flat to a fraction of a wavelength across the aperture of the beam, which is particularly challenging at x-ray wavelengths. Nonetheless, through the clever use of cut monolithic crystals, Bonse and Hart[11] have demonstrated the use of interferometric techniques at x-ray wavelengths using natural

---

[||] The phrase *fringe pattern* refers to the alternating bright and dark bands caused by positive and negative interference between the two waves.

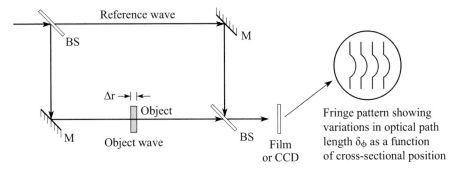

Figure 3.3 A Mach–Zehnder interferometer for measuring phase shift.

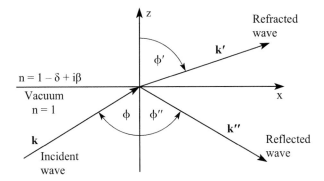

Figure 3.4 Interface geometry for incident, reflected, and refracted waves. The *plane of incidence* is defined as containing the incident wave vector **k** and the surface normal z_0.

crystals, which provide high reflection at the Bragg angle. Interferometry in the soft x-ray and EUV regions is more challenging because of the higher degree of absorption at these lower photon energies. However, sufficiently thin and flat substrates are being developed which are capable of splitting the beam without significantly affecting the reflected wavefront. Note that in the interferometer illustrated in Figure 3.3, the two wavefronts are recombined in a manner that maintains their original spatial orientation, thus requiring minimal coherence properties** for the formation of a high-contrast fringe pattern.

3.3   Reflection and Refraction at an Interface

To consider the phenomena of reflection and refraction we introduce Figure 3.4, which illustrates the incident, reflected, and refracted waves at a material interface. Note that all angles are measured from the surface normal (z-axis), and that propagation effects such

** The subjects of spatial and temporal coherence are discussed in Chapter 4.

as phase velocity and attenuation are contained in $\mathbf{k}$, the propagation vector, through the complex refractive index, e.g., as in Eq. (3.16):

$$k = |\mathbf{k}| = \frac{\omega}{c}(1 - \delta + i\beta)$$

where we assume that ω is real and that all waves have the same time dependence, $e^{-i\omega t}$, because they are driven by the incident wave. We consider a plane wave incident from the vacuum side and write the incident wave as

$$\mathbf{E} = \mathbf{E}_0 e^{-i(\omega t - \mathbf{k} \cdot \mathbf{r})} \tag{3.30a}$$

the *refracted wave* as

$$\mathbf{E}' = \mathbf{E}'_0 e^{-i(\omega t - \mathbf{k}' \cdot \mathbf{r})} \tag{3.30b}$$

and the *reflected wave* as

$$\mathbf{E}'' = \mathbf{E}''_0 e^{-i(\omega t - \mathbf{k}'' \cdot \mathbf{r})} \tag{3.30c}$$

where the subscript zero denotes the vector field amplitudes at the interface position $\mathbf{r} = 0$. Because the incident and reflected waves propagate in the same medium (vacuum), they experience the same refractive index ($n = 1$). Thus for the same oscillating frequency (ω), from Eq. (3.16), one can write

$$|\mathbf{k}| = |\mathbf{k}'| = \frac{\omega}{c} \tag{3.31}$$

At the interface ($z = 0$) where these three waves meet, the fields must obey certain boundary conditions[1-5] in order to satisfy Maxwell's equations.

Condition 1: Field components of $\mathbf{E}$ and $\mathbf{H}$ parallel to the interface must be continuous. For the geometry shown in Figure 3.4, where $\mathbf{z}_0$ is a unit vector normal to the interface, the boundary conditions in the absence of surface currents are

$$\mathbf{z}_0 \times (\mathbf{E}_0 + \mathbf{E}''_0) = \mathbf{z}_0 \times \mathbf{E}'_0 \tag{3.32a}$$

and

$$\mathbf{z}_0 \times (\mathbf{H}_0 + \mathbf{H}''_0) = \mathbf{z}_0 \times \mathbf{H}'_0 \tag{3.32b}$$

Condition 2: Field components of $\mathbf{D}$ and $\mathbf{B}$ normal (perpendicular) to the interface must be continuous. For the geometry in Figure 3.4 the boundary conditions in the absence of surface charge are

$$\mathbf{z}_0 \cdot (\mathbf{D}_0 + \mathbf{D}''_0) = \mathbf{z}_0 \cdot \mathbf{D}'_0 \tag{3.32c}$$

and

$$\mathbf{z}_0 \cdot (\mathbf{B}_0 + \mathbf{B}''_0) = \mathbf{z}_0 \cdot \mathbf{B}'_0 \tag{3.32d}$$

We consider the incident wave vector $\mathbf{k}$ to lie in the x, z-plane, and refer to this as the *plane of incidence*, as it contains both $\mathbf{k}$ and the surface normal $\mathbf{z}_0$. At this point we

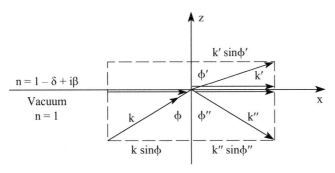

Figure 3.5 Parallel components of the wave vectors.

make no assumption as to the polarization of the incident wave, that is, to the direction of **E** with respect to this plane. If the parallel field components are to be continuous everywhere along the interface as required by Eqs. (3.32a) and (3.32b), then the phase and amplitude variations for all waves must be identical along the interface. This requires that the x-direction components of the wave vectors in Eqs. (3.30a–c) satisfy the condition

$$(\mathbf{k} \cdot \mathbf{x}_0 = \mathbf{k}' \cdot \mathbf{x}_0 = \mathbf{k}'' \cdot \mathbf{x}_0) \quad \text{at } z = 0 \tag{3.33}$$

Since **k** has no y-component by our orientation of axes, neither can $\mathbf{k}'$ or $\mathbf{k}''$. Thus all three vectors must lie in the plane of incidence defined by **k** and the surface normal direction $\mathbf{z}_0$. If the phase and amplitude factors are to match along the interface in the x-direction, all waves must have equal k_x components. In this way one ensures that the boundary condition is met everywhere along the interface if it is met at any one point. Following this argument Eq. (3.33) requires that

$$k_x = k'_x = k''_x \tag{3.34a}$$

or in terms of the angles shown in Figure 3.5,

$$k \sin \phi = k' \sin \phi' = k'' \sin \phi'' \tag{3.34b}$$

Since k and k'' propagate in vacuum, they are real and equal in magnitude as observed in Eq. (3.31); thus from (3.34b) we can solve for ϕ''

$$\sin \phi'' = \sin \phi \tag{3.35a}$$

or

$$\phi'' = \phi \tag{3.35b}$$

which states that *the angle of reflection equals the angle of incidence*. Considering the refracted wave $\mathbf{k}'$, Eq. (3.34b) permits us to write

$$k \sin \phi = k' \sin \phi' \tag{3.36}$$

Since both waves must oscillate at the same frequency (ω), we can write, by using Eq. (3.15), that

$$\omega = kc = k'c/n$$

or

$$k' = kn = \frac{\omega}{c}(1 - \delta + i\beta) \qquad (3.37)$$

indicating that the propagation vector in the medium is complex, representing both phase variation and amplitude decay as the wave propagates, as seen previously in Eqs. (3.16) and (3.17). Equation (3.36) can now be rewritten as *Snell's law* (1621):

$$\boxed{\sin \phi' = \frac{\sin \phi}{n}} \qquad (3.38)$$

which formally describes the refractive turning of a wave as it enters a uniform, isotropic medium of complex refractive index n. The fact that n is complex implies that $\sin \phi'$ is also complex for real incidence angle ϕ. Thus both the wave vector k' and the refracted angle ϕ', in the medium, have real and imaginary components, giving a somewhat more complicated representation of refraction and propagation.

Snell's law (3.38) is valid over a wide range of wavelengths and photon energies. It is widely used in lens designs at visible wavelengths, and to describe such interesting phenomena as total internal reflection of visible light within the denser medium at water–air and glass–air interfaces. We will use Snell's law here to describe the near-total reflection of short wavelength radiation at glancing incidence to a material surface.

3.4   Total External Reflection of X-Rays and EUV Radiation

For most angles of incidence the reflection coefficient for soft x-rays and extreme ultraviolet radiation is very small, as we will see in a following section of this chapter. This is because of the fact that the refractive index is very close to unity so that there is little change in field amplitudes across the interface. However, there is an important exception for radiation incident at a glancing angle to the material surface, far from the surface normal. We will see that in this case, radiation of any polarization experiences near total reflection. This *total external reflection* is widely used in experiments involving radiation transport, deflection, focusing, and filtering at x-ray and EUV wavelengths. Like its visible light counterpart of total internal reflection (commonly observed in fish tanks and swimming pools, and used for turning visible laser beams within glass prisms where the refractive index is greater than unity), the x-ray effect can be understood in large measure on the basis of Snell's law, Eq. (3.38). Snell's law indicates that visible light will be bent towards the surface normal ($\phi' < \phi$) when entering a medium of greater refractive index (n typically greater than 1.5 for glass or water at visible wavelengths). For EUV and x-rays, however, with the real part of the refractive index slightly less than unity, Snell's law indicates that the radiation is refracted in a direction slightly further from the surface normal. Inspection of Eq. (3.38) shows that for n slightly less than unity, $\sin \phi'$ is slightly larger than $\sin\phi$. Thus for near-glancing incidence (ϕ near $\pi/2$) the refraction angle ϕ' can equal $\pi/2$, indicating that to first order the refracted wave does not penetrate into the material, but rather propagates along the interface. In

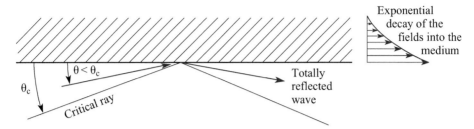

Figure 3.6 Glancing incidence radiation and total external reflection.

short order we will investigate the dependence of these fields on the parameters of the problem: the wavelength λ, incidence angle ϕ, and refractive index components δ and β. First, however, we consider the simplified problem with β approaching zero, which permits us to understand the basic phenomenon of total external reflection and quantify the critical angle with minimal mathematical complexity. The general effect is illustrated in Figure 3.6.

Considering Snell's law for a refractive index of $n \simeq 1 - \delta$, where for the moment we assume that β approaches zero, one has

$$\sin \phi' = \frac{\sin \phi}{1 - \delta} \tag{3.39}$$

Thus the refracted wave is at an angle ϕ', somewhat further from the surface normal than ϕ because of the $1 - \delta$ factor. As ϕ approaches $\pi/2$ it is evident that $\sin \phi'$ approaches unity somewhat faster. The limiting condition occurs at a *critical angle of incidence*, $\phi = \phi_c$, where $\phi' = \pi/2$, so that $\sin \phi' = 1$ and from Eq. (3.39)

$$\sin \phi_c = 1 - \delta \tag{3.40}$$

This is the condition for total external reflection; the incident x-rays do not penetrate the medium, but rather propagate along the interface at an angle $\phi' = \pi/2$. The angle for which this condition is just met is given by Eq. (3.40). Since $\delta \ll 1$ for x-rays, the phenomenon occurs only for glancing angles where ϕ is near 90°. Thus it is convenient to introduce the complementary angle θ, measured from the interface as shown in Figure 3.6, where

$$\theta + \phi = 90°$$

The critical angle condition (3.40) then becomes

$$\sin(90° - \theta_c) = 1 - \delta$$

or

$$\cos \theta_c = 1 - \delta$$

Since $\delta \ll 1$ for x-rays, $\cos \theta_c$ is near unity, θ_c is very small, and we may make the small angle approximation

$$1 - \frac{\theta_c^2}{2} + \cdots = 1 - \delta$$

which has the solution

$$\theta_c = \sqrt{2\delta}$$

(3.41)

as the *critical angle for total external reflection of x-rays and extreme ultraviolet radiation*, a result first obtained by Compton[6] in 1922.

Since the real part of the refractive index can be written as [Eq. (3.13a)]

$$\delta = \frac{n_a r_e \lambda^2 f_1^0(\lambda)}{2\pi}$$

we have, to first order,

$$\theta_c = \sqrt{2\delta} = \sqrt{\frac{n_a r_e \lambda^2 f_1^0(\lambda)}{\pi}}$$

(3.42a)

Because the atomic density n_a, in atoms per unit volume, varies only slowly among the natural elements, the major functional dependencies of the critical angle are

$$\theta_c \propto \lambda \sqrt{Z}$$

(3.42b)

where we have used the fact that to first order f_1^0 is approximated by Z, although as we have seen f_1^0 is also a complicated function of wavelength (photon energy) for each element. To obtain a conveniently large critical angle, Eq. (3.42b) suggests use of a relatively long wavelength and a higher Z material. We will see that other factors enter, such as the absorption β, specific absorption edges available with differing materials, and the availability of certain elements in convenient form for laboratory use. We will return to this subject, with illustrations for a variety of materials, later in this chapter.

As a specific example of the critical angle for total external reflection we consider a carbon mirror with incident radiation of 0.4 nm wavelength. We have previously observed, in the paragraph below Eqs. (3.13a, b), that for this case $\delta = 4.90 \times 10^{-5}$, and thus $\theta_c = \sqrt{2\delta} \simeq 10^{-2}$ rad, or about 0.6° from the surface. At longer wavelengths the critical angles are larger and at shorter wavelengths the critical angles are smaller, well below 1° for hard x-rays. Use of higher Z materials, such as nickel or a gold coating, increases the critical angle. Note that the smallness of these angles is not only inconvenient, but results in very small collection solid angles for use in imaging and spectroscopic experiments in EUV/x-ray astronomy and plasma diagnostics. The application of total external reflection to x-ray optics is discussed further in Chapter 10.

The above model of total external reflection is incomplete in that it does not include the effect of finite β. Since a portion of the field extends into the lossy medium, even if only in an evanescent manner, losses are incurred and total reflection is not achieved. In the following section we calculate the reflection coefficients for radiation incident on an interface at arbitrary angle of incidence for finite δ and β. In a later section we investigate further the nature of field penetration near the critical angle for finite β.

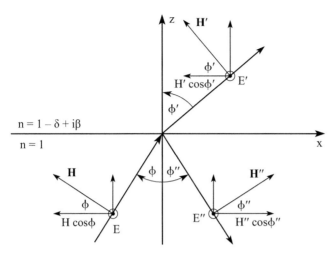

Figure 3.7 Field components for perpendicular (s) polarization, where **E** is polarized in the y-direction (into the paper as shown here), perpendicular to the x, z-plane of incidence.

3.5 Reflection Coefficients at an Interface

Returning to Figure 3.4, which shows the geometry of incident, reflected, and refracted waves at an interface, we use the various wave amplitudes given in Eqs. (3.30a–c) and the boundary conditions [Eq. (3.32a–d)] to determine the field amplitudes in both regions, and from these determine the reflectivity. Incident radiation of any polarization (linear, circular, elliptical) can be described in terms of two orthogonal polarizations with appropriate amplitudes and phase angle between the two. For the purpose of analysis, it is convenient to decompose incident radiation into two geometries, one with the incident electric field **E** perpendicular to the plane of incidence (containing **k** and $\mathbf{z}_0$), and one with **E** parallel to that plane. These orientations are often referred to as s- and p-polarizations, respectively, following the German words for perpendicular (*senkrecht*) and parallel (*parallele*). Any incident wave, polarized or not, can be represented in terms of these two polarizations. We can determine the refracted and reflected wave amplitudes at the interface, $\mathbf{E}_0'$ and $\mathbf{E}_0''$, by applying the boundary conditions given by Eqs. (3.32a) and (3.32b). We treat the two possible field orientations separately.

3.5.1 E_0 Perpendicular to the Plane of Incidence

For the case of the incident electric field polarized perpendicular to the plane of incidence, in the y-direction for our choice of axes (see Figure 3.7), application of Eq. (3.32a) is relatively simple. The scalar field amplitudes at the interface (subscript zero) must satisfy the condition

$$E_0 + E_0'' = E_0' \tag{3.43}$$

while continuity of the magnetic field [Eq. 3.32b], upon inspection of Figure 3.7, requires that at the interface

$$H_0 \cos\phi - H_0'' \cos\phi = H_0' \cos\phi' \tag{3.44}$$

Equations (3.43) and (3.44) can be combined to solve for the fields by recalling that for plane waves propagating in a medium of refractive index n, **E**, and **H** are related by Eq. (3.18):

$$\mathbf{H}(\mathbf{r}, t) = n\sqrt{\frac{\epsilon_0}{\mu_0}}\mathbf{k}_0 \times \mathbf{E}(\mathbf{r}, t)$$

or, more conveniently, the amplitudes are related by

$$H = n\sqrt{\frac{\epsilon_0}{\mu_0}}E$$

where the field orientations are described by Eq. (3.18) with $\mathbf{E}_0 = E_0\mathbf{y}_0$, $\mathbf{E}_0' = E_0'\mathbf{y}_0$, and $\mathbf{E}_0'' = E_0''\mathbf{y}_0$. Equation (3.44) can now be expressed, after eliminating magnetic fields, as

$$\sqrt{\frac{\epsilon_0}{\mu_0}}E_0 \cos\phi - \sqrt{\frac{\epsilon_0}{\mu_0}}E_0'' \cos\phi = n\sqrt{\frac{\epsilon_0}{\mu_0}}E_0' \cos\phi'$$

or

$$(E_0 - E_0'') \cos\phi = nE_0' \cos\phi' \tag{3.45}$$

From Snell's law [Eq. (3.38)] we know the relation between ϕ and ϕ':

$$\sin\phi' = \frac{\sin\phi}{n}$$

Thus we have three equations [(3.38), (3.43), and (3.45)] in five unknowns (E_0, E_0', E_0'', ϕ, and ϕ'), which can be solved by treating two (E_0 and ϕ) as independent variables describing the incident radiation. We can now solve for the refracted and reflected field amplitudes, E_0' and E_0'', in terms of the incident wave parameters E_0 and ϕ.

Combining Eqs. (3.43) and (3.45), we have

$$(E_0 - E_0'') \cos\phi = n(E_0 + E_0'') \cos\phi'$$

Combining terms according to incident and reflected fields, one has

$$(\cos\phi - n\cos\phi')E_0 = (\cos\phi + n\cos\phi')E_0''$$

so that the ratio of field amplitudes is

$$\frac{E_0''}{E_0} = \frac{\cos\phi - n\cos\phi'}{\cos\phi + n\cos\phi'}$$

From Snell's law,

$$\cos\phi' = \sqrt{1 - \sin^2\phi'} = \sqrt{1 - \frac{\sin^2\phi}{n^2}}$$

or

$$n \cos \phi' = \sqrt{n^2 - \sin^2 \phi}$$

The ratio of field amplitudes is then

$$\frac{E_0''}{E_0} = \frac{\cos \phi - \sqrt{n^2 - \sin^2 \phi}}{\cos \phi + \sqrt{n^2 - \sin^2 \phi}} \tag{3.46}$$

The refracted field E_0' can then be determined from Eq. (3.43):

$$\frac{E_0'}{E_0} = 1 + \frac{E_0''}{E_0} = 1 + \frac{\cos \phi - \sqrt{n^2 - \sin^2 \phi}}{\cos \phi + \sqrt{n^2 - \sin^2 \phi}}$$

or

$$\frac{E_0'}{E_0} = \frac{2 \cos \phi}{\cos \phi + \sqrt{n^2 - \sin^2 \phi}} \tag{3.47}$$

Thus we have the refracted and reflected field amplitudes for the case of perpendicular polarization. The reflectivity R, defined as the ratio of reflected to incident intensity (at the surface), is determined, with the use of Eq. (3.19), to be

$$R = \frac{\bar{I}''}{\bar{I}''} \equiv \frac{|\bar{\mathbf{S}}''|}{|\bar{\mathbf{S}}|} = \frac{\frac{1}{2} \mathrm{Re} \left(\mathbf{E}_0'' \times \mathbf{H}_0''^* \right)}{\frac{1}{2} \mathrm{Re} \left(\mathbf{E}_0 \times \mathbf{H}_0^* \right)} \tag{3.48}$$

With $n = 1$ for both incident and reflected waves,

$$R = \frac{|E_0''|^2}{|E_0|^2}$$

which with Eq. (3.46) becomes, for the case of perpendicular (s) polarization,

$$R_s = \frac{\left| \cos \phi - \sqrt{n^2 - \sin^2 \phi} \right|^2}{\left| \cos \phi + \sqrt{n^2 - \sin^2 \phi} \right|^2} \tag{3.49}$$

where n is complex. Knowing the values of δ and β as functions of photon energy, one can now calculate the reflectivity, at an arbitrary angle of incidence, for s-polarized radiation. Alternatively, angle-dependent reflectivity can be used to experimentally determine values of δ and β, that is, determine the refractive index at short wavelengths. Soufli and Gullikson[9] have used this technique, with tunable synchrotron radiation and specially prepared (smooth, unoxidized) surfaces, to measure the optical constants (δ and β) of silicon in the photon energy range from 50 eV to 180 eV, extending above and below the L-edges.

Two cases of particular interest are normal incidence reflection ($\phi = 0$) and glancing incidence reflection ($\phi \geq \phi_c$, $\theta \leq \theta_c$). For normal incidence ($\phi = 0$) one has

$$R_{s,\perp} = \frac{|1 - n|^2}{|1 + n|^2} = \frac{(1 - n)(1 - n^*)}{(1 + n)(1 + n^*)}$$

For $n = 1 - \delta + i\beta$

$$R_{s,\perp} = \frac{(\delta - i\beta)(\delta + i\beta)}{(2 - \delta + i\beta)(2 - \delta - i\beta)} = \frac{\delta^2 + \beta^2}{(2 - \delta)^2 + \beta^2}$$

which for $\delta \ll 1$ and $\beta \ll 1$ gives the *reflectivity for x-rays and EUV radiation at normal incidence* ($\phi = 0$) as

$$\boxed{R_{s,\perp} \simeq \frac{\delta^2 + \beta^2}{4}} \qquad (3.50)$$

which shows that the reflection is indeed very small for x-rays incident normally on a single interface. Similar results follow from Eq. (3.49) for all angles except those at glancing incidence.

The case of glancing incidence at or below the critical angle ($\theta \leq \theta_c$) can also be considered by examining Eq. (3.49). Using the definitions

$$\theta = 90° - \phi \leq \theta_c$$

where

$$\theta_c = \sqrt{2\delta} \ll 1$$

and where for glancing incidence

$$\cos\phi = \sin\theta \simeq \theta$$

we have to a high degree of accuracy

$$\sin^2 \phi = 1 - \cos^2 \phi = 1 - \sin^2 \theta \simeq 1 - \theta^2$$

Noting further that for $n = 1 - \delta + i\beta$

$$n^2 = (1 - \delta)^2 + 2i\beta(1 - \delta) - \beta^2$$

we see that the reflectivity [Eq. (3.49)] for glancing incidence of perpendicularly polarized x-rays can be written to a good approximation as

$$R_{s,\theta} = \frac{\left|\theta - \sqrt{(1 - \delta)^2 - \beta^2 + 2i\beta(1 - \delta) - (1 - \theta^2)}\right|^2}{\left|\theta + \sqrt{(1 - \delta)^2 - \beta^2 + 2i\beta(1 - \delta) - (1 - \theta^2)}\right|^2} \qquad (\theta \ll 1)$$

Dropping second-order terms of order δ^2, β^2, $\delta\beta$,

$$R_{s,\theta} = \frac{\left|\theta - \sqrt{\theta^2 - 2\delta + 2i\beta}\right|^2}{\left|\theta + \sqrt{\theta^2 - 2\delta + 2i\beta}\right|^2}$$

Recalling that $\theta_c = \sqrt{2\delta}$,

$$R_{s,\theta} = \frac{\left|\theta - \sqrt{(\theta^2 - \theta_c^2) + 2i\beta}\right|^2}{\left|\theta + \sqrt{(\theta^2 - \theta_c^2) + 2i\beta}\right|^2} \qquad (\theta \ll 1) \tag{3.51a}$$

Absolute values can be taken to obtain a purely real coefficient of reflection by first expressing the square root of the complex quantity in terms of a sum of real and imaginary components. This is accomplished by introducing the quantity $\sqrt{a + ib}$ such that

$$R_{s,\theta} = \frac{\left|\theta - \sqrt{a + ib}\right|^2}{\left|\theta + \sqrt{a + ib}\right|^2} \tag{3.51b}$$

where

$$a = \theta^2 - \theta_c^2 \tag{3.51c}$$

$$b = 2\beta \tag{3.51d}$$

and from Appendix D

$$\sqrt{a + ib} = \frac{1}{\sqrt{2}}\left[\sqrt{(a^2 + b^2)^{1/2} + a} + i\sqrt{(a^2 + b^2)^{1/2} - a}\right] = A + iB \tag{3.51e}$$

so that complex conjugates are readily identified and the reflectivity for glancing incidence can be written relatively simply, following Compton and Allison,[6] and Parratt,[12] as

$$R_{s,\theta} = \left|\frac{\theta - (A + iB)}{\theta + (A + iB)}\right|^2 = \frac{(\theta - A)^2 + B^2}{(\theta + A)^2 + B^2} \tag{3.52a}$$

where now

$$A = \sqrt{\frac{(a^2 + b^2)^{1/2} + a}{2}} \tag{3.52b}$$

$$B = \sqrt{\frac{(a^2 + b^2)^{1/2} - a}{2}} \tag{3.52c}$$

And where

$$a = \theta^2 - \theta_c^2 = \theta^2 - 2\delta \tag{3.52d}$$

$$b = 2\beta \tag{3.52e}$$

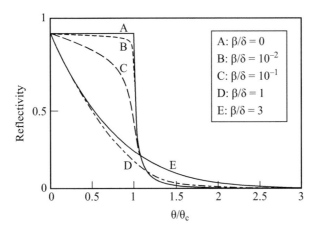

Figure 3.8 Reflectivity curves from Eq. (3.52a) as a function of the parameter β/δ for radiation incident from vacuum upon an idealized material interface. Finite absorption (β) causes a rounding of the otherwise sharp angular dependence of reflectance at the critical angle. These results apply to both perpendicular (s) and parallel (p) polarization.

which, although convenient in form, reveals a somewhat complicated dependence of the reflection coefficient on θ, δ, and β near glancing incidence. Numerical solutions are shown in Figure 3.8 for various values of the parameter β/δ. Analytic expressions are readily obtained for two special cases. Just at the critical angle $\theta = \theta_c$, one has $a = 0$ and one obtains

$$R_{s,\theta_c} = \frac{1 - \frac{\sqrt{2\delta\beta}}{\delta+\beta}}{1 + \frac{\sqrt{2\delta\beta}}{\delta+\beta}} \tag{3.53}$$

which is unity for $\beta/\delta = 0$, 0.20 for $\beta/\delta = \frac{1}{2}$, and 0.17 for $\beta/\delta = 1$. For $\theta = 0$, Eq. (3.52a) is unity for all values of δ and β.

3.5.2 E_0 Parallel to the Plane of Incidence

The second decomposition of incident polarization is that in which the electric field vector lies in the x, z-plane of incidence, as shown in Figure 3.9, and is referred to as parallel (p) polarization (in both English and German). Applying the boundary conditions as in the previous case, one now obtains

$$\frac{E_0''}{E_0} = \frac{n\cos\phi - \cos\phi'}{n\cos\phi + \cos\phi'}$$

Noting that

$$\cos\phi' = \sqrt{1 - \sin^2\phi'} = \sqrt{1 - \frac{\sin^2\phi}{n^2}}$$

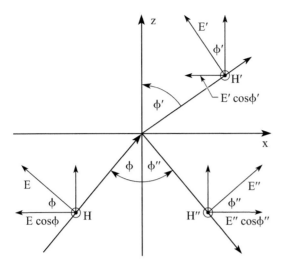

Figure 3.9 Field components for parallel (p) polarization, where **E** lies in the x, z-plane of incidence.

this becomes

$$\frac{E_0''}{E_0} = \frac{n\cos\phi - \frac{1}{n}\sqrt{n^2 - \sin^2\phi}}{n\cos\phi + \frac{1}{n}\sqrt{n^2 - \sin^2\phi}}$$

or

$$\frac{E_0''}{E_0} = \frac{n^2\cos\phi - \sqrt{n^2 - \sin^2\phi}}{n^2\cos\phi + \sqrt{n^2 - \sin^2\phi}} \tag{3.54}$$

for the reflected field amplitude. The refracted electric field is then determined from the boundary condition

$$nE_0' = E_0 + E_0''$$

to be

$$\frac{E_0'}{E_0} = \frac{1}{n}\left[1 + \frac{n^2\cos\phi - \sqrt{n^2 - \sin^2\phi}}{n^2\cos\phi + \sqrt{n^2 - \sin^2\phi}}\right]$$

or

$$\frac{E_0'}{E_0} = \frac{2n\cos\phi}{n^2\cos\phi + \sqrt{n^2 - \sin^2\phi}} \tag{3.55}$$

which we note are similar in form, but slightly different than the equations for reflected and refracted fields in the case of perpendicular polarization [Eqs. (3.46) and (3.47)].

The reflectivity for parallel polarization is determined from Eq. (3.54) to be

$$R_p = \left| \frac{E_0''}{E_0} \right|^2 = \frac{\left| n^2 \cos\phi - \sqrt{n^2 - \sin^2\phi} \right|^2}{\left| n^2 \cos\phi + \sqrt{n^2 - \sin^2\phi} \right|^2} \qquad (3.56)$$

which we note is different than Eq. (3.49) for the case of perpendicular polarization.

It is interesting that in several important special cases both polarizations give the same result. For normal incidence, Eq. (3.56) for p-polarization reduces to

$$R_{p,\perp} = \frac{(n-1)(n^*-1)}{(n+1)(n^*+1)} \simeq \frac{\delta^2 + \beta^2}{4} \qquad (3.57)$$

which is identical to Eq. (3.50) for s-polarization at $\phi = 0$, as it should be, since the two polarizations are physically indistinguishable at normal incidence. For glancing incidence Eq. (3.56) also reduces to the same result obtained previously for perpendicular polarization, given as Eq. (3.52) and plotted in Figure 3.8.

While Figure 3.8 is very instructive, showing the effect of finite β on the shape of an idealized critical angle reflection ($\beta = 0$), more practical results can also be obtained through the use of Eq. (3.52) for real materials at various photon energies. For instance, in laboratory studies it is often interesting to know the reflectivity vs. photon energy for a given mirror material at various angles of incidence near the critical angle.

To calculate such curves one must know the values of δ and β for the material of interest, across the photon energies of interest, and use them in Eq. (3.52a) for the incident angles relevant to the experiment. Values of δ and β have been tabulated by Henke, Gullikson, and colleagues[7–9]; examples are given in Appendix C for selected materials. The resultant reflectivity curves are interesting in that they relate to real materials and include the effects of absorption edges, which can be used to enhance angular cutoffs at a given photon energy, and also can include the effects of oxidation and other multi-element effects. Figure 3.10 illustrates glancing incidence reflection curves for carbon, aluminum, aluminum oxide, and gold. Results for other materials are given in Chapter 10 and Refs. 7 and 8.

As the reflectivity curves show, mirrors at glancing incidence do not reflect well at higher photon energies, and thus can be used as *low pass filters*. This contrasts with thin transmitting foils, which are highly absorptive at low photon energies and thus serve as *high pass filters*. In combination, a mirror–filter pair can provide a moderate resolution *notch filter*, which blocks low- and high-energy photons, but passes a central pass band of relative spectral bandwidth $E/\Delta E$ of 3–5, depending on the degree to which absorption edges can be used to enhance the sharpness of reflection or transmission curves. Figure 3.11 illustrates the idea of a notch filter based on an idealized mirror whose cutoff angle is matched to the K-absorption edge of a transmission filter. Such mirror–filter pairs are quite convenient to use, and are often utilized in imaging and transport applications involving broadband sources of radiation.

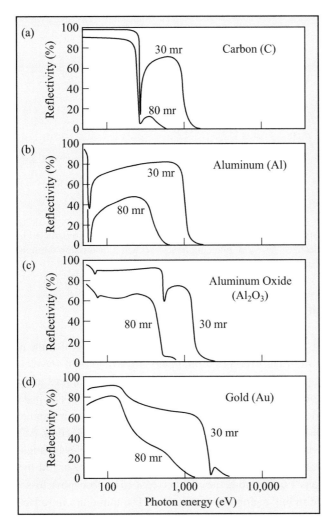

Figure 3.10 Glancing incidence reflectivity vs. photon energy for incidence angles (θ) of 30 mrad and 80 mrad, for materials of (a) carbon, (b) aluminum, (c) aluminum oxide, and (d) gold. The results follow from Eq. (3.52a), for both parallel and perpendicular polarization, as a function of θ. Values of δ and β are from the tabulations of Henke *et al.*[7] and Gullikson[8]. Note how the combination of cutoff angle and absorption edge can be used to enhance the sharpness of the reflectivity curve, which is useful when the material is used as a low pass filter. Recall that 1 mrad $= 0.0573° = 3.44$ arcmin (see Appendix A).

Another, more specialized option for spectral filtering involves the failure of total external reflectance in the vicinity of certain absorption edges. For EUV radiation there is strong anomalous dispersion near the $L_{2,3}$-edges of the third period elements, such as Al and Si, which results in a sign reversal of f_1^0, and thus δ. This results in a sharp transmission window, making it possible to exploit the effect to create narrow spectral bandwidth filters.[13]

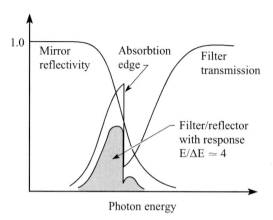

Figure 3.11 A moderate resolution spectral bandpass, or *notch filter*, is constructed by sequential reflection from an idealized mirror of glancing incidence, and transmission through a foil whose absorption edge is matched to the decline in mirror reflectivity.

3.6   Brewster's Angle

An effect unique to parallel polarization occurs at an angle for which the numerator of Eq. (3.56) is zero, or at least reaches a minimum. At visible light wavelengths, where β/δ is extremely small, this phenomenon has important applications and is known as Brewster's angle, or the polarizing angle (see Refs. 1, 4, and 5). For instance, visible light laser rods often have their ends cut at Brewster's angle to minimize intracavity reflective losses and to provide a mechanism for polarization selection. At EUV and x-ray wavelengths, where reflectivities are already very small at relevant angles (near 45°, as we shall see shortly), the effect is less dramatic but still very useful in polarization sensitive applications. Furthermore, the effect is reduced by absorptive losses (β), which are more important in this region of the spectrum. The minimum in reflectivity occurs when the numerator in Eq. (3.56) satisfies the condition

$$n^2 \cos \phi_B = \sqrt{n^2 - \sin^2 \phi_B} \qquad (3.58)$$

Squaring both sides, collecting like terms involving ϕ_B, and factoring, one has

$$n^2(n^2 - 1) = (n^4 - 1) \sin^2 \phi_B$$

or

$$\sin \phi_B = \frac{n}{\sqrt{n^2 + 1}}$$

This permits us to construct the diagram in Figure 3.12, from which we see that the

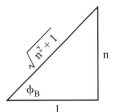

Figure 3.12 A diagram for understanding Brewster's angle.

condition for a minimum in the reflection coefficient, for parallel polarized radiation, occurs at an angle given by[††]

$$\tan \phi_B = n \tag{3.59}$$

Since n is complex, Eq. (3.59) does not yield a real angle ϕ_B for which $R_p(\phi)$ would be zero; rather, a minimum is achieved. From Eq. (3.56) we expand the reflectivity $R_p(\phi)$ in the small parameters δ and β, then find the minimum by setting the derivative with respect to ϕ to zero. Doing this, we find that the Brewster minimum occurs at a real angle of incidence given by

$$\tan \phi_B = 1 - \delta$$

which for $\delta \ll 1$ corresponds to an angle of incidence ϕ_B slightly less than $45°$ from the surface normal. Taking a Taylor expansion of $\tan \phi_B$ about $\pi/4$, one finds that Brewster's angle, or, if one prefers, the polarizing angle, is given by

$$\boxed{\phi_B \simeq \frac{\pi}{4} - \frac{\delta}{2}} \tag{3.60}$$

This result has an interesting physical interpretation. At Brewster's angle the refracted wave is turned in just such a manner that the refracted wave vector $\mathbf{k}'$ is at a right angle to the reflected wave vector $\mathbf{k}''$, as illustrated in Figure 3.13. Note that with parallel polarization the refracted electric field $\mathbf{E}'$ is then coaligned with the reflected wave vector $\mathbf{k}''$ – a condition in which it cannot generate a reflected wave. In this situation the atoms at the interface respond to the impressed field $\mathbf{E}_0'$, oscillating in a direction parallel to $\mathbf{k}''$, each radiating a $\sin^2 \theta$ pattern which is zero in the reflected direction and thus producing no reflected field component E_0''. Figure 3.14 shows an example of reflection coefficients for both parallel and perpendicular polarized radiation incident on a tungsten surface at a wavelength of 4.48 nm.[14] Note the sharp reflectivity dip at an incidence angle just below $45°$ for parallel polarization (dashed line).

[††] For visible and near-IR light, where solid state laser host materials typically have a refractive index near $n = 1.5$, Brewster's angle is about $56°$ from the surface normal, as often seen as the end cut of laser rods.

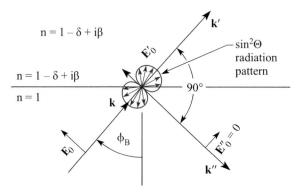

Figure 3.13 The special angle, Brewster's angle, ϕ_B, that results in minimal or no refracted wave for parallel polarization when the refracted wave direction is aligned with the null in the dipole scattering pattern.

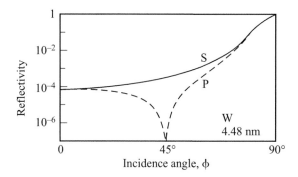

Figure 3.14 Reflectivity versus angle for parallel (dashed line) and perpendicular polarized radiation (solid line) incident on a tungsten surface at a wavelength of 4.48 nm (from J. H. Underwood[14]).

3.7   Field Penetration into a Lossy Medium Near the Critical Angle

In an earlier section we considered the nature of the refracted wave incident upon a lossless medium ($\beta = 0$) at glancing incidence. It was determined that for angles of incidence θ less than a critical angle $\theta_c = \sqrt{2\delta}$, the refracted wave propagated along the interface, with no energy flow into the material. However, in the preceding section where we considered reflectivity from a material with finite δ and β, we found that even at glancing incidence finite values of β have a significant effect on reflectivity, as seen in Eq. (3.52) and Figure 3.8. These results raise questions as to the nature of the wave at the interface, the penetration depth of the fields, and the flow of power across the interface when finite losses are considered. Because of finite absorptive losses in the medium, there must be some energy flow into the medium, even for $\theta < \theta_c$. This is different from the idealized case ($\beta = 0$) considered earlier, in which the refracted wave was found to propagate at a real angle $\phi' = \pi/2$, just along and parallel to the surface. Now with finite β, we must permit the refracted wave vector $\mathbf{k}'$ to have a finite real component $\mathbf{k}'_{zr}$

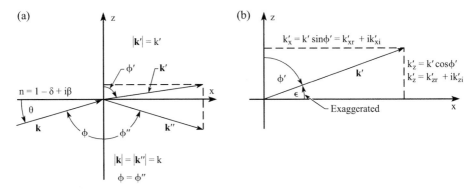

Figure 3.15 (a) Wave vectors for radiation incident on a lossy medium near the critical angle. (b) Real and imaginary components of the refracted wave vector for a complex angle ϕ'. The very small angle ϵ is exaggerated for clarity of the vector components.

into the medium, as illustrated in Figure 3.15. Thus we expect that even for $\theta < \theta_c$, ϕ' will have a solution slightly less than $\pi/2$.

We start our analysis, as we did for the ideal case ($\beta = 0$), with Snell's law [Eq. (3.38)], but now for a complex refractive index

$$\sin\phi' = \frac{\sin\phi}{n}$$

where

$$n = 1 - \delta + i\beta$$

Since n is complex, ϕ' will be complex for real angle of incidence ϕ. Since δ and β are very small for x-rays, we may write

$$\sin\phi' = (1 + \delta - i\beta)\sin\phi \qquad (3.61)$$

and assume

$$\phi' = \phi'_r - i\phi'_i \qquad (3.62)$$

where we have taken a negative imaginary component for consistency with the sign of β in Eq. (3.61). Upon inspection of Figure 3.15 we see that the propagation vector $\mathbf{k}'$ in the medium has components

$$k'_x = k'\sin\phi'$$

and

$$k'_z = k'\cos\phi'$$

where from Eq. (3.37)

$$k' = \frac{\omega}{c}(1 - \delta + i\beta)$$

so that

$$k'_x = \frac{\omega}{c}(1 - \delta + i\beta)\sin\phi' \tag{3.63a}$$

and

$$k'_z = \frac{\omega}{c}(1 - \delta + i\beta)\cos\phi' \tag{3.63b}$$

Recognizing that k'_x and k'_z are in general complex, we may rewrite the above in terms of real and imaginary components

$$k'_{xr} + ik'_{xi} = \frac{\omega}{c}(1 - \delta + i\beta)\sin\phi' \tag{3.64a}$$

and

$$k'_{zr} + ik'_{zi} = \frac{\omega}{c}(1 - \delta + i\beta)\cos\phi' \tag{3.64b}$$

For a complex angle $\phi' = \phi'_r - i\phi'_i$, as in Eq. (3.62), one has the trigonometric identities (see Appendix D.5)

$$\sin\phi' = \sin\phi'_r \cosh\phi'_i - i\cos\phi'_r \sinh\phi'_i \tag{3.65a}$$

and

$$\cos\phi' = \cos\phi'_r \cosh\phi'_i + i\sin\phi'_r \sinh\phi'_i \tag{3.65b}$$

The real and imaginary parts of the wave vector components can now be identified by combining Eqs. (3.64a, b) and (3.65a, b), viz.,

$$k'_{xr} + ik'_{xi} = \frac{\omega}{c}(1 - \delta + i\beta)[\sin\phi'_r \cosh\phi'_i - i\cos\phi'_r \sinh\phi'_i] \tag{3.66a}$$

and

$$k'_{zr} + ik'_{zi} = \frac{\omega}{c}(1 - \delta + i\beta)[\cos\phi'_r \cosh\phi'_i + i\sin\phi'_r \sinh\phi'_i] \tag{3.66b}$$

For $\beta \ll 1$, so that $\phi'_i \ll 1$, one can make the approximations

$$\sinh\phi'_i \simeq \phi'_i \tag{3.67a}$$

and

$$\cosh\phi'_i \simeq 1 + \frac{\phi'^2_i}{2} \tag{3.67b}$$

For angles ϕ near the critical angle we can again take the approximation

$$\sin\phi = \cos\theta \simeq 1 - \frac{\theta^2}{2} \tag{3.67c}$$

For angles near critical, with a refracted wave propagating very close to the interface of a lossy medium, we can take the solution for ϕ' in which the real part is very close to but slightly less than $\pi/2$, such that

$$\phi'_r = \frac{\pi}{2} - \epsilon, \quad \epsilon \ll 1 \qquad (3.68)$$

This permits the refracted wave to propagate very close to the interface, but with a small real component of k'_z so that energy can propagate across the interface and thus maintain steady state fields in the presence of finite absorptive losses. In this limit we can make the additional approximations

$$\sin \phi'_r = \cos \epsilon \simeq 1 - \frac{\epsilon^2}{2} \qquad (3.69a)$$

and

$$\cos \phi'_r = \sin \epsilon \simeq \epsilon \qquad (3.69b)$$

Collecting these various angular approximations [Eqs. (3.67a–c), (3.68), and (3.69a,b)] the expressions for real and imaginary wave vector components in the lossy medium [Eqs. (3.66a) and (3.66b)] become

$$k'_{xr} + ik'_{xi} = \frac{\omega}{c}(1 - \delta + i\beta)\left[\left(1 - \frac{\epsilon^2}{2}\right)\left(1 + \frac{\phi'^2_i}{2}\right) - i\epsilon\phi'_i\right] \qquad (3.70a)$$

and

$$k'_{zr} + ik'_{zi} = \frac{\omega}{c}(1 - \delta + i\beta)\left[\epsilon\left(1 + \frac{\phi'^2_i}{2}\right) + i\left(1 - \frac{\epsilon^2}{2}\right)\phi'_i\right] \qquad (3.70b)$$

Thus for k'_x, along the interface,

$$k'_{xr} = \frac{\omega}{c}(1 - \delta)\left(1 - \frac{\epsilon^2}{2}\right)\left(1 + \frac{\phi'^2_i}{2}\right) + \frac{\omega}{c}\beta\epsilon\phi'_i \qquad (3.71a)$$

and

$$k'_{xi} = \frac{\omega}{c}\beta\left(1 - \frac{\epsilon^2}{2}\right)\left(1 + \frac{\phi'^2_i}{2}\right) - \frac{\omega}{c}(1 - \delta)\epsilon\phi'_i \qquad (3.71b)$$

and for k'_z, in a direction perpendicular to the interface,

$$k'_{zr} = \frac{\omega}{c}(1 - \delta)\epsilon\left(1 + \frac{\phi'^2_i}{2}\right) - \frac{\omega}{c}\beta\left(1 - \frac{\epsilon^2}{2}\right)\phi'_i \qquad (3.72a)$$

and

$$k'_{zi} = \frac{\omega}{c}\beta\epsilon\left(1 + \frac{\phi'^2_i}{2}\right) + \frac{\omega}{c}(1 - \delta)\left(1 - \frac{\epsilon^2}{2}\right)\phi'_i \qquad (3.72b)$$

If we now apply the boundary condition $k'_x = k_x$, ensuring continuous transverse field components along the interface [Eq. (3.34a)], then we must set $k'_{xr} = k_{xr} = (\omega/c)\left(1 - \theta^2/2\right)$ in Eq. (3.71a) and $k'_{xi} = 0$ in Eq. (3.71b). From the latter we obtain

$$\beta\left(1 - \frac{\epsilon^2}{2}\right)\left(1 + \frac{\phi'^2_i}{2}\right) = (1 - \delta)\,\epsilon\phi'_i$$

Dropping second-order terms ϵ^2 and ϕ'^2_i as being small, and noting that $\delta \ll 1$, we have to first order

$$\phi'_i \simeq \frac{\beta}{\epsilon} \tag{3.73}$$

while the condition on k'_{xr} gives

$$\frac{\omega}{c}\left(1 - \frac{\theta^2}{2}\right) \simeq \frac{\omega}{c}(1 - \delta)\left(1 - \frac{\epsilon^2}{2}\right)\left(1 + \frac{\phi'^2_i}{2}\right) + \frac{\omega}{c}\beta\epsilon\phi'_i$$

$$\left(1 - \frac{\theta^2}{2}\right) \simeq (1 - \delta)\left(1 - \frac{\epsilon^2}{2}\right)\left(1 + \frac{\beta^2}{2\epsilon^2}\right) + \beta^2$$

Thus to first order, with $\beta^2 \ll \delta$,

$$\left(1 + \delta - \frac{\theta^2}{2}\right) \simeq \left(1 - \frac{\epsilon^2}{2}\right)\left(1 + \frac{\beta^2}{2\epsilon^2}\right)$$

$$\delta - \frac{\theta^2}{2} \simeq \frac{\epsilon^2}{2} + \frac{\beta^2}{2\epsilon^2}$$

which we can write in quadratic form

$$\epsilon^4 - (\theta^2 - 2\delta)\epsilon^2 - \beta^2 = 0 \tag{3.74}$$

which has solution for the (real) angular decrement ϵ (from $\pi/2$)

$$\epsilon^2 = \frac{\sqrt{\left(\theta^2 - 2\delta\right)^2 + 4\beta^2} + (\theta^2 - 2\delta)}{2} \tag{3.75}$$

and where we have taken the positive square root to ensure a real solution for the angle ϵ.

Thus general solutions for the complex refraction angle ϕ', at glancing incidence, are obtained from Eqs. (3.68) and (3.73) to be

$$\phi'_r = \frac{\pi}{2} - \epsilon = \frac{\pi}{2} - \left[\frac{\sqrt{\left(\theta^2 - 2\delta\right)^2 + 4\beta^2} + (\theta^2 - 2\delta)}{2}\right]^{1/2} \tag{3.76a}$$

$$\phi'_i = \frac{\beta}{\epsilon} = \frac{\sqrt{2}\beta}{\left[\sqrt{\left(\theta^2 - 2\delta\right)^2 + 4\beta^2} + (\theta^2 - 2\delta)\right]^{1/2}} \tag{3.76b}$$

We can now determine the various wave vector components for a refracted wave near the critical angle in a lossy medium. With the approximations θ^2, δ, and β all much less than unity, the wave vector components in the medium, Eqs. (3.71a,b) and (3.72a,b), reduce to

$$k'_{xr} \simeq \frac{\omega}{c}\left(1 - \frac{\theta^2}{2}\right) \tag{3.77a}$$

$$k'_{xi} = 0 \tag{3.77b}$$

$$k'_{zr} \simeq \frac{\omega}{c}\epsilon \tag{3.77c}$$

$$k'_{zi} \simeq \frac{\omega}{c}\frac{\beta}{\epsilon} \tag{3.77d}$$

The first two are set by the boundary conditions. The third, Eq. (3.77c), simply states that the real wave vector, which is related to power flow, propagates at an angle ϵ, slightly non-parallel to the interface. Lastly, Eq. (3.77d) shows that field decay into the medium is proportional to β/ϵ, which must be carefully analyzed for the various angles of interest.

Two cases of special interest offer insights into the nature of wave propagation, the resultant fields, and their relation to energy loss in a lossy medium near the critical angle of incidence. The special cases are those of extreme glancing incidence, such that $\theta \ll \theta_c$, and incidence just at the critical angle $\theta = \theta_c$. Combining Eqs. (3.76a, b) and (3.77a–d) we have the following.

(1) *Glancing incidence*, $\theta^2 \ll \theta_c^2$, $\epsilon = \beta/(\theta_c^2 - \theta^2)^{1/2} \to \beta/\sqrt{2\delta}$:

$$\phi'_r \simeq \frac{\pi}{2} - \frac{\beta}{\left(\theta_c^2 - \theta^2\right)^{1/2}} \to \frac{\pi}{2} - \frac{\beta}{\sqrt{2\delta}} \tag{3.78a}$$

$$\phi'_i \simeq \left(\theta_c^2 - \theta^2\right)^{1/2} \to \sqrt{2\delta} \tag{3.78b}$$

$$k'_{xr} \simeq \frac{\omega}{c}\left(1 - \frac{\theta^2}{2}\right) \tag{3.78c}$$

$$k'_{xi} = 0 \tag{3.78d}$$

$$k'_{zr} \simeq \frac{\omega}{c}\frac{\beta}{\left(\theta_c^2 - \theta^2\right)^{1/2}} \to \frac{\omega\beta}{c\sqrt{2\delta}} \tag{3.78e}$$

$$k'_{zi} \simeq \frac{\omega}{c}\sqrt{\left(\theta_c^2 - \theta^2\right)} \to \frac{\omega}{c}\sqrt{2\delta} \tag{3.78f}$$

where the arrows indicate limiting values as θ goes to zero.

(2) *Critical angle, $\theta = \theta_c$, $\epsilon_c \simeq \beta^{1/2}$:*

$$\phi'_r \simeq \frac{\pi}{2} - \beta^{1/2} \tag{3.79a}$$

$$\phi'_i \simeq \beta^{1/2} \tag{3.79b}$$

$$k'_{xr} \simeq \frac{\omega}{c}\left(1 - \frac{\theta^2}{2}\right) = \frac{\omega}{c}(1 - \delta) \tag{3.79c}$$

$$k'_{xi} = 0 \tag{3.79d}$$

$$k'_{zr} \simeq \frac{\omega}{c}\beta^{1/2} \tag{3.79e}$$

$$k'_{zi} \simeq \frac{\omega}{c}\beta^{1/2} \tag{3.79f}$$

In both special cases above, the solutions indicate an evanescent wave propagating very nearly parallel to the interface, with field amplitudes decaying with distance z into the material. The small angle ϵ by which the wave vector is non-parallel to the interface is zero only in the lossless limit, β equal to zero. It is interesting to note the differing field penetrations in the two cases, and relate these to reflectance curves previously seen in Figure 3.8. According to Eqs. (3.78a–f), at glancing incidence *the fields decay with an exponential dependence*

$$e^{-k'_{zi}z} \simeq \exp\left[-\left(\frac{\omega}{c}\sqrt{\theta_c^2 - \theta^2}\right)z\right] \rightarrow e^{-(2\pi\sqrt{2\delta}/\lambda)z} \tag{3.80a}$$

i.e., with a field penetration depth for θ near zero,

$$\boxed{z_0 \simeq \frac{\lambda}{2\sqrt{2}}\pi\delta^{1/2} \qquad (\theta \ll \theta_c)} \tag{3.80b}$$

This is to be compared with the case just at the critical angle, where according to Eqs. (3.79a–f)

$$e^{-k'_{zi}z} \simeq \exp\left[-\left(\frac{\omega}{c}\beta^{1/2}\right)z\right] = e^{-(2\pi\beta^{1/2}/\lambda)z} \tag{3.81a}$$

with a field penetration depth at the critical angle

$$\boxed{z_c \simeq \frac{\lambda}{2\pi\beta^{1/2}} \qquad (\theta = \theta_c)} \tag{3.81b}$$

Thus for $\beta < \delta$, the penetration depth at the critical angle will be greater than for incidence angles closer to the surface, leading to greater absorption of the wave and thus less reflectivity at the critical angle, as was seen in Figure 3.8. In an earlier section we calculated that for carbon at 4 Å wavelength, $\delta = 4.90 \times 10^{-5}$ and $\beta = 5.71 \times 10^{-7}$. Thus for a carbon mirror the critical angle is 9.9 mrad (0.57°) at 4 Å wavelength, and according to Eqs. (3.80b) and (3.81b) the penetration depth at near-zero glancing angle is 16λ, or 6.4 nm, with an increased penetration depth of 210 λ, or 84.0 nm at the critical angle. In both cases energy propagates across the interface accounting for the small absorptive loss, as indicated by the finite value of k'_{zr}, which we note goes to zero as β goes to zero.

The nature of energy flow into the lossy medium can be seen explicitly by considering the z-directed portion of the Poynting vector, $\bar{S}_z$, which we can deduce by taking the cross product of the appropriate fields. For convenience we consider the case of incident radiation with the electric field polarized in the y-direction. In addition to the primary component of magnetic field associated with the refracted plane wave calculated for a lossless medium far from the critical angle ($\theta \gg \theta_c$), we also expect a second-order magnetic field due to the non-uniform nature of the wave. This non-uniformity has contributions both due to absorption and due to the evanescent decay for incidence angles θ less than the critical angle. For this non-uniform wave, characterized by an evanescent electric field $E'_y(x, z)\mathbf{y}_0$ in the medium, we can calculate the magnetic fields from Faraday's law [Chapter 2, Eq. (2.2)]:

$$-\frac{\partial \mathbf{B}'}{\partial t} = \nabla \times \mathbf{E}'$$

In terms of vector components this can be written as

$$-\frac{\partial \mathbf{B}'}{\partial t} = \left(\frac{\partial}{\partial x}\mathbf{x}_0 + \frac{\partial}{\partial y}\mathbf{y}_0 + \frac{\partial}{\partial z}\mathbf{z}_0\right) \times E'_y(x, y)\mathbf{y}_0$$

or

$$-\frac{\partial \mathbf{B}'}{\partial t} = \underbrace{\frac{\partial E'_y}{\partial x}\mathbf{z}_0}_{B'_z} - \underbrace{\frac{\partial E'_y}{\partial z}\mathbf{x}_0}_{B'_x} \qquad (3.82)$$

In Eq. (3.82), above the first term on the right of the equality is B'_z, the first-order magnetic field associated with a uniform plane wave propagating in the x-direction. The second term on the right is B'_x, which includes the second-order magnetic fields due to absorption and below-critical evanescence, each of which give z-dependent variations, and also includes the z-direction component of plane wave propagation. Thus the second-order magnetic field, B'_x, which will cross with E'_y to give power flow in the z-direction (across the interface), corresponds to the $\mathbf{x}_0$-components of Eq. (3.82), viz.,

$$-\frac{\partial B'_x}{\partial t} = -\frac{\partial E'_y}{\partial z}$$

The time dependence for all fields is the same; thus $\partial/\partial t = -i\omega$. Furthermore, from Eqs. (3.77a–d) the z-dependence of $E'_y(x, z)$ is given by

$$E'_y(z) = E'_y(0)e^{i\omega\epsilon z/c}e^{-\omega\beta z/\epsilon c}$$

so that

$$B'_x(x, z) = -\frac{1}{i\omega}\left(\frac{i\omega\epsilon}{c} - \frac{\omega\beta}{\epsilon c}\right)E'_y(x, z)$$

$$B'_x(x, z) = -\left(\frac{\epsilon}{c} + i\frac{\beta}{\epsilon c}\right)E'_y(x, z)$$

For non-magnetic materials $\mu = \mu_0$ and the magnetic field vector is

$$H'_x(x, z)\mathbf{x}_0 = -\sqrt{\frac{\epsilon_0}{\mu_0}}\left(\epsilon + i\frac{\beta}{\epsilon}\right)E'_y(x, z)\mathbf{x}_0 \qquad (3.83)$$

where ϵ is the angle measured from the interface. With the magnitude of the average Poynting vector given by Eq. (3.19), the time-averaged power per unit area crossing the interface is given by

$$|\bar{\mathbf{S}}| = \tfrac{1}{2}\text{Re}\,(\mathbf{E} \times \mathbf{H}^*)$$

with z-component

$$\bar{S}'_z\mathbf{z}_0 = \tfrac{1}{2}\text{Re}(E'_y\mathbf{y}_0 \times H'^*_x\mathbf{x}_0)$$

so that

$$\bar{S}'_z = -\frac{1}{2}\text{Re}\left[(E'_y)\left(-\sqrt{\frac{\epsilon_0}{\mu_0}}\right)\left(\epsilon - i\frac{\beta}{\epsilon}\right)E'_y\right]$$

$$\boxed{\bar{S}'_z = \frac{1}{2}\sqrt{\frac{\epsilon_0}{\mu_0}}\epsilon|E'_y|^2} \qquad (3.84)$$

where the angle ϵ is given in Eq. (3.75). We see that the power flowing across the interface is directly proportional to ϵ, which has a complicated dependence on θ, δ, and β, but which goes to zero as β goes to zero. Tracing this term back through the mathematics, we observe that it is due to the non-parallelism of the refracted wave with respect to the surface, which arises because of the finite absorptivity β. The second term [for instance the $i\beta/\epsilon$ term in Eq. (3.83)], which is due to the evanescent nature of the fields, even when β is zero (β/ϵ is not zero in this limit), does not contribute to power flow across the interface, as is seen in the mathematical progression from Eq. (3.83) to Eq. (3.84). This term gives a measure of the stored energy in the evanescent fields.

In summary, we observe that for a lossy medium and a glancing angle of incidence θ less than the critical angle, the refracted wave is non-uniform, that is, the field amplitudes are no longer constant in planes transverse to the propagation direction. The wave now propagates at a slight angle (ϵ) to the interface, with second-order fields out of the plane. That is, the refracted wave is neither uniform nor plane. Power flows across the interface for non-zero absorptivity β. Details of the evanescent field decay, and power flow, depend on both δ and β, and of course the angle of incidence ϕ (or θ).

3.8   Determination of δ and β: The Kramers–Kronig Relations

In principle δ and β can be determined for all materials through measurements of absorption and phase shift, as discussed in the text leading to Eqs. (3.22) and (3.29). This is in fact how β is determined. However, determining δ proves to be more problematic as interferometry is not sufficiently advanced. Rather, the general approach is to return to Eqs. (3.13a,b), where δ and β are expressed in terms of real and imaginary parts of the complex atomic scattering factor, i.e.,

$$\delta = \frac{n_a r_e \lambda^2}{2\pi} f_1^0(\omega)$$

$$\beta = \frac{n_a r_e \lambda^2}{2\pi} f_2^0(\omega)$$

where from Eq. (2.79) the complex atomic scattering factor

$$f^0(\omega) = f_1^0(\omega) - i f_2^0(\omega)$$

is the ratio of the electric field strength scattered by an atom to that of a single free electron. Recall that the superscript zero refers to the limiting cases of either long wavelength or forward scattering [Eqs. (2.70)–(2.72)].

The determination of $f_2^0(\omega)$ is accomplished by measuring the absorption of radiation through thin foils (or gases) of an element of interest (C, O, ..., Si, ..., Au, ...) for a broad range of photon energies ($\hbar\omega$). This can be done by using broadly tunable synchrotron radiation (discussed in Chapter 5) and a suitable monochromator. From these measurements the macroscopic mass absorption coefficient $\mu(\omega)$ can be determined, as described in Chapter 1, or here in Eq. (3.24). The imaginary part of the atomic scattering factor is then determined from Eq. (3.26), viz.,

$$f_2^0(\omega) = \frac{A m_u}{2 r_e \lambda} \mu(\omega)$$

– a macroscopic-to-microscopic relationship, where for the element in question A is the number of atomic mass units, m_u is the atomic mass unit, r_e is the classical electron radius, λ is the wavelength in vacuum, and μ is the photon frequency (energy) dependent mass absorption coefficient.

The real part of the atomic scattering factor, $f_1^0(\omega)$, is then determined through mathematical relationships between f_1^0 and f_2^0, generally referred to as *Kramers–Kronig relations*,[15–18] first derived in 1927. For a broad class of physical problems – including damped electrical circuits, scattering, and refractive index – these relate the real and imaginary parts of the physical "response" to a "stimulus" in linear, stable, causal systems. If the system is causal, there is no response ($\mathbf{E}_{scatt}$) until there is a cause ($\mathbf{E}_{inc}$). The scattering of radiation is just such a problem. For the problem of interest here, we shall show that the real and imaginary parts of the complex atomic scattering factor are related by

$$f_1^0(\omega) = Z - \frac{2}{\pi} \mathcal{P}_C \int_0^\infty \frac{u f_2^0(u)}{u^2 - \omega^2}\, du \qquad (3.85a)$$

and

$$f_2^0 = \frac{2\omega}{\pi} \mathcal{P}_C \int_0^\infty \frac{f_1^0(u) - Z}{u^2 - \omega^2} \, du \tag{3.85b}$$

where f_1^0 is written as having a first-order term Z, the number of electrons per atom, and a departure therefrom due to the degree of binding, as discussed in Chapter 2. $\mathcal{P}_C$ indicates taking only the non-divergent Cauchy principal part of the integral. As seen in Appendix D, $\mathcal{P}_C(1/x)$ is defined by *Cauchy's Principal Value Theorem:*

$$\lim_{\epsilon \to 0} \frac{1}{x \mp i\epsilon} = \mathcal{P}_C(1/x) \pm i\pi\delta(x) \tag{3.86}$$

where the principal value $\mathcal{P}_C(1/x)$ refers to a function that behaves like $1/x$ everywhere except at $x = 0$, with the discontinuous behavior separated out and described by the Dirac delta function, $\delta(x)$, which is described in Appendix D.

The significance of Eqs. (3.85a) and (3.85b) is that knowledge of either the real or the imaginary part of the response function $f^0(\omega)$ across the full spectrum of frequencies is sufficient to determine the other. Thus if one can determine $f_2^0(\omega)$ through absorption measurements, across a sufficiently broad range of frequencies, that the integral converges, one can determine $f_1^0(\omega)$. In other words, measurement of β across a sufficiently broad photon energy range allows one to determine $f_1^0(\omega)$, and thus the real part of the refractive index decrement $\delta(\omega)$, by use of Eq. (3.85a). This is in fact the procedure used by Henke, Gullikson, and their colleagues[7] to deduce the values of f_1^0 and f_2^0 for all elements from hydrogen $(Z = 1)$ to uranium $(Z = 92)$, for photon energies extending from the extreme ultraviolet to the hard x-ray region of the spectrum. Experimental confirmations of these tables is of great interest, being pursued with several new types of interferometers, and with glancing incidence technologies based on best fits of Eq. (3.52) to reflections from very clean, homogeneous, and very flat surfaces.

To derive the Kramers–Kronig relations [Eqs. (3.85a) and (3.85b)] we begin with the expression for the atomic scattering factor [Eq. (2.72)]

$$f^0(\omega) = \sum_s \frac{g_s\omega^2}{\omega^2 - \omega_s^2 + i\gamma\omega}$$

where the oscillator strengths g_s in this semi-classical model[‡‡] sum to the total number of atomic electrons, Z, and where for small γ the poles lie in the lower half plane (LHP) at $\omega = \pm\omega_s - i\gamma/2$ (see Figure 3.16). Multiplying the numerator and denominator of $f^0(\omega)$ by the complex conjugate of the scattering factor, we obtain the real and imaginary parts

$$f_1^0(\omega) = \sum_s \frac{g_s\omega^2 \left(\omega^2 - \omega_s^2\right)}{\left(\omega^2 - \omega_s^2\right)^2 + \gamma^2\omega^2} \tag{3.87a}$$

[‡‡] The relationship between the semi-classical g_s and the quantum mechanical oscillator strength g_{nk} is discussed in Chapter 2, Section 2.7, Eq. (2.74).

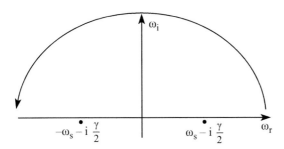

Figure 3.16 Representing $f^0(\omega)$ in the complex ω-plane. For $\gamma \ll \omega_s$ the function $f^0(\omega)$ has poles at $\pm\omega_s - i(\gamma/2)$ in the lower half plane, and is analytic in the upper half plane.

and

$$f_2^0(\omega) = \sum_s \frac{g_s \gamma \omega^3}{\left(\omega^2 - \omega_s^2\right)^2 + \gamma^2 \omega^2} \tag{3.87b}$$

where again

$$f^0(\omega) = f_1^0(\omega) - i f_2^0(\omega)$$

Using the Cauchy residue theorem[19] (Appendix D), we can represent the complex function $f^0(\omega)$, which has poles only in the lower half plane, in terms of a function $f(u)$ which is analytic in the upper half plane (UHP), viz.,

$$f^0(\omega) = \frac{1}{2\pi i} \oint \frac{f^0(u)}{u - \omega} \, du \tag{3.88a}$$

More conveniently, for a function that has a limiting value Z as ω approaches infinity, we can write this as

$$f^0(\omega) - Z = \frac{1}{2\pi i} \oint \frac{f^0(u) - Z}{u - \omega} \, du \tag{3.88b}$$

Recalling the normalization condition for g_s from Eq. (2.73), we can write

$$f^0(\omega) - Z = \sum_s \frac{g_s \omega^2}{\omega^2 - \omega_s^2 + i\gamma\omega} - \sum_s \frac{g_s \left(\omega^2 - \omega_s^2 + i\gamma\omega\right)}{\omega^2 - \omega_s^2 + i\gamma\omega}$$

or

$$f^0(\omega) - Z = \sum_s \frac{g_s \left(\omega_s^2 - i\gamma\omega\right)}{\omega^2 - \omega_s^2 + i\gamma\omega}$$

This sum goes to zero as ω approaches infinity, and thus does not contribute to the integral in Eq. (3.88b) along the semicircle of infinite radius in the UHP. Thus only the integral along the real axis remains, which can be written as

$$f^0(\omega) - Z = \frac{1}{2\pi i} \int_{-\infty}^{+\infty} \frac{f^0(u) - Z}{u - \omega} \, du$$

Using the principal value theorem, Eq. (3.86), this can be rewritten as

$$f^0(\omega) - Z = \frac{1}{2\pi i} \int_{-\infty}^{\infty} \left[\mathcal{P}_C \left(\frac{1}{u - \omega} \right) + \pi i \delta(u - \omega) \right] [f^0(u) - Z] \, du$$

or

$$f^0(\omega) - Z = \frac{1}{2\pi i} \mathcal{P}_C \int_{-\infty}^{\infty} \frac{f^0(u) - Z}{u - \omega} \, du + \frac{1}{2} [f^0(\omega) - Z]$$

Combining like terms,

$$f^0(\omega) - Z = \frac{1}{\pi i} \mathcal{P}_C \int_{-\infty}^{\infty} \frac{f^0(u) - Z}{u - \omega} \, du$$

Recalling that $f^0(\omega) = f_1^0(\omega) - i f_2^0(\omega)$, we can equate real and imaginary components to obtain

$$f_1^0(\omega) - Z = -\frac{1}{\pi} \mathcal{P}_C \int_0^{\infty} \frac{f_2^0(u)}{u - \omega} \, du \qquad (3.89a)$$

and

$$f_2^0(\omega) = -\frac{1}{\pi} \mathcal{P}_C \int_0^{\infty} \frac{f_1^0(u) - Z}{u - \omega} \, du \qquad (3.89b)$$

where the integration is along the real axis, from minus infinity to plus infinity. To rewrite these equations in terms of only positive frequencies, we divide Eq. (3.89a) into two parts as follows:

$$f_1^0(\omega) - Z = -\frac{1}{\pi} \left[\mathcal{P}_C \int_{-\infty}^0 \frac{f_2^0(u)}{u - \omega} \, du + \mathcal{P}_C \int_0^{\infty} \frac{f_2^0(u)}{u - \omega} \, du \right]$$

Replacing u by $-u$ in the first integral, and noting from Eq. (3.87b) that $f_2^0(-u) = -f_2^0(u)$ and that in general reversing the limits of integration causes a sign change, the integration can be rewritten in terms of positive frequencies only as

$$f_1^0(\omega) - Z = -\frac{1}{\pi} \mathcal{P}_C \int_0^{\infty} \left[\frac{1}{u + \omega} + \frac{1}{u - \omega} \right] f_2^0(u) \, du$$

or more concisely as

$$\boxed{f_1^0(\omega) - Z = -\frac{2}{\pi} \mathcal{P}_C \int_0^{\infty} \frac{u f_2^0(u)}{u^2 - \omega^2} \, du} \qquad (3.85a)$$

which gives a solution for $f_1^0(\omega)$, and thus δ, at some specific frequency ω in terms of an integral of $f_2^0(\omega)$, or equivalently β, over all real positive frequencies – a result we stated without proof at the beginning of this section. In similar fashion we can separate Eq. (3.89b) into integrals from minus infinity to zero and from zero to infinity, make the substitution $-u$ for u in the first integral, and observe that in this case according to Eq. (3.87a) we have $f_1^0(-u) = f_1^0(u)$, and that

$$\frac{1}{u - \omega} - \frac{1}{u + \omega} = \frac{u + \omega}{u^2 - \omega^2} - \frac{u - \omega}{u^2 - \omega^2} = \frac{2\omega}{u^2 - \omega^2}$$

With these substitutions Eq. (3.89b) becomes

$$f_2^0(\omega) = \frac{2\omega}{\pi} \mathcal{P}_C \int_0^\infty \frac{f_1^0(u) - Z}{u^2 - \omega^2} \, du \qquad (3.85b)$$

showing that f_2^0, and thus β, could in complementary fashion be obtained by an integration of $f_1^0(u)$, if it were δ that was more easily measured. Equations (3.85a) and (3.85b), the Kramers–Kronig relations,[14–16] provide the desired integral relationship between the real and imaginary parts of the atomic scattering factor, f_1^0 and f_2^0, or equivalently between δ and β. Since it is easier to determine β for a wide range of photon energies (frequencies) through absorption measurements, this provides a technique for numerically determining values of $\delta(\omega)$.

Henke, Gullikson, and colleagues[7–9] have compiled absorption data (f_2^0) for all elements from hydrogen to uranium, for photon energies extending from 10 eV to 30 keV, and from these data have computed and tabulated values of f_1^0 from 30 eV to 30 keV. Sample tabulations of f_1^0 and f_2^0 are given for some common elements in Appendix C. Recall that we have used the superscript zero to emphasize the simplification of the atomic scattering factor [see Eqs. (2.70)–(2.77)] when the exponent $\delta \mathbf{k} \cdot \delta \mathbf{r}_s$ goes to zero, i.e., in either the long wavelength or the forward scattering limit. As a consequence the reader must make the identifications of f_1^0 and f_2^0 here, with f_1 and f_2 in Refs. 7 and 8. Furthermore, the sign on f_2^0 is negative because of our choice regarding $e^{-i\omega t}$, as discussed earlier in this chapter.

3.9 Enhanced Reflectivity from Periodic Structures

In a medium of uniform refractive index, of infinite extent, there is no scattering. Scattering arises from variations in refractive index, within a material or at its boundary with another material (or vacuum). We have considered the scattering from isolated free and atomically bound electrons. With a large number of such scattering centers (electrons, atoms) per unit wavelength, the scattering in any direction is canceled by interference with that from another scattering center, a distance $\lambda/2$ away, along the direction of observation. Residual scattering occurs only because of fluctuations from the mean. The sky would be blue – or any other color – if it were not for density fluctuations in the atmosphere that redirect light away from its initial path. For short wavelengths this requires further consideration, but basically one observes scattering only to the degree that there are departures from the average density along a given line of sight. This occurs at an interface (for example, between a material and vacuum or between two materials) because the density of scatterers is different. In such cases one can apply the techniques described in Chapter 3 and solve for the resultant reflection in terms of changes in the refractive index at the interface – essentially uncompensated changes in scattering density or strength.

We begin this study by considering the scattering of radiation by a one-dimensional sinusoidal density profile $n_a(z)$, uniform across the x, y-plane, as shown in Figure 3.17.

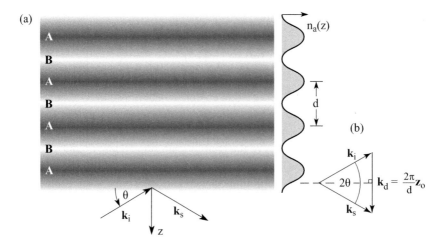

Figure 3.17 (a) Scattering of incident radiation of wave vector $\mathbf{k}_i$ by a one-dimensional sinusoidal density distribution of scattering centers (atoms or electrons) characterized by charge density $n_a(z)$ and periodicity d. (b) The associated charge density wave vector, $\mathbf{k}_d = (2\pi/d)\mathbf{z}_0$, points in the z-direction, where $\mathbf{z}_0$ is a unit vector. The scattered wave vector $\mathbf{k}_s$ is obtained by adding wave vectors as described in Eqs. (3.94a,b).

Considering Maxwell's equations, scattered fields are generated by induced bound electron currents that can be described for small angle, forward scattering in terms of the complex atomic scattering factor $f^0(\omega)$, as seen earlier in Chapter 2, Eq. (2.72). The induced current can then be written as

$$\mathbf{J}(\mathbf{r}, t) = -ef^0(\omega)n_a(\mathbf{r}, t)\mathbf{v}(\mathbf{r}, t) \tag{3.90}$$

where $-e$ is the electron charge, n_a is the density in atoms per unit volume, $\mathbf{v}$ is the oscillatory velocity that would be experienced by a single free electron (dominated by the electric field $\mathbf{E}_i$ of the incident field), and $f^0(\omega)$ is the frequency dependent factor that takes into account the effective number of bound electrons oscillating within each atom. Performing a space-time Fourier transform, or simply representing the three waves (incident, scattered, and density) in terms of amplitudes and exponential factors, one obtains an expression for the induced current that drives the scattering process,

$$\mathbf{J}_{\text{scatt}}e^{-i(\omega_s t - \mathbf{k}_s \cdot \mathbf{r})} = -ef^0(\omega_i)n_a e^{-i(\omega_d t - \mathbf{k}_d \cdot \mathbf{r})}\frac{-e\mathbf{E}_i}{-i\omega_i m}e^{-i(\omega_i t - \mathbf{k}_i \cdot \mathbf{r})} \tag{3.91}$$

where the subscript d denotes a density wave, and where the oscillating velocity $\mathbf{v}$ is related to the incident electric field by the equation of motion $md\mathbf{v}/dt = -e\mathbf{E}_i$, as in Chapter 2. Matching amplitudes and phase for all $\mathbf{r}$, t, one obtains the general scattering relationships

$$\mathbf{J}_s = \frac{ie^2 n_a f^0(\omega_i)}{\omega_i m}\mathbf{E}_i \tag{3.92}$$

$$\omega_s = \omega_i + \omega_d \tag{3.93a}$$

and

$$\mathbf{k}_s = \mathbf{k}_i + \mathbf{k}_d \qquad (3.93b)$$

Equations (3.93a) and (3.93b), multiplied by $\hbar$, provide familiar expressions for conservation of energy and momentum in the scattering process, i.e.,

$$\boxed{\hbar\omega_s = \hbar\omega_i + \hbar\omega_d} \qquad (3.94a)$$

$$\boxed{\hbar\mathbf{k}_s = \hbar\mathbf{k}_i + \hbar\mathbf{k}_d} \qquad (3.94b)$$

These general scattering relations can be applied to a wide variety of phenomena, including Raman and Brillouin scattering, which we consider in Chapter 8 for the case of propagating plasma waves. Here we are interested in the stationary density distribution of atoms, as illustrated in Figure 3.17. In this case the density "wave" does not move, so that ω_d is zero and $\omega_s = \omega_i$. In this case the magnitudes of the incident and scattered wave vectors must be equal, $|\mathbf{k}_s| = |\mathbf{k}_i| = 2\pi/\lambda$, so that the scattering diagram representing Eqs. (3.94a,b) must be isosceles, as shown in Figure 3.17(b). Taking $\sin\theta$ in one half of the isosceles triangle, one obtains

$$\sin\theta = \frac{k_d/2}{k_i}$$

or

$$\lambda = 2d\sin\theta \qquad (3.95a)$$

which we recognize as the first-order ($m = 1$) of *Bragg's Law* – a rather general relationship for scattering or diffraction from periodic structures, including crystal planes, multilayer interference coatings, and plasma waves. It was first fully described and demonstrated in 1913 by son and father Nobel Laureates W.L. and W.H. Bragg.[20–22] If the density distribution $n_a(z)$ is not a simple sinusoid, as would be the case in a natural crystal, the distribution can be Fourier decomposed into harmonic components of period d/m. In an angular (θ) scan of incident wave vector $k_i = 2\pi/\lambda$, the various Fourier components will generate successive scattering peaks, corresponding to wavenumbers $k_{d/m} = 2\pi/(d/m)$, yielding the more general form of Bragg's Law, or Bragg's condition

$$\boxed{m\lambda = 2d\sin\theta} \qquad (3.95b)$$

where θ is measured from the crystal planes and the diffraction orders are $m = 1, 2, 3, \ldots$ At this point the terminology used to describe this process generally evolves from use of the word "scattering" to "diffraction," or even "reflection," implying that the observed scattering has coalesced into a rather well-defined angular pattern, characteristic of the orderliness of the "diffracting medium." Use of the word "reflection" also becomes appropriate as the fraction of redirected energy evolves from weak,

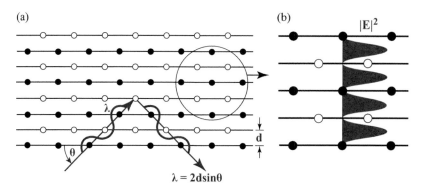

Figure 3.18 Diffraction of x-rays from a near-perfect natural crystal is illustrated. Scattering from the many atoms add in phase for a given wavelength, resulting in very strong diffraction at the Bragg angle, and cancellations at all other angles. The Braggs envisioned this as reflections from the many crystal planes, adding in phase at the angle named after them. The choice of dark and open circles is to suggest the presence of sodium and chlorine atoms in a NaCl crystal, much as was used in an early Bragg experiment.[22] The enlarged diagram in (b) shows the standing wave electric field interference pattern which is formed when near-equal intensity incident and diffracted radiation overlap throughout the crystal when the Bragg condition is met. The resultant null of the composite electric field, which occurs across each crystal plane, is essential to achieving high reflectivity.

broad angular scattering to strong, well-directed diffraction. For well-collimated hard x-rays, incident upon a near-perfect natural crystal, a reflectivity approaching unity is possible[23–25]. For example, a thin silicon crystal can reflect better than 95% of incident 10 keV x-rays of small divergence angle undulator or free electron laser radiation when oriented to diffract from its Si(111) planes. A diagram suggesting the first x-ray diffraction experiment with rocksalt,[20] the large scale mineral version of common salt (NaCl), is shown in Figure 3.18(a).

Interference between the incident and reflected waves within an atomic crystal plays a critical role in the achievement of very high reflectance for hard x-rays in the range of Ångström and sub-Ångström wavelengths, as illustrated in Figure 3.18(b). Just at the Bragg angle, θ_B satisfying Eq. (3.95b), the incoming and outgoing waves interfere forming a standing wave in the z-direction, with a null electric field located at the atomic planes. In such crystals the separation of atomic planes is set by the more distant valence electrons, which are relatively ineffective at absorbing hard x-rays, while the more absorbing K-shell orbitals are very tightly confined to regions near the respective nuclei, at radii of order a_0/Z, where a_0 is the Bohr radius and Z is the nuclear charge (e.g., see Figure 1.12). As all the atoms are located at null-field atomic planes they experience very minimal incident field. As a result, with properly controlled illumination of low angular divergence, the incident and reflected fields in atomic crystals scatter off the more numerous L-, M-, and N-shell outer electrons, while experiencing little absorption by the tightly bound K-shell electrons. This has a very interesting manifestation in the Borrmann effect[26] in which hard x-rays incident on a thin crystal are absorbed at incidence angles unequal to that set by Bragg's Law. But the radiation appears in transmission for incidence at the Bragg angle where interference within the crystal sets up the null field described above and greatly reduces absorption. For angles different than the Bragg angle the interference pattern shifts such that there is no longer a null at the crystal planes, resulting in strong absorption by the now exposed K-shell electrons.

There is a slight refraction of the x-ray beam as it enters the crystal, resulting in a refractive correction to Bragg's Law:

$$m\lambda = 2d \sin\theta \left(1 - \frac{\bar{\delta}}{\sin^2\theta}\right) \tag{3.96}$$

where $\bar{\delta}$ is the density weighted real part of the refractive index. This correction played an important role in the early development of x-ray diffraction, resolving annoying but repeatable anomalies observed as the precision of angular measurements improved.[27, 28] For detailed discussions of x-ray diffraction from crystals see the traditional texts of R.W. James[29] and Cullity and Stock[29], and the very current text by Als-Nielsen and McMorrow[30].

Multilayer interference coatings, consisting of a sequence of alternately high-Z and low-Z layered materials, also reflect EUV and x-rays according to Bragg's Law. They are commonly referred to as multilayer mirrors, or simply "multilayers," and sometimes as artificial crystals. A basic diagram is shown in Figure 3.19(a), and a side view TEM image¶¶ of a thinned, high-quality[31, 32] molybdenum/silicon (Mo/Si) multilayer coating is shown in Figure 3.19(b). The layers are generally amorphous in nature, with the atoms randomly positioned. Alternating layers are of different materials, typically one high Z and one low Z, so that there is a contrast in the refractive indexes experienced by x-rays. With sharp, planar interfaces between layers this leads to a well-defined but relatively weak reflection at each surface. With many interfaces properly placed for a given wavelength and periodicity, d, the weak reflections will add in phase at the Bragg angle. The resultant reflected electric field will be proportional to the number of interfaces, twice the number of periods (N), thus producing a strong reflection proportional to $4N^2$, at least for modest depths before absorption limits the process. As with natural crystals, interference of the incident and reflected waves will lead to a standing wave pattern within the crystal. In the multilayer case the nulls will be centered along the planes of the more absorptive high-Z layers. However, because these layers have a finite thickness the reduction in absorption is not as effective as with natural crystals, but still plays an essential role in achieving high reflectivity. To minimize absorption the high-Z layers are made as thin as possible, as a fraction of the two-material period (d-spacing), while still maintaining sharp interfaces. This maintains the number of interfaces contributing to the total reflection while minimizing absorptive losses. Maintaining sharp, flat interfaces, with uniform density within the layers, is particularly challenging on the nanometer scale required for EUV and x-ray wavelengths.

¶¶ Transmission Electron Microscope (TEM).

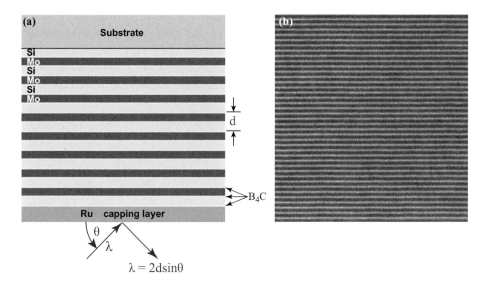

Figure 3.19 (a) A multilayer interference coating, also known as a multilayer mirror, in this example having alternate layers of molybdenum and silicon. Generally the thickness of the higher-Z material is less than that of the lower-Z material. The mirror shown was designed for use at a wavelength of 13.4 nm, just below the Si-L edge at 99 eV. Each period of 6.88 nm consists of a 4.14 nm Si layer and a 2.09 nm Mo layer, each covered by sub-half nm thick boron carbide (B_4C) diffusion barriers which limit interdiffusion during the coating process. There is also a ruthenium capping layer to isolate the multilayer coating structure from deleterious environmental effects such as penetrating oxidation. A mirror such as this has achieved a reflectivity of 70%.[29] (b) A transmission electron micrograph (TEM) side view of a thinned Mo/Si multilayer mirror showing the high-fidelity layer thicknesses and density uniformity required to achieve high reflectivity.[32] There are 40 layer pairs of 6.7 nm period each. Both figures are courtesy of S. Bajt, Lawrence Livermore National Laboratory, now at CFEL, DESY, Hamburg.

According to the Bragg condition, normal incidence reflection ($\theta = \pi/2$) is possible with a period $d = \lambda/2$. For the two-material period this implies an average layer thickness of $d/2 = \lambda/4$. In the EUV, with wavelengths of 10–50 nm, this corresponds to layer thicknesses of order a few nanometers (nm), which is possible. Mo/Si multilayer coatings with 50 layer pairs (periods) have achieved a near normal reflectance as high as 70%.[31] For shorter wavelengths normal incidence coatings are more challenging due to the required thinness of each layer.[33–35] This topic is discussed further in Chapter 10, Section 10.3. Multilayer mirrors are also successful at glancing incidence angles where they are typically used,[36] generally achieving high reflectivity at a few times the critical angle.

For hard x-rays, multilayer mirrors are used exclusively at small angles, a few times the critical angle, yet serve a very useful role by providing a modest degree of monochromatization and perhaps a doubling of incidence angle. The latter increases collection solid angle and numerical aperture, thus increasing flux and somewhat reducing off-axis optical aberrations. Both effects are very useful for hard x-ray fluorescence microprobes at synchrotron facilities.[37, 38] Multilayer coatings are also useful in medical applications,[39] and for astrophysical orbiting x-ray telescopes, the latter capable of forming images at photon energies from 8 keV to several hundred keV.[40–42] Further consideration of multilayer mirrors and their applications is presented in Chapter 10.

References

1. M. Born and E. Wolf, *Principles of Optics* (Cambridge University Press, 1999), Seventh Edition, Chapter VII.
2. J.D. Jackson, *Classical Electrodynamics* (Wiley, New York, 1998), Third Edition. See the discussion in Section 7.8 regarding group velocity in the vicinity of atomic resonances.
3. J.A. Stratton, *Electromagnetic Theory* (McGraw-Hill, New York, 1941).
4. G.R. Fowles, *Introduction tò Modern Optics* (Dover, New York, 1989).
5. E. Hecht, *Optics* (Addison-Wesley, Reading, 2002), Fourth Edition.
6. A.H. Compton and S.K. Allison, *X-Rays in Theory and Experiment* (Van Nostrand, New York, 1935).
7. B.L. Henke, E.M. Gullikson, and J.C. Davis, *Atomic and Nuclear Data Tables* **54**, 181–342 (1993).
8. E.M. Gullikson, www.cxro.LBL.gov/optical_constants .
9. R. Soufli and E.M. Gullikson, "Reflectance Measurements on Clean Surfaces for the Determination of Optical Constants of Silicon in the Extreme Ultraviolet-Soft X-Ray Region," *Appl. Optics* **36**, 5499 (1997); R. Soufli, "Optical Constants of Materials in the EUV/Soft X-Ray Region for Multilayer Mirror Applications," PhD thesis, Department of Electrical Engineering and Computer Sciences, University of California at Berkeley (1997).
10. M. Francon, *Optical Interferometry* (Academic, New York, 1966).
11. U. Bonse and M. Hart, "An X-Ray Interferometer," *Appl. Phys. Lett.* **6**, 155 (1965).
12. L.G. Parratt, "Surface Studies of Solids by Total Reflection of X-Rays," *Phys. Rev.* **95**, 359 (1954).
13. D.Y. Smith and J.H. Barkyoumb, "Sign Reversal of the Atomic Scattering Factor and Grazing Incidence Transmission at X-Ray Absorption Edges," *Phys. Rev. B* **41**, 11529 (1990).
14. J.H. Underwood, "Multilayer Mirrors for X-rays and the Extreme UV," *Optics News* **12**, 20 (Opt. Soc. Amer., Washington, DC, March 1996).
15. L.D. Landau and E.M. Lifshitz, *Electrodynamics of Continuous Media* (Addison-Wesley, New York, 1960), pp. 256–261.
16. J.E. Marsden and M.J. Hoffman, *Basic Complex Analysis* (Freeman, New York, 1973), Second Edition, p. 548.
17. F. Wooten, *Optical Properties of Solids* (Academic, New York, 1972).
18. R.W. Ditchburn, *Light* (Blackie, London, 1963), Second Edition, Appendix XIX.
19. E. Kreyszig, *Advanced Engineering Mathematics* (Wiley, New York, 1993), Seventh Edition, p. 770.

20. W.L. Bragg, "The Specular Reflection of X-rays," *Nature* **90**, No.2250, 410 (December 12,1912); W.L. Bragg, "The Diffraction of Short Electromagnetic Waves by a Crystal," *Proc. Cambridge Phil. Soc.* 17, 43 (1913), presented November 11, 1912.

21. W.H. Bragg, "X-rays and Crystals", *Nature* **90**, No.2256, 572 (January 23, 1913).

22. W.H. Bragg and W.L. Bragg, "The Reflection of X-rays by Crystals," *Proc. Royal Society (London)* **88**, No.605, 428 (July 1, 1913).

23. Y. Shvyd'ko, S. Stoupin, V. Blank and S. Terentyev, "Near 100% Bragg Reflectivity of X-Rays," *Nature Phot.* **5**, 539 (2011); Y. V. Shvyd'ko *et al.*, "High-Reflectivity High-Resolution X-ray Crystal Optics with Diamonds," *Nature Phys.* **6**, 196 (2010).

24. B.W. Batterman and D.H. Bilderback, "X-ray Monochromators and Mirrors," Chapter 4 in *Handbook of Synchrotron Radiation*, Vol. **3** (Elsevier Science, 1991), Edited by G. Brown and D.E. Moncton.

25. T. Matsushita and H. Hashizume, "X-Ray Monochromators," Chapter 4 in *Handbook on Synchrotron Radiation*, Vol. **1** (North Holland, 1983), Edited by E.E. Koch.

26. G. Borrmann, "Über Extinktionsdiagramme von Quarz," *Physikalische Zeitschrift* **XLII**, 157 (July 15, 1941); B.W. Batterman and H. Cole, "Dynamical Diffraction of X-Rays by Perfect Crystals," *Rev. Mod. Phys.* **36**(3), 681 (July 1964).

27. M. Siegbahn, *The Spectroscopy of X-rays* (Oxford University Press, London, 1925), translated by G.A. Lindsay; Section 12, pp. 21–29.

28. F.K. Richtmyer, E.H. Kennard and T. Lauritsen, *Introduction to Modern Physics* (McGraw-Hill, New York, 1955), Chapter 8, Section 161, pp. 392–393.

29. R.W. James, *The Optical Principles of the Diffraction of X-Rays* (Ox Bow Press, Woodbridge, CT, 1982); B.D. Cullity and S.R. Stock, *Elements of X-Ray Diffraction* (Prentice-Hall, NJ, 2001), Third Edition; B.E. Warren, *X-Ray Diffraction* (Dover, New York, 1990).

30. J. Als-Nielsen and D. McMorrow, *Elements of Modern X-Ray Physics* (Wiley, 2011), Second Edition.

31. S. Bajt, J.B. Alameda, T.W. Barbee *et al.*, "Improved Reflectance and Stability of Mo-Si Multilayers," *Opt. Eng.* **41**(8), 1797 (August 2002).

32. S. Bajt, D.G. Stearns and P.A. Kearny, "Investigation of the Amorphous-to-Crystalline Transition in Mo/Si Multilayers," *J. Appl. Phys.* **90**, 1017 (July 15, 2001).

33. M. Prasciolu, A.F.G. Leontowich, K.R. Beyerlein and S. Bajt, "Thermal Stability Studies of Short Period Sc/Cr and Sc/B$_4$C/Cr Multilayers," *Appl. Optics*, **53** (10), 2126 (April 1, 2014).

34. J.H. Underwood and T.W. Barbee, "Soft X-ray Imaging with a Normal Incidence Mirror," *Nature* **294**, 429 (December 3, 1981).

35. F. Eriksson, G.A. Johansson, H.M. Hertz *et al.*, "14.5% Near-Normal Incidence Reflectance of Cr/Sc X-ray Multilayer Mirrors for the Water Window," *Optics Lett.* **28**, 2494 (December 15, 2003).

36. M.A. MacDonald, F. Schäfers and A. Gaupp, "A Single W/B4C Transmission Multilayer for Polarization Analysis of Soft X-rays up to 1 keV," *Optics Expr.* **17**, 23290 (December 7, 2009).

37. J.H. Underwood, A.C. Thompson, Y. Wu and R.D. Giaque, "X-ray Microprobe Using Multilayer Mirrors," *Nucl. Instr. Meth. A* **266**, 296 (1988); J.H. Underwood, A.C. Thompson, J.B. Kortright, K.C. Chapman and D. Lunt, "Focusing X-rays to a 1-μm Spot Using Elastically Bent, Graded Multilayer Coated Mirrors," *Rev. Sci. Instrum.* 67,1 (1996).

38. K. Yamauchi, H. Mimura, T. Kimura *et al.*, "Single-Nanometer Focusing of Hard X-rays by Kirkpatrick-Baez Mirrors," *J. Phys.: Condens. Matter* **23**, 394206 (September 2011).

39. K.-H. Yoon, Y.-M. Kwon, B.-J. Choi *et al.*, "Monochromatic X-rays for Low-Dose Digital Mamography: Preliminary Results," *Investigative Radiology* **47**, 683 (December 2005).

40. F.A. Harrison *et al.*, "The Nuclear Spectroscopic Telescope Array (NuStar) High-Energy X-ray Mission," *Astrophys. J.* **770**, 103 (June 20, 2013).

41. D.L. Windt, S. Donguy, C.J. Hailey *et al.*, "W/SiC X-ray Multlayers Optomized for Use Above 100 keV," *Applied Optics* **42**, 2415 (May 1, 2003).

42. M. Fernández-Perea, M.-A. Descalle, R. Soufli *et al.*, "Physics of Reflective Optics for the Soft Gamma-Ray Photon Energy Range", *Phys. Rev. Lett.* **111**, 027404 (July 12, 2013).

Homework Problems

Homework problems for each chapter will be found at the website: www.cambridge.org/xrayeuv

4　Coherence at Short Wavelengths

Fourier transform pairs:　　　　Heisenberg Uncertainty Principle:

$$\Delta\tau\Delta\omega \geq 1/2 \quad (4.4a) \qquad\qquad \Delta E_{\mathrm{FWHM}}\Delta\tau_{\mathrm{FWHM}} \geq 1.825\ \mathrm{eV\cdot fs} \qquad (4.4b)$$

$$\Delta x\Delta k \geq 1/2 \quad (4.5a) \qquad\qquad \mathbf{\Delta x} \cdot \mathbf{\Delta p} \geq \hbar/2 \qquad\qquad (4.5b)$$

$$l_{\mathrm{coh}} = \lambda^2/(2\Delta\lambda)\ (\text{temporal (longitudinal) coherence}) \qquad (4.6)$$

$$d \cdot \theta = \lambda/2\pi\ (\text{spatial (transverse) coherence}) \qquad (4.7a)$$

$$\text{or} \qquad d_{\mathrm{FWHM}} \cdot 2\theta_{\mathrm{FWHM}} = \frac{2\ln 2}{\pi}\lambda = 0.4413\lambda \simeq \lambda/2 \qquad (4.7b)$$

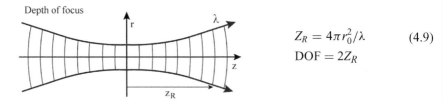

$$I/I_0 = \exp\left(-r^2/2r_0^2\right) \qquad (4.8)$$

Depth of focus

$$Z_R = 4\pi r_0^2/\lambda \qquad (4.9)$$
$$\mathrm{DOF} = 2Z_R$$

van Cittert–Zernike Relation

$$\mu_{\mathrm{OP}} = \frac{e^{-i\psi}\displaystyle\iint I(\xi, \eta)e^{ik(\xi\theta_x+\eta\theta_y)}\,d\xi\,d\eta}{\displaystyle\iint I(\xi, \eta)\,d\xi\,d\eta} \qquad (4.16)$$

Incoherently illuminated pinhole

$$\mu_{\mathrm{OP}}(\theta) = e^{-i\psi}\frac{2J_1(ka\theta)}{ka\theta} \qquad (4.24)$$

Whereas spatially and temporally coherent radiation is plentiful at visible wavelengths due to the availability of lasers, it is just becoming available at shorter wavelengths. In this chapter we review the concepts of spatial and temporal coherence, some applications that require radiation with these properties, and pinhole spatial filtering as a technique to generate spatially and temporally coherent radiation at extreme ultraviolet, soft x-ray, and x-ray wavelengths using largely incoherent sources. Techniques that generate highly coherent radiation at these wavelengths are discussed in Chapter 6 on free electron lasers and Chapter 9 on laser high harmonic generation.

4.1 Concepts of Spatial and Temporal Coherence

The ability to focus radiation to the smallest possible spot size, to propagate it great distances with minimal divergence, to encode wavefronts, and in general to form interference patterns, requires well-defined phase and amplitude variations of the fields throughout regions of interest. In general, simple phase distributions approaching those of plane or spherical waves are of greatest interest in those applications. Real laboratory sources, especially at very short wavelengths, generally radiate fields with more complex phase relationships that are well defined over only limited spatial and temporal scales. This brings us to the subject of coherence, its technical definition, and various convenient measures.

Coherence in our daily lives refers to a systematic connection or logical relationship between events, actions, or policies. In physics the word implies similar relationships among the complex field amplitudes used to describe electromagnetic radiation. Mathematically, one utilizes a *mutual coherence function*, Γ, as a measure of the degree to which the electric field at one point in space can be predicted, if known at some other point, as a function of their separation in space and time[1, 2]:

$$\Gamma_{12}(\tau) \equiv \langle E_1(t+\tau) E_2^*(t) \rangle \tag{4.1}$$

where in this scalar form E_1 and E_2 are the electric fields at points 1 and 2, and τ is the time delay. The angular brackets denote an expectation value, or a time average of the indicated product. It is often convenient to introduce a normalized *complex degree of coherence*, $\gamma_{12}(\tau)$, again in scalar form, as

$$\gamma_{12}(\tau) = \frac{\Gamma_{12}(\tau)}{\sqrt{\langle |E_1|^2 \rangle}\sqrt{\langle |E_2|^2 \rangle}} \tag{4.2}$$

where the normalizing factors in the denominator are related to the local intensities at the respective points, as was discussed in Chapter 2, Section 2.3. Thus, for example, in the case of a uniform plane wave, of very well-defined frequency, if the electric field is known at any given space-time point, it can be predicted everywhere else with certainty. As we quantify this later for real physical systems, we will consider this uniform plane wave as *coherent radiation*, meaning that $|\gamma_{12}(\tau)| = 1$ everywhere. The counter example would be one in which there were a large number of atoms moving randomly and radiating independently, at various frequencies, so that fields at the two separated points have almost no phase relationship. In this case the resultant degree of coherence approaches zero, and the fields are considered *incoherent*.

In many practical cases involving quasi-monochromatic x-ray sources it is possible to separate coherence properties in the forward, propagation direction, and in the transverse direction. For the analysis of such cases it is convenient to introduce a time-independent spatial coherence function between two points in the radiation field at a fixed time, viz.,

$$\mu_{12} = \frac{\langle E_1(t)E_2^*(t)\rangle}{\sqrt{\langle|E_1|^2\rangle}\sqrt{\langle|E_2|^2\rangle}} \tag{4.3}$$

which is known both as the normalized coherence factor and the complex degree of coherence.

One could write similar functions to describe amplitude and phase correlations in other physical systems. For a well-behaved water wave, for instance, one would expect the surface amplitude to be predictable over great distances, so that $|\gamma|$ would be near unity, implying a high degree of coherence over much of the observed field. On the other hand, the introduction of randomly thrown pebbles (or splashing ducks[3]) would create a jumble of uncorrelated disturbances, so that $|\gamma|$ would approach zero in the immediate vicinity, leading us to conclude that the fields in this vicinity were largely incoherent. Far from the pebbles, however, propagation effects permit the limited coherence to reveal itself, as we will see later in this chapter with consideration of the van Cittert–Zernike theorem.

In the theoretical limit of a point source oscillating at a single frequency for all time, from minus infinity to plus infinity, the radiated field quantities are perfectly correlated everywhere. That is, if one knew the electric field amplitude and phase at a given point and time, one would know these quantities at all points and for all time. In this limiting case the radiation field is said to be *coherent*. Real physical sources, however, are made up of spatially distributed radiators, and these radiate with a finite spectral bandwidth for a finite period of time. Consequently, well-defined phase relationships between field amplitudes are in practice restricted to a finite *region of coherence*.

To introduce the concept of a *coherence region*, we consider first the rather visual example of a marching band, as seen in Figure 4.1. The limit of full coherence corresponds to all the musicians in perfect step. However, in the presence of a strong wind, and perhaps some noisy spectators, not all musicians would clearly hear the band leader calling cadence. In this case those musicians close enough to hear clearly would remain in step, while those at further distances away would become increasingly out of step – resulting in a region of coherence near the band leader. The distance over which there is a reasonable expectation that the musicians were marching in step could be called a "coherence length." Figure 4.1 illustrates the near ideal case where members of the band are marching in step across the field of participants. Note that the coherence length need not be the same in all directions, for example being dependent on wind direction. The complete absence of cadence would result in uncorrelated stepping, a state of incoherence where $|\gamma|$ approaches zero for the smallest separations, and where the coherence

Figure 4.1 A marching band in which all members clearly hear the called cadence, steps coherently as they play their music. However, with a wind and noise from nearby spectators, band members at some distance might not hear the cadence and would become out of step, reducing the degree of coherence as a function of distance from the band leader. This introduces the concept of a coherence length, which might not be isotropic (not the same in all directions), for example depending on the wind's direction. Courtesy of the US Fife and Drum Corps (Pennsylvania Avenue inaugural parade, Washington DC, January 20, 2009).

length is essentially zero. In the following paragraphs we will attempt to provide measures of the distances over which electromagnetic fields can be expected to be well correlated, and thus useful for interference experiments as discussed in the first paragraph of this chapter.

Real sources are neither fully coherent nor fully incoherent, but rather are *partially coherent*.[1] In Figure 4.2(a) the point source radiates fields that are perfectly correlated, and thus coherently related everywhere. In Figure 4.2(b) a source of finite size and spectral bandwidth, restricted to radiate over a limited angular extent, generates fields with strong phase and amplitude correlation over only a limited extent. This brings us again to notions of "regions of coherence" and "coherence time": that is, spatial and temporal measures over which the fields are well correlated. In cases where there is a well-defined direction of propagation, it is convenient to decompose the region of coherence into orthogonal components, one in the direction of propagation and one transverse to it, as illustrated in Figure 4.3. Throughout the remainder of this chapter we will confine ourselves to the subject of partially coherent radiation in which there is a relatively well-defined direction of energy transport.

In exploring the limits of spatial and temporal coherence one considers the propagation of radiation from the source region to the far-field where experiments might

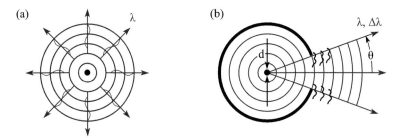

Figure 4.2 (a) Fully coherent radiation from a point source oscillator, which oscillates for all time. Note the circular or spherical nature of the outgoing waves. Knowing the phase at any position gives knowledge of the phase at all positions for all time. (b) Partially coherent radiation from a source of finite size, emission angle, and duration. Note that the outgoing radiation only approximates circular or spherical waves.

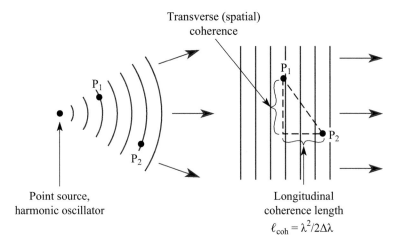

Figure 4.3 Transverse and longitudinal (temporal) coherence involve expectations of knowing the at some point, P_2, if one knows the phase at a different point, P_1. We define "coherence lengths" as distances over which such knowledge is within some defined expectation such as $1/e$ or ≥ 0.5.

be performed.[4] The resultant fields are described mathematically in terms of Fourier transforms in time and space.[1] For the time-frequency transform one finds a reciprocal relationship between the pulse duration, $\Delta\tau$, and the spectral bandwidth, $\Delta\omega$. For Gaussian distributions the resulting relationship is

$$\Delta\tau\,\Delta\omega \geq 1/2 \qquad\qquad (4.4a)$$

where $\Delta\tau$ and $\Delta\omega$ are single sided rms $(1/\sqrt{e})$ measures. That is, short pulses require a broad spectral bandwidth, while narrow spectral features require a longer pulse duration. Equation (4.4a) is closely related to Heisenberg's Uncertainty Principle,[5] famously setting limits within which the energy spread and temporal duration of quantum mechanical distributions can be described. Multiplying Eq. (4.4a) by $\hbar$ one obtains $\Delta E \cdot \Delta\tau \geq \hbar/2$, where $\Delta E = \hbar\Delta\omega$ is the rms bandwidth of photon energy required to generate an x-ray or EUV pulse of rms duration $\Delta\tau$. In convenient units $\hbar/2 = 0.3291$ eV·fs.

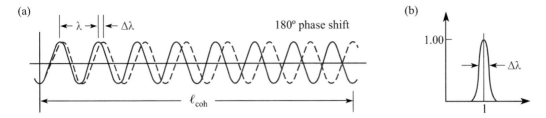

Figure 4.4 (a) Spectral bandwidth and (b) coherence length: destructive interference due to finite spectral bandwidth for radiation of wavelength λ and spectral bandwidth Δλ.

Writing Eq. (4.4a) as a product of two FWHM measures the constant in Eq. (4.5a) is multiplied by $(2.355)^2$, thus

$$\Delta E_{\text{FWHM}}\Delta\tau_{\text{FWHM}} \geq 1.825 \text{ eV}\cdot\text{fs} \tag{4.4b}$$

Complimentary Fourier relationships exist among spreads in spatial dimensions and spatial frequencies, or wavenumbers k. When written in one-dimensional the product is

$$\Delta x \Delta k \geq 1/2 \tag{4.5a}$$

For an uncertainty in vector momentum, $\Delta\mathbf{p} = \hbar\Delta\mathbf{k}$, this takes the form of Heisenberg's Uncertainty Principle for position and momentum[6]

$$\Delta\mathbf{x} \cdot \Delta\mathbf{p} \geq \hbar/2 \tag{4.5b}$$

We utilize this in a following paragraph when considering the limits of spatial coherence.

In the direction of propagation it is common to introduce a longitudinal, or *temporal*, coherence length l_{coh} over which phase relationships are maintained. For a source of bandwidth $\Delta\lambda$, one can define a coherence length

$$\boxed{l_{\text{coh}} = \frac{\lambda^2}{2\Delta\lambda}} \tag{4.6}$$

where $\Delta\lambda$ is the spectral width, as discussed by several authors.[1, 2, 4] The relationship between longitudinal coherence length (e.g., in the direction of propagation) and spectral bandwidth is illustrated in Figure 4.4. Here the coherence length is taken as the distance that results in two waves, of wavelength difference just equal to the bandwidth $\Delta\lambda$, becoming 180° out of phase. Over such a distance one would expect the waves emanating from a source of continuous spectral width to become largely uncorrelated, and thus not contribute significantly to a well-defined interference pattern. Equation (4.6) follows from Figure 4.4 on writing $l_{\text{coh}} = N_{\text{coh}}\lambda$ for the first wave and $l_{\text{coh}} = (N_{\text{coh}} - \frac{1}{2})(\lambda + \Delta\lambda)$ for the spectrally shifted wave, which executes one-half less oscillation (one-half fewer wavelengths) to travel the same distance, and then equating the two to solve for the "number of waves of coherence," $N_{\text{coh}} = \lambda/(2\Delta\lambda)$. Equation (4.6) is then obtained by multiplying N_{coh} by the wavelength, giving the coherence length for which radiation of bandwidth $\Delta\lambda$ becomes substantially dephased. In this manner one can describe the coherence length as the product of the "number of waves (or cycles) of coherence", $\lambda/\Delta\lambda$, times the wavelength. The numerical factor of $1/2$ appearing in Eq. (4.6) is somewhat arbitrary as obtained here, as it depends on the $180°$ dephasing criteria taken. Depending on the experiments to be performed alternate criteria might be selected. The numerical factors' dependence on spectral line shape is discussed by Goodman[2] in his Section 5.1.3. In the experimental formation of interference (fringe) patterns by amplitude dissection (e.g., using a beamsplitter) and recombination, as in interferometry[1] and holography,[4] it is essential that differences in propagation length be less than the coherence length; otherwise high-contrast interference patterns will not be obtained.

Transverse, or *spatial*, coherence is related to the finite source size and the characteristic emission (or observation) angle of the radiation. In this case one is interested in phase correlation in planes orthogonal to the direction of propagation. It is instructive to first associate spatial coherence with spherical waves in the limit of phase being perfectly correlated everywhere. In this first approximation spherical waves are concentric with constant phase across every spherical surface and with phase maxima separated by a wavelength in the outward propagation direction. Although somewhat restrictive, we consider the spherical case because it is common to our experience and yields a clear physical insight. We consider only a small angular portion of a spherical wave as it propagates in a relatively well-defined direction. With some appropriate spectral bandwidth, and thus finite coherence length, such a spherical wave could provide a reference wave for encoding complex wavefronts, as in interferometry or holography. Near-spherical waves can be focused to a spot size approaching finite wavelength limits, which we often refer to as "*diffraction limited*," meaning that the focal spot size is limited only by the finite wavelength and numerical aperture of the focusing lens. One application is a scanning transmission x-ray microscope (STXM) for which illumination by diffraction limited radiation is required to achieve the best spatial resolution. Other features of full spatial coherence include collimation with minimal divergence, which permits propagation of visible laser light to the moon and back for accurate metrology, or use in x-ray diffraction and protein crystallography with minimal angular blur.

Full spatial coherence, the situation in which phase is perfectly correlated at all points transverse to the propagation direction, can be achieved with a spherical wavefront, which we associate with a point source. We might then ask, "How small is a point source?" or more accurately, "How small must the source be to produce wavefronts suitable for our purpose?" and "How small must our undulator electron beam or x-ray laser aperture be in order to provide spatially coherent radiation?" We can obtain a simple estimate based on Heisenberg's Uncertainty Principle, previously cited as Eq. (4.5b)

$$\Delta\mathbf{x} \cdot \Delta\mathbf{p} \geq \hbar/2$$

Here $\Delta\mathbf{x}$ is the vector uncertainty in position and $\Delta\mathbf{p}$ the vector uncertainty in momentum, both being single-sided rms measures of Gaussian probability distributions.[5] With Eq. (4.5b), we can determine the smallest source size d resolvable with finite wavelength λ and observation half angle θ. For photons the momentum is $\hbar\mathbf{k}$, where the

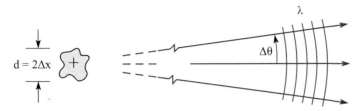

Figure 4.5 Spherical wavefronts and spatially coherent radiation are approached when the source size and far-field divergence angle are related to wavelength as indicated in Eq. (4.7a).

scalar wavenumber $|\mathbf{k}|$ is $2\pi/\lambda$. If the relative spectral bandwidth $\Delta\lambda/\lambda$, which is equal to $\Delta k/k$, is small, then the uncertainty in momentum, $\Delta\mathbf{p} = \hbar\Delta\mathbf{k}$, is due largely to the uncertainty in direction θ, so that for small angles $|\Delta\mathbf{p}| = \hbar k\Delta\theta$. Substituting into the uncertainty relation (4.5b)

$$\Delta x \cdot \hbar k\Delta\theta \geq \hbar/2$$

and noting that $k = 2\pi/\lambda$, one has

$$\Delta x \cdot \Delta\theta \geq \lambda/4\pi$$

Identifying the source diameter as $d = 2\Delta x$ and the divergence half angle θ with the uncertainty $\Delta\theta$, as illustrated in Figure 4.5, we obtain the limiting relationship[7–9]

$$d \cdot \theta = \lambda/2\pi \tag{4.7a}$$

where for convenience we have replaced $\Delta\theta$ by θ. Equation (4.7a) determines the smallest source size we can discern; that is, within the constraints of physical laws we would not be able to tell if our "point" source were any smaller. When this condition is met the implied spherical wavefronts appear to emanate from a near point source and reducing the lateral extend futher would have a minimal impact on the apparent size. We note that Eq. (4.7a) is written for rms measures of Gaussian intensity distributions (for θ, and twice so for d). For non-rms measures the numerical factor $(1/2\pi)$ will be different. For Gaussian intensity distributions measured in terms of a FWHM diameter (d) and a FWHM angle (2θ), the equivalent relation is

$$d_{\text{FWHM}} \cdot 2\theta_{\text{FWHM}} = \frac{2\ln 2}{\pi}\lambda = 0.4413\lambda \simeq \lambda/2 \tag{4.7b}$$

Radiation satisfying the equality (4.7a) is said to be *diffraction limited* – that is, limited in its propagation and focusability by its finite wavelength and the observation half angle (or numerical aperture $\sin\theta \cong \theta$). To generate a spatially coherent spherical wave at x-ray wavelengths one must develop a source whose characteristics approach the limiting values set by Eq. (4.7a).

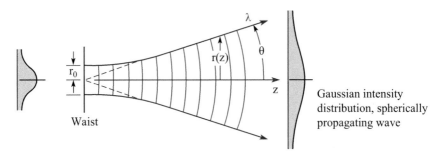

Figure 4.6 Propagation of a Gaussian beam with rms single-sided intensity radius r_0.

In some cases it is convenient to replace θ with an in-plane lateral coherence length at a propagation distance z from the source; e.g., if one defines $l_{\text{transverse}} \equiv z\theta$, one has an in-plane transverse coherence length

$$l_{\text{transverse}} = \frac{z\lambda}{2\pi d}$$

The primary relationship, however, is that of the space–angle formula given by Eq. (4.7a).

For comparison, a fully coherent laser radiating in a single transverse mode, TEM_{00}, satisfies this same condition, Eq. (4.7a) when the waist diameter d and far-field divergence half angle θ are written in terms of rms quantities, as illustrated in Figure 4.6. For a spherical wave propagating with a Gaussian intensity distribution, written as

$$I/I_0 = \exp(-r^2/2r_0^2) \tag{4.8a}$$

where r_0 is the $1/\sqrt{e}$ (rms) waist radius at the origin ($z = 0$), and where the intensity distribution grows with a $1/\sqrt{e}$ radius given by[6–8]

$$r(z) = r_0\sqrt{1 + \left(\frac{\lambda z}{4\pi r_0^2}\right)^2} \tag{4.8b}$$

Thus in the far field, where $z \gg 4\pi r_0^2/\lambda$, the $1/\sqrt{e}$ divergence half angle is

$$\theta \equiv \frac{r(z)}{z} = \frac{\lambda}{4\pi r_0}$$

With a waist diameter $d = 2r_0$, this TEM_{00} laser cavity mode exhibits a product of waist diameter and far-field divergence half angle (both in terms of $1/\sqrt{e}$ measures) given by

$$d \cdot \theta = \frac{\lambda}{2\pi}$$

as found previously in Eq. (4.7a) on the basis of Heisenberg's Uncertainty Principle. Note that the distance at which the cross-sectional area doubles provides a measure of the depth of field, or depth of focus (DOF). See Eq. (4.8b). This distance is known as the (single-sided) "Rayleigh range," typically expressed[6–9] for a Gaussian distribution in terms of the $1/e^2$ radius $\mathbf{r}_0 = 2r_0$ as

$$Z_R = 4\pi r_0^2/\lambda = \pi \mathbf{r}_0^2/\lambda \tag{4.9}$$

The double-sided depth of focus, $2Z_R$, is known as the "confocal parameter." The $1/e^2$ radius ($\mathbf{r}$) is of interest because 86.5% of a Gaussian beam's energy is contained within a cross-section of that radius. The various $1/\sqrt{e}$, FWHM, and $1/e^2$ radii are illustrated in the frontpiece of this chapter.

In summary, in this section we have developed two convenient relationships by which to gauge the coherence properties of a radiation field for the purpose of conducting phase sensitive interference experiments:

$$l_{\text{coh}} = \frac{\lambda^2}{2\Delta\lambda} \qquad \text{(temporal or longitudinal coherence)} \tag{4.6}$$

and

$$d \cdot \theta = \lambda/2\pi \qquad \text{(spatial or transverse coherence)} \tag{4.7a}$$

or equivalently for FWHM measures of the spatial coherence condition

$$d_{\text{FWHM}} \cdot 2\theta_{\text{FWHM}} = \frac{2\ln 2}{\pi}\lambda = 0.4413\lambda \simeq \lambda/2 \tag{4.7b}$$

In Chapter 5, Section 5.6, we will use these measures to determine what fraction of undulator power, or photon flux, is useful for experiments requiring spatially and temporally coherent radiation.

4.2 Spatial and Spectral Filtering

We concluded in the previous section that the limiting condition of spatially coherent radiation is a space–angle product [Eq. (4.7a)], or phase space* volume

$$d \cdot \theta = \lambda/2\pi$$

where d is a Gaussian $1/\sqrt{e}$ diameter and θ is the Gaussian half angle. All physical sources generate radiation of space–angle product larger than this, often considerably larger. At visible wavelengths, for instance, only lasers with intra-cavity mode control approach this limit, those operating in the so-called TEM_{00} mode.[6–8] The question here is: what if your source generates radiation into a larger phase space, largely incoherent in nature, but you wish to use it for phase sensitive experiments that require a

---

* This space–angle product is often referred to as a "phase space" volume. This derives from the study of dynamics, where particles are position–momentum phase space ($\Delta\mathbf{x}$, $\Delta\mathbf{p}$). For photons $p = \hbar\mathbf{k}$, and for nearly monochromatic radiation the interval in momentum $\Delta\mathbf{p} = \hbar\Delta\mathbf{k}$ becomes $\Delta\mathbf{p} = \hbar k\Delta\theta$, where $\Delta\theta$ is transverse to $\mathbf{k}$. Thus for nearly monochromatic photons the interval of position–momentum phase space becomes $\Delta\mathbf{x} \cdot \Delta\theta$, which has a scalar minimum given by Eq. (4.7a).

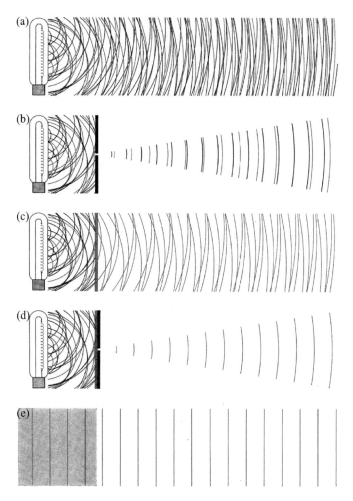

Figure 4.7 (a) An ordinary thermal light source. Spatial (b) and spectral (c) filtering is illustrated as a procedure to produce spatially and temporally coherent radiation, as in part (d), albeit at greatly diminished power. In the nomenclature used here, d would be the diameter of the pinhole and θ would be the divergence half angle, as seen in part (d), the latter set by the radiation emission characteristics, by a downstream acceptance aperture, or by a lens. Laser radiation is illustrated in (e), emitted with near-perfect phase coherence. Courtesy of A. Schawlow[9], Stanford University.

higher degree of coherence? Arthur Schawlow, in his article on lasers[9], introduces a very informative illustration to show how such radiation can be filtered, both spectrally and spatially, to obtain radiation of greatly improved coherence properties, albeit at the considerable loss of available power. The illustration is reproduced here in Figure 4.7. Shown is a typical thermal light bulb with an extended filament heated to a temperature such that many excited atoms randomly radiate a broad spectrum of white light – that is, radiate a continuum containing all colors of the spectrum visible to the human eye. The radiation is filtered in two ways. A pinhole is used [Figure 4.7(b)] to obtain spatially coherent radiation (over some angular extent), as set here by Eq. (4.7a). A color filter is used [Figure 4.7(c)] to narrow the spectral bandwidth, thus providing a degree of longitudinal coherence, as described here in Eq. (4.6). Combining both the pinhole and the filter, one obtains radiation that is both spatially and temporally coherent, as is seen in Figure 4.7(d), but with a power that is only a small fraction of the total power radiated by the light bulb.

The technique of spatial and spectral filtering is therefore important, and is now commonly used at synchrotron radiation facilities for a variety of experiments involving scanning x-ray microscopy, interferometry, coherent x-ray scattering, coherent diffractive imaging, ptychography, and holography. The use of spatial and temporal filtering of undulator radiation at powers sufficient to permit experiments at these very short wavelengths is discussed further in Chapter 5, Section 5.6. As Schawlow points out in his article, and illustrates in Figure 4.7(e), a visible light laser has the great advantage of providing these desired coherence properties, often with very long temporal coherence-length (very narrow $\Delta\lambda/\lambda$), without compromising available power.[†] This, however, is a much greater challenge at x-ray wavelengths, both because the energetics make lasing at high photon energy more difficult, and because the very short wavelengths place great demands on the achievement of substantial spatial coherence. Nonetheless, great progress has been made recently in the availability of coherent radiation with laser high harmonic generation (HHG) at EUV and soft x-ray wavelengths, and with free electron lasers (FELs) at EUV, soft and hard x-ray wavelengths. With both HHG and FELs essentially full spatial coherence is achieved, albeit with varying degrees of temporal coherence. Free electron lasers and their coherence properties are described in Chapter 6, while high harmonic generation and its coherence properties are described in Chapter 7.

4.3   Examples of Experiments that Require Coherence

As discussed in a preceding section, radiation from a real physical source can at best only approach the limits of full coherence due to both finite spectral width and finite physical extent. But it is possible to come very close, even with x-rays.[10] For example, if one wishes to focus radiation to the smallest possible spot size, at a given wavelength (λ) and lens numerical aperture (NA), the lens must be coherently illuminated, as illustrated in Figure 4.8. Such focusing is essential for the achievement of highest spatial resolution in a scanning x-ray microscope, a topic we take up in Chapter 10. The

---

[†] An advantage of visible light lasers is that full spatial coherence is easily achieved through inclusion of a mode-selecting pinhole within the cavity, permitting amplification of only spatially coherent radiation. Temporal coherence is generally very high due to the spectral narrowness of the quantum states involved, as well as further spectral line narrowing by the gain process itself, and perhaps the use of an intra-cavity interference filter known as an etalon. A description of a typical laser cavity, its components and their roles, is given in Chapter 9. On the other hand, lasers designed to generate very short pulses, in the femtosecond range, require very broad spectral bandwidth, such as the Ti:sapphire laser,[6] which is critical to current advances in high harmonic generation.

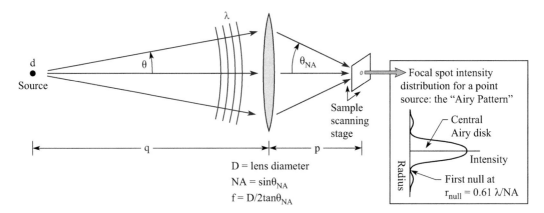

D = lens diameter

$NA = \sin\theta_{NA}$

$f = D/2\tan\theta_{NA}$

Figure 4.8 Diffraction limited focusing – that is, limited only by the finite wavelength and lens numerical aperture (NA) – requires a perfect lens and coherent illumination. The refractive lens shown is for illustration only. At x-ray wavelengths this would require diffractive or reflective optics, such as a Fresnel zone plate or a multilayer coated spherical mirror. In a scanning microscope a sample would be placed at the focus and raster scanned with a suitable translation stage while observing an appropriate signal such as transmitted x-rays, fluorescent emission of characteristic radiation, or photoelectrons. The spatial resolution of the measurement would be set by the focal spot size, assuming this is not compromised by lens imperfections, imperfect illumination, or scanning stage limitations.

advantage of scanning x-ray microscopy is that it is capable of achieving significantly smaller focal spots than are achievable with visible or ultraviolet radiation, with minimal radiation dose, and often combined with high spectral resolution in what is called spectromicroscopy. As a result, scanning x-ray microscopy has become a widely used tool for the study of environmental and biological materials, chemical fibers and material surfaces. To achieve the smallest possible focal spot size, for example with synchrotron radiation, spatial filtering is essential. The lens forms an image of the pinhole, now the source of spatially coherent radiation. This is referred to as *diffraction limited* focusing because the intensity distribution in the focal region is now limited by the finite wavelength and lens numerical aperture, rather than the actual source size. This is, of course, a limiting case. For a larger source size the image would simply be demagnified by the ratio $M = q/p$, where q is the source to lens distance, p is the lens to image distance, and these are related to the lens focal length f by the reciprocal thin lens equation $1/f = 1/p + 1/q$. In the diffraction limited case, however, the source size d is sufficiently small that the radiation intercepted by the lens (see θ in Figure 4.8) approximately satisfies the spatial coherence condition set by Eq. (4.7a). In this case the focal region intensity pattern approximate an Airy pattern[1, 11] with a focal region radius to the first null given by $0.61\lambda/NA$. The fact that this is a spherical wave illumination of the lens, rather than a plane wave illumination, simply moves the focal plane to a conjugate point determined by the thin lens equation for finite source distance q.

A second example in which coherence plays an important role is that of encoding complex wavefronts, as in holography.[4] To form an interference pattern in the plane of the recording medium, the reference wavefront and the wavefront scattered by the sample must maintain a time averaged phase relation, that is, the detected fields must at every point in this plane have a degree of coherence of order unity in order to form a recordable interference pattern. Choice of a spherical reference wavefront provides a clear mechanism for wavefront encoding and subsequent decoding, or reconstruction. Furthermore, to ensure high-contrast encoding (interference) it is essential that all path lengths from the source to the detector be equal to within a longitudinal coherence length, $l_{coh} = \lambda^2/(2\Delta\lambda)$. The path matching condition must be satisfied at every point in the detector plane. Having satisfied these conditions, and with sufficient coherent photon flux or power, a suitable interference pattern can be produced and recorded with an appropriate detector. In general, the interference pattern at x-ray wavelengths will be characterized by a very fine spatial scale, equal to the wavelength divided by the intersection angle of the two wavefronts.

X-ray holography has long provided an intellectual and experimental challenge, utilizing clever solutions to the requirements of spatial coherence for the recording of finely scaled x-ray interference patterns. Early examples that demonstrated off-axis holography with spatially and temporally filtered x-rays were described in the literature by Aoki, Kikuta, Kohra, and their colleagues[12, 13], and by others soon thereafter[14–17]. An illustration of a successful x-ray holographic experiment by Eisebitt, Lüning, Schlotter, and their colleagues[18] is shown in Figure 4.9. Circularly polarized soft x-rays at 1.59 nm wavelength (778 eV), emitted from an undulator at the BESSY II synchrotron facility in Berlin, were spatially filtered by a 20 μm diameter pinhole to coherently illuminate the mask and sample plane at a distance of 723 mm. The transverse coherence length in the mask plane was estimated to be about 9 μm. An upstream monochromator set the spectral bandwidth, and thus the longitudinal coherence length to 1.6 μm, sufficient to accomodate the path length differences associated with this narrow angle, Fourier transform geometry. The mask includes a 1.5 μm diameter cobalt/platinum magnetic sample as well as a second, 2.0 μm diameter, pinhole which provides the spherical reference wavefront for encoding the more complex object wavefront in this Fourier transform hologram geometry. High-contrast fringes were recorded as shown in Figure 4.9. Fringe patterns were recorded digitally at a distance of 315 mm downstream of the mask on a soft x-ray sensitive charge-coupled-device (CCD). The observed magnetic features of the Co/Pt multilayed sample compared well with those of the same object recorded by scanning transmission x-ray microscopy.[18] Further holographic experiments continue to extend this technique.[19] A complex reference wave technique described by Marchesini[20] promises to overcome the inefficiency of pinhole filtered Fourier holography. It employs a *uniformly redundant array* to generate a complex reference wave, essentially an array of reference beams, thus significantly rebalancing power in the object and reference waves. This larger area reference pattern could provide a considerably stronger reference wave than that obtained with a small pinhole, likely by several orders of magnitude, permitting more efficient recording and shorter exposure times. Techniques that permit femtosecond recording capabilities have been described by Chapman and his colleagues.[21] Consideration of spatially and temporally coherent power levels available for experiment such as these is discussed in detail in Chapter 5, Section 5.6. Broad

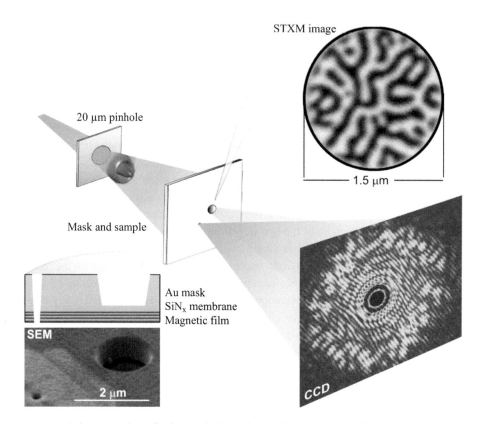

Figure 4.9 A demonstration of soft x-ray holography employing spatially filtered undulator radiation at a wavelength of 1.59 nm (778 eV) to probe the magnetic state of a cobalt–platinum multilayer sample at the Co-L_3 edge.[18] Courtesy of S. Eisebitt, BESSY/HZB and Technical University of Berlin.

reviews of coherent methods in the x-ray sciences are given by Vartanyants and Singer[22], and by Nugent.[23]

Another example of the use of coherent x-rays is the creation of x-ray standing waves at the surface of a mirror for the analysis of material properties. To create standing waves the radiation must be spatially and spectrally filtered to the degree required, typically with synchrotron radiation and the use of both a monochromator and an upstream pinhole (or crossed slits). The standing waves are created by interference between the incident and reflected monochromatic radiation, similar to the interference effects discussed in Chapter 3, Section 3.9. The interference pattern consists of a set of planes parallel to the mirror surface,[24] as seen in Figure 4.10, with a d-spacing satisfying the Bragg equation, Eq. (3.91a). The period can be varied by changing the wavelength or the angle of incidence. The number of periods within the interference pattern is proportional to the (inverse) relative spectral bandwidth, $\lambda/\Delta\lambda$.[25] By depositing stratified materials on the mirror surface, properties of the material can be probed as a function of height above the surface.[26–28] For example, by observing fluorescence emission as the angle of incidence is varied from zero to θ_c, the period of the pattern is varied

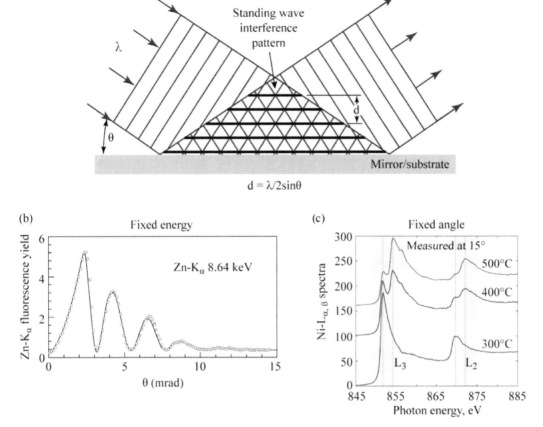

Figure 4.10 (a) The technique of x-ray standing wave analysis is based on the interference of incident and reflected waves just above the surface of a mirror.[24] To form a standing wave interference pattern with synchrotron radiation, the grazing incidence radiation must be spectrally and spatially filtered to provide the required degree of coherence. The interference pattern satisfies the Bragg condition. The properties of stratified material coated on the mirror surface can be analyzed by observing characteristic fluorescence emissions as a function of height from the surface as the angle of incidence is varied, causing peaks of the interference pattern to move through the sample. Both K- and L-shell fluorescence can be observed depending on incident photon energy and elements present. (b) An example is shown of Zn K_α emission at 8.64 keV as a function of incidence angle, obtained with 9.8 keV bending magnet radiation at CHESS.[24] Both (a) and (b) courtesy of M.J. Bedzyk, Cornell University (now at Northwestern University). (c) An example of NEXAFS data in the vicinity of the nickel (Ni) L_3 and L_2 edges, as a function of creation temperature, for a nickel coated silicon mirror with a BCN overlayer. The spectra were obtained at 15° incidence angle to enhance emission from the Ni–BCN interface, using 1060 eV radiation at the PTB undulator beamline at BESSY II.[27] Panel (c) courtesy of B. Pollakowski and B. Beckhoff, PTB.

and as a result the plannes of maximum intensity move vertically through the material. Both K- and L-shell fluourescence can be observed depending on the incident photon energy and the elemental composition of the material being studied. Once the planes of high concentration of a specific element are identified, Near-Edge X-ray Absorption

Structure (NEXAFS) spectroscopy can be used to determine chemical bonding information (chemical speciation) of that element at the identified planes. This is a powerful tool, for example for the study of buried nanostructures or the distribution of elemental and chemical content as a function of depth in stratified media.[26–28] Figure 4.10(a) shows the basic technique of interference between the incident and reflected waves, and the resultant stationary interference pattern with planes of constant phase parallel to the surface. The period of the planes, d, is set by the Bragg equation for wavelength λ and angle of incidence, θ. Shown as dark lines are several planes of high intensity which can be moved up or down in the vertical direction by varying either θ or λ. Shown in Figure 4.10(b) are experimental data, obtained with 9.8 keV bending magnet radiation at CHESS, of Zn K_α emission at 8.64 keV from a thin layer of zinc, supported above the mirror surface by a Langmuir–Blodgett thin film, as the angle of incidence is increased and the planes of high intensity periodically move through the Zn layer.[24] Shown in (c) is NEXAFS spectroscopic data near the L_3 and L_2 edges of nickel for a fixed angle of $15°$ and a photon energy of 1060 eV. This experiment was part of a study of chemical bonding at the interface between nickel and a boron–carbon–nitrogen (BCN) overcoating, each nominally 5 nm thick, and formed at various temperatures as indicated.[27] The incident photon energy was selected to be above the nominally 852–870 eV Ni-L edges. The angle of $15°$ was determined by fluorescence measurements to place a plane of maximum intensity at the Ni–BCN interface. Spectral shifts indicate a diffusion of Ni atoms into the BCN overcoat at the higher temperatures, moving some Ni atoms closer to the vacuum interface. Vertical data offsets are used for clarity. The measurements were made with the PTB undulator beamline at BESSY II.

4.4   The van Cittert–Zernike Theorem

In the previous sections of this chapter we have compared the phase-space of emitted radiation from an incoherent source with that from a nearly point source for which the radiated fields approach perfect correlation in the transverse plane and for which the phase-space is given by the limiting condition $d \cdot \theta = \lambda/2\pi$. In this section we consider the finite degree of spatial coherence that results when an extended incoherent source of quasi-monochromatic radiation is observed through the use of spatial and angular apertures. The complex degree of coherence in such a radiation field is described by the van Cittert–Zernike theorem.[1, 2, 29, 30] It is of particular interest to us as a means to predict the degree of spatial coherence that will result from pinhole spatial filtering of both undulator radiation and EUV/soft x-ray laser radiation, each of which is, to a large degree, spatially incoherent in nature. For the consideration of spatially coherent radiation one is primarily interested in the correlation of quasi-monochromatic fields in the limit that the time separation τ goes to zero. In this limit the normalized degree of coherence between fields at points 1 and 2 is given by

$$\mu_{12} = \frac{\langle E_1(t)E_2^*(t)\rangle}{\sqrt{\langle |E_1|^2\rangle}\sqrt{\langle |E_2|^2\rangle}} \tag{4.3}$$

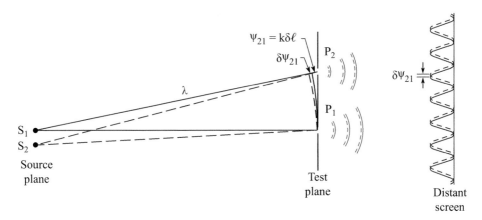

Figure 4.11 The persistence of interference fringes as observed in a distant plane is dependent on the spatial separation of two mutually incoherent quasi-monochromatic point sources, the angles, and the wavelength. The source S_1 illuminates small openings at points P_1 and P_2 in a *test plane*. For a single point source S_1 there is a fixed variation ψ_{21} between the two observation points. The resulting interference pattern is observed at a distant screen (solid line). Placing a second point source S_2 at a small distance from S_1 results in a second interference pattern, mutually incoherent with the first, which very nearly overlaps with the interference pattern due to S_1, but is shifted by a small amount, $\delta\psi_{21}$. This small phase increment is due to the separation between S_1 and S_2. Thus one sees that within some as yet to be determined limits on source size, observation geometry, and wavelength, mutually incoherent sources of emission can provide a highly coherent radiation field at a distant plane. This is the basis of the van Cittert–Zernike theorem.[24, 25]

with absolute values bounded by $0 \le |\mu_{12}| \le 1$. While generally referred to as the normalized degree of coherence, μ_{12} is also known as the complex coherence factor.

The van Cittert–Zernike theorem provides a very convenient method for calculating the degree of spatial coherence that can be derived from a collection of mutually incoherent but quasi-monochromatic radiators. That spatially coherent radiation can be obtained in any circumstance involving uncorrelated radiators may at first seem surprising. Figure 4.11 introduces the subject. Imagine first that in Figure 4.11 there is only the point source S_1, whose radiation at wavelength λ illuminates a mask with two small openings at points P_1 and P_2 in what we call the *test plane*, because we will use it to test the degree of coherence as a function of separation distance between the two observation points. Radiation will pass these two small openings, propagating to a distant screen where the two beams will overlap. Because the emission is quasi-monochromatic and from a point source (S_1), a well-defined interference pattern will be formed on the screen, as indicated by the solid sinusoidal pattern. Note that the exact location of maxima and minima in this self-interference pattern will depend on the phase difference of the paths from S_1 to P_1 and P_2, indicated as ψ_{21} in Figure 4.11. This phase difference is caused entirely by the geometry (and wavelength) and is therefore constant in time, so that the interference pattern (maxima and minima) is also constant in time. Now consider a second point source S_2 at a very small distance from the first (S_1). Its radiation

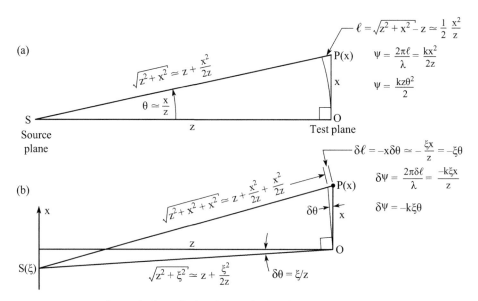

Figure 4.12 Propagation paths for radiation from a single point source to two observation points O and P at a distant observation plane. The separation distance z between the two parallel planes is large compared to the lateral source (ξ) and observation (x) distances in their respective planes ($z \gg \xi, x$). Shown in (a) are the differences in path length and the resulting phase difference ψ for a spherical wave propagating from a point source at the origin in the source plane to two points in the test (observation) plane, one at the origin O and one a distance x off axis at a point P(x). In (b) the path lengths and phase difference $\delta\psi$ are shown when the source point is displaced a distance ξ from the origin in the source plane.

paths are indicated by the dashed lines, resulting in a second stationary interference pattern at the screen, also shown as a dashed line in Figure 4.11. Because the two point radiators are uncorrelated, their fields do not combine to form a mixed (time averaged) interference pattern. Rather, they each form separate self-interference patterns that are quite similar, in fact displaced on the distant screen by only a small phase difference $\delta\psi_{21}$, which is evidently due to the small spatial separation between S_1 and S_2.

We thus begin to see the emerging picture. With some constraints on lateral source size, wavelength, and observation angles, it is possible to obtain rather well-defined interference patterns even with completely uncorrelated emissions. On the other hand, it is also possible to scramble the resultant interference pattern completely by observing a collection of such radiators whose positions in the source plane are sufficiently separated that the resultant phase shifts $\delta\psi_{21}$ are of order π rad or larger, so that the summed intensity patterns at the screen show no net modulation, indicating a total absence of coherence in the test plane. Following Born and Wolf[1], we will now detail the geometry of propagation paths involved and explore under what circumstances a finite degree of coherence can be obtained, and what the resultant degree of coherence will be.

In Figure 4.12(a) we revisit this geometry for the single point source S, showing propagation paths to two points in the test plane, the origin O, which we use to define an optical axis, and a point P a distance x away in the transverse plane. The distance z separating the parallel source and observation (test) planes is assumed very large compared to the lateral source and observation distances, so that the angle θ is very small. The phase difference between the two paths (ψ_{21} in Figure 4.10) is

$$\psi = \frac{2\pi l}{\lambda} = \frac{kx^2}{2z} = \frac{kz\theta^2}{2} \tag{4.10}$$

where l is the difference in geometrical path lengths, determined by comparing the sides of the triangle with equal radial distances measured from S. This difference gives the additional path length for a propagating wave. In Figure 4.12(b) we show the point source displaced a distance ξ in the source plane, and again calculate the various path lengths. With the source displaced a distance ξ we observe that a ξ-dependent increment of path difference is introduced, $\delta\psi$, where from the geometry

$$\delta l = -\xi x/z = -\xi\theta$$

with a corresponding increment in phase

$$\delta\psi = 2\pi\,\delta l/\lambda$$

or

$$\delta\psi = -k\xi x/z = -k\xi\theta \tag{4.11}$$

The negative sign indicates that negative values of the displacement ξ, as illustrated in Figure 4.12(b), produce a positive phase shift $\delta\psi$. These relations can be seen by forming a right angle from the path $\bar{S}\bar{O}$ to the line from S to P. The increment δl, due to displacement ξ, is also dependent on the distance x from O to P in the test plane.

The scalar electric field at a distance R from a point source of field E_ξ can be written as

$$E = \frac{E_\xi e^{ikR}}{R}$$

where the time dependence $e^{-i\omega t}$ and an arbitrary phase factor are suppressed. Thus the electric field at points O and P due to the source S at ξ can be written for $x, \xi \ll z$ as

$$E_O = \frac{E_\xi e^{ik(z+\xi^2/2z)}}{z+\xi^2/2z} \simeq \frac{E_\xi e^{ikz}e^{ik\xi^2/2z}}{z} \tag{4.12}$$

and

$$E_P = \frac{E_\xi \exp\left[ik\left(z+\frac{x^2}{2z}+\frac{\xi^2}{2z}-\frac{\xi x}{z}\right)\right]}{z+\frac{x^2}{2z}+\frac{\xi^2}{2z}-\frac{\xi x}{z}} \simeq \frac{E_\xi e^{ikz}e^{ik\xi^2/2z}e^{ikx^2/2z}e^{-ik\xi x/z}}{z}$$

or

$$E_P \simeq \frac{E_\xi e^{ikz}e^{ik\xi^2/2z}e^{i\psi}e^{-ik\xi\theta}}{z} \tag{4.13}$$

where ψ is given by Eq. (4.10) and where we observe that E_O and E_P contain two identical phase factors, e^{ikz} and $e^{ik\xi^2/2z}$. Thus for a single point source at ξ, the normalized degree of coherence for electric fields at a point O on the optic axis and a point P at distance x off axis can be obtained by combining Eqs. (4.12), (4.15a–d), and (4.16) to

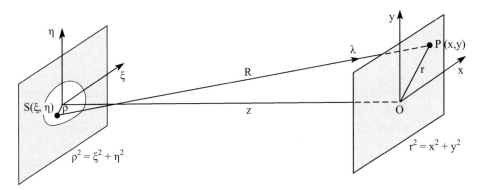

Figure 4.13 The geometry for deriving the van Cittert–Zernike theorem, which involves integration in the (ξ, η) source plane of the contributions from a continuum of quasi-monochromatic mutually incoherent sources of radiation to the fields at a distant observation plane at the origin O and the off-axis point P(x, y). Following M. Born and E. Wolf.[1]

form

$$\mu_{OP} = \frac{\langle E_O E_P^* \rangle}{\sqrt{\langle |E_O|^2 \rangle}\sqrt{\langle |E_P|^2 \rangle}} = e^{-i\psi} e^{ik\xi\theta} \qquad (4.14)$$

where for $z \gg x, \xi$ both normalizing fields can be approximated by

$$\sqrt{\langle |E_O|^2 \rangle} \simeq \sqrt{\langle |E_P|^2 \rangle} \simeq \frac{|E_\xi|}{z}$$

In Eq. (4.14) ψ is an x-dependent phase shift (for fixed z) due to the displacement of P from the axis, and the $ik\xi\theta$ factor is due to the ξ-dependent tilt of the wavefront, with a leverage due to x, where $\theta \simeq x/z$. Equation (4.14) is readily extended to three dimensions (ξ, η, z and x, y, z), as illustrated in Figure 4.13. The normalized correlation function becomes

$$\mu_{OP} = e^{-i\psi} e^{ik(\xi\theta_x + \eta\theta_y)} \qquad (4.15a)$$

where in the three-dimensional coordinate system

$$\psi = k\frac{x^2 + y^2}{2z} = kz\frac{\theta_x^2 + \theta_y^2}{2} \qquad (4.15b)$$

$$\theta_x = \frac{x}{z} \qquad (4.15c)$$

$$\theta_y = \frac{y}{z} \qquad (4.15d)$$

Again, the ψ factor is strictly due to the angular projection of the observation point P as measured from the optic axis, with no dependence on the source point location, while the $k(\xi\theta_x + \eta\theta_y)$ factor is now generalized to take account of the wavefront tilt due to both source point coordinates ξ and η.

   If we now assume that there is a distribution of quasi-monochromatic source points in the ξ, η-plane, all mutually incoherent, then we can sum their individual contributions to the fields at O and P in the x, y-plane. Integrating over the extended source as suggested in Figure 4.12, one obtains the complex degree of coherence by combining Eqs. (4.14) and (4.15a):

$$\mu_{OP} = \frac{e^{-i\psi} \iint |E(\xi, \eta)|^2 e^{ik(\xi\theta_x + \eta\theta_y)} \, d\xi \, d\eta}{\iint |E(\xi, \eta)|^2 \, d\xi \, d\eta}$$

or in terms of the source plane intensity distribution $I(\xi, \eta) \propto |E(\xi, \eta)|^2$,

$$\mu_{OP} = \frac{e^{-i\psi} \iint I(\xi, \eta) e^{ik(\xi\theta_x + \eta\theta_y)} \, d\xi \, d\eta}{\iint I(\xi, \eta) \, d\xi \, d\eta} \qquad (4.16)$$

This is the *van Cittert–Zernike theorem*[29, 30]. It states that the normalized degree of coherence for a distribution of uncorrelated quasi-monochromatic emissions, observed in a distant plane, is equal to the two-dimensional Fourier transform of the source intensity function. Again the phase $\psi = k(x^2 + y^2)/2z$ is purely geometrical, giving the predictable oscillation of phase as the observation point P is moved from the reference position O at the origin. Fourier transform pairs are given in Appendix D.6

Of particular interest to us is the axisymmetric case, as might be encountered in pinhole spatial filtering of undulator or EUV/soft x-ray laser radiation. For an axisymmetric geometry it is convenient to introduce cylindrical coordinates, (ρ, φ_S), in the (ξ, η) source plane, and (r, φ_P) in the (x, y) observation (test) plane. The geometrical details are clarified in Figure 4.14. The normalized degree of coherence can now be written as

$$\mu_{OP} = \frac{e^{-i\psi} \int_0^\infty \int_0^{2\pi} I(\rho, \phi_S) e^{ik(\rho\cos\phi_S \cdot \theta \cos\phi_P + \rho\sin\phi_S \cdot \theta \sin\phi_P)} \rho \, d\rho \, d\phi_S}{\int_0^\infty \int_0^{2\pi} I(\rho, \phi_S) \rho \, d\rho \, d\phi_S}$$

$$\mu_{OP} = \frac{e^{-i\psi} \int_0^\infty \int_0^{2\pi} I(\rho, \phi_S) e^{ik\rho\theta\cos(\phi_S - \phi_P)} \rho \, d\rho \, d\phi_S}{\int_0^\infty \int_0^{2\pi} I(\rho, \phi_S) \rho \, d\rho \, d\phi_S}$$

where $\cos(\varphi_S - \varphi_P) = \cos\varphi_S \cos\varphi_P + \sin\varphi_S \sin\varphi_P$. For the axisymmetric case, where $I(\rho, \varphi_S) = I(\rho)$, integration over the angle ϕ_s yields

$$\mu_{OP} = \frac{e^{-i\psi} \int_0^\infty I(\rho) J_0(k\rho\theta) \rho \, d\rho}{\int_0^\infty I(\rho) \rho \, d\rho} \qquad (4.17)$$

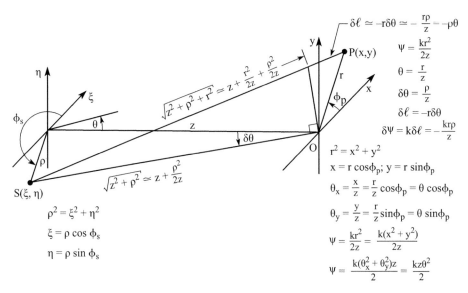

Figure 4.14 Geometry of the various propagation path lengths and phase differences (ψ) as a function of cylindrical coordinates (ρ, φ_S) in the source plane and (r, φ_P) in the observation (test) plane. The angle θ is measured from the optic axis to the point P.

where we have made the identification[31–33]

$$J_0(k\rho\theta) = \frac{1}{2\pi} \int_0^{2\pi} e^{ik\rho\theta\,\cos(\phi_S - \phi_P)}\, d\phi_S \qquad (4.18)$$

for a Bessel function[34] $J_0(\nu)$ of the first kind, order zero. In this axisymmetric geometry, the degree of coherence is described in terms of a Fourier–Bessel transform of the radial source function, sometimes described as a Hankel transform.[35, 36]

We next consider three axisymmetric cases of interest, (1) a point source, (2) a Gaussian intensity distribution, and (3) a uniformly but incoherently illuminated pinhole. The analysis of each follows.

(1) **A Point Source:** For a point source described by a normalized delta function[‡] in cylindrical coordinates,

$$I = I_0 \delta(\rho)/2\pi\rho \qquad (4.19)$$

the degree of coherence given by Eq. (4.17) is

$$\mu_{\mathrm{OP}} = \frac{e^{-i\psi} \displaystyle\int_0^\infty \delta(\rho) J_0(k\rho\theta)\, d\rho}{\displaystyle\int_0^\infty \delta(\rho)\, d\rho} = e^{-i\psi} J_0(0)$$

[‡] The normalization condition is determined by $\int_0^\infty \int_0^{2\pi} f(\rho,\phi)\rho\, d\rho\, d\phi = 2\pi \int_0^\infty f(\rho)\rho\, d\rho = 1$.

or

$$\mu_{OP} = e^{-i\psi} \tag{4.20a}$$

where $J_0(0) = 1$ and, from Figure 4.14, $\psi = kr^2/2z = kz\theta^2/2$. Thus for a true point source the normalized degree of coherence $|\mu_{OP}| = 1$, so that the radiated field is *fully coherent* in the distant (z) plane, with a *perfectly described phase variation* $\psi(r)$ as a function of off-axis position, i.e., the field variations in the observation or test plane are completely predictable, with a variation for distance z between the source and observation planes given by

$$\mu_{OP} = e^{-ikz\theta^2/2} \tag{4.20b}$$

with a θ^2 phase variation first encountered in Figure 4.12(a).

(2) **A Gaussian Intensity Distribution:** For an axisymmetric Gaussian intensity distribution of uncorrelated emitters with standard deviation a such that

$$I = I_0 e^{-\rho^2/2a^2} \tag{4.21}$$

the normalized degree of coherence [Eq. (8.20)] takes the form

$$\mu_{OP} = \frac{e^{-i\psi} \int_0^\infty e^{-\rho^2/2a^2} J_0(k\theta\rho)\rho \, d\rho}{\int_0^\infty e^{-\rho^2/2a^2} \rho \, d\rho}$$

These are standard integrals[37], which yield the result

$$\boxed{\mu_{OP} = e^{-i\psi} e^{-(ka\theta)^2/2}} \tag{4.22}$$

This degree of coherence for a Gaussian distributed source exhibits the same geometrical phase variation, $\psi = kz\theta^2/2$, as did the point source, now, however, with an additional Gaussian amplitude dependence as a function of θ. Note that for $ka\theta = 1/2$, corresponding to $d \cdot \theta = \lambda/2\pi$, the normalized degree of coherence is

$$|\mu_{OP}| = e^{-1/8} = 0.88$$

which is just 12% less than the maximum value of unity. Thus for a radiation source described spatially as a Gaussian with a standard deviation a, the far-field angular distribution is concomitantly determined to within one standard deviation θ, as described earlier in this chapter on the basis of uncertainty arguments, i.e., the quantities a and θ constitute an *uncertainty pair* or *transform pair*, more usually written as uncertainties in position and momentum $\Delta r \cdot \Delta p = \hbar/2$, where $\Delta r = a$ and $\Delta p = \hbar \Delta k = 2\pi \hbar \theta/\lambda$. We now see by use of the van Cittert–Zernike theorem that in the case where the uncertainty condition, written as $d \cdot \theta = \lambda/2\pi$ earlier in this chapter, is just met, the finite degree of coherence is not unity, but somewhat less at 0.88. That is to say, if we knew the electric field on axis, the expectation that we could predict the field at an off-axis angular position θ would be 0.88. Since the degree of coherence varies as θ^2 in this case the degree of coherence can be increased substantially, to $e^{-1/32} = 0.97$, by halving the observation angle to $ka\theta = 1/4$, albeit at a considerable loss of flux or power. On the other hand, the available photon flux could be increased substantially by somewhat compromising the degree of coherence, clearly a choice that depends on the experiment.

(3) **A Uniformly but Incoherently Illuminated Pinhole:** For a uniform circular disk of uncorrelated emitters, the equivalent of an incoherently illuminated pinhole, we can write the source function as

$$I(\rho) = \begin{cases} I_0 & \text{for } \rho \le a \\ 0 & \text{for } \rho > a \end{cases} \tag{4.23}$$

where $d = 2a$ is the pinhole diameter. The degree of coherence [Eq. (8.20)] then becomes

$$\mu_{OP} = \frac{e^{-i\psi} I_0 \displaystyle\int_0^a J_0(k\rho\theta)\rho \, d\rho}{I_0 \displaystyle\int_0^a \rho \, d\rho}$$

$$\mu_{OP} = \frac{8e^{-i\psi}}{d^2} \int_0^a J_0(k\rho\theta)\rho \, d\rho$$

which is of the standard integral form[38]

$$\int_0^1 x J_0(vx) \, dx = \frac{J_1(v)}{v}$$

where we have made the substitutions $\rho = ax$ and $k\rho\theta = vx$ so that $v = ka\theta$. Making these substitutions, one finds that for the *incoherently illuminated pinhole* the normalized degree of coherence is

$$\boxed{\mu_{OP}(\theta) = e^{-i\psi} \frac{2J_1(ka\theta)}{ka\theta}} \tag{4.24}$$

where $J_1(v)$ is a Bessel function of the first kind, of order one, and where again $\psi = kz\theta^2/2$. The absolute value of $2J_1(v)/v$ is plotted[34] in Figure 4.15. It has a maximum value of unity at $ka\theta = 0$, drops to 0.88 at $ka\theta = 1$, and to 0.5 at $ka\theta = 2.215$, which correspondes to $d \cdot \theta = 0.7051\lambda$. The function is zero, corresponding to complete incoherence (no correlation among the fields), for $ka\theta = 3.832$. This coherence null corresponds to $d \cdot \theta_{null} = 1.220\lambda$. These results are particularly interesting in a practical sense in that circular pinhole apertures, back-illuminated by essentially incoherent radiation, provide a particularly attractive method by which to obtain spatially coherent radiation at these very short wavelengths, as was stated without proof in Section 4.3. These techniques play a significant role in experiments with largely incoherent radiation, such as the use of undulator radiation as will be discussed in Chapter 5, Section 5.6. Tradeoffs between desired degree of coherence and available photon flux can now be made on

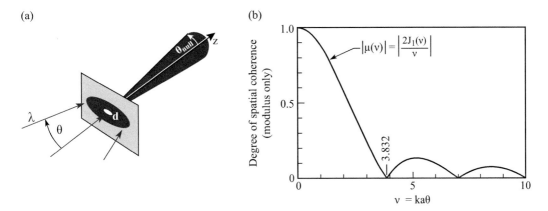

(a)

(b)

$$|\mu(v)| = \left|\frac{2J_1(v)}{v}\right|$$

Degree of spatial coherence
(modulus only)

$v = ka\theta$

Figure 4.15 (a) Geometry showing elliptically shaped, uniform but incoherent illumination at wavelength λ upon a pinhole of diameter $d = 2a$. (b) The degree of transverse coherence as observed at an angle θ in the far field of a circular pinhole. The modulus of the degree of spatial coherence $|\mu_{OP}|$ follows a $|2J_1(ka\theta)/(ka\theta)|$ behavior.

the basis of Eq. (4.24) and Figure 4.15. For a desired degree of transverse coherence in some distant (z) experimental plane, one can choose a pinhole diameter $d = 2a$, for the given wavelength λ, that sets a value for $ka\theta$, and thus the enclosed value of μ within a transverse dimension $2z\theta$, as shown in Figure 4.15(a). For example, if one selects a pinhole size such that $ka\theta = 1$, the degree of spatial coherence will be 0.88 at the edge of the θ-defined cone, and higher inside that cone. This would provide a high degree of coherence for many experiments but the pinhole size would be somewhat small, restricting the transmitted photon flux. By increasing the pinhole size such that $ka\theta = 2.215$, the transmitted coherent flux would increase by almost a factor of five while the degree of spatial coherence would fall to a still very useful value of $\mu = 0.5$ at the edge of the cone, and higher degrees of coherence within the cone. Selecting a larger pinhole diameter, such that $ka\theta$ approaches 3.832, will result in a significant reduction in the degree of spatial coherence.

4.5 Young's Double Slit Interference Technique

To measure spatial coherence one can perform an interference experiment by sampling the radiation passing through two pinholes at the points of interest, as shown in Figure 4.16. The technique is generally referred to as Young's double slit interference technique, after Thomas Young who in 1802 used it to demonstrate the wave nature of light[1, 2, 39] and measure visible wavelengths from red to violet. In modern practice, with sufficient available photon flux, pinhole pairs are used for better accuracy, as did Young on occasion. The source shown as an irregular spot of approximate diameter d in Figure 4.16 could be that of an undulator, a free electron laser, high harmonic radiation from high-intensity laser irradiation of a noble gas, an atomic EUV laser, or even a hot filament lightbulb as was illustrated in Figure 4.7. Radiation is allowed to fall

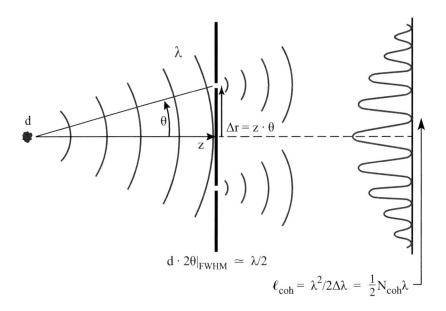

$$d \cdot 2\theta|_{\text{FWHM}} \simeq \lambda/2$$

$$\ell_{\text{coh}} = \lambda^2/2\Delta\lambda = \tfrac{1}{2}N_{\text{coh}}\lambda$$

Figure 4.16 The geometry of Young's double pinhole interference technique for measuring transverse coherence. If the source size and divergence satisfy Eq. (4.7b) for spatial coherence, $d \cdot \theta = \lambda/2\pi$, the fringe contrast (modulation) will be high. If the source size is larger, such that $d \cdot \theta > \lambda/2\pi$, the fringe contrast will be smaller, indicating a lower degree of coherence. The number of fringes observed on the screen to the right depends on the temporal coherence length and the difference in propagation paths from each pinhole to a common point on the screen. The number of fringes thus depends on the "number of waves of coherence," $N_{\text{coh}} = \lambda/\Delta\lambda$, at wavelength λ.

on the plane of interest, illuminating two small pinholes, shown here at angles $\pm\theta$. The transverse separation distance is $2\Delta r = 2z\theta$, where z is the distance from the source to the two-pinhole plane. The pinholes sample the degree of mutual coherence between the fields at the two points by allowing the transmitted radiation to form an interference, or "fringe," pattern on a distant screen, perhaps a charge coupled array detector (CCD). For a single point source radiator a self-interference pattern of unit contrast will form at the screen. If additional point source radiators of random phase are nearby, within a region such that $d \cdot \theta = \lambda/2\pi$, their self-interference patterns will overlap that of the first, such that a net fringe pattern forms with little degradation. We call this the persistence of fringes. If, on the other hand, additional radiators are located outside this region the resultant fringe pattern will be degraded, indicating a decreased degree of coherence. This corresponds to $d \cdot \theta > \lambda/2\pi$. For a given d and λ, as the pinholes are separated in sequential exposures (θ increased), the contrast will decrease, going to zero for $d \cdot \theta \gg \lambda/2\pi$, or $ka\theta \gg 1$. In this manner the degree of spatial coherence, $|\mu(\Delta r)|$, can be measured as a function of seperation distance across the transverse plane. Note that the number of fringes observed on the distant detector will be of order $\lambda/\Delta\lambda$, as the increasing mismatch of path lengths from the two pinholes to the screen exceeds the coherence length. When the various point source radiators are phase correlated additional considerations become important. That is discussed in the following section.

In 1957, Thompson and Wolf[40] published high-quality interference patterns obtained using Young's technique to study two-beam interference of spatially filtered, partially coherent visible light at a wavelength of 579 nm, the yellow doublet of a mercury discharge lamp. Shown in Figure 4.16 are more recent two-pinhole interference patterns obtained by Chang[41] using partially coherent undulator radiation at 13.4 nm wavelength (see Chapter 5). The pinhole pairs were located in a distant plane where there was a 60:1 demagnified image of the undulator output. Recorded interference patterns, formed with overlapping Airy patterns from two 420 nm diameter pinholes at separation distances of 3, 4, 6, and 9 μm, are shown in Figure 4.17. The interference patterns were recorded on a one inch square, 512 by 512 pixels, EUV sensitive CCD camera at a distance of 26 cm. Intensity modulation of the recorded patterns are also shown. The fringe contrast is observed to decrease with increasing pinhole separation providing a measure of the transverse coherence length in the plane of the pinholes. Further two-pinhole studies of partially coherent undulator radiation at 2.48 nm are discussed in Chapter 5, Section 5.6, and illustrated in Figure 5.32. A discussion of spatial coherence characterization using non-redundant arrays (NRAs) is given by Skopintsev, Vartanyants, and colleagues.[42]

4.6   Spatial and Temporal Coherence, and True Phase Coherence

In preceding sections we have discussed spatial and temporal filtering as a method to obtain a high degree of coherence with *uncorrelated radiators* of random phase. To achieve spatial coherence a geometry was chosen that permits radiation only from electrons within a restricted phase space $(d \cdot \theta)$ to propagate to the distant screen, or to an experimental station. The technique of spatial filtering was illustrated in Figure 4.7, and a technique for measuring the achieved degree of spatial coherence was illustrated in Figure 4.16. When we consider *radiators of correlated phase*, fundamental differences are observed. These effects are common with bound electrons involved in stimulated emission within the cavity of a multipass visible light laser, and now with free electrons which, through feedback from their own radiated fields, bunch to form a correlated electron wave moving within a long FEL undulator. The subtle but important phase difference between correlated and uncorrelated radiators is illustrated in Figure 4.18, for N total radiators in each case. In both (a) and (b) the radiation is spatially and temporally coherent, being sufficiently confined laterally to satisfy Eqs. (4.7a), and having a temporal coherence set by Eq. (4.6) that depends on just the wavelength λ and the spectral bandwidth, $\Delta\lambda$. However, the phase coherence as detected at some point in the far-field is quite different. The radiators in (a) are spatially and temporally uncorrelated, resulting in electric fields that contain a random phase. As a result the time average of $|E|^2$ contains no cross terms involving the fields of different radiators. Thus in this uncorrelated case the phase of the radiated electric field is random and the radiated power, which is proportional to $|E|^2$, is simply N times that of a single radiator. On the other hand, in

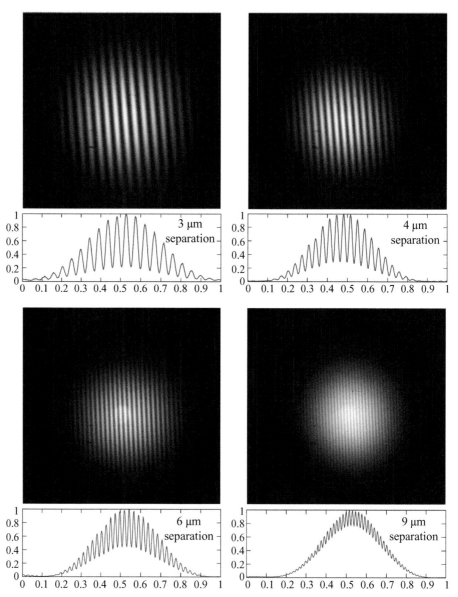

Figure 4.17 The recorded interference pattern[41] at 13.4 nm wavelength of two 420 nm diameter pinholes at separation distances of 3, 4, 6, and 9 μm as part of experiments to measure the partial coherence of undulator radiation in a distant, 60:1 demagnified image plane. Normalized intensity modulations through the center of the interference pattern are shown in each case. Plots of modulation vs. pinhole seperation indicate a lateral coherence length of 6.3 μm in the horizontal direction and 7.4 μm in the vertical direction. Courtesy of C. Chang, EECS University of California, Berkeley, and Lawrence Berkeley National Laboratory.

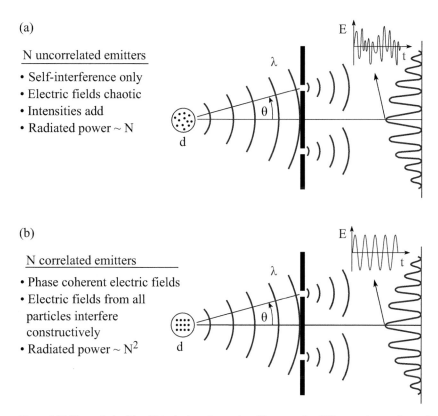

(a)

N uncorrelated emitters

- Self-interference only
- Electric fields chaotic
- Intensities add
- Radiated power ~ N

(b)

N correlated emitters

- Phase coherent electric fields
- Electric fields from all particles interfere constructively
- Radiated power ~ N^2

Figure 4.18 Young's double slit technique is used to illustrate the difference in resultant phase, and its consequences, for two cases involving a large number (N) of radiators (electrons). In both cases the radiation is spatially coherent ($d \cdot \theta = \lambda/2\pi$) and with the same coherence length ($l_{coh} = \lambda^2/\Delta\lambda$) but in one case with emitters of uncorrelated phase and in the other with phase correlated emitters. In the case of uncorrelated emitters the composite radiated electric field is of random phase, intensities add, and the total power radiated is proportional to N. In the case of phase correlated emitters the electric fields add constructively, the composite radiated electric field is characterized by a well-defined phase, and the total radiated power scales as N^2.

case (b), all the radiation is truly phase correlated at the chosen wavelength such that the electric fields from all radiators interfere constructively, i.e., sum coherently, in the far-field. This results in a net electric field of continuous, well-defined phase. Furthermore, since the electric fields of all radiators in this example add constructively, the radiated power is N^2 times that of a single radiator. For x-rays the number of electrons radiating coherently may be in the millions or billions, thus enormous power gains are achievable.

4.7   Diffraction of Radiation by a Coherently Illuminated Pinhole Aperture

In contradistinction to the previous section which dealt with partially coherent radiation, in this section our interest turns to the diffraction of spatially coherent short-wavelength

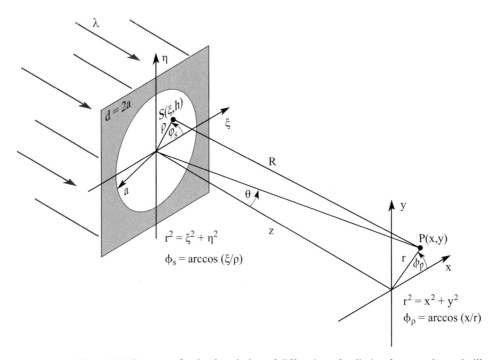

Figure 4.19 Geometry for the description of diffraction of radiation from a coherently illuminated aperture in a plane (ξ, η) to an observation point (x, y). R is the distance from a source point $S(\xi, \eta)$ in the aperture plane to an observation point $P(x, y)$. The two planes are parallel, separated by a distance z. The aperture illustrated here is a circle of diameter $d = 2a$. The angle θ is measured from the origin of the ξ, η-plane to the observation point P.

radiation by pinholes and other structures, such as zone plates, which are much used in this short-wavelength region of the electromagnetic spectrum. Coherent diffraction is a basic electromagnetic phenomenon, well known and much utilized at lower wavelengths. There is a long history of mathematical development on the theory of diffraction that successfully predicts physical observations. Most notable is the scalar theory of diffraction developed by Kirchhoff in 1882, extending the earlier work of Huygens (1690) and Fresnel (1818), and leading to what is now known as the *Fresnel–Kirchhoff diffraction formula*,[1, 2, 8, 11] which for small angle scattering in the near-forward direction ($\theta \simeq \lambda/d \ll 1$, where d is a characteristic dimension), can be written as[2]

$$E(x, y) = \frac{-i}{\lambda} \iint \frac{E(\xi, \eta) e^{ikR}}{R} \, d\zeta \, d\eta \qquad (4.25)$$

where $k = 2\pi/\lambda$, $E(x, y)$ is the electric field observed at a distant point $P(x, y)$, $E(\xi, \eta)$ is the incident field as a function of position in the aperture plane ($z = 0$), and R is the distance from each source point $S(\xi, \eta)$ to the observation point $P(x, y)$, as illustrated in Figure 4.19. Basically this states that the field detected in a distant plane is obtained by summing the contributions from every point in the aperture plane, allowing for its propagation distance R and phase e^{ikR}, as if each point were a secondary source of radiation. As we will see shortly, the finite aperture introduces unbalanced contributions near boundaries, leading to interference effects specific to the geometry and clearly dependent on the wavelength.

From the geometry of Figure 4.19 we see that

$$R = \sqrt{z^2 + (x - \xi)^2 + (y - \eta)^2}$$

For the case where $z \gg x, y$ and $z \gg \xi, \eta$, this becomes

$$R \simeq z + \frac{x^2}{2z} + \frac{y^2}{2z} + \frac{\xi^2}{2z} + \frac{\eta^2}{2z} - \frac{\xi x}{z} - \frac{\eta y}{z} \tag{4.26}$$

so that

$$E(x, y) = \frac{-i e^{ikz} e^{ik(x^2 + y^2)/2z}}{\lambda z} \iint E(\xi, \eta) e^{ik(\xi^2 + \eta^2)/2z} \, e^{-ik(\xi\theta_x + \eta\theta_y)} \, d\xi \, d\eta \tag{4.27}$$

where $\theta_x = x/z = r \cos \phi_p / z = \theta \cos \phi_p$ and $\theta_y = y/z = r \sin \phi_p / z = \theta \sin \phi_p$.

For a small pinhole of radius a, such that $ka^2/2z \ll 1$,[¶] assuming uniform plane wave illumination such that $E(\xi, \eta) = E_0$, and that the exponent term $k(\xi^2 + \eta^2)/2z$ can be neglected as second order compared to the $\xi\theta_x$ and $\eta\theta_y$ terms, Eq. (4.27) simplifies to

$$E(x, y) = \frac{-i E_0 e^{ikz} e^{ik(x^2 + y^2)/2z}}{\lambda z} \iint e^{-ik(\xi\theta_x + \eta\theta_y)} \, d\xi \, d\eta \tag{4.28}$$

Converting to polar coordinates as shown in Figure 4.19 for this axisymmetric case, where $(\xi, \eta) \to (\rho, \phi_s)$ and $(x, y) \to (r, \phi_p)$, the integral reduces to

$$E(r, \theta) = \frac{-2\pi i E_0 e^{ikz} e^{ikr^2/2z}}{\lambda z} \int_0^a J_0(k\rho\theta)\rho \, d\rho \tag{4.29}$$

where we have replaced r by θ, using $r = \theta z$ for final z, and where[§] the ϕ_s-integral has led to the identification[31-34]

$$J_0(k\rho\theta) = \frac{1}{2\pi} \int_0^{2\pi} e^{ik\rho\theta \cos(\phi_s - \phi_p)} \, d\phi_s \tag{4.30}$$

where $J_0(v)$ is a Bessel function of the first kind, of zero order. The integral in Eq. (4.29) is of the standard form[33]

$$\int_0^1 x J_0(vx) \, dx = \frac{J_1(v)}{v} \tag{4.31}$$

where we have made the substitutes $\rho = ax$ and $k\rho\theta = vx$, so that $v = ka\theta$. The field diffracted by a small pinhole, in the small angle (θ) approximation, is therefore obtained from Eqs. (4.29) and (4.31) to be

$$E(\theta) = \frac{-2\pi i a^2 E_0 e^{ikz} e^{ikz\theta^2/2}}{\lambda z} \cdot \frac{J_1(ka\theta)}{ka\theta} \tag{4.32}$$

[¶] This is the *far-field* approximation, valid for $z \gg \pi d^2/4\lambda$, where $d = 2a$ is the pinhole diameter.

[§] Note that this is mathematically identical to treatment of the propagation of partial coherence earlier in this Chapter, leading to Eqs. (4.16) and (4.17).

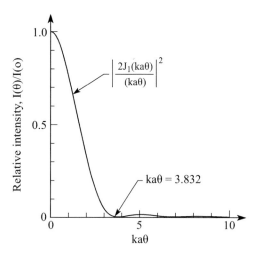

Figure 4.20 The normalized far-field intensity pattern resulting from the diffraction of a uniform plane wave by a circular aperture. $I(\theta)$ is the angular distribution of intensity in a distant (z) plane, $I(0)$ is the axial intensity at a distance z from the pinhole, a is the pinhole radius, $k = 2\pi/\lambda$ is the wavenumber, θ is measured from the axis of symmetry, and $J_1(v)$ is a Bessel function of the first kind, order one. This is known as an Airy pattern[1, 11] after George Airy who first described its functional dependence in 1835.

The corresponding intensity distribution in the far field of the pinhole is then given by

$$\frac{I(\theta)}{I_0} = \left(\frac{ka^2}{2z}\right)^2 \left|\frac{2J_1(ka\theta)}{ka\theta}\right|^2 \tag{4.33}$$

where $I_0 = \sqrt{\epsilon_0/\mu_0}|E_0|^2$ is the illumination intensity at the pinhole. Note that Eq. (4.33) can be written in terms of the radial coordinate in the observation plane (x, y) through the substitution $\theta = r/z$, for fixed z. Equation (4.33) gives the functional dependence of the far-field diffracted intensity pattern for a coherently illuminated (monochromatic plane wave) pinhole. The normalized far-field intensity pattern is illustrated in Figure 4.20.

The angular dependence is dominated by the quantity $|2J_1(ka\theta)/ka\theta|^2$, often referred to as an Airy pattern[1, 2, 11] after George Airy, who first described this functional dependence in 1835. The function $|2J_1(v)/v|^2$ is unity for $v = 0$, declines to zero at $v = 3.832$ and then oscillates with successively smaller maxima and minima for increasing values[1] of v. The central lobe of this angular diffraction pattern, sometimes called the Airy disk, is bounded by the first Airy null at $ka\theta = 3.832$, which corresponds to a first null diffraction half angle

$$\theta_{\text{null}} = \frac{0.610\lambda}{a} = \frac{1.22\lambda}{d} \tag{4.34}$$

where $d = 2a$ is the pinhole diameter. Recall that this result is valid in the far field where $z \gg \pi a^2/\lambda$, so that the radius r of this null, given by $r = z\theta$, is much greater than

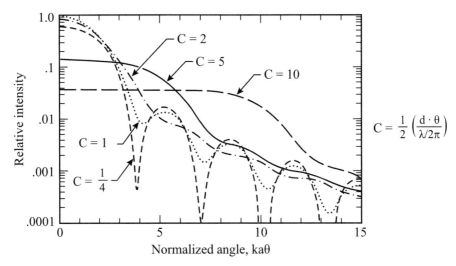

Figure 4.21 Diffraction from a circular aperture with varying degrees of partially coherent illumination.[43] The partially coherent illumination is characterized by the normalized constant $C = 0.5[d \cdot \theta/(\lambda/2\pi)]$, where in our terminology the source is characterized by diameter d, half angle θ, and wavelength λ. For $C = \frac{1}{4}$, corresponding to $d \cdot \theta = \lambda/4\pi$, the illumination is highly coherent and the diffraction pattern exhibits sharp angular features, approaching those described by Eq. 4.34. For a decreased degree of spatial coherence, $C = 1$, corresponding to $d \cdot \theta = \lambda/\pi$, the sharp features of the diffraction pattern are less pronounced. Further reductions of the degree of spatial coherence across the aperture result in a loss of the sharp angular features. Courtesy of B. Thompson, University of Rochester.

the pinhole radius a, i.e., $r = z\theta \gg 2a$. Thus the far-field Airy pattern is propagating outward from the pinhole with a half angle proportional to λ/a as in Eq. (4.34), with an Airy null of ever larger radius $r = 0.610\lambda z/a$. Furthermore, we observe from Eq. (4.33) that the on-axis intensity in the far field of the pinhole decreases in proportion to $I(0) = I_0(ka^2/2z)^2$, or

$$I(0) = \pi\varepsilon\left(\frac{a}{\lambda}\right)^2\left(\frac{1}{z}\right)^2 \tag{4.35}$$

where I_0 is the incident plane wave intensity at the pinhole, where we have defined $\varepsilon = \pi a^2 I_0$ as the total energy passing through the pinhole, and where $2J_1(0)/(0)$ is unity on axis, as can be shown with use of L'Hospital's rule. Thus the on-axis intensity of the far-field diffraction pattern decreases with the inverse square of the distance z as we would expect, with a proportionality $(a/\lambda)^2$ as we also would expect on a solid angle basis with a divergence half angle given by Eq. (4.34). Approximately 84% of the incident energy ε lies within the first Airy null.[1]

In 1966 Shore, Thompson, and Whitney[43, 2] published a very useful study of pinhole diffraction by radiation having varying degrees of spatial coherence across the pinhole. Their calculated diffraction results for partially coherent illumination of a circular aperture are shown in Figure 4.21.

$$\mu_{op}(\theta) = e^{-i\psi} \frac{2J_1(ka\theta)}{ka\theta} \qquad (4.24)$$

$$E(\theta) = \frac{-2\pi i a^2 E_0 e^{ikz} e^{ikz\theta^2/2}}{\lambda z} \frac{J_1(ka\theta)}{ka\theta} \qquad (4.32)$$

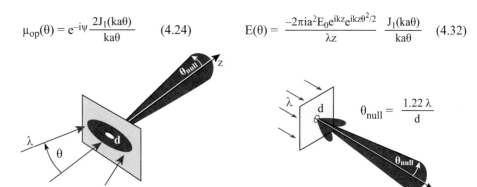

Uniformly but incoherently illuminated pinhole

Diffraction of coherent radiation by a pinhole

$$\theta_{null} = \frac{1.22\,\lambda}{d}$$

Figure 4.22 Similar propagation characteristics of mutual coherence and diffracted fields are illustrated for a circulat aperture of diameter $d = 2a$. (a) The diffraction of uniform, *incoherent radiation* by a pinhole results in a far-field pattern of high *spatial coherence* on axis, falling to lower values off axis with a $2J_1(ka\theta)/(ka\theta)$ functionality. (b) The diffraction of *coherent radiation* by a pinhole results in a far-field *electric field* pattern of the same $ka\theta$ functionality. The square of this function, the coherently diffracted intensity, is known as the Airy pattern, as discussed in Section 4.7 and illustrated in Figure 4.20. Comparison suggested by S. Kevan, U. Oregon.

4.8　Similarities between the Diffraction of Partially Coherent and Coherent Radiation by a Circular Aperture

The mathematical developments in Sections 4.4 and 4.5 are strikingly similar, but physically dissimilar, and deserve comment. Both developments involve applications of incident radiation on a pinhole, with an axial symmetry of radius a and wavelength λ, that naturally involve Bessel functions of argument $ka\theta$, where $k = 2\pi/\lambda$. Both involve Fourier transforms which describe propagation to the far-field. For an *incoherently illuminated pinhole, the degree of spatial coherence*, μ, is found to propagate to the far-field with a $J_1(ka\theta)/(ka\theta)$ dependence [Eq. (4.24)]. For the *coherently illuminated pinhole* the *electric field* is similarly found to propagate to the far-field with a $J_1(ka\theta)/(ka\theta)$ dependence [Eq. (4.32)]. In mathematical terminology they have the same propagator. Note, however, that for the coherently illuminated pinhole the far-field intensity distribution, which scales as the square of electric field, is proportional to $|J_1(ka\theta)/(ka\theta)|^2$, and is known as an Airy pattern. Physically, the first case describes the propagation of measurable spatial coherence far from the pinhole, while the second case describes the measurable intensity of radiation as it propagates far from the pinhole. In each case the first null occurs at $ka\theta = 3.832$, but otherwise the values are different, one being the square of the other (compare Figures 4.15(b) and 4.20). Figure 4.22 illustrates the two physically different applications involving incoherent and coherent illumination of circular apertures which result in similar mathematical forms for physically different quantities.

References

1. M. Born and E. Wolf, *Principles of Optics* (Cambridge University Press, 1999); Section 7.3.1 for a discussion of Young's double-pinhole technique, Section 7.58 for a discussion of bandwidths and coherence length; Section 8.5.2 for a discussion of the Airy pattern; and Chapter 10 for a broad discussion of coherence and partial coherence, correlation functions, and the van Cittert–Zernike theorem. Also see L. Mandel and E. Wolf, *Optical Coherence and Quantum Optics* (Cambridge University Press, 1995).
2. J.W. Goodman, *Statistical Optics* (Wiley, New York, 1985); Chapter 3 for a clear and concise exposition of diffraction theory, pp. 207–210 regarding the van Cittert–Zernike theorem, and pp. 215–218 regarding Young's double-pinhole technique.
3. W.H. Knox, M. Alonso and E. Wolf, "Spatial Coherence from Ducks," *Physics Today* (March 2010).
4. R.J. Collier, C. Burkhardt and L.H. Lin, *Optical Holography* (Academic Press, New York, 1985).
5. P.A. Tipler and R.A. Llewellyn, *Modern Physics* (Freeman, New York, 2012), Section 5.5.; also see R.N. Bracewell, *The Fourier Transform and Its Applications* (McGraw-Hill, New York, 1986), Sixth Edition, p. 160.
6. W.T. Sifvast, *Laser Fundamentals* (Cambridge University Press, 2004); Second Edition.
7. A.E. Siegman, *Lasers* (University Science, Mill Valley, CA, 1986), Chapters 1, 14, and 17, and Eq. 17–5, recast here for Eq.(4.8) in terms of rms quantities. For Rayleigh range and depth of focus considerations, Eq.(4.9), standard $1/e^2$ description used.
8. G.R. Fowles, *Introduction to Modern Optics* (Dover, New York, 1989), Chapters 3 and 9.
9. A. Schawlow, "Laser Light," *Scientific American* **219**, 120 (September1968).
10. D. Attwood, K. Halbach and K.-J. Kim, "Tunable Coherent X-Rays," *Science* **228**, 1265 (June 14, 1985).
11. E. Hecht, *Optics* (Addison-Wesley, San Francisco, 2002), Fourth Edition, Chapters 10–12; see Sec. 10.2.5 and 10.2.5 regarding diffraction by a circular aperture and the resultant Airy pattern.
12. S. Kikuta, S. Aoki, S. Kosaki and K. Kohra, "X-ray Holography of Lenseless Fourier-Transform Type," *Opt. Commun.* **5**, 86 (1972).
13. S. Aoki, Y. Ichihara, and S. Kikuta, "X-ray Hologram Obtained by Using Synchrotron Radiation," *Jpn. J. Appl. Phys.* **11**, 1857 (1972).
14. B. Reuter and H. Mahr, "Experiments with Fourier Transform Holograms Using 4.48 nm X-rays," *J. Physics E: Scientific Instrum.* **9**, 746 (1976).
15. A.M. Kondratenko and A.N. Skrinsky, "Use of Radiation of Electron Storage Rings in X-ray Holography of Objects," *Opt. Spectrosc.* **42**, 189 (February 1077); "X-Ray Holography of Microobjects", Institute for Nuclear Physics, Novosibirsk, preprint 76–405 (1976).
16. A.M. Kondratenko and A.N. Skrinsky, "X-Ray Holographic Microscopy," *Avtometriya* **2**, 3 (Novosibirsk, March–April 1977).
17. I. McNulty, J. Kirz, C. Jaconsen, M.R. Howells and D.P. Kern, "High-Resolution Imaging by Fourier Transform X-ray Holography," *Science* **256**, 1009 (May 15, 1992).
18. S. Eisebitt, J. Lüning, W.F. Schlotter *et al.*, "Lenseless Imaging of Magnetic nanostructures by X-ray Spectro-Holography," *Nature* **432**, 885 (December 16, 2004); J. Geilhufe *et al.*, "Monolithic Focused Reference Beam X-ray Holography," *Nature Commun.* **5**, 3008 (January 7, 2014); E. Guehrs, M. Fohler, S. Frömmel *et al.*, "Mask-Based Dual-Axis Tomoholography Using Soft X-Rays," *New J. Physics* **17** (October 2015).

19. W.F. Schlotter *et al.*, "Multiple Reference Fourier Transform Holography with Soft X-rays," *Appl. Phys. Lett.* **89**, 163112 (2006).

20. S. Marchesini, S. Boutet, A.E. Sakdinawat *et al.*, "Massively Parallel X-ray Holography," *Nature Photonics* **2**, 560 (September 2008); S. Eisebitt, "X-Ray Holography: The Hole Story," *Nature Photonics* **2**, 529 (September 2008).

21. H. N. Chapman, S. P. Hau-Riege, M. J. Bogan *et al.*, "Femtosecond Time-Delay X-Ray Holography," *Nature* **448**, 676 (2007).

22. I.A. Vartanyants and A. Singer, "Coherence Properties of 3-rd Generation Synchrotron Sources and Free Electron Lasers", Chapter 1 in *Handbook on Synchrotron Radiation and Free-Electron Lasers* (Springer, Heidelberg, 2015), E. Jaeschke, S. Khan, J.R. Schneider and J.B. Hastings, Editors.

23. K. Nugent, "Coherent Methods in the X-Ray Sciences," *Advances in Physics* **59**, 1–99 (2010).

24. M.J. Bedzyk, G.M. Bommarito and J.S. Schildkraut, "X-ray Standing Waves at a Reflecting Mirror Surface," *Phys. Rev. Lett.* **62**, 1376 (March 20, 1989).

25. M.K. Tiwari, G. Das and M.J. Bedzyk, "X-ray Standing Wave Analysis of Nanostructures using Partially Coherent Radiation," *Appl. Phys. Lett.* **107**, 103104 (September 7, 2015).

26. B. Beckhoff, "Reference-Free X-ray Spectrometry Based on Metrology Using Synchrotron Radiation," *J. Analyt. Atomic Spectrom.* **23**, 845 (2008).

27. B. Pollakowski, P. Hoffmann, M. Kosinova *et al.*, "Nondestructive and Nonpreparative Chemical Nanometrometry of Internal Material Interfaces at Tunable High Information Depths," *Analyt. Chem.* **85**, 193 (2013).

28. C. Becker, M. Pagels, C. Zachäus *et al.*, "Chemical Speciation at Buried Interfaces in High-Tempature Processed Polycrystalline Silicon Thin-Film Solar Cells on ZnO:Al," *J. Appl. Phys.* **113**, 044519 (2013).

29. P.H. van Cittert, "Die Wahrscheinliche Schwingungsverteilung in Einer von Einer Lichtquelle Direkt oder Mittels Einer Linse Beleuchteten Ebene" [The Probable Distribution of Vibrations in a Plane Illuminated by a Light Source Either Directly or Through a Lens], *Physica* **1**, 210 (1934); see *Coherence and Fluctuations of Light* (*1850–1966*) (SPIE, Bellingham, WA, 1990), L. Mandel and E. Wolf, Editors, p. 1.

30. F. Zernike, "The Concept of Degree of Coherence and its Application to Optical Problems," *Physica* **5**, 785 (1938); see *Coherence and Fluctuations of Light 1850–1966* (SPIE, Bellingham, WA, 1990), L. Mandel and E. Wolf, Editors, p. 100.

31. G.N. Watson, *A Treatise on the Theory of Bessel Functions* (Cambridge University Press, 1944), p. 20.

32. I.S. Gradshteyn and I.M. Ryzhik, *Table of Integrals, Series and Products* (Academic Press, New York, 1994), Fifth Edition, Section 8.411, No. 1, p. 961.

33. F.W. Oliver, "Bessel Functions of Integer Order," p. 360, No. 9.1.21, in *Handbook of Mathematical Functions* (Dover, New York, 1972), M. Abramowitz and I. Stegun, Editors.

34. E. Kreyszig, *Advanced Engineering Mathematics* (Wiley, New York, 1993), Seventh Edition, pp. 225 and A97; p. 834 shows the function.

35. R.N. Bracewell, *The Fourier Transform and Its Applications* (McGraw-Hill, New York, 1978), Second Edition, Chapter 12.

36. I.N. Sneddon, *Fourier Transforms* (Dover, New York, 1995).

37. Reference 32, p. 382, Section 3.460, No. 3, and p. 738, Section 6.632, No. 4.

38. Reference 32, p. 707, Section 6.561, No. 5.

39. T. Young, *A Course of Lectures on Natural Philosophy and the Mechanical Arts*, Volume 1, Lecture XXXIX, "On the Nature of Light and Colours" (J. Johnson, London, 2007),

pp. 464–465 and 776–777; A. Robinson, *The Last Man Who Knew Everything* (Pearson Education, Penquin Books, Pi Press, 2006).

40. B.J. Thompson and E. Wolf, "Two-Beam Interference with Partially Coherent Light," *J. Opt. Soc. Amer.* **47**, 898 (October 1957).

41. C. Chang, "Coherence Techniques at Extreme Ultraviolet Wavelengths", PhD thesis, EECS, University of California, Berkeley, 2002; C. Chang, P. Naulleau, E. Anderson and D. Attwood, "Spatial Coherence Characterization of Undulator Radiation," *Optics Commun.* **182**, 25 (August 1, 2000).

42. P. Skopintsev, A. Singer, J. Bach *et al.*, "Characterization of Spatial Coherence of Synchrotron radiation with Non-Redundant Arrays of Apertures," *J. Synchr. Rad.* **21** (4), 722 (July 2014).

43. R.A. Shore, B.J. Thompson, and R.E. Whitney, "Diffraction by Apertures Illuminated with Partially Coherent Light," *J. Opt. Soc. Amer.* **56**, 733 (June 1966).

Homework Problems

Homework problems for each chapter will be found at the website:
www.cambridge.org/xrayeuv

5 Synchrotron Radiation

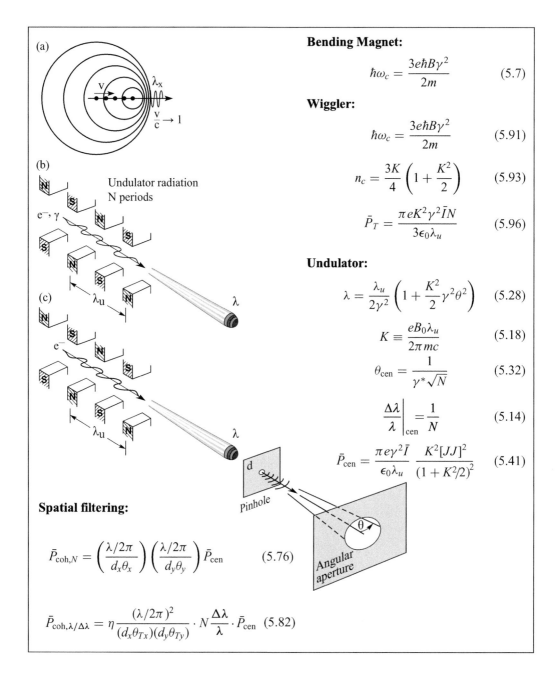

(a)

λ_x

$\dfrac{v}{c} \to 1$

(b) Undulator radiation N periods

e^-, γ

λ

(c)

λ_u

e^-

λ_u

λ

d

Pinhole

θ

Angular aperture

Bending Magnet:

$$\hbar\omega_c = \frac{3e\hbar B\gamma^2}{2m} \qquad (5.7)$$

Wiggler:

$$\hbar\omega_c = \frac{3e\hbar B\gamma^2}{2m} \qquad (5.91)$$

$$n_c = \frac{3K}{4}\left(1 + \frac{K^2}{2}\right) \qquad (5.93)$$

$$\bar{P}_T = \frac{\pi e K^2 \gamma^2 \bar{I} N}{3\epsilon_0 \lambda_u} \qquad (5.96)$$

Undulator:

$$\lambda = \frac{\lambda_u}{2\gamma^2}\left(1 + \frac{K^2}{2}\gamma^2\theta^2\right) \qquad (5.28)$$

$$K \equiv \frac{eB_0\lambda_u}{2\pi mc} \qquad (5.18)$$

$$\theta_{\text{cen}} = \frac{1}{\gamma^* \sqrt{N}} \qquad (5.32)$$

$$\left.\frac{\Delta\lambda}{\lambda}\right|_{\text{cen}} = \frac{1}{N} \qquad (5.14)$$

$$\bar{P}_{\text{cen}} = \frac{\pi e \gamma^2 \bar{I}}{\epsilon_0 \lambda_u} \frac{K^2[JJ]^2}{(1 + K^2/2)^2} \qquad (5.41)$$

Spatial filtering:

$$\bar{P}_{\text{coh},N} = \left(\frac{\lambda/2\pi}{d_x\theta_x}\right)\left(\frac{\lambda/2\pi}{d_y\theta_y}\right)\bar{P}_{\text{cen}} \qquad (5.76)$$

$$\bar{P}_{\text{coh},\lambda/\Delta\lambda} = \eta \frac{(\lambda/2\pi)^2}{(d_x\theta_{Tx})(d_y\theta_{Ty})}\cdot N\frac{\Delta\lambda}{\lambda}\cdot\bar{P}_{\text{cen}} \qquad (5.82)$$

In this chapter we briefly review the central features of synchrotron radiation, beginning with estimates of radiated photon energies and angular divergence based on the application of well-known results from the theory of relativity and Heisenberg's uncertainty principle. For bending magnet radiation, formulae describing photon flux as a function of angle and photon energy are summarized in a convenient handbook style. Undulator radiation, generated by ultra-relativistic electrons traversing a periodic magnet structure, is calculated in detail. The approach taken makes maximal use of the well-known classical results of dipole radiation. This is accomplished by solving the electron equation of motion in the laboratory frame of reference, then making a Lorentz transformation to the frame of reference moving with the average electron velocity. In this frame of reference the motion is non-relativistic, yielding the well-known $\sin^2 \Theta$ angular dependence of radiated power per unit solid angle. These results are then Lorentz transformed back to the laboratory (sample) frame of reference. A central radiation cone, defined as containing a $1/N$ relative spectral bandwidth, is shown to correspond to an angular half width of $1/\gamma\sqrt{N}$, where N is the number of magnet periods and γ is the Lorentz factor. Power radiated in the central cone is readily calculated from the dipole formula. Calculations of spectral brightness follow in a straightforward manner. Spatial coherence at x-ray wavelengths can be obtained using pinhole spatial filtering techniques, as described in Chapter 4. Curves of spatially and temporally coherent power vs. photon energy are here obtained for both soft and hard x-ray undulator facilities. Wiggler radiation, the strong field extension of undulator radiation, is shown to be dominated by a large number of harmonics that merge to a continuum at high photon energy. The spectral shape of wiggler radiation is similar to that of bending magnetic radiation, but shifted to higher photon energy (by the higher magnetic fields) and to increased ($2N$) photon flux.

5.1   Introduction

It is well known that an accelerated charged particle, such as one traveling on a curved trajectory, will emit radiation (see Figure 5.1, for example). When moving at relativistic speeds, this radiation is emitted as a narrow cone tangent to the path of the particle.[1–5] Synchrotron radiation is generated when relativistic electrons (or positrons) are accelerated (undergo a change of direction) in a magnetic field, as seen for example in Figure 5.2. There are three types of magnetic structures commonly used to produce synchrotron radiation: bending magnets, undulators, and wigglers. Bending magnets cause a single curved trajectory as pictured in Figure 5.2. The result is a fan of radiation around the bend. Undulators are periodic magnetic structures with relatively weak magnetic fields. The periodicity causes the electron to experience a harmonic oscillation as it moves in the axial direction, resulting in a motion characterized by small angular excursions called undulations,[2, 3] as shown in Figure 5.3. The relatively weak magnetic fields cause the amplitude of the undulation to be small. Hence, the resultant radiation cone is narrow. In combination with a tightly confined electron beam, this leads to radiation with small angular divergence and relatively narrow spectral width, properties we

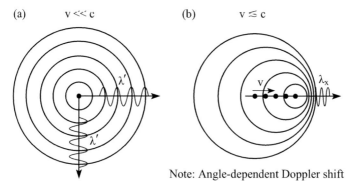

Note: Angle-dependent Doppler shift

Figure 5.1 Radiation from an oscillating charge moving at (a) a non-relativistic and (b) a relativistic speed. Shorter wavelengths are observed on axis due to the Doppler effect which reduces the separation of succeeding phase fronts. Indeed, as v approaches c, as in (b), the phase front separations (λ) are compressed by many orders of magnitude. The wavelengths of radiation emitted by ultra-relativistic electrons traversing periodic magnet structures ("undulators") at modern storage rings are shortened by an order of 10^8 or 10^9, from centimeter magnet periods to nanometer or sub-nanometer wavelengths. Because the Doppler effect is strongly angle dependent, the shortest wavelengths are confined to a very narrow forward cone. Following J. Madey.

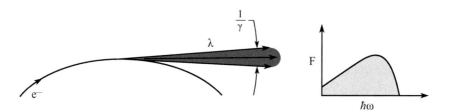

Figure 5.2 Bending magnet radiation occurs when a relativistic electron travels in a uniform magnetic field, executing a circular motion with acceleration directed toward the center. The radiation is directed tangentially outward in a narrow radiation cone, giving the appearance of a sweeping "searchlight." The radiation spectrum is very broad, analogous to a "white light" x-ray light bulb. The emission angle is typically $1/\gamma$, where γ is the Lorentz contraction factor.

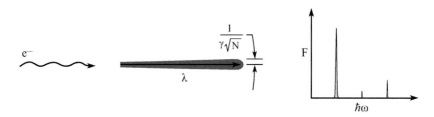

Figure 5.3 Undulator radiation is generated as a highly relativistic electron traverses a periodic magnetic field. In the undulator limit, the magnetic field is relatively weak and the resultant angular excursions of the electron are smaller than the angular width of the natural radiation cone, $1/\gamma$, normally associated with synchrotron radiation. The frequency spread of undulator radiation can be very narrow, and the radiation can be extremely bright and partially coherent, under certain circumstances. The characteristic emission angle is narrowed by a factor $\sqrt{N}$, where N is the number of magnetic periods.

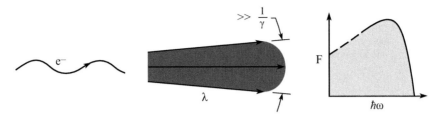

Figure 5.4 Wiggler radiation is also generated from a periodic magnet structure, but in the strong magnetic field limit where in at least one plane the angular excursions are significantly greater than the natural $(1/\gamma)$ radiation cone. Because accelerations are stronger in this limit, the radiation generated peaks at higher photon energies and is more abundant (higher photon flux and more power). The radiation spectrum is very broad, similar in shape to that of the bending magnet, but shifted to higher photon energies.

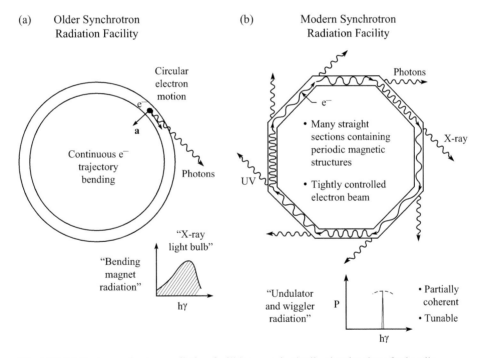

Figure 5.5 (a) Early synchrotron radiation facilities were basically circular rings for bending magnet radiation, although some have been retrofitted with periodic magnetic structures (undulators or wigglers). They generally have an electron beam of relatively large cross-section and angular divergence. (b) Modern storage rings are dedicated to broad scientific use and optimized for high spectral brightness through the inclusion of many long straight sections for undulators and wigglers, as well as very tightly confined (spatial and angular extent) electron beams. Bending magnet radiation is also generated in turning from one straight section to the next (not shown).

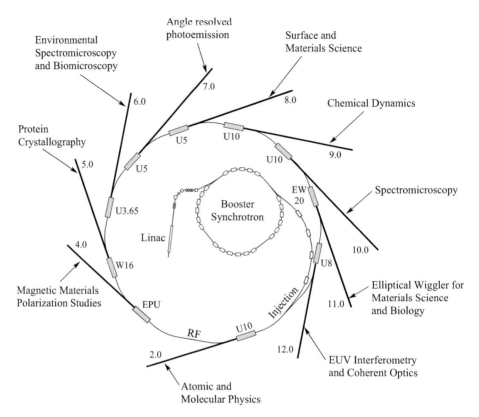

Figure 5.6 This sketch of an electron storage ring optimized for soft x-ray radiation shows a linear accelerator (linac) and booster synchrotron that bring electrons up to an energy matched to storage ring magnet settings, an injection system, which directs electrons into the ring, and a radio frequency (rf) generator to replenish the energy lost to synchrotron radiation as the electrons pass bending magnets, undulators, and wigglers. Straight sections for undulators and wigglers direct energy into beamlines and end sections for various scientific studies. Bending magnet radiation beamlines, located between straight sections, are not shown.

generally associate with the coherence properties of lasers. Wigglers are a strong magnetic field version of undulators. Owing to the stronger fields, the oscillation amplitude and concomitant radiated power is larger. The radiation cone is broader in both space and angle. The radiation spectrum is similar to that of bending magnets, but characterized by a larger photon flux and a shift to harder x-rays (shorter wavelengths), as seen in Figure 5.4. Although more power is radiated, wiggler radiation is less bright largely due to the substantially increased radiation cone.

Historically, synchrotron radiation was first observed as energy loss in electron storage rings. Logically, the first synchrotron radiation sources for general scientific use were simple parasitic beam ports utilizing otherwise lost radiation at existing storage rings. Over time, however, sources have been constructed for dedicated use as synchrotron radiation facilities (*second generation* facilities). The newest synchrotron facilities (*third generation* facilities) are composed of many straight sections specially optimized to produce high brightness undulator radiation. Figure 5.5 illustrates yesterday's and today's synchrotron radiation facilities.

Figure 5.6 provides a simple schematic of a synchrotron radiation source. The relativistic electrons are injected into the ring from a linear accelerator and (energy) booster synchrotron. Various magnetic lenses keep the electrons traveling along the desired trajectory. Synchrotron radiation is produced as the electrons pass through the bending magnets, undulators, and wigglers. Electron beam energy lost to synchrotron radiation is replenished with a radiofrequency accelerator (a cavity with an axial electric field oscillating at the frequency of arrival of sequential electron bunches). Typical parameters characterizing synchrotron radiation from two modern storage rings, one optimized for the generation of soft x-rays and one optimized for the generation of hard x-rays, are given in Table 5.1

5.2 Characteristics of Bending Magnet Radiation

In this introductory section, we wish to use simple arguments to show why one expects to see radiation at x-ray wavelengths. The arguments are based on an estimate of the time duration of the observed radiation signal and an application of Heisenberg's uncertainty principle for photon energy. Bending magnet radiation is sometimes described as a sweeping "searchlight," analogous to the headlight of a toy train on a circular track. This searchlight effect is a general manifestation associated with radiation from relativistic particles undergoing acceleration. An electron experiencing radial acceleration as it travels around a circle emits radiation through a broad angular pattern – as seen in its frame of reference. However, angular patterns are very much compressed upon Lorentz transformation from one frame of reference (that moving with the electron) to another (the laboratory frame of the observer) when the relative motion is highly relativistic. In Appendix F it is shown that angles measured from the direction of motion are related by

$$\tan \theta = \frac{\sin \theta'}{\gamma(\beta + \cos \theta')} \qquad (5.1)$$

where θ' is observed in the frame of reference moving with the electron, θ is in the laboratory frame, $\beta \equiv v/c$ (where v is the relative velocity between frames and c is the velocity of light), and $\gamma \equiv 1/(1 - v^2/c^2)^{1/2}$. For highly relativistic electrons β approaches unity, and $\gamma \gg 1$. Thus for arbitrarily large emission angles θ', in the electron frame, the radiation is folded into a narrow forward radiation cone of half angle

$$\boxed{\theta \simeq \frac{1}{2\gamma}} \qquad (5.2)$$

leading to the description of synchrotron radiation as being concentrated in a narrow "searchlight beam."

Table 5.1 Typical parameters for synchrotron radiation at two complementary storage ring facilities. Both rings are optimized for small electron phase space (emittance) and the use of multiple straight sections for undulators and wigglers. Bending magnet radiation is obtained as the electron beam turns from one straight section to the next. The two facilities are complementary in that one is optimized for soft x-rays while the other is optimized for hard x-rays. The Advanced Light Source (ALS) is operated by Lawrence Berkeley National Laboratory in California. The Advanced Photon Source (APS) is operated by Argonne National Laboratory in Illinois. Parameters for other facilities around the world can be found through the website given in reference 6. Both rings anticipate upgrades to near circular beams, which along with other improvements will increase spectral brightness and coherent flux by factors of several hundred. See Section 5.4.6, Figure 5.24, and references therein.

Facility	ALS	APS
Electron energy	1.90 GeV	7.00 GeV
γ	3720	13 700
Current (mA)	500	100
Circumference (m)	197	1100
RF frequency (MHz)	500	352
Pulse duration (psec, FWHM)	60	79
Bending Magnet Radiation:		
Bending magnet field (T)	1.27	0.599
Critical photon energy (keV)	3.05	19.5
Critical photon wavelength	0.407 nm	0.0636 nm (0.636 Å)
Bending magnet sources	24	35
Undulator Radiation:		
Number of straight sections	12	40
Undulator period (typical) (cm)	5.00	3.00
Number of periods	89	69
Photon energy ($K = 1, n = 1$)	457 eV	10.3 keV
Photon wavelength ($K = 1, n = 1$)	2.71 nm	0.120 nm (1.20 Å)
Tuning range ($n = 1$)	2.0–5.4 nm	0.9–1.8 Å
Tuning range ($n = 3$)	0.68–1.8 nm	0.5–1.2 Å
Central cone half-angle ($K = 1, \gamma^*$)	35 μrad	11 μrad
Power in central cone ($K = 1, n = 1$) (W)	2.9	13.1
Flux in central cone (ph/s; $1/N$ BW)	3.9×10^{16}	1.1×10^{16}
σ_x, σ_y (μm)	251, 8.3	275, 8.6
σ_x', σ_y' (μrad)	9.7, 4.8	11.4, 3.4
Brightness ($K = 1, n = 1$)[a] [(photons/s)/mm$^2 \cdot$ mrad$^2 \cdot$ (0.1% BW)]	6.4×10^{19}	4.3×10^{19}
Total power ($K = 1$, all n, all θ; W)	86	304
Other undulator periods (cm)	3.0, 7.0, 8.0, 10.0	2.3, 2.7, 3.3, 5.5
Wiggler Radiation:		
Wiggler period (typical) (cm)	11.4	8.5
Number of periods	29	28
Magnetic field (maximum) (T)	1.9	1.0
K (maximum)	20	7.9
Critical photon energy (keV)	4.6	33
Critical photon wavelength (nm)	0.27	0.038 (0.38 Å)
Total power (max. K) (kW)	0.97	7.4

[a] Using Eq. (5.65).

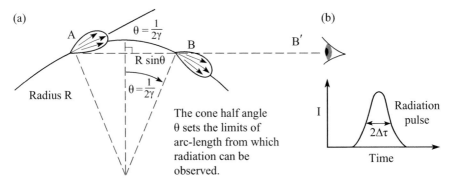

Figure 5.7 (a) A schematic of bending magnet radiation illustrating the "searchlight" effect, similar to that of the headlight of a train on a circular track, which is a general feature of radiation by highly relativistic electrons. (b) The time width of the observed radiation pulse is determined by transit time differences between radiation and electrons between points A and B. The uncertainty relationship between pulse duration and minimal spread of photon energy indicates that a broad range of photon energies, extending to the x-ray region, is to be expected. (Following Hofmann.[2])

As the electron traverses a curved path, radiation is emitted tangentially in a narrow radiation cone of half width $\theta \cong 1/2\gamma$, as seen in Figure 5.7. For electrons circulating in a ring, we can estimate the photon energies and wavelengths radiated using simple arguments based on Heisenberg's Uncertainty Principle[7], $\Delta E \cdot \Delta \tau \geq \hbar/2$, where $\Delta \tau$ is the (rms) time duration during which one detects radiation, and ΔE is the uncertainty (rms spread) in observed photon energies. We begin by estimating the detected pulse duration, $2\Delta \tau$, of radiation emitted by a short bunch of electrons following a circular trajectory of radius R. We estimate the time extent of the observed signal by considering a detector at point B or equivalently further to the right at B'. As the electron comes within an angle $\theta \cong 1/2\gamma$ of the horizon at point A, the detector will be in the path of emitted photons. These photons will be detected after a transit time of the light, τ_r. The signal will continue until the electron reaches point B, beyond which the radiation cone has turned too far to permit reception by our detector. The electron will reach point B after a transit time around the bend, τ_e. The pulse width, $\Delta \tau$, shown in Figure 5.7(b) is the difference between these two transit times, i.e., the detector detects radiation after a time τ_r, and stops detecting radiation at τ_e. Following this outline, we see that

$$2\Delta\tau = \tau_e - \tau_r$$

$$2\Delta\tau = \frac{\text{arc length}}{\text{v}} - \frac{\text{radiation path}}{c}$$

$$2\Delta\tau \simeq \frac{R \cdot 2\theta}{\text{v}} - \frac{2R\sin\theta}{c}$$

Noting that $\theta \cong 1/2\gamma$, making a small angle approximation for $\sin\theta$, and substituting $\text{v} = \beta c$, one obtains

$$2\Delta\tau \simeq \frac{R}{\gamma\text{v}} - \frac{R}{\gamma c} = \frac{R}{\gamma}\left(\frac{1}{\text{v}} - \frac{1}{c}\right)$$

Writing $v = \beta c$, one has

$$2\Delta\tau \simeq \frac{R}{\gamma\beta c}(1-\beta)$$

Noting that

$$\gamma \equiv \frac{1}{\sqrt{1-\beta^2}}$$

$$\gamma^2 = \frac{1}{1-\beta^2} = \frac{1}{(1-\beta)(1+\beta)}$$

and thus for $\beta = v/c$ approaching unity

$$1-\beta \simeq \frac{1}{2\gamma^2} \tag{5.3}$$

the expression for the duration of the radiation pulse becomes

$$2\Delta\tau \simeq \frac{R}{2c\gamma^3} \tag{5.4a}$$

This can be expressed as an anticipated photon energy spread through the use of Heisenberg's Uncertainty Principle[7] and an expression for the radius of curvature R. From the uncertainty principle,

$$\Delta E \cdot \Delta\tau \geq \hbar/2$$

Combining this with the expression in Eq. (5.4a) for the pulse duration, we see that the photons will have an rms energy spread of order*

$$\Delta E \geq \frac{2\hbar c\gamma^3}{R} \tag{5.4b}$$

To better appreciate the photon energies implied by Eq. (5.4b) it is useful to replace the electron radius of curvature R with an expression involving γ and the magnetic field. For electrons crossing a perpendicular magnetic field, as in a bending magnet, the relativistically correct form of the equation of motion can be written as

$$\mathbf{F} = \frac{d\mathbf{p}}{dt} = -e\mathbf{v} \times \mathbf{B}$$

where $\mathbf{p} = \gamma m\mathbf{v}$ is the momentum[7], m is the electron rest mass, γ is the Lorentz factor, $\mathbf{v}$ is the velocity, and $\mathbf{B}$ is the magnetic flux density. For electron motion in a uniform magnetic field, the electron energy and thus γ is a constant, so that only the direction of $\mathbf{v}$ changes, not its magnitude. To see this we write the rate of change of electron energy as

$$\frac{dE_e}{dt} = \mathbf{v} \cdot \mathbf{F} = \underbrace{-e\mathbf{v} \cdot (\mathbf{v} \times \mathbf{B})}_{\equiv 0}$$

---

* Similar arguments are given in J.D. Jackson (Ref. 8), First Edition, pp. 475–477.

which is zero by vector identity (see Appendix D.1). Thus the electron energy, which can be written[6] as γmc^2, is a constant, viz.,

$$\frac{dE_e}{dt} = \frac{d}{dt}(\gamma mc^2) = 0$$

Thus γ, and therefore the scalar magnitude v of the velocity, are both constant. The equation of motion can be rewritten as

$$\gamma m \frac{d\mathbf{v}}{dt} = -e\mathbf{v} \times \mathbf{B}$$

Since the magnitude of $\mathbf{v}$ is constant, the magnitude of the acceleration is also constant, equal to $evB/\gamma m$, in a plane perpendicular to $\mathbf{B}$. This corresponds to motion along a circle, with centripetal acceleration v^2/R, so that the scalar form of the equation of motion becomes

$$\gamma m \left(-\frac{v^2}{R}\right) = -evB$$

Solving for the radius of curvature, we have

$$R = \frac{\gamma mv}{eB}$$

or for highly relativistic electrons

$$R \simeq \frac{\gamma mc}{eB}$$

Using this in Eq. (5.4b), the rms spread of photon energies for bending magnet radiation becomes

$$\Delta E \geq \frac{2e\hbar B\gamma^2}{m} \tag{5.4c}$$

which we note depends on the electron charge to mass ratio, e/m, and the product $B\gamma^2$. If we substitute values for e, $\hbar$, and m, Eq. (5.4c) indicates photon energies in the keV range (nanometer wavelengths) for typical values of γ and B found in modern storage rings, e.g., γ of several thousand and B of 1T or more. For highly relativistic electrons it is convenient to express the total electron energy in terms of γ and the electron rest energy, mc^2, as[1]

$$\gamma = \frac{E_e}{mc^2} = 1957 \, E_e[\text{GeV}] \tag{5.5}$$

where on the right side we have used the fact that the electron rest energy is 0.5110 MeV, and by use of square brackets expressed the electron energy E_e in GeV.

The description of expected photon energy spread obtained above, Eq. (5.4c), is based on relatively simple arguments involving Heisenberg's uncertainty principle. It is valuable in that it provides a measure of the expected photon energies radiated by accelerated charges moving at relativistic speeds, and gives a functional dependence in terms of $B\gamma^2$. The numerical factor (2) obtained by this argument is, however, somewhat arbitrary in that it depends on the angular distribution of radiation embodied in our assumption that $\theta \simeq 1/2\gamma$. A more precise description of the photon energy

Table 5.2 Sample values of the functions $G_1(y)$ and $H_2(y)$, where $y = \omega/\omega_c$. Following Green.[9]

y	$G_1(y)$	$H_2(y)$
0.0001	9.959×10^{-2}	6.271×10^{-3}
0.0010	2.131×10^{-1}	2.910×10^{-2}
0.0100	4.450×10^{-1}	1.348×10^{-1}
0.1000	8.182×10^{-1}	6.025×10^{-3}
0.3000	9.177×10^{-1}	1.111×10^{0}
0.5000	8.708×10^{-1}	1.356×10^{0}
0.7000	7.879×10^{-1}	1.458×10^{0}
1.000	6.514×10^{-1}	1.454×10^{0}
1.500	4.506×10^{-1}	1.250×10^{0}
2.000	3.016×10^{-1}	9.780×10^{-1}
3.000	1.286×10^{-1}	5.195×10^{-1}
4.000	5.283×10^{-2}	2.493×10^{-1}
5.000	2.125×10^{-2}	1.131×10^{-1}
7.000	3.308×10^{-3}	2.107×10^{-2}
10.00	1.922×10^{-4}	1.478×10^{-3}

distribution, obtained by a rigorous solution of Maxwell's equations for a relativistic electron in a uniform magnetic field, introduces instead a factor of $3/2$ and a more useful definition of ΔE. The results are somewhat complex, involving modified Bessel functions of the second kind.[1–3] Defining θ as the in-plane observation angle for radiation from relativistic electrons traveling in a circular path, and ψ as the out-of-plane (vertical) angle, Kim[1] shows that the photon flux F_B for bending magnet radiation is given on axis by

$$
\left. \frac{d^3\overline{F_B}}{d\theta\, d\psi\, d\omega/\omega} \right|_{\psi=0}
$$
$$
= 1.33 \times 10^{13} E_e^2\,[\text{GeV}]\bar{I}[\text{A}]H_2(E/E_c)\frac{\text{photons/s}}{\text{mrad}^2 \cdot (0.1\%\,\text{BW})} \tag{5.6}
$$

where the electron energy E_e is expressed in GeV, the average current $\bar{I}$ in amperes, the units of relative spectral bandwidth $d\omega/\omega$ are expressed non-dimensionally as a factor of 10^{-3}, or 0.1% BW, and the function

$$
H_2(y) = y^2 K_{2/3}^2(y/2)
$$

is a modified Bessel function, tabulated in Table 5.2 and shown graphically in Figure 5.8. The ratio E/E_c is the photon energy normalized with respect to a *critical photon energy*

$$
E_c = \hbar\omega_c = \frac{3e\hbar B\gamma^2}{2m} \tag{5.7a}
$$

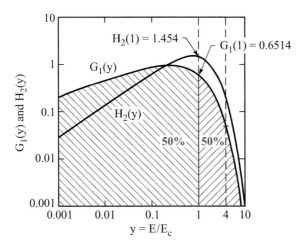

Figure 5.8 The functions $H_2(y)$, representing on-axis photon flux from a bending magnet, and $G_1(y)$, representing the vertically integrated photon flux, as functions of photon energy normalized to the critical photon energy. Half the radiated power is in photons of energy greater than E_c, and half in photons of energy less than E_c (following Kim[1]). Note that for a photon energy of $4E_c$ the photon flux is reduced a factor of about 10 from its value at E_c.

The critical photon energy is that for which half the radiated power is in higher energy photons and half is in lower energy photons. As such it provides a primary parameter for characterizing bending magnet radiation.

Equation (5.7a) can be rewritten in practical units as[1]

$$E_c[\text{keV}] = 0.6650 E_e^2[\text{GeV}]B[\text{T}] \tag{5.7b}$$

where the critical photon energy is in keV, the electron beam energy is given in GeV, and the magnetic field in teslas. The corresponding *critical wavelength* is

$$\lambda_c = \frac{4\pi mc}{3eB\gamma^2} \tag{5.7c}$$

which can be written in practical units of nanometers, GeV, and Tesla as

$$\lambda_c[\text{nm}] = \frac{1.864}{E_e^2[\text{GeV}]B[\text{T}]} \tag{5.7d}$$

Note that the critical photon energy given in Eq. (5.7a) is well within the range of photon energies estimated by Eq. (5.4c) on the basis of relativistic angular transformations and Heisenberg uncertainty arguments.

The critical photon energy is in fact a very useful parameter for characterizing synchrotron radiation from relativistic electrons as they traverse the fields of a bending magnet. For example, of two new storage rings operating in the United States, the Advanced Light Source (ALS) at Lawrence Berkeley National Laboratory in California, with a beam energy of 1.9 GeV and a bending magnet field strength of 1.27 T, has a critical photon energy of 3.1 keV and a critical wavelength of 0.41 nm (4.1 Å), while the Advanced Photon Source (APS) at Argonne National Laboratory in Illinois, with a beam

Table 5.3 Measures of angular divergence of
bending magnet radiation in the vertical plane,
as a function of normalized photon energy.
Single sided rms and full width at half maximum
(FWHM) measures are given. Following Kim.[1]

E/E_c	σ'_ψ (rms)	FWHM
0.01	$5.0/\gamma$	$12/\gamma$
0.03	$3.3/\gamma$	$7.8/\gamma$
0.1	$2.0/\gamma$	$4.7/\gamma$
0.3	$1.2/\gamma$	$2.8/\gamma$
1	$0.64/\gamma$	$1.5/\gamma$
3	$0.37/\gamma$	$0.9/\gamma$
10	$0.18/\gamma$	$0.4/\gamma$

energy of 7.0 GeV and a bending magnet field strength of 0.60 T, has a critical photon
energy of 20 keV and a critical wavelength of 0.064 nm (0.64 Å).

Typical parameters characterizing synchrotron radiation from these two representative facilities are presented in Table 5.1. Between the two they cover a broad region of
the electromagnetic spectrum. In fact, inspection of Figure 5.8 shows that on axis the
photon flux decreases by only a factor of 10 at a photon energy equal to $4E_c$. For many
experiments this significantly extends the useful range of bending magnet radiation, for
instance to 12 keV at the ALS, and to 80 keV at the APS. Further enhancements using
strong field periodic wigglers are also possible. Wiggler radiation is described at the end
of this chapter.

On occasion it is convenient to know the bending magnet photon flux per unit horizontal angle θ, integrating out the vertical plane ψ-dependence. In this case Kim[3] finds
that the radiated photon flux, in units of photons per second per milliradian per 0.1%
relative spectral bandwidth, is given by

$$\boxed{\frac{d^2 F_B}{d\theta \, d\omega/\omega} = 2.46 \times 10^{13} E_e [\text{GeV}] \bar{I}[\text{A}] G_1(E/E_c) \frac{\text{photons/s}}{\text{mrad} \cdot (0.1\% \, \text{BW})}} \qquad (5.8)$$

where the function

$$G_1(y) = y \int_y^\infty K_{5/3}(y') \, dy'$$

is also shown graphically in Figure 5.8. Note that by the definition of E_c, the integrals
of $G_1(y)$ from zero to one and from one to infinity are equal, as suggested in Figure 5.8.
Table 5.2 gives some specific values of the functions $H_2(\omega/\omega_c)$ and $G_1(\omega/\omega_c)$.

Note that bending magnet radiation is linearly polarized when viewed in the horizontal plane of acceleration. When viewed outside this plane, bending magnet radiation
is elliptically polarized. The out of plane photon flux, decomposed into horizontal and
vertical polarization components, is given by Kim.[1] Kim also introduces a convenient
measure of angular divergence[1] in the vertical plane, σ'_ψ, for bending magnet radiation.

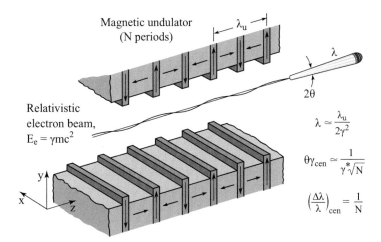

Figure 5.9 Illustration of narrow cone undulator radiation that is generated by electrons traversing a periodic magnet structure. The zeroth-order velocity of the electron beam is in the z-direction, the periodic magnetic field points in the y-direction, and the first order electron oscillation velocity is in the x-direction.

This divergence angle varies with normalized photon energy, E/E_c. Fitted to a Gaussian angular distribution, the rms half angle in the vertical plane is $0.64/\gamma$ at $E/E_c = 1$. Full width at half maximum (FWHM) measures are larger by a factor of 2.35. Sample values are given in Table 5.3 for sample values of E/E_c.

Since the acceleration of electrons is confined to the horizontal plane (for vertical bending magnet fields), the electric field of the resultant radiation will be linearly polarized in that plane. The general polarization properties of bending magnet radiation for arbitrary angles of observations are discussed in Ref.1.

5.3   Characteristics of Undulator Radiation

An electron traversing a periodic magnet structure[10] of moderate field strength will undergo a small amplitude oscillation and therefore radiate. If the electron's angular excursions are small compared to the natural radiation width, $\theta_e < 1/2\gamma$, the device is referred to as an *undulator* (see Figure 5.9). The resultant radiation is greatly reduced in wavelength, λ, from that of the magnet period, λ_u. We will see shortly that Lorentz contraction and relativistic Doppler shift lead to a reduction in the radiated wavelength by a factor of $2\gamma^2$. As γ can easily be several thousand, undulator periods measured in centimeters lead to observed x-ray wavelengths measured in angstroms.

While discussing undulator radiation, we will find it convenient to consider the radiation in several frames of reference. Many of the calculations will be done in the reference frame moving with the electron. We will then transform the results to the rest frame of the laboratory via Lorentz transformations (see Refs. 7 and 11, or Appendix F,

Lorentz Space-Time Transformations). The following is a brief introduction to undulator radiation. A more detailed discussion will follow in subsequent sections.

In the frame moving with the electron, the electron "sees" a periodic magnet structure moving toward it with a relativistically (Lorentz) contracted period, λ', given by

$$P_T = \frac{1.90 \times 10^{-6}(\text{W})K^2\gamma^2 NI(\text{A})}{\lambda_u(\text{cm})} \tag{5.9}$$

where $\gamma \equiv 1/\sqrt{(1 - \text{v}^2/c^2)}$, v is the relative velocity, and c is the velocity of light in vacuum, as discussed in Appendix F. Owing to the periodic magnet, the electron experiences an oscillation and consequently radiates. In the frame moving with the electron this problem is that of the classical *radiating dipole*, a point charge oscillating with an amplitude much smaller than the radiated wavelength. The frequency of this emitted radiation, in the reference frame of the electron, is

$$f' = \frac{c}{\lambda'} = \frac{c\gamma}{\lambda_u}$$

To the observer in the fixed laboratory reference frame, the radiation wavelength is further reduced by Doppler shifting. The Doppler shift is dependent on the relative velocity and therefore is dependent on the observation angle θ, as can be deduced from Figure 5.1. The shortest wavelength is observed on axis. The relativistic form of the Doppler frequency formula is [see Appendix F, Eq. (F.8b)]

$$f = \frac{f'}{\gamma(1 - \beta\cos\theta)} = \frac{c}{\lambda_u(1 - \beta\cos\theta)} \tag{5.10}$$

where $\beta \equiv \text{v}/c$ and θ is the observation angle measured from the direction of motion. Let us first analyze the observed frequency on axis. Here $\theta = 0$, $\cos\theta = 1$, and

$$f = \frac{c}{\lambda_u(1 - \beta)}$$

As noted in Eq. (5.3), for $\beta \simeq 1$ we have $1 - \beta \cong 1/2\gamma^2$. Therefore, the observed radiation frequency on axis is

$$f = \frac{2\gamma^2 c}{\lambda_u}$$

and the observed wavelength on axis is

$$\lambda = \frac{c}{f} = \frac{\lambda_u}{2\gamma^2} \tag{5.11}$$

Note that the observed wavelength, λ, is relativistically contracted by a factor $2\gamma^2$ from the period of the undulator. Again using the ALS as an example, with a 1.9 GeV electron energy, $\gamma \cong 3700$ [see Eq. (5.5)]; thus $2\gamma^2 \cong 2.8 \times 10^7$. If the undulator period is $\lambda_u = 5.0$ cm, the resultant on-axis radiation will be relativistically shifted to an observed wavelength of order

$$\lambda \simeq \frac{5.0\,\text{cm}}{2.8 \times 10^7} \simeq 1.8\,\text{nm}$$

Thus the periodic magnet generates radiation peaked in the soft x-ray region of the electromagnetic spectrum.

If we wish to consider Doppler shifts at small angles off axis ($\theta \neq 0$), we can return to Eq. (5.10) and use the small angle approximation. The Taylor expansion for small angles is $\cos\theta = 1 - \theta^2/2 + \cdots$; therefore,

$$f = \frac{\frac{c}{\lambda_u}}{1 - \beta\left(1 - \frac{\theta^2}{2} + \cdots\right)} = \frac{\frac{c}{\lambda_u}}{1 - \beta + \frac{\beta\theta^2}{2} + \cdots} = \frac{\frac{c}{(1-\beta)\lambda_u}}{1 + \frac{\beta\theta^2}{2(1-\beta)}}$$

Since $\beta \cong 1$ and by Eq. (5.3) $1 - \beta \simeq 1/2\gamma^2$, one has

$$f = \frac{\frac{2\gamma^2 c}{\lambda_u}}{1 + \frac{2\gamma^2\theta^2}{2} - \cdots} = \frac{2c\gamma^2}{\lambda_u(1 + \gamma^2\theta^2)}$$

In terms of the observed wavelength $\lambda = c/f$, one has to first order

$$\lambda = \frac{\lambda_u}{2\gamma^2}(1 + \gamma^2\theta^2) \tag{5.12}$$

We again see the $2\gamma^2$ contraction on axis, but now with the off-axis radiation having a wavelength increased by a factor $(1 + \gamma^2\theta^2)$. Hence, to observe the narrow bandwidth characteristic of this relativistic harmonic oscillator, it is necessary to select only near-axis radiation.

As we will see explicitly in a following section, the magnetically induced undulation causes the electron to follow a somewhat longer path length as it traverses the undulator. Thus, the mean axial velocity is reduced, resulting in a modified Doppler shift and therefore somewhat longer wavelengths than indicated by Eq. (5.12), and a broader radiation cone as well.

5.3.1 Undulator Radiation Pattern

As we saw in Chapter 2, Eqs. (2.25)–(2.33), an oscillating electron of charge $-e$ undergoing an acceleration $\mathbf{a}$ will radiate electromagnetic waves characterized by an electric field[11]

$$E(\mathbf{r}, t) = \frac{ea(t - r/c)}{4\pi\epsilon_0 c^2 r} \sin\Theta$$

and an orthogonal magnetic field

$$H(\mathbf{r}, t) = \frac{ea(t - r/c)}{4\pi cr} \sin\Theta$$

where $t - r/c$ is the retarded time (delayed arrival at distance r), and Θ is the angle between the direction of acceleration ($\mathbf{a}$) and the propagation direction ($\mathbf{k}_0$). Because the electric and magnetic fields are orthogonal, their cross product gives a Poynting vector $\mathbf{S}$ (power per unit area) of

$$\mathbf{S} = \mathbf{E} \times \mathbf{H} = \left[\frac{e^2 a^2 \sin^2\Theta}{16\pi^2\epsilon_0 c^3 r^2}\right]\mathbf{k}_0$$

(a) 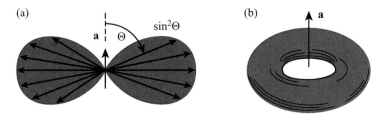 (b)

Figure 5.10 Illustration of an oscillating charge and the resultant radiation pattern. Note that there is no radiation in the direction of acceleration, giving the radiation pattern a doughnut-like appearance.

(a) 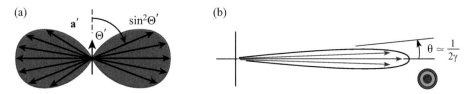 (b)

Figure 5.11 (a) Illustration of the radiation pattern of an oscillating electron in the frame of reference moving with the average electron speed. (b) Illustration of the radiation pattern of a highly relativistic electron as observed in the laboratory frame of reference.

The radiated power per unit solid angle is [Chapter 2, Eq. (2.34)]

$$\frac{dP}{d\Omega} = r^2 |\mathbf{S}| = \frac{e^2 a^2}{16\pi^2 \epsilon_0 c^3} \sin^2 \Theta$$

Hence, the radiation pattern has a toroidal $\sin^2 \Theta$ shape, because there is no radiation in the acceleration direction ($\Theta = 0$), as illustrated in Figure 5.10.

For an undulating electron, undergoing simple oscillations in its own reference frame (γ), one obtains the same radiation pattern. However, the radiation pattern as observed in the laboratory frame is relativistically contracted into a narrow radiation cone (the so-called searchlight effect) as shown in Figure 5.11(b). Considering the symmetry of the problem, it is convenient to work with a polar coordinate system measured from the z-axis. For instance, in the plane defined by the electron acceleration (**a**) and the z-axis, the factor $\sin^2 \Theta'$ becomes $\cos^2 \theta'$, θ' being the polar angle measured away from the z-axis in the primed coordinate system. In this primed electron frame of reference the radiation pattern has a half-intensity angle at $\cos^2 \theta' = \frac{1}{2}$ or $\theta' = 45°$. According to Eq. (5.1), this corresponds to an angle in the unprimed laboratory (observer) frame of reference of $\theta \cong 1/2\gamma$. Returning to the example of a 1.9 GeV electron ($\gamma \simeq 3700$), in this case traversing a periodic magnet structure, one anticipates that radiated x-rays will largely be confined to a cone of half angle 140 μrad. As we will see in the following paragraphs, further cone narrowing can be obtained in the case of undulator radiation.

5.3.2   The Central Radiation Cone

The spectrum of radiation in the two reference frames is shown in Figure 5.12(a) and (b). Figure 5.12(a) shows the narrow spectral width in the electron frame, set by the

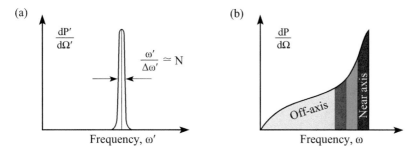

Figure 5.12 (a) The radiation spectrum as seen in the frame of reference moving with the electron is narrow with a relative spectral bandwidth of order $1/N$, where N is the number of oscillation periods. (b) In the laboratory frame of reference, the wavelengths are shorter, but the spectrum is broader due to off-axis Doppler effects (following Hofmann[2]).

harmonic oscillation for a fixed number of periods N. This is essentially a frequency–time (Laplace) transform. For example, the ALS has undulators of 5.0 cm period, with a length of 89 periods, so that one can expect $\Delta\omega'/\omega' = \Delta\lambda'/\lambda'$ of order 0.01. Note, however, that upon transformation to the laboratory frame of reference, off-axis Doppler effects will broaden this considerably. Figure 5.12(b) illustrates the Doppler shifted spectrum that results when the $\sin^2\Theta$ dipole radiation pattern is transformed according to Eqs. (5.1) and (5.12).

Recall that we have determined the undulator equation (5.12) in the laboratory frame, viz.,

$$\lambda \simeq \frac{\lambda_u}{2\gamma^2}(1 + \gamma^2\theta^2)$$

and have also noted that the radiation is primarily contained in a narrow cone of half angle $\theta = 1/2\gamma$. The corresponding spectral width within this cone can thus be estimated by taking the difference of Eq. (5.12) for two angles. Taking the wavelength as λ on axis ($\theta = 0$), and $\lambda + \Delta\lambda$ off axis at angle θ, then taking ratios, one obtains

$$\frac{\Delta\lambda}{\lambda} \simeq \gamma^2\theta^2 \tag{5.13}$$

where Eq. (5.13) shows how the wavelength increases as one observes the radiation off axis. Note that for radiation within the cone of half angle $\theta \cong 1/2\gamma$ the relative spectral bandwidth given by Eq. (5.13) is $1/4$; thus the cone of half-intensity half angle encloses a relative spectral bandwidth of about 25%. Use of aperture spectral filtering is illustrated in Figure 5.13. Often, further spectral narrowing is desired, for instance, when probing in the vicinity of sharp atomic resonance features. In such cases, a monochromator of some type is employed that acts as a narrow bandpass filter. In the case of radiation from a single electron or a tightly constrained bunch of electrons, modest spectral filtering (as narrow as $1/N$) can be obtained with a simple small-angle selecting aperture (pinhole). In this limit, we will see that angular width and spectral width are closely connected. The interrelationship is shown in Figure 5.14.

Further cone narrowing can be appreciated by considering the undulator equation for two angular positions, one on axis and one at angle θ, as we did previously in Eq. (5.13).

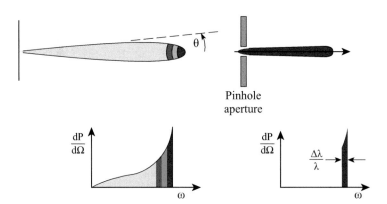

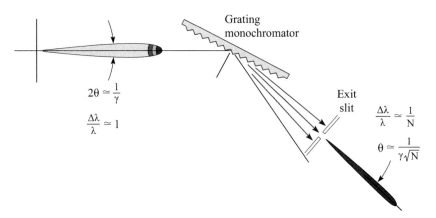

Figure 5.13 The spectrum of undulator radiation in the laboratory frame of reference before and after selecting an angular cone near the axis. With a sufficiently small electron beam phase space (size–angle product) this can provide a simple mechanism for monochromatization.

Figure 5.14 Illustration of a grating monochromator as used to pass a narrow band of soft x-ray undulator radiation corresponding to a "natural" spectral width of $1/N$, and the concomitant cone narrowing to $1/\gamma\sqrt{N}$ that occurs when radiated by a tightly constrained electron beam. A similar band pass can be achieved with hard x-rays using a crystal monochromator.

If one sets the monochromator for a "natural" bandwidth $\Delta\lambda/\lambda$, set by the number of oscillation periods, N, then one obtains the condition

$$\frac{\Delta\lambda}{\lambda} = \frac{1}{N} \tag{5.14}$$

which, when combined with Eq. (5.13), indicates that narrower bandwidth radiation occurs in a *concomitantly narrower "central" radiation cone* of half width

$$\theta_{\text{cen}} \simeq \frac{1}{\gamma\sqrt{N}} \tag{5.15}$$

This narrow undulator radiation cone implies an emission solid angle reduced by a factor $1/N$. These factors become very important when considering brightness and coherence (see Chapter 4). The above analysis is for a single electron. For these results to hold for an electron beam with many electrons, it is necessary that all electrons in the bunch be contained within an angular variance of less than $1/\gamma\sqrt{N}$. This angular constraint on the electron beam is referred to as the *undulator condition*. Again considering 1.9 GeV electrons, with $\gamma \cong 3720$ and $N \cong 100$, one expects the 1% spectral bandwidth radiation to be confined within a cone of angular half width $\theta_{cen} \cong 35$ μrad. A correction to this formula is given in Eq. (5.32) for the general case where $\gamma \to \gamma^*$, accounting for slowed electron velocity in the axial direction in the presence of finite undulator magnetic fields.

5.4 Undulator Radiation: Calculations of Radiated Power, Brightness, and Harmonics

Having introduced the basic features of undulator radiation, we now wish to solve the problem by considering the equations of motion for a highly relativistic electron traversing a periodic magnetic field. In the laboratory frame, the electron experiences only the static, albeit periodic, magnetic field for small K. Hence, the laboratory is a convenient reference frame for the calculation. After calculating electron trajectories in the laboratory frame, we will transform to the frame of reference moving with the average electron motion (γ). Our next step will be to calculate the radiated fields in the electron frame where we have simple harmonic motion (dipole radiation). We will see a multiplicity of harmonics, $n\omega$, of this radiation. Finally, we will transform the radiated fields to the laboratory frame. The approach follows that of Hofmann.[2]

5.4.1 The Undulator Equation

The Lorentz force, which describes the rate of change of momentum experienced by a charge in the presence of electric and magnetic fields, can be described in any frame of reference by

$$\frac{d\mathbf{p}}{dt} = q(\mathbf{E} + \mathbf{v} \times \mathbf{B}) \tag{5.16}$$

where $\mathbf{p} = \gamma m\mathbf{v}$ is the momentum, q is the charge, $\mathbf{v}$ is the velocity, and $\mathbf{E}$ and $\mathbf{B}$ are the electric field vector and magnetic flux density, determined through Maxwell's equations. The problem we are considering is dominated by the applied dc magnetic field associated with a periodic magnet structure (undulator), as illustrated in Figures 5.9 and 5.15. There are no applied electric fields. Further, we consider the radiated electromagnetic fields due to the undulator radiation to be relatively weak relative to the effects of the dc magnetic field, in the sense that the radiated fields have a negligible effect on the various electron motions. To this level of approximation, we take $\mathbf{E} \cong 0$ in Eq. (5.16). Note that this would not be the case in a sufficiently long undulator. In fact, the effect

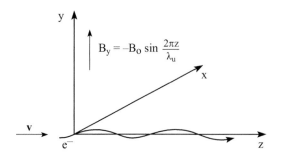

Figure 5.15 Electron motion in a periodic magnetic field.

of the radiated fields leads to free electron laser (FEL) action in long undulators with high current and a tightly controlled electron beam, as is discussed in Chapter 6. For the present case of a modest length undulator, with relatively weak radiated fields, the momentum equation is approximated by

$$\frac{d\mathbf{p}}{dt} = -e(\mathbf{v} \times \mathbf{B})$$

For the undulator case with relatively weak radiated fields (pre-FEL action), we take the approximations $E \cong 0$ and $\mathbf{B}_y = -\mathbf{B}_0 \sin(2\pi z/\lambda_u)$ plus a negligible radiation field. Additionally, taking to first order $\mathbf{v} \cong \mathbf{v}_z$, the vector components in the x-direction give

$$m\gamma \frac{d\mathbf{v}_x}{dt} = +e\mathbf{v}_z B_y$$

$$m\gamma \frac{d\mathbf{v}_x}{dt} = -e\frac{dz}{dt} \cdot B_0 \sin\left(\frac{2\pi z}{\lambda_u}\right) \qquad (0 \leq z \leq N\lambda_u)$$

Now we can solve for the transverse oscillation $\mathbf{v}_x$. This gives rise to the primary source of undulator radiation. To first order, we will find $\mathbf{v}_x$ as a function of axial position z. Continuing the algebra,

$$m\gamma \, d\mathbf{v}_x = -e \, dz \, B_0 \sin\left(\frac{2\pi z}{\lambda_u}\right)$$

Integrating both sides gives

$$m\gamma \mathbf{v}_x = -eB_0 \frac{\lambda_u}{2\pi} \int \sin\left(\frac{2\pi z}{\lambda_u}\right) \cdot d\left(\frac{2\pi z}{\lambda_u}\right)$$

or

$$m\gamma \mathbf{v}_x = \frac{eB_0 \lambda_u}{2\pi} \cos\left(\frac{2\pi z}{\lambda_u}\right) \qquad (5.17)$$

This is an exact solution of the simplified equation of motion, but note that z is not a linear function of time. That is, $\mathbf{v}_z$ is not constant, but rather involves oscillations itself. Hence, terms of the form $\sin(\ldots \sin)$ will appear, giving rise to harmonics.

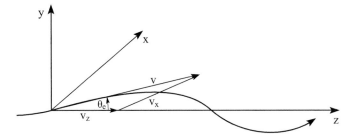

Figure 5.16 Electron angular excursions are harmonic, with maximum excursion K/γ. For $K < 1$ the angular excursions are within the natural radiation cone ($1/2\gamma$), leading to interesting interference effects that are manifested in cone narrowing, higher spectral brightness, and in some cases partial coherence. The case of small angular excursions, $K \le 1$, is referred to as the undulator limit. For $K \gg 1$ such interference effects are not possible. The limit $K \gg 1$ is referred to as the wiggler limit. The scales here are exaggerated in the x-direction for clarity of presentation.

It is convenient to write Eq. (5.17) in terms of the dimensionless undulator parameter for a periodic magnet as[12]

$$K \equiv \frac{eB_0\lambda_u}{2\pi mc} \qquad (5.18a)$$

or, in bracketed convenient units,[1]

$$K = 0.9337B_0[\text{T}]\lambda_u[\text{cm}] \qquad (5.18b)$$

The electron's transverse velocity can then be written as [cos]

$$v_x = \frac{Kc}{\gamma}\cos\left(\frac{2\pi z}{\lambda_u}\right) \qquad (5.19)$$

Note that the angle the electron motion makes with the z-axis is a sine function bounded by $\pm K/\gamma$, i.e.,

$$\tan\theta_e = \frac{v_x}{v_z} \simeq \frac{K}{\gamma}\cos\left(\frac{2\pi z}{\lambda_u}\right) \qquad (5.20)$$

In this context K is also referred to as the *magnetic deflection parameter*. Note that to good approximation we have taken $v_z \cong c$. Thus the maximum excursion angle (see Figure 5.16) is

$$|\theta_{e,\,\text{max}}| \simeq \frac{K}{\gamma} \qquad (5.21)$$

This is the root of differences between undulator radiation and wiggler radiation. Recall that the characteristic half angle for emission of radiation is $\theta_{\text{rad}} \cong 1/2\gamma$. Thus, for magnet strength characterized by $K \le 1$, the electron angular excursions lie within the radiation cone. This is the undulator case where interesting interference effects can occur, narrow bandwidths result, and narrower radiation cones are obtained.

For the strong field wiggler, $K \gg 1$, interference opportunities are lost because the radiation from various segments of an oscillation are widely separated in angle and therefore do not overlap in space after some propagation distance. Nonetheless, other valuable attributes appear. Primarily, wiggler radiation provides a $2N$ increase in radiated power and a broad shift to higher photon energies. We will discuss both cases ($K \leq 1, K \gg 1$) further.

Recall that Eq. (5.19) is not that of a simple time harmonic, because $z = z(t)$ is only approximately equal to ct. To see this explicitly, we recall that γ is constant in a magnetic field; thus for motion in the x, z-plane ($v_y = 0$),

$$\gamma \equiv \frac{1}{\sqrt{1 - \frac{v^2}{c^2}}} = \frac{1}{\sqrt{1 - \frac{v_x^2 + v_z^2}{c^2}}}$$

Thus,

$$\frac{v_z^2}{c^2} = 1 - \frac{1}{\gamma^2} - \frac{v_x^2}{c^2} \tag{5.22}$$

Knowing v_x from Eq. (5.19), we can solve for v_z:

$$\frac{v_z^2}{c^2} = 1 - \frac{1}{\gamma^2} - \frac{K^2}{\gamma^2} \cos^2 \left(\frac{2\pi z}{\lambda_u} \right)$$

To first order in the small parameter K/γ:

$$\frac{v_z}{c} = 1 - \frac{1}{2\gamma^2} - \frac{K^2}{2\gamma^2} \cos^2 \left(\frac{2\pi z}{\lambda_u} \right) \tag{5.23}$$

where $k_u = 2\pi / \lambda_u$ and $\cos^2 k_u z = \frac{1}{2}(1 + \cos 2k_u z)$, and thus

$$\frac{v_z}{c} = 1 - \frac{1 + K^2/2}{2\gamma^2} - \frac{K^2}{4\gamma^2} \cos 2k_u z \tag{5.24}$$

Hence, the velocity (z-direction) has a reduced average component and a component oscillating at twice the magnet spatial frequency. By averaging over an integer number of periods, we can determine the average axial velocity, which plays a major role in the relativistic transformations. The average axial velocity for finite K is then given by

$$\frac{\bar{v}_z}{c} = 1 - \frac{1 + K^2/2}{2\gamma^2} \tag{5.25}$$

where the contribution from the $\cos^2 k_u z$ term averages to zero for each period. From this, we can define an effective axial value of the relativistic factor,

$$\boxed{\gamma^* \equiv \frac{\gamma}{\sqrt{1 + K^2/2}}} \tag{5.26}$$

where the asterisk (*) refers to the reduction of the relativistic contraction factor by an amount $\sqrt{1 + K^2/2}$. Hence Eq. (5.25) can be rewritten as

$$\frac{\bar{v}_z}{c} = 1 - \frac{1}{2\gamma^{*2}} \tag{5.27}$$

As a consequence, the observed wavelength in the laboratory frame of reference is modified from that given in Eq. (5.12), now taking the form

$$\lambda = \frac{\lambda_u}{2\gamma^{*2}}(1 + \gamma^{*2}\theta^2)$$

that is, the Lorentz contraction and relativistic Doppler shift now involve γ^* rather than γ. Expanding γ^* according to Eq. (5.26), one has

$$\lambda = \frac{\lambda_u}{2\gamma^2}\left(1 + \frac{K^2}{2}\right)\left(1 + \frac{\gamma^2}{1 + K^2/2}\theta^2\right)$$

or

$$\boxed{\lambda = \frac{\lambda_u}{2\gamma^2}\left(1 + \frac{K^2}{2}\gamma^2\theta^2\right)} \tag{5.28}$$

where we recall that $K = eB_0\lambda_u/2\pi mc$. Equation (5.28) is the *undulator equation*, which describes the generation of short (x-ray) wavelengths through the factor $\lambda_u/2\gamma^2$, magnetic tuning through $K^2/2$, and off-axis wavelength variations through $\gamma^2\theta^2$. Note that wavelength tuning through variations of K requires changing the magnet gap. This is more desirable than γ-tuning, as it affects only the desired experimental station on a multi-undulator storage ring (see Figure 5.5). In practical units the wavelength λ and corresponding photon energy $E = 2\pi\hbar c/\lambda$ are given by

$$\lambda[\text{nm}] = \frac{1.360\lambda_u[\text{cm}]\left(1 + \frac{K^2}{2} + \gamma^2\theta^2\right)}{E_e^2[\text{GeV}]} \tag{5.29a}$$

and

$$E[\text{keV}] = \frac{0.9496 E_e^2[\text{GeV}]}{\lambda_u[\text{cm}]\left(1 + \frac{K^2}{2} + \gamma^2\theta^2\right)} \tag{5.29b}$$

where the bracketed photon energy, E, *beam energy*, E_e, and *undulator period*, λ_u, are to be given in the indicated units.[1]

5.4.2 Comments on Undulator Harmonics

In addition to modifying the observed wavelength of the fundamental [as given by Eq. (5.28)], the effect of transverse oscillations introduces higher harmonics into the motion. We will see that the harmonic amplitudes scale as K^n, where n is the harmonic number. These higher harmonics of the radiation will occur at frequencies $n\omega_1$ and wavelengths λ_1/n. Because short wavelengths are difficult to generate, harmonics are of great interest, especially since they are a natural consequence of the motion. Harmonics are frequently used to extend the photon energy range of a given undulator or facility.

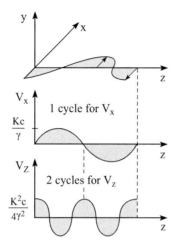

Figure 5.17 Illustration of the first and second harmonic motions of the electron.

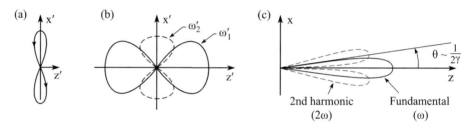

Figure 5.18 (a) Illustration of the figure eight electron motion in the frame of reference moving with the average electron velocity, and the resultant radiation patterns at the fundamental and second harmonic frequencies in both (b) the frame of reference moving with the electron and (c) the laboratory frame of reference.

We begin by considering second harmonic motion. From Eq. (5.24) we have

$$\frac{v_z}{c} = 1 - \frac{1 + K^2/2}{2\gamma^2} - \frac{K^2}{4\gamma^2} \cos\left(2 \cdot \frac{2\pi z}{\lambda_u}\right)$$

This expression displays both the decreased axial velocity and an axial velocity modulation at twice the fundamental frequency. This is referred to as a second harmonic of the motion and is illustrated in Figure 5.17. If the first-order (fundamental) motion leads to radiation at frequency ω_1' in the electron frame, then the axial harmonic will radiate at $\omega_2' = 2\omega_1'$; hence, it is called *second harmonic* radiation. Note that the magnitude of the second harmonic term scales as K^2, and also that the second harmonic oscillations of the electron are at right angles to the fundamental oscillations. That is, the fundamental radiation results from oscillations in the x-direction, while the second harmonic, and other even harmonics, result from oscillations in the z-direction. As a result, the polarizations will be different, and will be affected differently by transformation to the laboratory frame. Figure 5.18 illustrates the radiation patterns of the fundamental and second harmonics.

For higher harmonics the observed wavelengths will be governed by an extension of the undulator equation:

$$\lambda_n = \frac{\lambda_u}{2\gamma^2 n}\left(1 + \frac{K^2}{2}\gamma^2\theta^2\right)$$

(5.30)

Owing to the increased number of cycles, the relative spectral bandwidth is also improved, viz.,

$$\left(\frac{\Delta\lambda}{\lambda}\right)_n = \frac{1}{nN}$$

(5.31)

where n is the harmonic number and N is the number of magnetic periods.

From Figure 5.18, we see that the even harmonics radiate a pattern that peaks off axis and has zero intensity on axis. As a consequence, the even harmonics tend to be relatively weak on axis and, upon transformation to the laboratory frame, radiate into a hollow cone of radiation. The odd harmonics ($n = 1, 3, 5\ldots$) radiate on axis with a narrow spectrum and into a narrow forward cone. Hence, they are quite interesting as sources of high brightness and partially coherent x-rays. We will return to this subject in Section 5.6.

5.4.3 Power Radiated in the Central Radiation Cone

The undulator equation (5.28) tells us the wavelength of radiation as a function of magnet period λ_u, magnet deflection parameter K, electron energy γ (in rest energy units), and polar angle of observation θ. Now we would like to calculate the amount of power radiated. A natural and interesting choice is to calculate the power radiated into the central radiation cone, of half angle θ_{cen}, which we can identify with a relative spectral bandwidth $\lambda/\Delta\lambda \cong N$, where N is the number of magnetic periods and thus the number of oscillations the electron executes in traversing the undulator. This has a natural appeal, common to our experience with other physical phenomena involving oscillators, gratings, etc., which we embody mathematically in our time–frequency and space–angle transformations. The choice of a central radiation cone containing the $1/N$ relative spectral bandwidth is also interesting because applications of undulator radiation generally involve the use of narrow bandwidth, quasi-monochromatic radiation, and the $1/N$ bandwidth is as small[†] as one can obtain without use of a monochromator.

In Section 5.3.2 we used a simplified version of the undulator equation to introduce the concept of a central radiation cone, finding that for a bandwidth $1/N$ the cone half angle is $1/\gamma\sqrt{N}$. Having reconsidered electron motion in an undulator of finite K (Section 5.4.1), we can now follow the same arguments using the corrected undulator equation (5.28), viz.,

$$\lambda = \frac{\lambda_u}{2\gamma^2}\left(1 + \frac{K^2}{2} + \gamma^2\theta^2\right)$$

[†] In fact the $1/N$ value is idealistic in that in practice one utilizes radiation from electrons having some angular divergence due to the slightly varying trajectories within the bunch, as described by σ'. Later in this chapter we describe the additional spectral broadening as a function of σ' and θ_{cen}.

Writing this equation twice, once for a wavelength λ_0 corresponding to $\theta = 0$, and once for an off-axis angle θ_{cen} such that it encompasses a full bandwidth $\Delta\lambda$, subtracting the two equations and normalizing (as was done in Section 5.3.2, but now for finite K), one obtains a corrected formula for the central radiation cone

$$\theta_{\mathrm{cen}} = \frac{1}{\gamma^*\sqrt{N}} = \frac{\sqrt{1 + K^2/2}}{\gamma\sqrt{N}} \qquad (5.32)$$

of a single electron, containing a relative spectral bandwidth $\Delta\lambda/\lambda = 1/N$, where $\gamma^* = \gamma/\sqrt{1 + K^2/2}$, as defined earlier in Eq. (5.26). Thus for finite K there are not only longer wavelengths at each angle, but also an enlargement of the central radiation cone. We can trace both effects to the reduced average axial velocity of the electron for finite K, and thus to reduced effects of the angle dependent relativistic Doppler shift. A further discussion of spectral bandwidth is presented in Section 5.4.4.

Our task now is to calculate the power radiated within the central cone, at the fundamental frequency only. In later sections we will calculate other details, including the total power radiated. Our approach will be to use our knowledge of classical dipole radiation, as considered earlier in Chapter 2. We might ask how this can be done in a situation involving highly relativistic motion. The technique is to transfer the calculation to the frame of reference moving with the average electron velocity. In this frame of reference the electron motion is non-relativistic, at least for modest K, and the oscillation amplitude is small compared to the wavelength (in the frame of reference in which the calculation is made), as it should be for the dipole approximation to be valid. Having the desired power calculations, the results are then Lorentz transformed back to the laboratory (observer) frame of reference using straightforward but relativistically correct angular relationships given in Appendix F. This procedure gives us maximum leverage on the use of classical radiation results, and provides very valuable insights to the most important properties of undulator radiation. The process is outlined in Table 5.4.

Following the procedure outlined in Table 5.4, the electron velocity in the laboratory frame of reference has been derived, from Newton's second law of motion, as Eq. (5.19),

$$v_x = \frac{Kc}{\gamma}\cos\frac{2\pi z}{\lambda_u}$$

which we can write as

$$v_x = \frac{Kc}{\gamma}\cos k_u z$$

To obtain the acceleration we need v_x as a function of time. To first order we assume that $z \simeq \bar{v}_z t = \beta^* ct$, where $\bar{v}_z$ is the average electron velocity in the z-direction and β^* is very close to unity. The velocity can then be written as

$$v_x \simeq \frac{Kc}{\gamma}\cos k_u \beta^* ct = \frac{Kc}{\gamma}\cos\omega_u t$$

Table 5.4 An outline of the procedure for calculating power radiated by relativistic electrons traversing a periodic undulator. Electron motion is determined in the laboratory frame of reference. A Lorentz transformation to the frame of reference moving with the average electron velocity permits the use of classical dipole radiation (Chapter 2), as the electron motion is non-relativistic in this frame. The dipole radiation results are then Lorentz transformed back to the laboratory frame of reference.

x, z, t laboratory frame of reference	x', z', t' frame of reference moving with the average velocity of the electron

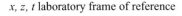

$\lambda_u'= \dfrac{\lambda_u}{\gamma^*}$

Lorentz transformation

x', z', t' motion
$a'(t')$ acceleration

Determine x, z, t motion:

$$\frac{d\mathbf{p}}{dt} = -e\,(\mathbf{E} + \mathbf{v} \times \mathbf{B})$$

Dipole radiation:

$$m\gamma\frac{dv_x}{dt} = -e\frac{dz}{dt}B_0\sin\frac{2\pi z}{\lambda_u}$$

$$\frac{dP'}{d\Omega'} = \frac{e^2\,a'^2\sin^2\Theta'}{16\pi^2\epsilon_0 c^3}$$

$$v_x(t);\ a_x(t) = \cdots$$
$$v_z(t);\ a_z(t) = \cdots$$

$$\frac{dP'}{d\Omega'} = \frac{e^2\,c\gamma^2}{4\epsilon_0\lambda_u^2}\frac{K^2}{(1+K^2/2)^2}(1-\sin^2\theta'\cos^2\phi')\sin^2\omega'_u t'$$

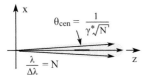

Lorentz transformation

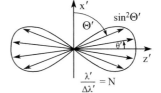

$$\frac{dP}{d\Omega} = 8\gamma^{*2}\frac{dP'}{d\Omega'}$$

$$\frac{d\bar{P}}{d\Omega} = \frac{e^2\,c\gamma^4}{\epsilon_0\lambda_u^2}\frac{K^2}{(1+K^2/2)^3}\quad\begin{array}{l}K\le 1\\ \theta\le\theta_{cen}\end{array}$$

$$\Delta\Omega_{cen} = \pi\theta_{cen}^2 = \pi/\gamma^{*2}N$$

$$\bar{P}_{cen} = \frac{\pi e^2\,c\gamma^2}{\epsilon_0\lambda_u^2 N}\frac{K^2}{(1+K^2/2)^2}$$

N_e uncorrelated electrons:

$$N_e = \bar{I}L\,/ec,\ L = N\lambda_u$$

$$\bar{P}_{cen} = \frac{\pi e\,\gamma^2\bar{I}}{\epsilon_0\lambda_u}\frac{K^2}{(1+K^2/2)^2}$$

where $\omega_u = k_u\beta^*c \cong k_u c$. Integrating once with respect to time t, we have the first-order oscillatory motion $x(t)$:

$$x \simeq \frac{K}{k_u\gamma}\sin\omega_u t$$

The Lorentz transformations from the (x, t) laboratory frame of reference to the (x', t') frame of reference moving with the average electron velocity ($\beta^* c$ or γ^*) are given in Appendix F, Eqs. (F.1b) and (F.1c), as

$$t = \gamma^* \left(t' + \frac{\beta^* z'}{c} \right) \simeq \gamma^* \left(t' + \frac{z'}{c} \right)$$

$$x = x'$$

(non-relativistic motion transverse to z for $K \leq 1$). Thus in the electron frame of

$$x' \simeq \frac{K}{k_u \gamma} \sin \omega_u \gamma^* \left(t' + \frac{z'}{c} \right)$$

where z' represents the small axial excursions about the average position in the reference frame moving with the electron. This is an important term, which we will see later provides a coupling of energy to higher harmonics. For small values of K, however, this term's contribution to the fundamental motion is minimal. Thus to a fair degree of accuracy we can write

$$x' \simeq \frac{K}{k_u \gamma} \sin \omega_u \gamma^* t'$$

Recognizing the Lorentz shifted frequency ω'_u, this becomes

$$x' \simeq \frac{K}{k_u \gamma} \sin \omega'_u t'$$

Taking the second derivative with respect to t', we have

$$a'_x \simeq -\frac{K \omega'^2_u}{k_u \gamma} \sin \omega'_u t'$$

where $a'_x \equiv d^2 x'/dt'^2$. Noting that $\omega'_u = \gamma^* k_u c = \gamma k_u c/(1 + K^2/2)^{1/2}$, one has the desired electron acceleration in the (x', t') moving frame of reference:

$$a'_x \simeq -\frac{2\pi c^2 \gamma}{\lambda_u} \frac{K}{(1 + k^2/2)} \sin \omega'_u t' \tag{5.33}$$

This acceleration can now be used in the dipole radiation formula [Chapter 2, Eq. (2.34)]

$$\frac{dP'}{d\Omega'} = \frac{e^2 a'^2 \sin^2 \Theta'}{16\pi^2 \epsilon_0 c^3}$$

where a' is the instantaneous electron acceleration, Θ' is the angle of observation measured from the direction of acceleration, and we have assumed that the amplitude of oscillation is small compared to the radiated wavelength (in the frame of reference where the calculation is made). Using Eq. (5.33) for the electron acceleration in the moving frame of reference, and averaging over one full cycle of the motion, we obtain the average power radiated per unit solid angle to be

$$\frac{d\bar{P}'}{d\Omega} = \frac{e^2 c \gamma^2}{8\epsilon_0 \lambda_u^2} \frac{K^2}{(1 + K^2/2)^2} \sin^2 \Theta' \tag{5.34}$$

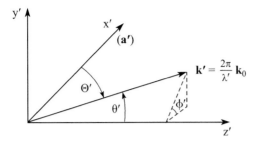

Figure 5.19 Illustration of the polar coordinate system (r', θ', ϕ') where $d\Omega' = \sin \theta' \, d\theta' \, d\phi'$, and where the coordinate system is oriented to have the polar axis $(\theta' = 0)$ oriented along the z'-axis of symmetry for undulator radiation.

where time averaging $\sin^2 \omega'_u t'$ over a full cycle (or N full cycles) has introduced a factor of one half. The $\sin^2 \Theta'$ factor can be set to unity, as only radiation in the vicinity of $\Theta' \cong \pi/2$ contributes to the central radiation cone in the laboratory frame. The angular factors will be discussed in detail in the next section. However, for clarity in understanding the approximation to various angular factors, we introduce the coordinate system shown in Figure 5.19. We recall that the Lorentz transformation to the laboratory frame will concentrate the radiation pattern into a narrow cone about the z-axis. It is sensible to organize our angular measurements about this natural symmetry axis. Hence, we introduce a polar coordinate system (r', θ', ϕ') where r' is the polar axis oriented collinear to the z-axis, θ' is the polar angle (0 to π), and ϕ' is measured from the x'-axis in the x', y'-plane(0 to 2π). For the fundamental at ω'_u, with acceleration α'_x, Θ' is measured from the x'-axis as shown in Figure 5.19.

In a polar coordinate system, the angle between the two vectors $\mathbf{k}'$ and $\mathbf{a}'$ (wave propagation direction and acceleration direction) is given by (see Appendix D)

$$\cos \Theta' = \cos \theta'_k \cos \theta'_a + \sin \theta'_k \sin \theta'_a \cos(\phi'_k - \phi'_a) \tag{5.35}$$

For the fundamental radiation at ω'_u, $\theta'_a = \pi/2$, and $\phi'_a = 0$. Equation (5.35) then simplifies to

$$\cos \Theta' = \sin \theta'_k \cos \phi'_k$$

so that for the radiated power,

$$\sin^2 \Theta' = 1 - \sin^2 \theta'_k \cos^2 \phi'_k \tag{5.36}$$

In what follows we drop the subscript k for convenience. The approximation that $\sin^2 \Theta' \cong 1$ can then be understood by examining the magnitude of the $\sin^2 \phi' \cos^2 \phi'$ term for angles that will transform to angles $\theta \leq \theta_{\text{cen}}$ in the laboratory frame. From Appendix F, Eq. (F.14), the polar angles in the two frames of reference are related by

$$\sin \theta' = \frac{2\gamma^*\theta}{1 + \gamma^{*2}\theta^2}$$

where both θ' and θ are measured from the z-axis in their respective frames of reference. For a central cone of half angle $\theta = 1/\gamma^*\sqrt{N}$, the corresponding angle in the primed reference frame is $\sin \theta' \sim 2/\sqrt{N}$, so that $\theta' \simeq 2/\sqrt{N}$ for large N (of order 100). For such small angles it is clear that, for all values of ϕ, $\sin^2 \Theta' \cong 1 - \theta'^2/2 \cong 1 - 2/N$, thus permitting, for large N, the first-order approximation $\sin^2 \Theta' \simeq 1$ in Eqs. (5.34) and (5.36).

Equation (5.34) above gives us the power radiated per unit solid angle, for angles near the z-axis, in the frame of reference moving with the average electron velocity. Following the procedure outlined in Table 5.4, we now want to transform this result back to the laboratory frame of reference. To do so we need a relativistically correct relation between $dP'/d\Omega'$ and $dP/d\Omega$. In the next section we will show that the desired relationship is

$$\frac{dP}{d\Omega} = \frac{8\gamma^{*2}}{(1 + \gamma^{*2}\theta^2)^3} \frac{dP'}{d\Omega'}$$

which for small angles within the central radiation cone reduces to

$$\frac{dP}{d\Omega} \simeq 8\gamma^{*2} \frac{dP'}{d\Omega'}$$

Thus using Eq. (5.34) with the approximation $\sin^2 \Theta' = 1$, the average power radiated per unit solid angle, as observed in the laboratory frame of reference, is

$$\left.\frac{dP'}{d\Omega}\right|_{e^-} \simeq \frac{e^2 c \gamma^4}{\epsilon_0 \lambda_u^2} \frac{K^2}{(1 + K^2/2)^3} \qquad (K \le 1, \; \theta \le \theta_{\text{cen}}) \qquad (5.37)$$

where the subscript e^- reminds us that this is for a single electron. To obtain power radiated we simply multiply by the element of solid angle associated with the central radiation cone, viz.,

$$\Delta\Omega_{\text{cen}} = \int_0^{2\pi} \int_0^{1/\gamma^*\sqrt{N}} \sin\theta \, d\theta \, d\phi = \frac{\pi}{(\gamma^*\sqrt{N})^2}$$

and thus conclude that, as observed in the laboratory frame of reference, the average power radiated into the central cone, for a single electron, is given by

$$\bar{P}_{\text{cen}}|_{e^-} \simeq \frac{\pi e^2 c \gamma^2}{\epsilon_0 \lambda_u^2 N} \frac{K^2}{(1 + K^2/2)^2} \qquad (5.38)$$

with an associated bandwidth of $\Delta\lambda/\lambda = 1/N$ and a radiation cone half angle of $1/\gamma^*\sqrt{N}$. This result is generally valid for $K \le 1$. We observe that the power radiated is proportional to $(K\gamma/\lambda_u)^2$, due to the dependence on electron acceleration squared, and inversely proportional to N. While the inverse dependence on N may at first seem surprising, it can be understood as a combination of increased power (N), combined with a decreased solid angle $(1/N)$ of the central radiation cone and a decreased central cone bandwidth $(1/N)$. The additional factor involving K in the denominator is associated with the reduced acceleration as the electron's axial motion is decreased. Recall that the wavelength of this radiation is given by the undulator equation (5.28).

An important extension of this result is to the practical case of multi-electron bunches traversing the undulator, in which case the radiated power is much greater and thus

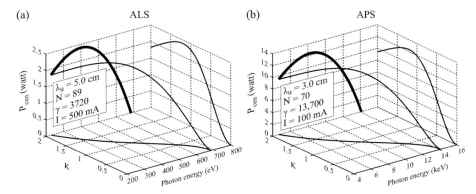

Figure 5.20 Power radiated into the central radiation cone ($\theta_{cen} = 1/\gamma^*\sqrt{N}$, $1/N$ relative spectral bandwidth, $n = 1$ only) as a function of photon energy and dimensionless undulator parameter K, for typical soft x-ray and hard x-ray undulators. Harmonic power is not included ($n = 1$ only).

much more valuable for laboratory applications. We show in the next section that for an average electron current $\bar{I}$ in the storage ring, the number of electrons radiating within an undulator, averaged over a long time period compared to the electron bunch structure, is equal to

$$N_e = \bar{I}L/ec$$

where $L = N\lambda_u$ is the length of the undulator. If the motion of the various electrons is uncorrelated, the radiated fields due to differing electrons will have no special relationship, and as a result the radiated power will increase proportionally to N_e, the number of electrons. Were the electron motions well correlated, as in an electron wave (sometimes called microbunching), this would lead to phase correlated electric and magnetic fields – as in a free electron laser (FEL). In such a case the fields of the N_e electrons would add in phase and since radiated power is proportional to E^2, far greater power could be radiated, perhaps N_e^2 times greater than for a single electron. For the uncorrelated case, generally understood as *undulator radiation*, the intensities rather than the fields add and one simply multiplies Eq. (5.38) by $N_e = \bar{I}L/ec = \bar{I}N_u/ec$ to obtain

$$\bar{P}_{cen} \simeq \frac{\pi e\gamma^2 \bar{I}}{\epsilon_0 \lambda_u} \frac{K^2}{(1 + K^2/2)^2} \tag{5.39}$$

for the average power radiated by electrons of average current $\bar{I}$, at the fundamental frequency ($n = 1$), within a relative spectral bandwidth $\Delta\lambda/\lambda = 1/N$, and into a central radiation cone of half angle $\theta_{cen} = 1/\gamma^*\sqrt{N} = \sqrt{1 + K^2/2}/\gamma\sqrt{N}$. Detailed spectral shapes and increases to the central radiation cone caused by random electron motions (divergence) within the electron beam are discussed in Section 5.4.5.

We give two examples of the use of this formula, involving soft x-ray and hard x-ray undulators at the ALS and APS, displayed here in Figure 5.20, and previously cited in Table 5.1. For a typical soft x-ray case ($\gamma = 3720$, $\lambda_u = 5.00$ cm, $N = 89$, and $\bar{I} = 500$ mA), power of order 2–5 W is radiated in first order into a half angle of about

35 μrad, within a wavelength region tunable from 2 nm to 5 nm (250 eV to 600 eV photon energy), and within a relative spectral bandwidth of approximately 1.1%. For a typical hard x-ray undulator at the APS ($\gamma = 13{,}700$, $\lambda_u = 3.00$ cm, $N = 70$, and $\bar{I} = 100$ mA), power of order 10–20 W is radiated into first order in an 11 μrad half angle cone, at wavelengths from about 1 to 2.5 Å (5 keV to 13 keV photon energy), and within a relative spectral bandwidth of approximately 1.4%. Tuning curves for these two undulators are presented in Figure 5.20(a) and (b), respectively, illustrating photon energy and power in the central cone as a function of the dimensionless undulator parameter K. At the lowest values of K there is little transverse acceleration and thus little power radiated. Because the electrons move through the undulator relatively fast at low K-values, the N oscillations are executed more rapidly, resulting in higher frequency radiation, higher photon energies, and shorter wavelength. For small K, power grows as K^2, peaking according to Eq. (5.39) at $K = \sqrt{2}$. At such high K-values, however, the coupling to higher harmonics becomes very efficient, and the accuracy of our results, which are valid for $K < 1$, requires further attention.

The formulation of power radiated in the central cone [Eq. (5.39)] is based in part on a small perturbation analysis in which we have assumed small K operation. These approximations were made in the development of a simplified expression for the electron acceleration [Eq. (5.33)] in the moving frame of reference, which is subsequently squared and used in the dipole radiation formula. To obtain Eq. (5.33) we started in the laboratory frame with the electron velocity [Eq. (5.19)] $v_x = (Kc/\gamma)\cos k_u z$ and assumed, to first order in K, that $z \cong \beta^* ct$, thus neglecting higher order harmonic motions, which scale as K^2, K^3, etc. Having made the transformation to the moving reference frame, where $x' \simeq -(K/K_u\gamma)\cos\omega_u\,\gamma^*(t' + z'/c)$, we again assumed on the basis of K^n scaling that the z'-term, which is associated with harmonic oscillations about the average electron trajectory, could be neglected. Now, however, observing that central cone power scales as K^2 with a peak just above $K = 1$ according to Eq. (5.39), we have a great interest in understanding the accuracy of the low K results as K approaches and exceeds unity. This knowledge will be of great value in planning experiments.

Kim[1] has analyzed undulator radiation in a very complete manner, accounting for all harmonics and accurate for all K. Comparing his results Ref. 1, Eq. (4.44) with that of Eq. (5.39) here, there is an additional multiplicative factor, $[JJ]$, associated with the transfer of power from the fundamental ($n = 1$) to the harmonics ($n > 1$) and given by the difference of Bessel functions of the first kind, orders zero and one,

$$[JJ]^2 = [J_0(x) - J_1(x)]^2 \tag{5.40a}$$

where $x = K^2/4(1 + K^2/2)$ and[‡]

$$J_n(x) = \sum_{s=0}^{\infty} \frac{(-1)^s}{s!(n+s)!}\left(\frac{x}{2}\right)^{n+2s}$$

[‡] E. Kreyszig, *Advanced Engineering Mathematics* (Wiley, 1993), Seventh Edition, p. 227.

Table 5.5 The multiplicative correction factor $[JJ]^2$ that enables the low K analytic formulation of undulator central cone power [Eq. (5.39)] to be extended to higher K-values.

K	$x = \frac{K^2}{4(1+K^2/2)}$	$J_0(x)$	$J_1(x)$	$[JJ]^2$
0	0	1.0000	0	1.000
0.5	0.0556	0.9992	0.0278	0.944
1.0	0.1667	0.9931	0.0831	0.828
$\sqrt{2}$	0.2500	0.9844	0.1240	0.740
1.5	0.2647	0.9826	0.1312	0.725
2.0	0.3333	0.9724	0.1643	0.653
2.5	0.3788	0.9644	0.1860	0.606

For small values of x (zero to $\frac{1}{3}$, for $0 \leq K \leq 2$) the relevant expansions are

$$J_0(x) = 1 - \left(\frac{x}{2}\right)^2 + \frac{1}{(2)^2}\left(\frac{x}{2}\right)^4 - \cdots$$

and

$$J_1(x) = \frac{x}{2} - \frac{1}{2}\left(\frac{x}{2}\right)^3 + \cdots$$

For small K the multiplicative factor $[JJ]^2$ can be approximated by

$$[JJ]^2 = 1 - x - \frac{x^2}{4} + \frac{3x^3}{8} + \cdots \tag{5.40b}$$

The factor $[JJ]^2$ is tabulated in Table 5.5 for selected values of K. Thus Eq. (5.39) overestimates the fundamental power radiated in the central cone by 6.6% for $K = 1/2$, and 17% for $K = 1$. These overestimates are incurred through the omission of second order and higher terms in K, which would have had the effect of reducing the first order electron velocity and acceleration, and thus of concomitant radiated power. Nonetheless, this analytic formulation [Eq. (5.39)] has provided a valuable tool for understanding the most important features of undulator radiation, with very simply interpreted physical insights.

A modified version of Eq. (5.39), which extends its applicability to higher K-values while preserving its simple form, is obtained by including the $[JJ]$ finite K corrective factor, so that the power in the central cone ($n = 1$, $1/N$ relative spectral bandwidth, finite K) becomes

$$\bar{P}_{\text{cen}} = \frac{\pi e \gamma^2 \bar{I}}{\epsilon_0 \lambda_u} \frac{K^2 [JJ]^2}{(1+K^2/2)^2} \tag{5.41a}$$

where one sometimes introduces a more compact abbreviation, as is often the case in the free electron laser community,

$$\hat{K} = K \cdot [JJ] \tag{5.41b}$$

The finite K corrective undulator parameter $[JJ]$ accounts for reductions of the first-order transverse velocity, v_x, due to the generation of higher order longitudinal and transverse harmonics, all of which draw energy from the first order fields and thus from radiated power in the fundamental ($n = 1$). Note that the $[JJ]^2$ factor affects only the power radiated in the fundamental and thus appears only in the numerator of Eq. (5.41a).

In practical units the central cone power can be written as[1]

$$\bar{P}_{\text{cen}} = (5.69 \times 10^{-6}\ \text{W}) \frac{\gamma^2 \bar{I}[A]}{\lambda_u[\text{cm}]} \frac{K^2[JJ]^2}{(1 + K^2/2)^2} \tag{5.41c}$$

Note that for the two undulators cited, Eq. (5.41c) indicates that for $K = 1$ the 5.0 cm undulator at the ALS radiates 2.9 W in the central cone at 2.7 nm wavelength (460 eV), while the 3.0 cm undulator at the APS radiates about 13 W within the central cone at 1.2 Å (10 keV).

The power radiated in the central cone can be written explicitly as a function of photon energy through use of the undulator equation (5.28), which relates K^2 to frequency. From the undulator equation, $f = 2\gamma^2 c/\lambda_u(1 + K^2/2)$ on axis. Thus $\hbar\omega = 4\pi\hbar\gamma^2 c/\lambda_u(1 + K^2/2) = \hbar\omega_0/(1 + K^2/2)$, where $\hbar\omega_0$ is defined as the photon energy on axis for the limiting case $K = 0$. From this one can solve for both $1 + K^2/2$ and K^2 and thus describe the radiated power in terms of photon energy, rather than K, as

$$\boxed{\bar{P}_{\text{cen}} = \frac{2\pi e\gamma^2 \bar{I}}{\epsilon_0 \lambda_u} \cdot \frac{\hbar\omega}{\hbar\omega_0}\left(1 - \frac{\hbar\omega}{\hbar\omega_0}\right)[JJ]^2} \tag{5.41d}$$

where, in terms of $\hbar\omega/\hbar\omega_0$, the multiplicative factor $[JJ]^2$ is given by

$$[JJ(\hbar\omega/\hbar\omega_0)]^2 \simeq \frac{7}{16} + \frac{5}{8}\frac{\hbar\omega}{\hbar\omega_0} - \frac{1}{16}\left(\frac{\hbar\omega}{\hbar\omega_0}\right)^2 + \cdots \tag{5.41e}$$

In numerical form this becomes

$$\bar{P}_{\text{cen}} = (1.14 \times 10^{-5}\ \text{W})\frac{\gamma^2 \bar{I}[A]}{\lambda_u[\text{cm}]} \cdot \frac{\hbar\omega}{\hbar\omega_0}\left(1 - \frac{\hbar\omega}{\hbar\omega_0}\right)[JJ(\hbar\omega/\hbar\omega_0)]^2 \tag{5.41f}$$

where $\hbar\omega_0 = 4\pi\hbar c\gamma^2/\lambda_u$ has the value 686 eV for $\gamma = 3720$ and $\lambda_u = 5.00$ cm, and the value 14.1 keV for $\gamma = 13\ 700$ and $\lambda_u = 3.30$ cm.

5.4.4   Power as a Function of Angle and Total Radiated Power

In this section we return to the calculation of power radiated per unit solid angle, for small K, in this case keeping the angular dependence. Again we follow the procedure outlined in Table 5.4. At the end of the section we integrate over all angles to obtain the total power radiated at the fundamental ($n = 1$) frequency. In the previous section we employed dipole radiation in the frame of reference moving with the electron,

$$\frac{dP'}{d\Omega'} = \frac{e^2 a'^2 \sin^2 \Theta'}{16\pi^2 \epsilon_0 c^3}$$

along with the first-order acceleration of an electron traversing an undulator [Eq. (5.33)],

$$a'_x \simeq -\frac{2\pi c^2 \gamma}{\lambda_u} \frac{K}{(1+k^2/2)} \sin \omega'_u t'$$

where $\omega'_u = \gamma^* \omega_u = 2\pi \gamma^* \beta^* c / \lambda_u \simeq 2\pi \gamma^* c / \lambda_u$, and where $a'_x \equiv d^2 x' / dt'^2$, to obtain the average power radiated per unit solid angle in the electron frame of reference [Eq. (5.34)], corrected for power reduced to higher harmonics by including the $[JJ]^2$ factor, one has

$$\frac{\overline{dP'}}{d\Omega'} = \frac{e^2 c \gamma^2}{8\epsilon_0 \lambda_u^2} \frac{K^2 [JJ]^2}{(1+K^2/2)^2} \sin^2 \Theta'$$

where averaging over a full cycle of the motion has introduced a factor of one-half. Thus, comparing with Eq. (5.34), here we have kept the factor $\sin^2 \Theta'$.

As illustrated in Figure 5.19 and described in Eq. (5.36), the factor $\sin^2 \Theta'$ can be written in terms of the polar angles

$$\sin^2 \Theta' = 1 - \sin^2 \theta' \cos^2 \phi'$$

where θ' is the polar angle (0 to π) measured from the z-axis and ϕ' is the azimuthal angle (0 to 2π) measured from the x'-axis in the x', y'-plane. The radiated power in the fundamental is therefore

$$\frac{d\bar{P}'}{d\Omega'} = \frac{e^2 c \gamma^2}{8\epsilon_0 \lambda_u^2} \frac{K^2 [JJ]^2}{(1+K^2/2)^2} (1 - \sin^2 \theta' \cos^2 \phi') \tag{5.42}$$

To transform this to the laboratory frame of reference we make use of angular relationships obtained in Appendix F as Eqs. (F.9)–(F.11):

$$\sin \theta' = \frac{\sin \theta}{\gamma^* (1 - \beta^* \cos \theta)}$$

$$\cos \theta' = \frac{\cos \theta - \beta^*}{1 - \beta^* \cos \theta}$$

$$\tan \theta' = \frac{\sin \theta}{\gamma^* (\cos \theta - \beta^*)}$$

and

$$\sin \theta = \frac{\sin \theta'}{\gamma^* (1 + \beta^* \cos \theta')}$$

$$\cos \theta = \frac{\cos \theta' + \beta^*}{1 + \beta^* \cos \theta'}$$

$$\tan \theta = \frac{\sin \theta'}{\gamma^* (\cos \theta' + \beta^*)}$$

In the highly relativistic case, where $\gamma^* \gg 1$, $\beta^* \simeq 1$, these take the approximate forms

$$\sin\theta' \simeq \frac{2\gamma^*\theta}{1 + \gamma^{*2}\theta^2} \tag{5.43a}$$

$$\cos\theta' \simeq \frac{1 - \gamma^{*2}\theta^2}{1 + \gamma^{*2}\theta^2} \tag{5.43b}$$

$$\tan\theta' \simeq \frac{2\gamma^*\theta}{1 - \gamma^{*2}\theta^2} \tag{5.43c}$$

Since the angles ϕ and ϕ' lie in planes perpendicular to the relativistic motion, we have

$$\phi' = \phi$$

Using the angular relations, Eq. (5.43), for $\gamma^* \gg 1$, the angular radiation pattern in Eq. (5.42) becomes

$$1 - \sin^2\theta'\cos^2\phi' = \frac{1 + 2\gamma^{*2}\theta^2(1 - 2\cos^2\phi) + \gamma^{*4}\theta^4}{(1 + \gamma^{*2}\theta^2)^2} \tag{5.44}$$

Similarly, the element of solid angle $d\Omega'$ can be rewritten as

$$d\Omega' \simeq \sin\theta' d\theta' d\phi' \simeq \frac{4\gamma^{*2}}{(1 + \gamma^{*2}\theta^2)^2} \cdot \theta\, d\theta\, d\phi$$

or

$$d\Omega' \simeq \frac{4\gamma^{*2}}{(1 + \gamma^{*2}\theta^2)^2} d\Omega \tag{5.45}$$

where we recognize that for small angles $d\Omega = \theta\, d\theta\, d\phi$.

To complete the transformation of power per unit solid angle in the moving (primed) frame of reference, $d\bar{P}'(\theta', \phi')/d\Omega'$, to the laboratory frame of reference, $d\bar{P}(\theta, \phi)/d\Omega$, we are left to consider the relationship between the radiated power, P' and P, as observed in the two frames of reference. To do this we consider the emission of a finite number of photons, $\mathcal{N}'$, during a time interval $\Delta t'$, as seen in the moving frame of reference where all photons have the same energy $\hbar\omega'$, independent of emission angle – a property of dipole radiation. These same photons, discretely counted in identical number, $\mathcal{N} = \mathcal{N}'$, in the laboratory frame of reference, are observed there in a time interval Δt, with an angle-dependent photon energy [due to the angle-dependent Doppler shift, Appendix F, Eq. (F.8b)] given by

$$\hbar\omega = \frac{\hbar\omega'}{\gamma^*(1 - \beta^*\cos\theta)}$$

Since the Lorentz transformation forces all angles θ' to very small angles of order $\theta \simeq O(1/\gamma^*)$ in the laboratory frame, we can approximate the Doppler shift by

$$\hbar\omega \simeq \frac{2\gamma^*}{1 + \gamma^{*2}\theta^2}\hbar\omega'$$

By noting that time intervals in the two frames of reference are related by Appendix F, Eq. (F.13), as

$$\Delta t = \gamma^* \Delta t'$$

we can likewise relate the incremental radiated power in the two frames of reference,

$$\Delta P' = \frac{\mathcal{N}' \hbar \omega'}{\Delta t'}$$

and

$$\Delta P = \frac{\mathcal{N} \hbar \omega}{\Delta t}$$

Rewriting the expression for ΔP and then substituting relationships in terms of primed quantities for $\mathcal{N}$, ω, and Δt, one obtains

$$\Delta P = \frac{\mathcal{N} \hbar \omega}{\Delta t} = \frac{\mathcal{N}' \left(\frac{2\gamma^*}{1 + \gamma^{*2}\theta^2} \right) \hbar \omega'}{\gamma^* \Delta t'}$$

$$\Delta P = \frac{2}{1 + \gamma^{*2}\theta^2} \frac{\mathcal{N}' \hbar \omega'}{\Delta t'}$$

Recognizing the quantity on the right as $\Delta P'$, and writing this in differential form, one has

$$dP = \frac{2}{1 + \gamma^{*2}\theta^2} dP'$$

Using the relation given in Eq. (5.45) between $d\Omega'$ and $d\Omega$, the relationship for power per unit solid angle between the two reference frames becomes

$$\frac{dP}{d\Omega} = \frac{8\gamma^{*2}}{(1 + \gamma^{*2}\theta^2)^3} \frac{dP'}{d\Omega'} \tag{5.46}$$

Combining Eqs. (5.42), (5.44), and (5.46), one obtains the average power radiated per unit solid angle at the fundamental frequency ($n = 1$), as observed in the laboratory frame of reference:

$$\left. \frac{d\bar{P}}{d\Omega} \right|_{e^-} = \frac{e^2 c K^2 [JJ]^2 \gamma^4}{\epsilon_0 \lambda_u^2 (1 + K^2/2)^3} \left[\frac{1 + 2\gamma^{*2}\theta^2 (1 - 2\cos^2 \phi) + \gamma^{*4}\theta^4}{(1 + \gamma^{*2}\theta^2)^5} \right] \tag{5.47}$$

This result is for a single electron. A more useful result would be the power radiated by an electron bunch in which the individual motions within the bunch are random. In this case the radiated fields due to different electrons are uncorrelated and the average power radiated is a simple sum of the radiated power from individual electrons; that is, we sum intensities, not fields. For the moment let us consider the electron bunch to be sufficiently constrained in spatial and angular extent that the angular dependencies are to first order as given by Eq. (5.47). We will see in the next section that this requires that the extent of random angular deviation within the electron bunch be limited to values $\sigma' < 1/\gamma^* \sqrt{N}$, where σ' is the rms measure of width of the electron angular distribution function about the z-axis.

To generalize Eq. (5.47) to the many electron case we must determine how many electrons, on average, are radiating from within the undulator at any given time. It is convenient to do this in terms of the current in the storage ring. Current is defined as the charge per unit time crossing a given plane. For electrons of velocity v the magnitude of the current can be written as

$$I = evn_l$$

where n_l is the number of electrons per unit length in the direction of motion. For a magnet structure (undulator) of length L containing on average N_e electrons in its entire length, each traveling with a velocity v $\simeq$ c, the average current is

$$\bar{I} = \frac{ecN_e}{L}$$

so that the total number of electrons radiating within the magnet structure at a given time is

$$N_e = \frac{\bar{I}L}{ec} \tag{5.48}$$

which was cited without proof in the previous section. The average power radiated per unit solid angle by a distribution of relativistic electrons of average current I is then obtained, following the same arguments regarding uncorrelated motions which led to Eq. (5.39), by simply multiplying the single electron result [Eq. (5.47)] by the average number of electrons within the undulator [Eq. (5.48)] to obtain

$$\frac{d\bar{P}}{d\Omega} = N_e \left.\frac{d\bar{P}}{d\Omega}\right|_{e^-} = \frac{eN\gamma^4\bar{I}}{\epsilon_0\lambda_u} \cdot \frac{K^2[JJ]^2}{(1+K^2/2)^3} \underbrace{\left[\frac{1 + 2\gamma^{*2}\theta^2(1-2\cos^2\phi) + \gamma^{*4}\theta^4}{(1+\gamma^{*2}\theta^2)^5}\right]}_{F(\theta,\phi)} \tag{5.49}$$

where we have used the fact that the undulator length L is equal to $N\lambda_u$. This is a significant result for undulator radiation ($K < 1$). Note that the angular function $F(\theta, \phi)$ in Eq. (5.49) is unity on axis, is approximately $1/3$ for an angle $\theta = 1/2\gamma^*$, and goes rapidly to zero for $\theta > 1/\gamma^*$. Thus we again see the generally anticipated searchlight effect for synchrotron radiation, this time with an explicit power dependence on angle.

For small amplitude oscillations, e.g., $K \leq 1$, the K^2 dependence reflects the acceleration (a^2) dependence on magnetic field. For $K > 1$ the power radiated in the fundamental begins to decline as the strong magnetic field couples energy into successively higher harmonics, thus beginning an evolution towards wiggler radiation. The γ^4 dependence reflects the relativistic photon energy shift (γ^2) and the ever narrowing emission solid angle ($1/\gamma^2$). The angular distribution given in Eq. (5.49) can be used to provide a small (several percent) correction to the previously derived power in the central radiation cone.

In the derivation of Eqs. (5.38) and (5.39) we assumed that the power per unit solid angle, $dP/d\Omega$, was to first order independent of θ for $\theta \leq \theta_{cen} = 1/\gamma^*\sqrt{N}$, where typically $N = 100$. We see, however, in Eq. (5.49) that there are several terms involving $\gamma^{*2}\theta^2$ that reach values of order $1/N$ within the central cone. Thus for improved accuracy in predicting undulator performance, attention should be paid to these angular factors. Towards this end, numerical integration programs such as that described by Walker and Diviacco,[13] are an important complement to the simplified analytic formulation. These computer codes also permit the inclusion of finite electron beam size, angular divergence, and energy spread, as is discussed in the next section.

Our next task is to calculate the undulator power radiated at the fundamental wavelength to all angles and all wavelengths, which is achieved by integrating Eq. (5.49) over all solid angles $d\Omega = \sin\theta\, d\theta\, d\phi$. Recalling that the integrand falls off rapidly for angles beyond $1/\gamma^*$, we can take $d\Omega \simeq \theta\, d\theta\, d\phi$, and proceed beyond that with an exact integration. Note that ϕ appears in only one term, which integrates to zero, viz.,

$$\int_0^{2\pi} (1 - 2\cos^2\phi)\,d\phi = -\frac{1}{2}\int_0^{2\pi}\cos 2\phi \cdot d\,(2\phi) = -\frac{1}{2}\sin 2\phi\Big|_0^{2\pi} = 0$$

In all other terms there is no ϕ-dependence, so that the ϕ-integration gives a simple 2π factor. The remaining integration is performed (see Appendix F) by introducing $u = \gamma^*\theta$ and $x = 1 + u^2$, then integrating the resultant polynomial over $dx = 2u\,du$ to obtain

$$\int_0^\pi \int_0^{2\pi} \frac{1 + \gamma^{*4}\theta^4}{(1 + \gamma^{*2}\theta^2)^5}\theta\, d\theta\, d\phi = \frac{\pi}{3\gamma^{*2}}$$

The total radiated power in the fundamental $(n = 1)$, to all angles and wavelengths is then

$$\boxed{\bar{P}_{T,1} = \frac{\pi e\gamma^2 \bar{I}N}{3\epsilon_0\lambda_u} \cdot \frac{K^2[JJ]^2}{(1 + K^2/2)^2}} \tag{5.50a}$$

where we have made the identification $L = N\lambda_u$. Note that the total power is larger by $N/3$ than that in the central cone [Eq. (5.39)]. Rewriting in practical units, the total power in the fundamental for is

$$\bar{P}_{T,1}\,[\mathrm{W}] = \frac{(1.90 \times 10^{-6}\,\mathrm{W})\gamma^2 N\bar{I}[\mathrm{A}]}{\lambda_u[\mathrm{cm}]} \cdot \frac{K^2[JJ]^2}{(1 + K^2/2)^2} \tag{5.50b}$$

where $\bar{I}$ and λ_u are in units as indicated. We note that for the two undulators cited in Table 5.1, the soft x-ray undulator radiates 86 W into the fundamental at all angles, while the hard x-ray undulator generates 304 W in the fundamental. The subject of harmonic motion and power radiated at harmonic frequencies is discussed in later sections.

5.4.5 Spectral Bandwidth of Undulator Radiation With Finite Electron Beam Parameters

For an undulator of N periods each electron oscillates through N cycles of its motion and thus radiates a wavetrain consisting of N well-defined cycles of the electric field as illustrated in Figure 5.21(a). The Fourier transform of this waveform (Appendix D.6) gives the spectral content of the fields, and is a $(\sin x)/x$, or sinc x, function, whereas used here $x = N\pi u$, $u = \Delta\omega/\omega_0$, and $\Delta\omega = \omega - \omega_0$ is the frequency shift away from

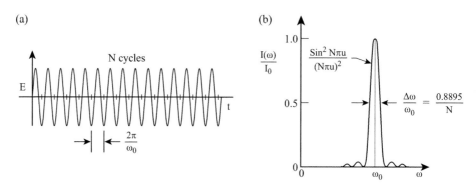

Figure 5.21 (a) Radiated wavetrain from a single electron traversing an undulator, as detected in the laboratory frame of reference, and (b) the corresponding spectral distribution function, where $u = \Delta\omega/\omega_0$ and $\Delta\omega = \omega - \omega_0$.

the central maximum at ω_0, as observed at a given angle and K-value. The intensity observed in the laboratory frame is proportional to the square of the electric field, so that in normalized form

$$\frac{I(\omega)}{I_0} = \frac{\sin^2\left(N\pi\,\Delta\omega/\omega_0\right)}{\left(N\pi\,\Delta\omega/\omega_0\right)^2} \tag{5.51}$$

For large N the major contribution to the fundamental occurs for small values of $\Delta\omega/\omega_0$. This is a commonly encountered function, tabulated[14] as $\text{sinc}^2 x$, which is normalized to unity with a full width at half maximum (FWHM) of approximately $1/N$, centered at ω_0, as illustrated in Figure 5.21(b).

For undulator radiation the central frequency ω_0 is equal to $2\pi c/\lambda_0$, which we determined earlier (without the use of a subscript zero) to be given by [Eq. (5.28)]

$$\lambda = \frac{\lambda_u}{2\gamma^2}\left[1 + \frac{K^2}{2} + \gamma^2\theta^2\right]$$

Thus the central maximum of the spectral distribution function, Eq. (5.51), occurs at a frequency (subscript zero suppressed)

$$\omega = \frac{4\pi c\gamma^2}{\lambda_u\left(1 + \frac{K^2}{2} + \gamma^2\theta^2\right)} \tag{5.52}$$

which we recall is a function of the observation angle θ in the laboratory frame. Thus for undulator radiation from a single electron moving along the undulator (z) axis, we expect to see radiation centered at a photon energy $\hbar\omega_0$ given as a function of angle by Eq. (5.52), with a spectral distribution of intensity given by Eq. (5.51), and having a relative spectral bandwidth of approximately $1/N$. If the acceptance cone for this observation, $\Delta\theta$ at θ, is finite, the detected frequency or energy bandwidth may be broader.

In a previous section we used a general knowledge of Fourier transform pairs to predict that undulator radiation would have a "natural" spectral bandwidth of $1/N$ and,

(a)

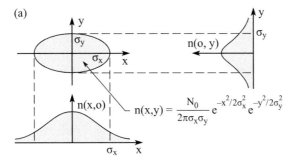

(b)

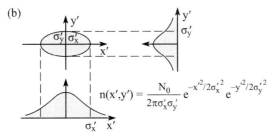

Figure 5.22 Cross-sectional view of an electron beam of elliptical cross-section with Gaussian distribution of density in both (a) spatial and (b) angular coordinates, where $x' = dx/dz$ and $y' = dy/dz$.

on the basis of this, defined a *central radiation cone* with an inclusive half angle of [Eq. (5.32)]

$$\theta_{\text{cen}} = \frac{1}{\gamma^* \sqrt{N}}$$

Beyond the natural line width due to the finite number N of oscillations, further spectral broadening can be incurred with the passage of many electrons through the undulator in a bunch of finite size, divergence, and energy spread. Parameters describing electron beam size and divergence[¶] are illustrated in Figure 5.22. If there is an electron energy spread within the bunch, $\Delta E_e / E_e = \Delta\gamma/\gamma$, there will be a corresponding photon energy

---

[¶] In an electron storage ring the beam size and divergence, σ and σ', are described in terms of a phase space product, known as the *emittance*, and a β-function, a parameter characterizing the magnet's structure (*lattice*) that confines the electrons within the ring. The phase space volume is the region of a six-dimensional position–momentum, or position–angle, space that encloses all particles. For particles with a Gaussian distribution of rms spatial and angular measures, σ and σ', respectively, accelerator physicists refer to the electron beam phase space volume ϵ as the emittance, where $\epsilon = \sigma\sigma'$ is generally written separately for the horizontal and vertical planes, i.e., $\epsilon_h = \sigma_x \sigma_x'$ and $\epsilon_v = \sigma_y \sigma_y'$. The rms electron beam parameters are given by $\sigma_{x,y} = \sqrt{\epsilon_{x,y}\beta_{x,y}}$ and $\sigma_{x,y}' = \sqrt{\epsilon_{x,y}/\beta_{x,y}}$, where the emittances ϵ_x and ϵ_y are fixed for a given storage ring, while βx and βy have different values around the ring depending on local magnetic fields. The ratio ϵ_y/ϵ_x is referred to historically as the coupling ratio. Typically the coupling ratio is of order 1%, but upgrades to couplings of order 100% are underway[18, 19]. For further discussion of emittance see the article by Cornacchia[15] and the texts by Kim, Huang, and Lindberg[5], and Schmüser, Dohlus, Rossbach, and Behrens[4].

spread according to Eq. (5.52), given by

$$\frac{\Delta E}{E} = \frac{2\Delta\gamma}{\gamma} \tag{5.53}$$

where the factor of two is due to the squared relationship between photon energy and electron energy. Since typical energy spreads in modern storage rings are of order $\Delta\gamma/\gamma = 10^{-3}$, this is generally negligible in the fundamental for undulators of about 100 periods, although it can be an observable factor for higher harmonic (n) radiation where the effective number of cycles in the observed radiation is nN.

A more significant effect is that due to random angular motion within the bunch. As a result some electrons traverse the magnet structure not along or parallel to the z-axis, but at a small angle α thereto. These electrons undergo the same number N of oscillations, but experience a somewhat longer period, and further, the observed radiation is affected by a non-axial relativistic Doppler shift. The net result is a Doppler-dominated energy shift, always to lower photon energy (longer wavelength), given by[§]

$$\frac{\Delta E}{E} = \gamma^{*2}\alpha^2 \tag{5.54}$$

This term is important for electron trajectories such that the angle α causes a photon energy shift $\Delta E/E$ of order $1/N$. For a collection of electrons passing through the undulator in a bunch, maintenance of the sharp single-electron spectral features requires that the rms angular divergence σ' cause a spectral broadening less than $1/N$. From Eq. (5.54), with α replaced by σ', and $\Delta E/E < 1/N$, one obtains the *undulator condition* restricting electron beam divergence, as anticipated in earlier sections:

$$\sigma'^2 \ll \frac{1}{\gamma^{*2}N} \tag{5.55a}$$

or

$$\boxed{\sigma'^2 \ll \theta_{\text{cen}}^2} \tag{5.55b}$$

A far-field computation of the predicted spectral distribution of undulator radiation for various values of electron beam divergence[16] is presented in Figure 5.23. Note that this is a summation of $\text{sinc}^2 N\pi u$ individual electron spectral functions, shifted to lower photon energies ΔE for an assumed Gaussian distribution of electron trajectories at an angle α to the z-axis. The example is chosen with $N = 89$, for a natural bandwidth of about 1.1%, and a (single electron) central cone of half angle 35 μrad. The rms electron angular distributions (σ'), ranging from zero to 70 μrad, illustrate the transition from a $\sin^2 N\pi u$ behavior for many parallel electrons, to an extended low photon energy (red shifted) tail when the rms values are comparable to the characteristic central cone angle.

As these curves illustrate, when σ' is comparable to the single electron value of θ_{cen}, the observed on-axis spectrum is spread to varying degrees. This requires some adjustment to our characterization of the observed radiation. For instance, we calculated in an

---

[§] Write the undulator equation (5.28) for an increased period $\lambda u/\cos\alpha$, and for an observation angle $\theta = \alpha$; then compare this with the $\alpha = 0$ case and normalize to form $\Delta\lambda/\lambda$ and $\Delta E/E$.

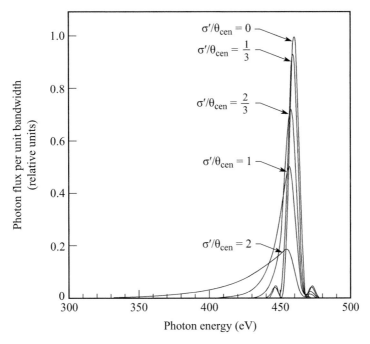

Figure 5.23 Spectral line shape and relative spectral photon flux of undulator radiation as a function of electron beam angular divergence σ' divided by the central cone half angle θ_{cen}, for a soft x-ray undulator having $\gamma = 3720$, $\lambda_u = 5.00$ cm, $N = 89$, and $K = 1$. Following Kitamura.[16]

earlier section the power radiated in the central cone. Clearly that power is not confined within the central cone of a single electron, but rather is spread to a larger angle by random electron trajectories within the beam. As a first approximation we can define a *total* central radiation cone θ_T as a near-Gaussian square root of the sum of squares. As storage ring electron beams are generally elliptical in cross-section, as was illustrated in Figure 5.22, we choose to define horizontal and vertical values, θ_{Tx} and θ_{Ty}, as

$$\theta_{Tx} = \sqrt{\theta_{\text{cen}}^2 + \sigma_x'^2} \qquad (5.56a)$$

and

$$\theta_{Ty} = \sqrt{\theta_{\text{cen}}^2 + \sigma_y'^2} \qquad (5.56b)$$

To appreciate the magnitude of these corrections we take as an example the ALS 5 cm period undulator operating at $K = 1$, which has a single electron central radiation cone half angle (θ_{cen}) of 35 µrad. For the ALS $\sigma_x' = 9.7$ µrad and $\sigma_y' = 4.8$ µrad, so that the total central cone half angles are $\theta_{Tx} = 36$ µrad and $\theta_{Ty} = 35$ µrad. Adding angles in quadrature, as in Eqs. (5.56a,b), is correct for Gaussian distributions, but only a convenient approximation for broadening of the central radiation cone. Nonetheless the model is simple, with a clear physical concept, and gives an easily obtained estimate of the degree to which single-electron calculations of power and divergence are modified by beam divergence. More accurate estimates require numerical studies, which include finite angular acceptance, real measures of electron angular distribution, and other storage-ring–beamline characteristics.[13]

Further spectral broadening can be incurred due to the finite electron bunch size, as was illustrated in Figure 5.22(a) in terms of a Gaussian radial measure σ, with subscripts indicating differing widths in the various coordinate directions. Although spatial effects can be quite large in older storage rings with large beam sizes, in modern storage rings spatial effects are typically somewhat smaller than those due to angular divergence. The effect of spatial broadening is due to the fact that electrons travelling parallel to the z-axis, but displaced at some lateral coordinates (x, y), will be observed at some finite angle, and thus again be Doppler shifted to longer wavelengths. As typical values of σ are of order 100 μm in modern storage rings, and observations (experimental chambers) are typically 10 m or more downstream, angular measures due to source size are of order 10 μrad, and thus cause less of a spectral broadening effect than that due to electron beam divergence (σ') as estimated by Eqs. (5.56a,b).

5.4.6   Spectral Brightness of Undulator Radiation

As measures of the radiation emitted by electrons traversing a periodic magnet undulator we have calculated the power, photon flux, and power per unit solid angle. Another important measure is brightness, or in fact spectral brightness.[‖] Brightness is defined here as radiated power per unit area and per unit solid angle at the source, or equivalently the photon flux per unit area and per unit solid angle. Spectral brightness is the brightness per unit relative spectral bandwidth, i.e., the brightness contained within a relative spectral bandwidth ($\Delta\lambda/\lambda$ or $\Delta\omega/\omega$) of interest. Brightness has an important conceptual role, as it is a conserved quantity in perfect optical systems. That is, in a lossless unaberrated optical system the brightness is equal in the source and image planes. For instance, in a simple imaging system the size magnification is matched by an equal angular demagnification, so that the size–angle product is fixed. The area–solid-angle product is therefore also equal in the object and image planes. It is thus an important quantity in designing microscopes, microprobes, and other imaging systems. Furthermore, it has very interesting wavelength limits when considering experiments that utilize the partially coherent nature of undulator radiation, as discussed in Chapter 4.

To first order one can define *brightness* as the power ΔP radiated from an area ΔA into a solid angle $\Delta\Omega$ as

$$B = \frac{\Delta P}{\Delta A \cdot \Delta\Omega} \qquad (5.57)$$

[‖] Brightness is the preferred term in optics, rather than brilliance. Refer to Born and Wolf[17], pp. 194 and 201.

and the *spectral brightness* as that portion of the brightness lying within a relative spectral bandwidth $\Delta\omega/\omega$ as

$$B_{\Delta\omega/\omega} = \frac{\Delta P}{\Delta A \cdot \Delta\Omega \cdot \Delta\omega/\omega} \qquad (5.58)$$

To specialize this to the case of undulator radiation we can use the previously calculated power in the central radiation cone, P_{cen}, which was defined as having a relative spectral bandwidth (BW) of $\Delta\omega/\omega = 1/N$ and a radiation cone of half angle θ_{cen}, which in the presence of an elliptically divergent electron beam becomes elliptical itself with half angles θ_{Tx} and θ_{Ty} as defined in Eq. (5.56). In the synchrotron community the tradition is to define spectral brightness in terms of photon flux (photons per unit time), rather than power, and furthermore to express the result in terms of a relative spectral bandwidth of 10^{-3}, often written as 0.1% BW. To accommodate this tradition we introduce the average photon flux within the central cone, $\bar{F}_{\text{cen}}$, which we define as the radiated power divided by the energy per photon, viz.,

$$\bar{F}_{\text{cen}} = \frac{\bar{P}_{\text{cen}}}{\hbar\omega/\text{photon}} \qquad (5.59)$$

Defining the *undulator spectral brightness in terms of photon flux* within the central cone, one has

$$\bar{B}_{\Delta\omega/\omega} = \frac{\bar{F}_{\text{cen}}}{\Delta A \cdot \Delta\Omega \cdot N^{-1}} \qquad (5.60a)$$

To write this in terms of a 0.1% bandwidth, rather than $1/N$, we multiply numerator and denominator by a unitless factor 10^{-3} to obtain

$$\bar{B}_{0.1\%\,\text{BW}} = \frac{\bar{F}_{\text{cen}} \cdot (N/1000)}{\Delta A \cdot \Delta\Omega \cdot (0.1\%\,\text{BW})} \qquad (5.60b)$$

The factor $N/1000$, which appears in the numerator, takes account of the fact that with a choice of photon flux within a unit relative bandwidth less than $1/N$, only a portion of the flux within the central cone is utilized, e.g., for $N = 100$ only $N/1000 = 10^{-1}$ of the central cone flux is within a relative bandwidth of 0.1%.

If the radiation emits from a source of elliptical cross-section, having a Gaussian distribution of density across both horizontal (x) and vertical (y) coordinates, the photon flux per unit area can be written as[**]

$$\frac{d\bar{F}}{dA} = \frac{\bar{F}_{\text{cen}}}{2\pi\sigma_x\sigma_y} e^{-x^2/2\sigma_x^2} e^{-y^2/2\sigma_y^2} \qquad (5.61)$$

where the spatially integrated distribution is normalized to $\bar{F}_{\text{cen}}$, the total flux within a $1/N$ relative spectral bandwidth without divergence. For the assumed Gaussian spatial distribution the on-axis value of photon flux per unit area is $\bar{F}_{\text{cen}}/2\pi\sigma_x\sigma_y$, where $\pi\sigma_x\sigma_y$ is the area of a cross-sectional ellipse of semi-major and semi-minor axes σ_x and σ_y.

[**] This is confirmed by integrating dF/dA over all x and y, and noting (Appendix D) that $\int_0^\infty e^{-a^2x^2} dx = \sqrt{\pi}/2a$. Note that if the vertical beam phase space is near or smaller than the diffraction limit at the wavelength and acceptance angle, a correction $[\sigma_y^2 + \lambda/4\pi\theta_{\text{cen}}^2]^{1/2}$ is required.

To complete the brightness calculation an expression is needed for $dF/d\Omega$ within the central cone. Various forms of this function will be appropriate, depending on the relative measures of $\sigma'_{x,y}$ and θ_{cen}. If the undulator condition (5.55a, b) is well satisfied, so that $\sigma'_{x,y} \ll \theta_{cen}$, the central cone will be rather well defined in terms of both its angular definition and spectrum. The spectrum will approximate the limiting case illustrated in Figure 5.23, while the cone half angle will be only slightly larger than θ_{cen}, which we can approximate by θ_{Tx} and θ_{Ty}, in the respective planes, as given in Eqs. (5.56a, b). In this case the central cone solid angle will be only slightly elliptical, and well approximated by $\Delta\Omega_{cen} = \pi\theta_{Tx}\theta_{Ty}$, so that within this cone

$$\frac{d\bar{F}}{d\Omega} = \frac{\bar{F}_{cen}}{\pi\theta_{Tx}\theta_{Ty}} \tag{5.62}$$

which is the expression we will use in Eqs. (5.60a, b), as it captures our sense of the ideal circumstances for observing undulator radiation. In other cases, however, where the undulator condition is not well satisfied, such as $\sigma'_{x,y} \simeq \theta_{cen}$, the concept of a well-defined central cone is somewhat diminished. As we observed in Figure 5.23, this leads to a broader emission spectrum of reduced spectral intensity. In such a case the angular distribution of central cone photon flux will also be spread. We distinguish this from the central cone arguments developed in this chapter, for which to first order the various electrons travel parallel to the z-axis and radiate a nearly uniform angular pattern out to θ_{cen}. Rather, the Gaussian distribution argument follows from a convolution of the many single-electron radiation patterns (θ_{cen}) with an angular distribution function whose measures are σ'_x and σ'_y. This leads to a smoother angular distribution, which can in some cases be fitted with a Gaussian distribution, particularly when the values of σ'_x and σ'_y approach that of θ_{cen}. This is the type of argument that led to the definitions of θ_{Tx} and θ_{Ty} as total central cone angles in their respective planes, as defined in Eqs. (5.56a) and (5.56b). In this case the angular distribution of radiation within a narrow bandwidth would be approximated by a Gaussian distribution of the form

$$\frac{d\bar{F}}{d\Omega} = \frac{\bar{F}_{cen}}{2\pi\theta_{Tx}\theta_{Ty}} e^{-(x')^2/2\theta_{Tx}^2} e^{-(y')^2/2\theta_{Ty}^2}$$

where x' is an abbreviation for the radiation angle in the x, z-plane measured from the z-axis, y' is an abbreviation for the radiation angle in the y, z-plane, and integration over all angles would be normalized to $\bar{F}_{cen}/2\pi\theta_{Tx}\theta_{Ty}$. Keeping to the spirit of a well-defined central radiation cone, with the undulator condition well satisfied, we will use the formulation of photon flux angular distribution given by Eq. (5.62). Where the condition $\sigma'_{x,y} \ll \theta_{cen}$ is not well satisfied, computational techniques will be very useful.

Combining expressions for a Gaussian spatial distribution [Eq. (5.61)] with the central cone angular distribution given by [Eq. (5.62)], the on-axis photon flux per unit area and per unit solid angle is

$$\frac{d^2\bar{F}}{dA\,d\Omega} = \frac{\bar{F}_{cen}}{2\pi^2\sigma_x\sigma_y\theta_{Tx}\theta_{Ty}} \tag{5.63}$$

so that the on-axis spectral brightness follows from Eq. (5.60b) as

$$\bar{B}_{0.1\%\mathrm{BW}}(0) = \frac{\bar{F}_{\mathrm{cen}} \cdot (N/1000)}{2\pi^2 \sigma_x \sigma_y \theta_{Tx} \theta_{Ty}(0.1\%\,\mathrm{BW})} \tag{5.64}$$

where $\bar{F}_{\mathrm{cen}}$ is given in Eq. (5.59) and where the zero in parentheses refers to an on-axis value with respect to both position and angle. Combining this with Eqs. (5.41b) and (5.59) for $\bar{P}_{\mathrm{cen}}$ and $\bar{F}_{\mathrm{cen}}$, for the special case of 0.1% bandwidth,

$$
\begin{aligned}
\bar{B}_{0.1\%\,\mathrm{BW}}(0) &= \frac{7.25 \times 10^6 \gamma^2 N^2 \bar{I}\,[\mathrm{A}]}{\sigma_x\,[\mathrm{mm}]\,\sigma_y\,[\mathrm{mm}]\left(1 + \frac{\sigma_x'^2}{\theta_{\mathrm{cen}}^2}\right)^{1/2}\left(1 + \frac{\sigma_y'^2}{\theta_{\mathrm{cen}}^2}\right)^{1/2}} \\
&\quad \times \frac{K^2[JJ]^2}{\left(1 + \frac{K^2}{2}\right)^2} \frac{\mathrm{photons/s}}{\mathrm{mm}^2\,\mathrm{mrad}^2\,(0.1\%\,\mathrm{BW})}
\end{aligned}
\tag{5.65}
$$

where σ_x and σ_y are in units of millimetres, $\bar{I}$ is in amperes, and $[JJ]^2$, defined by Eq. (5.40a), is the parameter that corrects for reduced power in the fundamental ($n = 1$) due to higher harmonic electron motion which diverts some radiated power to higher harmonics. This formulation is most accurate when the electron beam divergences σ_x' and σ_y' are less than θ_{cen} so that the undulator condition is well satisfied. For large electron beam divergence, such that $\sigma_{x,y}' \simeq \theta_{\mathrm{cen}}$, the undulator condition is not satisfied, so that the spectral content is significantly broadened, as illustrated in Figure 5.23, and Eq. (5.65) overestimates spectral brightness by a factor approaching two. In this case it is best to utilize numerical simulations, although the analytic formulation will continue to give useful insights. Sample values of spectral brightness for soft and hard x-ray undulators are given in Table 5.1, and general trends are shown in Figure 5.24, where brightness values for bending magnet radiation and wiggler radiation are shown for comparison.

5.4.7 Time Structure

The electron beam in a storage ring is not a continuous stream, but rather a highly modulated density function consisting of axial bunches. The spacing of these bunches is set by the radio frequency (rf) used to restore power to electrons, once each turn around the storage ring, to compensate for power lost to synchrotron radiation. The rf is fed to a microwave cavity operating in a mode with an axial electric field, synchronized so that slower electrons receive a small acceleration, while faster electrons experience a small deceleration. In this manner a sequence of potential wells is set up that tends to trap available electrons into a series of *buckets* that travel around the ring at the speed of light, with a bunch-to-bunch separation equal to the rf wavelength. Figure 5.25 illustrates electron bunch structure in the ALS storage ring for a multibunch operation in which all of the buckets are filled. By proper timing of electron injection into the ring any sequence of filled buckets can be obtained. A less common, but regularly used mode

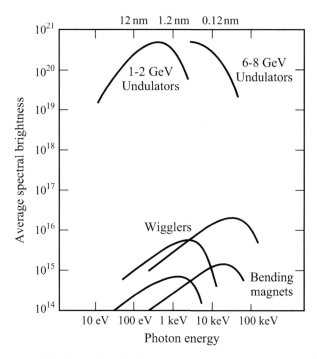

Figure 5.24 General trends of spectral brightness for undulator radiation, wiggler radiation, and bending magnet radiation, showing the complementary nature of soft x-ray (1–2 GeV) and hard x-ray (6–8 GeV) storage ring facilities. Units as in Eq. (5.65). High spectral brightness is particularly useful for experiments involving microscopy, partial coherence, diffraction from small crystalline samples, and other studies which generally benefit from radiation of minimal divergence emanating from a small source size. According to M. Eriksson of Lund University and MAXLAB, improved electron beam lattices are expected to significantly reduce horizontal beam size and divergence (σ and σ') in planned upgrades at several synchrotron radiation facilities, which in turn will lead to higher spectral brightness and coherent flux.[18, 19] The reduced horizontal beam sizes, likely of order 10^2, will greatly reduce data collection times for all x-ray imaging experiments. The upgrades are referred to as "diffraction limited storage rings (DLSRs)."

of operation employs only two filled bunches, which is convenient for use with time-of-flight measurements of chemical reaction products in photodissociation products.

For some experiments involving time-of-flight detection of photofragments (chemical dynamics) very few buckets contain electrons. More generally, most buckets are filled, leaving some sequence unfilled to counteract beam propagation instabilities that are thought to involve ion trapping within the ring. For the case illustrated in Figure 5.25 a 500 MHz rf produces a bunch-to-bunch separation of 60 cm, which accommodates a 328 bucket axial charge distribution within the 197 m storage ring circumference set by the beam energy and bending magnet fields. In the case illustrated, 288 of the 328 buckets are filled. Each of the electron bunches has a near-Gaussian pulse shape with a nominal 60 ps FWHM duration, corresponding to an axial charge distribution of $\sigma_z = 7.7$ mm rms.

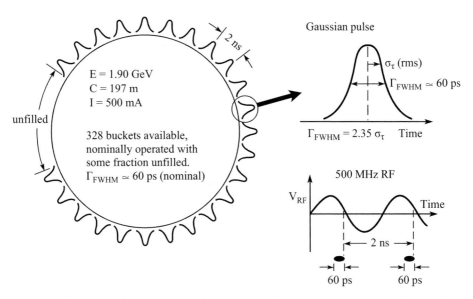

Figure 5.25 Illustration showing how the time structure of synchrotron radiation is related to rf power replenishment and resultant electron bunch structure (axial electron density modulation) for a typical storage ring. In the example shown here the radio frequency (rf) and storage ring circumference are matched for a 60 cm peak to peak structure of 328 axial buckets, of which a pre-selected fraction are filled with electrons. At the speed of light, 60 cm corresponds to a 2 ns x-ray pulse separation. The individual pulse duration, set by the rf voltage and beam dynamics, is nominally 60 ps FWHM, corresponding to a Gaussian axial bunch length $\sigma_z = 7.7$ mm rms.

In many of the examples considered throughout this chapter, the average power, photon flux, etc., were calculated on a time-average basis (indicated by a bar over the assigned symbol, as in $\bar{P}_{cen}$). To calculate peak powers ($\hat{P}$) one must know the time structure, that is, the pulse duration and separation. For instance, in the case illustrated in Figure 5.25, where a sequence of 60 ps FWHM Gaussian pulses occurs every 2.0 ns, the ratio of peak to average power, $\hat{P}/\bar{F}$, is 33. Thus the peak power, photon flux, and spectral brightness values are higher than the time-averaged values cited by a factor of 33. The APS typically operates with nominal 79 ps FWHM pulses with a pulse-to-pulse separation of 153 ns, so that the ratio of peak to average power is about 2000.

5.4.8 Polarization Properties of Undulator Radiation

For electron motion directed along the z-axis of an undulator (see Figures 5.8 and 5.17) with periodic magnetic fields oriented in the vertical (y) direction, electron oscillations are in the horizontal (x) direction for low K-values. This generates radiation at a fundamental ($n = 1$) wavelength given by the so-called undulator equation (5.28), having fields within the central cone polarized with the electric field in the horizontal (x) direction. For somewhat higher K-values, of order one, second and third harmonics become important, as was illustrated in Figure 5.17 and is discussed further in Section 5.5, which follows. For these modest K-values the second harmonic motion consists of a

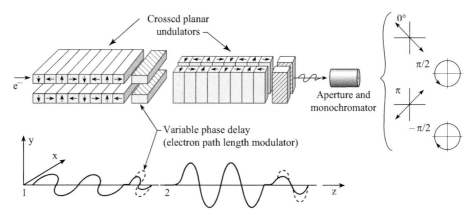

Figure 5.26 Schematic of a pair of crossed undulators that can be used to generate linearly polarized, circularly polarized, or elliptically polarized radiation across a broad range of EUV and soft x-ray energies. (Courtesy of K.-J. Kim.[21])

velocity modulation in the z-direction so that in the primed frame of reference, moving with the mean electron motion, the second harmonic radiation (ω_2') is polarized in the z'-direction, with maximum intensity off axis at $\theta' = \pi/2$, as shown in Figure 5.18(b). This radiation is then transformed to the laboratory frame of reference as an axisymmetric second harmonic radiation cone, with a peak intensity off axis having a radial electric field polarization, and an intensity null on axis. Thus by moving the point of observation circularly around the second harmonic peak $0 \leq \phi \leq 2\pi$, at an angle of $1/\gamma$ off axis, the polarization detected will vary smoothly from vertical to horizontal. Third harmonic radiation ($n = 3$) is similar to the fundamental, being horizontally polarized for angles of observation near the central (z) axis. The subject of off-axis polarization for arbitrary K-values is discussed by Kim.[1]

Specialized magnet structures for the generation of arbitrarily polarized undulator radiation are of great interest for probing magnetic materials, helical structures, and other samples with polarization dependent properties.[20] In general the greatest freedom in varying both polarization state and wavelength is valued. Toward this end efforts have been made to construct magnet structures that can provide orthogonal linear polarizations, left or right circularly polarized radiation, and elliptically polarized radiation. Figure 5.26 shows a pair of axially separated crossed undulators.[21] As electrons sequentially traverse the two undulators, radiation is generated with orthogonal polarizations. In each magnet structure the emitted phase is related to the motion of the electron. By controlling the electron transit time between the two undulators through the use of a variable-field magnet, the relative phase of the radiated field amplitudes ($\mathbf{E}$) can be selected. Thus linearly polarized radiation can be obtained with field orientations set to $\pm\pi/4$ from the vertical. By modulating the electron transit time between undulators one can switch between these two orthogonal polarizations. Similarly one can generate left or right circular polarization through proper phasing. The wavelength is controlled independently by setting the magnet gap. A pair of crossed undulators such as these has

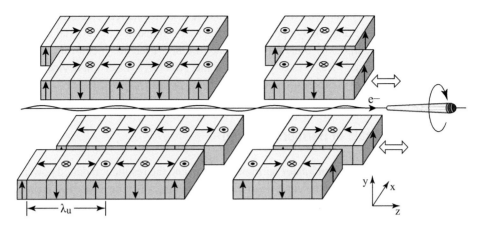

Figure 5.27 Schematic view of an APPLE II magnetic structure for generating variably polarized undulator radiation. (a) Isometric view showing two fixed rows (upper right and lower left) and two axially movable rows (upper left and lower right), shown displaced forward by one quarter period. (b) End-on view in a midplane of the magnet structure in a configuration that produces both horizontal and vertical magnetic fields on axis, which will result in a helical electron motion and the concomitant emission of elliptically polarized radiation. Courtesy of S. Sasaki.[25]

been installed at the BESSY Facility in Berlin for the study of spin resolved photo-emission at soft x-ray and EUV wavelengths.[22] As noted earlier in Eq. (5.55a), optimum undulator performance is obtained with small electron beam divergence, $\sigma' < 1/\gamma\sqrt{N}$, which limits devices such as this to modern storage rings.

A number of other magnet structures have been designed with various additional attributes. Onuki and colleagues[23] have designed and built a system in which two orthogonal undulators are placed side by side so that the electrons experience a net magnetic force due to the combined fields at each point along their trajectory. By phase shifting one undulator magnet set (say the horizontal field set) by a quarter period, that is, moving it axially by a distance $\pm\lambda_u/4$, one can induce left or right handed helical motion of the electrons as they traverse the magnet structure. This has the sometimes advantage of avoiding harmonic radiation.[12] Di Viacco and Walker, then at Sincrotrone Trieste[24] in Italy, have designed planar magnet structures that generate circularly polarized radiation.

S. Sasaki and colleagues in Japan have developed the APPLE II[††] undulator concept for the generation of linear and circular and elliptical polarized radiation based on the use of four rows of periodic magnet arrays[25–29], as illustrated in Figure 5.27. In this structure all the magnets are above or below the electron orbit plane, thus avoiding horizontal gap–electron-beam interactions. In this device the four rows of periodic magnets are arranged with two rows above and two rows below the electron orbit plane. In both the upper and lower magnet planes one row of magnets is fixed and one is movable (in the axial direction). The two movable magnet sets, upper front and lower back in Figure 5.27, move together. In combination, the four magnet structures create

---

[††] Advanced Planar Polarized Light Emitter II (APPLE II).

an on-axis magnetic field that can induce linear or helical electron motion as the position of the diagonally opposed moveable set is translated in the axial direction. Versions of this elliptically polarizing undulator are now used at synchrotron radiation facilities worldwide.[27] Reviews of periodic magnet structures (insertion devices) for third generation x-ray facilities are presented in Refs. 28–32.

5.5   The Scale of Harmonic Motion

To understand undulator harmonic motion and the resultant radiated power, it is instructive to continue the analysis of electron dynamics for the case of K approaching unity. We will transform to the frame of reference moving with the average electron velocity $\bar{v}_z$. In this frame, the motion is a simple harmonic. Therefore, we can utilize well-known results for radiation from an oscillating electron, as summarized earlier in this chapter.

To begin the analysis, we recall the earlier result, Eq. (5.19):

$$v_x = \frac{Kc}{\gamma} \cos\left(\frac{2\pi z}{\lambda_u}\right)$$

To determine the oscillation amplitude and accelerations we must find $v_x(t)$, rather than $v_x(z)$. To do so, we utilize our knowledge of $z = z(t)$ and do the necessary transformations to the frame of reference moving with the average electron motion. In this frame of reference, in the low-K undulator regime, the motion is described in terms of a few simply related harmonic motions. We recall Eq. (5.23a):

$$v_z \simeq c\left[1 - \frac{1}{2\gamma^2} - \frac{K^2}{2\gamma^2}\cos^2\frac{2\pi z}{\lambda_u}\right]$$

Recalling that $\cos^2 k_u z = (1 + \cos 2k_u z)/2$, where $k_u z = 2\pi z/\lambda_u$ is the wavenumber, we rewrite the above as previously in Eq. (5.23b):

$$v_z = c\left[1 - \frac{1 + K^2/2}{2\gamma^2} - \frac{K^2}{4\gamma^2}\cos(2k_u z)\right]$$

Noting that $z = \int v_z \, dt$, we have

$$z(t) = c\underbrace{\left(1 - \frac{1}{2\gamma^{*2}}\right)}_{\beta^*}t - \frac{cK^2}{4\gamma^2}\int \cos[2k_u z(t)]\,dt$$

where $z(t)$ appears twice. Noting that the cosine term is bounded, and approximating $z(t) \simeq \beta^* ct$ in the integral, this gives, to first order,

$$z - c\beta^* t \simeq -\frac{cK^2}{4\gamma^2 \cdot 2k_u c\beta^*}\sin 2k_u \beta^* ct \qquad (5.66)$$

where we have used the following definitions:

$$\gamma = \frac{1}{\sqrt{1 - \beta^2}}$$

and

$$\gamma^* = \frac{1}{\sqrt{1 - \beta^{*2}}} = \frac{\gamma}{\sqrt{1 + K^2/2}}$$

Therefore, for β approaching unity,

$$\beta^* = 1 - \frac{1}{2\gamma^{*2}} \tag{5.67a}$$

$$\beta^* = 1 - \frac{1 + K^2/2}{2\gamma^2} \tag{5.67b}$$

Substituting $\omega_u = \beta^* k_u c$ and rearranging terms, Eq. (5.66) becomes

$$z - \beta^* ct \simeq -\frac{K^2}{8\gamma^2 k_u \beta^*} \sin 2\omega_u t$$

The Lorentz transformation, as described in Appendix F, can be used to transform this equation from the laboratory frame of reference (z, t) to the electron frame of reference (z', t').

Using the transformation [Appendix F, Eq. (F.1)], the z'-motion takes the form

$$\frac{z'}{\gamma^*} \simeq -\frac{K^2}{8\gamma^2 k_u \beta^*} \sin 2\omega_u \left[\gamma^* \left(t' + \frac{\beta^* z'}{c} \right) \right]$$

Multiplying through by γ^*, noting that $\omega_u = k_u c$, and taking $\beta^* \simeq 1$, one has

$$z' \simeq -\frac{K^2}{8\gamma k_u (1 + K^2/2)^{1/2}} \sin \left(2\omega_u \gamma^* t' + 2k_u \gamma^* z' \right)$$

For convenience, we define

$$K^* \equiv \frac{K}{(1 + K^2/2)^{1/2}} \tag{5.68}$$

where we note that for large K, K^* approaches $\sqrt{2}$. We also note that because of the Lorentz contraction $k_u' = \gamma^* k_u$, with corresponding frequency $\omega_u' = \gamma^* \omega_u$. Thus z' can be

$$z'(t') = -\frac{K^{*2}}{8k_u'} \sin(2\omega_u' t' + 2k_u' z') \tag{5.69}$$

where we observe the complexity of this equation in that z' appears both on the left and in a phase term on the right. A numerical solution of this equation is illustrated in Figure 5.28, showing the basic second harmonic motion for a relatively low K-value (two full cycles of motion in a time period λ_u'/c), with an observable asymmetry for $K = 1$, where the sine function peaks early in the first half cycle and late in the second half cycle. This asymmetric behaviour indicates the growing presence of still higher harmonics as K grows from very small values to unity and beyond.

Before proceeding to the consideration of higher K-values, we first examine the low-K case somewhat further for a better insight into the motion and amplitude scaling in this

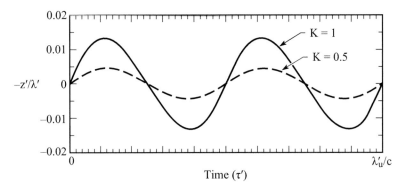

Figure 5.28 Normalized axial motion $z'(t')$ for an electron traversing a low K undulator, as seen in the moving (γ^*) frame of reference. Note the double oscillation for a single period of the magnet structure, and the distortion for higher K as a $4\omega'$ contribution pulls the positive amplitude peaks earlier in time and pushes the negative peaks later in time.

important parameter regime. Since z', as described in Eq. (5.69), is bounded by the sine function to an oscillation amplitude $z' \simeq K^{*2}/8k'_u$, the second phase term, $2k'_u z'$, never exceeds $K^{*2}/4$. This is in contrast with the first phase term, $2\omega'_u t'$, which dominates the oscillations for small K^*. In this limit the lowest-order solution for $z'(t')$ is

$$z'(t') \simeq -\frac{K^{*2}}{8k'_u} \sin 2\omega'_u t' \tag{5.70}$$

which displays a double-frequency oscillation in the axial direction, with an amplitude that scales as K^{*2}. Note that the validity of this approximation, for instance $K^{*2}/4 < 0.1$, corresponds to $K < 0.7$, while $K^{*2}/4 < 0.2$ corresponds to $K < 1.1$. As the axial "sloshing amplitude," the z' motion serves as the amplitude for second harmonic radiation, and also as the coupling mechanism between fundamental and higher harmonics. Note that the axial oscillation velocity, $v'_z = dz'/dt'$, is bounded by $K^{*2}c/4$.

To determine the third harmonic of the motion, we return to the calculation of $x(t)$, where, as we found in Section 5.4.3 [in the development of Eq. (5.33)],

$$x \simeq \frac{K}{\gamma k_u} \sin \omega_u t$$

Again using the Lorentz transformations $x = x'$ and $t = \gamma^*(t' + \beta^* z'/c) \simeq \gamma^*(t' + z'/c)$

$$x' \simeq \frac{K}{k_u \gamma} \sin \omega_u \gamma^* \left(t' + \frac{z'}{c}\right)$$

Using Eq. (5.70) for $z'(t')$, and noting that the Doppler-shifted wavenumber and frequency are $k'_u = \gamma^* k_u$, and $\omega'_u = \gamma^* \omega_u$, the x'-motion becomes

$$x'(t') \simeq \frac{K^*}{k'_u} \sin \left(\omega'_u t' + \frac{K^{*2}}{8} \sin 2\omega'_u t'\right) \tag{5.71}$$

If we let $\epsilon = K^{*2}/8$, Eq. (5.71) becomes

$$x'(t') \simeq \frac{K^*}{k'_u} \sin(\omega'_u t' + \epsilon \sin 2\omega'_u t')$$

Using the trigonometric identity (Appendix D.3) that $\sin(\alpha + \beta) = \sin\alpha \cos\beta + \cos\alpha \sin\beta$, the expression for $x'(t')$ can be expanded as

$$x'(t') \simeq \frac{K^*}{k'_u}[\sin\omega'_u t' \cos(\epsilon \sin 2\omega'_u t')] + \frac{K^*}{k'_u}[\cos\omega'_u t' \sin(\epsilon \sin 2\omega'_u t')]$$

For $\epsilon \ll 1$ this becomes

$$x'(t') \simeq \frac{K^*}{k'_u}[\sin\omega'_u t' + \epsilon \cos\omega'_u t' \sin 2\omega'_u t']$$

The third harmonic component is obtained from the last term by using the trigonometric identity (Appendix D.3) $\cos\alpha \sin\beta = [\sin(\alpha + \beta) - \sin(\alpha - \beta)]/2$. The expression for $x'(t')$ is then

$$x'(t') \simeq \frac{K^*}{k'_u}\left[\sin\omega'_u t' + \frac{\epsilon}{2}\sin\omega'_u t' + \frac{\epsilon}{2}\sin 3\omega'_u t'\right]$$

For $\epsilon = K^{*2}/8 \ll 1$ the dominant terms are

$$x'(t') \simeq \frac{1}{k'_u}\left[K^* \sin\omega'_u t' + \frac{K^{*3}}{16}\sin 3\omega'_u t'\right] \tag{5.72}$$

The first term describes the fundamental oscillation as the electron traverses the periodic magnet structure, while the second term describes the third harmonic component of the motion. Thus both first and third harmonic radiations will have the same (x') polarization.

For the low-K undulator case the lowest-order harmonic motions are described by Eqs. (5.70) and (5.72), which we restate below in terms of the fundamental wavelength in the moving reference frame, $\lambda'_u = 2\pi/k'_u$, as

$$\frac{x'(t')}{\lambda'_u} \simeq \frac{1}{2\pi}\left[K^* \sin\omega'_u t' + \frac{K^{*3}}{16}\sin 3\omega'_u t'\right] \tag{5.72}$$

$$\frac{z'(t')}{\lambda'_u} \simeq -\frac{1}{2\pi}\left[\frac{K^{*2}}{8}\sin 2\omega'_u t'\right] \tag{5.70}$$

We observe that the various harmonic amplitudes scale as K^{*n}, and the acceleration as $n^2 K^{*n}$, so that harmonic motion can be expected to grow very rapidly as K approaches and exceeds unity. Since the power radiated scales as the square of acceleration the harmonic power scales as $n^4 K^{*2n}$. Furthermore, the harmonic central radiation cone narrows to $1/\gamma^*\sqrt{nN}$ owing to the larger number of oscillations (nN), so that the useful solid angle is $\pi/nN\gamma^{*2}$. The power radiated in a given harmonic then scales as

$$\frac{dP'}{d\Omega'} \propto n^3 K^{*2n}$$

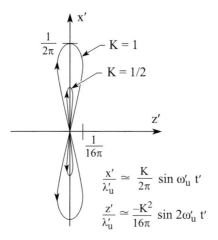

$$\frac{x'}{\lambda'_u} \simeq \frac{K}{2\pi} \sin \omega'_u t'$$

$$\frac{z'}{\lambda'_u} \simeq \frac{-K^2}{16\pi} \sin 2\omega'_u t'$$

Figure 5.29 The x', z'-motion of an electron in the moving (γ^*) frame of reference, as it traverses a low-K undulator.

Thus the radiated power grows very quickly with increasing K. Note, however, that the growth scales with K^*, so that for a given harmonic there is a built-in saturation for higher values of K. Rather, as we will see in Section 5.7 on high K wiggler radiation, the radiation evolves to ever higher harmonics.

From Eqs. (5.70) and (5.72), the harmonic amplitudes, scaled to their respective wavelengths (λ'_u/n), can be written as

$$\frac{|x'_{\omega'_u}|}{\lambda_u} \le \frac{K^*}{2\pi} \tag{5.73}$$

$$\frac{|z'_{2\omega'_u}|}{\lambda'_u/2} \le \frac{1}{4} \frac{K^{*2}}{2\pi} \tag{5.74}$$

$$\frac{|x'_{3\omega'_u}|}{\lambda'_u/3} \le \frac{3}{16} \frac{K^{*3}}{2\pi} \tag{5.75}$$

We see that, for small K, the dipole approximation is well satisfied in the moving frame of reference, as assumed when calculating radiated power. For modest K-values the motion in the electron reference frame approximates a figure eight. This can be seen by combining the fundamental and second harmonic terms in Eqs. (5.70) and (5.72), viz., for small K

$$\frac{x'(t')}{\lambda'_u} \simeq \frac{K^*}{2\pi} \sin \omega'_u t'$$

and

$$\frac{z'(t')}{\lambda'_u} \simeq -\frac{K^{*2}}{16\pi} \sin 2\omega'_u t'$$

so that the electron executes two z' oscillations for each x' oscillation, as illustrated in

Figure 5.29. Note that the relative width of the figure eight grows with increasing K, as x' is proportional to K, while the z'-motion is proportional to K^2.

In Section 5.7 we consider the more complex transition from undulator to wiggler radiation as K exceeds unity, leading to a domination by higher harmonics. In this limit the harmonics are closely spaced, and practical effects lead to their appearance as a smoothed continuum at very high photon energies.

5.6   Spatial and Spectral Filtering of Undulator Radiation

As discussed in Chapter 4 it is possible to obtain spatially and spectrally coherent radiation from an extended source of broad spectrum through the combined use of a monochromator and a pinhole spatial filter. This is particularly effective when the product of source size and divergence is not too much larger than the coherence limit at relevant wavelengths. This is the case for undulator radiation at modern synchrotron radiation facilities. The resultant radiation is valuable for scanning soft x-ray microscopy and hard x-ray fluorescence microprobes, both of which achieve their best spatial resolution with spatially coherent radiation, and for all experiments involving interference and holography.[33–39] The secret to success in this spatial filtering process is that the electron beam cross-section and divergence must be sufficiently small, so that a fair fraction of the radiated flux is able to pass through a pinhole–aperture combination for which $d \cdot \theta = \lambda/2\pi$, as described previously in Eq. (4.7a). Figure 5.30(a) depicts an undulator and one form of a spatial filter.

In general, for synchrotron radiation, the phase space of the central radiation cone is larger than the limiting condition required for spatial coherence, Eq. (4.7a). That is, if we take a typical electron beam diameter of 100 µm and a typical central cone half angle of 30 µrad, the product $d \cdot \theta$ is 3 nm, generally much greater than $\lambda/2\pi$. Thus for experiments that require spatial coherence the *pinhole spatial filter* is used to narrow, or filter, the phase-space of transmitted radiation, much as was illustrated in Figure 4.8. Filtering to $d \cdot \theta = \lambda/2\pi$ requires the use of both a small pinhole (d) as shown, and some limitation on θ, such that the product is equal to $\lambda/2\pi$. For example, one could accept the full central cone (θ_{cen}) and choose an appropriate pinhole diameter $d = \lambda/2\pi\theta_{\mathrm{cen}}$. Alternatively, one could use a downstream angular aperture of acceptance angle $\theta < \theta_{\mathrm{cen}}$, and choose d accordingly. To calculate the spatially coherent power transmitted by the pinhole spatial filter, one must consider the phase-space of the emitted radiation in both the vertical (y–z) and horizontal (x–z) planes, as the condition $d \cdot \theta = \lambda/2\pi$ must be satisfied for both. If the electron beam is elliptical, as illustrated in Figure 5.22, with major and minor diameters $d_x = 2\sigma_x$ and $d_y = 2\sigma_y$, and if the central radiation cone is also somewhat elliptical due to differences in the horizontal and vertical electron beam divergences, then so too are the characteristic half angles[‡‡] θ_x and θ_y, and then the phase-space volume containing the emitted power within the radiation cone, $\bar{P}_{\mathrm{cen}}$ of Eq. (5.41), will be $(d_x \theta_x)(d_y \theta_y)$. The pinhole spatial filter must reduce both

---

[‡‡] In Eqs. (5.56a,b) these were described as the "total" central cone half angles $\theta_{Tx} = \sqrt{\theta_{\mathrm{cen}}^2 + \sigma_x'^2}$ and $\theta_{Ty} = \sqrt{\theta_{\mathrm{cen}}^2 + \sigma_y'^2}$, where σ_x' and σ_y' are the respective measures of electron beam divergence in the two planes.

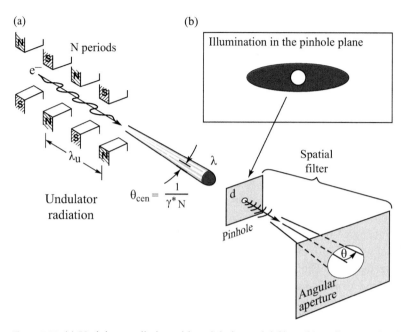

Figure 5.30 (a) Undulator radiation with a pinhole spatial filter. (b) A diagram showing a highly elliptical photon beam as projected or imaged to the plane of the pinhole. Generally the eccentricity of the ellipse at present facilities is of order 20:1; however, new electron beam lattices[18, 19] promise reductions to approximately 5:1. This will result in higher brightness photon beams and better utilization of available photons for imaging and other axisymmetric experiments.

$d_x \, \theta_x$ and $d_y \, \theta_y$ to $\lambda/2\pi$. The transmitted *spatially coherent power*[7] will therefore be reduced proportionally to a value

$$\bar{P}_{\text{coh},N} = \left(\frac{\lambda/2\pi}{d_x\theta_x} \right) \left(\frac{\lambda/2\pi}{d_y\theta_y} \right) \bar{P}_{\text{cen}} \qquad (5.76)$$

where $d = 2\sigma$ and $\theta = \theta_T$. The horizontal (x) and vertical (y) phase-space filter factors are written separately to remind us that each alone has a maximum value of unity. In much of what follows we will assume that in both planes $d \cdot \theta > \lambda/2\pi$, permitting some simplifications to the formulae.¶

We recall from Eq. (5.41a) that

$$\bar{P}_{\text{cen}} = \frac{\pi e \gamma^2 \bar{I}}{\epsilon_0 \lambda_u} \cdot \frac{K^2 [JJ]^2}{(1 + K^2/2)^2} \qquad (5.77)$$

where $\bar{I}$ is the average current, λ_u is the undulator period, and $[JJ]$ is a finite-K correction factor that accounts for reduced power radiated in the fundamental $(n = 1)$ due to the

---

¶ The phase-space assumption $d \cdot \theta > \lambda/2\pi$ is generally valid for the undulator radiation, but is near its diffraction limit in the vertical plane at third generation facilities and should be accounted for.[6]

emergence of higher harmonic electron beam oscillations and concomitant radiation emitted in higher harmonics [see Eq. (5.40a)].

Rather than plot radiated power versus K, it is often more useful for experimental planning to plot radiated power versus photon energy. Towards that end we note the wavelength dependence of coherent power; further we note that $\bar{P}_{\text{cen}}$ contains a factor $K^2/(1 + K^2/2)^2$ that is related to wavelength through the undulator equation [Eq. (5.28)]

$$\lambda = \frac{\lambda_u}{2\gamma^2}\left(1 + \frac{K^2}{2} + \gamma^2\theta^2\right)$$

For on-axis radiation ($\theta = 0$) one has

$$\lambda = \frac{\lambda_u}{2\gamma^2}\left(1 + \frac{K^2}{2}\right)$$

or more conveniently, in terms of photon energy ($\hbar\omega = 2\pi\hbar c/\lambda$),

$$\hbar\omega = \frac{\hbar\omega_0}{1 + K^2/2} \tag{5.78a}$$

where

$$\hbar\omega_0 \equiv 4\pi\hbar c\gamma^2/\lambda_u \tag{5.78b}$$

is the highest photon energy that can be radiated in the fundamental ($n = 1$) by a given undulator, and that corresponds to the limiting case $K = 0$. With some algebraic manipulation one can show that the three wavelength-dependent factors, λ^2 due to the coherent phase-space constraint, K^2 due to the transverse electron acceleration, and $(1 + K^2/2)^2$ due to the decreased axial velocity ($\gamma^* = \gamma/\sqrt{1 + K^2/2}$) for finite K, combine to give a photon energy dependence $(\hbar\omega_0 - \hbar\omega)/\hbar\omega$, so that the spatially coherent power [Eq. (5.76)] for an undulator of N cycles takes the form

$$\boxed{\bar{P}_{\text{coh},N} = \frac{e\lambda_u\bar{I}}{8\pi\epsilon_0 d_x d_y \theta_x \theta_y \gamma^2}\left(\frac{\hbar\omega_0}{\hbar\omega} - 1\right)[JJ]^2} \tag{5.79}$$

where in terms of photon energy the finite-K correction factor [Eq. 5.41(d)] can be rewritten as

$$[JJ]^2 = \frac{7}{16} + \frac{5}{8}\frac{\hbar\omega}{\hbar\omega_0} - \frac{1}{16}\left(\frac{\hbar\omega}{\hbar\omega_0}\right)^2 + \cdots \tag{5.80}$$

Note that for magnetic tuning of the undulator through a range $0 \le K \le 2$, the photon energy is varied by a factor of three, where now the factor $[JJ]^2$ varies from 1 to 0.65 over this range. For the case where the undulator condition $\sigma_{x,y}'^2 \ll \theta_{\text{cen}}^2$ is well satisfied, the product $\theta_x\theta_y$ can be approximated as

$$\theta_x\theta_y \simeq \theta_{\text{cen}}^2 = \frac{1 + K^2/2}{\gamma^2 N}$$

which by Eq. (5.78a) becomes $\theta_x \theta_y \simeq \hbar\omega_0 / \hbar\omega\gamma^2 N$. The spatially coherent power in this important special case then takes the form

$$\bar{P}_{\text{coh},N} = \frac{e\lambda_u \bar{I} N}{8\pi\,\epsilon_0 d_x d_y}\left(1 - \frac{\hbar\omega}{\hbar\omega_0}\right)[JJ]^2 \qquad (\sigma'^2 \ll \theta_{\text{cen}}^2) \tag{5.81}$$

On a global scale, coherent power scales roughly as λ, when corrected for differences in average electron current at various facilities. From Eq. (5.76) one might expect coherent power to scale as λ^2; however, higher values of γ are required to reach shorter wavelengths, and power in the central cone scales as γ^2/λ_u, or as $1/\lambda$, thus giving a net scaling proportional to λ. This too is modified further by beamline efficiencies which are typically higher for hard x-ray than for soft x-rays. Note that because of the duty cycle of the synchrotron facilities (typically 60ps FWHM Gaussian pulses every 2.0 ns at the ALS), the peak power can be considerably higher than the average power, for example a factor of about 30 at the ALS, and a factor of about 2000 at the APS. See Table 5.1 for details.

For many experiments it is also desirable to narrow the spectral bandwidth, either because improved spectral resolution is required to probe atomic or molecular states, because a chromatically sensitive zone plate focusing lens requires a relatively narrow spectral bandwidth (narrower than one divided by the number of zones), or because a longer longitudinal coherence length is required for high contrast interferometric or holographic fringe formation. The radiation must then be spectrally filtered by a monochromator (not shown in Figure 5.30) to further narrow the relative spectral bandwidth to a suitable value of $\Delta\lambda/\lambda$, thus increasing the longitudinal coherence length from a value of $N\lambda/2$ to a greater length $l_{\text{coh}} = \lambda^2/(2\,\Delta\lambda)$. For example, if monochromatization to a value $\lambda/\Delta\lambda = 10^3$ were desired, the longitudinal coherence length would become $l_{\text{coh}} = 10^3\lambda/2$. This of course is accomplished at a reduction in spatially coherent power. By filtering from $\Delta\lambda/\lambda = 1/N$ to $\Delta\lambda/\lambda = 1/10^3$, the transmitted power is necessarily reduced by a multiplicative factor $(\Delta\lambda/\lambda)/(1/N)$, or $N/10^3$ in the example cited. Furthermore, there will be an *insertion loss* due to the finite efficiency of the monochromator, including such factors as the grating or crystal efficiency, finite mirror reflectivities, etc. If we collect these factors into an inclusive *beamline efficiency* η, then the available coherent power can be written as

$$\bar{P}_{\text{coh},\lambda/\Delta\lambda} = \underbrace{\eta}_{\substack{\text{beamline} \\ \text{efficiency}}} \underbrace{\frac{(\lambda/2\pi)^2}{(d_x\theta_{Tx})(d_y\theta_{Ty})}}_{\substack{\text{spatial} \\ \text{filtering}}} \cdot \underbrace{N\frac{\Delta\lambda}{\lambda}}_{\substack{\text{spectral} \\ \text{filtering}}} \cdot \bar{P}_{\text{cen}} \tag{5.82}$$

which can be rewritten following the logic that led to Eq. (5.79) as

$$\bar{P}_{\text{coh},\lambda/\Delta\lambda} = \frac{e\lambda_u \bar{I}(\eta N\Delta\lambda/\lambda)}{8\pi\,\epsilon_0 d_x d_y \theta_x \theta_y \gamma^2}\left(\frac{\hbar\omega_0}{\hbar\omega} - 1\right)[JJ]^2 \tag{5.83}$$

where $\lambda/\Delta\lambda$ is the relative spectral bandwidth, N is the number of undulator periods, η is the beamline efficiency (insertion loss), $\hbar\omega_0 = 4\pi \, c\hbar\gamma^2/\lambda_u$ is the highest photon energy achievable with the fundamental ($n = 1$) of a given undulator in the limit $K = 0$, and $[JJ]^2$ is the finite-K correction factor given in Eq. (5.80). To emphasize the penalty paid for this further monochromatization we have bracketed the quantity $\eta N \, \Delta\lambda/\lambda$, a numerical factor, often much less than unity, that represents the loss of power incurred through monochromatization.

In the case where the undulator condition is well satisfied $\left(\sigma'_{x,y} \ll \theta_{\mathrm{cen}}\right)$, such that $\theta_x \, \theta_y \cong (1 + K^2/2)/N\gamma^2$, the expression for coherent power takes the form

$$\bar{P}_{\mathrm{coh},\lambda/\Delta\lambda} = \frac{e\lambda_u \bar{I}\eta \, (\Delta\lambda/\lambda \,) \, N^2}{8\pi \, \epsilon_0 d_x d_y} \cdot \left(1 - \frac{\hbar\omega}{\hbar\omega_0}\right) [JJ]^2 \quad \left(\sigma'^2 \ll \theta_{\mathrm{cen}}^2\right) \qquad (5.84)$$

which we note scales as N^2 in this limit. This expression is quite accurate for low emittance[§§] electron beams at modern synchrotron facilities.

An example of a beamline designed for soft x-ray science is shown in Figure 5.31(a). It utilizes the third harmonic ($n = 3$) of the undulator, nominally tunable from photon energies of 200 eV to 1 keV. It employs a grazing incidence varied line-space grating monochromator[40, 41] as is appropriate for use at these wavelengths. The first optical element is a retractable, water cooled, plane gold mirror (M_0) set at an incidence angle of 2°. When inserted, this mirror directs radiation to the two coherent soft x-ray branchlines, as shown. Following this first mirror are a set of curved mirrors, the first of which (M_2) focuses radiation vertically to the horizontal exit slit of the monochromator, then both horizontally ($M_{4a,b}$) and again vertically ($M_{5a,b}$) to the entrance planes of the two end-stations (experimental chambers). Mirrors $M_{4a,b}$ serve double roles as they direct radiation to the respective chambers when inserted, and project an image of the undulator exit plane to the entrance plane of the respective chambers. In combination, mirrors M_2, M_4, and M_5 form a 14:1 reduced image of the source at the Coherent Soft X-ray Optics end-station, in both horizontal and vertical directions, and an 8:1 reduced image at the Coherent Soft X-ray Scattering end-station[42–44]. This combination of grazing incidence mirrors, focusing sequentially in orthogonal planes, is known as a Kirkpatrick–Baez (KB) imaging system, and is discussed further in Chapter 10. Coherent power tuning curves are shown in Figure 5.31(b). These assume use of pinhole spatial filtering for full spatial coherence, and a monochromator set for a relative spectral bandwidth (passband) of $\lambda/\Delta\lambda = 10^3$. An estimated soft x-ray beamline efficiency of $\eta_3 = 0.5\%$ accounts for the loss of photons due to finite reflectivity of the mirrors and finite diffraction efficiency of the grating. As an example, Figure 5.31(b) indicates a predicted coherent power of 5 μW within a $1/1000$ spectral bandwidth when tuned to a photon energy of 500 eV.

Spatial filtering of the soft x-rays is accomplished with a pinhole aperture placed at the entrances of the respective end-stations, centered on the demagnified, highly elliptical image of the largely incoherent undulator source.[51] The pinhole size is determined

---

[§§] The phrase "low emittance" refers to an electron beam of small phase–space product $\epsilon = \sigma\sigma'$.

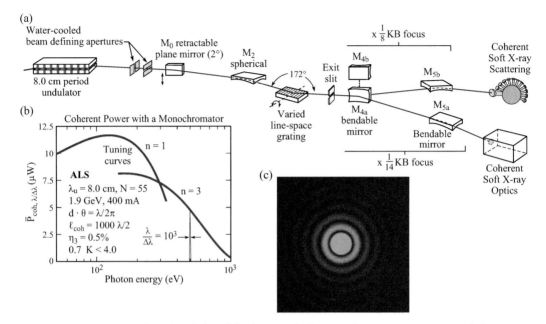

Figure 5.31 A beamline designed for the use of coherent soft x-rays at Lawrence Berkeley National Laboratory's Advanced Light Source (ALS). The upper end-station in (a) is used for time-resolved soft x-ray scattering from magnetic nanostructures,[42–44] while the lower endstation is used to study the performance of spatial filtering and the development of novel soft x-ray optical techniques.[45–52] The branchlines receive third-order harmonic undulator radiation ($n = 3$); tunable from 200 eV to 1 keV[52]. The branchlines incorporate a varied line space grating and horizontal exit slit as the soft x-ray monochromator. Curved total external reflection mirrors project reduced size images of the exit plane of the undulator to the entrance planes of the endstations, at reductions of 8:1 and 14:1, respectively. A 2.5 µm diameter pinhole placed at the entrance plane of the coherent soft x-ray optics chamber transmits radiation with a high degree of spatial coherence. Broadly tunable, spatially and temporally coherent power as a function of photon energy is presented in (b) for a relative spectral bandwidth $\lambda/\Delta\lambda = 1000$ and an assumed beamline efficiency of 0.5%. The pinhole generates high quality Airy patterns such as shown in (c) for 500 eV (2.48 nm, $K = 1.78$) radiation. Courtesy of K. Rosfjord and Y. Liu.[51]

using the van Cittert–Zernike theorem by considering the projected spatial coherence patch at the chamber entrance plane [Eq. (4.16)]. For a wavelength of 2.48 nm (500 eV), the undulator parameters described in Figure 5.31(b), and σ' values for the ALS at that time, the third harmonic central cone angle is 34 µrad, and the total central cone angles are 34 µrad vertically and 41 µrad horizontally. For the Coherent Soft X-ray Optics branchline, with an optical demagnification $M = 1/14$, the predicted beam size was 2.3 µm by 37 µm FWHM based on ALS σ values at the time. The beam parameters were about twice diffraction limited in the vertical and about twenty times diffraction limited in the horizontal at this wavelength. The measured beam size, however, was 9.4 µm by 60 µm FWHM, apparently enlarged by optical aberrations and defocus. The van Cittert–Zernike theorem predicts that for a double-Gaussian elliptical beam size at the source, the coherence cone for $\mu_{coh} = 0.5$ is also elliptical, with rms half angles given by $\theta_{v,h} = 0.187\lambda/\sigma_{v,h}$. This gives rms coherence angles of 34 µrad vertically and

2.2 μrad horizontally, resulting in an elliptical coherence patch at the entrance of the end station. Based on these coherence angles and beam sizes, and assuming ideal demagnification optics, one would expect the elliptical coherence patch for $\mu_{coh} = 0.5$ to have major and minor diameters in the demagnified plane of 2.3 μm (v) and 1.9 μm (h), at a wavelength of 2.48 nm. Thus with ideal optics one might choose a 2 μm diameter pinhole for the spatial filter and this degree of partial coherence. With the significantly larger measured beam size, however, a pinhole diameter of 2.5 μm was selected, which was expected to provide a reasonable compromise between degree of coherence across the pinhole and available photon flux. With the pinhole illuminated with a high degree of spatial coherence, the resultant radiation was expected to be that of an Airy pattern.║║

A recorded pinhole diffraction pattern[51] obtained under these conditions is shown in Figure 5.31(c). This fine example of an Airy pattern, which displays several well-defined rings, was obtained at the relatively short wavelength of 2.48 nm. The clear rings confirm a high degree of spatially coherent illumination across the pinhole. The diffraction pattern was recorded with a soft x-ray sensitive CCD detector*** at a distance of 1.44 m beyond the pinhole, with an exposure time of 200 ms. The measured photon flux within the central Airy cone, was approximately a factor of four below the predictions of Figure 5.31(b). This is likely caused by the enlarged size of the photon beam, perhaps some uncertainty in the beam parameters, and perhaps reduced reflectance of several mirrors due to carbon contamination of the surfaces.

To confirm the spatial coherence properties in the plane of the pinhole, Young's "double-slit" interference experiments were performed[51] with two pinholes of varied separations in both the vertical and horizontal directions [see Chapter 4, Section 4.5]. Measured interference patterns at 500 eV (2.48 nm) are shown in Figure 5.32 for vertical pinhole separations of 2, 6, and 8 μm vertically, and for horizontal separations of 1, 4, and 8 μm. For close separations the pinholes are within areas of high spatial coherence ($\mu \to 1$), resulting in a high degree of modulation. For larger separations the degree of coherence is reduced and the modulation of the interference patterns is decreased accordingly. Graphs of the resultant normalized degree of coherence, μ_{coh}, indicate transverse coherence lengths of 5.4 μm in the vertical direction (d) and 4.5 μm in the horizontal direction (h), each about twice what was anticipated based on idealized optics, but consistent with the enlarged size of the beam. As expected,††† measured transverse coherence lengths were somewhat shorter at the decreased wavelengths associated with 600, 700, and 800 eV photon energies.

Coherence experiments at hard x-ray wavelengths offer a wide range of new scientific opportunities.[53–58] The hard x-ray Coherent Optics Beamline (34-ID-C) for nanoscale imaging and scattering[59–64] at Argonne National Laboratory's Advanced Photon Source is illustrated in Figure 5.33(a). The facility operates at 7 GeV electron beam energy

---

║║ The Airy pattern, which results from diffraction of spatially coherent radiation by a circular aperture, is proportional to $[2J_1(\nu)/\nu]^2$ and is shown in Figure 4.20.

*** A $1'' \times 1''$, 1024×1024 pixel array, back-thinned charge-couple device (CCD) having single photon detection sensitivity at these wavelengths.

††† In the limit of full spatial coherence, $d \cdot \theta = \lambda/2\pi$ for Gaussians, the coherence requirements on $d \cdot \theta$ are more demanding for shorter wavelengths.

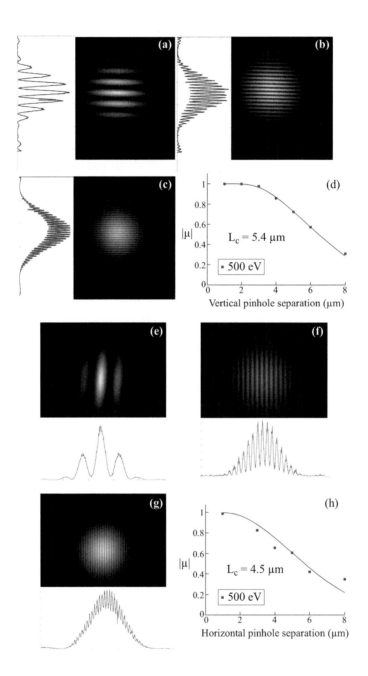

Figure 5.32 Measured two-pinhole interference patterns[51] at a photon energy of 500 eV (2.48 nm) and their respective line outs for vertical separations of (a) 2 μm, (b) 6 μm, and (c) 8 μm. In (d) the resultant normalized degree of spatial coherence versus pinhole separation, indicates a transverse coherence length of 5.4 μm in the vertical direction. Results for two-pinhole horizontal separations are shown in (e) 1 μm, (f) 4 μm, and (g) 8 μm. In (h) the normalized degree of spatial coherence versus pinhole separation indicates a transverse coherence length of 4.5 μm in the horizontal direction. All pinholes for this study were of 450 nm diameter. Courtesy of K. Rosfjord, EECS, and Y.W. Liu, AS&T, UC Berkeley.

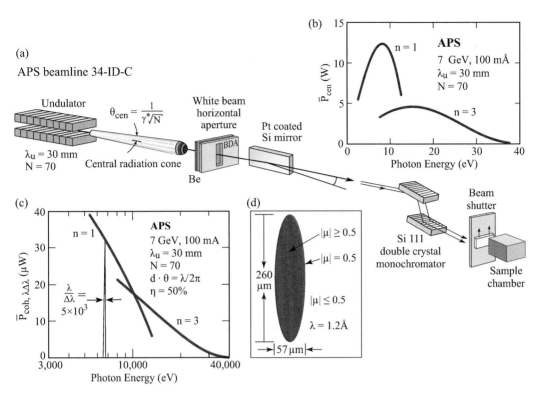

Figure 5.33 The hard x-ray Coherent Optics Beamline (34-ID-C) for nanoscale imaging and scattering[59-61] at Argonne National Laboratory's Advanced Photon Source is illustrated in (a). The electron beam energy is 7 GeV ($\gamma = 13\,700$) and the average current is 100 mA. The undulator contains 70 magnet periods of period $\lambda_u = 30$ mm. A platinum coated silicon mirror deflects light into the beamline and a Si 111 double-crystal monochromator efficiently provides spectral filtering to a resolution as high as 10^4, within an angular acceptance of 26 µrad for 10 keV photon energy. The combination of an upstream beryllium filter and a water-cooled horizontal slit (white beam aperture) remove unwanted power at longer wavelengths. The 10.8 µrad central radiation cone (θ_{cen}) is spread by electron beam divergences (σ') to θ_T values of 16 and 11 µrad in the horizontal and vertical directions. Further beam parameters are given in Table 5.1. Radiated power in the central cone is shown as a function of photon energy in (b) for a natural bandwidth of $1/70$. Values as high as 15 W are predicted for a photon energy of 7.8 keV (1.6 Å). Broadly tunable, spatially and temporally coherent power is shown in (c) for a typical operating spectral bandwidth of $\lambda/\Delta\lambda = 5000$. As an example, a coherent power of 14 µW, or 8.5×10^9 ph/s, is predicted for a photon energy of 10.3 keV (1.2 Å wavelength), assuming a beamline efficiency of 50%. The elliptical region of coherence, set in the vertical direction by the small beam size and in the horizontal direction by the vertical slit width, is shown in (d). The vertical slit aperture is located 27.3 m from the exit plane of the undulator and the sample chamber is located 50.6 m from the undulator exit plane. Beamline diagram (a) courtesy of R. Harder, APS, Argonne National Laboratory.

($\gamma = 13\ 700$), in this case with a 30 mm period undulator of 70 periods, radiating up to 15 W of average power in the central radiation cone, with a tunable photon energy range extending from 7 keV to 14 keV in first order, and to 25 keV in third order, as seen in (b). A platinum coated plane mirror (a) is used to deflect radiation into the branchline. A silicon double crystal monochromator (Si 111) is used to set the spectral bandpass, typically to $\lambda/\Delta\lambda = 5000$, with a Bragg angle of 11.4° and a Darwin width of 26.1 μrad or a photon energy of 10.3 keV (1.20 Å). A water-cooled, 50 μm wide, horizontal slit (white beam aperture) and a beryllium filter are used to remove unwanted power at longer wavelengths. The central radiation cone has a half-angle of 10.7 μrad, which is increased by electron beam divergences (σ') of 3.4 and 8.6 μrad, to total angles $\theta_{Th} = 16$ μrad and $\theta_{Tv} = 11$ μrad, in the vertical and horizontal directions, respectively (see beam parameters in Table 5.1). Note that the approximation $\sigma'/\theta_{cen} \ll 1$ does not hold in the horizontal direction, thus the radiation cone is not particularly sharp in that direction. Calculated spatially coherent power is shown in (c) for a spectral bandwidth of 1/5000. At 10.3 keV, for example, the spatially and temporally coherent power is predicted to be 14 μW, or 8.5×10^9 ph/s. Because a slit is used for spatial filtering, rather than a pinhole, the spatial coherence properties are not axisymmetric. Consequently, to obtain a reasonable approximation to the degree of spatial coherence in the sample plane, one must make separate calculations in the vertical and horizontal directions. Spatial coherence in the vertical direction is set by the small beam size in that direction, and the wavelength, while in the horizontal direction spatial coherence is largely set by the horizontal beam aperture (slit) and the wavelength. For a wavelength of 1.2 Å, a one-dimensional version of the van Cittert–Zernike theorem for a Gaussian intensity distribution [Eq. (4.25)] predicts a normalized degree of spatial coherence of $\mu_{coh} \geq 0.5$ within a 240 μm wide vertical region at the entrance to the experimental chamber. In the horizontal direction, where the undulator beam is much wider and spatial coherence is set by incoherent illumination of the 50 μm wide slit, a one-dimensional version of the van Cittert–Zernike theorem predicts $\mu_{coh} \geq 0.5$ within a 57 μm wide horizontal region. A sketch of the anticipated spatial coherence ellipse is shown in Figure 5.33(d). The dimensions of the elliptical coherence patch are proportional to the wavelength and thus are larger for longer wavelengths. The horizontal dimension of the coherence ellipse is inversely proportional to the slit width and thus can be increased by decreasing the slit width.[62–64] Samples smaller than the $|\mu| = 0.5$ ellipse experience a higher degree of spatial coherence. For some experiments a KB mirror pair (Chapter 10) is used to concentrate the coherent photon flux to a smaller region in the sample plane, and to isolate individual objects within a sample containing many such objects.

Coherent power at several synchrotron radiation facilities, including the APS (see M. Borlund et al.[19]), is expected to be increased by planned storage ring upgrades which will result in a reduced horizontal emittance, including reductions of both σ_h and σ'_h. The upgrades will use a gentler magnetic turning of the electron beam between straight sections in the horizontal plane.[18, 19] With weaker bending magnetic fields there will be less horizontal momentum kick to the radiating electrons, and thus reduced transverse electron beam size and divergence.

Explanatory Box Describing Estimates of Spatial Coherence for an Asymmetric Hard X-Ray Beamline

Vertical direction set by small σ_v:

One-dimensional version of Eq. (4.16) for a Gaussian yields

$$|\mu_{OP}| = e^{-(k\sigma_v\theta_y)^2/2}$$

$|\mu_{OP}| = 0.5$ for $(k\sigma_v\theta_y)^2/2 = 0.693$

$\theta_y = 0.187\lambda/\sigma_v$

For $\sigma_v = 8.6$ μm, $\lambda = 1.20$ Å,

$z = 50.6$ m

$\theta_y = 2.6$ μrad

$2\Delta y = 2(50.6$ m$)(2.6$ μrad$)$

$= 260$ μm

Thus a spatial coherence $|\mu_{OP}| \geq 0.5$ exists for vertical separations within 240 μm.

Horizontal direction set by narrow slit d:

One-dimensional version of Eq. (4.16) for a unit step function of width d yields

$$|\mu_{OP}| = \mathrm{sinc}\left(\frac{k\theta_x d}{2}\right)$$

where $|\mu_{OP}| = 1$ for $\theta_x = 0$, and where

$|\mu_{OP}| = 0$ for $\frac{k\theta_x d}{2} = \pi$, or $\theta_{x,\mathrm{null}} = \lambda/d$.

For $|\mu_{OP}| = 0.500$, $\theta_x = 0.605\,\theta_{x,\mathrm{null}}$.

For $|\mu_{OP}| = 0.5$, $d = 50$ μm, $\lambda = 1.20$ Å and $z = 19.5$ m, $2\Delta x = 2(19.5$ m$)(1.45$ μrad$) = 56$ μm.

Thus a spatial coherence $|\mu_{OP}| \geq 0.5$ exists for horizontal separations within 57 μm.

5.7   The Transition from Undulator to Wiggler Radiation

We have seen in the preceding sections that as K increases toward unity the radiated power in higher harmonics grows rapidly. Indeed, for $K \gg 1$ analysis shows the emergence of a large number of ever stronger harmonics, extending to ever higher photon energies. Figure 5.34 shows the development of this comb-like harmonic structure for increasing values of the magnetic deflection parameter K. The curves shown are for the limiting case in which all the electrons are contained within a vanishingly small phase space volume, $\epsilon = \pi\sigma\sigma' \to 0$, with the radiation observed in a very small acceptance cone of half angle $\Delta\theta \to 0$. Note that in these limits the even harmonics, which are zero on axis, do not appear. Although the harmonic spikes seen in Figure 5.34 would be broadened for finite emittance, a more important factor in the case of high K wiggler radiation is the observation cone angle, $\Delta\theta$. With the radiated energy spread into so many harmonics, and into a fan of angular width K/γ, it is natural to utilize the available photons by accepting a larger radiation cone. In this case the relative spectral bandwidth of each harmonic expands in frequency from values of order $1/N$ to values dominated by the $\gamma^2\theta^2$ term in the undulator equation (5.28). The harmonic spikes then merge into a quasi-continuum, much like that seen earlier for bending magnet radiation, with interesting consequences.

The broadening of harmonics and their merger into a smooth continuum can be understood better if we consider the various mechanisms involved and estimate the effect of each. We again follow the path of considering a small phase-space electron beam traversing a periodic magnet structure of N periods, and initially consider a near-axial observation of radiation within a relatively narrow acceptance cone. In this case the even

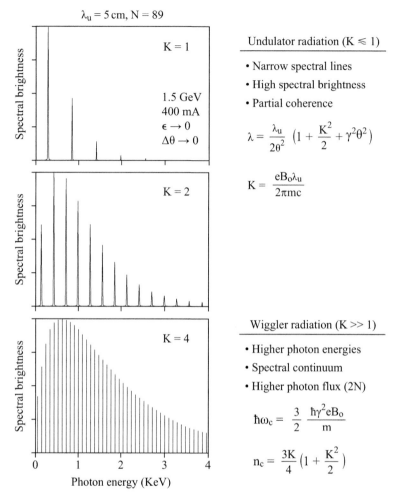

Figure 5.34 The transition from undulator radiation ($K \le 1$) to wiggler radiation ($K \gg 1$) is illustrated. The emergence of bright and more numerous harmonics is evident as the magnetic strength (K) is increased. Spectral brightness is indicated on axis in the limits of very small electron beam phase space (ϵ) and very small angular acceptance ($\Delta\theta$). The comb-like spectrum eventually merges to a continuum for finite emittance and acceptance angle in the wiggler limit, $K \gg 1$ (K.J. Kim, unpublished).

harmonics are, at least initially, not observed. Through a little algebraic manipulation of the undulator equation for harmonic n [see Eq. (5.30)] one can show that the separation between harmonics on axis (only odd harmonics appear) is given by

$$\Delta f_{\Delta n=2} = 2 \left(\frac{2c\gamma^{*2}}{\lambda_u} \right) = 2f_1 \qquad (5.85)$$

where f_1 is the frequency of the fundamental and the subscript $\Delta n = 2$ reminds us that the separation is between harmonics n and $n + 2$. Since the relative bandwidth

[Eq. (5.31)] for the nth harmonic of an N-period undulator in terms of frequency is

$$\frac{\Delta f_{n,N}}{f_n} = \frac{1}{nN}$$

the absolute bandwidth can be written as

$$\Delta f_{n,N} = \frac{f_n}{nN} = \frac{1}{N} \cdot f_1 \tag{5.86}$$

The bandwidth of radiation collected within an angular aperture 2θ (radiation cone of half angle θ) is, for the nth harmonic [an extension of Eq. (5.13) for finite K]

$$\Delta f_{n,\gamma*\theta} = \gamma^{*2}\theta^2 f_n = n\gamma^{*2}\theta^2 f_1 \tag{5.87}$$

Finally, we note that an individual electron traveling somewhat off axis at an angle α will experience a somewhat longer period, and further, the observed radiation will be affected by the equivalent of an off-axis Doppler shift. The net result is a shifted frequency [see Eq. (5.54)]

$$\Delta f_{n,\alpha} = n\gamma^{*2}\alpha^2 f_1 \tag{5.88a}$$

Note that for an electron storage ring there are many electrons in each bunch with random angular excursions described statistically by σ'. Thus the frequency spread due to random Doppler shifts [Eq. (5.88a)] becomes

$$\Delta f_{n,\alpha} = n\gamma^{*2}\sigma'^2 f_1 \tag{5.88b}$$

With substantial off-axis motion even harmonics are also observed on axis so that the harmonic separation, Eq. (5.85), is replaced by $\Delta f_{\Delta n=1} = f_1$. Combining this with Eq. (5.88b) one obtains the conditions for harmonic merging due to finite electron beam divergence

$$n_m\gamma^{*2}\sigma'^2 \geq 1 \tag{5.89}$$

Based on these estimates, Figure 5.35 sketches the expected frequency spectrum. The sketch is for a modest value of K where the various effects can still be distinguished. Note that it is presented in frequency space; conversion to photon energy requires multiplication by Planck's constant (h). The solid lines, corresponding to a small acceptance angle $\Delta\theta$, show a series of harmonics at frequencies n times the fundamental $(2c\gamma^{*2}/\lambda_u)$, with only the odd harmonics observed on axis. According to Eq. (5.85) they are separated by an interval $2f_1$, and according to Eq. (5.86) they each have a width $1/N$ times the fundamental.

A very important consideration is that of finite angular acceptance at moderate to high K operation. To understand this merging to a continuum for finite acceptance angle θ, we note that even harmonics will be observed off axis, and thus again replace Eq. (5.85) by $\Delta f_{\Delta n=1} = f_1$. Combining this with Eq. (5.76) we obtain the condition for harmonic merging due to a finite acceptance angle

$$n_m\gamma^{*2}\theta_m^2 = 1 \tag{5.90a}$$

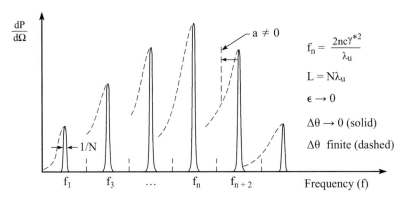

Figure 5.35 The effect of finite acceptance angle ($\Delta\theta$) is shown by dashed lines superposed on the spectrum of harmonic radiation for zero acceptance angle (solid lines). A small electron beam phase space ($\epsilon = 0$) is assumed in both cases. The solid lines also shift to lower frequencies (lower photon energies hf) for electron trajectories at angle α to the z-axis. For sufficiently large acceptance angle and magnetic field strength the harmonics merge into a smooth continuum. However, for finite K and $\Delta\theta$ the spectrum may continue to display many sharp peaks and valleys, particularly in the lower photon energy region.

Thus the condition for harmonic merging within a finite acceptance angle θ is

$$n \geq \frac{1}{\gamma^{*2}\theta^2} \qquad (5.90b)$$

Equation (5.90) suggests an evolution to a quasi-continuum spectrum, particularly for the higher harmonic region of the spectrum where much of the radiated energy resides.

To better understand the evolution from a spectrum characterized by sharply peaked harmonics to a relatively smooth continuum similar to that from a bending magnet, it is useful to consider further the relationship between harmonic number (n) and K. In an earlier section on bending magnet radiation we discussed the critical photon energy [Eq. (5.7a)]:

$$E_c = \hbar\omega_c = \frac{3e\hbar B\gamma^2}{2m}$$

for which half of the radiated energy appears at higher photon energy, and half at lower photon energy. Here on the other hand we have been discussing the generation of a comb-like structure of very high harmonics from a wiggler. A useful notion is to combine the two and introduce the concept of a critical harmonic number for which half the radiated energy is in higher harmonics, and half is in lower harmonics. A few simple substitutions provide the desired result. Rewriting the critical energy in terms of the non-dimensional undulator parameter, also referred to as the magnetic deflection parameter [Eq. (5.18a)]

$$K = \frac{eB_0\lambda_u}{2\pi mc}$$

the critical photon energy becomes

$$\hbar\omega_c = \frac{3\pi c\hbar\gamma^2 K}{\lambda_u} \tag{5.91}$$

For a periodic structure the harmonic frequencies, from the harmonic undulator equation (5.30), are

$$\omega_n = \frac{2\pi c}{\lambda_n} = \frac{4\pi nc\gamma^2}{\lambda_u(1 + K^2/2)} \tag{5.92}$$

Equating Eqs. (5.91) and (5.92), we can define a *critical harmonic number, n_c*, viz.,

$$\omega_c \equiv \omega_{n,c} = n_c\omega_1$$

thus

$$\frac{3\pi c\gamma^2 K}{\lambda_u} = \frac{4\pi c\gamma^2}{\lambda_u(1 + K^2/2)} \cdot n_c$$

or

$$\boxed{n_c = \frac{3K}{4}\left(1 + \frac{K^2}{2}\right)} \tag{5.93}$$

which for large K takes the form

$$n_c \simeq \frac{3K^3}{8}, \qquad K \gg 1 \tag{5.94}$$

The 29-period wiggler at the 1.9 GeV ALS, as indicated in Table 5.1, has a peak magnetic flux density on axis of $B_0 = 1.9$ T, a period of 11.4 cm, a critical photon energy of 4.6 keV (0.27 nm wavelength), and a deflection parameter at peak field of $K = 20$, with a corresponding critical harmonic number n_c in excess of 3000.

Under the circumstances it is interesting to reconsider the condition for harmonic merging given by Eq. (5.90a). For very large K the condition for spectral merging of the harmonics becomes

$$\theta_m = \frac{\sqrt{1 + K^2/2}}{\gamma\sqrt{n/2}} \simeq \frac{K}{\gamma\sqrt{2n}} \tag{5.95a}$$

or for n_c

$$\theta_{m,c} \simeq \frac{1.2}{\gamma\sqrt{K}} \tag{5.95b}$$

For the example cited, with $K = 32$, this corresponds to a collection angle of $\theta_{m,c} \simeq 0.2/\gamma$, or only 50 μrad for $\gamma = 3914$. Indeed, with a horizontal radiation fan of $\pm K/\gamma$, one might use a significantly larger acceptance angle to collect the radiation, thus ensuring spectral merging for very low harmonics, well below n_c.

The picture that emerges for very large K operation, where the spectral energy density shifts to very high harmonics, is one of spectrally isolated lower harmonics, a merger into a quasi-continuum well below the critical harmonic, and finally a relatively smooth

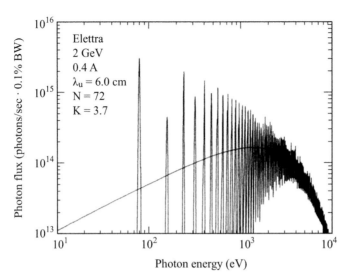

Figure 5.36 Calculated photon flux near the central axis for a 6 cm periodic magnet structure operating at an intermediate K-value of 3.7. The calculation is performed for the third generation 2 GeV synchrotron radiation facility Elettra in Trieste, Italy.[13, 25] The critical harmonic number, n_c is 22 for this K-value. The acceptance half angle is 0.1 mrad. Harmonic merging occurs for this acceptance angle at $n_m \simeq 51$, courtesy of R.P. Walker, then at Elettra, now at Diamond.

continuum for $n > n_c$. The high photon energy portion of the spectrum is similar to that of a bending magnet, but intensified by a factor $2N$ (two peak field locations per period), and shifted to a higher critical photon energy because the peak wiggler field is much greater than that of a bending magnet.[‡‡‡] An example of a detailed calculation is presented in Figure 5.36 for a 6 cm period magnet structure, operating at $K = 3.7$, in the 2 GeV storage ring Elettra in Trieste, Italy.[13] In this example $n_c = 22$, and with an acceptance angle of 0.1 mrad, harmonic merging occurs at $n \simeq 51$. This is an interesting illustration of the evolution from undulator to wiggler radiation. For significantly higher K and larger acceptance angle, harmonic merging occurs for $n \ll n_c$. This corresponds to the wiggler limit, which is discussed in the following section.

5.8   Wiggler Power and Flux

We have approached the subject of radiation from relativistic electrons traversing periodic magnet structures from a small K theory of undulator radiation. While this has clear advantages, the approximations made are not valid for large K wiggler radiation. The advantages of small K theory are that one is able to borrow significant results from well-known classical dipole radiation, transform the results to the laboratory frame of reference, and obtain simple analytic expressions and a clear physical model for the major

---

[‡‡‡] In a circular orbit the beam energy and magnetic field are matched so as to maintain a path of constant radius and thus avoid hitting the vacuum wall. In a periodic structure the magnetic fields can be considerably larger, as alternate poles redirect the beam, keeping it near the vacuum chamber axis at all times.

features of undulator radiation, including observable wavelengths, bandwidth, polarization, power scaling, and the emergence of harmonics. For wiggler radiation with $K \gg 1$ the motion becomes significantly more complex – relativistic even in the electron (γ^*) frame of reference – and the acceleration becomes very strong at the extremes of its off-axis excursions. Indeed, the accelerations become so strong at the extrema that the radiation appears not to come from a near-point-source oscillator on axis, but rather from an alternating sequence of two points at the extremes of motion in a highly non-sinusoidal trajectory. Fourier analysis of this radiation would clearly involve very sharp temporal gradients (time derivatives), and thus lead to very high harmonics of the basic periodic motion. This causes a shift to higher photon energies, and thus shorter wavelengths. Because the accelerations are greater (the time derivatives are sharper), there is also a substantial increase in radiated power. The angular width, however, is increased in the horizontal plane to a value K/γ, and the apparent source size is increased, particularly if observed from a position somewhat off axis, thus significantly reducing overall brightness.

A discussion of the relevant physics, including analytic and numerical solutions for arbitrary K, is given by Kim.[1] For example, the total radiated average power in *all* harmonics, integrated over all angles and wavelengths, for arbitrary K, is given by

$$\bar{P}_T = \frac{\pi e K^2 \gamma^2 \bar{I} N}{3\epsilon_0 \lambda_u} \qquad (5.96a)$$

Note that this is similar to our earlier result for $P_{T,1}$, the total power radiated to all angles in the fundamental ($n = 1$ only), Eq. (5.50a), except that the factor $(1 + K^2/2)^2$ is absent from Eq. (5.96a). This factor accounts for the power radiated to *all harmonics*. In practical units the total power radiated to all harmonics, for arbitrary K, is given by[1]

$$\bar{P}_T[W] = \frac{(1.90 \times 10^{-6}) K^2 \gamma^2 N \bar{I}[A]}{\lambda_u[\text{cm}]} \qquad (5.96b)$$

Typical values for radiated power from high K wigglers are given in Table 5.1. Both radiate kW-level x-ray power, with critical photon energies of 4.6 keV and 33 keV for the soft x-ray and hard x-ray facilities, respectively. Expressions for the photon flux from a wiggler, in the limits $K \gg 1$ and $n \gg 1$, are similar to those for bending magnet radiation, but increased by a factor of $2N$ due to the strong acceleration that occurs twice in each period at the peaks of magnetic field. For instance the on-axis photon flux per unit solid angle, per unit relative spectral bandwidth, is given by Kim[1]

$$\left.\frac{d^2\bar{F}}{d\Omega d\omega/\omega}\right|_0 = 2.65 \times 10^{13} N E_e^2[\text{GeV}]\bar{I}[A]H_2(E/E_c)\frac{\text{photons/s}}{\text{mrad}^2(0.1\%\,\text{BW})} \qquad (5.97)$$

where $H_2(y)$ is illustrated in Figure 5.7 and partially tabulated in Table 5.2. Relative spectra for both a bending magnet and a wiggler are presented in Figure 5.37. Note the $2N$ times larger photon flux for the wiggler, and the shift to higher photon energies ($\hbar\omega_c$). Expressions for off-axis photon flux in the two orthogonal polarizations are given by Kim.[1]

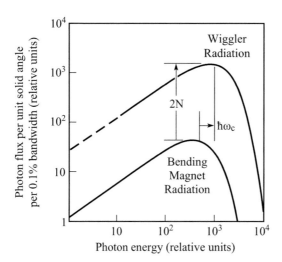

Figure 5.37 Comparison of on-axis photon flux per unit solid angle and per unit relative spectral bandwidth for bending magnet radiation and wiggler magnet radiation from the same storage ring. The shape of the curves is basically the same, $H_2(\omega/\omega_c)$. The photon flux from the wiggler is $2N$ higher in the case illustrated, and shifted $2\times$ to higher photon energy by the critical energy $\hbar\omega_c$. The wiggler curve below the critical photon energy (shown dashed here) has a complex harmonic content that may not be smooth, depending on the acceptance angle and electron beam divergence (see text).

To calculate the photon flux or power radiated by a wiggler within a given spectral bandwidth we can begin with Eq. (5.8) for bending magnet radiation with the dependence on vertical angle (ψ) already integrated out. To adapt this to wiggler radiation we multiply by $2N$ to obtained the photon flux per unit horizontal angle θ, and per unit relative spectral bandwidth

$$\frac{d^2\bar{F}}{d\theta d\omega/\omega} = 4.92 \times 10^{13} N E_e[\text{GeV}]\bar{I}[\text{A}]G_1(E/E_c)\frac{\text{photons/s}}{\text{mrad} \cdot (0.1\%\,\text{BW})} \quad (5.98)$$

where the various parameters are described below Eq. (5.8), and where the function $G_1(E/E_c)$ is tabulated in Table 5.2 and shown graphically in Figure 5.8. As we have seen, wiggler radiation in the horizontal plane is dominated by the electron trajectory [Eq. (5.20)] with angular deflection limits $\pm K/\gamma$. This results in an angular photon flux dependence[1] $d\bar{F}/d\theta = \bar{F}_0\sqrt{1 - (\gamma\theta/K)^2}$, which has a characteristic angular extent $(2\theta)_{\text{FWHM}} = \sqrt{3}K/\gamma$. With this θ-dependence an angular integration of Eq. (5.98) leads to a multiplicative factor of $1.57K/\gamma$, giving a wiggler radiated photon flux per unit relative spectral bandwidth of [1]

$$\bar{F} = 3.94 \times 10^{13} N K \bar{I}[\text{A}]G_1(E/E_c)\frac{\text{photons/s}}{(0.1\%\,\text{BW})} \quad (5.99\text{a})$$

where we have used Eq. (5.5) to replace γ by E_e, and where square brackets read "in units of." This equation can be written alternatively in terms of the magnetic flux density B_0 and the wiggler length $L = N\lambda_w$, using Eq. (5.18b) to express K in terms of $B_0[\text{T}]$

and λ_ω [cm], as[1]

$$\bar{F} = 3.68 \times 10^{15} L[\mathrm{m}] B_0[\mathrm{T}] \bar{I}[\mathrm{A}] G_1(E/E_c) \frac{\mathrm{photons/s}}{(0.1\% \, \mathrm{BW})} \qquad (5.99b)$$

where $E = \hbar\omega$ is the photon energy, and $E_c[\mathrm{keV}] = 0.665 E_e^2[\mathrm{GeV}] B_0[\mathrm{T}]$. Inspection of Eq. (5.99b) indicates that there is a linear dependence on wiggler length, peak magnetic flux density, and current as expected. To first order there is no dependence on electron beam energy, only a very slow dependence through E_c in $G_1(E/E_c)$.

References

1. K.-J. Kim, "Characteristics of Synchrotron Radiation," pp. 565–632 in *Physics of Particle Accelerators*, AIP Conference Proceedings 184 (Amer. Inst. of Physics, New York, 1989), M. Month and M. Dienes, Editors; also see K.-J. Kim, "Characteristics of Synchrotron Radiation," *X-Ray Data Booklet* (Center for X-Ray Optics, Lawrence Berkeley National Laboratory, 1999), Second Edition; K.-J. Kim, "Optical and Power Characteristics of Synchrotron Radiation Sources," *Opt. Engr.* **34**, 342 (1995).
2. A. Hofmann, *The Physics of Synchrotron Radiation* (Cambridge University Press, 2004); also "Quasi-Monochromatic Radiation from Undulators," *Nucl. Instrum. and Meth.* **152**, 17 (1978); and "Theory of Synchrotron Radiation," Stanford Synchrotron Radiation Laboratory report ACD-Note 38.
3. P.A. Duke, *Synchrotron Radiation* (Oxford University Press, 2000).
4. P. Schmüser, M. Dohlus. J. Rossbach and C. Behrens, *Free-Electron Lasers in the Ultraviolet and X-ray Regime* (Springer, Heidelberg, 2014).
5. K.-J. Kim, Z. Huang and R. Lindberg, *Synchrotron Radiation and Free Electron Lasers: Principles of Coherent X-Ray Generation* (Cambridge University Press, 2017), to be published.
6. Beam parameters for modern synchrotrons worldwide can be found through the website http: tandfonline.com; As an example of a modern, low emittance storage ring, see S.-L. Chang and C.-T. Chen, "From Taiwan Light Source to Taiwan Photon Source", *Synchrotron Radiation News* **26** (4), 32 (2013).
7. P.A. Tipler and R.A. Llewelyn, *Modern Physics* (Freeman, New York, 2012), Sixth Edition.
8. J.D. Jackson, *Classical Electrodynamics* (Wiley, New York, 1999), Third Edition, Chapter 14; also see First Edition (1962), Section 14.4.
9. G.K. Green, Brookhaven National Laboratory, Upton, NY, Report 50522 (also in Report 50595, Vol. II), 1977.
10. K. Halbach, "Physical and Optical Properties of Rare Earth Cobalt Magnets," *Nucl. Instrum. Meth.* **187**, 109 (1981); also "Permanent Magnet Undulators," *J. Phys. (Paris)*, **44**, Colloq.C1, Suppl. 2, 211 (1983); and N.A. Vinokurov and E.B. Levichev, "Undulators and Wigglers for Production of Radiation and Other Applications," *Uspekhi Fizicheskikh Nauk* **185** (90), 917–939 (September 2015).
11. R.B. Leighton, *Principles of Modern Physics* (McGraw-Hill, New York, 1959), Chapter 12.
12. B.M. Kincaid, "A Short-Period Helical Wiggler as an Improved Source of Synchrotron Radiation," *J. Appl. Phys.* **48**, 2684 (1977); also see M.R. Howells and B.M. Kincaid, "The Properties of Undulator Radiation," pp. 315–359 in *New Directions in Research with Third Generation Soft X-Ray Synchrotron Radiation Sources* (Kluwer, Dordrecht, 1994), A.S. Schlachter and F.J. Wuilleumier, Editors.

13. R.P. Walker and B. Diviacco, "A Computer Program for Calculating Undulator Radiation and Spectral, Angular, Polarization, and Power Density Properties," *Rev. Sci. Instrum.* **63**, 392 (1992); also see R.P. Walker, "Interference Effects in Undulator and Wiggler Radiation Sources," *Nucl. Instr. Methods A* **335**, 328 (1993).

14. E. Hecht, *Optics* (Addison-Wesley, San Francisco, 2002), Fourth Edition, pp. 50, 313, and 653.

15. M. Cornacchia, "Lattices," Chapter 2, pp. 30–58 in *Synchrotron Radiation Sources: A Primer* (World Science, Singapore, 1994), H. Winick, Editor.

16. H. Kitamura, "Future of Synchrotron Radiation," *Kasokuki Kagaku (Accelerator Sci.) 1*, 45 (Ionics Publishing, Tokyo, 1986).

17. M. Born and E. Wolf, *Principles of Optics* (Cambridge Univ. Press, New York, 1999), Sections 4.8.1 and 4.8.3, pp. 194 and 201, which deal with photometry and brightness.

18. M. Eriksson, J. Friso van der Veen and C. Quitman, "Diffraction-Limited Storage Rings – a Window to the Science of Tomorrow," *J. Synchr. Rad.* **21**, 837 (September 2014).

19. P.F. Tavares *et al.*, "The MAX IV Storage Ring Project", ibid, 862; M. Borlund *et al.*, "Lattice Design Challenges for Fourth-Generation Storage-Ring Light Sources", ibid, 912; J. Susini *et al.*, "New Challenges in Beamline Instrumentation for the ESRF Upgrade Programme Phase II", ibid, 986; R. Hettel, "DLSR Design and Plans: an International Overview", ibid, 843; C. Steier *et al.*, "Towards a Diffraction Limited Upgrade of the ALS," *Proc. IPAC 2016*, Busan (May 2016).

20. R.D. Schlueter, "Wiggler and Undulator Insertion Devices," Chapter 14, p. 377 in *Synchrotron Radiation Sources: A Primer* (World Science, Singapore, 1994), H. Winick, Editor.

21. K.-J. Kim, "A Synchrotron Radiation Source with Arbitrarily Adjustable Elliptical Polarization," *Nucl. Instrum. and Meth. A* **219**, 425 (1984).

22. J. Bahrdt, A. Gaupp, W. Gudat *et al.*, "Circularly Polarized Synchrotron Radiation from the Crossed Undulator at BESSY," *Rev. Sci. Instrum.* **63**, 339 (1992); R. David, P. Stoppmanns, S.-W. Yu *et al.*, "Circularly Polarized Undulator Radiation from the New Double Crossed Undulator Beamline at BESSY and its First Use for Spin Resolved Auger Electron Emission Spectroscopy," *Nucl. Instrum. and Meth. A* **343**, 650 (1994).

23. H. Onuki, "Elliptically Polarized Synchrotron Radiation Source with Crossed and Retarded Magnetic Fields," *Nucl. Instrum. Meth. A* **246**, 94 (1986); H. Onuki, N. Saito and T. Saito, "Undulator Generating any Kind of Elliptically Polarized Radiation," *Appl. Phys. Lett.* **52**, 173 (1988).

24. B. Di Viacco and R.P. Walker, "Fields and Trajectories in Some New Types of Permanent Magnet Helical Undulators," *Nucl. Instrum. Meth. A* **292**, 517 (1990).

25. S. Sasaki, "Analysis for a Planar Variably-Polarizing Undulator," *Nucl. Instrum. and Meth. A* **347**, 83 (1994); S. Sasaki, K. Kakunori, T. Takada *et al.*, "Design of a New Type of Planar Undulator for Generating Variably Polarized Radiation," *Nucl. Instrum. Meth. A* **331**, 763 (1993).

26. H. Kitamura, "Present Status of SPring-8 Insertion Devices," *J. Synchr. Rad.* **5**, 184 (1998).

27. A. Agui *et al.*, "First Operation of Circular Dichroism Measurements with Periodic Photon-Helicity Switching by a Variably Polarizing Undulator at BL23SU at SPring-8," *Rev. Sci. Instrum.* **72**, 3191 (August 2001); T. Nakatani, A. Agui, H. Aoyagi *et al.*, "Scheme for Precise Correction of Orbit Variation Caused by Dipole Error Field of Insertion Device," *Rev. Sci. Instrum.* **76**, 055105 (2005).

28. *Undulators, Wigglers and their Applications* (Taylor and Francis, London, 2003), edited by Hideo Onuki and Pascal Elleaume.

29. J.A. Clarke, *The Science and Technology of Undulators and Wigglers* (Oxford, 2004).

30. *Synchrotron Light Sources and Free Electron Lasers* (Springer, Heidelberg, 2015), edited by E. Jaescke, S. Khan, J.R. Schneider and J.B. Hastings, ISBN 978-3-319-04507-8.

31. E. Gluskin, "APS Insertion Devices: Recent Developments and Results," *J. Synchr. Rad.* **5**, 189 (1998).

32. G.N. Kulipanov, A.N. Skrinsky and N.A. Vinokurov, "Synchrotron Light Sources and Recent Developments of Accelerator Technology," *J. Synchr. Rad.* **5**, 176 (1998).

33. D.T. Attwood, K. Halbach and K.-J. Kim, "Tunable Coherent Radiation," *Science* **228**, 1265 (June 14, 1985).

34. A.M. Kondratenko and A.N. Skrinsky, "Use of Radiation of Electron Storage Rings in X-Ray Holography of Objects," *Opt. Spectrosc. (USSR)* **42**, 189 (February 1977); p. 338 in the original Russian journal.

35. D. Attwood, K.-J. Kim, N. Wang, and N. Iskander, "Partially Coherent Radiation at X-Ray Wavelengths," *J. de Phys. (Paris)* **47**, C6–203, Suppl. 10 (October 1986).

36. R. Coïsson, "Spatial Coherence of Synchrotron Radiation," *Appl. Opt.* **34**, 904 (1995); K. Fezzaa, F. Comin, S. Marchesini, R. Cöisson and M. Belakhovsky, "X-Ray Interferometry at ESRF Using Two Coherent Beams from Fresnel Mirrors," *J. X-Ray Sci. Technol.* **7**, 12 (1997).

37. D.L. Abernathy, G. Grübel, S. Bauer *et al.*, "Small-Angle X-Ray Scattering Using Coherent Undulator Radiation at ESRF," *J. Synchr. Rad.* **5**, 37 (1998).

38. Y. Takayama, R.Z. Tai, T. Hatano *et al.*, "Measurement of the Coherence of Synchrotron Radiation," *J. Synchr. Rad.* **5**, 456 (1998); Y. Takayama *et al.*, "Relationship between Spatial Coherence of Synchrotron Radiation and Emittance," *J. Synchr. Rad.* **5**, 1187 (1998).

39. K.A. Nugent, "Coherent Methods in the X-ray Sciences," *Advances in Physics* **59**, 1–99 (2010).

40. The monochromator illustrated in Figure 8.11 is similar to that described in J. Underwood, E. Gullikson, M. Koike *et al.*, "Calibration and Standards Beamline 6.3.2 at the Advanced Light Source," *Rev. Sci. Instrum.* **67**(9), 1 (September 1996); also R. Beguiristain, "Thermal Distortion Effects on Beamline Design for High Flux Synchrotron Radiation," PhD Thesis, Nuclear Engineering Department, University of California at Berkeley, October 1997.

41. M. Hettrick, J. Underwood, P. Batson and M. Eckart, "Resolving Power of 35,000 in the Extreme Ultraviolet Employing a Grazing Incidence Spectrometer," *Appl. Opt.* **27**, 200 (January 15, 1988), and references therein.

42. K. Chesnel, J.J. Turner, M. Pfeifer and S.D. Kevan, "Probing Complex Materials with Coherent Soft X-rays," *Applied Physics A* **92**, 431 (2008).

43. J. J. Turner, K. J. Thomas, J. P. Hill *et al.*, "Orbital Domain Dynamics in a Doped Manganite," *New J. Phys.* **10**, 053023 (2008).

44. S.-W. Chen, H. Guo, K. A. Seu *et al.*, "Jamming Behavior of Dynamics in a Spiral Antiferromagnetic System," *Phys. Rev. Lett.* **110**, 217201 (2013).

45. C. Chang, "Diffractive Optical Elements Based on Fourier Optical Techniques: A New Class of Optics for Extreme Ultraviolet and Soft S-ray wavelengths," *Applied Optics* **41**, 7384 (December 10, 2002); "Coherence Techniques at EUV Wavelengths", PhD Thesis, EECS, UC Berkeley (2002).

46. C. Chang, A.E. Sakdinawat, P. Fischer, E.H. Anderson and D. Attwood, "Single-Element Objective Lens for Soft X-ray Differential Interference Contrast Microscopy," *Optics Lett.* **31**, 1564 (2006).

47. A.E. Sakdinawat and Y.W. Liu, "Soft X-ray Microscopy Using Spiral Zone Plates," *Optics Lett.* **32**, 2635 (2007).
48. A.E. Sakdinawat and Y.W. Liu, "Phase Contrast Soft X-ray Microscopy Using Zernike Zone Plates," *Optics Express* **16**, 1559 (2008).
49. A.E. Sakdinawat, "Extended Depth of Field Soft X-ray Microscopy", unpublished (2009).
50. A.E. Sakdinawat, "Contrast and Resolution Enhancement Techniques for Soft X-ray Microscopy", PhD Thesis, Bioengineering, UC Berkeley (2008).
51. K.M. Rosfjord, Y.W. Liu and D.T. Attwood, "Tunable Coherent Soft X-Rays," *IEEE J. Select. Topics Quant. Electr.* **10**, 1405 (November/December 2004).
52. K.M. Rosfjord, "Tunable Coherent Radiation at Soft X-ray Wavelengths: Generation and Interferometric Applications", PhD Thesis, EECS, UC Berkeley (2004).
53. D. Patterson, B.E. Allman, P.J. McMahon *et al.*, "Spatial Coherence Measurement of X-ray Undulator Radiation," *Optics Commun.* **195**, 79 (August 1, 2001); I.K. Robinson, I.A. Vartanyants, G.J. Williams and M.A. Pitney, "Reconstruction of the Shapes of Gold Nanocrystals Using Coherent X-ray Diffraction," *Phys. Rev. Lett.* **87** (19), 195505 (November 5, 2001).
54. J.J.A. Lin, D. Patterson, A.G. Peele *et al.*, "Measurement of the Spatial Coherence Function of Undulator Radiation using a Phase mask," *Phys. Rev. Lett.* **90**, 074801 (February 21, 2003).
55. C.Q. Tran, A.G. Peele, A. Roberts *et al.*, "Synchrotron Beam Coherence: A Spatially resolved Measurement," *Optics Lett.* **30**, 204 (January 15, 2005).
56. C.Q. Tran, A.G. Peele, D. Patterson *et al.*, "Phase Space Density Measurement of Interfering X-rays", *J. Electr. Spectrosc. Rel. Phenom.* **144**, 947 (March 7, 2005).
57. C.Q. Tran, A.G. Peele, A. Roberts *et al.*, "X-ray Imaging: A Generalized Approach using Phase-Space Tomography," *JOSA* **22**, 1691 (August 2005).
58. M.A. Pfeiffer, G.J. Williams, I.A. Vartanyants, R. Harder and I.K. Robinson, "Three-Dimensional Mapping of a Deformation Field Inside a Nanocrystal," *Nature* **442**, 63 (July 6, 2006).
59. S.J. Leake, M.C. Newton, R. Harder and I.K. Robinson, "Longitudinal Coherence Function in X-ray Imaging of Crystals," *Optics Expr.* **17**, 15853 (August 31, 2009).
60. S.J. Leake, R. Harder and I.K. Robinson, "Coherent Diffractive Imaging of Solid State Reactions in Zinc Oxide Crystals," *New J. Phys.* **13**, 113009 (2011).
61. W. Cha, N.C. Jeong, S. Song *et al.*, "Core-Shell Strain Structure of Zeolite Microcrystals," *Nature Mat.* **12**, 729 (July 7, 2013).
62. I. Robinson and R. Harder, "Coherent x-ray Diffraction Imaging of Strain at the Nanoscale," *Nature Mat.* **8**, 291 (April 2009).
63. J.N. Clark, X. Huang, R. Harder and I.K. Robinson, "High-Resolution Three-Dimensional Partially Coherent Diffraction Imaging," *Nature Commun.* **3**, 993 (August 7, 2012).
64. J.N. Clark, X. Huang, R.J. Harder and I.K. Robinson, "Dynamic Imaging Using Ptychography," *Phys. Rev. Lett.* **112**, 113901 (March 21, 2014).

Homework Problems

Homework problems for each chapter will be found at the website:
www.cambridge.org/xrayeuv

6   X-Ray and EUV Free Electron Lasers

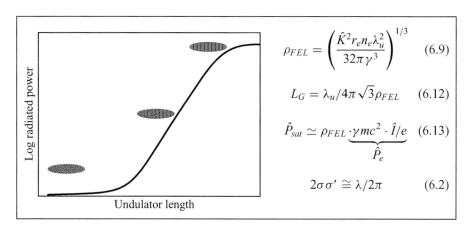

$$\rho_{FEL} = \left(\frac{\hat{K}^2 r_e n_e \lambda_u^2}{32\pi\gamma^3}\right)^{1/3} \quad (6.9)$$

$$L_G = \lambda_u/4\pi\sqrt{3}\rho_{FEL} \quad (6.12)$$

$$\hat{P}_{sat} \simeq \rho_{FEL} \cdot \underbrace{\gamma mc^2 \cdot \hat{I}/e}_{\hat{P}_e} \quad (6.13)$$

$$2\sigma\sigma' \cong \lambda/2\pi \quad (6.2)$$

In this chapter, the physics of x-ray free electron lasing is described, building on knowledge of undulator radiation acquired in the previous chapter. For the FEL, undulator magnet structures are extended from lengths of meters to order 100 m, electron beam size and divergence are reduced to diffraction limited dimensions, and electron currents are increased significantly, resulting in the generation of markedly increased electric fields. Through the Lorentz force, now with non-negligible electric fields, the uncorrelated, random motion of electrons within the bunch evolve to well-correlated electron waves in a physical process known as SASE, self-amplified spontaneous emission. This evolution starts from random statistical fluctuations of density within the electron bunch and evolves with passage through the long undulator to an electron wave, or microbunch,* of several hundred to one thousand cycles. The small phase-space of the electron beam results in constructive interference of the radiated electric fields which then grow collectively in proportion to N_e, the number[†] of electrons participating in the wave. As radiated power scales with the square of the electric field, thus as N_e^2 for these correlated electrons, the radiated power grows rapidly. With feedback from increasing fields driving ever increased electron density modulation, and thus ever stronger fields, the radiated power grows exponentially with traversed undulator distance. At saturation this can lead to radiated powers approaching terawatts (TW) in 10–100 femtosecond

---

* Use of the word microbunch in the broader FEL community refers to a single, sub-wavelength region of increased electron density, but here we use it in the somewhat broader sense of an extended region of many periods of modulated electron density, forming a correlated electron wave packet of period equal to the radiated x-ray wavelength, i.e., an electron wave.

† Recall that as in previous chapters, capital N_e refers to a number of electrons, while lower case n_e refers to a density of electrons. Lower case n without a subscript refers to refractive index, as in Chapter 3.

(fs) pulses, each with a high degree of spatial coherence. Because many microbunches can coexist in the longer electron beam, with slightly varying spectral characteristics, the overall spectrum can be rather broad, especially for probing the electronic states of matter. Thus it is of great interest to develop techniques for narrowing the spectrum. Narrower spectral properties may be obtained with an external laser used as a coherent "seed," if it is intense enough to dominate the spontaneous "noise" of undulator radiation. This has been demonstrated at EUV wavelengths. At x-ray wavelengths "self-seeding" techniques, a form of spectral filtering, are more appropriate, at least at this time.

In this chapter references to pioneering work[1-21], review papers[22-26] and dedicated x-ray FEL textbooks[27-30] are given. Of particular interest are papers where John Madey, as a graduate student, first described[1,2] a lasing process based on free electrons rather than quantized electronic states; papers[7,8] where Evgeny Saldin and his colleague A.M. Kondratenko correctly calculated the gain for coherent, short wavelength radiation that could be generated from noise by a high current, relativistic beam traversing a periodic magnetic undulator; and contributions by Claudio Pellegrini, Rudolfo Bonifacio and their colleagues who showed that the physics of the FEL process could be understood as an unstable wave growing out of statistical noise to a well-defined electron wave through to saturation, without short wavelength limits, and with all major features of the growth process characterized by a single parameter[10], ρ_{FEL}. Further references to pioneering contributions include papers by Kwang-Je Kim where a microscopic Maxwell–Klimontovich formalism was used to track the emergence of electron correlations, and the concomitant evolution of spatially and temporally coherent radiated power through to saturation[14], and papers by Li-Hua Yu and colleagues[70,75] furthering the understanding and experimental demonstration of high gain harmonic generation (HGHG) techniques for the efficient extension of free electron lasing to FEL wavelengths.

6.1   The Free Electron Laser

The free electron laser (FEL) is very closely related to undulator radiation, using basically the same technologies, but with finer control of the relativistic electron bunch (reduced size and divergence), use of a much longer undulator and a much higher peak current. The radiated wavelengths are identical to those given by the undulation equation (5.28)

$$\lambda = \frac{\lambda_u}{2\gamma^2} \left(1 + \frac{K^2}{2} + \gamma^2\theta^2\right) \qquad (6.1)$$

where λ is the radiation wavelength, λ_u is the undulator period, γ the Lorentz contraction factor, $\gamma \equiv 1/(1 - v^2/c^2)^{1/2}$, $K \equiv eB_o\lambda_u/2\pi mc$ is the non-dimensional undulator parameter[31] for a periodic magnet, B_o is the peak undulator magnetic density on axis, and θ is the radiation angle measured from the axis in the forward direction [Figure 5.15 and Eq. (5.18)]. The undulator central radiation cone, $\theta_{cen} = 1/\gamma\sqrt{N}$, contains a $1/N$ spectral bandwidth centered on λ, where N is the number of undulator periods. The emergent FEL radiation, when compared to undulator radiation, provides far higher

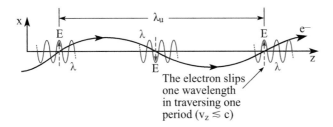

The electron slips
one wavelength
in traversing one
period $(v_z \lesssim c)$

Figure 6.1 As electrons and a radiated wave traverse one period of a magnet structure, the wave advances one wavelength due to its slightly higher velocity. The electrons thus "slip" exactly one wavelength per magnet period. This results in the coherent addition of generated fields in the forward direction. Note that the wavelengths shown here are not to scale. For an x-ray FEL there are of order 10^8 cycles (λ) per magnet period (λ_u).

peak power, much shorter pulses (due to shorter electron bunches), a high degree of spatial coherence and, concomitantly, a much smaller radiation angle, $\theta_{cen} \to \theta_{coh}$, as discussed in Section 6.5. Essential to the FEL action are tight control of the electron beam such that the product of its lateral size and angular divergence (its "phase space product")* is approximately equal to the radiation wavelength

$$2\sigma\sigma' \cong \lambda/2\pi \qquad (6.2)$$

where σ is the rms electron beam radius and σ' is the rms electron beam angular divergence, as illustrated earlier in Chapter 5, Figure 5.22, and discussed with respect to spatial coherence in Section 4.1, Eq. (4.5), where an rms beam diameter $d = 2\sigma$ was used. With this well-controlled electron beam size and divergence the emitted radiation from the various electrons will be capable of constructive interference, essentially adding in phase, depending on the orderliness of the electron positions and motion. This offers the possibility of large increases in radiated electric field, and by its square, very large increases in radiated power. The key is to transform the random positions and motion of the various electrons into a well-defined electron wave of high modulation.

While the undulator Eq. (6.1) was derived in Chapter 5 based on Lorentz contraction of the periodic magnet structure length by the passing relativistic electrons, and subsequent Doppler shift of radiation, the free electron laser community arrives at the same equation based on a different but very useful concept known as the "slip condition," illustrated here in Figure 6.1.

The electrons and the radiated wave (the x-rays) travel together as they move through the magnet structure at about the same axial velocity, $v = (1 - 1/2\gamma^{*2})c$ for the electrons[‡] and c for the radiated wave [see Chapter 5, Eqs. (5.26) and (5.27)]. The electrons execute one cycle of sinusoidal motion in traversing one period of the magnet structure. According to the "slip condition" the wave advances one wavelength *with respect to the electrons* as it propagates across the same path. Thus the electron's "slippage" behind one cycle of phase (one wavelength) with respect to the co-propagating wave. This permits continuous constructive interference as both electrons

[‡] $\gamma^* \equiv \gamma \Big/ \sqrt{1 + \dfrac{K^2}{2}}$

and wave execute one phase cycle per magnet period. For the generation of intense radiation, with the electrons slightly slower, the condition for constructive interference is

$$\Delta v \cdot \Delta t = \lambda$$

$$(c - \bar{v}_z)(\lambda_u/c) = \lambda$$

where Δv is the difference in wave and electron velocities and Δt is the transit time. From the equations of motion in Chapter 5, Eq. (5.25), the axial electron velocity averaged over one period is

$$\frac{\bar{v}_z}{c} = 1 - \frac{1 + K^2/2}{2\gamma^2} = 1 - \frac{1}{2\gamma^{*2}}$$

Combining these two equations, one recognizes the on-axis ($\theta = 0$) undulator equation

$$\lambda = \frac{\lambda_u}{2\gamma^2}\left(1 + \frac{K^2}{2}\right)$$

The physical insight of the slippage condition is very useful in understanding both coherent addition of fields in the forward direction and continuous modification of transverse electron velocities due to wave–particle coupling. The latter leads to modified electron trajectories and eventually to "microbunching" of the electrons, a critical feature of free electron lasing.

The very long undulators characteristic of FELs, of order 100 m for x-rays rather than meters at synchrotron facilities (see Table 6.2 later in this chapter), lead to much stronger radiated fields, sufficient to impose a wavelength-scale microbunching of the electrons into a well-defined electron wave whose modulation increases in amplitude as the bunch moves through the long magnet structure. As more electrons are induced to participate in the evolving electron wave, the constructively summed electric fields grow rapidly. This coherent addition of radiated fields leads to an ever stronger Lorentz electric force, further increasing the electron wave modulation, and in turn further increasing the enhanced radiation fields. This process leads to an exponentiation of electron bunching, that is an increase in electron wave amplitude, coupled to an exponential growth of the radiated fields and power. This feedback driven exponential growth of electron bunching and radiated fields constitutes the essential physics of the free electron laser. As we shall see in the coming sections of this chapter, current x-ray free electron lasers routinely generate 100 gigawatts (GW) of radiated power, typically in femtosecond duration pulses, and are spatially coherent to a high degree. Their spectral bandwidth and degree of temporal coherence are typically measured in hundreds of waves, but that property is rapidly evolving as new techniques for temporally coherent seeding and "self-seeding" are explored. Note that the FEL properties of high peak power, good spatial coherence, and femtosecond duration pulses, are complementary to those of synchrotron radiation facilities, for example opening new scientific opportunities for dynamical studies of atomic, molecular and solid state systems, the use of new coherent scattering and imaging techniques, and studies of high field-matter interactions, all with the advantages of short x-ray wavelengths, and with photon energies tunable to the distinctive atomic resonances and absorption edges of essentially all atoms.

6.2 Evolution from Undulator Radiation to Free Electron Lasing

Our approach to understanding the x-ray free electron lasing process is to begin with the knowledge of undulator radiation as described in Chapter 5, Section 5.4, and as illustrated in Figure 5.8. Here we consider just a single electron bunch, typically containing of order 10^9 electrons, randomly distributed as the bunch enters the periodic magnet structure. The electrons have a mean energy[¶] $E_e = \gamma mc^2$ and momentum $p = \gamma mv$, where m is the electron rest mass and c is the velocity of light in vacuum. The motion of these relativistic electrons is described by the Lorentz force

$$\mathbf{F} = \frac{d\mathbf{p}}{dt} = -e(\mathbf{E} + \mathbf{v} \times \mathbf{B}) \tag{6.3}$$

where $\mathbf{E}$ is the electric field and $\mathbf{B}$ is the magnetic flux density. The undulator provides a strong, time-independent, periodic magnetic field which can be represented by

$$\mathbf{B} = -B_0 \mathbf{y}_0 \sin(2\pi z/\lambda_u) \tag{6.4}$$

where λ_u is the undulator period, $\mathbf{y}_0$ is the unit vector, z is the axial coordinate of the magnet structure and also the zeroth-order direction of the entering electron beam. With no applied electric field the electron motion follows, to first order, a $\mathbf{v} \times \mathbf{B}$ driven oscillation in the x-direction, described by [see Chapter 5, Eqs. (5.18) and (5.19)]

$$v_x = \frac{Kc}{\gamma}\cos\left(\frac{2\pi z}{\lambda_u}\right) \tag{6.5}$$

where

$$K \equiv \frac{eB_0\lambda_u}{2\pi mc} \tag{6.6}$$

is the non-dimensional undulator parameter.[31] These oscillations generate the undulator radiation fields, associated power and photon flux, as was described in Eq. (5.39). In the relatively short undulators described for synchrotron radiation, and the first few meters of the otherwise similar FEL undulator, the radiated fields are not strong enough to significantly contribute to the Lorentz force and thus have little effect on motions within the electron bunch. That is to say, there is not a significant feedback from the radiated fields to the electrons. Indeed, although the undulator radiated fields, their associated power and photon flux, are very valuable for scientific experimentation at synchrotron radiation facilities, these fields have no discernible effect on the electron bunch properties. However, for the much longer FEL magnet structure, and higher beam current, this is not the case. Rather, the undulator is now sufficiently long that the radiated electric field does indeed affect the electron motion, through its contribution to the Lorentz

¶ To avoid confusion, the electron energy always has a subscript e.

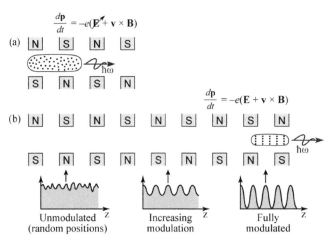

Figure 6.2 (a) A short undulator traversed by an electron bunch with uncorrelated, random positions, radiating uncorrelated fields such that intensities add, and power scales with N_e, the total number of electrons in the bunch. The electric field is negligible. (b) A SASE free electron laser (FEL) in which electrons are "microbunched" by their own radiated fields, to increasing modulation depth, as they traverse the long undulator, evolving to an orderly, strongly correlated electron wave such that radiated electric fields now add, and radiated power scales as N_e^2.

force (Eq. 6.3), and eventually has a significant effect upon the spatial distribution of electrons within the bunch. The general concept is illustrated in Figure 6.2, where the effect of the radiated fields, back on the electrons in the long undulator, is shown to cause a redistribution of the random electron positions into an electron wave within the bunch. The modulation depth of this electron wave increases as the bunch moves through the magnet structure, and leads to substantially increased fields and radiated power due to the constructive interference permitted by the orderliness (correlation) of the electron positions and the relative phases of their emissions.

To understand the induced bunching we begin with a simple model of an initially uniform, one-dimensional electron distribution entering the undulator, as illustrated in Figure 6.2(b). The electrons each oscillate as they traverse the magnet structure, as described by Eqs. (6.1)–(6.6), and thus radiate. There is a transverse coupling between the radiated electric field E_x and the electron's transverse velocity, v_x, the latter set by its motion in the undulator's magnetic field, as described by Eq. (6.5). This leads to an energy exchange between the fields and the particles, described by the rate of change of electron energy

$$\frac{dE_e}{dt} = \mathbf{v} \cdot \mathbf{F} \tag{6.7}$$

where $E_e = \gamma mc^2$ is the total electron energy, m is the rest mass, and $\mathbf{F}$ is the Lorentz force, given by Eq. (6.3), so that

$$mc^2 \frac{d\gamma}{dt} = -eE_x v_x \tag{6.8}$$

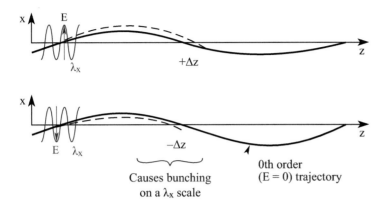

Figure 6.3 Depending on each electron's position with respect to both its transverse motion within the magnet structure and the phase of the co-propagating electromagnetic wave, its transverse velocity will either be increased or decreased, affecting its time of return to the z-axis, and thus affecting local charge density. This causes an electron density modulation on a wavelength scale, so-called "microbunching."

where the $\mathbf{v} \cdot (\mathbf{v} \times \mathbf{B})$ is zero by vector identity [Appendix D, Eq. (D.9)]. Thus the energy exchange between particles $(d\gamma/dt)$ and fields $(E_x v_x)$ is explicit. This energy exchange depends directly on the relative phases of the wave's electric field (E_x) and the electron's position in the magnet structure, as the electron's oscillation amplitude v_x depends on its "phase" with respect to the magnet structure, z/λ_u, as in Eq. (6.5). Thus some electrons gain energy and some lose energy. These changes in electron velocity, some positive and some negative depending on their relative position with respect to the phase of the radiated electric field, lead to changes in their respective trajectories and cause microbunching. Figure 6.3 shows the broad trajectory (solid black line) of electrons oscillating as they traverse one period of an undulator. Also shown are a few cycles of a co-propagating electromagnetic wave that satisfies the undulator equation (6.1) for γ and the magnet period λ_u. Depending on the position of a given electron relative to the co-propagating wave, it will be accelerated, or decelerated, adding to or subtracting from its velocity due to the magnet structure. If the electron receives an increase in velocity it will travel a little farther in the x-direction before returning towards the z-axis and thus take a little longer before crossing the z-axis. Similarly, a deceleration by the electric field (see the arrow down) will decrease that electron's x-coordinate excursion, so that it will arrive sooner at the z-axis. Thus electrons begin to develop a reduced charge density in some axial positions and an increased charge density in others. This is the initiation of microbunching within the larger electron bunch, which continues as the electrons move further through the long undulator, slowly increasing the number of electrons participating in the electron wave formation, while decreasing the background level of those uniformly or randomly distributed. As this orderliness (electron density modulation) increases, as suggested in Figure 6.2(b), ever more electrons radiate in phase with each other, the radiated electric field grows with the number of participating electrons, $\tilde{N}_e$, and thus the radiated power grows with $\tilde{N}_e^2$ [Chapter 2, Section 2.3, and Eq. (2.31)].

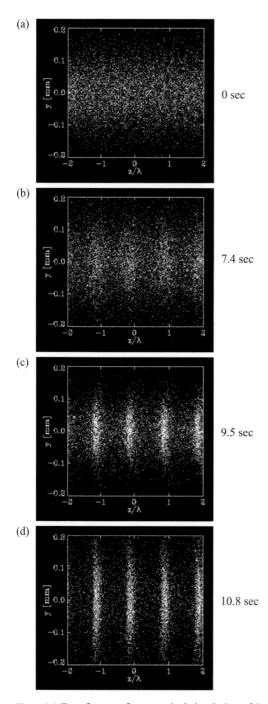

(a)   0 sec

(b)   7.4 sec

(c)   9.5 sec

(d)   10.8 sec

Figure 6.4 Four frames of a numerical simulation of SASE free electron lasing. The simulation movie shows evolution from (a) an initially random electron distribution; to (b) the emergence of microbunching; to (c) a highly modulated beam well into the region of exponential growth; and to (d) essentially at saturation with most electrons participating in microbunching, with a period equal to the lasing wavelength, λ, which satisfies the undulator ("resonance condition") equation, Eq. (6.1). Parameters: $\lambda = 109$ nm, $E_e = 233$ MeV ($\gamma = 456$), $\hat{I} = 500$ A, $\lambda_u = 27.3$ mm, $K = 1.17$, $N = 494$, $L_u = N\lambda_u = 13.5$ m is the undulator length; TESLA (TTF), precursor to FLASH at DESY, Hamburg, Germany. Courtesy of Sven Reiche,[32] Paul Scherrer Institute, Villigen, Switzerland.

This positive feedback, between the increased electric field and increased electron wave amplitude modulation, continues throughout the length of the magnet structure until a high percentage of the beam electrons are organized into a well-defined electron wave. For this constructive interference of the radiated fields to occur it is necessary that the electron bunch be contained laterally within a phase-space equal to that of diffraction limited radiation at the wavelength λ, as discussed earlier and described analytically by Eq. (6.2). As a consequence, the radiation not only grows rapidly in power but also evolves to a high degree of spatial coherence.§ This continuing cycle of positive feedback between induced microbunching and increased radiation field leads to an exponentiation of spatially coherent radiated power, a process known as "free electron lasing"[1, 10]. Unlike conventional lasers (see Chapter 9), which involve stimulated emission of radiation through excitation of quantized atomic states, the free electron laser is classical, involving a continuous buildup of the "inverted population" (the electron wave) from "noise" (the initial undulator radiation from fluctuations of the randomly distributed electrons). The mathematics of this lasing process[27, 28] are presented in the following Section. Results of a numerical simulation[32] of the FEL microbunching process are presented in Figure 6.4, where one observes the emergence of a fully modulated electron microbunch in the presence of growing short wavelength radiation. The achievement of full modulation is referred to as "saturation" in that the highest degree of electron modulation has been achieved and passage through further magnetic periods will not be useful. At this distance the FEL is at maximum radiated power with a high degree of spatial coherence. Figure 6.5 illustrates the general features of radiated power versus traversed undulator distance. For short distances radiated energy grows linearly with distance, as described for undulator radiation in Chapter 5, Eqs. (5.39) and (5.41). As the radiated fields begin to cause microbunching within the electron beam, the radiated fields themselves begin to grow more rapidly than linear.

This continued feedback process leads to an ever increasing growth of electron density modulation and, concomitantly, an exponential growth in radiated power, eventually reaching a saturated state of highest peak power and spatial coherence. This form of free electron lasing, in which the process starts from the "spontaneous emission" of undulator radiation, dominated by particularly strong components of random density fluctuations, and evolves to fully coherent, high peak power, is known as "self-amplified spontaneous emission," or SASE.[10] While SASE has provided the primary path towards gigawatt x-ray free electron lasing, with coherent, femtosecond duration radiation, it has the disadvantage that many microbunches may form independently within the longer electron bunch, with slightly different wavelengths and phases. Efforts to avoid or minimize this involve several techniques. One potential technique is that of coherent seeding with a high harmonic of a conventional, femtosecond duration laser (see Chapter 7). This requires a laser harmonic wavelength matched to the undulator,

---

§ The evolution from chaos to coherence is a general feature found in many areas of science and culture, such as the evolution of coherent clapping in concerts and soccer (football) matches. See Z. Neda, *Nature* **403**, 849 (February 24, 2000), and H. Fountain, "Making Order Out of Chaos When the Crowd Goes Wild," *NY Times* (March 7, 2000).

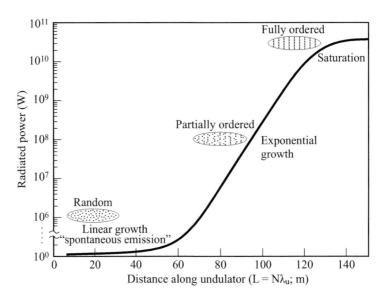

Figure 6.5 A generic power gain curve for an x-ray free electron laser showing no density modulation within the electron bunch during the linear "undulator radiation" region, followed by the emergence of density modulation and exponential growth of radiated power, and finally saturation characterized by nearly full electron density modulation, maximum radiated power and essentially full spatial coherence. Courtesy of K.-J. Kim (Argonne National Laboratory and U. Chicago).

and with sufficient intensity to dominate spontaneous emission in the first sections of the undulator, a combination which is possible in the EUV, but not likely at x-ray wavelengths in the near future. An alternative technique being developed for x-ray wavelengths is known as "self-seeding". This involves the placement of a monochromator within the photon beam path, passing only radiation within a narrow spectral window.[||] In this self-seeding technique the monochromator is placed at an axial position where the exit beam intensity is still sufficient to dominate the lasing process, ideally allowing further amplification of only radiation within the narrowed spectral bandwidth. In this sense, the monochromator plays a role similar to that of a longitudinal mode selector, or Fabry–Perot etalon in a conventional visible light laser (see Chapter 9, Figure 9.6 and supporting text).

Regarding polarization control, note that undulator radiation (see Section 5.4.8) and SASE FEL lasing can be generated for both linear and circularly polarized radiation. Note too that it is possible to use most of the undulator length to establish high modulation electron waves (microbunches) in a linear polarization mode, but then send this two-dimensional planar electron wave into a final sector of circularly polarized undulators, which will serve as the "radiator," in this case emitting circularly polarized x-rays of desired helicity [see Chapter 5, Section 5.4.8].

[||] The electron beam is magnetically deflected through a separate path and realigned with the radiation at an axial position beyond the monochromator.

6.3   The FEL Equations and Characteristic Parameters

In the previous sections the physics of free electron lasing has been described qualita-
tively. In this section, the theory of free electron lasing is developed from basic equations
describing relativistic particle dynamics, wave propagation and growth.[9, 22, 26, 27] These
include equations for the rate of change of particle momentum (the Lorentz force equa-
tion), rate of change of particle energy ($\mathbf{v} \cdot \mathbf{F}$), and the wave equation. Combining these
equations results in a third-order partial differential equation for the electric field, for
which the solution involves a third-order dispersion relation with three solutions, one of
which is a wave with exponential growth – the lasing mode described in Sections 6.1
and 6.2. This analysis provides expressions for wave growth and radiated power in terms
of the physical parameters of the FEL: γ, λ_u, K, $\hat{I}$, and n_e, where $\hat{I}$ is the peak current
associated with the short, femtosecond duration, electron bunch, and n_e is the number of
electrons per unit volume. A particularly important non-dimensional parameter which
emerges from the mathematical analysis is the "FEL parameter", first introduced in the
paper by Bonifacio, Pellegrini and Narducci[11]

$$\rho_{FEL} = \left(\frac{\hat{K}^2 r_e n_e \lambda_u^2}{32\pi \gamma^3} \right)^{1/3} \tag{6.9}$$

where $\hat{K}$ is an effective undulator parameter which accounts for energy transferred to
higher undulator harmonics,

$$\hat{K} \equiv K \cdot [JJ] \tag{6.10}$$

where K was defined below Eq. (6.1), $[JJ] \equiv [J_0(x) - J_1(x)]$, $x = K^2/(4 + 2K^2)$, and
where $J_0(x)$ and $J_1(x)$ are Bessel functions of the first kind, orders zero and one, respectively [see Chapter 5, Eqs. (5.40) and (5.41)]. Of significant interest to the study of lasers
is the distance L_G which the wave must travel to realize an e-folding gain of power, i.e.,

$$\frac{\hat{P}}{\hat{P}_0} = e^{z/L_g} \tag{6.11}$$

It will be seen later in this section that for the free electron laser the idealized one-
dimensional gain length, L_G, is given by

$$L_G = \lambda_u/4\pi \sqrt{3} \rho_{FEL} \tag{6.12}$$

Further analysis will show that the peak radiated power at saturation is given approximately by

$$\hat{P}_{\text{sat}} \cong \rho_{FEL} \cdot \underbrace{\gamma mc^2 \cdot \hat{I}/e}_{\hat{P}_e} \tag{6.13}$$

where $\hat{P}$ and $\hat{I}$ refer to the peak (largest) values in the respective time-dependent functions, as describe previously in Section 5.4.7. Thus we see that the FEL parameter,
ρ_{FEL}, plays a major role in characterizing critical measures of FEL performance. For

the SLAC x-ray FEL, the Linear Coherent Light Source (LCLS),[33–36] to be considered in a later section, the energy per particle (γmc^2) is 13.6 GeV, the peak content ($\hat{I}$) is 3 kA, the units "e" cancel in GeV/e, so that the electron beam power, $\hat{P}_e$, is about 40 TW. The FEL parameter, ρ_{FEL}, is estimated to be approximately 5×10^{-4}, so that according to Eq. (6.13) a radiated peak x-ray power, $\hat{P}$, of about 20 GW is anticipated. The ideal one-dimensional gain length for the early LCLS experiments, with an undulator period λ_u of 30 mm and a wavelength of 1.5 Å, is expected according to Eq. (6.12) to be approximately $L_G \cong 3$ m, of a total 112 m length of periodic magnetic undulator available. These projected measures of performance will be compared to experimental results in Section 6.5.

To begin a theoretical analysis of the FEL process a particular model and specific goals must be established as there are quite a few practical matters to be considered.[23, 27, 28] For example, is the electron beam mono-energetic, or does it have some energy spread $\Delta\gamma$; are three-dimensional effects such as the cross-sectional overlap between the electrons and photons to be considered, accounting for both the finite emittance (phase-space product) of the electrons, and the diffraction of radiation from the finite-sized electron beam**; are only initial (linear) growth rates to be considered or gain lengths well into the region of high modulation microbunching; is the lasing process to be driven by a strong monochromatic wave, or by a competition among undulator radiation "modes" of random phase and slightly varying wavelengths ("spectral spikes") emitted by statistical variations of density fluctuations in the beam; indeed is the incoming beam of electrons assumed uniform or random in position, velocity, and angle with respect to the z-axis?

For the purposes of this text, a satisfactory achievement would be to demonstrate exponential gain in a simple, idealized, one-dimensional FEL system, with a quantitative identification of the gain length, leaving the two-dimensional, polychromatic and multi-energetic aspects to the literature and to dedicated FEL texts.[27, 28] With that motivation the analysis begins with the assumption of a transversely uniform, mono-energetic electron beam and a strong co-propagating, monochromatic wave, all collectively satisfying the undulator equation (6.1). The co-propagating wave could be a strong seed wave, or a single dominant wave emerging statistically from the undulator's "spontaneous emission," to be approximated here as being monochromatic and of well-defined phase. The former case would be characterized as a cold beam, high gain, one-dimensional, FEL amplifier, while the latter case is an example of SASE.

6.3.1   The FEL Particle Equations for a One-Dimensional, High Gain FEL

The quantitative analysis of this idealized one-dimensional, high gain FEL amplifier begins with the particle dynamics. The electron velocity is dominated by the strong,

---

** Simulations show that losses due to diffraction from the small diameter source are compensated by the very high near-axis gain whereas off-axis diffracted radiation (higher-order modes) is not amplified. This very helpful effect is known as "gain guiding."

periodic magnetic field, but is modified by the electromagnetic wave as it grows. Assuming a periodic magnetic field given by Eq. (6.4):

$$\mathbf{B} = -B_0 \mathbf{y}_0 \sin(2\pi z/\lambda_u)$$

where $\mathbf{y}_0$ is the unit vector in the y-direction, λ_u is the undulator period, and z defines the axial direction of the electrons. The zeroth-order electron oscillatory motion, in the weak electron field limit, was determined from the Lorentz force equation to be (Eqs. (5.19) and (6.5))

$$v_x = \frac{Kc}{\gamma} \cos k_u z$$

where $k_u = 2\pi/\lambda$. With a strong co-propagating electromagnetic wave

$$\mathbf{E} = E_0 \mathbf{x}_0 \cos(kz - \omega t) \tag{6.14}$$

where E_0 is the initial electric field amplitude, $\mathbf{x}_0$ is the unit vector in the x-direction, $k = 2\pi/\lambda_u$, $\omega = kc$, and where the energy exchange driven by the electric field, as seen earlier in Eq. (6.7),

$$\frac{dE_e}{dt} = \mathbf{v} \cdot \mathbf{F}$$

where $E_e = \gamma mc^2$ is the total electron energy, m is the rest mass, and $\mathbf{F}$ is the Lorentz force

$$\mathbf{F} = -e(\mathbf{E} + \mathbf{v} \times \mathbf{B})$$

As mc^2 is a constant, the rate of energy exchange can be written as

$$mc^2 \frac{d\gamma}{dt} = -e\mathbf{v} \cdot (\mathbf{E} + \mathbf{v} \times \mathbf{B}) \tag{6.15a}$$

or

$$mc^2 \frac{d\gamma}{dt} = -e\mathbf{v} \cdot \mathbf{E} \tag{6.15b}$$

where $\mathbf{v} \cdot (\mathbf{v} \times \mathbf{B})$ is identically zero by vector theorem (Appendix D.1). Combining Eqs. (6.15b), (6.5), and (6.14), the energy coupling between electrons and the wave can be written as

$$mc^2 \frac{d\gamma}{dt} = -\frac{eE_0 Kc}{\gamma} \cos(k_u z) \cdot \cos(kz - \omega t) \tag{6.16}$$

Using the trigonometric relation $\cos A \cos B = \frac{1}{2}[\cos(A + B) + \cos(A - B)]$, Appendix D.3, one has

$$\frac{d\gamma}{dt} = -\frac{eE_0 Kc}{2mc^2\gamma} \left\{ \cos \underbrace{[(ku + k)z - \omega t]}_{\text{slow, } \theta_s \ (6.17b)} + \cos \underbrace{[(ku - k)z + \omega t]}_{\text{fast, } \theta_f \ (6.17c)} \right\} \tag{6.17a}$$

where the first term, θ_s, is slowly varying over many undulator periods, allowing energy transfer, while the second term, θ_f, varies too rapidly, two cycles per undulator period, to contribute significantly to energy transfer. From the algebra of Appendix G we see that

$$\frac{d\theta_s}{dt} = 2\omega_u \eta \tag{6.18}$$

and

$$\frac{d\theta_f}{dt} = 2\omega_u \tag{6.19}$$

where we have introduced the relative electron energy

$$\eta \equiv \frac{\gamma - \gamma_0}{\gamma_0} \tag{6.20}$$

where γ_0 is the initial beam energy, γ its evolving energy, and where $\eta \ll 1$. Note that $d\gamma/dt = \gamma_0 \, d\eta/dt$, so that the energy equation takes the form

$$\frac{d\eta}{dt} = -\frac{eE_0 K}{2mc\gamma^2} \cos\theta_s \tag{6.21}$$

where only the slowly varying phase term contributes. To a high degree of accuracy the average velocity in the axial direction is very nearly c (to within one part in $2\gamma^2$), so that $z \simeq ct$, and the equation for electron position with respect to the phase of the wave, θ_s, and normalized energy becomes

$$\frac{d\theta_s}{dz} = 2k_u \eta \tag{6.22}$$

$$\frac{d\eta}{dt} = -\frac{eE_0 K}{2mc\gamma^2} \cos\theta_s \tag{6.23}$$

Here $\theta_s \equiv (ku + k)z - \omega t$. These are the *coupled equations of motion*, sometimes referred to as the "*pendulum equations*,"[5] which describe the oscillation between electron energy (η) and position (relative-phase, θ_s) in the combined fields of the wave and the undulator. To ease comparisons with the widely known "pendulum equations," authors[5, 27, 28] often introduce a $\pi/2$ phase shift, thus transforming $\cos\theta_s$ to a new $\sin\theta_s'$.

6.3.2   One-Dimensional Equation for a Growing Wave

The electron equations of motion ((6.22) and (6.23)) are coupled to the amplitude of the electric field. In a high gain FEL the wave amplitude will grow with propagation as part of the lasing process. For one-dimensional propagation in the z-direction,[††] an idealized but important model, the wave equation can be written for an electric field polarized in the x-direction as

$$\left[\frac{\partial^2}{\partial t^2} - c^2 \frac{\partial^2}{\partial z^2} \right] E_x(z, t) = -\frac{1}{\varepsilon_0} \frac{\partial J_x(z, t)}{\partial t} \tag{6.24}$$

[††] As discussed in the literature,[15, 23] and further at the end of this section, this assumes a sufficiently small electron beam and negligible lateral diffraction.

where $c^2 = 1/\mu_0\varepsilon_0$, and where as in Eq. (3.1) the charge density term has cancelled with the longitudinal component of the current density for the propagation of strictly transverse waves [see Chapter 3, Section 3.1]. The current density is given by [see Eq. (2.10)]

$$J_x(z, t) = -en_e(z, t)v_x(z, t) \tag{6.25}$$

Where $n_e(z)$ is the electron density, whose microbunching is critical to the FEL process, and v_x is the electron velocity. The dominant seed wave, whose growth we wish to study, was described earlier in Eq. (6.14) as a wave of constant amplitude E_0, but now we wish to consider growth of this seed wave in the FEL amplifier, and thus introduce the complex electric field amplitude $\tilde{E}_x$, where the superscript tilde indicates a complex quantity. Separating the rapidly varying phase from the slowly growing, z-dependent field amplitude, we now write for amplification of this monochromatic, one-dimensional wave[‡‡]

$$E_x(z, t) = \tilde{E}_x(z)e^{-i(\omega t - kz)} \tag{6.26}$$

where $\tilde{E}_x(0) = E_0$. The full wave equation (6.24) can now be simplified for the case of a one-dimensional, slowly growing wave, $\tilde{E}_x$, as shown in Appendix G.2, viz.,

$$\frac{\partial \tilde{E}_x}{\partial z} = -i\frac{\mu_0}{2k}e^{i(\omega t - kz)}\frac{\partial J_x(z, t)}{\partial t} \tag{6.27}$$

where all the high-frequency components have been moved to the right side of the equation, where they will all drop out in the following "slice" integration. To determine the slow growth of the wave, we perform an average over a small but finite number of cycles[¶], in a time interval $\Delta = 2\pi n/\omega$, where n is an integer. This has the effect of averaging out the high-frequency variations. The time-averaged gradient of the slowly varying electric field is then

$$\left\langle\frac{d\tilde{E}_x}{dz}\right\rangle_\Delta = -i\frac{\mu_0}{2k}\int_{t-\frac{\Delta}{2}}^{t+\frac{\Delta}{2}} \underbrace{e^{i(\omega t' - kz)}}_{w}\underbrace{\frac{\partial J_x(z, t')}{\partial t'}dt'}_{du} \tag{6.28}$$

Performing an integration by parts, $\int w\,du = uw - \int u\,dw$,

$$\left\langle\frac{d\tilde{E}_x}{dz}\right\rangle_\Delta = -i\frac{\mu_0}{2k}\left\{\underbrace{\left[J_x(z, t')e^{i(\omega t' - kz)}\right]_{t-\frac{\Delta}{2}}^{t+\frac{\Delta}{2}}}_{0} - i\omega\int_{t-\frac{\Delta}{2}}^{t+\frac{\Delta}{2}} J_x(z, t')e^{i(\omega t' - kz)}dt'\right\}$$

$$\left\langle\frac{d\tilde{E}_x}{dz}\right\rangle_\Delta = -\frac{\mu_0 c}{2}\left\langle J_x(z, t)e^{i(\omega t - kz)}\right\rangle_\Delta \tag{6.29}$$

[‡‡] For the SASE process, which involves a spread of frequencies, this monochromatic representation would not be appropriate. One could, however, treat the SASE case in an analogous manner in terms of an emerging dominant wave of finite bandwidth, and then analyze the SASE case in terms of its Fourier components, as in Ref. 27, section 3.5.

[¶] A "slice" averaging technique suggested by Z. Huang, Ref. 28, Section 3.4.1.

where the brackets indicate quantities averaged across a time-slice Δ. The current density, from Eq. (6.25), is

$$J_x(z, t) = -en_e(z, t)v_x(z, t)$$

Based on the physical process of field-induced microbunching discussed earlier in this chapter, we anticipate the formation of a periodic electron wave whose period is the same as the seed wave, whose amplitude is directly related to that of the slowly growing seed wave, $\tilde{E}_x$, and whose rate of growth is closely tied to the slow phase, θ_s, defined in Eq. (6.17b). This model suggests an electron density with high-frequency components in space and time which follow that of the seed wave, and would be removed in the Δ-slice averaging, and a slowly varying amplitude, $n_e(z)$, which contributes to growth of the seed wave over many periods.

The transverse velocity, which contributes to the current $J_x(z, t)$, should according to the Lorentz force equation [Eq. (6.3)] be the *sum* of two components, one driven by the electric field of the growing seed wave, $\tilde{E}_x$, and one driven by the undulator's $\mathbf{v}_x \times \mathbf{B}_0$. However, the undulator term is far greater, so that in this summation we can neglect the contribution of the seed wave. Where these two components *multiply*, as in the product term $n_e\mathbf{v}$ in J_x, and $v_x\tilde{E}_x$ in the energy exchange equations [Eqs. (6.16) and (6.23)], the amplitudes and phase variations of both the seed and the undulator roles are preserved, each contributing to the beat frequencies, θ_s and θ_f, as well as to the growth rates of microbunching and the seed wave. With these understandings, the undulator-dominated velocity in the current term, from Eq. (6.5), is

$$v_x = \frac{Kc}{\gamma}\cos\left(\frac{2\pi z}{\lambda_u}\right)$$

Then Eq. (6.29) becomes

$$\left\langle \frac{\partial \tilde{E}_x}{\partial z} \right\rangle_\Delta = \left\langle \frac{eK}{2\varepsilon_0\gamma}n_e(z)\cos(k_u z)e^{i(\omega t - kz)} \right\rangle_\Delta \tag{6.30}$$

Evaluation of $\cos k_u z$ multiplied by $\exp[i(\omega t - kz)]$ in Eq. (6.30) can be treated as in Appendix G; in this case expanding the exponential into cosine and sine, using double angle trigonometric relations, and then recognizing θ_s and θ_f (Eqs. (6.17b, c)). Doing so, one obtains

$$(\cos k_u z)e^{i(\omega t - kz)} = \frac{1}{2}\{(\cos\theta_s + \cos\theta_f) - i(\sin\theta_s - \sin\theta_f)\}$$

The fast contributions, θ_f, go to zero when averaged over the slice interval Δ, so that the slowly varying gradient becomes

$$\frac{d\tilde{E}_x}{dz} = \frac{eK}{4\varepsilon_0\gamma}n_e(z)\langle\cos\theta_s - i\sin\theta_s\rangle_\Delta$$

where we have dropped the explicit averaging brackets for the electric field gradient and electron density modulation as these are understood to be slowly varying, and have replaced $\partial/\partial z$ with d/dz acting on the electric field as the high-frequency time dependence has been removed. The slowly varying gradient of electric field can now be written in exponential form as

$$\frac{d\tilde{E}_x}{dz} = \frac{eK}{4\varepsilon_0\gamma}n_e(z)\langle e^{-i\theta_s}\rangle_\Delta \tag{6.31}$$

where the subscript Δ indicates an average over an integer number of cycles.

Combined with the particle equations derived earlier, Eqs. (6.22) and (6.23), one has three coupled first-order equations in three variables, $\tilde{E}_x$, η and θ_s, and where E_0 has been replaced by the slowly growing field amplitude, $\tilde{E}_x$. The three are repeated here for convenience as

$$\frac{d\theta_s}{dz} = 2k_u\eta \tag{6.32}$$

$$\frac{d\eta}{dz} = -\frac{e\hat{K}}{2mc^2\gamma^2}\tilde{E}_x\cos\theta_s \tag{6.33}$$

$$\frac{d\tilde{E}_x}{dz} = \frac{e\hat{K}}{4\varepsilon_0\gamma}n_e e^{-i\theta_s} \tag{6.34}$$

where the slice averages over Δ are understood for the three slowly varying quantities $\tilde{E}_x$, n_e, and θ_s, and where, as in Eq. (6.10), we have introduced $\hat{K} = \hat{K} = K[JJ]$, the effective undulator parameter which accounts for the reductions of the first-order transverse velocity due to the appearance of longitudinal and higher-order transverse harmonics, all of which draw energy from the first-order fields and radiated power, as described in Chapter 5, Eqs. (5.40) and (5.41). These three equations provide a basis for describing the one-dimensional, high gain FEL amplification of a monochromatic seed wave interacting with a uniform, monoenergetic electron beam through to saturation. Note that the one-dimensional (1D) approximation fails in axisymmetic geometry when 3D effects become noticeable.[15, 23] Thus validity of the 1D model requires that the electron beam phase space, $\sigma\sigma'$, be less than $\lambda/4\pi$, essentially a phase matching condition between the electrons and diffraction limited photons, and that transverse diffraction be negligible, which requires that the gain length be smaller than the Rayleigh length, $L_g < Z_R$. The process is non-linear, as can be seen by the product terms $\tilde{E}_x\cos\theta_s$ and $n_e e^{-i\theta_s}$ in Eqs. (6.33) and (6.34), and thus not analytically solvable without further simplifications. Numerical simulations of these three coupled equations, and similar but more general equations (finite electron beam energy spread, 3D, SASE, etc.), have provided valuable insights to the FEL process through saturation[32, 54, 55, 162]. Analytic solutions, which provide broad, functional insights to the FEL process and to its most important parameters and relationships are possible, but require linearization. An analytic solution is described in the following section.

6.3.3   The Cubic FEL Dispersion Relation

In this section, we continue development of equations to describe free electron lasing quantitatively, utilizing an idealized, but very useful, one-dimensional model of a cold, uniform, and highly relativistic electron beam interacting with a strong, monochromatic electromagnetic wave of x-ray wavelength. To obtain an analytic solution we now limit the evolving process to a modestly modulated beam, whose amplitude increases in a coupled manner with the ever increasing x-ray/EUV wave amplitude and radiated power.

By following the FEL process for a distance of only modest wave growth, where θ_s and η remain small, and modulation of n_e is still modest, the three first-order, coupled FEL equations [(6.32), (6.33), and (6.34)] can be linearized and combined into a single third-order differential equation for the slowly varying field amplitude, $\tilde{E}_x$. From this a cubic dispersion relation can be identified,[9, 22, 26, 27] with three solutions, one of which represents an unstable wave characterized by an exponential growth of the radiation field. The linearization process then proceeds by approximating $\cos\theta_s \simeq 1$ in Eq. (6.33), $\exp(-i\theta_s) \simeq 1 - i\theta_s$ in Eq. (6.34), and treating n_e as relatively constant, so that the three coupled equations, now numbered as *linearized* (ℓ), are

$$\frac{d\theta_s}{dz} = 2k_u\eta \qquad\qquad (6.32\ell)$$

$$\frac{d\eta}{dz} = -\frac{e\hat{K}}{2mc^2\gamma^2}\tilde{E}_x \qquad\qquad (6.33\ell)$$

$$\frac{d\tilde{E}_x}{dz} = \frac{e\hat{K}}{4\varepsilon_0\gamma}n_e\,e^{-i\theta_s} \qquad\qquad (6.34\ell)$$

Taking d^2/dz^2 of Eq. (6.34ℓ), replacing the $d^2\theta_s/dz^2$ factor which appears with d/dz of Eq. (6.32ℓ), and then replacing the $d\eta/dz$ by Eq. (6.33ℓ), one obtains the third order differential equation for the evolving electric field amplitude, $\tilde{E}_x$

$$\frac{d^3\tilde{E}_x}{dz^3} - i\Gamma^3\tilde{E}_x = 0 \qquad\qquad (6.35)$$

where

$$\Gamma = \left[\frac{\mu_0\hat{K}^2e^2n_ek_u}{4\gamma^3m}\right]^{1/3} \qquad\qquad (6.36)$$

known as the gain parameter, is the product of coefficients in Eqs. (6.32)–(6.34), n_e is the average electron density, and $k_u = 2\pi/\lambda_u$ is the wavenumber associated with the undulator period. Assuming solutions of the form $\tilde{E}_x = E_0 e^{\mu z}$, one obtains the third-order dispersion relation

$$\mu^3 = i\Gamma^3 \qquad\qquad (6.37)$$

with three solutions[§§]

$$\mu_1 = -i\Gamma, \quad \mu_2 = (i - \sqrt{3})\Gamma/2, \quad \mu_3 = (i + \sqrt{3})\Gamma/2 \tag{6.38}$$

The first solution corresponds to a constant amplitude, purely oscillatory mode, the second to a damped oscillation, and the third to a wave of exponentially growing amplitude, i.e.,

$$\tilde{E}_x = E_0 e^{i(\Gamma/2)z} e^{(\sqrt{3}\Gamma/2)z} \tag{6.39}$$

with a corresponding power growth

$$\hat{P}(z) \propto |\tilde{E}_x|^2 \propto e^{\sqrt{3}\Gamma z}$$

or in terms of a *power gain length*[||||], L_G, (see Eq. 6.11)

$$\hat{P}(z) \propto E_0^2 e^{z/L_G} \tag{6.40}$$

where

$$L_G = \frac{1}{\sqrt{3}\Gamma} \tag{6.41}$$

Note that in terms of the FEL parameter, ρ_{FEL}, Eq. (6.9), the gain length can be written as

$$L_G = \frac{\lambda_u}{4\sqrt{3}\pi \rho_{FEL}} \tag{6.42}$$

where this is for an idealized one-dimensional FEL amplifier model. Finite electron beam energy spread, σ_γ, three-dimensional effects, including overlap of the electron beam cross-section with that of the x-ray beam, alignment and co-propagation of the two, and diffraction from the finite size x-ray emission area, to name a few, will tend to extend values of the power gain length beyond that given in Eq. (6.42). More complete developments of the third-order dispersion relation and its solutions are given in the texts of Schmüser, Dohlus, Rossbach, and Behrens[27] and of Kim, Huang, and Lindberg.[28]

Two other important quantities that characterize an FEL are its saturation power, $\hat{P}_{\text{sat}}$ and the relative spectral bandwidth, $\Delta\omega/\omega$, or $\Delta\lambda/\lambda$ (the ratios are equal for narrow bandwidths). In the preceding analysis an idealized electron beam of zero energy width, $\sigma_\gamma = 0$, was assumed. With finite energy width an important energy spread parameter σ_γ/ρ_{FEL} comes into play. Numerical solutions[32] show that σ_γ/ρ_{FEL} must be well below unity for the exponential gain to approximate that given in Eq.(6.42). For larger electron energy spread the radiation spectrum will be broadened, which concomitantly involves a smearing of the induced electron wave (the microbunching), thus affecting the coherent

---

[§§] The solution μ_3, with a positive real part corresponding to exponential growth, quickly dominates the oscillatory solution, μ_1, and the decaying wave, μ_2.[10, 23, 27, 28]

[||||] Note that the power gain length is half that of the field gain length because power scales as electric field squared, and thus exponentiates more quickly.

addition of fields and reducing radiated power. Thus a requirement for high FEL gain is that

$$\frac{\sigma_\gamma}{\rho_{FEL}} \ll 1 \tag{6.43}$$

Maximum power is obtained at saturation [see Fig. (6.7)], where the electron beam approaches full modulation and radiated fields add coherently for maximum benefit. The peak saturated power is given approximately by[11]

$$\hat{P}_{\text{sat}} \simeq \rho_{FEL} \cdot \underbrace{\gamma mc^2 \cdot \hat{I}/e}_{\hat{P}_e} \tag{6.44}$$

where γmc^2 is the energy per electron in the beam and $\hat{I}/e$ is the number of electrons passing per second, the product of the two equaling the electron beam energy, $\hat{P}_e$. This equation was discussed earlier in text associated with Eq. (6.13), including a numerical example. Here we see that the FEL parameter, ρ_{FEL}, represents an efficiency factor for energy transfer from the electron beam to the photons, and thus can be described as the ratio of radiated power at saturation to power of the electron beam.

The relative spectral bandwidth, $\Delta\omega/\omega$, is a very important property of any laser. For an x-ray FEL it is especially critical as many scientific studies involve the probing of very well-defined atomic states, typically defined to parts per 10^3 or 10^4. The FEL spectral bandwidth will depend on whether the radiation evolves from an injected narrow bandwidth coherent seed wave, or from statistical noise as in SASE. For a seed wave we expect the coherence properties of the FEL, including the spectral characteristics, to largely follow those of the seed.*** For the SASE process, however, the bandwidth evolves from noise, initially with spectral widths of order $1/N$, as we saw for undulator radiation in Chapter 5, Eq. (5.14), then varying with distance as the microbunches evolve, finally reaching saturation where energy transfer to the photons leads to a reduction in γ, which sets a limit to the narrowness of the emission spectrum. Indeed it is possible that many microbunches, with slightly different center frequencies, will co-exist within the longer electron bunch, all competing for dominance, as suggested in Figure 6.6.

These separate microbunches will differ somewhat in frequency, within the $1/N$ bandwidth of the initial undulator radiation noise. Typically the SASE induced electron microbunches consist of several hundred x-ray or EUV/soft x-ray wavelengths, about 60 nm long for a 400 cycle, 1.5 Å wavelength FEL. The electron bunch, however, may be 20 μm long, so that several hundred microbunches could co-exist, each with slightly different wavelengths within the overall $1/N$ spectral bandwidth. Data from LCLS will show this explicitly in a later section. Kim[14, 15] has used the coupled Maxwell–Klimontovich equations (see Chapter 8, Section 8.4) to analyze the emergent intensity, spectral, and coherence properties of SASE FELs, including finite electron beam energy spread and detuning. His studies reproduce undulator radiation results for short distances (see Fig. 6.5), with incoherent emission within a spectral bandwidth of

---

*** Allowing for some differences due to beam overlap, finite electron beam energy spread, σ_γ, head-to-tail or other electron beam energy variations.

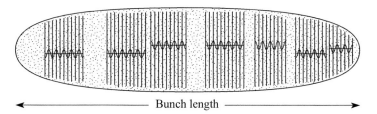

Figure 6.6 Many microbunches can co-exist within an electron bunch, perhaps as many as several hundred, depending on the length of the electron bunch and the number of waves within each of the microbunches. While each microbunch will have a high degree of spatial coherence due to the small electron phase-space and coherent addition of fields, the emission spectrum will be broad due to the presence of many random spectral spikes, all of wavelength and phase within the initial $1/N$ undulator "noise" bandwidth. At full modulation the N_{ej} electrons within each microbunch will radiate coherently, proportional to N_{ej}^2. As the various microbunches evolve independently from noise, the phases of their electron motions are uncorrelated, shifting with time due to differing central wavelengths, and thus their contributions to total FEL power will add linearly. With n_j microbunches, the total radiated power will then be proportional to the sum of $n_j \cdot N_{ej}^2$. For 300 microbunches, averaging 3×10^6 electrons each, 10^9 total, this yields a net gain of 3×10^6 relative to zero gain undulator radiation.

order $1/N$, narrowing in the high gain region of exponential growth, and then returning to a $1/N$ bandwidth at saturation as increased energy spread $(\Delta\gamma)$, caused by significant energy transfer, ends the exponential growth process. As the electrons traverse a distance $z = N\lambda_u$, where $N \rightarrow N_{sat}$ and $N_{sat} \simeq 1/\rho_{FEL}$, the SASE FEL rms spectral bandwidth[29] $(\sigma\omega/\omega)$ evolves to approximately ρFEL, and the FWHM relative spectral bandwidth evolves *at saturation* to

$$\left.\frac{\Delta\omega}{\omega}\right|_{\substack{sat \\ \text{FWHM}}} \cong 2.35\rho_{FEL} \tag{6.45a}$$

The somewhat more convenient inverse spectral bandwidth is

$$\left.\frac{\omega}{\Delta\omega}\right|_{\substack{sat \\ \text{FWHM}}} \cong \frac{1}{2.35\rho_{FEL}} \tag{6.45b}$$

Recall that Eqs. (6.42) and (6.45a, b) are for the case of a cold, mono-energetic electron beam, neglecting two-dimensional effects. For real x-ray FEL experiments the gain lengths are expected to be somewhat longer, and the spectral bandwidths broader, due to the finite initial electron beam energy spread $\sigma_{\gamma 0}$, and possible "head-to-tail" beam energy variation. In many cases the experimental bandwidths are twice those given in Eqs. (6.45a, b).

6.4   First FEL Lasing Experiments at EUV and X-Ray Wavelengths

Free electron lasers at short wavelengths burst onto the scene in 2000 with word that the TESLA FEL at DESY Lab[37] had achieved SASE lasing at 109 nm wavelength (11 eV)

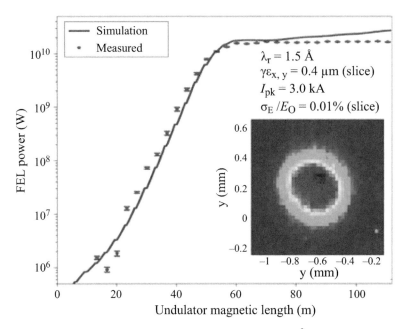

Figure 6.7 The LCLS gain curve based on SASE lasing at 1.5 Å wavelength. A gain length L_G of about 3.5 m is measured, with saturation at 60 m. Further parameters are given in Table 1. The insert shows an image of the x-ray beam after 50 m of additional propagation beyond that where saturation is achieved[33] (courtesy of P. Emma, SLAC).

with a laser gain of 3000, and with all the characteristics of a SASE FEL. Shortly there-after, the same team demonstrated saturation[38, 39] at GW power levels, with 30–100 fs pulses, leading to many pioneering scientific studies.[40, 41] Later, with the upgrade to "FLASH",[†††] the team published FEL lasing results at 13.7 nm (90.5 eV), 10 GW peak power, pulse durations as short as 10 fs, and significant harmonic output (0.6%) at 4.6 nm, and beyond.[42]

As startling as the above EUV FEL demonstrations were, the experiments at Stanford University's SLAC National Acceleration Laboratory were even more so. After nurturing and aligning the SLAC linac, a gift from the down-sizing high-energy particle physics program there, high gain, high power SASE lasing at 1.5 Å wavelength (8.3 keV photon energy) was achieved in a single day.[33] This was the dramatic birth of the Linac Coherent Light Source (LCLS). Although the physics of SASE lasing had been clearly demonstrated at TELSA and FLASH, at LCLS the 1.5 Å wavelength was a hundred times shorter, and high gain lasing required co-axial alignment of the undulator magnet sections to very tight tolerances to assure electron beam and photon beam overlap, a major challenge with an electron beam diameter of about 100 μm FWHM and an x-ray propagation path of 132 m (112 m of active magnetic material). Assuming the beam must be aligned to a tenth of the beam size, this corresponds to a required beam align-ment accuracy of order 0.1 μrad. Within a few weeks full saturation was achieved at

---

[†††] Free electron LASer at Hamburg (FLASH), Deutches Elektronen-Synchroton Laboratory (DESY).

Table 6.1 Nominal parameters for the first experiments at SLAC's LCLS x-ray free electron laser.[33] The parameters σ and σ' are the rms electron beam radius and divergence, σ_γ is the rms spread of γ, and thus of beam energy, σ_z is the rms beam length, $\varepsilon_{\text{slice}}$ is the beam emittance (phase-space product) in the microbunched region, Q is the total electric charge in the bunch, and R_{ep} is the repetition rate. Note that the electron beam phase-space is about 25% larger than that of the photons. The measured gain length, 3.5 m, is longer than the 3.0 m ideal 1D value. Note too, that the Rayleigh range [Chapter 4, Eq. (4.9)], Z_R, with a $1/e^2$ photon beam radius roughly equal to twice that of the electrons, $r_0 = 2r_0 \cong 2\sigma \cong$ 80 μm, is many times that of the gain length, L_G. The measured x-ray power vs. undulator length traversed[33] is shown in Fig. 6.7.

Electron beam	Photons
$E_e = 13.6$ GeV	$\lambda = 1.5$ Å (8.3 keV)
$\gamma = 26{,}600$	$\varepsilon_{\text{coh}} = \lambda/4\pi = 12$ pm · rad
$\hat{I} = 3.4$ kA	$\hat{F} = 0.84 \times 10^{12}$ ph/pulse
$\varepsilon_{\text{slice}} = \sigma\sigma' = 15$ pm · rad	$\hat{E} = 1.1$ mJ/pulse
$Q = 0.25$ nC ($\sim 10^9$ e⁻s)	$\Delta\tau_{\text{ph}} = 70$ fsec FWHM
$R_{\text{ep}} = 120$ Hz	$\lambda/\Delta\lambda = 200\text{–}500$ FWHM
$\sigma_\gamma/\gamma = 1 \times 10^{-4}$	$\hat{P} = 15$ GW
$2.35\sigma = 95$ μm FWHM	
$2.35\sigma' = 0.87$ μr FWHM	**FEL parameters**
$2.35\sigma_z = 21$ μm FWHM	
(70 fsec @ 300 nm/fsec)	$\rho_{FEL} = \left(\dfrac{\hat{K}^2\, r_e n_e \lambda_u{}^2}{32\pi\gamma^3} \right)^{1/3} = 4.7 \times 10^{-4}$
Undulator	$\hat{P}_{\text{sat}} \simeq \rho_{FEL}\cdot\gamma mc^2 \cdot \hat{I}/e = 25$ GW
$\lambda_u = 3$ cm	
$g = 6.8$ mm (fixed)	$L_G = 3.5$ m
$K = 3.5$	$L_{\text{sat}} = 60$ m (of 112 m)
$N = 33 \times 113 = 3733$	
$L_{\text{magnetic}} = 3733 \times 3$ cm $= 112$ m	$z_R = 4\pi r_0^2/\lambda = \pi r_0^2/\lambda = 140$ m

1.5 Å. A gain curve for these early lasing experiments is shown in Figure 6.7, with an insert showing a cross-sectional image of the x-ray beam. A summary of electron beam, undulator, and photon beam parameters for these first LCLS experiments is shown in Table 6.1.

It is interesting now to compare the measured LCLS values in Table 6.1 with the formulas considered earlier in this chapter. The values of λ, γ, and K match the resonance condition given by Eq. (6.1). The measured values of σ and σ' are a match to the wavelength limited phase-space set by Eq. (6.2) to within 25%. The anticipated high degree of spatial coherence was confirmed in later experiments, as will be described in the following section. The radiated power curve, Figure 6.7, closely follows SASE models, showing a region of linear growth for short distances, exponential growth, and finally saturation, except that Figure 6.7 provides real data for the gain length, saturated power, and distance to achieve saturated power.[33] Based on the beam current and measured beam cross-section values, σ_x and σ_y, a value of n_e is obtained, and with this

Eq. (6.36) yields a value for the FEL parameter $\rho_{FEL} = 4.7 \times 10^{-4}$. With this value of ρ_{FEL}, the idealized 1D power gain length, L_G, is about 3 m, which compares well with the experimental value of 3.5 m obtained from the data in Figure 6.7 (3D, finite σ_γ, etc.). According to Eq. (6.44), the idealized saturated power, $\hat{P}_{\text{sat}}$, is predicted to be about 22 GW, while the measured value presented in Table 6.1 is 15 GW, corresponding to a photon energy of 1.1 mJ at 1.5 Å, in a 70 fs pulse, and thus a flux of nearly 10^{12} photons/pulse. The inverse relative spectral bandwidth, Eq. (6.45b), was estimated in the previous section to be $(\omega/\Delta\omega)_{FWHM} \cong 900$ for the ideal one-dimensional, cold electron beam at saturation, while experimental values in Table 6.1 are 200–500 FWHM. Possible explanations for this difference in spectral width include finite electron energy spread within the electron beam, and varied head to tail energy of electrons within the bunch.

Characterization and control of the overall spectral bandwidth, as in Eq. (6.45), is essential. The spectral content of SASE driven radiation evolves from noise and is expected to be highly structured and statistical in nature, not only on a shot-to-shot basis but also within a single radiated pulse. It is anticipated that under these conditions hundreds of microbunches will evolve independently out of the noise, and compete for dominance within a single electron bunch, each contributing its own spectral spikes, as was suggested earlier in Figure 6.6. Experimental hard x-ray spectra can be obtained with crystal spectrographs.[‡‡‡][43] Data obtained at LCLS with crystal spectrographs operating at an 8.3 keV photon energy[44] are shown in Figure 6.8. Two complementary crystal spectrographs are shown, one that provides a broad spectrum of the entire breadth of emitted photon energies, and one that selects a limited region within that broad range for display at very high spectral resolution, both recorded simultaneously on a single-shot basis. The geometry of these measurements is shown later in Figure 6.10, along with sample spectra across the full FEL emission band as well as high spectral resolution spectra within a narrow band, for the same single radiation pulse. The spectrographs were located 192 m downstream of the last undulator.

The statistical nature of the SASE process is clearly evident in the many spectral spikes seen in Figure 6.8(b), and in the high shot-to-shot variability. Clearly there are many random temporal modes, originating from electron density fluctuations in the early "spontaneous" emissions (Figure 6.7) of the SASE lasing process, each competing for dominance in the FEL. The thin Si (333) spectrograph has a spectral resolving capability of 6.4×10^4, permitting spectral spikes as close as 0.2 eV to be clearly identified, as seen in Figure 6.8(d). These experiments were conducted in a low-charge LCLS operating mode, with a bunch charge of about 150 pC and optimal bunch compression, so that the pulse durations for these experiments were likely 25 fs or less. Following these first experiments, LCLS began operations in the soft x-ray region,[45–48] from about 250 eV to 2 keV, including measured pulse durations as short as 2.6 fs.[49]

[‡‡‡] A "spectrograph" disperses radiation of different wavelengths (photon energies) so that the full spectrum can be displayed. A "monochromator" passes a narrow spectral band for further use in experiments.

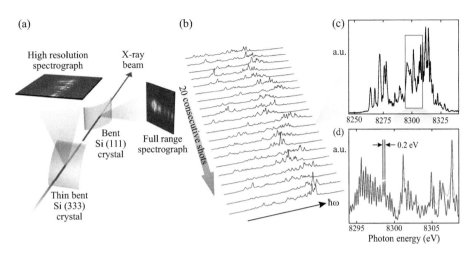

Figure 6.8 A schematic diagram (a) showing the use of two crystals to simultaneously disperse and record hard x-ray FEL spectra at LCLS on a single shot basis.[44] The crystals are located 192 m beyond the last undulator. The bent Si (111) spectrograph displays the full FEL emission spectra, as shown in (b) for 20 consecutive shots, and in (c) for single shot. The thin (83% transmissive) Si (333) crystal provides very high resolution spectra (d) within a narrower band (defined by the red rectangle in (c)) for this same, nominally 25 fs, x-ray pulse. Courtesy of D. Zhu, J. Hastings and Y. Feng, SLAC.

6.5 Spatial and Temporal Coherence of X-Ray FEL Radiation

The subject of spatial and temporal coherence of x-ray radiation was considered in Chapter 4, where concepts were clarified and limiting relationships were developed in terms of properties of the radiator. These relationships were used in Chapter 5, Section 5.6, to show how undulator radiation at x-ray wavelengths could be spatially and temporally filtered to obtain considerable degrees of coherence, albeit at a significant loss of photon flux, or power. It was also noted that despite high degrees of spatial and temporal coherence, true phase coherence was not obtained. In this section we address these same issues for x-ray FELs, pointing out substantial advantages with regard to coherent power, typically increased from fractions of a mW ($\overline{P}_{und}$) to tens or even hundreds of GW ($\hat{P}_{FEL}$), essentially true phase-coherence vs. random phase, and femtosecond pulse durations vs. nominally one second exposure times for undulator experiments.

The condition for spatial coherence, essentially that the radiation appears to be emitted by a point source equivalent at the given wavelength, is that the radiation phase-space area, equal to the product of the source diameter and radiation half angle, be equal to the wavelength within a constant multiplier dependent on the measures taken, e.g., rms or FWHM (see Chapter 4, title page). For instance, in terms of rms quantities for a source characterized by Gaussian spatial and angular distributions, the condition for full spatial coherence is [Eq. (4.7a), renumbered here]

$$d \cdot \theta = \lambda / 2\pi \qquad (6.46a)$$

where the accelerator community, which generally prefers rms quantities, would write $d = 2\sigma$ and $\theta = \sigma'$, so that Eq. (6.46a) would appear as $\sigma\sigma' = \lambda/4\pi$, as quoted earlier in Eq. (6.2). Alternatively, this same phase space constraint can be expressed in terms of FWHM quantities, more popular in the traditional atomic laser community, as

$$d \cdot 2\theta|_{\text{FWHM}} = \frac{2 ln 2}{\pi} \lambda = 0.4413\lambda$$

or for convenience

$$d \cdot 2\theta|_{\text{FWHM}} \simeq \frac{\lambda}{2} \qquad (6.47b)$$

A measure of the temporal coherence length (longitudinal direction of propagation) was determined to be (Eq. 4.6, renumbered here)

$$l_{\text{coh}} = \frac{\lambda^2}{2\Delta\lambda} \qquad (6.47)$$

where $\Delta\lambda$ is the spectral FWHM and the constant two depends on definition, spectral shape[50] and the nature of its experimental use, as discussed in Section 4.1. We will return to the subject of temporal coherence length and its dependence on spectral bandwidth towards the end of this section.

How then is spatial coherence measured? A property of coherent radiation is the ability to form modulated interference patterns ("fringe patterns") utilizing all emissions from the source. Assuming the radiation is sufficiently monochromatic ("quasi-monochromatic"), it should be possible to divide the radiation into two parts, then recombine them and form a fringe pattern. A spatially coherent source will produce fringes with 100% modulation. A partially coherent source, one whose phase-space is a bit too large, will result in a decreased modulation, and an essentially incoherent source, one whose phase-space areas is far too large (spatial extent too large, divergence too large, Eqs. (6.46a, b) not satisfied) will produce a fringe pattern of near zero modulation. A classic technique for measuring spatial coherence is Young's double-slit interferometer,[50] shown here in Figure 6.9, for several source characteristics relevant to this discussion. Young used this in his demonstration of the interference of light in 1802, supporting a wave nature not fully accepted at that time[50, 51]. The diagrams in Figure 6.9 show Young's geometry in four situations. The first, as Young used it with light focused to a small aperture of diameter d, and slits separated by half angles θ, with a fringe pattern displayed on a distant screen. Full (100%) modulation is obtained if d, θ, and λ satisfy Eqs. (6.46a, b), with an appropriate coefficient for the aperture shape.

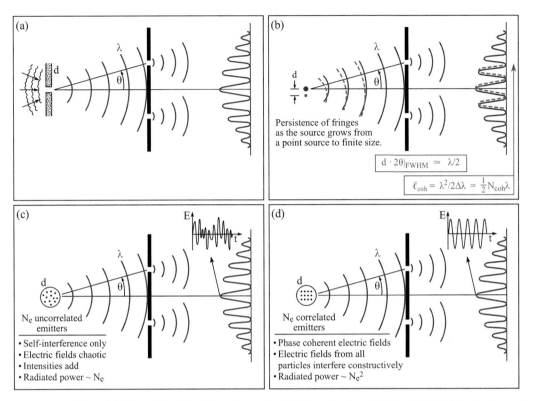

Figure 6.9 Young's double-slit interferometer for (a) an incoherently illuminated pinhole, where d, θ, and λ satisfy the spatial coherence requirement of Eqs. (6.46a, b); (b) a situation where several point radiators of random, uncorrelated positions are located within a region of dimension d that also satisfies Eqs. (6.46a, b), such that their individual fringe intensity patterns overlap, or nearly so, such that there is a "persistence of fringes" even with uncorrelated positions and motion; (c) a collection of N_e uncorrelated emitters all within a region of diameter d satisfying Eqs. (6.46a, b), producing a very high degree of fringe intensity modulation, even though the phases of the various electric fields are chaotic, and as a result radiated power scales as N_e; (d) an altogether different situation where the N_e emitters are perfectly arranged into a periodic pattern of wavelength λ such that the electrical fields add constructively, and the radiated power scales as N_e^2. Parts (c) and (d) remind us of the FEL process discussed in the text and illustrated in Figure 6.5. The phase coherence of (d) would emerge with SASE for a single microbunch, or for an FEL with a phase coherent seed.

As the pinhole size d is increased the fringe modulation is decreased, eventually going to zero for a large phase-space source, such as the sun or a conventional light bulb. The number of fringes observed is dependent on the coherence length [Eq. (6.47)], as was discussed in Chapter 4. For a point-source radiator that itself satisfies Eq. (6.46), the aperture in Figure 6.9(a) is not necessary, full modulation occurs unaided. Figure 6.9(b) shows a situation where two point sources, uncorrelated in phase, and separated by a small distance d, each generate separate high-modulation interference patterns, but the two are angle shifted by the geometry, resulting in a reduced composite modulation (reduced "fringe visibility"). With sufficient separation the crests of one will match the valleys of the other, significantly reducing visibility. Similarly, one could imagine a large number of phase independent point sources extending a distance greater than d, and again requiring an aperture. This is the technique of "spatial filtering" discussed in Chapter 4, Section 4.2, and illustrated for undulator radiation in Chapter 5, Section 5.6. Unlike an FEL, undulator radiation typically is characterized by a phase-space much larger than that given by Eqs. (6.46a, b). The diagrams particularly relevant to the evolution from incoherent undulator radiation to coherent free electron lasing are shown in Figures 6.9(c) and (d). Figure 6.9(c) shows a collection of N_e uncorrelated radiators, suggestive of the tight phase-space but short FEL undulator distances characteristic of the "spontaneous emission" region in Figure 6.5. Each of the radiators separately produces its own highly modulated interference pattern, and if all are contained within a phase-space defined by Eqs. (6.46a, b), the sum of intensities (fields do not add due to the random, uncorrelated motions) will overlap and a high modulation will be obtained. Note, however, that while the intensity patterns add, the electric fields are chaotic. As a result, the radiated power is proportional to N_e in Figure 6.9(d) and an altogether different situation is depicted. Now similar to the high gain, near-saturation FEL of Figure 6.5, the particles are well aligned by the SASE process into strongly correlated positions (an electron wave of well-defined phase) such that the radiated electric fields add constructively in the forward direction, and radiated power thus scales as N_e^2. As the number of electrons in an x-ray FEL microbunch may be of order 10^6 (Table 6.1, $Q = 0.25$ nC, 10^9 electrons but hundreds of independent microbunches, as illustrated in Figure 6.6), this orderliness of the radiators results in a significant power gain. Theoretical[52, 53] and numerical studies[54, 55] of the coherence properties of FELs, and factors affecting the degree of coherence, are discussed in the literature.

Note, too, that since the electron bunch diameter d, or 2σ, is fixed throughout its traversal through the long undulator, this implies that the half-angle of emission decreases from $\theta = 1/\gamma^*\sqrt{N}$, characteristic of undulator radiation for short magnet structures of N periods [Chapter 5, Eqs. (5.15) and (5.32)], to a value of $\theta_{coh} = \lambda/2\pi d$, or $\sigma'_{coh} = \lambda/4\pi\sigma$, for long distances, a very noticeable decrease. The determination that indeed SASE FEL lasing has occurred is then based on a very large increase of radiated power, an observed decrease in the emission angle, and a confirmation of spatial coherence by a Young's double-slit measurement. Young's double-slit technique has been used to measure the degree of spatial coherence at both LCLS[56, 57] and FLASH.[58, 59]

The spatial coherence of LCLS was measured, using the Young's double-slit technique described above,[56] plus some combinational patterns[57] which permit a more efficient and more complete coherence characterization. The measurements were made on a single shot basis during soft x-ray operation at 780 eV (1.59 nm wavelength), an operational mode that followed operations at 8 keV and 2 keV. The experiments were conducted in what is called a "diffract and destroy mode"[40, 41] in which the double-slits, double-pinholes, or other coherence measuring patterns were destroyed upon irradiation, but "slowly" due to their finite mass, so that diffraction of the femtosecond duration pulse was completed before the structure disassembled. Figure 6.10 shows sample data obtained with three different pinhole separations (thus three values of the angle θ)

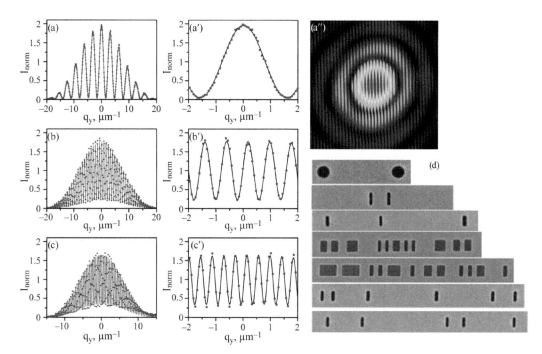

Figure 6.10 Interference patterns obtained at LCLS,[56] using 780 eV (1.59 nm wavelength) radiation[45–47] and three different pinhole separations: (a), and expanded scale (a'), obtained with a pair of 340 nm diameter pinholes at 2 μm separation, and a two-dimensional record of the interference pattern in (a''); (b) and (b'), pinholes at 8 μm separation; (c) and (c') with 500 nm diameter pinholes at 15 μm separation, corresponding to half angles of 1.2 μrad, 4.8 μrad, and 9.0 μrad, respectively. An SEM image of the test pattern array[57] is shown in (d), where one sees double pinholes, double and triple (off-centered) slits, and other patterns useful for characterizing the degree of spatial coherence. Portions (a) to (c) courtesy of I. Vartanyants, DESY.

and an SEM[¶¶¶] image of the nano-fabricated structure containing double-pinhole pairs of varying separation, double slits, triple slits, and non-redundant array patterns, all in multiple copies for sequential single-pulse diffract-and-destroy experiments.

These spatial coherence measurements at LCLS utilized a pair of Kirkpatrick–Baez glancing incidence mirrors (see Chapter 10, Section 10.2) to illuminate regions of the diffractive test pattern containing selected pinhole pairs, with focal spots of approximately 6 μm FWHM in the horizontal and 17 μm FWHM in the vertical direction. The mirrors were located 140 m beyond the last undulator, and the CCD[§§§] detector was 0.8 m beyond that. The irradiated regions were destroyed in a single shot and the test pattern then moved for the next measurement. For these soft x-ray SASE experiments, LCLS was operated at 4.3 GeV and 1.5 kA, with a 0.25 nC bunch charge. The electron beam size and divergence were $\sigma = 20$ μm and $\sigma' = 8.5$ μrad. Note that this product

---

[¶¶¶] Scanning Electron Microscope (SEM).
[§§§] A 2024 × 2024 pixel charge-coupled device (CCD).

corresponds to an electron beam emittance (phase-space area) that is about 1.4 times larger than that of spatially coherent radiation at 1.59 nm wavelength 780 eV. The active magnet length was 44 m, corresponding to 1469 periods, 3 cm each. The gain length was estimated to be 1.5 m, and the saturation length about 25 m. A typical energy of 1 mJ, in a 300 fs pulse, corresponds to peak radiated powers of order 3 GW, about 10^{13} ph/pulse. The recorded interference patterns showed some variability, possibly due to illumination non-uniformities, or perhaps due to pulse-to-pulse mode content variability. Analysis of interference patterns with the highest visibility indicates that for those pulses the measured spatial coherence across the beam was very close to the full beam width, and that although several modes appear to be present, 78% of the beam energy was contained in a single dominant TEM_{00} transverse mode. From some of the high-visibility interference patterns, those that displayed many cycles across the full CCD detector, it was possible to do a spectral analysis which indicated a relative spectral bandwidth, $\Delta\omega/\omega$, of $1/200$ FWHM, a value consistent with numerical simulations, but spectrally broader than that given by Eq. (6.45) where, for soft x-ray experiments near 800 eV, $\rho_{FEL} \simeq 1 \times 10^{-3}$. Further understanding of the coherence properties of LCLS is aided by very helpful numerical simulations[54, 55] which show that spatial coherence is maximized just before saturation. For hard x-rays at LCLS, essentially full spatial coherence has also been confirmed, in this case using a speckle interference technique.[60]

6.6   Seeded and Self-Seeded FELs

In many scientific studies the goal is to study femtosecond and attosecond electron dynamics in various atomic, molecular, and solid state systems. This is typically to be done by observing well-defined spectral features. Towards this end it is of great interest to control the FEL spectral bandwidth and phase as well as possible. Among methods to control the spectral, spatial, and temporal coherence of FEL pulses are the use of intense seed pulses of high coherence, and "self-seeding" in which the radiation spectrum is filtered mid-way through the lasing process, then amplified further at reduced bandwidth. Seeding requires an intense, fully coherent, external input pulse at the beginning of the undulator, synchronous with arrival of an electron bunch of similar pulse duration and repetition rate, with a full lateral overlap among the two, and with the coherent electric field of the seed laser sufficiently strong to clearly dominate spontaneous emissions in driving the FEL process. Plans for coherent seeding generally involve using an intense harmonic of a femtosecond duration atomic laser. "Self-seeding" of hard x-ray FELs to date have involved use of a forward scattering, single-crystal monochromator inserted at a distance within the FEL where the continuing, monochromatized x-ray beam is still sufficiently intense to dominate spontaneous emissions generated in that region.

The viability of coherent seeding with an intense, external laser harmonic is very much wavelength dependent. The present laser of choice is titanium:sapphire[61, 62] ($Ti:Al_2O_3$), which has a very broad spectral output centered at 780–800 nm, which can support 10–20 fs pulses in a highly coherent TEM_{00} mode (see Chapters 8 and 9), with per pulse energies of order 30 mJ at 10–20 Hz repetition rate. The first few harmonics, in the ultraviolet region (780 nm/3 = 260 nm), can be generated very efficiently in transmissive non-linear crystals.[63] Higher odd harmonics, in the region of $h = 31$–61 (approximately 13–25 nm; 49–97 eV) are efficiently generated when focused to intensities of order 10^{14} W/cm^2 in inert gases such as neon and argon, resulting in harmonic combs (many harmonics; see Chapter 7) with energies per harmonic of several nJ/pulse in the EUV, dropping off sharply beyond 100 eV. Single EUV harmonics can then be selected, directed, and focused to overlap with FEL electron bunches. The technique has been demonstrated in several EUV FEL laboratories.[64–69] Successful use requires a seed intensity of order 100 times that of the undulator spontaneous emission,||||| within the first few gain lengths. This gives a factor of 10 stronger electric field driving the FEL amplification process. This is necessary to overcome the largest noise fluctuations. Transport from the inert gas EUV generation site to the electron bunch can be an issue depending on the number and efficiency of required mirrors (see Chapter 10). Seeding beyond the EUV, in the soft x-ray region is difficult to implement due to the rapidly declining energy per pulse achievable at these shorter wavelengths (see Chapter 7). Rather, for soft x-rays, coherent seeding is supplemented by further frequency multiplication (higher harmonic photon energy) using what is known as a "high gain harmonic generation" (HGHG) which, as the name implies, involves the generation of high undulator harmonics.[70–75] This combination of techniques has been demonstrated at the FERMI**** EUV/soft x-ray FEL in Trieste, Italy. Starting with the third harmonic of Ti:sapphire at 260 nm (780 nm/3) and then two stages of HGHG, tunable, phase-coherent radiation at wavelengths as short as 4 nm have been achieved[76–81]. The FLASH facility in Hamburg, Germany, which currently operates in the EUV/soft x-ray region[37–39] is considering an upgrade to FLASH II, including an option for an EUV HHG harmonic followed by the HGHG technique.[82] For hard x-rays there is no clear path at this time to coherent seeding, but "self-seeding" techniques,[83–86] which essentially consist of mid-path spectral filtering, provide considerable value in reducing spectral bandwidth to values better matched to many scientific applications, while still achieving high power at saturation. The technique does not, however, provide temporal phase control.

Results of the first hard x-ray self-seeding experiments[85, 86], conducted at the LCLS, are shown in Figure 6.11. The experiments were performed at 8.3 keV (1.5 Å) in a low-charge (40 pC) operating mode, generating nominally 10 fs pulses. Only half of the available undulator sections were used (U3–U15, of 33 total) before insertion of a single diamond crystal monochromator operating in a forward Bragg diffraction mode. Transmission of the 8.3 keV photons through the crystal involves a delay of 19 fs, which is matched by use of a short, horizontally deflecting delay line for the electrons (a "chicane"). This allows the electrons to bypass the crystal and provides temporal overlap between electrons and photons as they are rejoined in the second set of undulators.

||||| Based on Eq. (5.39), but interpreted for peak power, $\hat{P}_{cen}$, and peak current, $\hat{I}$. Density fluctuations, and therefore current fluctuations which drive the SASE process, suggest taking a safety factor of at least five. An additional safety factor may be required for particular experiments to assure sufficient dominance over amplified noise fluctuations to provide acceptable contrast relative to the broader, spectrally integrated "background" signal.

**** Honoring physics Nobel Laureate, Enrico Fermi; Free Electron LaseR for Multidisciplinary Investigations.

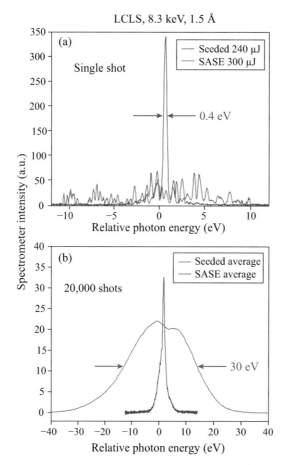

Figure 6.11 LCLS hard x-ray spectra (8.3 keV, 1.5 Å) compared for "self-seeded" and unseeded SASE FEL lasing,[85] on (a) single-pulse basis, and (b) averaging over 20 000 pulses. The "self-seeding" mode of operation includes a transmission x-ray crystal monochromator about half way through (~60 m) the full length of LCLS periodic magnet structures. Note that with monochromatization by self-seeding the single pulse FEL bandwidth is narrowed by a factor of about 50, from 20 eV to 0.4 eV. The narrowed relative spectral bandwidth of 2×10^4 stretches the temporal coherence length to about 2 μm. Courtesy of P. Emma, SLAC.

This has the additional advantage of clearing short period density modulations within the bunch, thus providing a relatively uniform density of electrons for "self-seeding" by the narrowed bandwidth 8.3 keV radiation. The monochromator is located at a distance of about 60 m, corresponding in this case to a traversed (active) undulator distance of 44 m (1469 periods at 3 cm each). Beyond the monochromator the filtered beam traveled through another 13 undulator segments (U17–U29), also 44 m of active magnet structure. Output energies as high as 240 μJ/pulse (~2 × 10^11 ph/pulse; 20 GW peak power) were achieved with the crystal (SASE plus self-seeded). The surviving SASE signal is spread across a broad spectral bandwidth, typically 20 eV FWHM, for a single pulse, while the self-seeded signal is spectrally confined to a bandwidth of just

0.4 eV FWHM, thus a considerably higher photon flux per eV. Of the broad 30 eV band-width observed over many pulses, as seen in Figure 6.11(b), about 20 eV is due to the spectral bandwidth of any given pulse, while the remainder is due to pulse-to-pulse elec-tron beam energy variations. The relative spectral bandwidth at 8.3 keV thus narrows to about $1/20\,000$ FWHM. Figure 6.11(a) shows a comparison of SASE and self-seeded spectra for single pulses of similar energy. The spectral advantage for atomic probing is significant. The temporal coherence length is extended to about 2 μm. The spectra were obtained with the same Si (333) thin bent crystal spectrometer shown earlier in Figure 6.8. The rms intensity jitter on a pulse-to-pulse basis, however, was about 50%. Figure 6.13(b) shows data averaged over 20 000 pulses, in this case obtained with the broader spectral range Si (111) crystal described in Figure 6.8. The 0.4 eV vs. 20–30 eV FWHM spectral advantage is evident. When continued to saturation in future experi-ments with added undulator sections, a saturated peak power is expected, with reduced pulse-to-pulse intensity variations. Additional options for hard x-ray self-seeding are considered in the literature.[87–90]

As mentioned earlier in this section, direct seeding with an external coherent laser pulse offers the possibility of imprinting the full phase coherence of an external laser on the emergent FEL pulse at much shorter wavelength, and with significantly improved pulse energy stability. Typically one hopes that this can be done with a high harmonic of an intense femtosecond duration pulse (HHG, Chapter 7) focused to a waist diameter approximately equal to, or perhaps somewhat larger[54] than, that of the electron beam. While it is anticipated that this may eventually be accomplished with an EUV harmonic of a femtosecond duration titanium sapphire ($Ti_2O_3 : Al_2O_3$), to date there has been only limited success. Rather, the FERMI EUV FEL in Trieste employs as a seed the third harmonic of a Ti:sapphire laser, generated very efficiently at 260 nm in a solid state crystal,[63] to drive the high gain, high harmonic (HGHG) process.[70–78] The latter utilizes undulator harmonics coupled to the FEL lasing process, to reach very high har-monics with good spatial and temporal coherence properties, and much improved inten-sity statistics. Undulator harmonics, used here in the HGHG multiplication process, are discussed in Chapter 5, Sections 5.5 and 5.6.

The seeded HGHG process[††††] used at FERMI involves injection of intense, coherent laser light (260 nm at FERMI) into a short undulator (the "modulator") at moderately high K-value (3–8; 100 nm period), overlapping with the nominal 1.24 GeV ($\gamma = 2430$) electron beam and resonantly driving strong transverse velocity oscillations [Eq. (6.1) at 260 nm]. The electron beam next traverses a dispersion section where the velocity modulation evolves to an axial density modulation, as we have seen earlier for the basic FEL process. Because the resonant interaction in the "modulator" is at rather high K (7.8) there are many high harmonics present, and also in the dispersion section, where energy modulations evolve to density modulations. Recall that at high K-values the

---

[††††] A precursor to what is now known as the HGHG process was first suggested by Bonifacio.[73] Later the concept was significantly extended by Yu and colleagues at Brookhaven National Laboratory, includ-ing experiments with CO_2 laser injection at 10.6 μm wavelength and later with Ti:sapphire at 800 nm wavelengths, the latter producing high-power FEL lasing at 266 nm.

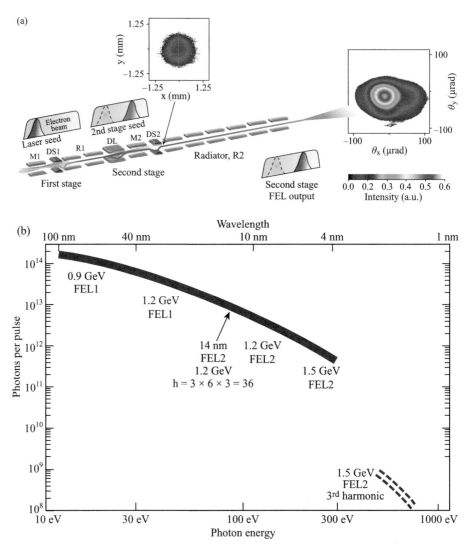

Figure 6.12 Component geometry and photon energy tuning curves for the FERMI free-electron laser.[76–81] (a) A diagram[77] of FERMI's FEL-2 showing a coherent seed pulse at 260 nm plus two stages of HGHG. Components include modulators M, dispersion sections DS, and radiators R. A "fresh" region of the somewhat longer electron bunch is used for the the second stage of HGHG. Images of the beam and its divergence are shown at different stages of the FEL. (b) Broad tunability, from 100 nm to 4 nm, is achieved by using one or two sectors of HGHG (FERMI FEL-1 or FERMI FEL-2), combined with magnetic undulator K-tuning and varied electron beam energies from 0.9 to 1.5 Gev. Harmonics as high as 195 (3 × 13 × 5) have been generated, corresponding to a wavelength of 4.0 nm (310 eV), with 20 μJ per puse, or 4 × 10¹¹ photons/pulse. Dashed lines indicate expectations for future harmonic radiation. Courtesy of L. Giannessi and M. Svandrlik[77], FERMI, Trieste.

number of undulator harmonics scales as K^3 [see Chapter 5, Eq. (5.82)]. The modulated electron bunch, with considerable harmonic content, then enters a second, long undulator, the so-called "radiator" ($\lambda_u = 55$ mm, $N = 262$) where it is K-tuned for selection of a particular harmonic, for example $K = 3.45$ for resonant enhancement [Eq. (6.1)] and growth of the $h = 24$ harmonic (3×8) at 32.5 nm wavelength. The undulator used is an Apple type II,[91, 92] capable of generating either linear or circular polarization depending on the selected configuration of the periodic magnets [see Chapter 5, Section 5.4.8]. At the twentyfourth harmonic (32.5 nm wavelength) FEL amplification has generated peak powers of several hundred MW, with both planar and circular polarization, in nominally 85 fs pulses,[76] with measured gain lengths of order 2 m, approaching saturation with both polarizations. Importantly, shot-to-shot pulse energy variability was observed to have a standard deviation of only 10%, and a normalized central wavelength stable to about 1.5 parts in 10^4 FWHM, a great improvement when compared to the unseeded SASE process. Young's double slit experiments showed a high degree of spatial coherence,[76, 79, 81] while a high-resolution monochromator confirmed spectral stability and a measured bandwidth of 47 meV at 38.5 eV photon energy (32.5 nm wavelength) for a relative spectral bandwidth of 1.2×10^{-3} FWHM,[76] which is about equal to $\rho_{FEL} = 1.3 \times 10^{-3}$ for these experiments. In later experiments, referred to as FERMI FEL-2, the same 260 nm seed is used in a double FEL HGHG cascade technique,[77] with sequential modulators and radiators, to further extend the harmonic process. A schematic diagram and broad tuning curves are shown in Figure 6.12. This cascaded technique has extended the double HGHG, coherent seed FEL amplification process to 4.0 nm (310 eV, $h = 3 \times 13 \times 5 = 195$) with an electron beam energy of 1.5 GeV. The special advantage of both spatial and longitudinal coherence has permitted the development of coherently controlled, two-color photoemission studies to asec phase accuracy.[78]

Ultimate control of spatial and temporal coherence properties of an x-ray FEL including temporal phase control, will come with the introduction of cavity optics in an FEL oscillator. Efforts towards this end have begun[93–98] for hard x-rays based on the formation of an x-ray optical cavity using high reflectivity, near normal incidence crystal optics at opposite ends of an active amplification e-beam/FEL, operating at wavelengths of order 0.5–1 Å, with limited tunability. Lasing would be limited to a very narrow spectral bandwidth, of order $\Delta\omega/\omega \simeq 10^{-4}$, with tight phase control. Options for a soft x-ray FEL oscillator are also considered.[99, 100]

6.7   Pump-Probe Capabilities

"Two-color" (two photon energies) pump-probe capabilities provide a powerful tool for studying the dynamic response of systems in which the first pulse perturbs or changes the system and the second records the reaction as a function of time, set by the delay between the two. The technique promises to be a powerful tool in both the EUV and x-ray spectral regions where primary, well-defined resonances exist for all atoms. Femtosecond dynamical studies of atomic and molecular systems are just becoming

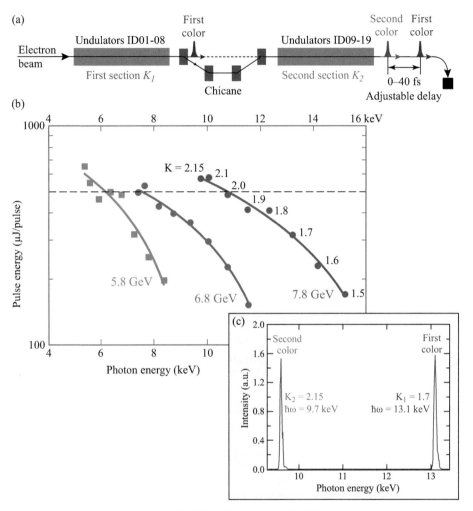

Figure 6.13 Broad photon energy availability, with up to a 30% difference between the two time-separable pulses, is available at SACLA.[101] (a) Schematic layout of the SACLA undulators configured for two-color pump-probe experiments showing a first section of eight periodic magnet modules tuned to K_1, followed by a chicane to deflect and delay the electron bunch before entering the remaining 11 modules tuned to K_2. By the undulator Eq. (6.1), K_1 and K_2 generate different photon energies at the chosen electron beam energy (γ). (b) Single color photon energies for three distinct electron beam operating energies. (c) An example of two widely separated photon energies generated with equal intensity at an electron beam energy of 7.8 GeV. Time delay between the two photon beams, shown at 9.7 and 13.1 keV, can be varied from zero to 40 fs, set by the selected geometry of the chicane. Pulse energies above 250 μJ per pulse can be achieved at all electron beam energies. Courtesy of T. Hara and H. Tanaka, SACLA.

available, with extensions to attosecond dynamics likely in the near future, opening significantly new research opportunities in areas such as core level atomic state dynamics, ultrafast chemistry, and solid state physics. Two-color pump-probe capabilities presently exist in the EUV at FERMI[78, 80], and for x-rays at LCLS[49] and SACLA.[101]

A two-color, hard x-ray pump-probe capability,[101] with both broad tunability and broad "color" separation has become available at the SACLA[‡‡‡‡] FEL facility [102–109] in Hyogo Prefecture, Japan, contiguous to the SPring-8 synchrotron radiation facility. This unique capability provides broad tunability, from 5–15 keV photon energies, delivered in sub-10 fs pulses at 60 Hz, with variable time-delays from 0 to 40 fs.

As illustrated in Figure 6.13(a), the undulator modules are divided into two groups separated by a variable delay electron chicane.[101] The first group of variable gap periodic magnet modules is set at a selected undulator parameter K_1, and the second group at a different value, K_2. At the selected electron beam energy (5.7–8.3 GeV; $\gamma = 11\,000$–16 000), these K-values generate nominally 7 fs pulses of photon energies extending from 5 to 15 keV, capable of spanning an energy spread of about 30% at any given beam energy, as seen in Figure 6.13(b). An example is shown in Figure 6.13(c) for a beam energy of 7.8 GeV. With the first group of undulator modules set at $K_1 = 1.7$, photons of energy 13.1 keV are generated. With the second group of modules set at $K_2 = 2.15$, photons of energy 9.7 keV are generated. The time delay between the two pulses is independently controlled by selection of electron path length in the electron chicane, adjustable from 0 to 40 fs with sub-femtosecond accuracy. The number of undulator modules in the two groups can be adjusted to control the relative intensities of the two pulses. Generally the second group is larger in number to compensate for increasing electron beam energy spread (σ_γ), which extends the FEL gain length. Pulse energies of up to 0.5 mJ ($\sim 4 \times 10^{11}$ photons/pulse at 8 keV) are obtained. Angular separations are also possible, providing additional flexibility for sample illumination.

An example of new science enabled by the two-color capability at SACLA involves the generation of atomic x-ray lasing[¶¶¶¶] on the copper $K_{\alpha 1}$, $K_{\alpha 2}$ doublet lines at wavelengths of 1.541 Å and 1.544 Å. In this experiment[110] a 7.0 GeV electron bunch traverses two sets of undulators, as in Figure 6.13, but without use of the chicane. The first string of eight SACLA undulators are magnetically tuned at $K = 1.93$ to radiate at 9.05 keV. This photon energy is just above the copper K-absorption edge at 8.979 keV, ideal for removal of $n = 1$ core electrons and thus serving as a "pump," creating a population inversion in each affected atom (see Chapter 1, Figure 1.11 and Appendix B.1). The FEL photons are focused to spot sizes of 50–120 nm FWHM on a 20 μm thick copper foil placed beyond the last undulator. Focusing at these uniquely small spot sizes is accomplished with a newly developed two-stage Kirkpatrick–Baez mirror set[111] (KB system, four mirrors total, see Chapter 10). With a nominally 0.03 mJ, 7 fs, x-ray pulse at 9.05 keV, focused to spot sizes of 50–120 nm diameter,[111] illumination intensities as high as 10^{20} W/cm² are achieved, with an incident photon flux of 2×10^{10} photons/pulse. This number of photons is sufficient to not only remove all $n = 1$ electrons from the Cu atoms within a conventional absorption length of about 4 μm, but is sufficient to continue ionizing Cu atoms (from $n = 1$) as the photons penetrate ever deeper into the increasingly transparent, already bleached foil.[112] This creates a cylinder of population inverted Cu atoms as small as 120 nm in diameter and 20 μm

---

‡‡‡‡ SPring-8 Angstrom Compact free electron LAser (SACLA).
¶¶¶¶ See Chapter 9, Extreme Ultraviolet and Soft X-Ray Lasers, and Reference 129 in this chapter.

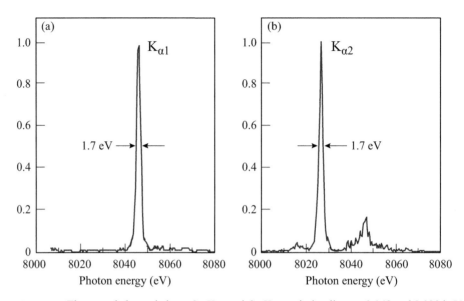

Figure 6.14 The recorded atomic laser Cu $K_{\alpha 1}$ and Cu $K_{\alpha 2}$ emission lines at 8.048 and 8.028 keV, are shown in (a) and (b), respectively.[110] The corresponding wavelengths are 1.541 and 1.544 Å, respectively. The measured bandwidths were 1.7 eV in both cases. Numerical simulations suggest the lines were of 3–4 fs duration. Two sequential FEL pulses were used in a pump and seed sequence. The pump pulse at 9.05 keV, just above the 8.979 keV K-absorption edge of copper, arrived first to generate a volume of population inverted Cu atoms. The second pulse, at 8.05 keV, arrived at the copper sample just 1 fs later. This "seed" pulse was then amplified by stimulated emission as it traversed the region of population inverted Cu atoms. The intense pump and seed radiation were focused to diameters of 50–120 nm FWHM by a special four-mirror Kirkpatrick–Baez reflective mirror pair,[111] corresponding to intensities of 10^{19} to10^{20} W/cm^2. This stripped most Cu atoms in the irradiated area of their core $n = 1$ electrons, leaving the atoms in an inverted population state. The intensity of the beam, and number of photons, were sufficient to create a cylinder-like region of population inverted Cu atoms with a length to diameter aspect ratio of 170:1. The inverted population was then amplified by the 8.05 keV FEL seed pulse in the long direction of the cylinder, which set the resultant divergence angle. These are the first ångström (Å) wavelength, narrow linewidth, atomic inner-shell x-ray laser lines amplified by stimulated emission of radiation. Courtesy of H. Yoned, U. Electro-Communications, Tokyo, and Y. Inubushi and M. Yabashi, SACLA.

deep, corresponding to an aspect ratio of 170:1. Just as the population of inverted atoms might begin to fluoresce in all directions, with $n = 2$ to $n = 1$ transitions to the Cu 8.028/8.048 keV doublet, a second FEL pulse, undulator tuned to $K = 2.11$ for radiation at 8.04 keV, arrives and dominates radiation within the excited Cu cylinder through stimulated emission. This second pulse, which seeds the stimulated emission, arrives ~1 fs after the first, delayed only by the velocity difference between electrons and radiation in the first set of undulators. This short arrival time delay of the seed pulse is critical as the natural life time of the $n = 2$ to 1 transition is only about 1 fs. The cylinder of population inverted Cu atoms provides a preferential path for the stimulated emission of radiation and thus defines the divergence angle of the Cu$_{k\alpha}$ laser emission. As discussed

Table 6.2 Nominal parameters and capabilities of EUV and x-ray FELs. The range of values is typically broader than shown, and frequently upgraded. For example, pulse durations are now pushing to 10 fs and less, with improved spectral definition. Several of the facilities are planning extensive upgrades to improve both operating parameters and user access. An upgraded version of this table will be maintained at the book's website http://www.cambridge.org/xrayeuv

Parameters	FLASH FEL (Hamburg 2005)	Fermi (Trieste, 2010)	LCLS (Stanford, 2009)	SACLA (Hyogo, 2011)	Swiss FEL (PSI, 2017)	PAL (Pohang, 2016)	EU XFEL (Hamburg, 2016)
E_e	1.25 GeV	1.5 GeV	13.6 GeV	8 GeV	5.8 GeV	10 GeV	17.5 GeV
γ	2450	2900	26600	15700	11300	19600	35000
$\hat{I}$	1.3 kA	300 A	3.4 kA	9 kA	3 kA	3 kA	5 kA
λ_u	27.3 mm	55 mm	30 mm	18 mm	15 mm	26 mm	40 mm
N	989	216	3733	4986	3200	3456	4375
L_u	27 m	14 m	112 m	90 m	48 m	90 m	175 m
$\hbar\omega$	30–300 eV (4–40 nm)	20–300 eV (4–60 nm)	250 eV–10 keV (1.2–50 Å)	4.5–15 keV (0.8–2.8 Å)	2–12 keV (1–6 Å)	2–12 keV (1–6 Å)	3–25 keV (0.05–4 Å)
$\lambda/\Delta\lambda_{FWHM}$	100	1000	200–500	200–400	200–500	200–500	1000
$\Delta\tau_{FWHM}$	50 fsec	85 fsec	<10–70 fsec	<10 fsec	2–20 fsec	5–60 fsec	100 fsec
$\tilde{\mathcal{F}}$ (ph/pulse)	3×10^{12}	5×10^{12}	2×10^{12}	4×10^{11}	10^{11}	5×10^{11}	10^{12}
rep rate	1 MHz @ 10 Hz	10 Hz	120 Hz	60 Hz	100 Hz	60 Hz	2.7 kHz @ 5 Hz
$\hat{P}$	1 GW	1 GW	30–60 GW	60 GW	3 GW	15 GW	20 GW
L	260 m	200 m	2 km	710 m	730 m	1.1 km	3.4 km
Polarization	Linear	Variable	Linear	Linear	Linear	Linear	Linear
Mode	SASE (3ω Ti: sapphire)	Seeded	SASE & self-seeded	SASE & self-seeded	SASE & self-seeded	SASE & self-seeded	SASE

in Chapter 9, this is the process of atomic lasing, *light amplification by stimulated emission of radiation*, in this case an FEL pumped atomic inner-shell x-ray laser at 8 keV. Spectra are recorded with a flat Si (111) dispersive crystal spectrometer.[44] Examples of the resultant Cu K_α emission[110] are shown in Figure 6.14. The measured linewidth in each case is 1.7 eV. Numerical simulations, constrained by the measured data, indicate pulse durations of 3–4 fs. Both lines can be generated separately, depending on spectral spikes within the 40 eV bandwidth of the FEL seed pulse. These data represent the first achievement of hard x-ray inner-shell lasing based on narrow linewidth atomic transitions,[110] and are of further significance in that the radiated pulses are nominally only a few fsec duration, and at a 1 Hz repetition rate. A broad range of emerging scientific applications, including unique non-linear studies, can be expected to follow using this high-intensity, two-color x-ray capability.

6.8 Current FEL Facilities, Parameters, and Capabilities

A brief summary of EUV and x-ray free electron laser facilities is presented in Table 6.2. Early references to FLASH,[113, 36–39] to LCLS,[33–36] to FERMI,[114, 76, 78] and to SACLA[102, 104] are in the literature. Important parameters include the electron beam energy and γ, the peak current, undulator parameters (λ_u, N, and active magnet structure length L_u), the photon energy (and wavelength) range covered, typical photon flux (photons/pulse), peak radiated power, $\hat{P}$, pulse duration ($\Delta\tau$), repetition rate, polarization options, and mode of operation (SASE or seeded; with self-seeded understood as an option for SASE facilities). The European x-ray FEL (EU XFEL)[115] is expected to reach a higher photon energy. The Swiss FEL (SwissFEL),[116] with a planned beam energy of 5.8 GeV, nominal photon energy range of 200 eV to 15 keV, and a high repetition rate, will have multiple user stations. The PAL FEL facility in Pohang is also well underway.[117] All three are expected to be completed in 2016.

6.9   Scientific Applications of Coherent, Intense, and Short Duration (fs/as) EUV and X-Ray FEL Radiation

The dramatic appearance of x-ray FEL user facilities, after decades of analysis and planning, brings significant new capabilities and opportunities to the scientific community. The spatially coherent x-rays, now at gigawatt power levels and femtosecond pulse durations open opportunities for dynamical studies with femtosecond (fs) and likely attosecond (as) pulses tuned to the primary energy resonances of atoms in molecular, biological, and solid state materials. The range of scientific areas affected include atomic, molecular and optical (AMO) sciences,[118–129] clusters,[130] surface science,[131–132] solid state physics,[133–136] magnetism,[137–140] and nanoscale diffractive imaging and protein crystallography.[141–151] Here we sample just a few. A relevant and important technique, first demonstrated at FLASH,[40, 41, 152, 153] involves what is now known as "diffract-before-destroy,"[142] in which the x-ray interaction with the sample is completed in femtoseconds, before the inertia of the finite mass atoms permits them to move from their initial positions, before the sample is completely destroyed. In this manner the sample can be imaged, or otherwise probed, before it disassembles.

Figure 6.15 shows the recorded diffraction pattern (a) from a sub-micron test pattern irradiated by an intense 62nm wavelength, 25 fs duration EUV pulse from FLASH, and (b) the successfully reconstructed image of stick figures, using the technique known as "coherent diffractive imaging,"[40, 79, 143, 146, 154–157] (CDI; see Chapter 11). The technique has broad interest as it has the potential for imaging nanoscale samples without the limitations of available lenses, assuming a convergence of reconstruction techniques with available degrees of spatially coherent illumination. Significant opportunities also exist for efficient, high resolution massively parallel x-ray holography.[158]

An early example of femtosecond dynamic probing of atoms[118, 119] is illustrated in Figure 6.16, based on 800 eV to 2 keV experiments at LCLS, where the intense, tuned x-rays were essentially able to effectively remove core K-shell electrons, producing "hollowed out" neon atoms and, concomitantly, an x-ray induced transparency. Indeed for photon energies of 2 keV, and at intensities of order 10^{18} W/cm^2 (10^5 x-ray photons/Å^2), it is possible to fully strip the neon atoms to Ne^{+10}.

Other AMO related research addresses the dynamics of charge redistribution following multiple ionizations of poly-atomic molecules, double core-hole formation in small molecules, pump-probe dynamical studies,[123] and the creation of FEL pumped, very

(a)

(b)

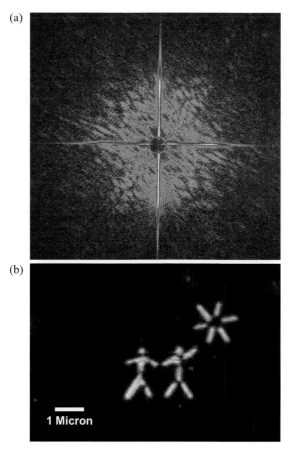

1 Micron

Figure 6.15 An early demonstration of the femtosecond "diffract-before-destroy" technique.[40, 41] (a) The recorded diffraction pattern, at 62 nm wavelength and 25 fs duration, of a sub-micron test pattern, obtained in early experiments at FLASH. (b) The reconstructed image of stick figures using the reconstruction technique known as CDI, "coherent diffractive imaging." Courtesy of H. Chapman, CFEL, DESY and U. Hamburg.

narrow linewidth, L- and K-shell atomic x-ray lasers.[129, 110] With x-ray focal intensities[111] now of order 10^{16} W/cm^2 to 10^{20} W/cm^2 there is great interest in highly non-linear x-ray experiments such as x-ray frequency doubling,[159] two-photon Compton scattering,[160] and others, as well as four-wave mixing techniques such as visible/IR with either EUV or x-ray FEL radiation. Each of these techniques has served as a powerful tool at longer wavelengths, now demonstrated with the atomic selectivity of spectrally tuned EUV[161] and x-ray[162] radiation.

A scientific opportunity of significant interest is that x-ray protein nanocrystallography, an effort already well underway[141–151] to determine the structure of proteins for which larger crystals cannot be grown. If these small crystals are exposed to intense synchrotron radiation, damage is incurred before a complete diffraction data set can be obtained (larger crystals have the advantage of manyfold parallel addition of

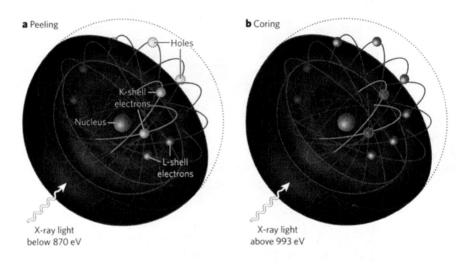

Figure 6.16 Intense, fsec duration x-rays tuned to appropriate K or L-edges of neon can completely strip the atoms of all electrons, producing N_e^{+10}, or selectively remove just the ls electrons, producing the "hollowed out", N_e^{+2} atoms.[118] Figure[119] is courtesy of J. Wark, University of Oxford.

intensities). With femtosecond x-ray pulses, the proposition is that the nanocrystals will survive intact until the diffraction process is complete, another application of "diffraction before destruction."[142] Because the diffraction pattern from a single nanocrystal is necessarily small, a technique is needed for sequentially exposing a large number of identical crystals.

Figure 6.17 illustrates a technique for sequential, high repetition rate, femtosecond x-ray protein nanocrystallography[141] in which a stream of identical crystals were passed in front of 1.8 keV (0.69 nm), 70 fs duration, x-rays at LCLS. The nanocrystals consisted of a membrane protein complex known as "*photosystem-1.*" The individual nanocrystals ranged in size from 200 nm to 2 μm, and were delivered at a repetition rate of 30 Hz. The delivery system consisted of a 4 μm diameter water jet, well matched to the KB focused x-ray beam (see Chapter 10) of nominally 7 μm diameter. The choice of photosystem-1, an important membrane protein complex, centered on its availability in much larger sizes, suitable for conventional synchrotron crystallography, so that there would exist an accurate data set with which to compare the new nanocrystallography technique. The geometry of these first results is shown in Figure 6.17(a); recorded diffraction patterns for 15 000 nanocrystals are shown in (b), and a calculated electron density map from the data is shown in (c). Comparisons of this reconstruction with electron density maps obtained with larger crystals and synchrotron radiation is favorable. Further evaluation of the technique, and extensions to a wider class of nanocrystals, is underway.

Future scientific opportunities are plentiful for EUV, soft x-ray, and hard x-ray free electron lasers with their high power, femtosecond pulses, and spatial coherence. Opportunities will increase as capabilities improve for the control of emission spectra.

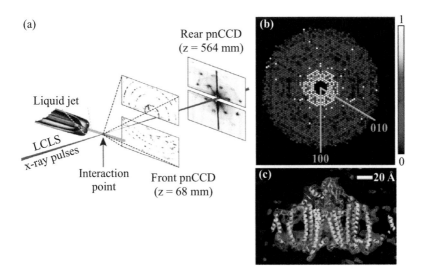

Figure 6.17 Femtosecond x-ray protein nanocrystallography is demonstrated with 1.8 keV, 70 fs duration x-ray pulses at LCLS.[141] (a) A water jet delivery system that provides 30 nanocrystals per second. A recorded diffraction pattern (b) obtained for 15 000 sequential nanocrystals, and (c) the calculated electron density distribution based on the recorded diffraction pattern in (b). Courtesy of H. Chapman, CFEL, DESY and U. Hamburg.

Dynamical studies with femtosecond, and eventually attosecond, pulses are likely to play a major role in scientific advances based on the short wavelength FELs and appropriately advanced optical techniques. The temporal probing of atomic, molecular, and solid state electron dynamics are likely to provide particularly rich opportunities. Studies of valence electrons, typically held with a few eV to tens of eV binding energies, are presently being studied with laser high harmonic generation (HHG) techniques, already reaching well into the attosecond region (see Chapter 7). Studies involving core electron dynamics, typically characterized by hundreds to tens of thousands of eV binding energies, are clearly the domain of x-ray free electron lasers. Currently the shortest x-ray FEL pulses are 3–7 fs or less,[49, 104, 108] but the accelerator technology for shorter pulses is rapidly evolving[163] and theoretical limits are distant. For bandwidth limited pulses, having a Gaussian spectral distribution (no chirp), the Heisenberg Uncertainty Principle[164] for photon energy bandwidth and pulse duration, stated here in terms of FWHM measures, is [Chapter 4, Eq.(4.4b)]

$$\Delta E_{\mathrm{FWHM}} \cdot \Delta t_{\mathrm{FWHM}} \geq 1.825 \, \mathrm{eV} \cdot \mathrm{fsec} \qquad (6.48)$$

For example, a bandwidth limited Gaussian pulse of 100 as requires a spectral bandwidth of about 18 eV. For 8 keV photons this corresponds to a relative spectral bandwidth of approximately 1/400, definitely within the range of possibilities. Considerable literature is emerging with regard to the best methods for generating attosecond duration x-ray pulses[163–172].

References

1. J.M.J. Madey, "Stimulated Emission of Bremsstrahlung in a Periodic Magnetic Field," *J. Appl. Phys.* **42**, 1906 (April, 1971); J.M.J. Madey, PhD thesis, Stanford University (November 1970).

2. J.M.J. Madey, "Stimulated Emission of Radiation in Periodically Deflected Electron Beam," *United States Patent*; 3,822,410 (July 2, 1974; filed May 8, 1972).

3. L.R. Elias, W.M. Fairbank, J.M.J. Madey, H.A. Schwettman and T.I. Smith, "Observation of Stimulated Emission of Radiation by Relativistic Electrons in a Spatially Periodic Transverse Magnetic Field," *Phys. Rev. Lett.* **36**, 717 (1976).

4. D.A.G. Deacon, L.R. Elias, J.M.J. Madey *et al.*, "First Operation of a Free-Electron Laser," *Phys. Rev. Lett.* **38**, 892 (1977).

5. W.B. Colson, "One-Body Electron Dynamics in a Free Electron Laser," *Phys. Lett. A* **64**, 190 (1977).

6. J.M.J. Madey, "Invention of the Free Electron Laser," *Rev. Accel. Sci. Techn.* **3**, 7. 1 (2010).

7. J.M.J. Madey, "Wilson Prize Article:From Vacuum Tubes to Lasers and Back Again", *Phys. Rev. Spec. Topics* **17**, 074901 (July 2014).

8. A.M. Kondratenko and E.L. Saldin, "Generation of Coherent Radiation by a Relativistic-Electron Beam in an Undulator," *Sov. Phys. Dokl.* **24**, 986 (1979).

9. A.M. Kondratenko and E.L. Saldin, "Generation of Coherent Radiation by a Relativistic Electron Beam in an Ondulator," *Particle Accel.* **10**, 207 (1980).

10. Ya.S. Derbenev, A.M. Kondratenko and E.L. Saldin, "On the Possibility of using a Free Electron Laser for Polarization of Electrons in a Storage Ring," *Nucl. Instr. Meth.* **193**, 415 (1982).

11. R. Bonifacio, C. Pellegrini and L.M. Narducci, "Collective Instabilities and High-Gain Regime in a Free Electron Laser," *Optics Commun.* **50**, 373 (1984).

12. J.B. Murphy, C. Pellegrini and R. Bonifacio, "Collective Instability of a Free Electron Laser Including Space Charge and Harmonics," *Optics Commun.* **53**, 197 (1985).

13. J.B. Murphy and C. Pellegrini, "Generation of High-Intensity Coherent Radiation in the Soft-X-Ray and Vacuum-Ultraviolet Region," *J. Opt. Soc. Am. B* **2**, 259 (1985).

14. K.J. Kim, J.J. Bisognano, A.A. Garren, K. Halbach and J.M. Peterson, "Issues in Storage-Ring Design for Operation of High-Gain FEL," *Nucl. Instr. Meth. A* **239**, 54 (1985).

15. K.-J. Kim, "Three-Dimensional Analysis of Coherent Amplification and Self-Amplified Spontaneous Emission in Free-Electron Lasers," *Phys. Rev. Lett.* **57**, 1871 (1986); K.-J. Kim, "An Analysis of Self-Amplified Spontaneous Emission," *Nucl. Instr. Meth. A* **250**, 396 (1986).

16. J.-M. Wang and L.-H. Yu, "A Transient Analysis of a Bunched Beam Free Electron Laser," *Nucl. Instr. Meth. A* **250**, 484 (1986).

17. L.-H. Yu, S. Krinsky and R.L. Gluckstern, Calculation of Universal Scaling Function for Free-Electron-Laser Gain," *Phys. Rev. Lett.* **64**, 3011 (June 1990).

18. C. Pellegrini, J. Rosenzweig, H.D. Nuhn *et al.*, "A 2 to 4 nm High Power FEL on the SLAC Linac," *Nucl. Instr. Meth. A* **331**, 223 (1993).

19. R. Bonifacio, L. De Salvo, P. Pierini, N. Piovella and C. Pellegrini, "Spectrum, Temporal Structure, and Fluctuations in a High-Gain Free-Electron Laser Starting from Noise," *Phys. Rev. Lett.* **73**, 70 (1994).

20. M. Xie, "Exact and Variational Solutions of 3D Eigenmodes in High Gain FELs," *Nucl. Instr. Meth. A* **445**, 59 (2000).

21. S.V. Milton, E. Gluskin *et al.*, "Exponential Gain and Saturation of a Self-Amplified Spontaneous Emission Free-Electron Laser," *Science* **292**, 2037 (2001).

22. C. Pellegrini, "The History of X-Ray Free-Electron Lasers," *Eur. Phys. J. H* **37**, 65 (2012).

23. Z. Huang and K.-J. Kim, "Review of X-Ray Free-Electron Laser Theory," *Phys. Rev. Spec. Topics–Accel. Beams* **10**, 034801 (2007).

24. B.W.J. McNeil and N.R. Thompson, "X-Ray Free-Electron Lasers," *Nature Photon.* **4**, 814 (2010).

25. C. Pellegrini and S. Reiche, "The Development of X-Ray Free-Electron Lasers," *IEEE J. Select Topics QE* **10**, 1393 (2004).

26. W.B. Colson and A.M. Sessler, "Free Electron Lasers," *Ann. Rev. Nucl. Part. Sci.* **35**, 25 (1985).

27. P. Schmüser, M. Dohlus, J. Rossbach and C. Behrens, *Free-Electron Lasers in the Ultraviolet and X-Ray Regime: Physical Principles, Experimental Results and Technical Realization* (Springer, Berlin, 2014).

28. K.-J. Kim, Z. Huang and R. Lindberg, *Synchrotron Radiation and Free Electron Lasers: Principles of Coherent X-ray Generation* (Cambridge University Press, to be published 2016).

29. Z. Huang and P. Schmüser, "Free Electron Lasers", p. 229 in *Handbook of Accelerator Physics and Engineering* (Second Edition, World Scientific Publishing, Singapore, 2013), A.W. Chao, K.H. Mess, M. Tigner and F. Zimmermann, Editors.

30. E.L. Saldin, E.A. Schneidmiller and M.V. Yurkov, *The Physics of Free Electron Lasers* (Springer-Verlag, Berlin, 2000).

31. B.M. Kincaid, "A Short-Period Helical Wiggler as an Improved Source of Synchrotron Radiation," *J. Appl. Phys.* **48**, 2684 (1977).

32. S. Reiche, "Numerical Studies for Single Pass, High Gain Free Electron Lasers," PhD Thesis (University of Hamburg, DESY report 2000–012 (2000)); also CERN School on Free-Electron Lasers (Darmstadt, Germany, 2009).

33. P. Emma, R. Akre, J. Arthur *et al.*, "First Lasing and Operation of an Ångstrom-Wavelength Free-Electron Laser," *Nature Photon.* **4**, 641 (2010).

34. LCLS Design Study Group, "LCLS Design Study Report," SLAC R-521 (April 1998); W.E. White, A. Robert and M. Dunne, "The Linac Coherent Light Source," *J. Synchrotron Rad.* **22**, 472 (May 2015).

35. C. Pellegrini, J. Rosenzweig, G. Travish *et al.*, "The SLAC Soft X-ray High Power FEL," *Nucl. Instr. Meth. A* **341**, 326 (1994).

36. H.D. Nuhn and J. Rossbach, "LINAC-Based Short Wavelength FELs: The Challenges to be Overcome to Produce the Ultimate X-Ray Source – The X-Ray Laser," *J. Synchr. Rad.* **13**, 18 (2000).

37. J. Andruszkow *et al.*, "First Observation of Self-Amplified Spontaneous Emission in a Free-Electron Laser at 109 nm Wavelength," *Phys. Rev. Lett.* **85**, 3825 (2000).

38. V. Ayvazyan *et al.*, "Generation of GW Radiation Pulses from a VUV Free-Electron Laser Operating in the Femtosecond Regime," *Phys. Rev. Lett.* **88**, 104802 (2002).

39. V. Ayvazyan *et al.*, "First Operation of a Free-Electron Laser Generating GW Power Radiation at 32 nm Wavelength," *Eur. Phys. J. D* **37**, 297 (2006).

40. H.N. Chapman, A. Barty, M.J. Bogan *et al.*, "Femtosecond Diffractive Imaging with a Soft-X-Ray Free-Electron Laser," *Nature Phys.* **2**, 839 (2006).

41. H.N. Chapman, S.P. Hau-Riege, M.J. Bogan *et al.*, "Femtosecond Time-Delay X-Ray Holography," *Nature* **448**, 676 (2007).

42. W. Ackermann *et al.*, "Operation of a Free-Electron Laser from the Extreme Ultraviolet to the Water Window," *Nature Photon.* **1**, 336 (2007).

43. M. Yabashi, J. B. Hastings, M.S. Zolotorev *et al.*, "Single-Shot Spectrometry for X-Ray Free-Electron Lasers," *Phys. Rev. Lett.* **97**, 084802 (2006).

44. D. Zhu, M. Cammarata, J.M. Feldkamp *et al.*, "A Single-Shot Transmissive Spectrometer for Hard X-Ray Free Electron Lasers," *Appl. Phys. Lett.* **101**, 034103 (2012).

45. P. Heimann *et al.*, "Linac Coherent Light Source Soft X-Ray Materials Science Instrument Optical Design and Monochromator Commissioning," *Rev. Sci. Instr.* **82**, 093104 (2011); S. Moeller *et al.*, "Photon Beamlines and Diagnostics at LCLS," *Nucl. Instr. Meth. A* **635**, S6 (2011); R. Soufli, M. Fernández-Perea, S.L. Baker *et al.*, "Development and Calibration of Mirrors and Gratings for the Soft X-ray Materials Science Beamline at the Linac Coherent Light Source Free-Electron Laser," *Appl. Optics* **51**(12), 2118 (April 20, 2012).

46. W.F. Schlotter *et al.*, "The Soft X-Ray Instrument for Materials Studies at the Linac Coherent Light Source X-Ray Free-Electron Laser," *Rev. Sci. Instr.* **83**, 043107 (2012); G.L. Dakovski *et al.*, "The Soft X-ray Research Instrument at the Linac Coherent Light Source," *J. Synchrotron Rad.* **22**, 498 (May 2015).

47. J. Galayda, J. Arthur, D. Ratner and W. White, "X-Ray Free-Electron Lasers – Present and Future Capabilities (Invited)," *JOSA* **27** B106 (2010).

48. A.A. Lutman, R. Coffee, Y. Ding *et al.*, "Experimental Demonstration of Femtosecond Two-Color X-Ray Free-Electron Lasers," *Phys. Rev. Lett.* **110**, 134801 (March 2013); A. Marinelli *et al.*, "High Intensity Double Pulse X-Ray Free-Electron Laser," *Nature Commun.* **6**, 7369 (March 6, 2015).

49. C. Behrens *et al.*, "Few-Femtosecond Time-Resolved Measurements of X-Ray Free-Electron Lasers," *Nature Commun.* **5**, 4762 (April 30, 2014); W. Heml *et al.*, "Measuring the Temporal Structure of Few-Femtosecond Free-Electron Laser X-Ray Pulses Directly in the Time Domain," *Nature Photon.* **8**, 950 (December 2014); M.P. Minitti, J.M. Budarz, A. Kirrander *et al.*, "Imaging Molecular Motion: Femtosecond X-Ray Scattering of an Electrocycle Chemical Reaction," *Phys. Rev. Lett.* 114, 255501 (June 26, 2015).

50. M. Born and E. Wolf, *Principles of Optics* (Cambridge University Press, 1999), Seventh Edition, pp. xxvii, 290–299.

51. T. Young, *A Course of Lectures on Natural Philosophy and the Mechanical Arts*, Volume 1, Lecture XXXIX, "On the Nature of Light and Colours" (J. Johnson, London, 2007), pp.464–465 and 776–777; A. Robinson, *The Last Man Who Knew Everything* (Pearson Educ., Pi Press, 2006).

52. I.A. Vartanyants and A. Singer, "Coherence Properties of Hard X-Ray Synchrotron Sources and X-Ray Free-Electron Lasers," *New J. Phys.* **12**, 035004 (2010).

53. E.L. Saldin, E.A. Schneidmiller and M.V. Yurkov, "Statistical and Coherence Properties of Radiation from X-Ray Free-Electron Lasers," *New J. Phys.* **12**, 035010 (2010).

54. S. Reiche, "Coherence Properties of the LCLS X-Ray Beam," Proc. PAC07 (Albuquerque, New Mexico, USA, 2007), pp. 1272.

55. Y. Ding and Z. Huang, "Transverse-Coherence Properties of the FEL at the LCLS," Proc. FEL2010 (Malmö, Sweden, 2010), p. 51.

56. I.A. Vartanyants, A. Singer, A.P. Mancuso *et al.*, "Coherence Properties of Individual Femtosecond Pulses of an X-Ray Free-Electron Laser," *Phys. Rev. Lett.* **107**, 144801 (2011).

57. A. Sakdinawat, "Nanostructured Patterns for FEL Coherence Measurements" (unpublished, 2010).

58. A. Singer, I.A. Vartanyants, M. Kuhlmann *et al.*, "Transverse-Coherence Properties of the Free-Electron-Laser FLASH at DESY," *Phys. Rev. Lett.* **101**, 254801 (2008).

59. A. Singer *et al.*, "Spatial and Temporal Coherence Properties of Single Free-Electron Laser Pulses," *Optics Expr.* **20**, 17480 (2012).

60. C. Gutt, P. Wochner, B. Fischer *et al.*, "Single Shot Spatial and Temporal Coherence Properties of the SLAC Linac Coherent Light Source in the Hard X-Ray Regime," *Phys. Rev. Lett.* **108**, 024801 (2012).

61. W. Sibbett, A.A. Lagatsky and C.T.A. Brown, "The development and application of femtosecond laser systems," *Optics Expr.* **20**, 6989 (2012).

62. D.E. Spence, P.N. Kean and W. Sibbett, "60-fsec Pulse Generation from a Self-Mode-Locked Ti:sapphire Laser," *Opt. Lett.* **16**, 42 (1991).

63. R.W. Boyd, *Nonlinear Optics* (Academic Press, Third Edition, 2008).

64. D. Garzella, T. Hara, B. Carré *et al.*, "Using VUV High-Order Harmonics Generated in Gas as a Seed for Single Pass FEL," *Nucl. Instr. Meth. A* **528**, 502 (2004).

65. M. Gullans, J.S. Wurtele, G. Penn and A.A. Zholents, "Performance Study of a Soft X-Ray Harmonic Generation FEL Seeded with an EUV Laser Pulse," *Optics Commun.* **274**, 167 (2007).

66. T. Watanabe, X.J. Wang, J.B. Murphy *et al.*, "Experimental Characterization of Superradiance in a Single-Pass High-Gain Laser-Seeded Free-Electron Laser Amplifier," *Phys. Rev. Lett.* **98**, 034802 (2007).

67. G. Lambert, T. Hara, D. Garzella *et al.*, "Injection of Harmonics Generated in Gas in a Free-Electron Laser Providing Intense and Coherent Extreme-Ultraviolet Light," *Nature Phys.* **4**, 296 (2008).

68. T. Togashi, E.J. Takahashi, K. Midorikawa *et al.*, "Extreme Ultraviolet Free Electron Laser Seeded with High-Order Harmonic of Ti:sapphire Laser," *Optics Expr.* **19**, 317 (2011).

69. S. Ackermann, A. Azima, S. Bajt *et al.*, "Generation of Coherent 19- and 38-nm Radiation at a Free-Electron Laser Directly Seeded at 38 nm," *Phys. Rev. Lett.* **111**, 114801 (2013).

70. L.-H. Yu, M. Babzien, I. Ben-Zvi *et al.*, "High-Gain Harmonic-Generation Free-Electron Laser," *Science* **289**, 932 (2000).

71. S. Krinsky and L. H. Yu, "Output Power in Guided Modes for Amplified Spontaneous Emission in a Single-Pass Free-Electron Laser," *Phys. Rev. A* **35**, 3406 (1987).

72. S. Krinsky, "Transient Analysis of Free-Electron Lasers with Discrete Radiators," *Phys. Rev. E* **59**, 1171 (1999).

73. R. Bonifacio, L. De Salvo Souza, P. Pierini and E.T. Scharlemann, "Generation of XUV Light by Resonant Frequency Tripling in a Two-Wiggler FEL Amplifier," *Nucl. Instr. Meth. A* **296**, 787 (1990).

74. L.H. Yu, "Generation of Intense UV Radiation by Subharmonically Seeded Single-Pass Free-Electron Lasers," *Phys. Rev. A* **44**, 5178 (1991).

75. L.H. Yu, L. DiMauro, A. Doyuran *et al.*, "First Ultraviolet High-Gain Harmonic-Generation Free-Electron Laser," *Phys. Rev. Lett.* **91**, 074801 (2003).

76. E. Allaria *et al.*, "Highly Coherent and Stable Pulses from the FERMI Seeded Free-Electron Laser in the Extreme Ultraviolet," *Nature Photon.* **6**, 699 (2012).

77. L. Giannessi *et al.*, "First Lasing of FERMI-2 and FERMI-1 Recent Results", *Proc. FEL2012*, **13** (Nara, August 2012); E. Allaria *et al.*, "Two-Stage Seeded Soft-X-ray Free-Electron Laser," *Nature Phononics* **7**, 913 (November 2013); M. Svandrlik *et al.*, *Proc. FEL2014*, TuP 085 (Basel, August 2014).

78. K. Prince *et al.*, "Coherent Control with a Short-Wavelength Free-Electron Laser," *Nature Photon.* **10**, 176 (March 2016).

79. F. Capotondi, E. Pedersoli, N. Mahne *et al.*, "Coherent Imaging Using Seeded Free-Electron Laser Pulses with Variable Polarization: First Results and Research Opportunities," *Rev. Sci. Instr.* **84**, 051301 (May, 2013) (invited).

80. G. De Ninno, B. Mahieu, E. Allaria, L. Giannessi and S. Spampinati, "Chirped Seeded Free-Electron Lasers: Self-Standing Light Sources for Two-Color Pump-Probe Experiments," *Phys. Rev. Lett.* **110**, 064801 (2013).

81. E. Pedersoli *et al.*, "Multipurpose Modular Experimental Station for the DiProI Beamline of Fermi@Elettra Free Electron Laser," *Rev. Sci. Instr.* **82**, 043711 (2011).

82. J. Bödewadt and C. Lechner, "Results and Perspectives on the FEL Seeding Activities at FLASH," (Proc. FEL 2013).

83. J. Feldhaus, E.L. Saldin, J.R. Schneider, E.A. Schneidmiller and M.V. Yurkov, "Possible Application of X-Ray Optical Elements for Reducing the Spectral Bandwidth of an X-ray SASE FEL," *Optics Commun.* **140**, 341 (1997).

84. G. Geloni, V. Kocharyan and E. Saldin, "A Novel Self-Seeding Scheme for Hard X-Ray FELs," *J. Modern Optics* **58**, 1391 (2011).

85. J. Amann, W. Berg, V. Blank *et al.*, "Demonstration of Self-Seeding in a Hard-X-Ray Free-Electron Laser," *Nature Photon.* **6**, 693 (2012).

86. M. Yabashi and T. Tanaka, "Self-Seeded FEL Emits Hard X-Rays," *Nature Photon.* **6**, 648 (2012).

87. G. Stupakov, "Using the Beam-Echo Effect for Generation of Short-Wavelength Radiation," *Phys. Rev. Lett.* **102**, 074801 (2009).

88. D. Xiang and G. Stupakov, "Echo-enabled harmonic generation free electron laser," *Phys. Rev. Spec. Topics–Accel. Beams* **12**, 030702 (2009).

89. D. Xiang, E. Colby, M. Dunning *et al.*, "Demonstration of the Echo-Enabled Harmonic Generation Technique for Short-Wavelength Seeded Free Electron Lasers," *Phys. Rev. Lett.* **105**, 114801 (2010).

90. Z.T. Zhao *et al.*, "First Lasing of an Echo-Enabled Harmonic Generation Free-Electron Laser," *Nature Photon.* **6**, 360 (2012).

91. S. Sasaki, "Analyses for a Planar Variably-Polarizing Undulator," *Nucl. Instr. Meth. A* **347**, 83 (1994).

92. H. Kitamura, "Recent Trends of Insertion-Device Technology for X-Ray Sources," *J. Synchrotron Rad.* **7**, 121 (2000).

93 R. Colella and A. Luccio, "Proposal for a Free Electron Laser in the X-Ray Region," *Optics Commun.* **50**, 41 (1984).

94. Z. Huang and R.D. Ruth, "Fully Coherent X-Ray Pulses from a Regenerative-Amplifier Free-Electron Laser", *Phys. Rev. Lett.* **96**, 144801 (April 14, 2006).

95. K.-J. Kim, Yu. Shvyd'ko and S. Reiche, "A Proposal for an X-Ray Free-Electron Laser Oscillator with an Energy-Recovery Linac," *Phys. Rev. Lett.* **100**, 244802 (2008).

96. K.-J. Kim and Yu.V. Shvyd'ko, "Tunable Optical Cavity for an X-Ray Free-Electron-Laser Oscillator," *Phys. Rev. Spec. Topics-Accel. Beams* **12**, 030703 (2009).

97. K.J. Kim, Yu.V. Shvyd'ko and R.R. Lindberg, "An X-Ray Free-Electron Laser Oscillator for Record High Spectral Purity, Brightness, and Stability," *Synchr. Rad. News* **25**, 25 (2012).

98. R.R. Lindberg and Yu.V. Shvyd'ko, "Time Dependence of Bragg Forward Scattering and Self-Seeding of Hard X-Ray Free-Electron Lasers," *Phys. Rev. Spec. Topics-Accel. Beams* **15**, 050706 (2012).

99. J. Wurtele, P. Gandhi and X.-W. Gu, "Tunable Soft X-Ray Oscillators," *Proc. FEL Conf.* (Malmö, Sweden, 2010).

100. P. Gandhi, G. Penn, M. Reinsch, J.S. Wurtele and W.M. Fawley, "Oscillator Seeding of a High Gain Harmonic Generation Free Electron Laser in a Radiator-First Configuration," *Phys. Rev. Spec. Topics-Accel. Beams* **16**, 020703 (2013).

101. T. Hara, Y. Inubushi, T. Katayama *et al.*, "Two-Colour Hard X-Ray Free-Electron Laser with Wide Tunability," *Nature Commun.* **4**, 1 (December 5, 2013); T. Osaka *et al.*, *Optics Express* **24**(9), 9187 (May 2, 2016).

102. T. Shintake, H. Tanaka, T. Hara *et al.*, "A Compact Free-Electron Laser for Generating Coherent Radiation in the Extreme Ultraviolet Region," *Nature Photon.* **2**, 555 (2008).

103. B. McNeil, "Free-Electron Lasers: A Down-Sized Design," *Nature Photon.* **2**, 522 (2008).

104. T. Ishikawa, M. Yabashi *et al.*, "A Compact X-Ray Free-Electron Laser Emitting in the Sub-Ångström Region," *Nature Photon.* **6**, 540 (2012); M. Yabashi, H. Tanaka and T. Ishikawa, "Overview of the SACLA Facility," *J.Synchrotron Rad.* **22**, 477 (May 2015).

105. D. Pile, "First Light from SACLA," *Nature Photon.* **5**, 456 (2011).

106. Z. Huang and I. Lindau, "SACLA Hard-X-Ray Compact FEL," *Nature Photon.* **6**, 505 (2012).

107. M. Yabashi and H. Tanaka, "And Then There Were Two," *Nature Photon.* **6**, 566 (2012).

108. Y. Inubushi, K. Tono, T. Togashi *et al.*, "Determination of the Pulse Duration of an X-Ray Free Electron Laser Using Highly Resolved Single-Shot Spectra," *Phys. Rev. Lett.* **109**, 144801 (2012).

109. K. Tono, T. Togashi, Y. Inubushi *et al.*, "Beamline, Experimental Stations and Photon Beam Diagnostics for the Hard X-Ray Free Electron Laser of SACLA," *New J. Phys.* **15**, 083035 (2013).

110. H. Yoneda, Y. Inubushi, K. Nagamine *et al.*, "Atomic Inner-Shell Laser at 1.5-Ånsgström Wavelength Pumped by an X-ray Free-Electron Laser," *Nature* **524**, 446 (August 27, 2015); L. Young, "A Stable Narrow-Band X-Ray Laser," *Nature* **524**, 424 (August 27, 2015).

111. H. Yumoto *et al.*, "Focusing of X-Ray Free-Electron Laser Pulses with Reflective Optics," *Nature Photon.* **7**, 43 (2013); H. Mimura *et al.*, "Generation of 1020 W/cm² Hard X-ray laser Pulses with Two-Stage Reflective Focusing System," *Nature Commun.* **5**, 3539 (2014); K. Yamauchi, M. Yabashi, H. Ohashi, T. Koyama and T. Ishikawa, "Nanofocusing of X-ray Free-Electron Lasers by Grazing-Incidence Reflective Optics," *J. Synchrotron Rad.* **22**, 592 (May 2015).

112. H. Yoneda *et al.*, "Saturable Absorption of Intense Hard X-Rays in Iron," *Nature Commun.* **5**, 1038 (October 1, 2014).

113. J. Rossbach (for the TESLA FEL Study Group), "A VUV Free Electron Laser at the TESLA Test Facility at DESY," *Nucl. Instr. Meth. A* **375**, 269 (1996).

114. C. Rizzuto, C. Bocchett *et al.*, "FERMI at Elletra, Conceptual Design Report" (December 2007); E. Allaria *et al.*, "The FERMI Free Electron Laser," *J. Synchrotron Rad.* **22**, 485 (May 2015).

115. EU XFEL: The European X-Ray Free-Electron Laser (Technical Design Report), DESY, 2006–097, M. Altarelli *et al.* (Editors); Y. Li *et al.*, "Magnetic Measurement Techniques for the Large-Scale Production of Undulator Segments for the European XFEL," *Synchr. Rad. News* **28**(3), 23 (May/June 2015).

116. Ultrafast Phenomena at the Nanoscale: Science opportunities at the SwissFEL X-ray Laser, B.D. Patterson *et al.* (editors), Paul Scherrer Institute Report 09–10 (2009).

117. E.-S. Kim and M. Yoon, "Beam Dynamics in a 10-GeV Linear Accelerator for the X-Ray Free Electron Laser at PAL," *IEEE Trans. Nucl. Sci.* **56**, 3597 (2009); H.-S. Kang, private communication (March 2015).

118. L. Young, E.P. Kanter, B. Krässig *et al.*, "Femtosecond Electronic Response of Atoms to Ultra-Intense X-Rays," *Nature* **466**, 56 (2010); K.R. Ferguson *et al.*, "The Atomic, Molecular and Optical Science Instrument at the Linac Coherent Light Source," *J. Synchrotron Rad.* **22**, 492 (May 2015).

119. J. Wark, "X-Ray Laser Peels and Cores Atoms," *Nature* **466**, 35 (2010).

120. B. Nagler *et al.*, "Turning Solid Aluminium Transparent by Intense Soft X-ray Photoionization," *Nature Phys.* **5**, 693 (2009).

121. E.P. Kanter *et al.*, "Unveiling and Driving Hidden Resonances with High-Fluence, High-Intensity X-Ray Pulses," *Phys. Rev. Lett.* **107**, 233001 (2011).

122. H. Fukuzawa *et al.*, "Deep Inner-Shell Multiphoton Ionization by Intense X-Ray Free-Electron Laser Pulses," *Phys. Rev. Lett.* **110**, 173005 (2013).

123. K. Tamasaku, M. Nagasono, H. Iwayama *et al.*, "Double Core-Hole Creation by Sequential Attosecond Photoionization," *Phys. Rev. Lett.* **111**, 043001 (2013).

124. T. Katayama *et al.*, "Femtosecond X-Ray Absorption Spectroscopy with Hard X-Ray Free Electron Laser," *Appl. Phys. Lett.* **103**, 105 (2013).

125. V. Zhaunerchyk *et al.*, "Using Covariance Mapping to Investigate the Dynamics of Multi-Photon Ionization Processes of Ne Atoms Exposed to X-FEL Pulses," *J. Phys. B: At. Mol. Opt. Phys.* **46**, 164034 (2013).

126. B. Erk *et al.*, "Inner-Shell Multiple Ionization of Polyatomic Molecules with an Intense X-Ray Free-Electron Laser Studied by Coincident Ion Momentum Imaging," *J. Phys. B: At. Mol. Opt. Phys.* **46**, 164031 (2013).

127. M. Larsson, P. Salén *et al.*, "Double Core-Hole Formation in Small Molecules at the LCLS Free Electron Laser," *J. Phys. B: At. Mol. Opt. Phys.* **46**, 164030 (2013).

128. C. Bostedt *et al.*, "Ultra-Fast and Ultra-Intense X-Ray Sciences: First Results from the Linac Coherent Light Source Free-Electron Laser" (Invited Paper), *J. Phys. B: At. Mol. Opt. Phys.* **46**, 164003 (2013).

129. N. Rohringer, D. Ryan, R.A. London *et al.*, "Atomic Inner-Shell X-Ray Laser at 1.46 Nanometres Pumped by an X-Ray Free-Electron Laser," *Nature* **481**, 488 (2012).

130. H. Wabnitz *et al.*, "Multiple Ionization of Atom Clusters by Intense Soft X-Rays from a Free-Electron Laser," *Nature* **420**, 482 (2002).

131. M. Dell'Angela *et al.*, "Real-Time Observation of Surface Bond Breaking with an X-ray Laser," *Science* **339**, 1302 (2013).

132. M. Beyeet *et al.*, "Selective Ultrafast Probing of Transient Hot Chemisorbed and Precursor States of CO on Ru(0001)," *Phys. Rev. Lett.* **110**, 186101 (2013).

133. J.N. Clark, L. Beitra, G. Xiong *et al.*, "Ultrafast Three-Dimensional Imaging of Lattice Dynamics in Individual Gold Nanocrystals," *Science* **341**, 56 (2013).

134. Y.D. Chuang *et al.*, "Real-Time Manifestation of Strongly Coupled Spin and Charge Order Parameters in Stripe-Ordered $La_{1.75}Sr_{0.25}NiO_4$ Nickelate Crystals Using Time-Resolved Resonant X-Ray Diffraction," *Phys. Rev. Lett.* **110**, 127404 (2013).

135. M. Först *et al.*, "Displacive Lattice Excitation Through Nonlinear Phononics Viewed by Femtosecond X-Ray Diffraction," *Solid State Commun.* **169**, 24 (2013).

136. D. Milathianaki *et al.*, "Femtosecond Visualization of Lattice Dynamics in Shock-Compressed Matter," *Science* **342**, 220 (October 2013).

137. C. Gutt, S. Streit-Nierobisch, L.M. Stadler *et al.*, "Single-Pulse Resonant Magnetic Scattering Using a Soft X-Ray Free-Electron Laser," *Phys. Rev. B* **81**, 100401 (2010).

138. T. Wang *et al.*, "Femtosecond Single-Shot Imaging of Nanoscale Ferromagnetic Order in Co/Pd Multilayers Using Resonant X-Ray Holography," *Phys. Rev. Lett.* **108**, 267403 (2012).

139. C.E. Graves *et al.*, "Nanoscale Spin Reversal by Non-Local Angular Momentum Transfer Following Ultrafast Laser Excitation in Ferrimagnetic GdFeCo," *Nature Mat.* **12**, 293 (2013).

140. S. de Jong *et al.*, "Speed Limit of the Insulator–Metal Transition in Magnetite," *Nature Mat.* **12**, 882 (2013).

141. H.N. Chapman *et al.*, "Femtosecond X-Ray Protein Nanocrystallography," *Nature* **470**, 73 (2011); R.M. Wilson, "X-Rays from a Free-Electron Laser Resolve the Structures of Complex Biomolecules," *Physics Today* **64**, 13 (April 2011).

142. R. Neutze, R. Wouts, D. van der Spoel, E. Weckert and J. Hajdu, "Potential for Biomolecular Imaging with Femtosecond X-ray Pulses," *Nature* **406**, 752 (August 17, 2000).

143. M.M. Seibert *et al.*, "Single Mimivirus Particles Intercepted and Imaged with an X-Ray Laser," *Nature* **470**, 78 (2011); G. van der Schot *et al.*, "Imaging Single Cells in a Beam of Live Cyanobacteria with an X-ray laser," *Nature Commun.* **6**, 5704 (February 11, 2015); T. Eckberg *et al.*, "Three-Dimensional Reconstruction of the Giant Mimivirus Particle with an X-ray Free-Electron Laser," *Phys. Rev. Lett.* **114**, 098102 (March 6, 2015); T. Kimura *et al.*, "Imaging Live Cell in Micro-liquid Enclosure by X-ray Laser Diffraction," *Nature Commun.* **5**, 3052 (January 7, 2014).

144. A. Aquila *et al.*, "Time-Resolved Protein Nanocrystallography Using an X-Ray Free-Electron Laser," *Optics Expr.* **20**, 2706 (2012); M. Liang *et al.*, "The Coherent X-ray Imaging Instrument at the Linac Coherent Light Source," *J. Synchrotron Rad.* **22**, 514 (May 2014).

145. A. Barty, C. Caleman, H.N. Chapman *et al.*, "Self-Terminating Diffraction Gates Femtosecond X-Ray Nanocrystallography Measurements," *Nature Photon.* **6**, 35 (2012).

146. J.C.H. Spence, U. Weierstall and H. N. Chapman, "X-Ray Lasers for Structural and Dynamic Biology," *Rep. Progr. Phys.* **75**, 102601 (2012).

147. L. Redecke *et al.*, "Natively Inhibited Trypanosoma Brucei Cathepsin B Structure Determined by Using an X-ray Laser," *Science* **339**, 227 (2013).

148. J. Kern *et al.*, "Simultaneous Femtosecond X-ray Spectroscopy and Diffraction of Photosystem II at Room Temperature," *Science* **340**, 491 (2013).

149. C. Kupitz *et al.*, "Serial Time-Resolved Crystallography of Photosystem II Using a Femtosecond X-Ray laser," *Nature* **513**, 261 (September 11, 2014).

150. L.C. Johansson *et al.*, "Structure of a Photosynthetic Reaction Centre Determined by Serial Femtosecond Crystallography," *Nature Commun.* **4**, 2911 (2013).

151. T.R.M. Barends, L. Foucar, S. Botha *et al.*, "De novo Protein Crystal Structure Determination from X-Ray Free-Electron Laser Data," *Nature* **505**, 244 (2014); M. Suga *et al.*, "Native Structure of Photosystem II at 1.95 Å Resolution Viewed by Femtosecond X-Ray Pulses", *Nature* **517**, 99 (January 1, 2015).

152. S.P. Hau-Riege *et al.*, "Subnanometer-Scale Measurements of the Interaction of Ultrafast Soft X-Ray Free-Electron-Laser Pulses with Matter," *Phys. Rev. Lett.* **98**, 145502 (2007).

153. A. Barty, S. Boutet, M.J. Bogan *et al.*, "Ultrafast Single-Shot Diffraction Imaging of Nanoscale Dynamics," *Nature Photon.* **2**, 415–419 (2008).

154. A.P. Mancuso, O.M. Yefanov and I.A. Vartanyants, "Coherent Diffractive Imaging of Biological Samples at Synchrotron and Free Electron Laser Facilities," *J. Biotechn.* **149**, 229 (2010).

155. H.M. Quiney and K.A. Nugent, "Biomolecular Imaging and Electronic Damage Using X-Ray Free-Electron Lasers," *Nature Phys.* **7**, 142 (2011).

156. T. Kimura *et al.*, "Imaging live sell in Micro-Liquid Enclosure by X-ray Laser Diffraction," *Nature Commun.* **5**, 3052 (January 7, 2014).

157. G. van der Schot *et al.*, "Imaging Single Cells in a Beam of Live Cyanobacteria with an X-ray Laser," *Nature Commun.* **6**, 5704 (February 11, 2015).

158. S. Marchesini, S. Boutet, A.E. Sakdinawat *et al.*, "Massively Parallel X-Ray Holography," *Nature Photon.* **2**, 560 (2008); S. Marchesini *et al.*, "Coherent Diffracrive Imaging: Applications and Limitations," *Optics Express* **11**, 2344 (August 15, 2003).

159. S. Shwartz, M. Fuchs, J.B. Hastings *et al.*, "X-ray Second Harmonic Generation," *Phys. Rev. Lett.* **112**, 163901 (April 25, 2014).

160. M. Fuchs, M. Trigo, J. Chen *et al.*, "Anomolous Nonlinear X-ray Compton Scattering," *Nature Physics* **11**, 964 (November 2015).

161. F. Bencivenga, R. Cucini, F. Capotondi *et al.*, "Four-Wave Mixing Experiments with Extreme Ultraviolet Transient Gratings," *Nature* **520**, 205 (April 9, 2015).

162. T.E. Glover, D.M. Fritz, M. Cammarata *et al.*, "X-Ray and Optical Wave Mixing," *Nature* **488**, 603 (2012).

163. A. Zholents, "Next-Generation X-Ray Free-Electron Lasers," *IEEE J. Selected Topics Quant. Electr.*, **18**, 248 (2012).

164. P. Tipler and R. Llewellyn, *Modern Physics* (Freeman, 2012), Sixth Edition. Section 5.5, p. 213.

165. E.L. Saldin, E.A. Schneidmiller and M.V. Yurkov, "Scheme for Attophysics Experiments at an X-ray SASE FEL," *Optics Commun.* **212**, 377 (2002).

166. P. Emma, K. Bane, M. Cornacchia *et al.*, "Femtosecond and Subfemtosecond X-Ray Pulses from a Self-Amplified Spontaneous-Emission–Based Free-Electron Laser," *Phys. Rev. Lett.* **92**, 074801 (2004).

167. N.R. Thompson and B.W.J. McNeil, "Mode Locking in a Free-Electron Laser Amplifier," *Phys. Rev. Lett.* **100**, 203901 (2008).

168. S. Reiche, P. Musumeci, C. Pellegrini and J.B. Rosenzweig, "Development of Ultra-Short Pulse, Single Coherent Spike for SASE X-Ray FELs," *Nucl. Instr. Meth. A* **593**, 45 (2008).

169. B.W.J. McNeil and N.R. Thompson, "Cavity Resonator Free Electron Lasers as a Source of Stable Attosecond Pulses," *EPL (Europhysics Letters)* **96**, 54004 (2011).

170. D.J. Dunning, B.W.J. McNeil and N.R. Thompson, "Few-Cycle Pulse Generation in an X-Ray Free-Electron Laser," *Phys. Rev. Lett.* **110**, 104801 (2013).

171. T. Tanaka, "Proposal for a Pulse-Compression Scheme in X-Ray Free-Electron Lasers to Generate a Multiterawatt, Attosecond X-Ray Pulse," *Phys. Rev. Lett.* **110**, 084801 (2013).

172. C. Kealhofer, *et al.*, "All Optical Control and Metrology of Electron Pulses," *Science* **352**, 429 (April 22, 2016).

Homework Problems

Homework problems for each chapter will be found at the website:
www.cambridge.org/xrayeuv

7 Laser High Harmonic Generation

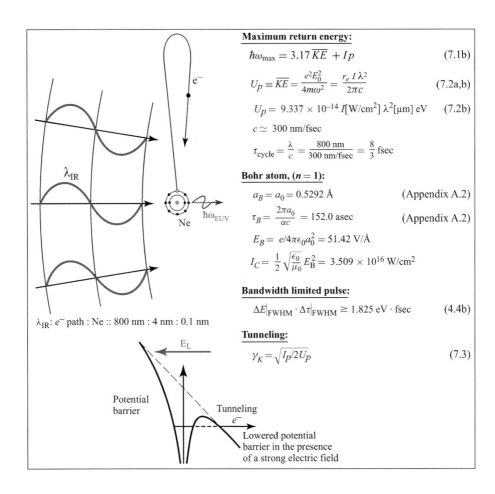

Maximum return energy:

$$\hbar\omega_{\max} = 3.17\,\overline{KE} + Ip \tag{7.1b}$$

$$U_p \equiv \overline{\overline{KE}} = \frac{e^2 E_0^2}{4m\omega^2} = \frac{r_e\,I\,\lambda^2}{2\pi c} \tag{7.2a,b}$$

$$U_p = 9.337 \times 10^{-14}\,I[\text{W/cm}^2]\,\lambda^2[\mu\text{m}]\ \text{eV} \tag{7.2b}$$

$$c \simeq 300\ \text{nm/fsec}$$

$$\tau_{\text{cycle}} = \frac{\lambda}{c} = \frac{800\ \text{nm}}{300\ \text{nm/fsec}} = \frac{8}{3}\ \text{fsec}$$

Bohr atom, ($n = 1$):

$$a_B = a_0 = 0.5292\ \text{Å} \tag{Appendix A.2}$$

$$\tau_B = \frac{2\pi a_0}{\alpha c} = 152.0\ \text{asec} \tag{Appendix A.2}$$

$$E_B = e/4\pi\epsilon_0 a_0^2 = 51.42\ \text{V/Å}$$

$$I_C = \frac{1}{2}\sqrt{\frac{\epsilon_0}{\mu_0}}\,E_B^2 = 3.509 \times 10^{16}\ \text{W/cm}^2$$

Bandwidth limited pulse:

$$\Delta E\big|_{\text{FWHM}} \cdot \Delta \tau\big|_{\text{FWHM}} \geq 1.825\ \text{eV} \cdot \text{fsec} \tag{4.4b}$$

Tunneling:

$$\gamma_K = \sqrt{Ip/2U_p} \tag{7.3}$$

λ_{IR}: e^- path : Ne :: 800 nm : 4 nm : 0.1 nm

Potential barrier

Tunneling e^-

Lowered potential barrier in the presence of a strong electric field

7.1 The Basic Processes of High Harmonic Generation (HHG)

In this chapter we explore the generation of very high harmonics of intense, femtosecond (fs) duration, near-infrared (IR) pulses. Typically this begins with 30 fs duration Ti: sapphire laser pulses at nominally 800 nm wavelength, focused into a nobel gas at intensities of order 10^{14} to 10^{15} W/cm², with correspondingly strong electric fields of order 10^9 V/cm, or 10 V/Å. Gas jets of He, Ne, Ar, Kr, or Xe, at a fraction of an atmosphere pressure, are most frequently used as the non-linear medium with propagation paths of several millimeters at high intensity. A semi-classical description of the HHG process, validated by quantum mechanical calculations, begins with the illumination of a single atom, such as neon. The electric field of the intense infrared pulse essentially pulls a bound electron from the atom by lowering the binding potential, permitting one of the bound electrons to tunnel through the barrier to the continuum. The free electron, beginning with zero initial energy, is then accelerated away from the parent atom (ion), gaining energy in the strong laser electric field, typically beginning its travel slightly beyond the peak of the laser field in a highly non-linear process. Just under a quarter cycle later the laser field direction reverses, the free electron is initially slowed, then reversed in direction and moves back towards the parent ion. The kinetic energy gained depends on the "time of birth" (emission time at the end of the tunneling process) within the incident electric field cycle. Maximum return kinetic energy corresponds to electron release times just beyond the peak of the pulse. Return kinetic energies of order 20–200 eV are common; higher depending on laser intensity and wavelength.

With linearly polarized laser light there is a finite probability that the returning electron will radiatively recombine with the parent ion, releasing energy equal to the kinetic energy gained plus the reversed ionization potential (for example, 21.6 eV for neon). The released energy appears in the form of high-energy photons extending well into the extreme ultraviolet (EUV) and soft x-ray regions of the electromagnetic spectrum. With a many cycle incident pulse (e.g., 30 fs, $\sim$ 11 cycles) the emitted radiation appears in harmonics of the driving laser, $\omega_h = h\omega$. The harmonic number, h, equals the IR laser wavelength divided by the EUV/SXR wavelength. Because the process is temporally symmetric, occuring twice per cycle, the harmonics are odd-order only. This process occurs, with finite probabilities, for all atoms in the illuminated region of the gas jet. Owing to "time of birth" variations among the many atoms there is a spread in the return electron energies, resulting in a similar spread of emitted photon energies.

The phase of the emitted radiation from all atoms, regardless of their positions, is synchronized to that of the driving IR laser pulse, with some variation due to the various electron return trajectories. As a consequence the coherence properties of the incident laser are largely impressed upon the high harmonic radiation. At all wavelengths the harmonic radiation adds constructively in the forward direction and is spatially coherent. The emitted harmonic pulses are of femtosecond duration, somewhat shorter than the incident infrared pulse due to the non-linear nature of the process. The temporal coherence is, however, dominated by the shorter wavelength and spectral bandwidth of the emitted radiation, resulting in a longitudinal coherence length shorter than the IR pulse length by a factor approximately equal to the harmonic number.

The intensity of the emitted radiation scales approximately as the square of the number of participating atoms in the interaction region defined by the overlap of the gas jet and the incident infrared radiation. This scaling is limited to some extent by curvature of the incident IR wavefront. This region is defined transversely by the incident beam size and longitudinally by the distance over which the emitted short wavelength radiation adds constructively. This limiting distance is affected by the free electrons created within the interaction region by non-recombining electrons, creating a plasma refractive index (see Chapter 8) that advances the phase of the infrared wave compared to that of the EUV/soft x-rays, the latter traveling at a phase velocity very close to the speed of light in vacuum, c. Neutral atoms also contribute to this effect. Beyond this propagation distance, typically a few millimeters, the newly created EUV radiation is out of phase with that created earlier, thus setting limits to the effective interaction volume. This "dephasing distance" depends on the gas pressure and laser intensity as these affect the free electron density and thus the mismatch between IR and EUV/SXR phase velocities.

The three processes, electron tunneling, acceleration in the laser field, and radiative recombination, result in the generation of a comb of intense, femtosecond duration, polarized, coherent, odd-order, EUV/soft x-ray high harmonics. Typical efficiencies are of order 10^{-5} to 10^{-8} to each harmonic from incident IR laser pulse energies of mJs, thus generating nJs per high harmonic per pulse, or μJ at lower harmonics. Repetition rates are typically 1– 0 kHz, but moving towards 100 KHz, and towards 100 MHz using intracavity techniques. Recent research and applications have pursued the use of nearly-single cycle incident IR pulses to generate attosecond duration EUV pulses which are well suited in both photon energy and pulse duration for probing electron dynamics in molecules, clusters, near surfaces, and in solids. The EUV/SXR pulses are well synchronized to the driving IR laser pulse, thus opening many possibilities for the use of "two-color" pump-probe techniques. The necessary equipment is sufficiently compact and affordable that HHG labs are being set up to pursue dynamical aspects of HHG enabled science at many forefront universities, industrial and national laboratories.

7.2 Electron Tunneling, Trajectories, Return Energies, and the Efficiency of High Harmonic Generation

The early history of high harmonic generation (HHG) started with experiments in which noble gas atoms were irradiated with near infrared (IR) laser pulses at high intensities, of order 10^{14} W/cm^2 to 10^{15} W/cm^2, resulting in the emission of a comb of odd-order harmonics extending well into the extreme ultraviolet (EUV) region of the electromagnetic spectrum. The general technique is illustrated in Figure 7.1. At sufficiently high intensities the peak electric field of the incident IR laser approaches the binding electric field of the nucleus. Early researchers[1–4] observed that the harmonics extended to photon energies equal to the kinetic energy (KE) gained by an electron in such a laser field after tunneling free from its parent atom, plus the binding energy that would be regained upon recombination with that particular atom, that is

$$\hbar\omega = KE + I_p \tag{7.1a}$$

with a maximum value

$$\hbar\omega_{\text{max}} = 3.17\,\overline{KE} + I_p \tag{7.1b}$$

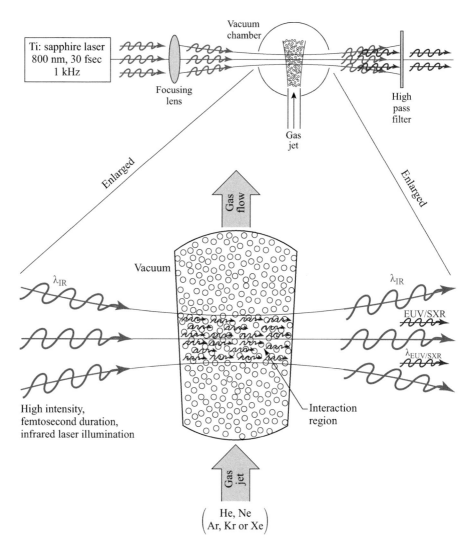

Figure 7.1 A schematic diagram illustrating the basic laboratory technique for generating high harmonic radiation. A femtosecond duration near infrared laser pulse is focused to high intensity, of order 10^{14} to 10^{15} W/cm^2, generally irradiating a volume of nobel atoms (He, Ne, Ar, Kr, or Xe) emerging from a gas jet, typically at gas pressures of a tenth of an atmosphere or less. In some cases a gas-filled cell or capillary tube is used instead. A filter is used to absorb the incident laser light while passing the EUV/soft x-ray radiation, which then goes on to various diagnostic test stations or to excite and/or probe scientific samples (not shown). The enlarged interaction region shows emission of EUV/soft x-ray harmonic radiation from some atoms, interfering constructively in the forward direction. Owing to the fractional probability of a released electron returning to its parent ion, a large number of ions and free electrons are also created in the interaction region, the latter creating plasma refractive effects that can limit the extent of the interaction region in the propagation direction, typically 100 μm wide and up to several millimeters long. Examples of spectra, to harmonic numbers h in excess of 100, observed in early HHG experiments are shown in Figure 7.2.

where $\hbar\omega_{\max}$ is the maximum photon energy observed, $\overline{KE}$ is the cycle averaged kinetic energy of a free electron in an oscillating electric field, $3.17\,\overline{KE}$ is the maximum kinetic energy that can be gained by a free electron as it is accelerated by the laser electric field, and I_p is the ionization potential of the atom, the energy required to move a bound electron from the ground state to the continuum. We will see later in this section that the return kinetic energy gained by an electron is strongly dependent on the time of liberation from the parent atom within the phase of the laser cycle ("time of birth"), which leads to the numerical factor 3.17 in Eq. (7.1b). The cycle averaged kinetic energy of a free electron, often referred to in the HHG community as the "ponderomotive energy" (or potential), is

$$\overline{KE} = U_p = e^2 E_0^2 / 4m\omega^2 \qquad (7.2a)$$

or in convenient units

$$\overline{KE} = U_p = r_e I \lambda^2 / 2\pi c = 9.337 \times 10^{-14} I[\text{W/cm}^2]\lambda^2[\mu\text{m}]\text{eV} \qquad (7.2b)$$

where r_e is the classical electron radius [see Chapter 2, Eq. (2.4)], E_0 is the electric field amplitude, ω is the radian frequency, λ is the wavelength, and I is the intensity of the incident laser radiation. Equation (7.2b) gives $\overline{KE}$ in units of eV when numerical values of I and λ are inserted in units indicated by the brackets.

The high-order harmonic spectra seen in Figure 7.2 can be explained in terms of a semi-classical model known as the "three step model,"[1, 2, 7] and its primary prediction that the harmonics extend in photon energies to the limit given in Eq. (7.1b). The model has been validated by quantum mechanical calculations,[8] which yields small corrections due to quantum mechanical aspects of electron tunneling and diffusion. In step 1 of this model the potential barrier of the atom is supressed by the presence of a strong laser electric field, of order but smaller than the shielded electric field due to the nucleus, allowing a single bound electron to tunnel through to the continuum. In step 2 the released electron, initially with zero kinetic energy, is accelerated away from the ion in the now dominant laser electric field (the residual ionic field is now comparatively weak). The free electron can then gain significant kinetic energy while returning to the ion in the reversed laser field, with greatest gained energy corresponding to release from the atom just beyond the peak of the pulse. In step 3 the electron can recombine, with some probability, with the parent ion releasing the acquired kinetic and ionization energies in the form of photons of equal energy. As the returning electrons can have different kinetic energies due to their different times of birth, so do the emitted photons. As seen in Figure 7.2, the emitted photon energies, even in early experiments, extended well into the EUV, to photon energies limited by Eq. (7.1b). In later experiments harmonics were reported reaching into the soft x-ray region,[9, 10] and weakly beyond that.

Step 1. The probability of an electron tunneling through the potential barrier of an atom in the presence of a strong laser electric field, step 1 in the HHG process, was described successfully by L.V. Keldysh[11] of the Lebedev Physical Institute, and later extended by Ammosov, Delone, and Kraĭnov[12] (ADK) of the Institute of General Physics, Moscow. The tunneling probability is exponential in nature[13–17], involving the atoms' ionization potential, I_p, and the laser electric field strength, E_0. These analyses

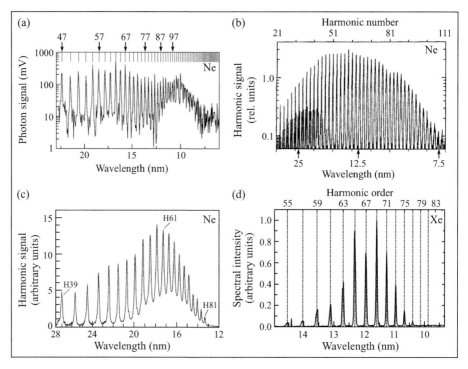

Figure 7.2 Very high order, odd harmonics extending to 7.8 nm wavelength (160 eV, $h = 135$) upon illumination of neon in a pulsed gas jet at a pressure of 40 Torr, by a 1.053 μm wavelength, 1 ps duration, 30 mJ pulse from a Nd-glass laser at an intensity of 1.5×10^{15} W/cm^2 (L'Huillier and Balcou[3]); (b) harmonic spectra extending to 7.3 nm wavelength (170 eV, $h = 109$) observed by Maclin, Kmetec, and Gordon[4] upon illumination of neon gas at 13 Torr in a 2.5 mm long tube by an 800 nm wavelength, 125 fs duration, 35 mJ pulse, from a Ti:sapphire laser at an intensity of 1.3×10^{15} W/cm^2. (Note that the additional, lower amplitude features seen at longer wavelengths are due to second order of the grating); (c) harmonic spectra to 13 nm wavelength (95 eV, $h = 81$) reported by Schulze et al.[5] upon illumination of a 1 mm wide jet of neon at 60 Torr with a 700 fs, 1.053 μm wavelength at 5×10^{14} W/cm^2 with confirmed polarization; (d) harmonics to 10.7 nm (116 eV, $h = 77$) observed by Pfeifer[6, 16] through 100 μm of xenon gas at 2 atmospheres pressure, and illuminated at 800 nm, 80 fs, and an intensity of 5×10^{14} W/cm^2. In all cases cited here the observed spectra, especially at the longer wavelengths, are affected by the use of absorption filters used to block the incident laser light, and by the spectral response of the monochromators.

invoke what is known as the "slow field approximation" in which the duration of a half cycle of the laser field is much longer than the electron tunneling time, and thus the field can be treated as relatively static, an assumption well matched to the HHG process. The governing non-dimensional parameter, known as the Keldysh parameter, γ_K, is written as

$$\gamma_K = \sqrt{I_p/2U_p} \tag{7.3}$$

where I_p is the binding potential due the electric field of the nucleus as felt by the bound electron (shielded by other electrons) and $U_p = \overline{KE} = e^2 E_0^2/4m\omega^2$ is the

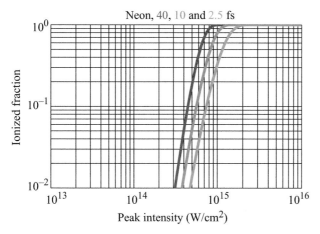

Figure 7.3 Calculated ionization probabilities for neon atoms as a function of laser intensity. The curves show the ionized fraction of neon atoms after interacting with linearly polarized, Gaussian laser pulses of various durations: 40 fs, 10 fs, and 2.5 fs. The curves have been calculated using cycle-averaged ADK tunnelling ionization rates (Ref. 13, Eq. 12). Only the dominant ionization channel ($m = 0$) was considered. The ionization potential for Ne was taken as 21.5645 eV. These curves follow the ADK model,[13] employing the slowly varying envelope approximation (SVEA). For the shortest pulse duration, the SVEA becomes increasingly invalid, and the actual electric field waveform, including the carrier envelope phase (CEP), needs to be taken into account. The 2.5 fs curve plotted here should thus be regarded as an approximate visualization, illustrating that shorter pulse durations permit higher peak intensities before the atoms are fully ionized. Additional curves for He, Ne, Ar, Kr, and Xe are found in Appendix H. Courtesy of C. Ott[18], Max Planck Institute for Nuclear Physics, Heidelberg.

ponderomotive potential [Eq. (7.2a)]. In the parameter range of interest to high harmonic generation, $\gamma_K \leq 1$, the electric field of the laser is smaller but comparable to the (shielded) nuclear electric field felt by the bound electron, and tunneling is the primary mechanism by which the electron is removed from the atom. U_p [Eq. (7.2b)] can be written in terms of the laser intensity, so that we expect an ionization rate that grows exponentially as the laser intensity is increased. The intensities in Figure 7.2 are consistent with a Kelydsh parameter of order 0.1 at the higher intensities, for which tunneling of electrons is possible and continues to rise with increasing intensity.

The highly efficient nature of tunneling is illustrated in Figure 7.3 which shows calculated ionization fractions for neon atoms as a function of laser intensity for various pulse durations. Note how sharply the ionization fraction rises with intensity, reflecting the exponential nature of the tunneling process. The curves were generated by C. Ott[18] following the ADK model.[12] Only the dominant ionization channel ($m = 0$) is considered. Similar curves for He, Ar, Kr, and Xe are presented in Appendix H. See comments in the Figure 7.3 caption. Experimental data exhibiting this trend of ionization vs. laser intensity were first reported in 1983 by L'Huillier *et al.*[19] for He, Ne, Ar, Kr, and Xe, gases typically used in HHG experiments. Further results helped to clarify the role of tunneling ionization.[20, 21]

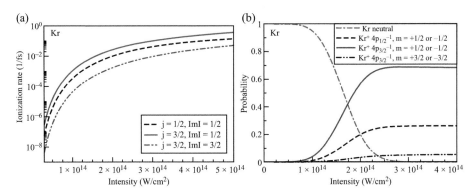

Figure 7.4 (a) Calculated ionization rates for the 4p orbitals of Kr and (b) the probability of neutral Kr atoms surviving by the end of the pulse, and the emergence of 4p-type states of Kr$^+$, both as a function of intensity for 800 nm, 40 fs duration laser pulses. In the legend, m is the projection quantum number and 4p$_j$$^{-1}$ refers to a hole in the 4p$_j$ orbital. Courtesy of R Santra,[22] Argonne National Laboratory; presently CFEL/DESY and U. Hamburg.

Calculated ionization rates for krypton as a function of laser intensity, explicitly showing contributions from various $n = 4$ orbitals, are given by Santra, Dunford, and Young[22] in Figure 7.4. The calculations are for laser pulses of 40 fs duration (15 cycles) and 800 nm wavelength. Also shown is the fraction of neutral Kr atoms remaining at the end of the pulse. Note that the intensity region for which the availability of neutral atoms declines rapidly is a very narrow window, 1–2×10^{14} W/cm^2, and that the density of neutral Kr atoms approaches zero beyond 3×10^{14} W/cm^2. These curves too demonstrate the very high efficiency and sharp intensity dependence of tunneling under typical HHG conditions.

Step 2. Following the release of an electron by tunneling, as described in step 1, the free electron is now accelerated in the electric field of the irradiating laser pulse.[2] The three step model assumes that the electron exits the atom at zero kinetic energy, and that the residual field of the ion is negligible, so that the free electron's motion is dominated by the laser field. As we will see in what follows, the energy gained by the electron in the laser field, and delivered upon return to the parent ion, depends greatly on its "time of birth" (exiting the tunneling process) relative to the phase of the driving laser electric field. The dynamics of the electron motion in step 2 are purely classical,[2] described by Newton's second law of motion, $\mathbf{F} = m\mathbf{a}$, where $\mathbf{F}$ is the Lorentz force on the free electron

$$m\mathbf{a} = -e[\mathbf{E} + \mathbf{v} \times \mathbf{B}] \tag{7.4}$$

In Chapter 2, Section 2.5, it was shown that the $\mathbf{v} \times \mathbf{B}$ term is of order v/c compared to $\mathbf{E}$, and thus negligible in the non-relativistic regime considered here. Assuming a linearly polarized field, $E = E_0 \cos \omega t$, the equation of motion for the oscillation amplitude is

$$a(t) = d^2x/dt^2 = -[eE_0/m] \cos \omega t \tag{7.5a}$$

Integrating twice one has

$$v(t) = -[eE_0/m\omega]\sin\omega t + v_0 \tag{7.5b}$$

and

$$x(t) = [eE_0/m\omega^2]\cos\omega t + v_0 t + x_0 \tag{7.5c}$$

where the constants v_0 and x_0 are determined by the initial conditions $v(t_0) = 0$ and $x(t_0) = 0$ at the electron's time of birth, $\omega t_0 = \varphi_0$, within the phase φ of the laser electric field. From Eq. (7.5b) one determines that $v_0 = [eE_0/m\omega]\sin\varphi_0$, and then from Eq. (7.5c), $x_0 = -[eE_0/m\omega^2](\cos\varphi_0 + \varphi_0\sin\varphi_0)$. The time dependent velocity of the electron is then

$$\Delta k = k_h - hk = \Delta k_e + \Delta k_a + \Delta k_{\text{other}} \tag{7.6a}$$

and the time dependent amplitude of the electron (away from the parent atom ion) is

$$x(t) = [eE_0/m\omega^2]\,[(\cos\varphi - \cos\varphi_0) + (\varphi - \varphi_0)\sin\varphi_0] \tag{7.6b}$$

where $\varphi = \omega t$. Upon return to the parent ion the electron's x-coordinate is again zero, $x(t_r) = 0$, at a phase $\varphi_r = \omega t_r$. Setting the right most bracket in Eq. (7.6b) to zero yields the relationship between initial and final times (phases) of the electron's motion within the laser field, viz.,

$$\varphi_r\sin\varphi_0 + \cos\varphi_r = \varphi_0\sin\varphi_0 + \cos\varphi_0 \tag{7.6c}$$

This equation can be solved numerically for the return time, or phase φ_r, as a function of the time of the electron's emergence from the tunneling process at phase φ_0. For release within the interval $0 \leq \varphi_0 \leq \pi/2$ real solutions exist for return to the parent ion at phase φ_r. Indeed, for release just at the peak of the electric field, $\varphi_0 = 0$, the return is at $\varphi_r = 2\pi$. In this special case the electron undergoes a simple oscillation away from, and then back to the ion, with fairly high cycle-averaged kinetic energy but, unfortunately, returning with zero kinetic energy and thus not contributing to HHG emission. High return kinetic energies occur, as we will see below, for electrons released just after the peak of the pulse, with a maximum return energy at $\varphi_0 \cong \pi/10$, or $18°$. For initial phases $\pi/2 \leq \varphi_0 \leq \pi$ there are no real solutions for φ_r; the electron never returns to the parent ion and thus cannot generate HHG photons. The process repeats every half cycle, with identical excursion amplitudes and return kinetic energies, but travelling in the opposite direction due to the reversed direction of the laser electric field. A second maximum return energy occurs at an initial phase of $198°$; $18°$ beyond the peak electric field at $\varphi = \pi$. Illustrative trajectories are shown for these three cases in Figure 7.5.

Next we consider electron trajectories calculated from Eqs. (7.6a–c) for various values of time of birth ($\varphi_0 = \omega t_0$) relative to the phase of the incident laser electric field. By way of example, trajectories are calculated which correspond to the irradiation of neon atoms by focused laser light at an intensity of 5.00×10^{14} W/cm^2 and a wavelength of 800 nm. The corresponding peak electric field, E_0, is 6.14×10^8 V/cm, well below the binding field of a neon $n = 2$ electron, thus well matched for efficient tunneling

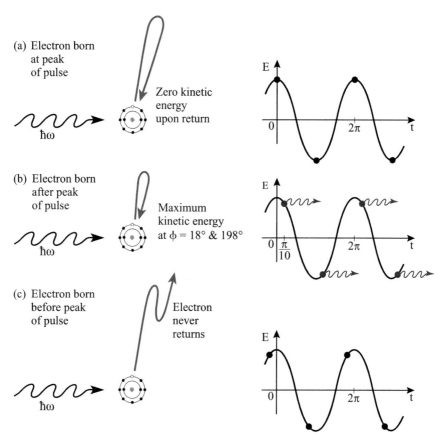

Figure 7.5 Electron trajectories are shown in (a) for an electron liberated from the tunneling process just at the peak of the pulse ($\varphi_0 = 0$), gaining significant energy during its excursion in the laser field but returning to the parent ion with zero kinetic energy; (b) an electron born just after the peak of the pulse ($0 \leq \varphi \leq \pi/2$) which returns to the parent ion with significant kinetic energy, with finite probability of being converted to an HHG photon; (c) an electron liberated before the peak of the pulse ($-\pi/2 \leq \varphi \leq 0$, or $\pi/2 \leq \varphi \leq \pi$) never returning to the parent ion and thus not contributing to the HHG process. Figure with suggestions from A. L'Huillier, Lund University.

with a Keldysh parameter of $\gamma_K \cong 0.6$. The electron trajectories for these parameters, described by Eq. (7.6b), are as follows

$$x(t) = (1.947\,\text{nm})\left[(\cos\varphi - \cos\varphi_0) + (\varphi - \varphi_0)\sin\varphi_0\right] \tag{7.7a}$$

The electron return energies are determined from Eq. (7.6a) as

$$E_e(t_r) = \frac{1}{2}mv^2(t_r) = \left[e^2 E_0^2/2m\omega^2\right](\sin\varphi_r - \sin\varphi_0)^2$$

or

$$E_e(t_r) = (59.76\,\text{eV})(\sin\varphi_r - \sin\varphi_0)^2 \tag{7.7b}$$

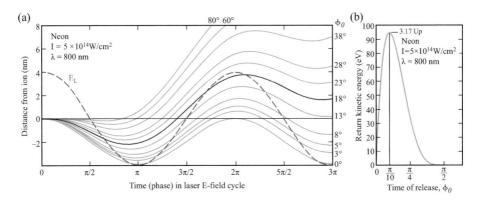

Figure 7.6 (a) Electron trajectories are shown for a variety of initial emission times, labeled to the right as initial phases, $\varphi_0 = \omega t_0$, within the cycle of the incident IR laser electric field, EL. The example is for neon atoms irradiated at an intensity of 5×10^{14} W/cm² and a wavelength of 800 nm. Of particular interest are the trajectories for $\varphi_0 = 0$, where the electron gains significant kinetic energy but returns with none, simply oscillating with a 4 nm total excursion, and the trajectory for $\varphi \cong \pi/10$, or 18°, which corresponds to a return with maximum kinetic energy, 94.8 eV in this example, as seen in (b). Note that electrons emerging from the tunneling process at initial phases extending from $\pi/2$ to π just before the peak of the pulse, never return to the parent ion and thus do not generate HHG photons (see Figure 7.5). The process repeats every half cycle, with identical excursion amplitudes and return kinetic energies, but travelling in the opposite direction due to the reversed direction of the laser electric field. A second maximum return kinetic energy occurs at a phase of 198°; 18° beyond the reversed peak field.

Note that from Eq. (7.7a) the trajectory for $\varphi_0 = 0°$ (birth at the peak of the laser electric field) corresponds to a simple electron oscillation with excursion distances of nearly 4 nm from the parent ion, gaining a maximum kinetic energy of near 60 eV, but returning to the ion with zero kinetic energy. This is illustrated in Figure 7.6 along with several other trajectories. As observed there, electrons which exit the tunneling process just after the peak of the pulse return at varying times and with broadly varying return kinetic energies. Of particular interest is the violet colored curve $\varphi_0 \cong \pi/10 = 18°$, just beyond the peak of the pulse, which corresponds to the highest return kinetic energy. For the parameters considered in this example, the maximum return kinetic energy is calculated from Eq. (7.7b) to be 3.17 $\overline{KE} = 94.8$ eV, where from Eq. (7.6c) $\varphi_0 = 18°$ and $\varphi_r = 252°$. Note that the cycle averaged* kinetic energy, $\overline{KE} = 29.88$ eV, demonstrating the 3.17 factor seen earlier in Eq. (7.1b). This multiplicative factor expresses the maximum possible return energy in terms of the average energy of an oscillating electron in the same field. Return energies are plotted in Figure 7.6(b) for initial phases from zero to greater than $\pi/2$, showing the breadth of return kinetic energies generated with monochromatic laser light. $\varphi \cong \pi/10$, or 18°, which corresponds to a return with maximum kinetic energy, 94.8 eV in this example, as seen in (b). Note that electrons emerging from the tunneling process at initial phases extending from $\pi/2$ to π

---

* This is half the 59.76 eV in Eq. (7.7b) due to the cycle average of $\sin^2 \omega t = 1/2$.

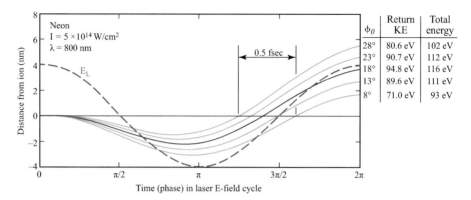

Figure 7.7 Representative electron trajectories are shown for various initial phases, with return kinetic energies (*KE*) displayed on the right side. Also shown are the total energies upon return, $KE + I_p$. These are chosen to illustrate the effect of bandwidth filtering as might be obtained with a mirror–filter pair[23, 24] or a multilayer coated mirror[25, 26] with bandwidth appropriate to a specific scientific experimentation. For the selected parameters, these values correspond to an approximately 8% relative spectral bandwidth around a peak of 116 eV (10.7 nm wavelength, harmonic number $h = 75$). The radiation will appear as harmonics within the emission bandwidth. Higher energies could be achieved with higher incident laser intensites and/or longer wavelengths photon (see Eqs. (7.1) and (7.2)), but likely with reduced HHG efficiency.

just before the peak of the pulse, never return to the parent ion and thus do not generate HHG photons (see Figure 7.5). The process repeats every half cycle, with identical excursion amplitudes and return kinetic energies, but travelling in the opposite direction due to the reversed direction of the laser electric field. A second maximum return kinetic energy occurs at a phase of 198°; 18° beyond the reversed peak field. Similarly, there is no return for trajectories beginning with initial phases $-\pi/2$ to 0; $3\pi/2$ to 2π; etc.

Return energies are explored further in Figure 7.7, where only electron trajectories which would contribute to emitted photon energies within a selected spectral bandpass are displayed. This could be achieved, for example, with a mirror–filter combination[23, 24] or a multilayer mirror[25, 26] (Chapter 10) used to narrow the emission spectrum for use in particular scientific experiments. Here curves are shown for initial phases of 8° to 28°, providing a bandwidth approximately centered on the maximum return energy. The return kinetic energies for each trajectory are shown to the right, as well as the total energies, $KE + I_p$, available for conversion to high harmonic radiation. Note that for the example chosen, the spread of photon energies constitutes an approximately 8% relative spectral bandwidth about the 116 eV (94.8 eV + 21.6 eV) maximum. Note too that the predicted burst of emitted photons occurs within a 0.5 fs window twice per laser cycle, thus creating a temporal train of EUV pulses of sub-femtosecond duration. As the emission bursts occur symmetrically in time, twice per cycle, a Fourier transform predicts only odd harmonics, as seen in the data of Figure 7.2. In a later section we will explore the generation of single attosecond HHG pulses driven by shorter, near single cycle, intense IR laser pulses. Harmonic photon energies higher than those seen in Figures 7.5

and 7.7 can be achieved with increased intensity and/or longer wavelength, as described by Eqs. (7.1a, b) and (7.2a,b), however, with decreased conversion efficiency from the IR to the EUV/SXR region. This will be discussed further in step 3, which follows.

While the maximum return kinetic energy scales directly with intensity, conversion efficiency may decrease due to excessive depletion of available neutrals, as was seen in Figure 7.4, and also due to the increased density of free electrons which affects the phase velocity of the IR pulse, causing further dephasing between the IR and generated harmonics, thus limiting the effective co-propagation path length. Return kinetic energies also scale with IR wavelength squared, so this too might provide considerable increases in emitted photon energies. However, the excursion distances from the parent ion also scale as wavelength squared, thus reducing the probability of recombination and significantly diminishing photon flux at these higher photon energies. These effects are discussed further in the following section.

Step 3. The third step in the HHG process is the radiative recombination of a returning electron to the ground state of its parent ion with the emission of a photon, which is possible with finite probability. For those electrons that do recombine to the ground state there is a release of energy in the form of a photon whose energy is equal to the gained kinetic energy plus the atom's ionization energy, as given by Eqs. (7.1a, b). Examples of anticipated photon energies, extending throughout the EUV region, were given in Figures 7.6 and 7.7 for neon irradiated at high intensity. The probability of electron recombination, however, is relatively small as a result of several factors. For one, half of the photoelectrons which emerge from tunneling never return to their parent ion, as illustrated in Figure 7.5. For those electrons released with initial phases from $-\pi/2$ to zero, and $\pi/2$ to π, just before the peak of the pulse, their trajectories take them far from the ion, never to return. For those released with initial phases of 0 to $\pi/2$, and π to $3\pi/2$, just after the peak of the pulse, they do return to the ion and thus can radiatively recombine. The relatively small probability of radiative recombination, however, is dominated by the long excursion path of the electron away from the ion (Eq. (7.6c)) and the three-dimensional divergent nature of the electron emission process due to quantum diffusion. The photoemission process is properly described as diffraction of the quantum mechanical electron wave packet as it emerges from the transversly confined ångström scale atom. Thus the electron moves away from the ion as a somewhat divergent electron wavepacket, with a fractional probability of recombination upon return. Discussions of the quantum mechanical nature of both tunneling and recombination are given in papers by Lewenstein et al.,[27], Ivanov et al.,[28] Brabec and Krausz,[14] Corkum and Krausz,[29] Krausz and Ivanov,[30] and Pfeifer, Spielmann, and Gerber.[16] Corkum and Krausz[29] estimate an electron wavepacket divergence of 5 Å/fs for an IR laser pulse of 800 nm wavelength and near 10^{15} W/cm^2 intensity. This results in a return wavepacket width (FWHM) of about 10 Å after an excursion of about 2 fs in the laser field. This increased width of the wavepacket, about 10 Å upon return, is to be compared to a 2p lateral atomic radius of order 0.3 Å, resulting in relative cross-sectional areas of order 10^{-3}, greatly limiting the probability of radiative recombination. Longitudinal wavepacket spreading further contributes to the diluted probability of radiative recombination.

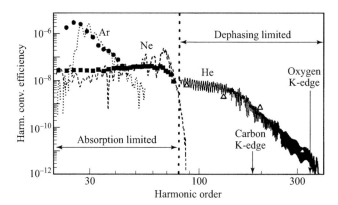

Figure 7.8 Measured energy conversion efficiency into individual harmonics is shown versus harmonic number.[14, 31, 32] Efficiencies of several times 10^{-6} to 10^{-8} are observed across much of the EUV region, then decreasing into the soft x-ray region. Values for Ar (solid circles), Ne (squares), and He (triangles), are shown at respective intensities, gas pressures, and effective gas propagation lengths as follows: Ar at 2×10^{15} W/cm^2, 230 Torr, 3 mm; Ne at 5×10^{14} to 2×10^{15} W/cm^2, 300 Torr, 3 mm; He at 4×10^{15} W/cm^2, 3000 Torr, 0.3 mm, all at approximately 800 nm wavelength and 7 fs duration. Experimental values follow the original publications.[31, 32] Calculated spectra, shown by dashed lines, are given by Brabec and Krausz.[14] Courtesy of M. Schnürer, Technical University of Vienna, and F. Krausz, Max Planck Institute for Quantum Optics and Ludwig Maxmillians University, Garching.

The combined effects of tunneling with an efficiency of order 10^{-1} to 10^{-2} and the low probability of radiative recombination, results in an overall efficiency of harmonic generation from the IR laser to a specific EUV harmonic of order 10^{-5} to 10^{-8} across the EUV spectrum, decreasing further towards soft x-ray wavelengths. Experiments reported in the literature[31, 32] for several gases, irradiated at approximately 800 nm wavelength, have been summarized in the review article by Brabec and Krausz,[14] and are shown here in Figure 7.8. For neon with an effective propagation path of 3 mm at a pressure of 300 Torr, irradiated with a 7 fs duration, 790 nm laser pulse at an intensity of 2×10^{15} W/cm^2, the data indicate a conversion efficiency of about 4×10^{-8} from the energy of the IR laser pulse to the $h = 65$ harmonic at 12 nm (100 eV). With a modern Ti:sapphire laser delivering as much as 30 mJ in 7 fs pulses at 10 kHz, this efficiency suggests that 1 nJ per 12 nm harmonic, or about 6×10^7 photons per individual pulse can be generated at 10 kHz. The role of absorption is addressed in the literature,[33] and here in a later section.

A further effect of the small probability of recombination is the creation of a sufficient density of ions and free electrons to affect the phase matching of IR and EUV/SXR beams. The free electron density produces a plasma refractive index effect which can limit the extent of the coherent interaction region (Figure 7.1). In this situation the free electron plasma imparts a high phase velocity to the IR laser wavefront such that it runs faster than the EUV wavefront, eventually generating EUV radiation that is out of phase with the EUV created earlier along the propagation path ("dephasing"). This effect is considered further in Section 7.4. The role of EUV absorption is also significant, even dominant in some cases, as discussed in Section 7.4.

Although there is great interest in utilizing longer IR laser wavelengths to achieve higher photon energies (Eqs. (7.1b) and (7.2b)), numerical simulations, which solve the time-dependent Schrödinger equation, indicate a strongly decreasing probability of radiative recombination with longer wavelengths,[34, 35] with a wavelength dependence in the range of $1/\lambda^4$ to $1/\lambda^6$, thus reducing the advantage of extending the laser driver to longer IR wavelengths to achieve high photon flux at SXR photon energies. Experiments with various wavelengths,[36, 37] from 500 nm 2.1 µm, confirm this strong wavelength dependence, with variances depending on specific experimental conditions.

7.3   Spatial and Temporal Coherence of High Harmonic Radiation

The spatial coherence of HHG radiation is largely imprinted from that of the fully coherent IR laser wavefront. This is true despite the randomness of atoms in the gas. Every participating atom across a uniform IR wavefront emits radiation in the forward direction with a fixed phase relative to the laser electric field, and thus to each other. As a consequence of this impressed spatial coherence, the electric fields for a given harmonic add constructively *in the forward direction*, generating very intense radiation, proportional to the square of the number of participating atoms in the interaction region. The number of participating atoms is defined as those within the overlap region of the gas jet and the incident laser field, as seen in Figure 7.1. Note, however, that radiation emitted in any other direction does not share this feature. For example, the phases of radiation in the reverse direction are dependent on the random atomic positions. As a consequence, the phases in the reverse direction are also random, and cancel when averaged over a sufficiently large number of atoms. This is also true for any lateral emissions, indeed for any direction except the forward direction. In these other directions, emission from any atom, radiating with any particular phase, will always be cancelled by that of another atom emitting with a π phase difference somehere along the same path. Because real experiments involve focused radiation, there will be both curvature and intensity variation across the IR wavefront, each of which will to some degree affect the phase fronts and coherence of the emitted harmonics. There will also be an intensity dependent generation of free electrons, due to the finite probability of not returning to the parent ion, and this too can affect and possibly distort harmonic wavefronts. These factors, departures from the idealized uniform wavefront, can have varying impact depending on focusing geometry and laser intensity, all of which will be discussed further below. The propagation distance for which the IR spatial coherence continues to be impressed on the harmonics, even for the idealized transversly uniform case, is limited by the relative phase velocities of the IR laser and the high harmonics, a topic of special importance discussed in the following section.

Following Chapter 4, Eq. (4.7a), radiation of full spatial coherence satisfies the limiting equation

$$d \cdot 2\theta = \lambda/2\pi \tag{7.8a}$$

where both $d/2$ and θ are rms measures of Gaussian distributions. Alternately, this condition can be expressed in terms of FWHM measures, as are commonly used in describing laser experiments, i.e., Eq. (4.7b), renumbered here as Eq. (7.8b)

$$d_{\text{FWHM}} \cdot 2\theta_{\text{FWHM}} = 0.4413\,\lambda \cong \lambda/2 \tag{7.8b}$$

where both d and 2θ are FWHM measures of Gaussian distributions of the source diameter and divergence angle. Because the laser intensity is less off axis we expect, based on discussions and equations in Section 7.2, step 2, that the spatial extent of the harmonic radiation, d, will be somewhat less than that of the IR laser pulse. Furthermore, since the harmonic wavelengths are significantly shorter than that of the IR laser, constructive interference will lead to a significantly narrower harmonic emission angle, 2θ, as can be deduced from Eq. (7.8). Confirmation of a high degree of spatial coherence of HHG radiation[38-40] has been demonstrated using Young's double-slit, or two pinhole interference technique,[41] as was described in Chapter 4, Section 4.5. The papers by Ditmire et al.[38, 39] include an extensive discussion of the HHG process with particular attention to coherence issues, including, for example, the non-uniform generation of excessive free electrons, which results in a distortion of the laser wavefront due to plasma refractive effects and a concomitant reduction to the degree of spatial coherence. Papers by Lyngå et al.,[42] Salières et al.,[43, 44] and Hergott et al.[45] discuss further aspects of the HHG process that affect the degree of coherence. An example of the two-pinhole interference technique as it is used to measure spatial coherence of HHG radiation is shown in Figure 7.9. The experiments were performed with a 5 kHz Ti:sapphire laser operating at 760 nm, with 0.8 mJ per 25 fs pulse, and focused to a laser intensity of about 2×10^{14} W/cm^2 in argon at 29 Torr. The spatial coherence was measured by recording fringe pattens simultaneously for harmonics $h = 17$, 19, and 21. Interference (fringe) patterns are shown in Figure 7.9 for separation distances of 142 μm, 242 μm, 384 μm, and 779 μm, which correspond to sampling 14%, 24%, 38%, and 78% of the EUV beam diameter. Analysis of the data indicates a high degree of spatial coherence across most of the EUV beam.

The temporal coherence of HHG radiation is significantly different than that of the IR laser. It is determined by the much shorter wavelength and the relatively broad spectral bandwidth. The longitudinal, or temporal, coherence length is defined as (Chapter 4, Eq. (4.6))

$$l_{\text{coh}} = \lambda^2/2\Delta\lambda \tag{7.9}$$

where λ is the harmonic wavelength and $\Delta\lambda$ is the spectral bandwidth, which for HHG is largely determined by electron time of birth differences within the phase of the IR laser pulse. These time of birth differences lead to variances of photoelectron return energies and excursion times, thus differences in photon energies and phase. The detected bandwidth can, of course, also be affected by spectral filters. For the example considered in Figure 7.7, with a central wavelength of 10.7 nm (116 eV) and a filter imposed 8% relative spectral bandwidth, the longitudinal coherence length is $l_{\text{coh}} = \lambda^2/2\Delta\lambda \cong 130$ nm. While this may seem relatively short, in fact it is sufficient for many interference experiments utilizing relatively shallow angles between the interfering wavefronts.

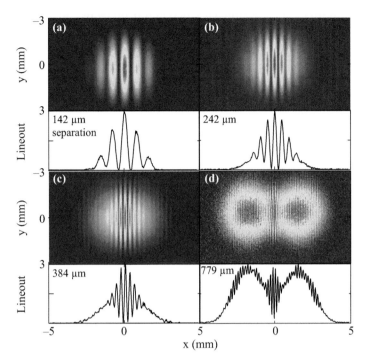

Figure 7.9 Interference patterns recorded on an EUV sensitive CCD detector for the measurement of spatial coherence of HHG radiation.[40] Three harmonics of 760 nm were simultaneously recorded, centered at harmonic number 19 (36 nm, 34 eV). Pulses of 25 fsec duration were focused to 2×10^{14} W/cm^2 in argon at 29 Torr. The EUV radiation is centered at 36 nm (34 eV), with a photon flux of about 2×10^{12} photons/s at 5 kHz (4×10^8 ph/pulse at 36 nm). Pairs of 50 μm diameter pinholes were used at separations of (a) 142 μm, (b) 242 μm, (c) 384 μm, and (d) 779 μm. Exposure times varied from 20 s to 240 s (100 000 to 1 200 000 laser shots). The high contrast fringe patterns reveal a high degree of spatial coherence across most of the EUV beam. Courtesy of R. Bartels, University of Colorado, Boulder, now at Colorado State University, Fort Collins, and Y. Liu, UC Berkeley, now at KLA-Tencor.

7.4   IR/EUV Dephasing and the Effective Propagation Path Length for HHG

For moderately focused IR laser pulses,[†] with a depth of focus greater than the gas–laser interaction distance, the high harmonics are radiated, to first order, with a fixed phase with respect to the driving IR laser pulse. In this manner the coherence properties of the laser are imprinted on the high harmonic radiation, as discussed in the previous section. Indeed, this phase coherence imprint continues plane by plane throughout the interaction region. However, owing to differences in phase velocities of the IR laser and high

---

[†] In Chapter 4, Section 4.1, the equations for a focused Gaussian beam are discussed in terms of $1/e^2$ radial measures. As an example, for an 800 nm wavelength beam focused to a 100 μm FWHM diameter, the $1/e^2$ radius is $\omega_0 = 85.1$ μm, and the Rayleigh range is $Z_R = \pi \omega_0^2 / \lambda = 28.5$ mm. The full ($\pm$) depth of focus is $2Z_R \cong 60$ mm, a value beyond the interaction lengths of many HHG experiments. For example, an 800 nm, 1 mJ, 30 fs pulse focused to 100 μm FWHM, would result in a focal plane intensity of about 4×1014 W/cm^2, typical of neon HHG experiments.

harmonic wavefronts, a phase difference develops between the two. At some distance, known as the dephasing distance, the harmonic radiation generated at the beginning and at the end of the gas–laser interaction region become out of phase with each other by a half-wave of the EUV radiation (180° or π radians), leading to destructive interference and thus a limiting distance for efficient HHG energy conversion. The differing phase velocities are related to the respective refractive indices[‡] n by $v_\varphi = c/n$, as was seen in Chapter 3, section 3.1. Thus the dephasing distance can be analyzed in terms of contributions to the respective IR and harmonic refractive indicies. The IR laser radiation is of considerably lower frequency than that of the high harmonics, by a factor h, and thus the two respond differently to the atoms, ions, and free electrons in the interaction region. The harmonics are at very high frequencies and propagate at phase velocities very close to the speed of light in vacuum. The IR laser pulse is strongly affected by the density of free electrons, which are present due to the high degree of non-recombination during the HHG collision process. The refractive index for a plasma of free electrons, n, was developed in Chapter 8, Physics of Hot Dense Plasmas, Section 8.4.7, Eq. (8.114b). For small electron density, $n_e/n_c \ll 1$

$$n = 1 - (n_e/2n_c) \tag{7.10a}$$

where n_e is the electron density, in units of electrons/cm^3, n_c is the critical electron density

$$n_c = \varepsilon_0 m \omega^2 / e^2 = \left(1.11 \times 10^{21} \text{ e/cm}^3\right) / \lambda^2 \text{ [}\mu\text{m]} \tag{7.10b}$$

and λ is bracketed above in units of microns. The critical electron density is that at which radiation at a given wavelength would be totally reflected by the plasma electrons, or said differently, when the radiation frequency equals that of the electron plasma oscillation frequency at $\omega = \omega_p$. The plasma frequency is given by

$$\omega_p^2 = n_e^2 / \varepsilon_0 m \tag{7.10c}$$

Thus from Eq. (7.10b), $n_c = 1.73 \times 10^{21}$ e/cm^3 for the commonly used 0.8 μm HHG wavelength. Because the free electron contribution to refractive index is less than unity it causes an increase in the phase velocity (c/n) of the IR laser pulse. This same free electron refractive effect is much smaller for the high harmonics because their frequencies are so much higher. Indeed, by Eq. (7.10b), the critical electron density is higher by the harmonic number squared, h^2, thus the free electron contribution to the harmonic refractive index is smaller by h^2, and negligible. Neutral atoms, and ions, contribute to the refractive index of both IR laser and harmonic radiation, but generally less than that of the free electrons, as discussed below.

The dephasing distance, L_π, is the distance required to produce a half-wave (π) relative phase shift at the harmonic wavelength due to differences between the phase velocities of the IR and harmonic waves. The difference in phase velocities, Δv_φ, is due to

---

[‡] Recall from chapters 2 and 3 that n without a subscript refers to refractive index, whereas use of a subscript implies a density, such as n_e or n_a, for electrons and atoms. In this chapter we break that tradition to introduce refractive indices for the IR laser and high harmonics as n_{ir} and n_h, respectively.

differences in the refractive indices at the two wavelengths, that is $\Delta v_\varphi = c/\Delta n$. The dephasing distance can be written as

$$L_\pi = (0.5\lambda/h)/\Delta n \qquad (7.11)$$

where λ is the IR laser wavelength in vacuum and $(0.5\lambda/h)$ corresponds to a π phase shift for the h harmonic. This would correspond to a situation in which the IR wavefront propagates through N cycles at wavelength λ, and the harmonic through $(hN + \frac{1}{2})$ cycles at wavelength λ_h. At the end of this path the high harmonic radiation generated at the beginning and at the end of the path would be a half wave out of phase. We can write these path lengths as $L_{ir} = N\lambda_{ir} = N\lambda/n_{ir}$ and $L_h = (hN + \frac{1}{2})\lambda_h = (hN + \frac{1}{2})\lambda/hn_h$, where $\lambda = 2\pi c/\omega$ is the vacuum IR wavelength and where we have introduced the IR and high harmonic refractive indices, n_{ir} and n_h. Setting the path lengths equal, $L_{ir} = L_h$, canceling the common λ, and solving for N, we have the number of cycles required for a half wave high harmonic phase difference,

$$N = \frac{1/2hn_h}{\frac{1}{n_{ir}} - \frac{1}{n_h}} \qquad (7.12a)$$

The dephasing distance, L_π, is equal to both L_{ir} and L_h, and thus we can write

$$L_\pi = L_{ir} = N\lambda_{ir} = \frac{\lambda_{ir}/2hn_h}{\left[\frac{1}{n_{ir}} - \frac{1}{n_h}\right]} \qquad (7.12b)$$

Expressing the refractive indicies as $n_{ir} = 1 - \alpha_e + \alpha_a$, and $n_h = 1 - \alpha_\delta$, where α_e, α_a, and α_δ are all positive and small ($\ll 1$) in the spectral range of interest here, where α_e is the free electron contribution to the IR refractive index, α_a is the atom (and ion) contribution to the IR refractive index, and α_δ is the atom (and ion) contribution to the high harmonic refractive index, the dephasing distance can be written as

$$L_\pi = \frac{\lambda_{ir}/2hn_{hh}}{[\alpha_e - \alpha_a - \alpha_\delta]} \qquad (7.13)$$

Identifying the difference between IR and harmonic refractive indices as

$$\Delta n = \alpha_e - \alpha_a - \alpha_\delta \qquad (7.14)$$

one obtains Eq. (7.11). Note that the free electron contribution to the fast IR phase velocity, α_e, is compensated by both the IR slowing due to the neutral atoms (and ions), α_a, and the speeding of the high harmonics due to the atoms (and ions), α_δ. The incremental contributions of the atoms changes sign from IR to high harmonic because the IR, for a photon energy of 1.55 eV, is well below the atomic resonances, while the harmonics are at photon energies well above the nearby resonances, which are of order 30–100 eV. See Chapter 3, Section 3.1, and Figure 3.1 for further discussion. As pointed out earlier, the free electron contribution to the high harmonic refractive index is smaller than α_e by $1/h^2$, and thus is not included in Eq. (7.14). Before putting values into the dephasing distance equation just derived, we note that in the field of non-linear optics,[46] particularly consideration of visible and infrared frequency mixing and harmonic upconversion in crystals at high laser intensity, the treatment of conversion efficiency and phase mismatch would be described in terms of a wavenumber mismatch

$$\Delta k = k_h - hk = \Delta k_e + \Delta k_a + \Delta k_{\text{other}} \tag{7.15a}$$

where $k = n\omega/c$ is the wavenumber at the fundamental frequency ω in a medium of refractive index n, k_h is the harmonic wavenumber at frequency $h\omega$ and refractive index at $h\omega$, and Δk_{other} could include additional phase shifting effects such as wavefront curvature due to tight focusing or waveguiding effects in capillary tubes, etc. In the traditional non-linear optics treatment[46, 47] the sensitivity of harmonic intensity to phase or wavenumber mismatch, would be proportional to $\text{sinc}^2(\Delta k \cdot L)$ where L is the propagation distance. The function $\text{sinc}^2(u)$ is unity for $u = 0$ (maximum conversion efficiency) and goes to zero[48] at $u = \Delta k \cdot L = \pi$. By our definition the π phase shift corresponds to a propagation distance L_π so that

$$\Delta k = \frac{\pi}{L_\pi} = \frac{2\pi h}{\lambda}(\alpha_e - \alpha_a - \alpha_\delta) \tag{7.15b}$$

Next we attempt to determine a set of values for gas pressure, wavelength, pulse duration, laser intensity, and percentage ionization that would combine to give a long dephasing distance, suitable for efficient HHG conversion efficiency, much as would be done in designing an experiment. A critical parameter will be the laser intensity, as this very sensitively sets the ionization fractions, as seen in Figure 7.3, and also to a large degree the free electron density. Indeed, we understand that the calculated ionization rates are for an idealized, transversly uniform intensity profile, while we know that this is not the case for typically Gaussian shaped focused laser beams, so that there will be some averaging involved and thus some fine tuning of intensity and pressure in actual experiments. To determine L_π from Eq. (7.13) values of the various refractive index increments are required. Towards that end we select a set of values for an imagined experiment. We choose neon at a pressure of 300 Torr ($\sim$0.4 atm), illuminated by an 800 nm wavelength, 40 fs pulse at an intensity of 3.2×10^{14} W/cm^2. At this intensity, according to Figure 7.3(d), the ionization fraction should be appoximately 1.1%. The atomic density can be determined from Loschmidt's number (Appendix A, Table A.2), $n_L = 2.69 \times 10^{19}$ atoms/cm^3 (the number of atoms per unit volume at standard temperature and pressure, STP), multiplying this by the ratio of gas pressure in Torr to the standard atmospheric pressure, 760 Torr. Thus the atomic density of neon is $n_a = 2.69 \times 10^{19}$ atoms/cm^3 times (300/760), giving a density of 1.06×10^{19} neon atoms/cm^3. With an ionization fraction of 1.1%, the free electron density is 1.17×10^{17} e/cm^3. With a critical electron density of 1.73×10^{21} e/cm^3, the IR refractive index increment due to free electrons, $\alpha_e = n_e/2n_c = 3.37 \times 10^{-5}$. The atomic contribution to the IR refractive index is given in the literature[49] as $\alpha_a = (6.70 \times 10^{-5})(300/760) = 2.64 \times 10^{-5}$ at 800 nm wavelength. For the high harmonic refractive indicies we follow Chapter 3, Eqs. (3.9) and (3.12), and utilize tabulated data by Henke, Gulikson, and Davis,[50, 51] available here for neon in Appendix C, Table C.1. For neon at this intensity the highest harmonic photon energy is 82.2 eV (15.1 nm, $h = 53$), well above the neon L-absorption edges, at 21.6–48.5 eV, and well below the neon K-aborption edge at 870 eV. From the tables we find that $f_1^0 = 5.20$ and $f_2^0 = 6.12$. From f_1^0 and the Chapter 3 equations, we calculate that $\alpha_\delta = \delta = n_a r_e \lambda^2 f_1^0 / 2\pi = 5.63 \times 10^{-6}$. The ions have essentially the same refractive index as the atoms at these photon energies, far from the absorption features. The net refractive index increment, Δn, Eq. (7.14), describing the differences in phase velocity for the IR and high harmonic wavefronts for this example is $\Delta n = \alpha_e - \alpha_a - \alpha_\delta = 3.37 \times 10^{-5} - 2.64 \times 10^{-5} - 0.56 \times 10^{-5} = 1.7 \times 10^{-6}$. The dephasing distance would then be $L_\pi = 4.4$ mm, a value similar to the distances quoted for the experiments described in Figure 7.2.

For the situation where Δn is small and the dephasing distance relatively large, one must consider the absorption of high harmonics by neutral gas atoms. For the parameters above we calculated $f_2^0 = 6.12$ for neon, which determines the imaginary component of the refractive index, $\beta = 6.63 \times 10^{-6}$, and an absorption length $l_{\text{abs}} = \lambda / 4\pi\beta = 0.18$ mm (Chapter 3, Eq. (3.22)). Thus for this example at 82 eV, the harmonic must propagate through several absorption lengths, greatly limiting the achievable conversion efficiency.[31–33] For higher harmonics with photon energies above 100 eV, necessarily generated at higher intensity, absorption rapidly decreases, as seen in the f_2^0 curve for neon in Appendix C, Table C.1. Using krypton instead of neon could significantly reduce absorption, as f_2^0 is smaller by a factor of ten in the vicinity of 80 eV (see Appendix C), potentially extending the absorption length to about 2 mm. But at the required higher intensity to reach beyond 100 eV, the fractional ionization in Kr would be much higher (see Appendix H), significantly reducing L_π. Tradeoffs among gas and gas pressure, intensity and pulse duration are discussed in the literature.[14, 31–33, 52]

An additional effect involves the sharply curved wavefront of a strongly focused laser beam. The sharp curvature cannot be followed by a diffraction limited high harmonic wavefront and thus introduces an off-axis phase mismatch. Additionally, the IR laser intensity typically has a Gaussian transverse profile, so that the ionization rate varies across the wavefront, generating a non-uniform free electron plasma. Similarly, substructure in the incident laser wavefront will lead to similar spatial variations in the high harmonic wavefront, further reducing the degree of spatial coherence.[39, 52] As a practical matter, these high-intensity laser effects can modify the local wavefront, its coherence across the interaction region and, additionally, can introduces a negative plasma lensing effect which can result in increased divergence of the IR laser light. This could decrease the IR laser intensity and further curve its wavefront beyond what spatially coherent EUV radiation can follow.[43]

An approach to extending the interaction region, and thus significantly increasing the radiated power, involves mechanisms which alternately turn on and off the HHG process in periodic cycles matched to the dephasing distance. This technique is referred to as *quasi-phase matching* in the literature of non-linear optics.[46] For example, it might consist of a sequence of gas jets[53] each of width L_π and separated by a vacuum region of width L_π, in a geometry reminiscent of a musician's flute. In this manner EUV/SXR radiation would only be generated in regions that constructively interfere, but not generated in regions which would destructively interfere. The gain would be proportional to the number of periods squared. Various forms of quasi-phase matching have been pursued by the University of Colorado, Boulder group, in one embodiement focusing

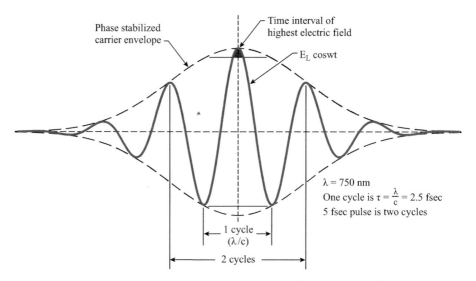

Figure 7.10 A very short, few cycle laser electric field (red) within a phase stabilized carrier envelope (CE). This is referred to as a cosine stabilized waveform in that the maximum electric field is matched to the maximum of the carrier envelope, at a centered phase $\varphi = 0$. Each laser cycle has a duration $\tau = \lambda/c$. The horizontal bars mark the highest electric field attained by the two lower laser peaks, which themselves represent the strongest electric field other than that of the central laser peak. The small shaded region at the top of the central laser peak highlights the unique region of strongest electric field, which will accelerate photoelectrons and return them to the parent ion with the highest kinetic energies, leading to the highest emitted photon energies.

the IR laser radiation into diameter modulated, hollow optical fibers,[54, 55] which causes a modulation of laser intensity and thus turns on and off generation of the high harmonics at alternating distances. This was accomplished with typical half-periods of 1 mm, depending on the value of L_π for particular experiments. They have also explored the use of counterpropagating beams,[56–58] which results in higher spatial frequency interference and can also be used to reduce high harmonic generation in alternating regions.

7.5   Attosecond Duration EUV/Soft X-Ray Pulses

There is great excitement in the scientific community regarding the generation of attosecond EUV and soft x-ray (SXR) radiation for the probing of electron dynamics in atoms, molecules, clusters, and solids. In previous sections HHG of femtosecond pulses was described, driven by many-cycle femtosecond duration IR laser pulses. In that case a similar train of odd numbered harmonic pulses is generated, each with a duration affected by the time spread and spectrum of return electron energies, but generally each having duration somewhat shorter than the driving IR laser pulses. In this section we discuss the extension to attosecond EUV/SXR pulses driven by the shortest possible IR laser pulses, which have durations of less than two cycles. Figure 7.10 shows the phase stabilized, carrier envelope, and laser electric field within that envelope, for what is essentially a two-cycle laser pulse, stabilized in such a manner that the maximum of the laser field coincides with the peak of its envelope in the interaction region. For a 750 nm wavelength laser the duration of a single cycle is $\lambda/c = 2.5$ fs, thus a two-cycle laser pulse has an envelope function of only 5 fs duration (FWHM of the intensity envelope). The laser electric field of a few-cycle pulse can be written in terms of an oscillation frequency ω, a phase φ, and an envelope function, such as the Gaussian shown below

$$E_L(t) = e^{-(t^2/2\tau_L^2)} \cos(\omega t + \varphi) \tag{7.16}$$

where t is time and φ is the envelope phase, both measured from the center of the pulse, and 2.35 τ_L is the FWHM of the intensity envelope. All values of φ are possible. The laser field shown in Figure 7.10 is a cosine function with $\varphi = 0$, and thus its peak electric field coincides with the peak of the carrier envelope function. As will be seen shortly, the phase stabilized cosine function offers a path to single sub-femtosecond pulses. A laser pulse stabilized with $\varphi = \pi/2$ has a sine function within the envelope, with an electric field of zero amplitude at the peak of the envelope ($\varphi = 0$), and symmetric peaks in opposite directions at $\varphi = \pm \pi/2$. The phase stabilized sine function offers a path to generating two sub-femtosecond EUV/SXR pulses separated by just a half IR cycle. What is particularly interesting about the cosine function is that its central ($\varphi = 0$) electric field is stronger than elsewhere in the envelope, and thus accelerates photoelectrons (generated near the peak of the previous half cycle) to the highest return energy for this pulse. Spectral filtering of these highest energy photons will result in a single, very short pulse.[59, 27] As the single pulse radiation is not repetitive it will not be harmonic in nature. The horizontal lines in Figure 7.10 mark the highest electric field strength in the pulse, other than the central peak. Attention is brought to the unique region of highest electric field by shading just above that line. The shaded region corresponds to a time interval where photoelectrons emitted near the peak of the previous half cycle (negative electric field) will experience the greatest acceleration, return to the parent ion with the highest kinetic energies, and upon recombination generate the highest energy photons in a single sub-femtosecond pulse. While the shaded region is very short, less than 0.5 fs, the resultant pulse duration for these highest energy photons will depend on other factors, such as the temporal spread of electrons due to path differences away from and returning to the parent ion.[28]

Baltuška, Krausz, and colleagues[59, 60, 27] have demonstrated phase stabilization with few-cycle cosine and sine functions as shown in Figure 7.11. Their emission data were obtained with a 2 mm wide column of neon gas at a pressure of 120 Torr, illuminated by single 750 nm pulses of 5 fs (two-cycle) duration, and an intensity of 7×10^{14} W/cm^2. Harmonic spectra for both cosine (a) and sine (b) stabilized carrier waveforms are shown for photon energies extending to 100 eV. Figure 7.11(a) shows harmonic emissions to about 88 eV, followed by a continuum extending from there to more than 100 eV. Also shown is a dashed violet colored line representing the bandpass of a thin zirconium filter used to absorb the laser fundamental and lower energy harmonics, transmitting only the higher energy continuum. This continuum of highest energy radiation is associated with the highest electric fields which are achieved only at the center

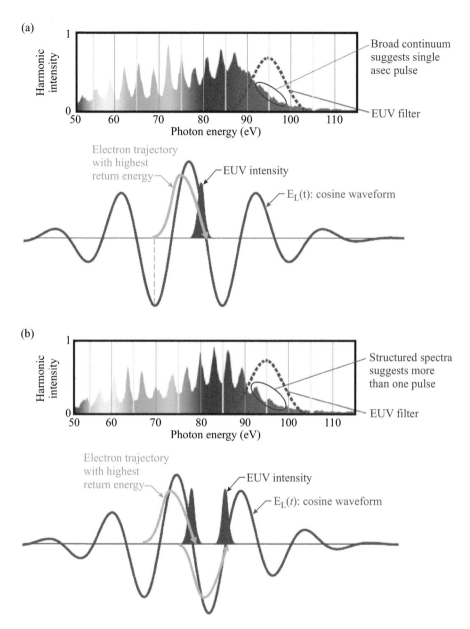

Figure 7.11 Harmonic spectra[30, 59] obtained with phase stabilized carrier envelopes of cosine and sine waveforms of 750 nm wavelength, 5 fs (two-cycle) duration. (a) Emission spectra with harmonics to about 88 eV and a continuum from there to greater than 100 eV, based on irradiation by a cosine stabilized wavefunction. The continuum spectra at the highest photon energies suggest these emissions are driven by a single, non-repetitive pulse. (b) Spectra of harmonics generated with a sine stabilized waveform. In this case the spectrum is significantly modulated in the high energy region, suggesting that in this case the highest energy emissions are generated by more than a single pulse. Courtesy of F. Krausz, Max Planck Institute for Quantum Optics and Ludwig Maxmillians University, Garching.

of a very short duration single pulse. Shown in red is the cosine waveform of the laser electric field that generated the spectrum in (a), plus an electron trajectory in green corresponding to the highest electron return energy. Note that this trajectory corresponds to an electron born just after the peak of the previous half cycle (negative direction electric field), but then accelerated by the force of the strongest electric field (positive half cycle) before recombining with the parent ion and emitting the sub-femtosecond pulse shown in violet. The violet color matching is meant to suggest a correspondance to the violet continuum in (a). Figure 7.11(b) shows harmonic spectra obtained with a sine stabilized carrier envelope of the same intensity which produces two sub-femtosecond pulses, one each driven by the two oppositely directed half cycles of equally strong electric field amplitude. For this case two electron trajectories are shown in green, one in each direction, resulting in the generation of two identical sub-femtosecond EUV pulses separated by a half laser IR cycle (1.25 fs). The question then is what is the duration of these very short emission bursts?

The measurement of sub-femtosecond EUV/SXR pulses is based on a "streaking" technique similar in concept to that used for recording laser and x-ray pulses[61-63] of picosecond temporal resolution, as was illustrated in Chapter 8, Physics of Hot-Dense Plasmas, Figure 8.24. In the picosecond case an external ramp voltage was used to streak a photon induced electron signal across a phosphor screen. In this case a very much faster ramped field is provided by the IR laser pulse itself, permitting temporal resolution extending well into the attosecond region.[27, 64-71] Figure 7.12 illustrates the technique[64] in which the just generated attosecond duration EUV/SXR radiation is focused onto a second gas jet, where it generates photoelectrons of energy, and in proportion to its own intensity, as a function of time. The focused EUV/SXR pulse is synchronously overlapped by a more modestly focused IR pulse of much larger spatial extent.

A result obtained using the attosecond streaking technique is shown in Figure 7.13 where the resultant streak spectrogram is shown for a single 80 as, nominally 80 eV pulse.[72] The experiment utilized a 3.3 fs FWHM duration (<1.5 cycles), 720 nm wavelength, and about 300 μJ per pulse, focused to an intensity of about 6×10^{14} W/cm^2 into a 2 mm thick neon gas jet at a pressure of about 230 Torr. The emission spectrum extended from 60eV to 110 eV, with a 28 eV FWHM bandwidth. A Mo/Si multilayer mirror was used to reduce the bandwidth to about 30 eV, centered at 75 eV. To obtain this broad bandwidth, the ML mirror was coated with only two periods, and consequently had a reflectivity of only 3.5%. To eliminate IR and lower harmonics from the 80 eV pulse a 0.3 μm thick Zr foil was also used. Within its spectral bandwidth, this single 80 as pulse contained approximately 0.5 nanojoules of energy, corresponding to a conversion efficiency of order 10^{-6}. The reduction of pulse duration from the 3.3 fs IR laser pulse to the 80 as EUV pulse corresponds to a factor of 40, characteristic of a truly non-linear process.

The recorded energy spectrogram was obtained by taking energy data in 126 steps across the 10 fs time window. A full data set takes about 30 minutes. The recorded streaked spectrogram carries with it both systematic and shot-to-shot variations of peak laser intensity, delay times, and spatial distributions of intensity, to name a few. These

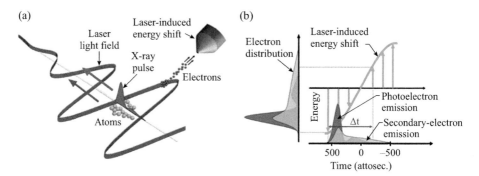

Figure 7.12 The technique of attosecond laser light streaking is illustrated following Kienberger and Krausz.[64] (a) Following generation of the attosecond EUV/SXR pulse in a first jet of atoms (not shown), the electric field of the one- to two-cycle IR laser pulse (shown in red) and the attosecond duration EUV/SXR pulse (shown in blue) move synchronously to the left. The EUV/SXR pulse is tightly focused to a second jet of atoms which it ionizes on a single photon basis, creating free electrons of energy different from the EUV/SXR photons only by the binding energy of the (second) gas jet atoms. The more modestly focused IR pulse provides an electric field strong enough to accelerate the free electrons, but not so strong as to affect the ionization process. The direction of the laser electric field is oriented so as to accelerate a fraction of the free electrons towards a time-of-flight electron energy analyzer. (b) The velocity of the free electrons is changed, increasing or decreasing their energies, by amounts depending on the relative phase of the EUV/SXR pulse with respect to the IR electric field. By varying the phase between the two, a series of "streak" records (time dependence stretched into a spatial dimension) is obtained, providing a history of the number of electrons generated, and their energies, as a function of time. The time delay between the IR electric field and EUV/SXR pulse is controlled by two concentric mirrors, one coated for the IR wavelength and one a multilayer coated optic matched to the EUV/SXR spectral band. For illustrative purposes, the electron emission history shown here in blue is more appropriate to a situation in which the EUV/SXR pulse illuminates a surface, where there would be a prompt photoemission signal followed by a slower secondary emission of straggling, scattered electrons, as shown in (b), rather than the jet of atoms suggested in (a). Courtesy of R. Kienberger, Technical University of Munich and Max Planck Institute for Quantum Optics, Garching.

are corrected with a FROG-like algorithm[73] (Frequency Resolved Optical Gating), which models these effects and provides a reconstructed version of the spectrogram. The FROG algorithm iteratively retrieves phase from the time-dependent photoelectron spectrum and, by doing so, retrieves the amplitude and phase of the generating attosecond pulse. After a successful FROG reconstruction one can obtain measures of systematic variations within the experimental system, such as time delay drifts and intensity fluctuations, by comparing the measured and reconstructed streak traces. Shown in Figure 7.13(b) is the FROG reconstructed attosecond streak spectrogram, showing shifted photoelectron energy in eV along the vertical axis and time delay in fs along the horizontal axis. The prominent features seen in Figure 7.13(b) reflect energy variations through several half periods of the IR laser electric field. Each vertical column of data gives the range of recorded electron energies at that particulat time. The signal integrated over any column is related to the intensity of the EUV/SXR pulse at

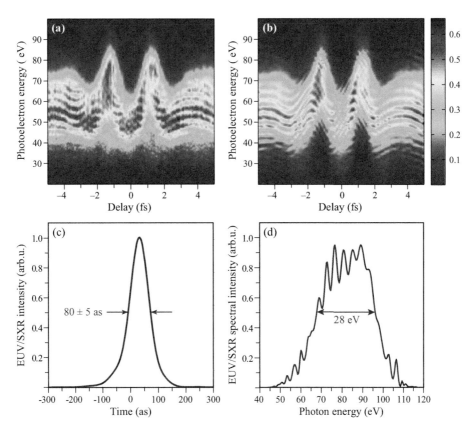

Figure 7.13 (a) A streaked spectrogram containing data relevant to the temporal intensity profile and photon energy spectrum of a nominally 80 eV pulse of 80 asec duration.[72] (b) FROG reconstructeded data.[73] (c) The temporal profile extracted from the spectrogram showing a duration of 80 ± 5 as FWHM. (d) The energy spectrum of the pulse showing a bandwidth of 28 eV centered at a photon energy of about 80 eV. The data was obtained by irradiating 230 Torr of neon with a 3.3 fs FWHM, 1.5 cycles, 720 nm wavelength IR laser pulse at an intensity of about 6×10^{14} W/cm^2. Courtesy of E. Goulielmakis, Max Planck Institute for Quantum Optics, Garching.

that time, and the recorded energy spread to that of the pulse. Figure 7.13(c) shows the retrieved EUV/SXR temporal intensity profile, with a duration of 80 ± 5 as FWHM, while Figure 7.13(d) shows the extracted energy spectrum of the EUV/SXR pulse. Note that while the 28 eV measured spectral bandwidth is clearly non-Gaussian, it is, in combination with the 80 as FWHM, close to the uncertainty limit for idealized Gaussian distributions

$$\Delta E_{\mathrm{FWHM}} \cdot \Delta \tau_{\mathrm{FWHM}} \geq 1.825\,\mathrm{eV} \cdot \mathrm{fs} \qquad (7.17)$$

where ΔE is the uncertainty in photon energy, in eV, and $\Delta \tau$ is the uncertainty, or pulse duration, in fs, as described in Chapter 4, Eq. (4.4b).

7.6   Attosecond Probing of Dynamical Processes in Atoms, Molecules, Nanoparticles, and Solids

The availability of attosecond duration EUV/SXR pulses has provided a unique new tool for the probing of electron dynamics in atoms, molecules, and solids. The pulses are not only very short, but the photon energies are well matched to the energy levels and resonances of electronic states in all forms of matter, and the EUV/SXR pulses are very well synchronized to the driving IR laser pulse, thus opening many possibilities for the use of "two-color" pump-probe techniques where one is used to perturb the system and the other to probe its response. The necessary equipment for such experiments is sufficiently compact and affordable that laboratories are being organized in more and more research institutes and universities, enabling an explosion in research and published papers. In this section we very briefly describe the nature of some recent experiments described in the literature. The reader, however, is advised to scan current literature for the latest developments as this is a rapidly evolving field.

To cite a few examples of asec probing we consider the following. In 2002 Drescher et al.[69] used the attosecond streak spectroscopy technique (750 nm, 7 fs) to measure the lifetime of a 3d inner shell vacancy in atomic Kr to be 7.9 fs. A 0.5 fs, nominally 100 eV pulse provided the excitation while the variable IR-EUV streak spectrograph traced the lifetime of an MNN Auger decay. In 2010 Schultze et al.[74] also using attosecond streak spectroscopy (750 nm, 3.3 fs), studied photoemission in atomic Ne, observing that emission of an electron emitted from a 2p orbital is delayed 21 as with respect to emission from a 2s orbital. The experiment employed a sub-200 as, nominally 100 eV excitation. In 2010 Goulielmakis, Loh, and their colleagues[75] used *attosecond transient EUV absorption spectroscopy* to probe the dynamics of valence electrons in ionized Kr. They used <150 as, spectrally broad EUV pulses in a geometry very similar to the attosecond streak spectroscopy discussed in the previous section, but instead measuring EUV photoabsorption. The geometry and initial results are shown in Figure 7.14. An intense IR pulse (750 nm, <4 fs FWHM, 3 kHz) was focused to about 7×10^{14} W/cm^2 in krypton gas at a density of $\sim 10^{18}$ atoms/cm^3 (~ 30 Torr). The absorption spectra were recorded with an EUV grating spectrometer, as shown in Figure 7.14(a). Independent positioning of the two-component, co-linear IR and EUV mirrors[72] was used to vary the time delay between IR excitation and absorption of the broadband (24 eV wide, centered at near 84 eV) attosecond radiation. Electrons were removed from 4p valence subshells of Kr, creating singly, doubly, and triply ionized Kr, as shown in Figure 7.14(b). An example of the absorption spectrum before and after ionization is shown in Figure 7.14(c). The ratio of these spectra was then inverted and normalized to plot relative absorptance, as seen in Figure 7.14(d), where absorption features of the various Kr ions are shown as a function of delay time. The time dependence of the recorded spectra reveals beating among intra-atomic wavepackets with a fairly high degeree of coherence.

Further exploration of coherent electron wavepacket using EUV attosecond pulses is being pursued by several groups. Klünder et al.[76] have used an interferometric

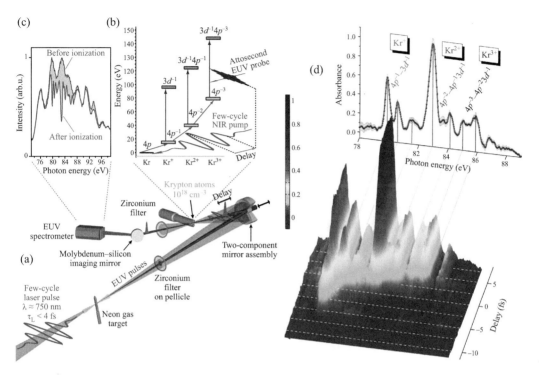

Figure 7.14 Intra-atomic electron motion, in the form of coherently excited electron wavepackets, is studied using attosecond transient EUV absorption spectroscopy.[75] The geometry of the experiment is shown in (a). In these pump-probe experiments Kr atoms are ionized to various degrees (b) by high intensity, 4 fs duration, 750 nm IR pulses and then probed by <150 as, spectrally broad EUV pulses (~24 eV wide, centered near 84 eV). EUV transmission spectra through the sample is shown in (c), then inverted and normalized as seen in (d) where absorption signatures of three ionization states are observed. The time-delay dependent data (300 as steps), displayed in color, reveal oscillations of coherently interacting electron wavepackets within single ions. Courtesy of Z.H. Loh, U. California, Berkeley, now National Technical University, Singapore, and E. Goulielmakis, Max-Planck Institute for Quantum Optics, Garching.

technique to study shell dependent photoemission times from 3s and 3p orbitals of argon with a temporal accuracy of ~50 as. Wirth *et al.*[77] have used EUV transient absorption spectroscopy with krypton atoms to better understand ionization dynamics and electronic coherence in single, double, and multiply charged ions. Ott *et al.*[78] have used EUV transient absorption spectroscopy to study and control two-electron wavepacket dynamics in helium atoms, as a model system to test and further develop three-body quantum dynamical theories. Månsson *et al.*[79] have used a train of attosecond pulses to study single-photon double-ionization in xenon as a means to evaluate two-electron wavepacket correlations in this inherently two-electron process and thus better understand competing ionization pathways. Studies of electron dynamics in molecules, such as N_2, CO_2, C_2H_2, are discussed by Smirnova *et al.*,[80] Niikura *et al.*,[81] and Neidel *et al.*,[82] respectively. Calegari *et al.*[83] have taken attosecond EUV absorption an additional step towards more complex, biological systems, studying charge dynamics in

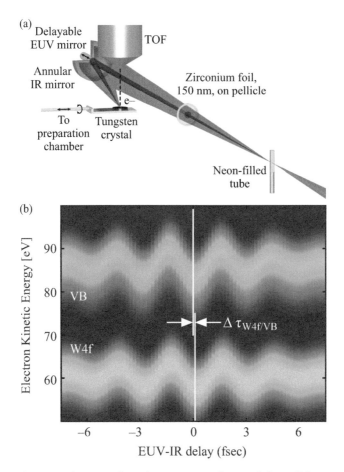

Figure 7.15 Attosecond streak spectrometry is extended to solid state systems by recording relative photoemission times from valence 4f and conduction band states of a tungsten crystal. Emission times from 4f orbitals are observed to be delayed by 99 as relative to those from valence states.[85] See text for comments regarding post-publication update to the data as shown here. Courtesy of R. Kienberger, Technical University of München and Max Planck Institute for Quantum Optics, Garching.

the large amino acid molecule phenylalaine by employing two-color photoionization. Lépine *et al.*[84] discuss the broader implications of attosecond techniques applied to molecular systems, including the manipulation of wavepacket coherence and the control of charge migration on asec and fsec time scales.

In 2007 Cavalieri *et al.*[85, 86] investigated attosecond dynamics in solid state systems, studying the electron photoemission from both localized core states and delocalized conduction band states in tungsten. A diagram of their experimental setup is seen in Figure 7.15(a), which utilizes an attosecond streak spectrometer similar to that described in Figure 7.12. A 750 nm, 5 fs, 0.4 mJ laser pulse is focused onto a 2 mm wide neon gas, generating a nominally 300 as FWHM duration pulse of EUV radiation having a 6 eV spectral bandwidth centered at ~91 eV. Axial positions of the concentric EUV multilayer mirror and surrounding annular IR mirror allow asec control of delay times between the two. The EUV and IR pulses are then focused onto a clean tungsten 110 crystal surface; the EUV is used to generate photoemitted electrons from the tungsten valence and conduction bands, and the IR laser electric field to accelerate those electrons towards the time-of-flight (TOF) electron spectrometer. The streaked spectrum of photoelectron energies is shown in Figure 7.15(b), where the range of emission kinetic energies permits identification of origin as tungsten 4f and conduction band states. The data shown here in (b) are from an unpublished post-publication experiment taking advantage of improved data analysis techniques.[85] The data show a 99 ($\pm$6) as emission delay of 4f orbital electrons relative to electrons from the conduction band. These experiments have extended the scientific application of attosecond streak techniques to processes in solid state systems.

In follow on experiments Neppl *et al.*[87, 88] have used attosecond streaking as a diagnostic to study not only the relative photoemission times of electons in solids, but also the subsequent transport of electron wavepackets through thin films of monolayer thicknesses. In these experiments the relative transport delays of electron wavepackets associated with tungsten 4f (-32.5 eV) and magnesium 2p (-50 eV) photoelectrons were measured. The photoelectrons were generated by 118 eV, 450 as EUV pulses in W(100) crystals and in thin overlaid magnesium coatings, and observed by sub-femtosecond IR streaked spectrometry. The Mg coatings had thicknesses from zero to four monolayers, approximately 0–10 Å. Relative emission times from the outermost surface were affected by transport variations through the Mg layers. Observed emission delay differences between the W 4f and Mg 2p varied from 80 as to 220 as for thicknesses of one to four monolayers, an average delay of just under 50 as per 2.7 Å monolayer. These experiments clearly establish the ability to measure physical processes on the scale of ångströms and attoseconds.

Further solid state experiments have been described by Schultze *et al.*[89] who performed experiments to understand how to control dielectric switching processes on attosecond and femtosecond time scales, as might be relevant to future electronic devices and information technologies. Using similar techniques as above, both IR laser and EUV pulses were moderately focused onto a thin SiO_2 film. The IR laser pulse exposed the sample to electric fields of several volts per ångström. The resultant electron responses were studied utilizing the 105 eV, 72 as EUV pulse to lift silicon L-shell electrons into the conduction band, and then recording modifications to the EUV absorption spectrum. The data suggest that it may be possible to vary conductivity over many orders of magnitude on these ultra-fast time scales. As tools for their studies they combined attosecond streak spectrometry, transient EUV absorption spectroscopy, and IR reflectivity. In a further study, Schultze *et al.*[90] used attosecond pulses tuned to the 99 eV silicon $L_{2,3}$ edge to observe a small reduction of the Si band gap as a result of excitations driven by few femtosecond IR pulses. They also identified electron tunneling as the primary process for valence to conduction band transitions. The study of HHG in solids presently attracts much attention.[91–93]

Indeed the ability to observe and control electron wavepackets on attosecond time scales, in ever more complicated systems is just entering its Golden Age.[94] The basic tools are available and improving. Furthermore, HHG photon flux levels are on a path to increase as femtosecond laser techniques achieve higher power[95] and higher repetition rates.[96] Interpretation will not be trivial. There is an interesting community-wide commentaryon the present status of attosecond electron dynamic studies, the potential of attosecond pump- attosecond probe experiments, the perturbative effects of those measurements, and the ability to properly interpret the experiments.[97]

References

1. J.L. Krause, K.J. Schaler and K.C. Kulander, "High-Order Harmonic Generation from Atoms and Ions in the High Intensity Regime," *Phys. Rev. Lett.* **68**, 3535 (June 15, 1992).
2. P.B. Corkum, "Plasma Perspective on Strong-Field Multiphoton Ionization," *Phys. Rev. Lett.* **71**, 1994 (September 27, 1993).
3. A. L'Huillier and Ph. Balcou, "High-Order Harmonic Generation in Rare Gases with a 1-ps 1053 nm Laser," *Phys. Rev. Lett.* **70**, 774 (February 8, 1993).
4. J.J. Macklin, J.D. Kmetec and C.L. Gordon III, "High-Order Harmonic Generation Using Intense Femtosecond Pulses," *Phys. Rev. Lett.* **70**, 766 (February 8, 1993).
5. D. Schulze, M. Dorr, G. Sommerer *et al.*, "Polarization of the 61st Harmonic from 1053-nm Laser Irradiation in Neon," *Phys. Rev. A* **57**, 3003 (1998).
6. T. Pfeifer, "Adaptive Control of Coherent Soft X-rays," PhD Dissertation, Universität Würzburg (2004).
7. T.F. Gallagher, "Above-Threshold Ionization in Low-Frequency Limit," *Phys. Rev. Lett.* **61**, 2304 (November 14, 1988).
8. M. Lewenstein, Ph. Balcou, M.Y. Ivanov, A. L'Huillier and P.B. Corkum, "Theory of High-Harmonic Generation by Low-Frequency Laser Fields," *Phys. Rev. A* **49** (3), 2117 (March 1994).
9. Ch. Spielmann, N.H. Burnett, S. Sartania *et al.*, "Generation of Coherent X-rays in the Water Window Using 5-Femtosecond Laser Pulses," *Science* **278**, 661 (October 24, 1997).
10. J. Zhou, J. Peatross, M.M. Murnane, H.C. Kapteyn and I.P. Christov, "Enhanced High-Harmonic Generation Using 25 fs Laser Pulses," *Phys. Rev. Lett.* **76**, 752 (1996).
11. L.V. Keldysh, "Ionization in the Field of a Strong Electromagnetic Wave," *Soviet Physics JETP* **20** (5), 1307 (May 1965).
12. M.V. Ammosov, N.B. Delone and V.P. Kraĭnov, "Tunnel Ionization of Complex Atoms and of Atomic Ions in an Alternating Electromagnetic Field," *Soviet Physics JETP* **64** (6), 1191 (December 1986).
13. A.S. Landsman, M. Weger, J. Maurer *et al.*, "Ultrafast Resolution of Tunneling Delay Time," *Optica* **1**, 343 (November 2014).
14. T. Brabec and F. Krausz, "Intense Few-Cycle Laser Fields: Frontiers of Nonlinear Optics," *Rev. Mod. Phys.* **72** (April, 2000).
15. G.L. Yudin and M.Yu. Ivanov, "Nonadiabatic Tunnel Ionization: Looking Inside a Laser Cycle," *Phy. Rev. A* **64**, 013409 (June 2001).
16. T. Pfeifer, C. Spielmann and G. Gerber, "Femtosecond X-ray Sciences," *Rep. Prog. Phys.* **69**, 443–505 (2006).
17. C.J. Joachain, N.J. Kylstra and R.M. Potvliege, *Atoms in Intense Fields* (Cambridge University Press, 2012), Chs. 1 and 6.

18. C. Ott, Max Planck Institute for Kern Physics, Heidelberg; private communication.

19. A. L'Huillier, L.A. Lompré, G. Mainfray and C. Manus, "Multiply Charged Ions Induced by Multiphoton Absorption in Race Gases at 0.53 μm," *Phys. Rev. A* **27** (5), 2503 (May1983).

20. S. Augst, D. Strickland, D.D. Meyerhofer, S.L. Chin, and J.H. Eberly, "Tunneling Ionization of Noble Gases in a High-Intensity Laser Field," *Phys. Rev. Lett.* **63**, 2212 (November 13, 1989).

21. T. Auguste, P. Monot, L.A. Lomparé, G. Mainfray, and C. Manus, "Multiply Charged Ions Produced in Noble Gases by a 1 ps Laser Pulse at λ = 1.053 μm," *J. Phys.B*, **25**, 4181 (1992).

22. R. Santra, R.W. Dunford and L. Young, "Spin-Orbit Effect on Strong-Field Ionization of Krypton," *Phys. Rev. A* **74**, 043403 (2006).

23. E.J. Takahashi, H. Hasegawa, Y. Nabekawa, and K. Midorikawa, "High-Throughput, High-Damage-Threshold Broadband Beam Splitter for High-Order Harmonics in the Extreme-Ultraviolet Region," *Optics Lett.* **29**, 507 (March 1, 2004).

24. E. Gustafsson, T. Ruchon, M. Swoboda *et al.*, "Broadband Attosecond Pulse Shaping," *Optics Lett.* **32**, 1353 (June 1, 2007).

25. A. Wonisch, U. Neuhäusler, N.M. Kabachnik *et al.*, "Design, Fabrication, and Analysis of Chirped Multilayer Mirrors for Reflection of Extreme-Ultraviolet Attosecond Pulses," *Applied Optics* **45**, 4147 (June 10, 2006).

26. C. Bourassin-Bouchet, S. de Rossi, J. Wang *et al.*, "Shaping of Single-Cycle Sub-50-Attosecond Pulses with Multilayer Mirrors," *New J. Phys.* **14**, 023043 (February 2012).

27. M. Lewenstein, Ph. Balcou, M.Yu. Ivanov, A. L'Huillier, and P.B. Corkum, "Theory of High Harmonic Generation by Low-Frequency Laser Fields," *Phys. Rev. A* **49** (3), 2117 (March 1994).

28. M.Yu. Ivanov, M. Spanner, and O. Smirnova, "Anatomy of Strong Field Ionization," *J. Mod. Optics* **52**, 165–184 (February 2005).

29. P. Corkum and F. Krausz, "Attosecond Science," *Nature Phys.* **3**, 381 (June 2007).

30. F. Krausz and M. Ivanov, "Attosecond Physics," *Rev. Mod. Phys.* **81**, 163–234 (January–March 2009).

31. M. Schnürer, Z. Cheng, M. Hentschel *et al.*, "Absorption-Limited Generation of Coherent Ultrashort Soft X-ray Pulses," *Phys. Rev. Lett.* **83**, 722 (July 26, 1999).

32. M. Schnürer, Z. Cheng, M. Hentschel *et al.*, "Few-Cycle-Driven XUV Laser Harmonics: Generation and Focusing," *Appl. Phys. B*, **70**, S227 Supplement (May 2000).

33. E. Constant *et al.*, "Optimizing High Harmonic Generation in Absorbing Gases: Model and Experiment," *Phys. Rev. Lett.* **82**, 1668 (February 22, 1999).

34. J. Tate, T. Auguste, H.G. Muller *et al.*, "Scaling of Wave-Packet Dynamics in an Intense Midinfrared Field," *Phys. Rev. Lett.* **98**, 013901 (January 5, 2007).

35. A.D. Shiner, C. Trallero-Herrero, N. Kajumba *et al.*, "Wavelength Scaling of High Harmonic Generation Efficiency," *Phys. Rev. Lett.* **103**, 073902 (August 14, 2009).

36. C.-J. Lai, G. Cirmi, K.-H. Hong *et al.*, "Wavelength Scaling of High Harmonic Generation Close to the Multiphoton Ionization Regime," *Phys. Rev. Lett.* **111**, 073901 (August 16, 2013).

37. K.-H. Hong, C.-J. Lai, J.P. Siqueira *et al.*, "Multi-mJ, kHz, 2.1 μm Optical Parametric Chirped-Pulse Amplifier and High-Flux Soft X-Ray High-Harmonic Generation," *Optics Lett.* **39**, 3145 (June 1, 2014).

38. T. Ditmire, E. Gumbrell, R. Smith *et al.*, "Spatial Coherence Measurement of Soft X-Ray Radiation Produced by High Order Harmonic Generation," *Phys. Rev. Lett.* **77**, 4756 (December 2, 1996).

39. T. Ditmire, J. Tisch, E. Gumbrell *et al.*, "Spatial Coherence of Short Wavelength High-Order Harmonic," *Appl. Phys. B*, **65**, 313 (1997).

40. R. Bartels, A. Paul, H. Green *et al.*, "Generation of Spatially Coherent Light at Extreme Ultraviolet Wavelengths," *Science* **297**, 376 (July 19, 2002).

41. M. Born and E. Wolf, *Principles of Optics* (Cambridge University Press, 1999), Seventh Edition.

42. C. Lynga, M. Gaarde, D. Deflin *et al.*, "Temporal Coherence of High-Order Coherence," *Phys. Rev. A*, **60**, 4823 (December 1999).

43. P. Salières, A. L'Huillier and M. Lewenstein, "Coherence Control of High-Order Harmonics," *Phys. Rev. Lett.*, **74**, 3776 (May 8, 1995).

44. P. Salières, B. Carré, L. Le Déroff *et al.*, "Feynman's Path-Integral Approach for Intense-Laser-Atom Interactions," *Science* **292**, 902 (May 4, 2001).

45. J.-F. Hergott, M. Kovacev, H. Merdji *et al.*, "Extreme-Ultraviolet High-Order Harmonic Pulses in the Microjoule Range," *Phys. Rev. A*, **66**, 021801(R) (2002).

46. R.W. Boyd, *Nonlinear Optics* (Academic Press, 2008), Third Edition; Phase Matching, Section 2.3; Quasi-Phase-Matching Section 2.4.

47. C.M. Heyl, "Scaling and Gating Attosecond Pulse generation," Doctoral Thesis, Department of Physics, Lund University (2014).

48. E. Hecht, *Optics* (Addison-Wesley, 2002); Fourth Edition; Table 1, p. 653.

49. C.W. Allen, *Astrophysical Quantities* (University of London, Athlone Press, 1963).

50. B.L. Henke, E.M. Gullikson and J.C. Davis, "X-Ray Interactions: Photoabsorption, Scattering, Transmission and Reflection at E = 50–30,000 eV, Z = 1–92," *Atomic Data and Nuclear Data Tables*, **54**, 181 (1993).

51. E.M. Gullikson, http://www.cxro.LBL.gov/optical_constants

52. T. Ditmire, J.K. Crane, H. Nguyen, L.B. DaSilva, and M.D. Perry, "Energy-Yield and Conversion-Efficiency Measurements of High-Order Harmonic Radiation," *Phys. Rev. A*, **51**, R902 (February 1995).

53. R. Kienberger and A. Scrizi, "Attosecond Pulses", in *Frequency-Resolved Optical Gating: The Measurement of Ultrashort Pulses* (Springer, 2000), edited by R. Trebino; http://www.frog.gatech.edu/prose.html

54. A. Paul, R.A. Bartels, R. Tobey *et al.*, "Quasi-phase-Matched Generation of Coherent Extreme Ultraviolet Light," *Nature* **421**, 51 (January 2, 2003).

55. E.A. Gibson, A. Paul, N. Wagner *et al.*, "Coherent Soft X-Ray Generation in the Water Window with Quasi-Phase Matching," *Science* **302**, 95 (October 3, 2003).

56. A.L. Lytle, X. Zhang, J. Peatross *et al.*, "Probe of High-Order Harmonic Generation in a Hollow Waveguide Geometry Using Counterpropagating Light," *Phys. Rev. Lett.* **98**, 123904 (March 23, 2007).

57. X. Zhang, A.L. Lytle, T. Popmintchev *et al.*, "Quasi-Phase-Matching and Quantum-Path Control of High-Harmonic Generation Using Counterpropagating Light," *Nature Phys.* **3**, 270 (April 2007).

58. O. Cohen, A.L. Lytle, X. Zhang, M.M. Murnane, and H.C. Kapteyn, "Optimizing Quasi-Phase-Matching of High Harmonic Generation Using Counterpropagating Pulse Trains," *Opt. Lett.* **32**, 2975 (October 15, 2007).

59. A. Baltuška, Th. Udem, M. Uiberacker *et al.*, "Attosecond Control of Electronic Processes by Intense Light Fields," *Nature* **421**, 611 (February 6, 2003).

60. S. Sartania, Z. Cheng, M. Lenzner *et al.*, "Generation of 0.1-TW 5-fs Optical Pulses at a 1-kHz Repetition Rate," *Opt. Lett.* **22**, 1562 (October 15, 1997).

61. D.J. Bradley, B. Liddy, and W.E. Sleat, "Direct Linear Measurement of Ultrashort Laser Pulses with a Picosecond Streak Camera," *Opt. Commun.* **2**, 391 (1971).

62. M.Ya. Schelev, M.C. Richardson, and A.J. Alcock, "Image Converter Streak Camera with Picosecond Resolution," *Appl. Phys. Lett.* **18**, 354 (1971).

63. D.T. Attwood, L.W. Coleman, M.J. Boyle *et al.*, "Space-Time Implosion Characteristics of Laser-Irradiated Fusion Targets," *Phys. Rev. Lett.* **38**, 282 (February 7, 1977); and references 60–62 in Chapter 7.

64. R. Kienberger and F. Krausz, "Attosecond Metrology Comes of Age," *Physica Script* **T110**, 32 (2004).

65. P. Agostini and L.F. DiMauro, "The Physics of Attosecond Light Pulses," *Rep. Progr. Phys.* **67**, 813–855 (May 2004).

66. M. Drescher, M. Hentschel, R. Kienberger *et al.*, "X-Ray Pulses Approaching the Attosecond Frontier," *Science* **291**, 1923 (March 9, 2001).

67. M. Hentschel, R. Kienberger, Ch. Spielmann *et al.*, "Attosecond Metrology," *Nature* **414**, 509 (November 29, 2001).

68. J. Itatani, F. Quéré, G.L. Yudin *et al.*, "Attosecond Streak Camera," *Phys. Rev. Lett.* **88**, 173903 (April 29, 2002).

69. M. Drescher, M. Hentschel, R. Kienberger *et al.*, "Time-Resolved Atomic Inner-Shell Spectroscopy," *Nature* **419**, 803 (October 24, 2002).

70. R. Kienberger, E. Goulielmakis, M. Uiberacker *et al.*, "Atomic Transient Recorder," *Nature* **427**, 817 (February 26, 2004).

71. E. Goulielmakis, M. Uiberacker, R. Kienberger *et al.*, "Direct Measurement of Light Waves," *Science* **305**, 1267 (August 27, 2004).

72. E. Goulielmakis, M. Schultze, M. Hofstetter *et al.*, "Single-Cycle Nonlinear Optics," *Science* **320**, 1614 (June 20, 2008); A.L. Aquila, "Development of Extreme Ultraviolet and Soft X-Ray Multilayer Optics for Scientific Studies with Femtosecond/Attosecond Sources," PhD Thesis, Applied Science and Technology, University of California, Berkeley, May 2009.

73. J. Gagnon and V.S. Yakovlev, "The Robustness of Attosecond Streaking Measurements," *Opt. Express* **17**, 17678 (September 28, 2009).

74. M. Schultze *et al.*, "Delay in Photoemission," *Science* **328**, 1658 (June 25, 2010); H. W. van der Hart, "When Does Photoemission Begin?" *Science* 328, 1645 (June 25, 2010).

75. E. Goulielmakis, Z.-H. Loh, A. Wirth *et al.*, "Real-Time Observation of Valence Electron Motion," *Nature* **466**, 739 (August 5, 2010).

76. K. Klünder, J.M. Dahlstrom, M. Gisselbrecht *et al.*, "Probing Single-Photon Ionization on the Attosecond Time Scale," *Phys. Rev. Lett.* **106**, 143002 (April 8, 2011).

77. A. Wirth, R. Santra, and E. Goulielmakis, "Real Time Tracing of Valence-Shell Electronic Coherences with Attosecond Transient Absorption Spectroscopy," *Chem. Physics* 414, 149 (2013).

78. C. Ott, A. Kaldun, L. Argenti *et al.*, "Reconstruction and Control of a Time-Dependent Two-Electron Wave packet," *Nature* **516**, 374 (December 18, 2014).

79. E.P. Månsson, D. Guénot, C.L. Arnold *et al.*, "Double Ionization Probed on the Attosecond Timescale," *Nature Phys.* **10**, 207 (March 2014).

80. O. Smirnova, Y. Mairesse, S. Patchkovskii *et al.*, "High Harmonic Interferometry of Multi-Electron Dynamics in Molecules," *Nature* **460**, 972 (August 20, 2009).

81. H. Niikura, H.J. Wörner, D.M. Villeneuve, and P.B. Corkum, "Probing the Spatial Structure of a Molecular Attosecond Electron Wave Packet Using Shaped Recollision Trajectories," *Phys. Rev. Lett.* **107**, 093004 (August 26, 2011).

82. Ch. Neidel, J. Klei, C.-H. Yang *et al.*, "Probing Time-Dependent Molecular Dipoles on the Attosecond Time Scale," *Phys. Rev. Lett.* **111**, 033001 (July 19, 2013).

83. F. Calegari, D. Ayuso, A. Trabattoni *et al.*, "Ultrafast Electron Dynamics in Phenylalanine Initiated by Attosecond Pulses," *Science* **346**, 336 (October 17, 2014).

84. F. Lépine, M.Y. Ivanov, and M.J.J. Vrakking, "Attosecond Molecular Dynamics: Fact or Fiction?" *Nature Photon.* **8**, 195 (March 2014).

85. A.L. Cavalieri, N. Müller, Th. Uphues *et al.*, "Attosecond Spectroscopy in Condensed Matter," *Nature* **449**, 1029 (October 25, 2007).

86. D.M. Villeneuve, "Attosecond at a Glance," *Nature* 449, 997 (October 25, 2007).

87. S. Neppi, R. Ernstorfer, E.M. Bothschafter *et al.*, "Attosecond Time-Resolved Photoemission from Core and Valence States of Magnetism," *Phys. Rev. Lett.* **109**, 087401 (August 24, 2012).

88. S. Neppl, R. Ernstorfer, A.L. Cavalieri *et al.*, "Direct Observation of Electron Propagation and Dielectric Screening on the Atomic Scale Length," *Nature* **517**, 342 (January 15, 2015).

89. M. Schultze, E.M. Bothschafter, A. Sommer *et al.*, "Controlling Dielectrics with the Electric Field of Light," *Nature* **493**, 75 (January 3, 2013).

90. M. Schultze, K. Ramasesha, C.D. Pemmaraju *et al.*, "Attosecond Band-Gap Dynamics in Silicon," *Science* **346**, 1348 (December 12, 2014).

91. T.T. Luu, M. Garg, S.Yu. Kruchinin *et al.*, "Extreme Ultraviolet High-Harmonic Spectroscopy of Solids," *Nature* **521**, 498 (May 28, 2015).

92. G. Vampa, T.J. Hammond, N. Thiré *et al.*, "Linking High Harmonics from Gases and Solids," *Nature* **522**, 462 (June 25, 2015).

93. M. Hohenleutner, F. Langer, O. Schubert *et al.*, "Real-Time Observation of Interfering Crystal Electrons in High-Harmonic Generation," *Nature* **523**, 572 (July 30, 2015).

94. O. Pedatzur, G. Orenstein, V. Serbinenko *et al.*, "Attosecond Tunnelling Interfereometry," *Nature Physics* **11**, 815 (October 2015).

95. O. Pronin, M. Seidel, F. Lücking *et al.*, "High-Power Multi-Gigahertz Source of Waveform-Stabilized Few-Cycle Light," *Nature Commun.* **6**, 7988 (May 5, 2015).

96. I. Pupeza, S. Holzberger, T. Eidam *et al.*, "Compact High-Repitition-Rate Source of Coherent 100 eV Radiation," *Nature Photonics* **7**, 608 (August 2013).

97. S.R. Leone *et al.*, "What Will it Take to Observe Processes in 'Real Time'?" *Nature Photon.* **8**, 162 (March 2014).

Homework Problems

Homework problems for each chapter will be found at the website:
www.cambridge.org/xrayeuv

8   Physics of Hot Dense Plasmas

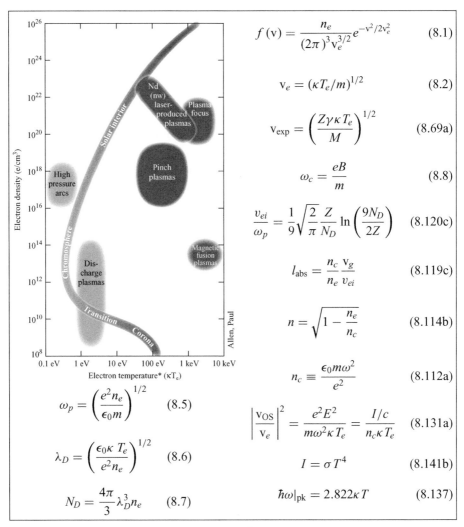

$$f(v) = \frac{n_e}{(2\pi)^3 v_e^{3/2}} e^{-v^2/2v_e^2} \tag{8.1}$$

$$v_e = (\kappa T_e/m)^{1/2} \tag{8.2}$$

$$v_{exp} = \left(\frac{Z\gamma\kappa T_e}{M}\right)^{1/2} \tag{8.69a}$$

$$\omega_c = \frac{eB}{m} \tag{8.8}$$

$$\frac{v_{ei}}{\omega_p} = \frac{1}{9}\sqrt{\frac{2}{\pi}}\frac{Z}{N_D}\ln\left(\frac{9N_D}{2Z}\right) \tag{8.120c}$$

$$l_{abs} = \frac{n_c}{n_e}\frac{v_g}{v_{ei}} \tag{8.119c}$$

$$n = \sqrt{1 - \frac{n_e}{n_c}} \tag{8.114b}$$

$$n_c \equiv \frac{\epsilon_0 m\omega^2}{e^2} \tag{8.112a}$$

$$\left|\frac{v_{OS}}{v_e}\right|^2 = \frac{e^2 E^2}{m\omega^2\kappa T_e} = \frac{I/c}{n_c\kappa T_e} \tag{8.131a}$$

$$I = \sigma T^4 \tag{8.141b}$$

$$\hbar\omega|_{pk} = 2.822\kappa T \tag{8.137}$$

$$\omega_p = \left(\frac{e^2 n_e}{\epsilon_0 m}\right)^{1/2} \tag{8.5}$$

$$\lambda_D = \left(\frac{\epsilon_0 \kappa T_e}{e^2 n_e}\right)^{1/2} \tag{8.6}$$

$$N_D = \frac{4\pi}{3}\lambda_D^3 n_e \tag{8.7}$$

* $\kappa T_e = 100$ eV corresponds to $T_e = 1.16 \times 10^6$ K

In this chapter the physics of hot dense plasmas is considered. Combining both high temperature and high density, such plasmas are particularly bright sources of EUV through soft x-ray radiation. In general the radiation consists of a broad spectral continuum, plus narrow line emission from the various ionization stages of those elements present. Such plasmas are found in the stars and, on a laboratory scale, at the focus of intense laser beams irradiating material surfaces.

8.1   Introduction

The study of hot dense plasmas constitutes a subset of plasma physics relevant to the generation of intense x-rays. Such plasmas are found in the stars and, on a smaller scale, at the focus of intense laser beams irradiating material surfaces (see chapter frontispiece). The physics of plasmas involves interaction between many charged particles on a microscopic scale through the electric and magnetic fields associated with their positions and velocities. Fortunately, this extremely complex *many body* problem often may be simplified by the consideration of macroscopic, collective interactions where the charges are described in terms of charge densities and currents. The study of plasmas is rich in interesting phenomena at both the linear and the non-linear level.[1–15] The term "non-linear" refers to various phenomena, such as wave growth or particle acceleration, which depend on some parameter, like temperature or density, in other than a linear manner, or on some combination of such parameters. Non-linear processes can involve frequency sums and differences, harmonics, and mixing phenomena.

Hot dense plasmas have a natural tendency to push at the threshold of many of these strong non-linear mixing processes. In order to generate EUV and soft x-ray radiation, the plasma must consist of particles at very high energies – of the order 100 eV to several keV if they are to radiate such energies during particle–particle interactions, since total energy must be conserved. This may also be understood in terms of blackbody radiation. The peak photon energy is related to the temperature of the radiating body, so that soft x-rays require radiators that are extremely hot. In addition, for the radiation to be intense, the emissions must come from a large number of particles in a small volume, perhaps approaching densities characteristic of solids. Thus the descriptive phrase "hot dense plasma."

Of course, these conditions are far from equilibrium in our $1/40$ eV world. As a consequence, such plasmas are inherently short lived. The high temperatures imply high velocities that cause the plasmas to rapidly expand and cool. For electron temperatures of 1 keV, a plasma of electrons and silicon ions that is electrically neutral overall will expand at a velocity of order 0.3 µm/ps (where $1\,\text{ps} = 10^{-12}$ s, and $1\,\text{µm} = 10^{-6}$ m). For comparison, the speed of light in these units is 300 µm/ps. Since a great deal of energy must be imparted to each particle, and there are so many particles per unit volume, these plasmas tend to be very small, on the order of 100 µm in dimension. Thus, a typical time for expansion is

$$\Delta t = \frac{\Delta x}{v} = \frac{100\,\mu\text{m}}{0.3\,\mu\text{m/ps}} \simeq 300\,\text{ps}$$

As a result the world of hot dense plasmas is generally one of microns (µm) and picoseconds. Clearly, there are detailed phenomena that occur on both shorter and longer time scales, but these simple arguments give a general idea of the domains involved.

This raises a question: How can we create a plasma quickly enough to deliver significant energy to a small volume in a short time? The primary technology with this capability is high peak power lasers, which can deliver single pulses with gigawatt to terawatt peak powers to spot sizes of characteristic dimension 100 μm in sub-nanosecond pulses. Sub-picosecond pulses may also be used, but these provide less input energy, and thus less energy radiates out of the resulting plasma. Such short pulse lasers are primarily employed in cases where extremely short time scales are required. On longer time scales, of order 10–100 ns, other technologies such as electrical discharges may be employed, but these generally involve larger volumes at lower densities and temperatures, tending to be better optimized for extreme ultraviolet radiation.

Hot dense plasmas are of interest in basic physics research because of the multitude of interesting phenomena that arise. Moreover, they are of technological and industrial importance in such research areas as laser fusion, EUV and soft x-ray lasers, lithography, and other areas well known for concentrated energy densities. Because of the high energy concentration, which implies high temperatures and pressures, these plasmas tend to involve rapid expansions and thus sharp gradients in density and other parameters. This introduces a fair degree of complexity into the description of plasma processes, requiring the use of several tools. Theoretical models are created that attempt to explain the behavior of these plasmas within limited parameter ranges, perhaps of density and temperature. They describe the system in purely analytic terms, and strive to find closed form mathematical descriptions of various phenomena. To deal with the wide variations of density, temperature, and field intensities, numerical simulations are employed.

Several types of computer modeling are used. On a small scale, the detailed motion of a finite number of particles is tracked, for instance, to study non-linear plasma motion in the presence of extremely high incident laser radiation. On a somewhat larger scale, fluid-like zoning techniques are used to follow energy and particle transport in the presence of sharp spatial and temporal gradients in the presence of localized heating.

Finally, it is essential that theory and simulations be compared with real experiments. Only in the laboratory (or the stars) can the plasma be studied in a rigorous manner, with all Mother Nature's interactions present and accounted for. However, to understand these experiments, they must be carefully considered and executed with appropriate diagnostic instrumentation. Indeed, because of the considerable complexity, a satisfactory interpretation of the experiments generally requires substantial use of both theory and simulations.

Of primary interest here is the resultant emission of radiation, particularly at short wavelengths. As hot dense plasmas are fully ionized, the radiation consists of a broad continuum of so-called *bremsstrahlung*,* due to free-electron–ion interactions, and narrow line emissions due to bound–bound transitions in the atoms (ions) of various charge states. A composite sketch of what such an emission spectrum might look like is shown in Figure 8.1.

---

* German word for "braking radiation."

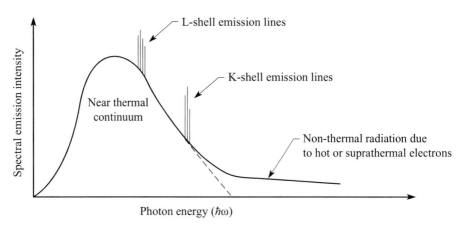

Figure 8.1 Line and continuum radiation from a hot dense plasma. The narrow emission lines are from ions of various ionization states.

The spectrum consists of a broad continuum, perhaps near-thermal in nature, with narrow atomic emission lines of characteristic L-shell and K-shell transitions. These atoms have generally lost several, perhaps many, electrons in collisions with energetic free plasma electrons, and thus radiate emission lines characteristic of ions of several ionization states. In addition, there may be a long tail of energetic x-rays emitted by hot, or *suprathermal*, electrons generated by non-linear wave–particle processes in the plasma.

8.2   Short and Long Range Interactions in Plasmas

Several basic processes occur in plasmas. Among these are short distance binary *collisions*, and longer distance many-particle interactions better described in terms of collective phenomena. The short distance collisions transfer energy from particle to particle, in a somewhat random fashion, thereby thermalizing the plasma and ionizing the atoms. The level of ionization (the number of electrons lost) is set by the electron temperature of the plasma and the atomic binding energies, a topic we discuss further in a later section. Generally, multiple ionization states are formed, each with its own characteristic emission lines, leading to a rich complex of lines, often useful for diagnostic purposes, providing information regarding the electron and ion temperatures and the density. Collisions also cause the plasma to radiate. When an electron collides with an ion, as illustrated in Figure 8.2, it is accelerated and therefore radiates.

If the electron comes very close to the ion (small value of the *impact parameter b* in Figure 8.2), the acceleration is strong, the deflection angle is large, and the radiated photon is of high energy. For a more distant interaction (larger impact parameter) the acceleration is weaker and the radiated photon is of lower energy. With a random interaction process involving many electrons of differing velocity and impact parameter, one can expect a rather smooth continuum spectrum, related to the plasma's electron velocity distribution, or more simply its temperature. By a *thermal* plasma we mean

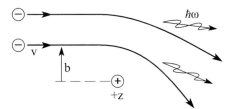

Figure 8.2 Bremsstrahlung occurs when a passing electron is accelerated by an ion, causing it to radiate. Because of the wide range of incident electron velocities and the range of distances of closest approach (*impact parameter b*) a broad continuum of radiation is generated in a plasma, with a spectrum closely related to the electron velocity distribution, or its characteristic temperature.

an idealized equilibrium plasma in which all species (electrons, ions, radiation) are characterized by a single temperature and appropriate energy or velocity distributions. In practice the different particles will be characterized by different temperatures that vary with space and time. Indeed, we may find that for a single species, such as electrons, the velocity distribution cannot be described by a single temperature. This then leads us to concepts such as *near-thermal radiation* and the use of two-temperature models involving a thermal component and a suprathermal component, as used in Figure 8.1. Nonetheless, it is very useful to consider the thermal limit. For electrons characterized by a single electron temperature[†] κT_e, the Maxwellian velocity distribution is[16]

$$f(\mathrm{v}) = \frac{n_e}{(2\pi)^3 \mathrm{v}_e^{3/2}} e^{-\mathrm{v}^2/2\mathrm{v}_e^2} \tag{8.1}$$

where

$$\mathrm{v}_e = (\kappa T_e/m)^{1/2} \tag{8.2}$$

is the root mean square thermal velocity, n_e is the electron density, and m is the electron mass. The closely related topic of blackbody thermal radiation is described in a later section.

Longer range interactions are also important in plasmas, and especially so in hot dense plasmas. These often take the form of plasma oscillations, collective waves that propagate naturally, much like sound waves or deep water waves in their respective media. Generally these are high-frequency waves associated with electrons, and lower-frequency waves associated with the heavier ions, referred to respectively as electron-acoustic and ion-acoustic waves. A sketch of a propagating electron-acoustic wave is shown in Figure 8.3.

It is longitudinal in nature and propagates at a very high phase velocity, as we will see in the theory section that follows. Indeed, at long plasma wavelengths this wave propagates at phase velocities much greater than the electron thermal velocity, as shown in Figure 8.3(b). Collective oscillations such as these are naturally damped at short wavelengths by wave–particle interactions. This occurs on a scale related to the Debye

---

[†] The Boltzmann constant is $\kappa = 8.6174 \times 10^{-5}$ eV/K (see Appendix A), so that when expressed in energy units a temperature of 100 eV corresponds to 1.16×10^6 Kelvin (K).

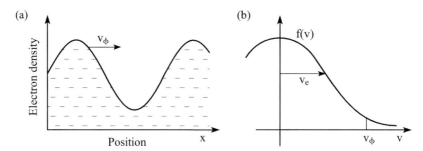

Figure 8.3 (a) An electron-acoustic wave, typically oscillating near the plasma frequency ω_p, propagates as an electron density modulation. Uniformly distributed ions (not shown) are too massive to participate in this relatively high frequency wave, but do provide overall charge neutrality. The wave is shown propagating to the right with a phase velocity v_ϕ. (b) The velocity distribution function $f(v)$ of individual electrons with a characteristic (rms) "thermal velocity," v_e. Electron-acoustic waves, as described in (a), generally propagate with a phase velocity v_ϕ significantly greater than the thermal velocity $v_e = (\kappa T_e/m)^{1/2}$.

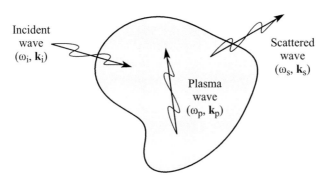

Figure 8.4 Three-wave mixing among natural modes of the plasma. In resonant mixing the three satisfy conservation of energy and momentum.

screening distance λ_D, which gives a measure of the transition between short and long range interactions in a plasma. It is discussed in the following section. Where circumstances conspire to cause such a wave to grow to large amplitude, the crest regions of high electron density can form a very high potential well in which individual electrons can be trapped and accelerated to enormous energies. These energetic suprathermal or "hot" electrons will eventually exhibit their presence through the ensuing bremsstrahlung process, giving off a high photon energy tail as suggested in Figure 8.1. Such high amplitude waves are indeed encountered in hot dense plasmas. They are driven to high amplitude by various non-linear processes involving wave–wave and wave–particle interactions, as suggested in Figure 8.4.

These non-linear processes are particularly strong in hot dense plasmas because the energy, time, and space scales require very high power densities, and thus high field amplitudes. These high field amplitudes (electric field, velocity, etc.) tend to force the plasma away from equilibrium, near-thermal states toward highly non-thermal states.

For instance, a high intensity focused laser pulse can excite plasma waves out of the noise of random particle motions (which are always present). As an example, in what is known as *stimulated Raman scattering* (SRS), an intense incoming electromagnetic (laser) wave rattles all electrons at a single frequency, with well-defined spatial periodicity ($\mathbf{k}$), and with high velocity. This can lead to a well-organized, large amplitude electron-acoustic wave. A portion of the incoming electromagnetic wave can be scattered collectively from the growing plasma wave, creating an outgoing scattered wave with appropriate Doppler frequency shift and directional change. This can lead to a *resonant* three-wave interaction in which the wave frequencies (ω) and wave vectors ($\mathbf{k}$) satisfy conservation of energy and momentum relations of the form

$$\hbar\omega_i = \hbar\omega_p + \hbar\omega_s \tag{8.3}$$

$$\hbar\mathbf{k}_i = \hbar\mathbf{k}_p + \hbar\mathbf{k}_s \tag{8.4}$$

where (ω_i, $\mathbf{k}_i$) characterizes the incoming or incident radiation, (ω_p, $\mathbf{k}_p$) characterizes the particular plasma wave involved, and (ω_s, $\mathbf{k}_s$) represents the outgoing scattered wave. With a very intense incident wave the plasma wave can be amplified to very high amplitude at the beat frequency ($\omega_p = \omega_i - \omega_s$) between the two electromagnetic waves, growing quickly out of the noise (of many plasma waves), and soon dominating the process. This can then seriously affect the observed emission spectrum as the excited plasma wave, now characterized by high fields (electric potential) and generally high velocities ($\omega_p/\mathbf{k}_p$), traps and accelerates individual electrons to very high velocities, to a large fraction of the speed of light in some cases, resulting in the emission of very high energy photons, as mentioned earlier. Where this process involves excitement of a high-frequency electron-acoustic wave, there is a substantial shift of the scattered wave frequency and the process is called stimulated Raman scattering – stimulated because the plasma wave's growth from noise is driven by the strong incident wave it eventually scatters and resonates with. Where the process involves the emergence of a high amplitude ion-acoustic wave, the scattered wave experiences only a small frequency shift and the process is referred to as *stimulated Brillouin scattering* (SBS).

As the generation of suprathermal processes necessarily takes energy away from thermal processes, and often is deleterious in its own right, the avoidance of such processes is often of great interest. Thresholds are thus well studied, and countermeasures, principally operating at lower intensities and high frequencies, are employed. In laser-produced plasmas this generally requires that the incident intensity I be kept below a threshold $I\lambda^2/\kappa T_e$, depending on the wavelength λ and temperature κT_e, which we will discuss in a later section.

8.3   Basic Parameters for Describing a Plasma

To further explore basic processes in a plasma it is necessary to develop an appropriate framework. This will necessarily involve important physical quantities such as the electron density n_e, the electron temperature T, and the magnetic induction $\mathbf{B}$. These in

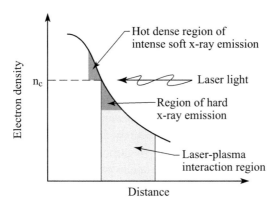

Figure 8.5 Intense laser light is absorbed in the region just below and at the critical electron density region. Energy transport to a thin region just beyond the critical electron density creates a region of intense soft x-ray emission. Harder x-rays tend to be generated in the higher temperature region just below the critical electron density. Laser light that reaches the critical surface of electron density, n_c, is reflected.

turn will lead to the natural introduction of characteristic parameters such as the plasma frequency ω_p, the Debye screening distance λ_D, the cyclotron frequency ω_c, and the collision mean free path l_{mfp}, among others. In general the plasmas are electrically neutral, so that the ion density differs from the electron density only by the average ionization state. Hot dense plasmas are effectively fully ionized, so that neutral atoms are of little consequence. The electron density n_e determines the electron plasma frequency

$$\omega_p = \left(\frac{e^2 n_e}{\epsilon_0 m} \right)^{1/2} \tag{8.5}$$

a natural frequency at which electrons tend to oscillate. In Eq. (8.5) e and m are the electron charge and mass, respectively, and ϵ_0 is the permittivity of free space.

In the next section we will see that electron-acoustic waves tend to oscillate at frequencies at or just above ω_p. Furthermore, we will see that electromagnetic waves can propagate in a plasma only if their frequency ω is greater than ω_p. For a plasma with an electron density gradient, as shown in Figure 8.5, the incident electromagnetic wave is totally reflected at the critical electron density n_c, where $\omega = \omega_p$. Except for a short exponential penetration depth, the wave is totally excluded from the region characterized by $\omega < \omega_p$.

For hot dense plasmas, particularly laser produced plasmas, the electron densities of interest are near-solid densities. For a common neodymium (Nd) laser of 1.06 μm wavelength, the laser–plasma interaction occurs near the critical electron density, $n_c = 1 \times 10^{21}$ e/cm^3, where the plasma frequency is just equal to the laser light radian frequency, about 1.8×10^{15} rad/s. The absorption and various scattering processes, some of which were discussed in the previous section, occur predominantly in the low density (10^{19} e/cm^3 to 10^{21} e/cm^3) region just below the critical region. Classical absorption occurs here, as the incident radiation causes the electrons to oscillate, giving them energy that is then lost in part to random collisions with ions, as was illustrated in Figure 8.2. X-ray emission tends to come predominantly from a thin, somewhat higher density region (10^{21} e/cm^3 to 10^{23} e/cm^3), just behind the critical layer, where energy has been transported by charged particles and radiation. This region characterized by high density and high temperature (see Figure 6.5), is ideal for intense x-ray generation.

A second important plasma parameter is the Debye screening distance

$$\lambda_D = \left(\frac{\epsilon_0 \kappa T_e}{e^2 n_e} \right)^{1/2} \tag{8.6}$$

– a distance beyond which individual charges tend to be screened by the presence of other nearby and mobile charges. For a 1 keV plasma at 10^{21} e/cm^3, the Deybe screening distance is about 7 nm. On spatial scales shorter than λ_D the presence and effects of individual charges are evident. On longer spatial scales the individual charges tend to be screened by neighboring charges, so that on this longer scale charged particle interactions tend to occur through collective motions, such as the electron-acoustic and ion-acoustic waves.

In the theoretical study of plasma waves, one finds that the dispersion relation $\omega(k)$ contains a damping term due to individual electron–wave interactions which, although negligible at long wavelengths, become very strong for plasma of wavenumber $k > 1/\lambda_D$. This natural decay of short wavelength plasma waves is known as Landau damping. It is an example of a plasma process whose understanding and theoretical description require not only use of fluid mechanical quantities, such as density and temperature, but also a detailed knowledge of the shape of the velocity distribution function $f(v)$, shown earlier in Figure 8.3(b). We will return to this subject in the next section.

What begins to emerge here is a more detailed understanding of the manner in which the Debye screening distance separates short range interactions from long range interactions in a plasma. In describing collective effects in plasma an important parameter is the number of electrons in the Debye sphere

$$N_D = \frac{4\pi}{3} \lambda_D^3 n_e \tag{8.7}$$

For $N_D \gg 1$, fluctuations in the microscopic fields are small, and as a result the collective description in terms of averaged field quantities is more effective.

A further important plasma parameter, associated with a directional or imposed magnetic induction **B**, is the electron cyclotron frequency

$$\omega_c = \frac{eB}{m} \tag{8.8}$$

with which electrons circle about the magnetic field. For a 1 MG field the electron cyclotron frequency is about 2×10^{13} rad/s. The electron cyclotron frequency is important for the understanding of energy transport in plasmas, as electrons tend to circle about magnetic field lines with a Larmor radius[‡] $r_L = v/\omega_c = mv/eB$, interrupted only by collisions. This tends to inhibit energy transport to density regions of potentially

---

[‡] The Larmor radius is 3.7 μm for a 100 eV electron in a 9 Tesla magnetic field.

intense x-ray emission. For instance, x-ray emission from the sun shows dark spots and bright coronal loops, which are clear evidence for the presence of strong magnetic fields and constrained charged particle transport. The cyclotron frequency can also play an important role in the collective plasma oscillations, at both low and high frequency, introducing strong dispersion and polarization effects. Cyclotron *resonances*, where the dispersion relation is flat, can play an important role in absorption of low frequency electromagnetic waves (microwaves) by magnetic fusion plasmas.

A fifth parameter, also important to energy transport by charged particles, is the electron–ion collision mean free path, l_{mfp}, which has various dependencies on basic plasma properties, but typically is proportional to $(\kappa T)^2 / n_i Z$, where n_i is the ion density and $+Ze$ is the average charge state. For a 1 keV electron in a 10^{20} ion/cm^3 plasma of $+10$ charge state, the mean free path is about 4 μm (that is, at the critical electron density 10^{21} e/cm^3). A 10 keV electron, well out on the velocity distribution curve, would have a similar mean free path at the $100n_c$ surface, which could be only a few microns away for a sharp density gradient in the supracritical region of a laser-produced plasma. Thus much of the absorbed energy could be stopped by classical collisions in a distance of only a few microns from the critical surface, causing this region to light up with intense (radiated power per unit area) x-ray emissions.

8.4   Microscopic, Kinetic, and Fluid Descriptions of a Plasma

The theoretical description of a plasma may take several forms with varying levels of detail.[1, 3–6] The most exact and complex model involves a description of the position and velocity of each and every particle in the plasma as a function of time. With so many particles present, this level of description is mathematically intractable and must be left to numerical simulations with small numbers of particles.

A simplification is obtained by averaging over a spatial volume containing a large number of particles. This leads to a kinetic description in terms of a more tractable velocity distribution function, which can still have a slow space and time dependence, but omits the details of any particular particle. Further simplification results from averaging over all velocities and describing the plasma in terms of fluid parameters such as density, temperature, and pressure. We begin our analysis at the microscopic level and work our way to the fluid description.

8.4.1   The Microscopic Description

A formal description of plasma dynamics suggested by Klimontovich[5] involves a microscopic distribution function describing the position and velocity of all particles in a six-dimensional velocity–position phase-space:

$$f(\mathbf{v}, \mathbf{r}; t) = \sum_{i=1}^{N} \delta[\mathbf{r} - \mathbf{r}_i(t)] \, \delta[\mathbf{v} - \mathbf{v}_i(t)] \qquad (8.9)$$

where the detailed motion of the i th point particle is described by $\mathbf{r}_i(t)$ and $\mathbf{v}_i(t)$. The distribution function is normalized to the total number of particles, N, by the phase space integral

$$\int_{\mathbf{r}} \int_{\mathbf{v}} f(\mathbf{v}, \mathbf{r}; t) d\mathbf{r} \, d\mathbf{v} = N \tag{8.10}$$

where we define the shorthand notation, for example in Cartesian coordinates

$$\delta(\mathbf{r}) \equiv \delta(x)\delta(y)\delta(z) \tag{8.11}$$

$$(\mathbf{v}) \equiv \delta(v_x)\delta(v_y)\delta(v_z) \tag{8.12}$$

and

$$d\mathbf{r} \equiv dx \, dy \, dz \tag{8.13}$$

$$d\mathbf{v} \equiv dv_x \, dv_y \, dv_z \tag{8.14}$$

where the properties of the Dirac delta function $\delta(x)$ are described in Appendix D.7.

The dynamics of the particle distribution can be determined by taking a partial derivative of $f(\mathbf{v}, \mathbf{r}; t)$ with respect to time:

$$\frac{\partial f}{\partial t} = \sum_i \left[\frac{\partial}{\partial t} \delta(\mathbf{r} - \mathbf{r}_i) \right] \delta(\mathbf{v} - \mathbf{v}_i) + \sum_i \left[\frac{\partial}{\partial t} \delta(\mathbf{v} - \mathbf{v}_i) \right] \delta(\mathbf{r} - \mathbf{r}_i)$$

The first bracketed quantity can be simplified by use of the chain rule for differentiation. For simplicity we first use a scalar, one-dimensional version, defining the functions $g(t) = r_i(t)$ and $f(g) = \delta(r - r_i(t)) = \delta(r - g(t))$, so that by the chain rule

$$\frac{\partial f(g)}{\partial t} = \frac{\partial g}{\partial t} \cdot \frac{\partial f}{\partial g}$$

or explicitly

$$\frac{\partial}{\partial t} \delta(r - r_i(t)) = \frac{\partial r_i}{\partial t} \frac{\partial \delta_i(r - r_i)}{\partial r_i}$$

Noting that for delta functions (see Appendix D.7) $(d/dx)(x - a) = -(d/da)(x - a)$, we can interchange differentials to obtain

$$\frac{\partial \delta(r - r_i)}{\partial r_i} = \frac{\partial \delta(r - r_i)}{\partial r}$$

and thus

$$\frac{\partial}{\partial t} \delta(r - r_i(t)) = -\frac{\partial r_i}{\partial t} \frac{\partial}{\partial r} \delta(r - r_i)$$

In its three-dimensional generalization this becomes

$$\frac{\partial}{\partial t} \delta(\mathbf{r} - \mathbf{r}_i(t)) = -\frac{\partial \mathbf{r}_i}{\partial t} \cdot \nabla \delta(\mathbf{r} - \mathbf{r}_i(t))$$

where we recognize the differential $\partial \mathbf{r}_i / \partial t = d\mathbf{r}_i/dt = \mathbf{v}_i$, so that

$$\frac{\partial}{\partial t}\delta(\mathbf{r} - \mathbf{r}_i(t)) = -\mathbf{v}_i \cdot \nabla\delta(\mathbf{r} - \mathbf{r}_i(t))$$

Likewise, for the second bracketed quantity one obtains

$$\frac{\partial}{\partial t}\delta(\mathbf{v} - \mathbf{v}_i(t)) = -\frac{\partial \mathbf{v}_i}{\partial t} \cdot \nabla_v\delta(\mathbf{v} - \mathbf{v}_i(t))$$

where use of the Lorentz force on each particle permits the substitution

$$\frac{\partial}{\partial t}\delta(\mathbf{v} - \mathbf{v}_i(t)) = -\frac{q_i}{m}(\mathbf{E} + \mathbf{v}_i \times \mathbf{B}) \cdot \nabla_v\delta(\mathbf{v} - \mathbf{v}_i(t))$$

Combining these results for the two bracketed quantities, we see that the microscopic particle distribution function $f(\mathbf{v}, \mathbf{r}; t)$ obeys the equation

$$\frac{\partial f}{\partial t} = \sum_i (-\mathbf{v}_i) \cdot \nabla[\delta(\mathbf{r} - \mathbf{r}_i)\delta(\mathbf{v} - \mathbf{v}_i)]$$
$$+ \sum_i -\frac{q_i}{m}(\mathbf{E} + \mathbf{v}_i \times \mathbf{B}) \cdot \nabla_v[\delta(\mathbf{r} - \mathbf{r}_i)\delta(\mathbf{v} - \mathbf{v}_i)]$$

By identifying $f(\mathbf{v}, \mathbf{r}; t)$ in each term above one obtains the *Klimontovich equation*

$$\boxed{\frac{\partial f}{\partial t} + \mathbf{v} \cdot \nabla f + \frac{q}{m}(\mathbf{E} + \mathbf{v} \times \mathbf{B}) \cdot \nabla_v f = 0} \qquad (8.15)$$

which describes the evolution of the microscopic distribution function, as a function of time, in phase space.

A self-consistent solution is required because of the interdependence of variables, namely, the velocity distribution function $f(\mathbf{v})$ depends on the electric and magnetic fields $\mathbf{E}$ and $\mathbf{B}$, which in turn depend on $f(\mathbf{v})$ through the charge density ρ and current $\mathbf{J}$ as they appear in Maxwell's equations. To find such a solution we note that the charge and current densities can be written in terms of the distribution functions $f_j(\mathbf{v}, \mathbf{r}; t)$, for each particle type present ($j = 1$ for electrons, $j = 2$ for ions) as

$$\rho_j(\mathbf{r}, t) = \sum_i q_j\delta(\mathbf{r} - \mathbf{r}_i) = \sum_i q_j \int \delta(\mathbf{r} - \mathbf{r}_i)\delta(\mathbf{v} - \mathbf{v}_i)d\mathbf{v} = q_j \int f_j(\mathbf{v}, \mathbf{r}; t)\,d\mathbf{v}$$

$$(8.16)$$

and

$$\mathbf{J}_j(\mathbf{r}, t) = \sum_i q_j\mathbf{v}_i\delta(\mathbf{r} - \mathbf{r}_i) = \sum_i q_j \int \mathbf{v}\delta(\mathbf{r} - \mathbf{r}_i)\delta(\mathbf{v} - \mathbf{v}_i)d\mathbf{v} \qquad (8.17)$$
$$= q_j \int \mathbf{v}f_j(\mathbf{v}, \mathbf{r}; t)\,d\mathbf{v}$$

where for electrons $q_1 = -e$ and the sum Σ_i is over all individual electrons, and for ions $q_2 = +Ze$ and the sum Σ_i is over all individual ions. The formal set of self-consistent field equations, which describe plasma dynamics at the microscopic level, are called the *Maxwell–Klimontovich equations*. They take the form [see Chapter 2, Eqs. (2.1)–(2.6)]

$$\nabla \times \mathbf{H} = \frac{\partial \mathbf{D}}{\partial t} + \sum_j q_j \int \mathbf{v} f_j(\mathbf{v}, \mathbf{r}; t) \, d\mathbf{v} \tag{8.18}$$

$$\nabla \times \mathbf{E} = -\frac{\partial \mathbf{B}}{\partial t} \tag{8.19}$$

$$\nabla \cdot \mathbf{D} = \sum_j q_j \int f_j \int f_j(\mathbf{v}, \mathbf{r}; t) \, d\mathbf{v} \tag{8.20}$$

$$\nabla \cdot \mathbf{B} = 0 \tag{8.21}$$

with constitutive relations

$$\mathbf{D} = \epsilon_0 \mathbf{E} \tag{8.22}$$

$$\mathbf{B} = \mu_0 \mathbf{H} \tag{8.23}$$

and for each particle of species (j)

$$\frac{\partial f_j}{\partial t} + \mathbf{v} \cdot \nabla f_j + \frac{q_j}{m_j} (\mathbf{E} + \mathbf{v} \times \mathbf{B}) \cdot \nabla_{\mathbf{v}} f_j = 0 \tag{8.24}$$

where the field equations are written for a multicomponent plasma (electrons and ions) with respective distribution functions $f_j(\mathbf{v}, \mathbf{r}; t)$, for species of charge q_j and mass m_j (see Ref. 2).

The Klimontovich description is a simple yet formal approach to the microscopic description of plasma phenomena. It postulates a distribution function, takes its time derivative, and writes it in a form that easily evolves into a reduced *kinetic theory*, which we consider in the next section.

8.4.2   The Kinetic Description

The microscopic description in terms of Klimontovich's density function is highly stochastic, varying rapidly over space and time, and involves details regarding too many individual particles for analytic treatment. A reduced description, averaged somewhat over a space containing a large number of particles, forms a more slowly varying distribution, $f(\mathbf{v}, \mathbf{r}; t)$, which contains no information regarding individual particles, but rather describes an average velocity distribution function $f(v)$, with a slow space-time dependence. We drop the subscript j here for convenience, but understand that this process must be followed separately for each species. Whereas Klimontovich's distribution function is discontinuous and stochastic, with wildly varying amplitude, the kinetic distribution function f is analytic. Integrating the Maxwell–Klimontovich equations over a spatial volume sufficient to include many particles – so that statistical fluctuations are not so wild – produces such a distribution.[3] We can write the distribution function in terms of a slowly varying part and a fluctuating part, as

$$f(\mathbf{v}, \mathbf{r}; t) = \bar{f}(\mathbf{v}, \mathbf{r}; t) + \tilde{f}(\mathbf{v}, \mathbf{r}; t)$$

with a similar description for the fields

$$\mathbf{E}(r, t) = \bar{\mathbf{E}}(\mathbf{r}, t) + \tilde{\mathbf{E}}(\mathbf{r}, t), \qquad \text{etc.}$$

Then substituting these into the Klimontovich equation [Eq. (8.15)], and averaging over a spatial scale sufficiently large to give a smoothed kinetic equation for the velocity distribution function, gives us a kinetic equation formally equivalent to the Boltzmann equation:

$$\frac{\partial f}{\partial t} + \mathbf{v} \cdot \nabla f + \frac{q}{m}(\mathbf{E} \times \mathbf{v} \times \mathbf{B}) \cdot \nabla_{\mathbf{v}} f(\mathbf{v}, \mathbf{r}; t) = -\frac{q}{m}\nabla_{\mathbf{v}} \cdot \overline{(\tilde{\mathbf{E}} + \mathbf{v} \times \tilde{\mathbf{B}})\tilde{f}} \qquad (8.25)$$

where for simplicity of notation we have dropped, on the left side of the equation, the overbars denoting slowly varying variables, i.e., $\bar{f}, \bar{E}$, and $\bar{B}$. The right side is a symbolic "collision" term, non-linear (because of the product terms) in fluctuations from the mean, which tends to bring the distribution function back toward an equilibrium. In situations not too distant from equilibrium, and not turbulent, the collision term may be small. In such a case one has the *collisionless Vlasov equation*, which is valuable in solving many plasma kinetic problems. Of course this must be solved self-consistently with similarly smoothed Maxwell's equations for the slowly varying quantities $\bar{\mathbf{E}}, \bar{\mathbf{H}}, \bar{\mathbf{B}}, \bar{\mathbf{D}}$, and $\bar{f}$. Substituting in Eqs. (8.18)–(8.24) quantities such as $\mathbf{E} = \bar{\mathbf{E}} + \tilde{\mathbf{E}}, \mathbf{H} = \bar{\mathbf{H}} + \tilde{\mathbf{H}}$, etc., retaining only the slowly varying quantities ($^-$), and then, for simplicity of notation, dropping the bars, we obtain the *collisionless Maxwell–Vlasov equations* for a plasma,

$$\nabla \times \mathbf{H} = \frac{\partial \mathbf{D}}{\partial t} + \sum_j q_j \int \mathbf{v} f_j(\mathbf{v}, \mathbf{r}; t)\, d\mathbf{v} \qquad (8.26)$$

$$\nabla \times \mathbf{E} = \frac{\partial \mathbf{B}}{\partial t} \qquad (8.27)$$

$$\nabla \cdot \mathbf{D} = \sum_j q_j \int f_j(\mathbf{v}, \mathbf{r}; t)\, d\mathbf{v} \qquad (8.28)$$

$$\nabla \cdot \mathbf{B} = 0 \qquad (8.29)$$

plus the constitutive relations in vacuum,

$$\mathbf{D} = \epsilon_0 \mathbf{E} \qquad (8.30)$$

$$\mathbf{B} = \mu_0 \mathbf{H} \qquad (8.31)$$

and the collisionless Vlasov equations for each species (electrons, ions),

$$
\frac{\partial f_j(\mathbf{v}, \mathbf{r}; t)}{\partial t} + \mathbf{v} \cdot \nabla f_j(\mathbf{v}, \mathbf{r}; t)
$$
$$
+ \frac{q_j}{m_j} (\mathbf{E} + \mathbf{v} \times \mathbf{B}) \cdot \nabla_{\mathbf{v}} f_i(\mathbf{v}, \mathbf{r}; t) = 0 \qquad (8.32)
$$

These equations are formally identical to the Maxwell–Klimontovich equations [Eqs. (8.18)–(8.24)], but contain significantly less detail, making them mathematically more tractable. Whereas the Maxwell–Klimontovich equations were written for an N-particle distribution function, containing detailed positions and velocities of all N particles, the Maxwell–Vlasov equations involved a simpler *kinetic* distribution function, which does not distinguish (or recognize) any individual particles. This kinetic distribution function is sometimes referred to in the literature as the "single particle" distribution function, emphasizing that it does not have discrete N-particle information.

Examples of phenomena that can be mathematically described by the use of a kinetic description are those of electron-acoustic wave amplification and collisionless decay, known commonly as Landau growth and damping. The distribution function for thermal electrons takes the form shown previously in Figure 8.3(b). For example, in the case of the electron-acoustic wave, which is a longitudinal wave of electrons in a uniform positive ion density background, a charge density modulation propagates in some direction as shown in Figure 8.3(a). Most of the electrons do not move with the wave, but rather oscillate in a nearly fixed position as the wave passes by, as in the case of the motion of molecules in a sound wave. However, as shown by the distribution function, some electrons travel at a velocity near that of the wave. Electrons that move slightly slower than the wave are pushed by the negative potential of the charge density peak; therefore, these electrons accelerate, taking energy from the wave. Likewise, electrons traveling slightly faster than the wave push the charge density peaks, thus decelerating themselves while imparting energy to the wave. In a thermal distribution, there are more electrons traveling slower than the phase velocity of the wave, v_ϕ, so the wave is damped.

On the other hand, if a beam of electrons with velocities slightly larger than the wave velocity (v_ϕ) is injected into the plasma, the new electron velocity distribution function, shown in Figure 8.6, causes wave growth. In this case the injected electrons tend to "push" the potential crests of the wave, transferring energy to the wave, and losing energy themselves as they merge into it. Mathematical solutions[4, 18] to Eqs. (8.26)–(8.32) show that the Landau damping or growth rate is related to the slope of $f(v)$ at $v = v_\phi$.

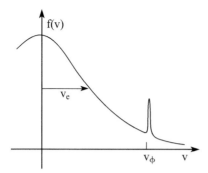

Figure 8.6 Velocity distribution function with an injected electron beam.

8.4.3   The Fluid Description

The Vlasov equation has a great deal of information about velocity distributions and how they evolve in space and time. In some problems velocity information is critical, but in some it is not needed and further simplification is possible. A set of fluid mechanical equations,[4] containing physical variables such as the particle density $n_j(\mathbf{r}, t)$ for each species, the average velocity $\bar{\mathbf{v}}_j(\mathbf{r}, t)$, partial pressure $P_j(\mathbf{r}, t)$, and others, can be developed directly from the kinetic Vlasov equation. These common fluid mechanical quantities correspond to so-called *velocity moments*, or velocity weighted integrals, of the kinetic velocity distribution function $f_j(\mathbf{v}, \mathbf{r}; t)$ with scalar, vector, and dyadic multipliers of the velocity. Specifically, for the electrons ($q_j = -e$) one defines

$$n_e(\mathbf{r}, t) = \int f(\mathbf{v}, \mathbf{r}; t)\, d\mathbf{v} \tag{8.33}$$

$$n_e \bar{\mathbf{v}}(\mathbf{r}, t) = \int \mathbf{v} f(\mathbf{v}, \mathbf{r}; t)\, d\mathbf{v} \tag{8.34}$$

where the subscript e has been suppressed for simplicity on both $\bar{\mathbf{v}}_j$ and $f_j(\mathbf{v}, \mathbf{r}; t)$. In terms of the random (thermal) component of the velocity, $\tilde{\mathbf{v}}$, that is, the departure from the average velocity $\bar{\mathbf{v}}$, higher velocity moments are

$$\bar{\bar{\mathbf{P}}}_e = n_e m \overline{\tilde{\mathbf{v}}\tilde{\mathbf{v}}} = m \int \tilde{\mathbf{v}}\tilde{\mathbf{v}} f(\mathbf{v}, \mathbf{r}; t)\, d\mathbf{v} \tag{8.35}$$

$$n_e U_e = \frac{1}{2} n_e m \overline{\tilde{v}^2} = \frac{1}{2} m \int \tilde{v}^2 f(\mathbf{v}, \mathbf{r}; t)\, d\mathbf{v} \tag{8.36}$$

and

$$\mathbf{Q}_e = \frac{1}{2} n_e m \overline{\tilde{v}^2 \tilde{\mathbf{v}}} = \frac{1}{2} m \int \tilde{v}^2 \tilde{\mathbf{v}} f(\mathbf{v}, \mathbf{r}; t)\, d\mathbf{v} \tag{8.37}$$

where $n_e(\mathbf{r}, t)$ is the local electron (particle) density, $\bar{\mathbf{v}}(\mathbf{r}; t)$ is the distribution weighted average velocity, the fluctuation component of fluid velocity is defined by $\mathbf{v} = \bar{\mathbf{v}} + \tilde{\mathbf{v}}$, $\bar{\bar{\mathbf{P}}}_e$ is the tensor electron pressure dyadic, U_e is the electron thermal (kinetic) energy density, and $\mathbf{Q}_e$ is the vector heat flux or thermal flux density carried by electrons. Note that the notation $\tilde{v}^2$ is understood to represent the scalar quantity $\tilde{\mathbf{v}} \cdot \tilde{\mathbf{v}}$ and that simple juxtaposition ($\tilde{\mathbf{v}}\tilde{\mathbf{v}}$) indicates a nine component tensor multiplication of two vectors, as described in Appendix D.1. As a specific example, the pressure dyadic $\bar{\bar{\mathbf{P}}}_e$ of Eq. (8.35), which involves the tensor velocity product $\tilde{\mathbf{v}}\tilde{\mathbf{v}}$, can be written out in terms of its components as

$$
\bar{\bar{\mathbf{P}}}_e = n_e m
\begin{bmatrix}
\overline{\tilde{v}_x^2}\mathbf{x}_0\mathbf{x}_0 & \overline{\tilde{v}_x\tilde{v}_y}\mathbf{x}_0\mathbf{y}_0 & \overline{\tilde{v}_x\tilde{v}_y}\mathbf{x}_0\mathbf{z}_0 \\
\overline{\tilde{v}_y\tilde{v}_x}\mathbf{y}_0\mathbf{x}_0 & \overline{\tilde{v}_x^2}\mathbf{y}_0\mathbf{y}_0 & \overline{\tilde{v}_y\tilde{v}_z}\mathbf{y}_0\mathbf{z}_0 \\
\overline{\tilde{v}_z\tilde{v}_x}\mathbf{z}_0\mathbf{x}_0 & \overline{\tilde{v}_z\tilde{v}_y}\mathbf{z}_0\mathbf{y}_0 & \overline{\tilde{v}_x^2}\mathbf{z}_0\mathbf{z}_0
\end{bmatrix}
\tag{8.38}
$$

where $\mathbf{v} = v_x\mathbf{x}_0 + v_y\mathbf{y}_0 + v_z\mathbf{z}_0$ in Cartesian coordinates, with unit vectors $\mathbf{x}_0, \mathbf{y}_0, \mathbf{z}_0$. The unit dyad has diagonal units equal to one, and zeros elsewhere, so that its dot product with any vector equals the vector, i.e., $\bar{\bar{\mathbf{1}}} \cdot \mathbf{v} = \mathbf{v}$. One can show (see Appendix D.1) that in general, for two vectors $\mathbf{A}$ and $\mathbf{B}$, $\nabla \cdot (\mathbf{A}\mathbf{B}) = (\mathbf{A} \cdot \nabla)\mathbf{B} + \mathbf{B}(\nabla \cdot \mathbf{A})$, a relationship we will find useful in following paragraphs.

The relevant equations for these fluid mechanical quantities ($n, \bar{\mathbf{v}}, \bar{\bar{\mathbf{P}}}$, etc.) can be obtained by multiplying the Vlasov equation by 1, $\mathbf{v}$, $\mathbf{vv}$, etc., and integrating over all velocities. These are the so called *moment* equations, as they each multiply the Vlasov equation by some quantity before integration. Assuming for the moment that the collision term is small, we start with the collisionless Vlasov equation [Eq. (8.32)] and integrate over velocity, viz.,

$$
\int \left[\frac{\partial f}{\partial t} + \mathbf{v} \cdot \nabla f + \frac{q}{m}(\mathbf{E} + \mathbf{v} \times \mathbf{B}) \cdot \nabla_v f \right] d\mathbf{v} = 0
\tag{8.39}
$$

where we have written this for a specific species j, for instance for the electrons or for a particular ionic species. Recall that $d\mathbf{v}$ is not a vector quantity, but rather shorthand notation for the volume element in velocity space, e.g., $d\mathbf{v} = dv_x \, dv_y \, dv_z$. Long range interactions between various species are included through the slowly varying electromagnetic fields. Short range interactions (collisions) that involve strong space-time variations of these fields are included, where appropriate, through the product of fluctuation term on the right-hand side of Eq. (8.25).

To perform the indicated integrations we note that the independent coordinates in this non-relativistic kinetic plasma are $\mathbf{v}$, $\mathbf{r}$, and t. Thus in the first term of Eq. (8.39) the time derivative passes through the velocity integral, leaving

$$
\frac{\partial}{\partial t} \int f(\mathbf{v}) \, d\mathbf{v} = \frac{\partial}{\partial t} n(\mathbf{r}, t)
$$

by the definition in Eq. (8.32). The second term of Eq. (8.39), involving $\mathbf{v} \cdot \nabla f$, can be integrated using the vector identity $\nabla \cdot (\phi \mathbf{A}) = \phi \nabla \cdot \mathbf{A} + \mathbf{A} \cdot \nabla \phi$, so that

$$
\mathbf{v} \cdot \nabla f = \nabla \cdot (f\mathbf{v}) - \underbrace{f \nabla \cdot \mathbf{v}}_{=0}
$$

The last term above is zero, as $\mathbf{v}$ is not a function of $\mathbf{r}$ and thus $\nabla \cdot \equiv 0$. The second term of Eq. (8.39) is then integrated as follows:

$$\int \mathbf{v} \cdot \nabla f(\mathbf{v}) \, d\mathbf{v} = \int \nabla \cdot f\mathbf{v} \, d\mathbf{v} = \nabla \cdot \int f(\mathbf{v})\mathbf{v} \, d\mathbf{v} \equiv \nabla \cdot n\bar{\mathbf{v}}$$

using the definition given in Eq. (8.33). The fluid equation now emerging from Eq. (8.39) becomes

$$\frac{\partial n}{\partial t} + \nabla \cdot (n\bar{\mathbf{v}}) + \frac{q}{m} \int (\mathbf{E} + \mathbf{v} \times \mathbf{B}) \cdot \nabla_{\mathrm{v}} f(\mathbf{v}) \, d\mathbf{v} = 0$$

With a few steps we will show that for this first fluid equation the Lorentz force term does not contribute; however, in higher-order fluid equations, to be considered next, it will contribute. Considering the remaining integral in Eq. (8.39), we again use the $\nabla \cdot (\phi\mathbf{A})$ vector relation, i.e.,

$$(\mathbf{E} + \mathbf{v} \times \mathbf{B}) \cdot \nabla_{\mathrm{v}} f(\mathbf{v}) = \nabla_{\mathrm{v}} \cdot [(\mathbf{E} + \mathbf{v} \times \mathbf{B})f] - \underbrace{f\nabla \cdot (\mathbf{E} + \mathbf{v} \times \mathbf{B})}_{=0}$$

where we note that in the term $\nabla_{\mathrm{v}} \cdot \mathbf{E}$, $\mathbf{E}$ is not a function of $\mathbf{v}$, and $\nabla_{\mathrm{v}} \cdot (\mathbf{v} \times \mathbf{B}) = (\nabla_{\mathrm{v}} \times \mathbf{v}) \cdot \mathbf{B} = 0$ since $\nabla_{\mathrm{v}} \times \mathbf{v} = 0$. The remaining term can then be integrated using Gauss's theorem for an arbitrary vector $\mathbf{A}$:

$$\int_{\mathrm{volume}} (\nabla \cdot \mathbf{A}) \, dV = \oint_{\mathrm{surface}} \mathbf{A} \cdot d\mathbf{S}$$

so that in velocity space

$$\int \nabla_{\mathrm{v}} \cdot [(\mathbf{E} + \mathbf{v} \times \mathbf{B})]f(\mathbf{v}) \, d\mathbf{v} = \int_{\substack{\mathrm{velocity} \\ \mathrm{space} \\ \mathrm{surface}}} [(\mathbf{E} + \mathbf{v} \times \mathbf{B})f(\mathbf{v})] \cdot d\mathbf{S}_{\mathrm{v}} = 0$$

where the integral is evaluated at a velocity space surface where $\mathbf{v}$ approaches infinity (or some other sufficiently large value not exceeding c). Taking $d\mathbf{S} = 4\pi v^2 \, d\Omega\mathbf{v}_0$, where $\mathbf{v}_0$ is a unit vector in the outgoing velocity direction, the $\mathbf{v} \times \mathbf{B}$ term is normal to $\mathbf{v}_0$ and thus does not contribute.

The remaining surface integral involving $\mathbf{E}f(\mathbf{v})$ does not contribute for the case where $f(\mathbf{v})$ goes to zero faster than $1/v^2$ for large v, a very reasonable assumption for any physical plasma. The resultant fluid equation, the *first moment* (lever arm) of the collisionless Vlasov equation for electrons, is then

$$\boxed{\frac{\partial n_e}{\partial t} + \nabla \cdot (n_e\bar{\mathbf{v}}) = 0} \tag{8.40}$$

which in fluid mechanics is referred to as the *continuity equation*, and expresses the conservation of particles of a certain type, in this case electrons.

The next fluid equation, involving conservation of momentum, takes on the form of Newton's second law of motion for a fluid. It is obtained by taking a second moment of the collisionless Vlasov equation [Eq. (8.32)], this time by multiplying all terms by the momentum $m\mathbf{v}$ and then integrating over all velocities to obtain

$$\int m\mathbf{v}\left[\frac{\partial f}{\partial t}+\mathbf{v}\cdot\nabla f-\frac{e}{m}(\mathbf{E}+\mathbf{v}\times\mathbf{B})\cdot\nabla_{\mathbf{v}}f\right]d\mathbf{v}=0 \qquad (8.41)$$

Noting again the fluid definitions of Eqs. (8.33)–(8.37), and the interchangeability of order among $\mathbf{r}$, t, and $\mathbf{v}$ derivatives, one has

$$m\frac{\partial}{\partial t}(n_e\bar{\mathbf{v}})+m\nabla\cdot\int\mathbf{v}\mathbf{v}f(\mathbf{v})\,d\mathbf{v}-e\int\mathbf{v}[(\mathbf{E}+\mathbf{v}\times\mathbf{B})\cdot\nabla_{\mathbf{v}}f]\,d\mathbf{v}=0$$

$$m\frac{\partial}{\partial t}(n_e\bar{\mathbf{v}})+m\nabla\cdot(n_e\overline{\mathbf{v}\mathbf{v}})-e\int\mathbf{v}[(\mathbf{E}+\mathbf{v}\times\mathbf{B})\cdot\nabla_{\mathbf{v}}f(\mathbf{v})]\,d\mathbf{v}=0$$

This begins to look like the desired fluid mechanical momentum equation if we expand the first two terms. If we write the velocity as the sum of a slowly varying component $\bar{\mathbf{v}}(\mathbf{r};t)$ and a faster fluctuating component $\tilde{\mathbf{v}}$, such that, $\mathbf{v}=\bar{\mathbf{v}}+\tilde{\mathbf{v}}$, then the product $\overline{\mathbf{v}\mathbf{v}}$ becomes

$$\overline{\mathbf{v}\mathbf{v}}=\overline{(\bar{\mathbf{v}}+\tilde{\mathbf{v}})(\bar{\mathbf{v}}+\tilde{\mathbf{v}})}=\bar{\mathbf{v}}\bar{\mathbf{v}}+2\underbrace{\overline{\bar{\mathbf{v}}\tilde{\mathbf{v}}}}_{=0}+\overline{\tilde{\mathbf{v}}\tilde{\mathbf{v}}}$$

where $\overline{\bar{\mathbf{v}}\tilde{\mathbf{v}}}=\bar{\mathbf{v}}\overline{\tilde{\mathbf{v}}}=0$, since $\overline{\tilde{\mathbf{v}}}\equiv 0$. The $\bar{\mathbf{v}}\bar{\mathbf{v}}$ term is a non-linear product of velocities that gives rise to interesting fluid mechanical properties, including aerodynamic flight. With expansion of the differential product terms, the second fluid equation (8.39) takes the form

$$mn_e\frac{\partial\bar{\mathbf{v}}}{\partial t}+m\bar{\mathbf{v}}\left[\frac{\partial n_e}{\partial t}+mn_e\bar{\mathbf{v}}\cdot\nabla\bar{\mathbf{v}}+m\bar{\mathbf{v}}\nabla\cdot\right]n\bar{\mathbf{v}}+m\nabla\cdot(n_e\overline{\tilde{\mathbf{v}}\tilde{\mathbf{v}}})$$

$$-e\int\mathbf{v}[(\mathbf{E}+\mathbf{v}\times\mathbf{B})\cdot\nabla_{\mathbf{v}}f[\mathbf{v}]]\,d\mathbf{v}=0$$

where the second and fourth terms cancel by the continuity equation [Eq. (8.40)], where the expansion of the $\nabla\cdot(n\,\overline{\mathbf{v}\mathbf{v}})$ term made use of the dyadic relation $\nabla\cdot(\mathbf{AB})=(\mathbf{A}\cdot\nabla)\mathbf{B}+\mathbf{B}(\nabla\cdot\mathbf{A})$, and where we recognize $mn_e\overline{\tilde{\mathbf{v}}\tilde{\mathbf{v}}}$ from Eq. (8.35) as the dyadic pressure $\bar{\bar{\mathbf{P}}}_e$. To simplify the remaining integral involving the Lorentz force term, we replace $-e(\mathbf{E}+\mathbf{v}\times\mathbf{B})$ by the force $\mathbf{F}(\mathbf{v},\mathbf{r};t)$. The remaining integral is then

$$\int\mathbf{v}[\mathbf{F}(\mathbf{v})\cdot\nabla_{\mathbf{v}}f(\mathbf{v})]\,d\mathbf{v}=\int\mathbf{v}\{\nabla_{\mathbf{v}}\cdot[\mathbf{F}(\mathbf{v})f(\mathbf{v})]-\underbrace{f(\mathbf{v})\nabla_{\mathbf{v}}\cdot\mathbf{F}}_{=0}\,d\mathbf{v}$$

where we have used the vector identity $\nabla\cdot(\phi\mathbf{A})=\phi\nabla\cdot\mathbf{A}+\mathbf{A}\cdot\nabla\phi$, and noted that $\nabla_{\mathbf{v}}\cdot\mathbf{F}=-e\nabla_{\mathbf{v}}\cdot(\mathbf{E}+\mathbf{v}\times\mathbf{B})=0$ (since $\mathbf{E}$ is not a function of $\mathbf{v}$) and $\nabla_{\mathbf{v}}\cdot\mathbf{v}\times\mathbf{B}=\nabla_{\mathbf{v}}\times\mathbf{v}\cdot\mathbf{B}=0$ (since $\nabla_{\mathbf{v}}\times\mathbf{v}=0$). Again using the dyadic expansion of $\nabla\cdot(\mathbf{AB})$, this time with $\mathbf{A}=\mathbf{F}(\mathbf{v})f(\mathbf{v})$ and $\mathbf{B}=\mathbf{v}$, the remaining integral becomes

$$\int\mathbf{v}\{\nabla_{\mathbf{v}}\cdot[\mathbf{F}(\mathbf{v})f(\mathbf{v})]\}\,d\mathbf{v}=\int\{\nabla_{\mathbf{v}}\cdot[\mathbf{F}(\mathbf{v})f(\mathbf{v})\mathbf{v}]-\mathbf{F}(\mathbf{v})f(\mathbf{v})\cdot\underbrace{\nabla_{\mathbf{v}}\mathbf{v}}_{1}\}\,d\mathbf{v}$$

The first term is again set equal to zero on the basis of Gauss's theorem in velocity space for the dyadic quantity $\mathbf{F} f(\mathbf{v})\mathbf{v}$, which requires a somewhat faster decay of $f(\mathbf{v})$ with large $\mathbf{v}$ than in the first moment equation (for the continuity equation). The second term, however, contributes in this case. Note that $-\mathbf{F}(\mathbf{v})f(\mathbf{v})\cdot\mathbf{1} = -\mathbf{F}(\mathbf{v})f(\mathbf{v}) = e(\mathbf{E} + \mathbf{v}\times\mathbf{B})f(\mathbf{v})$, so that the remaining integral in Eq. (8.41) becomes

$$-e\int [(\mathbf{E}+\mathbf{v}\times\mathbf{B})\cdot\nabla_v f(\mathbf{v})]\,\mathbf{v}\,d\mathbf{v} = e\int (\mathbf{E}+\mathbf{v}\times\mathbf{B})f(\mathbf{v})\,d\mathbf{v} = en_e(\mathbf{E}+\bar{\mathbf{v}}\times\mathbf{B})$$

where we have again used the definitions for $n_e(\mathbf{r},t)$ and $n_e(\mathbf{r},t)\bar{\mathbf{v}}(\mathbf{r};t)$ given by Eqs. (8.33) and (8.34). Combining all terms, we obtain the fluid mechanical equation expressing conservation of momentum for electrons:

$$m\underbrace{\left[\frac{\partial}{\partial t}+\bar{\mathbf{v}}\cdot\nabla\right]}_{D/Dt}\bar{\mathbf{v}} = -\frac{1}{n_e}\nabla\cdot\bar{\bar{\mathbf{P}}}_e - e(\mathbf{E}+\bar{\mathbf{v}}\times\mathbf{B}) \tag{8.42}$$

where $D/Dt = \partial/\partial t + \bar{\mathbf{v}}\cdot\nabla$ is the *substantial* derivative, a time derivative moving with the average velocity, and where $\bar{\bar{\mathbf{P}}}_e = mn_e\overline{\bar{\mathbf{v}}\bar{\mathbf{v}}}$ is the dyadic electron pressure, as defined in Eq. (8.35).

Equation (8.42) provides a mathematical description for the rate of change of momentum for a compressible fluid – it is essentially Newton's second law of motion where the unbalanced forces are due to a gradient in pressure along the fluid trajectory, and to the Lorentz force $-e(\mathbf{E}+\bar{\mathbf{v}}\times\mathbf{B})$ on the electrons (charged particles). In many cases, involving an isotropic distribution function, the dyadic pressure reduces to a scalar pressure P such that $\bar{\bar{\mathbf{P}}} = P\mathbf{1}$, and $\nabla\cdot\bar{\bar{\mathbf{P}}} = \nabla P$, yielding a more common form of the fluid mechanical *momentum equation* for electrons:

$$m\left(\frac{\partial}{\partial t}+\bar{\mathbf{v}}\cdot\nabla\right)\bar{\mathbf{v}} = -\frac{1}{n_e}\nabla P_e - e(\mathbf{E}+\bar{\mathbf{v}}\times\mathbf{B}) \tag{8.43}$$

The inclusion of viscosity, a frictional effect involving velocity differences (gradients) among adjoining regions of the same species, leads to an additional term in the momentum equation [Eq. (8.43)], which is then referred to in fluid mechanics as the *Navier–Stokes equation*.[17] In its most common form this involves the addition of a viscous force term $\mu\nabla^2\mathbf{v}$ to the right-hand side of Eq. (8.43). A discussion of the fluid transport equations, including viscosity (for uncharged particles), is given in Ref. 16.

Another commonly encountered situation in which an alternative form of Eq. (8.43) occurs is that in which short range collisions between the various species (electrons, ions, and neutrals) leads to a transfer of momentum among species, such as between electrons and ions, or between electrons and neutrals. For this to arise naturally would require an appropriate collision term on the right side of the kinetic Vlasov equation (8.25). Inclusion of such a collision term produces no additional term in the continuity equation (8.40), but a collision term of the form $-mv\bar{\mathbf{v}}$ will appear in Eq. (8.43), where v is an effective collision frequency for momentum transfer among different species due to short-range collisions.[16] In the absence of such viscosity and inter-particle collisions, the fluid equations are referred to as the *Euler equations*. Collecting the fluid equations (8.40) and (8.43) for each species j along with Maxwell's equations, we have the *Maxwell–Euler equations* that describe plasma dynamics on fluid level:

$$\nabla \times \mathbf{H} = \frac{\partial \mathbf{D}}{\partial t} + \sum_j q_j n_j \mathbf{v}_j \tag{8.44}$$

$$\nabla \times \mathbf{E} = -\frac{\partial \mathbf{B}}{\partial t} \tag{8.45}$$

$$\nabla \cdot \mathbf{D} = \sum_j q_j n_j \tag{8.46}$$

$$\nabla \cdot \mathbf{B} = 0 \tag{8.47}$$

with the constitutive relations

$$\mathbf{D} = \epsilon_0 \mathbf{E} \tag{8.48}$$

$$\mathbf{B} = \mu_0 \mathbf{H} \tag{8.49}$$

and for each particle species (j)

$$\frac{\partial n_j}{\partial t} + \nabla \cdot (n_j \mathbf{v}_j) = 0 \tag{8.50}$$

$$m \left(\frac{\partial}{\partial t} + \mathbf{v}_j \cdot \nabla \right) \mathbf{v}_j = -\frac{1}{n_j} \nabla P_j + q_j (\mathbf{E} + \mathbf{v}_j \times \mathbf{B}) \tag{8.51}$$

where for electrons $q = -e$ and for ions $q = +eZ$. Note that for simplicity we have dropped the overbar on the average velocities $\bar{\mathbf{v}}_j$ with the understanding that the simpler notation $\mathbf{v}_j$ now carries the connotation of a slowly varying function of space and time.

The Maxwell–Euler equations are an *independent* set in that the particles affect the electromagnetic fields, through the charge [Eq. (8.46)] and current [Eq. (8.44)] distributions, while the fluid equations are in turn affected by $\mathbf{E}$ and $\mathbf{B}$ [Eq. (8.38)]. The coupled set of equations as described here is *incomplete* in that for each species we have added five new fluid quantities (n_j, $\mathbf{v}_j$, and the scalar P_j), but only four new scalar equations, one from Eq. (8.40) and three from Eq. (8.43). An additional equation is required for each. To complete the set of coupled fluid–electromagnetic equations we must generate another equation – an energy related equation – by taking an additional moment of the Vlasov equation (8.32) for each species, this time by multiplying through by a scalar factor $mv^2/2$ and then proceeding with the velocity-space integrals. From Eq. (8.32) we form equations for each species:

$$\int \frac{mv^2}{2} \left[\frac{\partial f_j}{\partial t} + \mathbf{v} \cdot \nabla f_j + \frac{q_j}{m} (\mathbf{E} + \mathbf{v} \times \mathbf{B}) \cdot \nabla_v f_j \right] d\mathbf{v} = 0 \tag{8.52}$$

Integration techniques similar to those used to obtain Eqs. (8.40) and (8.43), yield[2] a *conservation of energy* equation:

$$n_j \left[\frac{\partial}{\partial t} + \mathbf{v} \cdot \nabla \right] U_j + (\bar{\bar{\mathbf{P}}}_j \cdot \nabla) \cdot \mathbf{v} + \nabla \cdot \mathbf{Q}_j = 0 \qquad (8.53)$$

where for each species U_j is the (random) thermal energy defined by Eq. (8.36), $\mathbf{Q}_j$ is the thermal flux vector defined by Eq. (8.37), and we recall that $\tilde{\mathbf{v}}^2 \equiv \tilde{\mathbf{v}} \cdot \tilde{\mathbf{v}}$.

For an isotropic plasma (no directional preference) with a symmetric distribution function $[f(\mathbf{v}) = f(-\mathbf{v})]$ the pressure dyad reduces to a scalar pressure times the unit dyad $P_j\mathbf{1}$, and the thermal energy flux $\mathbf{Q}_j$ is zero. The simplified *adiabatic* ($\mathbf{Q} = 0$) equation (8.53) is then

$$n_j \frac{DU_j}{Dt} + P_j \nabla \cdot \mathbf{v} = 0 \qquad (8.54)$$

where we use the substantial derivative $D/Dt = \partial/\partial t + \mathbf{v} \cdot \nabla$, and where we recall that there is a separate $\mathbf{v}_j$ for each species, but that we have suppressed the j for simplicity. Writing the continuity equation (8.40) in terms of the substantial derivative as

$$\nabla \cdot \mathbf{v} = -\frac{1}{n_j} \frac{Dn_j}{Dt}$$

the adiabatic energy equation (8.54) can be written as

$$n_j \frac{DU_j}{Dt} - \frac{P_j}{n_j} \frac{Dn_j}{Dt} = 0 \qquad (8.55)$$

As we shall see, this leads to a very simple relation between P and n, as needed to complete the Maxwell–Euler equation set.

For a fluid with three degrees of translational freedom, elementary kinetic theory tells us that the thermal energy U can be expressed in terms of a temperature T, for each species, by

$$U_j = \frac{1}{2} m_j \overline{\tilde{\mathbf{v}}^2} \equiv \frac{3}{2} \kappa T_j \qquad (8.56)$$

e.g., an energy of $\frac{1}{2}\kappa T$ per degree of freedom. From our definition of the pressure dyadic [Eq. (8.35)], for the symmetric and isotropic case,

$$\bar{\bar{\mathbf{P}}} = nm\overline{\tilde{\mathbf{v}}\tilde{\mathbf{v}}} = nm \left(\overline{\tilde{v}_x^2}\mathbf{x}_0\mathbf{x}_0 + \overline{\tilde{v}_y^2}\mathbf{y}_0\mathbf{y}_0 + \overline{\tilde{v}_z^2}\mathbf{z}_0\mathbf{z}_0 \right)$$

where $\overline{\tilde{v}_x^2} = \overline{\tilde{v}_y^2} = \overline{\tilde{v}_z^2} = \frac{1}{3}\overline{\tilde{v}^2}$ so that

$$\bar{\bar{\mathbf{P}}}_j = \frac{mn_j \overline{\tilde{v}^2}\mathbf{1}}{3} = P_j\mathbf{1} \qquad (8.57)$$

Combining Eqs. (8.56) and (8.57), we obtain the perfect gas relation for partial pressures,

$$P_j = n_j \kappa T_j \tag{8.58}$$

From Eq. (8.56) the thermal energy can now be written as

$$U_j = \frac{3}{2} \frac{P_j}{n_j} \tag{8.59}$$

From Eq. (8.55) we can write the adiabatic energy equation as

$$\frac{DU_j}{Dt} = \frac{P_j}{n_j^2} \frac{Dn_j}{Dt} = -P_j \frac{D}{Dt} \left(\frac{1}{n_j} \right)$$

so that with Eq. (8.59)

$$\frac{3}{2} \frac{D}{Dt} \left(\frac{P_j}{n_j} \right) = -P_j \frac{D}{Dt} \left(\frac{1}{n_j} \right)$$

Moving along a streamline, i.e., with D/Dt, the differential relation takes the form

$$\frac{3}{2} \left[\frac{dP_j}{n_j} + P_j d \left(\frac{1}{n_j} \right) \right] = -P_j d \left(\frac{1}{n_j} \right)$$

or

$$\frac{dP_j}{P_j} = \left(1 + \frac{2}{3} \right) \frac{dn_j}{n_j} = \gamma \frac{dn_j}{n_j} \tag{8.60a}$$

Integrating, one obtains the desired *adiabatic condition* between pressure and density for processes (wave motions, etc.) involving no heat transfer:

$$\frac{P_j}{P_{0j}} = \left(\frac{n_j}{n_{0j}} \right)^{\gamma} \tag{8.60b}$$

where P_{0j} and n_{0j} are background values, and $\gamma = 1 + 2/N$ is the thermodynamic *ratio of specific heats* for a system with N degrees of freedom. In this case $\gamma = \frac{5}{3}$, for three degrees of translational motion. For a diatomic molecule, one would have two additional degrees of rotational freedom, and one degree of vibrational freedom. This then completes the Maxwell–Euler equation set for a fluid mechanical description of an isotropic, collisionless plasma with a symmetric velocity distribution function.

Using Eqs. (8.60a, b), the pressure P can be eliminated from Eq. (8.43) so that there are an equal number of equations and unknowns. The coupled set of equations then permits a mathematical or computational description of fluid level plasma phenomena (particle transport, wave motion, radiation, etc.) in which the particle densities and currents determine the electromagnetic fields, and these fields *self-consistently* determine the particle densities and velocities. Descriptions of various plasma phenomena are presented in References 1–12.

The fluid model has an appropriate level of detail to describe some of the basic properties of wave propagation in plasmas. For instance, the propagation of both transverse and longitudinal waves is easily described, dispersion relations obtained, and some dominant wave–wave couplings identified. Wave mixing occurs as a result of the non-linear terms in the Maxwell–Euler equations: $n_j \mathbf{v}$, $\mathbf{v} \cdot \nabla \mathbf{v}$, and $\mathbf{v} \times \mathbf{B}$. For instance, if waves with frequencies ω_1 and ω_2 propagate in the plasma, then the density and velocity will have terms that vary in space and time as

$$e^{-i(\omega_1 t - \mathbf{k}_1 \cdot \mathbf{r})} \quad \text{and} \quad e^{-i(\omega_2 t - \mathbf{k}_2 \cdot \mathbf{r})}.$$

Multiplying these terms together results in a cross term with beat frequencies $\omega_3 = \omega_1 \pm \omega_2$. Thus, these terms can lead to scattering processes where two waves create a third, or the inverse process where a wave decays into two other waves. Examples of this are stimulated Raman scattering (SRS), stimulated Brillouin scattering (SBS), and the $2\omega_{pe}$ instability. These will be discussed later.

An example of something missing from the fluid treatment of a plasma is the collisionless Landau damping discussed earlier. Since the magnitude and sign of this damping process depend entirely upon the shape of the velocity distribution near the electron-acoustic wave phase velocity, it cannot be described by the fluid equations, in which all information about the velocity distribution has been integrated out. In general, any process that involves velocity specific wave–particle interactions cannot be described in the fluid description. However, in some cases the averaged consequences of these processes can be included, for instance by adding a collision term or damping rate to the fluid equations.

8.4.4   Plasma Expansion

An important characteristic of hot dense plasmas is the fact that they expand into vacuum with a speed determined by the temperature of the electrons (usually hotter than the ions) and the mass of the ions. The rate at which this occurs determines how fast energy must be supplied to the plasma if it is to reach a high temperature. The expansion rate of a plasma may be described in terms of the one-dimensional isothermal expansion of a hot fluid with two species: electrons and ions. This can be seen by examination of the conservation of mass and momentum equations (8.40) and (8.43) for both electrons and ions. Because the resultant expansion velocity is small (with a high ion mass), the electron momentum equation is dominated by the non-velocity terms, so that for a one-dimensional plasma of electron density n_e and electron pressure P_e

$$n_e e E = -\frac{\partial}{\partial x} P_e. \tag{8.61}$$

The one-dimensional continuity (8.40) and momentum (8.43) equations describing ions of density n_i, partial pressure P_i, charge $+Ze$ and mass M are

$$\frac{\partial n_i}{\partial t} + \frac{\partial}{\partial x}(n_i \mathbf{v}) = 0 \tag{8.62}$$

$$M n_i \left[\frac{\partial}{\partial t} + \mathbf{v}\frac{\partial}{\partial x} \right] \mathbf{v} = n_i Z e E - \frac{\partial}{\partial x} P_i. \tag{8.63}$$

The attraction of the electrons to the ions maintains an overall neutrality in the plasma, so that

$$n_e = Zn_i \tag{8.64}$$

The pressure terms in the electron and ion momentum equations (8.61) and (8.63) can be replaced with expressions involving the respective densities through use of the adiabatic energy condition, Eq. (8.60a), in the form

$$dP = \frac{\gamma P}{n} dn$$

Writing this separately for the electron and ion partial pressures, and using the perfect gas relation, Eq. (8.58), for both, the pressure gradient terms in Eqs. (8.61) and (8.63) become

$$\frac{\partial P_e}{\partial x} = \gamma \kappa T_e \frac{\partial n_e}{\partial x} \tag{8.65}$$

and

$$\frac{\partial P_i}{\partial x} = \gamma \kappa T_i \frac{\partial n_i}{\partial x} \tag{8.66}$$

The momentum equation for ions then becomes

$$Mn_i \left[\frac{\partial}{\partial t} + v \frac{\partial}{\partial x} \right] v = n_i Z \, eE - \gamma \kappa T_i \frac{\partial n_i}{\partial x} \tag{8.67}$$

Substituting for the electric field E from Eq. (8.61), which couples electron and ion motion, i.e.,

$$E = -\frac{1}{en_e} \frac{\partial P_e}{\partial x} = -\frac{\gamma \kappa T_e}{en_e} \frac{\partial n_e}{\partial x}$$

the ion momentum equation (8.67) becomes

$$Mn_i \left[\frac{\partial}{\partial t} + v \frac{\partial}{\partial x} \right] v = -(Z\gamma \kappa T_e + \gamma \kappa T_i) \frac{\partial n_i}{\partial x}$$

or for $T_e \gg T_i$

$$\left[\frac{\partial}{\partial t} + v \frac{\partial}{\partial x} \right] v = -v_{\exp}^2 \frac{1}{n_i} \frac{\partial n_i}{\partial x} \tag{8.68}$$

where we define an electron–ion thermal expansion velocity

$$v_{\exp} = \left(\frac{Z\gamma \kappa T_e}{M} \right)^{1/2} \tag{8.69a}$$

driven by the electron pressure (through $n_e \kappa T_e$), but limited by the inertia of the ions through their mass M. In practical units the expansion velocity can be expressed as[¶]

$$v_{\exp} = 0.28 \left(\frac{Z\kappa T_e}{M} \right)^{1/2} \, \mu\text{m/ps} \tag{8.69b}$$

where the ion charge Z is in units of ten, κT is in keV, the ion mass M is expressed in units of 20 times that of a proton, and γ is taken as $5/3$.

[¶] For comparison, the speed of light in vacuum, in these units, is $c \simeq 300 \, \mu\text{m/ps}$.

The ion continuity equation (8.62) can be written as

$$\left[\frac{\partial}{\partial t} + v\frac{\partial}{\partial x}\right] n_i + n_i\frac{\partial v}{\partial x} = 0 \tag{8.70}$$

As can be seen by substitution, a solution to the fluid equations (8.68) and (8.70) is[1]

$$v = v_{exp} + \frac{x}{t} \tag{8.71}$$

and

$$n_i = n_{i0}e^{-x/v_{exp}t} \tag{8.72}$$

Thus the plasma expands from an initial ion density n_{i0}, at a surface $x = 0$, with the electron–ion thermal velocity, v_{exp}. Examining the density function $n_i\,(x, t)$ we see that the density gradient length l_{exp}, due to the expansion, is given by

$$l_{exp} \equiv -n_i/(\partial n_i/\partial x) = v_{exp}t \tag{8.73}$$

and increases with time at a rate set by the expansion velocity v_{exp}. According to Eq. (8.69), a 1 keV plasma of Ne-like titanium ions with an average charge state of $Z = +12$ will expand at a velocity of approximately 0.20 μm/ps.

8.4.5   Electron-Acoustic Waves

The propagation of high frequency longitudinal waves in a plasma, known widely as electron-acoustic waves, and also known as electron-plasma waves or as Langmuir oscillations, is readily described on the basis of the Maxwell–Euler fluid equations (8.40) and (8.43)–(8.49). For high frequency waves the more massive ions are relatively immobile and simply provide a uniform, electrically neutralizing charge distribution. For longitudinal waves in which the field quantities n, v, P, E, etc., vary only in the wave propagation (**k**) direction, one need consider only the equations of continuity (8.40), momentum conservation (8.43), and Gauss's law (8.46), which for electrons are written as

$$\frac{\partial n_e}{\partial t} + \nabla \cdot (n_e\mathbf{v}) = 0$$

$$m\left(\frac{\partial}{\partial t} + \mathbf{v} \cdot \nabla\right)\mathbf{v} = -\frac{1}{n_e}\nabla P_e - e(\mathbf{E} + \mathbf{v} \times \mathbf{B})$$

$$\nabla \cdot \mathbf{E} = -en_e/\epsilon_0$$

where the uniform ion distribution does not contribute to the last equation.

The non-linear terms, involving products like $n_e\,\mathbf{v}$, $\mathbf{v} \cdot \nabla\mathbf{v}$, and $\mathbf{v} \times \mathbf{B}$, can be simplified through a *linearization* process in which one assumes that each field can be written as the sum of a background value and a small fluctuation therefrom. Thus we write

$$n_e = n_0 + \tilde{n}_e \tag{8.74a}$$

$$\mathbf{v} = \mathbf{v}_0 + \tilde{\mathbf{v}} \tag{8.74b}$$

$$\mathbf{E} = \mathbf{E}_0 + \tilde{\mathbf{E}} \tag{8.74c}$$

and so on. We then assume that the waves of interest are of small amplitude, such that $\tilde{n}_e/n_0 \ll 1$, etc. In the case of the velocity modulation we assume that there is no directed average motion, so that $\mathbf{v}_0 = 0$. Substituting Eqs. (8.74a–c) into the fluid equations and dropping all product of fluctuation terms as being of second order (very small), we have (dropping the tildes for fluctuating quantities)

$$\frac{\partial n_e}{\partial t} + n_0 \nabla \cdot \mathbf{v} + \underbrace{\mathbf{v}_0 \cdot \nabla n_e}_{=0} = 0 \tag{8.75}$$

$$m\frac{\partial \mathbf{v}}{\partial t} = -\frac{1}{n_0}\gamma \kappa T_e \nabla n_e - e\mathbf{E} \tag{8.76}$$

$$\nabla \cdot \mathbf{E} = -\frac{en_e}{\epsilon_0} \tag{8.77}$$

where the pressure gradient was handled as in Eq. (8.65). Taking $\partial/\partial t$ of Eq. (8.75),

$$\frac{\partial^2 n_e}{\partial t^2} + n_0 \frac{\partial}{\partial t}(\nabla \cdot \mathbf{v}) = 0$$

and the divergence $(\nabla \cdot)$ of Eq. (8.76),

$$m\frac{\partial}{\partial t}(\nabla \cdot \mathbf{v}) = -\frac{\gamma \kappa T_e}{n_0}\nabla^2 n_e - e\nabla \cdot \mathbf{E}$$

these can be combined to form a wave equation

$$\frac{\partial^2 n_e}{\partial t^2} - \frac{\gamma \kappa T_e}{m}\nabla^2 n_e - \frac{en_0}{m}\nabla \cdot \mathbf{E} = 0$$

Using Eqs. (8.46) and (8.48), this can be written as

$$\frac{\partial^2 n_e}{\partial t^2} + \frac{e^2 n_0}{\epsilon_0 m}n_e - \frac{\gamma \kappa T_e}{m}\nabla^2 n_e = 0$$

or

$$\left[\frac{\partial^2}{\partial t^2} + \omega_p^2 - a_e^2 \nabla^2\right]n_e(\mathbf{r}, t) = 0 \tag{8.78}$$

which we recognize as a longitudinal wave equation for electron density fluctuations, with *electron sound speed* a_e given by

$$a_e = \left(\frac{\gamma \kappa T_e}{m}\right)^{1/2} \tag{8.79}$$

where $\gamma = 1 + 2/N$, as described below Eq. (8.60b), and where, as we will understand shortly, the natural frequency of oscillation ω_p, known as the *plasma frequency*, is given by

$$\omega_p = \left(\frac{e^2 n_0}{\epsilon_0 m} \right)^{1/2} \tag{8.80}$$

for background electron density n_0. For the 1 keV electron temperature plasma considered earlier, the electron sound speed is $a_e \simeq 17$ μm/ps.

To better appreciate the plasma frequency, we consider an electron density wave of the form

$$n_e(\mathbf{r}, t) = n_e e^{-i(\omega t - \mathbf{k} \cdot \mathbf{r})} \tag{8.81}$$

where the wave is of frequency ω and wave vector $\mathbf{k}$, with scalar wavenumber $k = 2\pi/\lambda$. For a wave of this form the time and space differentials are replaced by (see Chapter 2, Section 2.2)

$$\frac{\partial}{\partial t} \to -i\omega \tag{8.82a}$$

and

$$\nabla \to -ik \tag{8.82b}$$

so that the wave equation (8.78) takes the form

$$\left[\omega^2 - \omega_p^2 - k^2 a_e^2 \right] n_e = 0 \tag{8.83}$$

where the exponential with time and space dependence is suppressed. Following the same procedures as in Chapter 2 for electromagnetic waves, we observe that Eq. (8.83) has solutions for finite n when the bracketed quantity is zero. This then is a natural oscillation or wave of the system, requiring, in principle, no driving term. Setting the bracketed quantity equal to zero yields the *dispersion relation for the electron-acoustic wave*,

$$\omega^2 = \omega_p^2 + k^2 a_e^2 \tag{8.84}$$

This tells us that for long period plasma waves, where k goes to zero, there is natural oscillation at the electron plasma frequency, $\omega \cong \omega_p$. For waves of finite k, in the range of $0 \le k \le \omega_p/a_e$, the frequency increases somewhat, to a value of $\sqrt{2}\omega_p$ at $k = \omega_p/a_e$, as shown in the dispersion diagram of Figure 8.7.

Here the frequency as a function of wavenumber is shown for naturally occurring waves in a plasma. The parameter $\omega_p/a_e = 1/\sqrt{\gamma}\lambda_D$ is approximately equal to one over the Debye screening distance, discussed earlier in Section 8.3, Eq. (8.6). For waves characterized by $k < k_D \equiv 1/\lambda_D$, the wavelength λ is greater than λ_D and the discreteness of individual charges within the plasma is not "seen" by the wave – they are screened.

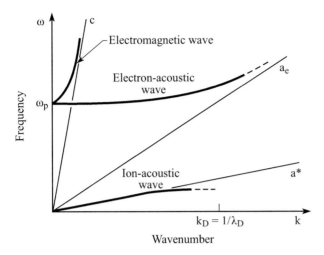

Figure 8.7 The dispersion diagram for naturally occurring waves in an isotropic plasma. Shown are the transverse electromagnetic wave with cutoff at the plasma frequency ω_p, a high frequency longitudinal wave called the electron-acoustic wave, and a low frequency longitudinal wave called the ion-acoustic wave.

In this region ($k < 1/\lambda_D$), the fluid model is quite accurate and the wave propagates as indicated. However, for $k > 1/\lambda_D$, the wavelength of the plasma wave is less than the Debye screening distance and the discreteness or individuality of charges should be apparent, i.e., *not* screened. As a consequence we can suspect that the fluid theory, which averages out (ignores) individual charge effects, might be inadequate.

Indeed, if one considers the propagation of the electron-acoustic wave based on a kinetic theory, as discussed in Section 8.4.2, one obtains the same basic dispersion relation [Eq. (6.84)], but the frequency is found to be complex, with an imaginary component ω_i corresponding to wave decay[1, 3, 4, 12] where for $\omega = \omega_r + i\,\omega_i$

$$\omega_i = \frac{\pi}{2}\frac{\omega_p^2 \omega_r}{k^2}\frac{\partial f}{\partial v}\bigg|_{v=\omega/k} \tag{8.85}$$

so that a negative slope for $\partial f / \partial v$ corresponds to damping. The expression for a three-dimensional Maxwellian distribution $f(v)$ is given by

$$f(v) = \frac{1}{(2\pi)^{3/2} v_e^3} e^{-v^2/2v_e^2} \tag{8.86a}$$

However, in Eq. (8.85), $f(v)$ may be considered to be a one-dimensional Maxwellian electron velocity distribution as follows:

$$f(v) = \frac{1}{\sqrt{2\pi}\,v_e} e^{-v^2/2v_e^2} \tag{8.86b}$$

where the electron thermal velocity is

$$v_e = (\kappa T_e/m)^{1/2} \tag{8.86c}$$

the damping term becomes

$$\frac{\omega_i}{\omega_r} = -\sqrt{\frac{\pi}{8}} \frac{\omega_p^2 \omega_r}{k^3 v_e^3} e^{-\frac{\omega_r^2}{2k^2 v_e^2}} \tag{8.87}$$

where from Eq. (8.84) $\omega_r^2 \simeq \omega_p^2(1 + k^2/k_D^2)$, and where $k_D = \omega_p/v_e$. For $k \ll k_D$ the exponential factor in Eq. (8.87) dominates, so that ω_i/ω_r goes to zero and damping is negligible. However, for larger k, near k_D, damping is very strong, with

$$\frac{\omega_i}{\omega_r} \simeq -\sqrt{\frac{\pi}{4}} \left(\frac{k_D}{k}\right)^3 e^{-(k_D/k)^2}$$

so that for $k = k_D$ the wave decays to a $1/e$ field amplitude in just a few oscillations. Decay of electron-acoustic waves is known as Landau damping,[18] § and is due to particle–wave interactions, particularly for electrons traveling in the wave direction with velocities approximately equal to the wave's phase velocity.

As was suggested earlier in Figure 8.3, the electron-acoustic wave consists of regions of high charge density that propagate at high phase velocity v_ϕ. From Eq. (8.84) we can now conclude, with a little algebraic manipulation, that

$$v_\phi = \frac{\omega}{k} = a_e\sqrt{1 + \frac{k_D^2}{k^2}} \tag{8.88}$$

so that in general $v_\phi > a_e$ for propagating waves ($k < k_D$). In this region ($v > a_e$) the velocity distribution (8.86a–c) falls very rapidly with increasing velocity (negative slope), so that in general there are more slow electrons ($v \leq v_\phi$) than fast electrons ($v \geq v_\phi$) interacting with the wave. Because the velocities of these *resonant* electrons are close to that of the wave, there is a relatively long interaction time in which energy can be exchanged between the electrons and the wave. Traveling with the wave, the somewhat faster electrons tend to "push" the potential crest of the wave, giving up energy as they merge with it. Somewhat slower electrons are dragged along, electrostatically, taking energy from the wave, As shown in Figure 8.8, the thermal velocity distribution has a negative slope for $v > a_e$, so that there are more slow electrons taking energy away from the wave than fast electrons contributing energy. Thus on balance, there is a net loss of energy for a plasma wave of high phase velocity. An interesting counter example occurs when an electron beam is injected into the plasma with a beam velocity ($v_b > a_e$), as was illustrated in Figure 8.6. In this case the electron velocity distribution has a positive slope (more fast electrons than slow electrons) near v_b, leading to wave growth. This beam-plasma instability is sometimes described as "inverse Landau damping."

8.4.6   Ion-Acoustic Waves

The fluid level plasma equations (8.40) and (8.43)–(8.49) also have a low frequency solution, a natural mode of oscillation at a frequency $\omega \ll \omega_p$. For this longitudinal wave the frequency is sufficiently low that the ions play a major role and, as we will

§ Named for the Russian scientist L.D. Landau.

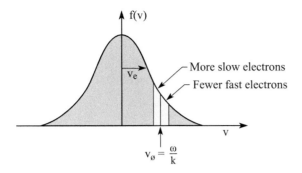

Figure 8.8 The electron velocity distribution [Eqs. (8.86a–c)] in a thermal plasma has more slow electrons than fast electrons in the vicinity of v_ϕ, the phase velocity electron-acoustic waves. This is leads to wave damping, especially for short period (high k) waves for which $v_\phi \geq v_e$. In this region there are many resonant electrons and the gradient in velocity is sharp, leading to strong damping of the wave. This process, called *Landau damping*, is a collisionless wave–particle interaction in which energy is transferred from the wave to the individual electrons.

see, the electrons also participate. At these very low frequencies the electron and ion charge densities are very closely coupled, a condition called *quasi-neutrality*. We again seek a linearized version of the fluid equations (8.40) and (8.43), through a perturbation analysis $n_i = n_{0i} + \tilde{n}_i$, $n_e = n_{0e} + \tilde{n}_e$, $\mathbf{v}_i = \mathbf{v}_{0i} + \tilde{\mathbf{v}}_i$, and $\mathbf{v}_e = \mathbf{v}_{0e} + \tilde{\mathbf{v}}_e$. With no average motion $\mathbf{v}_{0e} = \mathbf{v}_{0i} = 0$. For ions of charge Z and mass M, the condition of quasi-neutrality permits us to make the approximations $n_{0i} = n_{0e}/Z$, $\tilde{n}_i \simeq \tilde{n}_e/Z$, and $\tilde{\mathbf{v}}_e \simeq \tilde{\mathbf{v}}_i$. Furthermore, for small amplitude waves the process is adiabatic (no heat transfer), so that the pressure gradients can be replaced by density gradients [as seen previously in Eqs. (8.65) and (8.66)], viz.,

$$\nabla \tilde{P}_i = \gamma \kappa T_i \, \nabla \tilde{n}_i \tag{8.89}$$

and

$$\nabla \tilde{P}_e = \gamma \kappa T_e \, \nabla \tilde{n}_e \tag{8.90}$$

where the values of γ will depend on the nature of the wave and thus could be different for ions and electrons.

With these approximations the linearized fluid equations for the ions can be written as

$$\frac{\partial n_i}{\partial t} + \frac{n_0}{Z} \nabla \cdot \mathbf{v}_i = 0 \tag{8.91}$$

$$\frac{M n_0}{Z} \frac{\partial \mathbf{v}_i}{\partial t} = -\gamma \kappa T_i \, \nabla n_i + e n_0 \mathbf{E} \tag{8.92}$$

and the electron momentum equation, with $\mathbf{v}_e \simeq \mathbf{v}_i$ and $n_e \simeq Z n_i$, is

$$m n_0 \frac{\partial \mathbf{v}_i}{\partial t} = -Z \gamma \kappa T_e \, \nabla n_i - e n_0 \mathbf{E} \tag{8.93}$$

where we have dropped the tildes for simplicity (they appear on all but background quantities), have replaced $\mathbf{v}_e$ by $\mathbf{v}_i$ as discussed above, and have simplified the background charge densities by writing $n_{0e} = n_0$ and $n_{0i} = n_0/Z$. With these low frequency approximations the electron continuity equation offers no additional information. Adding the two momentum equations (8.91) and (8.92) to eliminate $\mathbf{E}$, and noting that $m \ll M/Z$, we have

$$\frac{Mn_0}{Z}\frac{\partial \mathbf{v}_i}{\partial t} = -(\gamma \kappa T_i + Z\gamma \kappa T_e)\nabla n_i \tag{8.94}$$

A wave equation can now be formed by taking $\partial / \partial t$ (8.91) and $\nabla \cdot$(8.94) to obtain

$$\frac{\partial^2 n_i}{\partial t^2} + \frac{n_0}{Z}\frac{\partial}{\partial t}(\nabla \cdot \mathbf{v}_i) = 0$$

$$\frac{Mn_0}{Z}\frac{\partial}{\partial t}(\nabla \cdot \mathbf{v}_i) = -(\gamma \kappa T_i + Z\gamma \kappa T_e)\nabla^2 n_i$$

which can be combined to form the *ion-acoustic wave equation*

$$\frac{\partial^2 n_i}{\partial t^2} - \left(\frac{\gamma \kappa T_i + Z\gamma \kappa T_e}{M}\right)\nabla^2 n_i = 0 \tag{8.95}$$

For the common case $T_e \gg T_i$, generally resulting from the fact that energy is delivered to the electrons and only indirectly transferred to the ions through collisions, the *ion-acoustic wave equation* can be written as

$$\boxed{\frac{\partial^2 n_i}{\partial t^2} - a^{*2}\nabla^2 n_i = 0} \tag{8.96}$$

where the wave propagates with a hybrid sound speed

$$\boxed{a^* = \sqrt{\frac{Z\gamma \kappa T_e}{M}}} \tag{8.97}$$

which has the characteristics of an electron temperature and an ion mass.[4] This low frequency plasma oscillation is thus seen to be driven by electron thermal energy, but with an inertia set by the more massive ions. Due to the quasi-neutrality at this low frequency, each electron must drag an equivalent (per unit charge) mass of M/Z.

Following our usual procedures, the dispersion relation for an ion-acoustic wave with $n_i = n_{i0}\exp[-i(\omega t - \mathbf{k} \cdot \mathbf{r})]$ follows from Eq. (8.96) as

$$\boxed{\omega = ka^*} \tag{8.98}$$

as illustrated by the lower branch in Figure 8.7. This dispersion relation indicates a linear relationship between k and ω, so that in fact it lacks dispersion – all frequencies propagate at the same phase velocity, a^*. A somewhat more refined analysis shows a rollover to lower phase velocities as k approaches the Debye wavenumber. Again Landau damping can be important, but in this case it is the ions that do the damping, and the degree of damping is a function of ZT_e/T_i due to the variation in sound speed. At the phase velocity of this wave, a^*, the electron velocity distribution is nearly flat [f' (v) $\simeq$ 0], so that the number of relatively fast and slow electrons is about equal, and there is little transfer of energy. For the ions, with $ZT_e/T_i \gg 1$, a^* is much greater than the ion thermal speed ($a_i = \sqrt{\gamma\kappa T_i/M}$) and thus the phase velocity of the ion-acoustic wave is so far out on the ion velocity distribution curve that $f_i (a^*)$ goes to zero and again there is little damping, i.e., although f_i'(v) is sharp, there are few ions in this velocity region. For $Z T_e/T_i$ approaching unity, a^* approaches a_i and ion damping becomes very strong for k near the Debye wavenumber.

8.4.7   Transverse Electromagnetic Waves in a Uniform Plasma

Transverse electromagnetic waves also propagate in a plasma, much like those considered in Chapter 2, but with a cutoff appearing at the plasma frequency $\omega \cong \omega_p$, which has important consequences for energy delivery in laser-produced plasmas. The Maxwell–Euler fluid equations give a very satisfactory description of these waves. At these high frequencies the ions are immobile, so that we need consider only the electrons in Eqs. (8.44)–(8.51). For transverse electromagnetic waves of relatively weak intensity these equations simplify considerably. The current term in Eq. (8.44), which in general is non-linear, simplifies in this case to $\mathbf{J} = -en_0 \mathbf{v}$, where n_0 is the background electron density, assumed in this weak field limit to be unmodulated by the passing electromagnetic wave. In this case Maxwell's equations (8.44)–(8.49) can be written as

$$\nabla \times \mathbf{H} = \epsilon_0 \frac{\partial \mathbf{E}}{\partial t} - en_0\mathbf{v} \tag{8.99}$$

$$\nabla \times \mathbf{E} = -\mu_0 \frac{\partial \mathbf{H}}{\partial t} \tag{8.100}$$

In Eqs. (8.50) and (8.51), which describe particle motion, all non-linear terms can be neglected in the weak field limit, so that with only electrons mobile, these become

$$\frac{\partial n_e}{\partial t} + n_0 \nabla \cdot \mathbf{v} = 0 \tag{8.101}$$

and

$$m\frac{\partial \mathbf{v}}{\partial t} = -\frac{\gamma\kappa T_e}{n_0}\nabla n_e - e\mathbf{E} \tag{8.102}$$

For transverse electromagnetic waves, as seen earlier in Chapters 2 and 3, only the transverse component of the current density $\mathbf{J}_T$ contributes to the wave, and thus only to the transverse component of $\mathbf{v}$. This is also the case for the plasma in the weak-field limit. Thus in Eqs. (8.101) and (8.102), the $\nabla \to i\mathbf{k}$ terms are longitudinal and do not contribute to the transverse motion. The remaining terms in Eq. (8.102) yield a simplified version of Newton's law, $\mathbf{F} = m\mathbf{a}$, for the electrons:

$$m\frac{\partial \mathbf{v}}{\partial t} = -e\mathbf{E} \tag{8.103}$$

We can now develop a wave equation by differentiating Eq. (8.99) with respect to time,

$$\nabla \times \frac{\partial \mathbf{H}}{\partial t} = \epsilon_0 \frac{\partial^2 \mathbf{E}}{\partial t^2} - e n_0 \frac{\partial \mathbf{v}}{\partial t} \tag{8.104}$$

and taking the curl of Eq. (8.100),

$$\nabla \times (\nabla \times \mathbf{E}) = -\mu_0 \nabla \times \frac{\partial \mathbf{H}}{\partial t} \tag{8.105}$$

We combine Eqs. (8.104) and (8.105) by eliminating $\nabla \times \partial \mathbf{H}/\partial t$, and use the vector relation $\nabla \times (\nabla \times \mathbf{E}) = \nabla(\nabla \cdot \mathbf{E}) - \nabla^2 \mathbf{E}$ (see Appendix D) to obtain

$$\nabla(\nabla \cdot \mathbf{E}) - \nabla^2 \mathbf{E} = -\mu_0 \epsilon_0 \frac{\partial^2 \mathbf{E}}{\partial t^2} + \mu_0 e n_0 \frac{\partial \mathbf{v}}{\partial t} \tag{8.106}$$

For transverse waves $\nabla \cdot \mathbf{E} = 0$ (recall from Chapter 2 that $\nabla \to i\mathbf{k}$). Furthermore, we can replace $\partial \mathbf{v}/\partial t$ with an expression involving $\mathbf{E}$ by use of Eq. (8.103), so that

$$\frac{\partial^2 \mathbf{E}}{\partial t^2} + \frac{e n_0}{\epsilon_0} \left(\frac{e \mathbf{E}}{m} \right) - \frac{1}{\epsilon_0 \mu_0} \nabla^2 \mathbf{E} = 0$$

Recognizing $c^2 = 1/\epsilon_0 \mu_0$ and $\omega_p^2 = e^2 n_0/\epsilon_0 m$ [Eq. (8.80)], we have the *wave equation for a transverse wave in a plasma*,

$$\boxed{\left(\frac{\partial^2}{\partial t^2} + \omega_p^2 - c^2 \nabla^2 \right) \mathbf{E}(\mathbf{r}, t) = 0} \tag{8.107}$$

For a plane wave of the form $\mathbf{E}(\mathbf{r}, t) = E_0\, e^{-i\,(\omega t\, -\mathbf{k}\cdot\mathbf{r})}$ Eq. (8.107) yields a dispersion relation

$$\boxed{\omega^2 = \omega_p^2 + k^2 c^2} \tag{8.108}$$

where we have essentially taken the indicated derivatives, or equivalently used the identification of Eqs. (8.82a) and (8.82b). The dispersion relation for waves propagating in plasma differs from that in vacuum by the appearance of the ω_p^2 term. In vacuum this term is zero, giving $\omega^2 = k^2 c^2$, or equivalently $f \lambda = c$, whereby waves of all frequencies propagate with the same phase velocity, c. According to Eq. (8.108) there is a cutoff frequency in the plasma at $\omega = \omega_p$. For $\omega < \omega_p$, the solution for k is imaginary, indicating that the wave cannot propagate in this *overdense* plasma. Rather the wave decays exponentially with wavenumber

$$k = \frac{\sqrt{\omega^2 - \omega_p^2}}{c} \tag{8.109}$$

or in the highly overdense limit $\omega^2 \ll \omega_p^2$

$$k = i \frac{\omega_p}{c} \tag{8.110}$$

which corresponds to a penetration depth l into the highly overdense plasma of

$$l = c/\omega_p \qquad (8.111)$$

The frequency for which $\omega = \omega_p$ is referred to as the *critical frequency*, and the corresponding electron density is defined as the *critical electron density*, n_c, where from Eq. (8.5)

$$n_c \equiv \frac{\epsilon_0 m \omega^2}{e^2} \qquad (8.112a)$$

or in terms of the wavelength (numerically in units of microns)

$$n_c = \frac{1.11 \times 10^{21} \text{ e/cm}^3}{\lambda^2 [\mu\text{m}]} \qquad (8.112b)$$

Thus for a Nd laser of wavelength 1.06 μm, the critical density is $n_c \simeq 1.00 \times 10^{21}$ e/cm^3. For frequency quadrupled light at 0.266 μm wavelength (ultraviolet) the critical electron density quadruples to 1.60×10^{22} e/cm^3, and for a CO_2 laser at 10.6 μm wavelength the critical density is 1.00×10^{19} e/cm^3.

Referring back to Figure 8.5, we can now better appreciate the role of the critical density region. There we see the laser light incident from the right on a plasma of sharply rising electron density. During passage through the *underdense* region of the plasma ($n < n_c$) the wave experiences classical absorption as it causes electrons to oscillate, some of which then lose their energy through electron–ion collisions, thermally heating the plasma in the process. This transfer of energy to the plasma increases in efficiency as the light wave propagates to higher densities. Eventually the laser light of frequency ω reaches the critical density n_c, beyond which it cannot propagate ($\omega_p > \omega$, k imaginary), and the wave is reflected back toward the vacuum and lost.

The overdense plasma is important in several other well-known situations. Astronauts and cosmonauts regularly experience a communication blackout during reentry into the earth's atmosphere as their capsules are engulfed in an overdense plasma created as it heat and ionizes atmospheric molecules, preventing the transmission or reception of shortwave or microwave signals. Common AM broadcasts, at frequencies around 1 MHz, are reflected from the earth's ionosphere[17] (10^5 e/cm^3 to 10^6 e/cm^3 at heights of 100 km to 400 km), often permitting distant reception at night when absorption is minimal. There are interesting daily and seasonal variations to this phenomenon, affected by cycles of ionizing radiation from the sun, longer charged particle lifetimes at higher altitudes, and increased collisional absorption during the day as the sun's energy warms the atmosphere below, causing it to expand outward into the lower regions of the ionosphere where 1 MHz radiation is typically reflected (or absorbed). Broadcast emissions in the 100 MHz region ($n_c \sim 10^8$ e/cm^3), typically used for FM, are of sufficiently high frequency that they propagate through the ionosphere and into outer space, which

Table 8.1 Electron density, plasma frequency, critical photon energy for $\omega_c = \omega_p$, and critical wavelength for electromagnetic radiation.

$n_e(\text{e/cm}^3)$	$\omega_p/2\pi$	$\hbar\omega_c(\text{eV})$	λ_c	Comments
1.00×10^6	8.94 MHz		33.5 m	Between AM and FM radio
1.00×10^{14}	89.4 GHz		3.35 mm	Microwaves
1.00×10^{19}			10.6 μm	CO_2 laser
1.00×10^{21}		1.17	1.06 μm	Nd laser
1.60×10^{22}		4.86	266 nm	4ω of Nd laser
4.60×10^{24}		80.0	15.5 nm	Ne-like Y laser

explains why the reflection phenomena experienced with AM radio do not occur for FM broadcasts. Critical parameters are given for representative values of n_e in Table 8.1.

For waves of frequency $\omega > \omega_p$ there is a real propagating wave in the plasma, with properties much like those considered in Chapters 2 and 3, except that now there is considerable dispersion (the phase velocity is not constant) and the refractive index, or dielectric constant, is different. The wave's dispersion relation [Eq. (8.108)] is shown in Figure 8.7 with phase velocity (Chapter 3, Section 3.2) $v_\phi = \omega/k$ approaching c, the phase velocity of light in vacuum, for $\omega \gg \omega_p$. From Eq. (8.108) the phase velocity of the wave is

$$v_\phi = \frac{\omega}{k} = \frac{c}{\sqrt{1 - \omega_p^2/\omega^2}} = \frac{c}{\sqrt{1 - n_e/n_c}} \tag{8.113a}$$

while the group velocity $\partial\omega/\partial k$, is given by

$$v_g = \frac{\partial\omega}{\partial k} = c\sqrt{1 - \frac{\omega_p^2}{\omega^2}} = c\sqrt{1 - \frac{n_e}{n_c}} \tag{8.113b}$$

We see that these velocities are not constant, but vary with ω_p/ω, or n_e/n_c. For low electron densities both the phase and group velocities approach c. However, for n_e/n_c approaching unity the phase velocity can be very large and the group velocity, with which we associate the transport of information, can be very small.[19-21] From Eq. (8.113a) we can see that the refractive index (see Chapter 3) of the plasma, $n \simeq c/v_\phi = ck/\omega$, is given by

$$n = \sqrt{1 - \frac{\omega_p^2}{\omega^2}} \tag{8.114a}$$

or equivalently

$$n = \sqrt{1 - \frac{n_e}{n_c}} \tag{8.114b}$$

This analysis is readily extended to include the effect of collisions between electrons, oscillating due to the transverse wave, and ions. By including a collision term,[4, 6, 16] the electron momentum transfer equation (8.103) becomes

$$m\frac{\partial \mathbf{v}}{\partial t} = -e\mathbf{E} - m v_{ei}\mathbf{v} \tag{8.115}$$

where the momentum transfer is proportional to the electron momentum $m\mathbf{v}$ and where v_{ei} is the electron–ion collision frequency. The electron velocity can now be written as

$$\mathbf{v} = -\frac{ie}{m(\omega + i\, v_{ei})}\mathbf{E} \tag{8.116}$$

and the dispersion relation (8.108) is modified, for $v_{ei} \ll \omega$, to

$$\omega^2 = \omega_p^2 \left(1 - i\frac{v_{ei}}{\omega}\right) + k^2 c^2 \tag{8.117}$$

If we set $\omega = \omega_r + i\omega_i$, substitute into Eq. (8.117), and solve separately for the real and imaginary parts, we find, for $v_{ei}/\omega \ll 1$, that the real part of the frequency satisfies

$$\omega_r^2 = \omega_p^2 + k^2 c^2 \tag{8.118a}$$

much as before, but we now have an imaginary component

$$\boxed{\omega_i \simeq -\frac{v_{ei}\omega_p^2}{2\omega^2} = -\frac{n_e}{2n_c}v_{ei}} \tag{8.118b}$$

where the negative sign indicates damping. Physically, as the electromagnetic wave propagates through the plasma, its electric field induces an oscillatory component to the velocity of all electrons, superposed on their otherwise random motion. As the electrons experience collision with ions, their energy of oscillation is converted to random energy, thus heating the electrons to a higher temperature. Thus there is a transfer of energy from the wave to the plasma, increasing the thermal energy of the plasma and decreasing the intensity of the wave. This linear damping mechanism, referred to as collisional damping or inverse bremsstrahlung, is very important for the creation and heating of laser-produced plasmas.[1, 23]

In this connection we note that for real ω, the dispersion relation (8.117) yields real and imaginary components of the magnitude of the propagation vector, $k = k_r + ik_i$, given by

$$k_r = \sqrt{\frac{\omega^2 - \omega_p^2}{c}} \tag{8.119a}$$

and

$$k_i = \frac{v_{ei}\omega_p^2}{2c\omega\sqrt{\omega^2 - \omega_p^2}} = \frac{v_{ei}\omega_p^2}{2v_g\omega^2} \tag{8.119b}$$

The attenuation length for $1/e$ intensity decay is a distance

$$l_{abs} = \frac{1}{2k_i} = \frac{\omega^2}{\omega_p^2} \frac{v_g}{v_{ei}}$$

In terms of electron densities, the absorption length for a transverse wave in a plasma is given by[1]

$$l_{abs} = \frac{n_c}{n_e} \frac{v_g}{v_{ei}} \tag{8.119c}$$

The collision frequency v_{ei} in a hot plasma is complicated, as it depends on many long range relatively weak interactions. There are, however, many of the these interactions, and the result can lead to a very strong absorption process,[22-24] as we shall see for the case of laser-produced plasmas of even relatively short density scale lengths, e.g., tens of wavelengths. The momentum transfer in collisions can be studied by considering Figure 8.2, where the distance b is called the impact parameter. The amount of momentum transfer clearly depends on the velocity of the electron v, the ion charge Z, and b. The change in momentum $\Delta p = m \Delta v$ is equal to the force experienced multiplied by the interaction time $\mathbf{F} \Delta t$. For an interaction time $\Delta t \cong 2b/v$ and a Coulomb force $e^2 Z / 4\pi\epsilon_0 b^2$, the resultant hyperbolic trajectory has a corresponding scalar momentum change[1]

$$m \Delta v \simeq \frac{e^2 Z}{4\pi\epsilon_0 b^2} \cdot \frac{2b}{v}$$

or

$$\Delta v \simeq \frac{e^2 Z}{2\pi\epsilon_0 m v b}$$

We see that the velocity change in a collision has an inverse dependence on both v and b. The time required to undergo a substantial momentum change such that $\Delta v_{rms} \sim v$ clearly depends on the range of values of v and b, and on the ion density. The reciprocal of this collision time is the effective electron–ion collision frequency v_{ei}.

Dawson and his colleagues[23, 24] have determined the collision frequency for a Maxwellian velocity distribution, as a function of density, temperature, and ion charge. According to Johnson and Dawson,[24]

$$v_{ei} = \frac{e^4 Z n_e \ln \Lambda}{3(2\pi)^{3/2}\epsilon_0^2 m^{1/2}(\kappa T_e)^{3/2}} \tag{8.120a}$$

or

$$\frac{v_{ei}}{\omega_p} = \frac{Z\omega_p^3 \ln \Lambda}{3(2\pi)^{3/2} n_e v_e^3} \tag{8.120b}$$

where v_e is the electron thermal velocity [Eq. (8.86b)], $v_e = (\kappa T_e/m)^{1/2}$, and Λ is the ratio b_{max}/b_{min} of the impact parameters corresponding to the Debye length (b_{max}), beyond which the individual ion is effectively screened and the classical distance of closest approach (b_{min}) without capture, the latter determined by equating the energy in the Coulomb field of the ion at closest approach, $e^2 Z/4\pi \epsilon_0 b_{min}$, to the average electron thermal energy $\frac{3}{2}mv_e^2$ (i.e., $\frac{1}{2}m\bar{v}^2 = \frac{3}{2}\kappa T_e = \frac{3}{2}mv_e^2$). Observing that the Debye length [Eq. (8.6)] can be written as $\lambda_D = v_e/\omega_p$, the electron collision frequency can be written as

$$\frac{\nu_{ei}}{\omega_p} = \frac{1}{9}\sqrt{\frac{2}{\pi}}\frac{Z}{N_D}\ln\left(\frac{9N_D}{2Z}\right) \tag{8.120c}$$

where $N_D = (4\pi/3)\lambda_D^3 n_e$ is the number of electrons in a Debye sphere, given in practical units as

$$N_D = \frac{4\pi}{3}\lambda_D^3 n_e = 1.7 \times 10^3 \frac{(\kappa T_e)^{3/2}}{n_e^{1/2}} \tag{8.121}$$

for κT_e in keV and n_e in units of 10^{21} e/cm^3.

For a plasma with $\kappa T_e = 1$ keV, created by a 1.06 μm laser, at half-critical density 0.5×10^{21} e/cm^3 and with $Z = +14$ (neon-like chromium, discussed later in this chapter) one has $N_D \cong 2.4 \times 10^3$ and $\nu_{ei}/\omega_p \cong 3.4 \times 10^{-3}$. At half-critical density, $\omega_p \cong 1.3 \times 10^{15}$ rad/s, so that $\nu_{ei} \cong 4.6 \times 10^{12}$/s. In this example the absorption length [Eq. (8.119c)] at this density is $l_{abs} \cong 93$ μm. In view of the density scale length [Eq. (8.73)] for an expanding plasma, this same chromium plasma would have an expansion velocity [Eq. (8.69)] of $v_{exp} \cong 0.21$ μm/ps and thus a density scale length $l_{exp} \cong 110$ μm after 500 psec of irradiation.

As this scale length is about equal to the absorption length, we expect a fairly significant collisional absorption of 1.06 μm light for nanosecond duration or longer pulses, especially in the region approaching the critical density. This can be seen more clearly by algebraically rearranging the parameters in Eqs. (8.119c) and (8.120a) so as to better illustrate the scaling of collisional absorption with density, temperature, and ion charge state:

$$l_{abs} \propto \frac{\sqrt{1 - n_e/n_c}(\kappa T_e)^{3/2}}{n_e^2 Z} \tag{8.122}$$

Thus in the example above the absorption scale length decreases substantially as the wave propagates into more dense plasma (above $n_c/2$), leading to rapidly increasing absorption. The advantage of short wavelength illumination is also clear. Using a harmonic of Nd at 2ω or 3ω (0.53 μm or 0.35 μm), where harmonic conversion can be done very efficiently, raises the critical density by a factor of four or nine, respectively, again leading to a substantial decrease of the absorption length in the subcritical density region. Furthermore, as the wave approaches the critical surface the group velocity goes to zero, again enhancing absorption as represented by the factor $\sqrt{1 - n_e/n_c}$ in Eq. (8.122).

Kruer has solved Maxwell's equations for a wave propagating into a one-dimensional expanding plasma with density profile $n/n_c = \exp(-1/Z)$. For the case of non-resonant (see next section) s-polarized light he obtains an analytic solution for the collisional absorption fraction[1]

$$f_{abs} = 1 - \exp\left[-\frac{(8v_{ei}^* \, l \cos^3 \theta)}{3c}\right]$$

where θ is the angle of incidence measured from the surface normal and v_{ei}^* is the value of v_{ei} at $n_e = n_c$, where much of the absorption occurs. The strong angular dependence is due to refraction in the plasma. For the 1 keV temperature chromium plasma considered above, with normal incidence irradiation ($\theta = 0$) at 1.06 μm wavelength, the wave will experience 80% absorption with a density scale length of about 30 μm, thus ensuring high absorption of nanosecond duration irradiation at modest intensity. For frequency-doubled Nd at 0.53 μm wavelength the required scale length for 80% absorption is only about 7 μm, permitting strong collisional absorption even with rather short duration modest scale length plasmas.

In a following section we will discuss the effect of high-intensity irradiation on expanding density profiles. In such cases the intense illumination can generate a radiation pressure that steepens the electron density profile and thus reduces the role of collisional absorption in the underdense plasma. In very steep density profiles, collision absorption is compromised, but the incident wave can approach very close to the critical region where, depending on the electric field polarization, an enhanced *resonance absorption* can become important. This is discussed in the following section.

8.4.8   Resonance Absorption

In very steep gradient plasmas the incident wave is very close to the critical density surface before it is reflected (or, more accurately in an expanding plasma, refracted away). Steep gradient plasmas are encountered with very short pulse irradiations where expansion is minimized, and in very high intensity illuminations where radiation pressure inhibits the expansion. For a one-dimensional plasma expanding from a planar surface with a density gradient ∇n_e and an incident wave vector $\mathbf{k}$ at an angle of incidence θ from the surface normal, the wave is refracted out of the plasma, reaching a highest electron density $n_c \cos^2 \theta$ at the turning point, beyond which k is imaginary, representing an evanescent or tunneling field.[25, 26] Depending on the polarization of the incident radiation, it is possible to directly excite plasma waves through the resonance $\omega = \omega_p$ at the critical density. For p-polarized radiation, with the electric field lying in the plane of incidence defined by $\mathbf{k}_i$ and the surface normal, the electric field at the turning point has a component in the direction of the gradient that tunnels into the critical region. As we saw in Chapters 2 and 3, Maxwell's equations must satisfy the condition $\nabla \cdot (\epsilon \mathbf{E}) = 0$ at an interface, where for a plasma $\epsilon = \epsilon_0 n^2 = \epsilon_0 (1 - n_e/n_c)$ and $n = c/v_\phi$ is the refractive index as given in Eq. (8.113a). One then has $\epsilon_0 \nabla \cdot [(1 - n_e/n_c)\mathbf{E}] = 0$, which shows that the tunneling field $\mathbf{E}$ will drive a resonant response at the critical surface $n = n_c$, strongly driving plasma oscillation at $\omega = \omega_p$. A solution of Maxwell's equations in

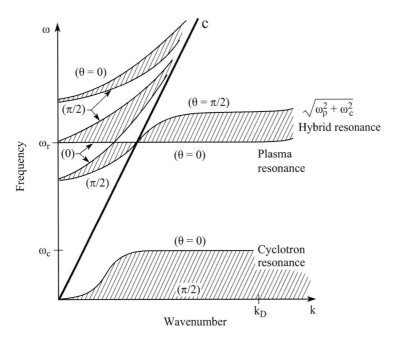

Figure 8.9 Dispersion diagram for plasma waves in the presence of an external magnetic field with $\omega_c < \omega_p$. Courtesy of N. Marcuvitz, New York University.[3]

the critical region[26] shows that the fraction of energy absorbed depends on the parameter $(k_i \, l)^{1/3} \sin \theta$, where l is the density scale length. The absorption peaks at about 50% when the parameter is equal to about 0.8. For normal incidence, $\theta = 0$, there is no axial component of electric field to drive the resonance, and for glancing incidence (large θ) the wave is refracted away from the critical surface, so that the tunneling field is weak.

8.4.9   Waves in a Magnetized Plasma

The presence of a static magnetic field $\mathbf{B}_0$ can substantially modify the nature of waves that propagate in a plasma, particularly if the cyclotron frequency $\omega_c = eB_0/m$ is comparable to or greater than the plasma frequency. Magnetized plasmas such as this are of great interest in astrophysics[27–30] and for the pursuit of fusion energy using magnetic confinement techniques. Not surprisingly, the orientation of the static magnetic field with respect to the propagation direction and electric field polarization is a significant factor. An example of dispersion curves for a magnetized plasma with $\omega_c < \omega_p$ is shown in Figure 8.9, for electromagnetic waves propagating along ($\theta = 0$) and perpendicular ($\theta = \pi/2$) to the static magnetic field direction ($\mathbf{B}_0$). Solutions for intermediate angles are shaded. The subject of magnetized plasma is beyond the scope of this text, but the interested reader will find substantial material in the plasma literature.[4, 6, 11, 14]

8.4.10   Non-Linear Processes in a Plasma

If we look back at the fluid level Maxwell–Euler equations, we see that product terms, which are inherently non-linear,[31] appear in several places. For convenience we repeat these equations here and box the terms that involve a product of field quantities:

$$\nabla \times \mathbf{H} = \epsilon_0 \frac{\partial \mathbf{E}}{\partial t} + \sum_j \boxed{(nq\mathbf{v})_j} \tag{8.123a}$$

$$\nabla \times \mathbf{E} = -\mu_0 \frac{\partial \mathbf{H}}{\partial t} \tag{8.123b}$$

$$\frac{\partial n_j}{\partial t} + \nabla \cdot (n_j \mathbf{v}) = 0 \tag{8.123c}$$

$$m_j n_j \left(\frac{\partial}{\partial t} + \mathbf{v} \cdot \nabla \right) \mathbf{v} = -\nabla P_j + \boxed{q_j n_j (\mathbf{E} + \mathbf{v} \times \mathbf{B})} \tag{8.123d}$$

where we have used $\mathbf{D} = \epsilon_0 \mathbf{E}$ and $\mathbf{B} = \mu_0 \mathbf{H}$, and where Eqs. (8.123c) and (8.123d) must be written for both electrons and ions ($j = 1$ and 2) with appropriate mass and charge. We see that in at least four places there is a boxed term involving a product of fields. These introduce the possibility for both non-linear growth and frequency mixing, and are known to play a major role in the development of non-thermal processes such as runaway suprathermal electrons and hard x-ray emission tails.

If we Fourier analyze one of these terms, we can gain some appreciation of the manner in which these processes operate, as well as some insight into the characteristic signatures we might look for. For instance, we can analyze the current term for electrons in Eq. (8.44),

$$\mathbf{J}(\mathbf{r}, t) = -e n_e(\mathbf{r}, t) \mathbf{v}(\mathbf{r}; t) \tag{8.124a}$$

by writing each field in terms of its respective wave component, viz.,

$$\mathbf{J} e^{-i(\omega_1 t - \mathbf{k}_1 \cdot \mathbf{r})} = -e n_e e^{-i(\omega_2 t - \mathbf{k}_2 \cdot \mathbf{r})} \mathbf{v} e^{-i(\omega_3 t - \mathbf{k}_3 \cdot \mathbf{r})} \tag{8.124b}$$

where a term by term match shows that for the amplitudes

$$\mathbf{J} = -e n_e \mathbf{v}$$

while for the frequencies and wavenumbers

$$\boxed{\omega_1 = \omega_2 \pm \omega_3} \tag{8.125a}$$

and

$$\boxed{\mathbf{k}_1 = \mathbf{k}_2 \pm \mathbf{k}_3} \tag{8.125b}$$

Equations (8.125a) and (8.125b) are sometimes referred to as *conservation of energy* and *conservation of momentum* equations for three-wave interactions, as is suggested by multiplying through by $\hbar$. Figure 8.10 captures the simplicity of this idea.

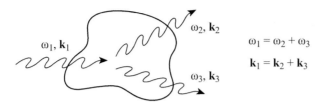

$$\omega_1 = \omega_2 + \omega_3$$
$$\mathbf{k}_1 = \mathbf{k}_2 + \mathbf{k}_3$$

Figure 8.10 A three-wave mixing process.

The process can in fact be either linear or non-linear. If a wave exists at a certain fixed amplitude, such as a density wave $n_e(\mathbf{r}, t)$, or even a fixed grating, then incoming waves ω_1, $\mathbf{k}_1$ scatter from it in linear fashion to a new frequency (for a moving wave) and wave vector ω_3, $\mathbf{k}_3$. This is a linear process: n is fixed, and the scattered field ω_3, $\mathbf{k}_3$ depends linearly on the incident field ω_1, $\mathbf{k}_1$. This specific example was the subject of Figure 8.4 in Section 8.2.

These processes can also be non-linear, and may apply to or grow out of any of the boxed non-linear terms in Eqs. (8.123a–d). For instance, it may occur that the incoming wave ω_1, $\mathbf{k}_1$ is very intense, and as a result, as it propagates into or through the plasma, it scatters from a spectrum of natural waves ω_2, $\mathbf{k}_2$ which pre-exist in the normal noise of random low level plasma oscillations. Of course ω_1, $\mathbf{k}_1$ will scatter from this spectrum of natural waves ω_2, $\mathbf{k}_2$, generating a spectrum of scattered waves ω_3, $\mathbf{k}_3$. Not all wave combinations will satisfy the conservation equations (8.125a, b) – matching of frequency and wave vectors is simply not guaranteed for three natural modes, each with its own dispersion relation. However, in some cases they may match, perhaps only at some special density.[32, 33] In those cases the three waves are said to be *in resonance*. The incoming wave scatters off waves in the noise. The new scattered wave grows in amplitude, and interferes with the incident wave at the beat frequency $\omega_3 - \omega_1$ and difference wave vector $\mathbf{k}_3 - \mathbf{k}_1$, causing the initial noise at $\omega_2 = \omega_3 - \omega_1$, $\mathbf{k}_2 = \mathbf{k}_3 - \mathbf{k}_1$ to grow in amplitude. This of course causes further scattering, and the process of growth and scattering continues. This process is called *stimulated scattering*. An incoming wave drives a plasma wave out of the noise and stimulates it to grow, by the very process of scattering from it in a resonant three-wave mixing process. Note that since these processes are resonant, they are sensitive to the background plasma parameters and to gradients of these quantities.

Figure 8.11 shows dispersion diagrams[7] for two such processes involving intense incident electromagnetic radiation: stimulated Brillouin scattering (SBS) and stimulated Raman scattering (SRS). In SBS the incident wave scatters from a low frequency ion-acoustic wave generating a scattered wave (ω_R, k_R) of slightly shifted frequency. In SRS the incident wave scatters from a high frequency electron-acoustic wave, generating a scattered wave (ω_R, ω_R) at a substantially shifted frequency.

The first is called Brillouin scattering because of the small scattered wave frequency shift, analogous to light scattering from acoustic waves in (neutral) gases. The second is called Raman scattering because of the large frequency shift reminiscent of light scattering from vibrational states in molecules. Both are called "stimulated" because

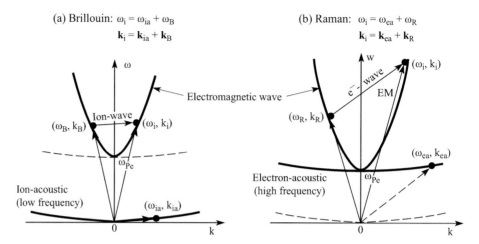

Figure 8.11 Dispersion diagrams for (a) Brillouin and (b) Raman scattering of incident electromagnetic radiation by ion-acoustic and electron-acoustic waves, respectively. The scattered wave experiences a relatively small frequency shift in the Brillouin case. Negative k-values indicate backscattered waves. Note that **k**-vector matching occurs for a strong incident wave participating in a three-wave process in which frequency and wave vector matching occur, i.e., when the ω, **k** for all three waves lie on the naturally occurring dispersion curves. Following H. Motz.[7]

the third wave – ion-acoustic in one case, electron-acoustic in the other – is caused to grow out of the noise through stimulation at the beat frequency. In each case the wave equations can be written with the non-linear terms appearing as driving terms on the right-hand side of the otherwise homogeneous (source-free) wave equation – now inhomogeneous. Whether the wave grows, and to what amplitude, is determined by the balance of loss processes (collisions or Landau damping, non-linear saturation processes, etc.) against the gain provided by the resonant beat frequency driver.

   To further illustrate this three-wave mixing, we consider the stimulated Raman scattering of Figure 8.11(b) in some additional detail. We consider backscattered radiation in which the incident frequency ω_i is somewhat greater that $2\omega_p$, so that both the scattered transverse wave frequency ω_R and the excited electron-acoustic wave frequency ω_{ea} are just slightly above ω_p as indicated in Figure 8.11. Because $\omega_i = 2\omega_p$, this corresponds to the quarter-critical density region, $n \cong n_c/4$. Since the scattered wave has $\omega_R \cong \omega_p$, it has a very small wave wavenumber in the plasma, $\Delta k \ll k_i/2$, smaller than it would have in vacuum. In order to resonantly match both frequencies and wavenumbers, as required by Eqs. (8.123a–d), the electron-acoustic wave must have a wavenumber $k_{ea} = k_i + \Delta k$, as shown in Figure 8.12. The frequency matching condition, $\omega_i = \omega_R + \omega_{ea}$, for waves with dispersion relations given by Eqs. (8.84) and (8.108) is given by

$$\omega_i = \omega_p\left[1 + \frac{(\Delta k)^2 c^2}{\omega_p^2}\right]^{1/2} + \omega_p\left[1 + \frac{(k_i + \Delta k)^2 a_e^2}{w_p^2}\right]^{1/2} \qquad (8.126)$$

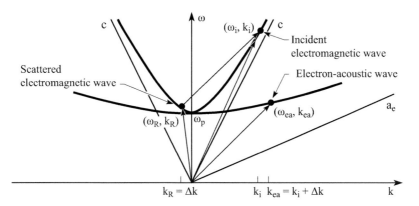

Figure 8.12 Three-wave mixing is illustrated for stimulated Raman backscattering at the quarter-critical electron density surface, $n_e \cong n_c/4$.

or for small Δk,

$$\omega_i = \omega_p + \omega_p \left(1 + \frac{k_i^2 a_e^2}{\omega_p^2} \right)^{1/2} \tag{8.127}$$

We can eliminate k_i through use of the transverse dispersion relation, Eq. (8.107), where

$$k_i = \left(\frac{\omega_i^2 - \omega_p^2}{c^2} \right)^{1/2} \simeq \left(\frac{4\omega_p^2 - \omega_p^2}{c^2} \right)^{1/2} \simeq \frac{\sqrt{3}\omega_p}{c} \tag{8.128}$$

The frequency matching condition (8.127) then becomes

$$\omega_i \simeq \omega_p + \underbrace{\omega_p \left(1 + \frac{3a_e^2}{c^2} \right)^{1/2}}_{\omega_{ea}} \tag{8.129}$$

For a 1 keV electron temperature, $a_e \simeq c/18$, so that the electron-acoustic frequency is $\omega_{ea} \simeq 1.005\omega_p$, and thus by Eq. (8.127) $\omega_i = 2.005\omega_p$. The three-wave mixing therefore occurs at an electron density in the vicinity of $n \simeq 0.249\, n_c$, that is, very close to the quarter-critical surface.

Note that an electron density gradient, as depicted in Figure 8.5, will limit this three-wave resonant mixing to a small region of the plasma. Also note that the phase velocity of the electron-acoustic wave is

$$v_\phi = \frac{\omega_{ea}}{k_{ea}} \simeq \frac{\omega_p}{\sqrt{3}\omega_p/c} \simeq \frac{c}{\sqrt{3}} \tag{8.130}$$

If this wave is driven to large amplitude by the high intensity three-wave resonance, it offers the possibility of trapping electrons within its high potential crests and accelerating them to velocities of $c/\sqrt{3}$, or to energies of order 100 keV. Turner and colleagues[34, 35] report experiments in which suprathermal electrons of order 100 keV energy are generated in a nominally 1 keV plasma by high intensity laser irradiation experiments in which SRS is identified as the non-linear acceleration mechanism. Once accelerated to high energies, these suprathermal electrons will eventually generate suprathermal x-rays as they collide with ions and nearby dense materials. In the following two subsections we consider intensity thresholds for non-linear processes, and numerical simulations of just such processes.

8.4.11   Threshold for Non-linear Processes

As we have seen in the previous subsection, the Maxwell–Euler equations (8.123a–d) are non-linear, with several product terms available for mode mixing and large amplitude growth. It is possible to determine the initial growth rate of such processes by expanding the fields in a power series, as was done earlier in the linearization process, but now carrying selected second order, or *product of fluctuation*, terms. The procedure is to keep linear terms to the left side of the equal signs, and treat the second order (non-linear) terms to the right as driving terms, much as in the treatment of inhomogeneous differential equations. At relatively small amplitude it is then possible to consider the product terms as exciting the natural (linear) modes to finite *initial* growth rates. Thresholds for the onset of non-linear growth of fields are then obtained by comparing these initial growth rates with natural wave decay rates such as collisional or collisionless (Landau) damping. For instance, in the stimulated Raman process just considered, one would determine what incident electric field **E**, or equivalently wave intensity I, would be required to overcome collisional and Landau damping of the electron-acoustic wave, and collisional damping (ν_{ei} of the two transverse waves). This would be a threshold intensity; below this intensity the waves naturally decay, and above it they grow.

We will not consider quasi-linear growth rates of plasma waves here, as we are generally confronted in hot dense plasmas, particularly in intense laser-produced plasmas, by growth rates that rapidly exceed these thresholds and for which further tools are required. These tools include numerical simulations in which a finite number of individual charged particles, constrained to a limited range of background density and temperature variations, are followed in the presence of high intensity laser illumination. Examples of such "particle in cell" calculations are described in the following section. By numerically studying the growth of waves and the acceleration of particles as a function of laser intensity, it is possible both to determine threshold values and to gain a better insight into the evolution of non-linear processes and their resultant field distributions.

Before proceeding to the numerical simulations, however, it is useful to develop some intuitive appreciation for the general nature of these stimulated processes and a likely order of magnitude estimate of threshold conditions. For instance, in the stimulated Raman process we can imagine a relatively weak electromagnetic wave, of frequency ω and electric field E, incident on a thermal plasma of electron density n_e and temperature κT_e. At low intensity the electron motions are dominated by random interactions (collisions) with other electrons and ions, both at short distances and through longer range wave motions. Superimposed on this random motion is a small sinusoidal velocity component, v_{os}. As the incident laser intensity increases, the oscillatory component of the velocity becomes more important and there evolves a distinctive *coherent* nature to the electron motions – phase locked in both space and time to the electric field of the transverse wave. Where the imposed frequency and wavenumber of the collective electron motion are a close match to a natural mode of the system – an electron-acoustic wave – we can expect the physics to change from one of random thermal processes to coherently driven wave motions. A measure of this transition is the intensity at which the imposed oscillation velocity v_{os} is comparable to a random thermal velocity, which we can take as v_e in Eq. (8.86c).

We can determine the oscillating component of velocity, v_{os}, as we did earlier using Eq. (8.103), which in scalar form gives

$$m\frac{\partial v_{os}}{\partial t} = -eE$$

or with imposed time dependence $e^{-i\omega t}$ is

$$v_{os} = -i\frac{eE}{m\omega}$$

We then compare this with a measure of the electron's random thermal motion, v_e from Eq. (8.86c):

$$v_e = \left(\frac{\kappa T_e}{m}\right)^{1/2}$$

We take the ratio and square it so that it represents a ratio of energies, that is, the ratio of electron energy in coherent oscillations to that in random motion. The result is

$$\left|\frac{v_{os}}{v_e}\right|^2 = \frac{e^2 E^2}{m\omega^2 \kappa T_e} = \frac{I/c}{n_c \kappa T_e} \tag{8.131a}$$

where the relationship between I and E^2 follows from Chapter 3, Eq. (3.20). For a plasma produced by a 1.06 μm wavelength Nd laser, with critical electron density $n_c = 1 \times 10^{21}$ e/cm^2 and an assumed (typical) electron temperature of 1 keV, the ratio of coherent to thermal electron energies is unity for a focused laser intensity $I = 4.7 \times 10^{15}$ W/cm^2. This is a commonly achieved value – and one such that the effects of non-linear processes are readily evident in the literature.[36] We can rewrite this ratio of energies in terms of common laboratory values as

$$\left|\frac{v_{os}}{v_e}\right|^2 = \frac{0.021 I \left[10^{14}\text{W/cm}^2\right] \lambda^2 [\mu\text{m}]}{\kappa T_e \,[\text{keV}]} \tag{8.131b}$$

where I is in units of 10^{14} W/cm^2, λ is in microns, and κT_e is in keV.

To avoid the excitation of non-thermal processes in a laser-produced plasma it is clearly advantageous to utilize low intensities and short wavelengths where possible. For instance, in laser-driven inertial fusion, where high intensities are essential, the use of short wavelengths through harmonic generation is common.[37] The achievable thermal temperature κT_e is also closely related to the incident intensity I, as we will see in a following section on blackbody radiation, so there too it may be more convenient to use a shorter wavelength. This is particularly true for the generation of thermal x-rays, but less so for extreme ultraviolet radiation where the requisite laser intensities are rather modest, of order 10^{12} W/cm^2.

For further discussion of these non-linear processes the reader is referred to the book by Kruer[1] and the article by Baldis et al.[33] in which they describe in detail several stimulated processes, their specific thresholds, and in particular, limits to growth rates in sharp gradient plasmas where energy and momentum matching can only be achieved over limited spatial dimensions. In the case of sharp gradients, the intensity–wavelength thresholds, $I\lambda^2$ in Eq. (8.131b), evolve to $I\lambda l$ thresholds, where l is the density scale length. For instance, see Table 2 of Baldis et al.[33] Experiments confirming several of these thresholds are reviewed by Drake.[38]

8.5   Numerical Simulations

The plasma theories considered in Section 8.4 are complicated in several ways. As we have seen, they are highly non-linear, the parameters vary sharply in space and in time, and they involve both fluid and kinetic details characteristic of long range and short range interactions. Linearized theories built around slowly varying background quantities are very useful for understanding the basic phenomena and obtaining initial growth rates for non-linear processes, but they are not adequate in themselves for the inherent complexities of hot dense plasmas. Numerical simulations offer important tools to address these complex phenomena. We discuss two methods in particular that address different aspects of this problem. One technique, called *particle in cell* calculations, follows the detailed kinetic motion of a finite number of particles in a background of limited space-time variations. The iterative process is illustrated in Figure 8.13. These techniques are particularly useful for studying particle kinetics and wave growth in highly non-linear laser plasma interactions. A second numerical technique, sometimes referred to as "hydrodynamic transport codes," utilizes zonal tracking of fluid properties such as density, temperature, and velocity in the presence of strong localized heating (laser energy deposition), including thermal and nonthermal energy transport among zones, with the use of energy bins as needed. These are generally called *Lagrangian techniques* in that they tend to follow identifiable mass regions as their positions and shape evolve in time.

8.5.1   Particle in Cell Simulations

In the particle in cell numerical simulations, a finite number of charged particles are tracked as their positions and velocity evolve in response to the self-consistent fields, electric and magnetic, that they themselves produce, as well as any applied fields. As an initial condition, a distribution of charged particle positions and velocities is selected, perhaps to represent a modest one-dimensional density ramp, with electron velocities

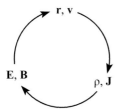

Figure 8.13 The basic cycle of a particle simulation code. Following W. Kruer.[1]

chosen to represent a selected electron temperature, and ions immobile. From these initial positions and velocities the charge distributions and currents are determined on a spatial scale (grid) sufficient to resolve collective motion. These are then used with Maxwell's equations to calculate the electric and magnetic fields generated. These fields, averaged over a suitable grid, are then used with the Lorentz force to determine changes in the position and velocity of all the particles. This constitutes one step in the simulation. The process is continued,[1, 10] as shown in Figure 8.13, with charge densities and currents re-determined for each step. The process is repeated through many cycles at small time intervals, sufficient to resolve the phenomena of interest, but typically on a time scale equal to a small fraction of $1/\omega_p$.

Figure 8.14 shows the results of a numerical simulation of stimulated Raman scattering by Forslund, Kindel, and Lindman,[39] in which the motion of 15 000 electrons, initially distributed in an electron density map extending from $n_c/8$ to $n_c/4$, with a thermal velocity distribution $v_e/c = 0.028$, is tracked as a function of time during irradiation by intense laser light (λ) that induces an oscillatory motion of the electrons at $v_{os}/c = 0.30$, thus just above the condition for $|v_{os}/v_e|^2 = 1$ in Eq. (8.122). The ions are held immobile. The electron density map and initial (thermal) velocity distribution are shown in Figure 8.14(a) and (b), respectively. The velocity distributions are shown later, after 1600 and 3000 oscillations, in Figure 8.14(c) and (d), displaying very large amplitude oscillations much larger than v_e, near the quarter-critical density. The amplitude of the scattered wave's electric field is shown in Figure 8.14(e), indicating an extremely sharp rise, with a growth period measured in tens of cycles of the incident radiation, essentially equal to the transit time across the density ramp. The power spectrum of scattered radiation shows a very strong component at $\omega = \omega_i/2$, as would be expected for Raman scattering, and many additional half-frequency harmonics due to the onset of further wave mixing in this very intense and highly non-linear interaction. In these calculations about 30% of the incident electromagnetic energy goes to the plasma oscillation.

Further simulations by Estabrook *et al.*[40] explore the electron heating due to the intense stimulated Raman process in a wide range of electron densities, with mobile ions, and with competition among non-linear processes. Figure 8.15 shows the heated electron energy distribution for a simulation in which an incident laser wave (λ) of intensity I, such that the ratio in Eqs. (8.131a, b), $|v_{os}/v_e|^2 = 0.53$, traverses a uniform plasma of electron density $n = n_c/10$ and length $L = 127\lambda$, and initial electron temperature 1 keV. The figure shows a cold (thermal) component and a heated electron component characterized by $\kappa\, T_{hot} \cong 13$ keV, due to the wave–particle interaction between electrons

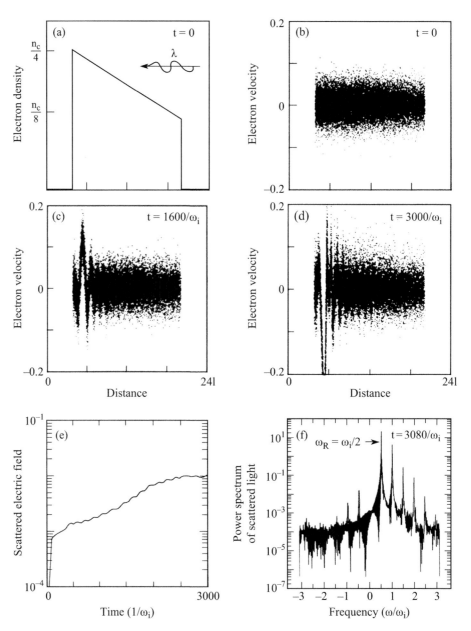

Figure 8.14 Numerical simulation[39] of stimulated Raman scattering from a plasma with an electron density ramp (a) just reaching $n_c/4$ in a distance of about 16 wavelengths of the incident laser radiation. The initial electron (thermal) temperature (b) is $v_e/c = 0.028$, and the incident laser intensity is such that $(v_{os}/v_e)^2 = 1.1$. The electron velocities as a function of position are shown at 1600 (c) and 3000 (d) cycles. The Raman scattered electromagnetic wave at $\omega_i/2$ and other half harmonics is shown in (e) and (f). Courtesy of D. Forslund, J. Kindel, and E. Lindman, Los Alamos National Laboratory.

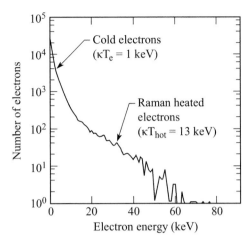

Figure 8.15 The electron energy distribution determined in a numerical simulation[40] of the stimulated Raman scattering process, for a uniform plasma of electron density $n = n_c/10$, plasma length of 127 electromagnetic wavelengths, initial electron temperature of 1 keV, and incident laser intensity I corresponding to $(v_{os}/v_e)^2 = 0.53$. The simulation shows a heated electron tail at $\kappa T_{hot} = 13$ keV, essentially corresponding to electrons of velocity equal to the phase velocity of the simulated electron-acoustic wave. Courtesy of K. Estabrook, W. Kruer, and B. Lasinski, Lawrence Livermore National Laboratory.

and the stimulated electron-acoustic wave. Further simulations show that, as expected from the arguments in Section 8.4.10, Eqs. (8.126) to (8.130), the energy of the heated electrons is largely dependent on the phase velocity of the Raman stimulated electron-acoustic wave (e.g., on κT_e and n_e or ω_p), and only weakly dependent on the incident wave intensity. Indeed, the simulations show that through control of the phase velocity of the electron-acoustic wave, largely the choice of n_e/n_c, it is possible to generate a Raman heated electron tail of energy up to 100 keV, as was suggested below Eq. (8.130).

8.5.2 Lagrangian Zonal Calculations of Plasma Mass and Energy Transport

Our description of the hot dense plasma includes sharp density profiles, rapid thermal expansion, and a host of non-linear terms, which in general cannot be linearized. Furthermore, we have seen that the energy distributions of particles and radiation can have substantial non-thermal components. Under these circumstances the mathematical problem of describing the plasma evolution is analytically intractable. Rather we again seek numerical solutions, in this case fluidlike Lagrangian computer codes in which the plasma mass is grouped into zones of specified mass, with appropriate fluid level parameters, whose evolution is followed as they absorb energy, increase in temperature and pressure, and expand. Basic parameters in each zone include density, temperature, pressure, vector velocity, multiple species, perhaps energy bins for electrons and photons, etc. As an example, Figure 8.16 shows a Lagrangian mesh calculated for a hot dense plasma as it expands from a laser-heated parylene (CH) disk. These

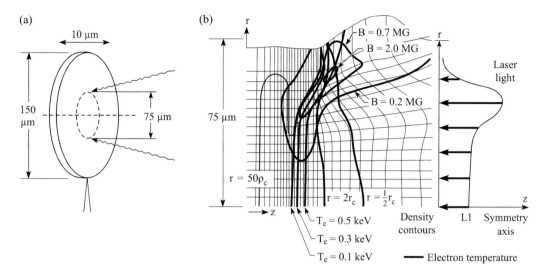

Figure 8.16 Results of a Lagrangian calculation of hot dense plasma expansion from a laser-heated parylene disk[42] (a) at the peak of a nominally 1×10^{14} W/cm^2, 150 ps FWHM pulse of 1.06 μm light. The non-uniform laser light intensity is shown to the right in (b). Shown in (b) are mass density contours corresponding to electron densities of $n_c/2$, $2n_c$, and $50n_c$, at the peak of the 0.5 ns laser pulse. Also shown are electron temperature contours for 0.5 keV, 0.3 keV, and 0.1 keV in the dense region of the plasma (higher in the low-density region). Various contours for calculated off-axis "dc" (varying only on the time scale of the expansion, not of the laser light) magnetic fields in megagauss are also shown. Courtesy of G. Dahlbacka, M. Mead, C. Max, and J. Thomson, Lawrence Livermore National Laboratory.

computations are done with the program LASNEX developed by Zimmerman and colleagues[41] at Lawrence Livermore National Laboratory.

Shown are the expanded zones at the peak of the laser pulse after each has absorbed prescribed amounts of laser energy and increased in temperature and pressure, thus exerting pressure on surrounding zones causing fluid motion.[42] Note that one side is shown for this axisymmetric problem. Larger zone size indicates lower density; thus we see directly the plasma expansion into vacuum at the right. Mass density contours corresponding to electron densities of $n_c/2$, $2n_c$, and $50n_c$ are shown, as are electron temperature contours for 0.1, 0.3, and 0.5 keV. Expansion velocities of the critical density surface in these numerical simulations, obtained by comparing zone positions at the two different times in the calculation, are typically 0.3 μm/ps, similar to the values estimated in Section 8.4.4 following Eqs. (8.69a, b). These Lagrangian computer codes are of great utility in understanding the complex fluid (plasma) dynamics, and also in the interpretation of experimental data, as the measurements generally involve finite space-time and spectral resolution in the presence of the sharp space-time gradients.

The simulation in Figure 8.16 was part of an effort to determine possible mechanisms for inhibiting energy transport from the sub-critical absorption region to the more dense supra-critical region. Two candidates considered were turbulent electron density fluctuations that would enhance electron scattering and thus inhibit energy transport, the other megagauss magnetic fields that might also inhibit energy transport to the dense plasma region. In the particular case examined in Figure 8.16 a somewhat annular laser light illumination pattern was used to generate strong lateral gradients, creating a $\nabla n_e \times \nabla T_e$ driven annular magnetic field, in this case reaching levels as high as 2 MG in a wrap-around (toroidal) field. Although not confirmed in early experiments, the subject of laser-generated megagauss magnetic fields continued to be of great interest.[43]

Through this example we see that numerical techniques can provide a very valuable tool for the study of complex hot dense plasmas, and also provide substantial assistance for interpreting experimental results that themselves are limited by finite spatial and temporal resolution, and thus not uniquely understandable without further information. These numerical simulation techniques provide a major capability for the planning and subsequent analysis of laser-fusion experiments,[41] as well as the emission characteristics from laser plasma sources.

8.6   Density Gradients: UV and EUV Probing of Plasmas

We have seen in preceding chapters that the electron density scale length plays an important role in determining which of several possible mechanisms dominates the absorption process. For a long scale, at moderate laser intensity, collisional absorption is very efficient, while for short scale length resonant absorption can also be important. For long scale lengths several non-linear processes can be important at high intensity. However, at high intensities the plasma density distribution can be strongly affected by radiation pressure, which we will discuss shortly. In any event it is important to measure the electron density scale length to better understand the competition among linear and non-linear mechanisms, and to quantitatively understand the dominant process. Toward that end we discuss measurements of electron density through the use of short wavelength interferometry in this section.

First we note that for high intensity radiation it is possible to generate sufficiently strong radiation pressure P_r, comparable to the plasma's electron thermal pressure P_e, such that the plasma is partially excluded from regions of otherwise high density.[1] The size of this effect can be estimated by considering the momentum transfer of absorbed and reflected photons near the region. For absorbed photons that deliver a momentum $\hbar k$, and reflected photons that have a change of momentum $2\hbar k$, one can readily show that the radiation pressure, expressed as the momentum change per unit area, is approximately

$$P_r = \frac{(2 - f_{\text{abs}})\,F\,\hbar k}{A}$$

where f_{abs} is the absorption fraction and F is the incident photon flux within an area A. With $k = \omega/c$ and intensity $I = \hbar \omega F / A$, the radiation pressure is

$$P_{\text{r}} = \frac{(2 - f_{\text{abs}})\,I}{c}$$

For a plasma of electron density n_e and temperature κT_e, the electron thermal pressure P_e is

$$P_e = n_e k T_e$$

so that the ratio is

$$\frac{P_r}{P_e} \simeq \frac{(2 - f_{\text{abs}})\, I}{c n_c \kappa T_e}$$

For a 1 keV plasma at 10^{21} e/cm^3 and an absorption fraction (near critical) of 0.5, the ratio is unity for a laser intensity of about 3×10^{15} W/cm^3. Thus even for some fraction of this intensity the radiation pressure can significantly affect the sub-critical electron density profile,[44] and thus the competition and effectiveness of the various linear and non-linear absorption processes.

It is possible to measure electron density distributions in the critical region using interferometric probing techniques. Generally the experiments employ a shorter wavelength probe to minimize refractive bending in the steep gradient plasma, and small targets so as to minimize the path length and thus the total turning angle. For plasma of a given electron density $n_e \cong n_c$, the refractive index n is given by Eq. (8.114b) as

$$n \simeq 1 - \frac{1}{2}\frac{n_e}{n_c}$$

where by Eqs. (8.112a, b)

$$n_c = \frac{1.11 \times 10^{21}\ \text{e/cm}^3}{\lambda[\mu m]^2}$$

for a probe of wavelength λ in microns. The first experiments[45, 46] to successfully probe the critical region of a Nd-laser irradiated target utilized a frequency quadrupled (4ω) probe of ultraviolet wavelength 266 nm, so that the plasma contribution to the refractive index was reduced by a factor of 16. The number of fringes N_F observed after propagating a distance L in a medium of refractive index n, and comparing with an equal path L in vacuum, is (see Chapter 3)

$$N_F = \frac{1}{\lambda_p} \int_0^L (1 - n)\, ds \tag{8.132}$$

where λ_p is the probe wavelength, ds is the incremental path length, and L is the total extent of propagation in the medium. For a region of relatively uniform density over a path length L, the number of fringes is then $N_F = n_e L / 2 n_c\, \lambda_p$, where n_c is the critical density for the probe at λ_p. For a laser heating pulse of wavelength λ and associated critical density n_c, the number of fringes is $N_F = \lambda_p L / 2\lambda^2$. To avoid the difficulty of too many fringes, perhaps too closely spaced to be optically resolved or so close that they are easily time smeared, it is clearly advantageous to choose a short probe wavelength λ_p and a short plasma propagation path L. If the axial gradient is such that the electron density falls to a value $1/e$ in a distance l, then the fringe separation distance[45] at critical is

$$\Delta z|_c = \frac{l}{N_F/e} \simeq \frac{5.4l\lambda^2}{\lambda_p L} \qquad (8.133a)$$

or in terms of the critical electron density [Eqs. (8.112a, b)]

$$\Delta z|_c \simeq 2.1 \times 10^2 \frac{m}{\mu_0 e^2} \cdot \frac{l}{\lambda_p L n_c} \qquad (8.133b)$$

which explicitly shows the functional dependence of required spatial resolution for short wavelength interferometry on the plasma parameters l, L, and n_c, for a probe wavelength λ_p. Clearly, to probe high electron densities in steep gradient plasmas *it is necessary to minimize the product $\lambda_p L$*. To probe the critical density associated with a laser heating pulse of wavelength $\lambda = 1.06$ µm, with a 4ω probe at wavelength $\lambda_p = 266$ nm, assuming a gradient length $\lambda = 1.5\lambda \cong 1.6$ µm and a plasma lateral extent of $L = 30$ µm requires a spatial resolution [Eqs. (8.133a, b)] of about 1.1 µm, which can be accomplished at 266 nm with a commercially available UV objective lens.

Figure 8.17 shows two such interferograms,[45] obtained using a holographic interferometer at 266 nm wavelength to probe plasma irradiated at 3×10^{14} W/cm² with a spherical shell target (a) and with a flat disk target (b). In both cases the electron density distribution was determined assuming axial symmetry. Both show clear effects of radiation pressure pushing plasma (partially) out of the near-critical interaction region. Assuming an electron temperature of 1 keV, the ratio of p_r to P_e is about 0.1 in these experiments. The spherical glass shell target experiment [Figure 8.17(a)] had the advantage of a short propagation path and thus was able to probe to just above the critical density. The target diameter was only 41 µm, and the lateral probing distance, equivalent to L, was shorter still. Note that the axial density profile shows a pronounced steepening just below critical, with a measured scale length $l \cong 1.6$ µm. This is quite similar to values seen in numerical simulations and is perhaps affected somewhat by the finite resolution of the UV objective lens, which is just under 1 µm at 266 nm wavelength. The flat disk targets are of larger diameter, that in Figure 8.17(b) having a 70 µm diameter, thus limiting the measurements to about half-critical density. Their analysis, however, is to first order simpler. All the targets in this series show relatively flat fringe patterns, which for an axisymmetric geometry clearly indicate the presence of a density depression,[46] or cavity, in the sub-critical region.

It is evident from Eqs. (8.133a, b) that to resolve fringes at high electron density, one key is to probe at shorter wavelength. Toward this end Da Silva and his colleagues[47] have developed a 15.5 nm probe beam based on EUV lasing, a subject discussed in Chapter 9, and have used it to probe relatively large scale plasmas at $L \cong 1$ mm and electron densities in excess of 10^{21} e/cm². Figure 8.18 shows (a) a fringe pattern obtained with a hot dense plasma expanding from a laser heated Mylar (CH) target, and (b) the measured electron density as a function of position away from the target surface. The target is

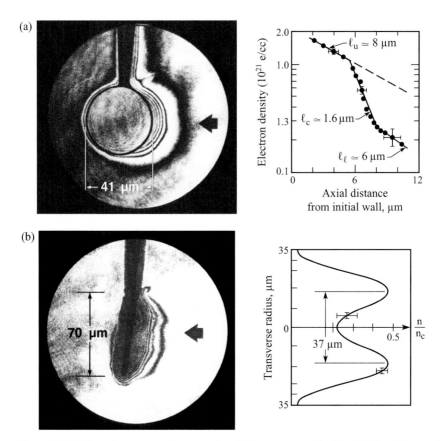

Figure 8.17 UV interferograms of laser produced plasmas, created with a 1.06 μm laser pulse, 30 ps FWHM, with nominal intensities of 3×10^{14} W/cm^2. The probe pulse wavelength is 266 nm, with a 15 ps FWHM duration, timed to arrive at the peak of the heating pulse. The interferograms are obtained using a holographic technique. The lack of temporal smearing in many such experiments suggests that the critical surface moves more slowly than predicted, perhaps due to radiation pressure.[45, 46]

irradiated with a 1 nsec duration rectangular pulse of 0.53 μm (frequency doubled Nd) laser light, at an intensity of 2.7×10^{13} W/cm^2, in a 0.7 mm diameter focal spot. The expanding plasma was integrated with a 350 ps duration, 15.5 nm probe pulse, arriving 1.1 ns after the arrival of the laser heating pulse. Analysis of the fringe pattern indicates a peak electron density of 3×10^{21} e/cm^3 and a scale length of $l \cong 40$ μm. This clearly represented a new capability of the study of electron density distributions in hot dense plasmas, extendible to densities of 10^{22} e/cm^3 in sub-millimeter plasmas. Indeed, the authors have used the 15.5 nm laser pulse to interferometrically probe colliding high density plasmas of interest to the inertial confinement fusion (ICF, or laser fusion) community.[47] Filevich and colleagues[48] have reported the study of plasma dynamics to near critical electron density using interferometry at 48.9 nm with a table top Ne-like argon laser (see Chapter 9, Section 9.5).

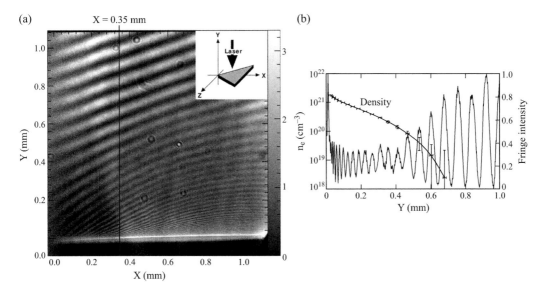

Figure 8.18 (a) An interferogram of an expanding CH plasma obtained with a 15.5 nm wavelength, 350 ps duration laser pulse. The plasma is produced by a 1 ns square pulse of 0.5 μm laser light at an intensity of 2.7×10^{13} W/cm^2 on a triangular shaped CH target. The probe pulse, from a neon-like yttrium laser, is of 300 ps duration, timed to arrive 1.1 ns after initiation of the heating pulse. (b) Measured electron density, to values above 10^{21} e/cm^2, are shown for a position 0.35 mm from the target surface. Courtesy of L. Da Silva,[47] Lawrence Livermore National Laboratory.

8.7   X-Ray Emission from a Hot Dense Plasma

That a plasma is both hot and dense, characterized by keV temperatures and near-solid densities, ensures that it will be an intense source of short wavelength radiation. As we have seen in the preceding sections, there is the possibility of both thermal and non-thermal processes taking place, from classical electron–ion collisions to highly non-linear three-wave mixing processes, each impressing its own signatures on the electron velocity distribution, and eventually on the observed emission spectra as the electrons collide with ions or nearby dense material. In the following subsections we consider various aspects of the emission process, with examples of line and continuum radiation.

Figure 8.19 shows data of the type we might expect to encounter.[49] It shows the radiated spectral energy density, in joules per keV bandwidth, radiated into 4π sr, for a gold disk target irradiated with 0.35 μm light (3ω of Nd) at an intensity of 5×10^{14} W/cm^2, in a 1 ns duration pulse. The emission spectrum shows a near-thermal continuum in the sub-kilovolt photon energy range, characteristic line spectra at a few keV (relatively broad M-band structure in this case), an exponentially falling spectrum of mixed free–bound and free–free bremsstrahlung radiation extending to photon energies of order 10 keV, and finally a suprathermal tail extending to 100 keV and beyond. In the follow-ing sections we consider separately general aspects of the thermal component, followed

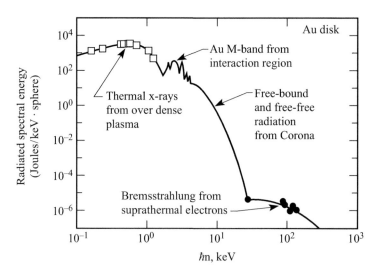

Figure 8.19 Typical x-ray spectra[49] from a laser irradiated high-Z target at moderate $I\lambda^2$. The data were obtained at Lawrence Livermore National Laboratory's Nova Laser Facility. Courtesy of R. Kauffman, Lawrence Livermore National Laboratory.

by various examples of line and continuum radiation in the sub-kilovolt and kilovolt photon energy ranges.

8.7.1   Continuum Radiation and Blackbody Spectra

We have seen in previous sections that hot dense plasmas are characterized by sharp spatial and temporal gradients, rapid expansion, and a variety of characteristic temperatures (T_e, T_i, T_{hot}, . . .). Such a plasma is clearly far from equilibrium. Nonetheless, a great fraction of the plasma energy is invested in a near-thermal distribution, and thus it is valuable to pause and consider the limiting case of *blackbody radiation*, that emitted by material characterized by a single temperature T that is in thermodynamic equilibrium with its surroundings.

In 1900 Max Planck, in an early contribution to the quantum theory of matter,[50] showed that if one assumed radiation to be emitted in discrete quanta of energy, with energy proportional to frequency, that the spectral energy density of radiation for such a body in equilibrium is[51, 52]

$$U_{\Delta\omega} = \frac{\hbar\omega^3/\pi^2 c^3}{e^{\hbar\omega/\kappa T} - 1} \tag{8.134a}$$

in units of energy per unit volume, per unit frequency interval $\Delta\omega$, at frequency ω, i.e., $\Delta^2 E/\Delta V \Delta\omega$. Expressing this in terms of relative spectral bandwidth ($\Delta\omega/\omega$), the spectral energy density $[\Delta^2 E/\Delta V(\Delta\omega/\omega)]$ becomes

$$U_{\Delta\omega/\omega} = \frac{\hbar\omega^4/\pi^2 c^3}{e^{\hbar\omega/\kappa T} - 1} \tag{8.134b}$$

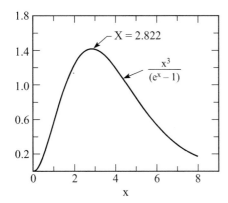

Figure 8.20 The Planck distribution function, of the form $x^3/(e^x-1)$ where $x = \hbar\omega/\kappa T$, for the spectral distribution of radiation from a body in equilibrium with its surroundings, a so-called *blackbody*. The function peaks at $x = 2.822$, or $\hbar\omega = 2.822\kappa T$, where κT is the temperature in eV.

Since the radiation is isotropic and propagating at the speed of light, we may write the spectral brightness as

$$B_{\Delta\omega/\omega} = \frac{cU_{\Delta\omega/\omega}}{4\pi} = \frac{\hbar\omega^4/4\pi^3 c^2}{e^{\hbar\omega/\kappa T}-1} \tag{8.135a}$$

now in units of energy per unit time, per unit area, per steradian, per unit relative spectral bandwidth $[\Delta^4 E/\Delta t\ \Delta A\ \Delta\Omega\ (\Delta\omega/\omega)]$. Observing that energy per unit time can be written in terms of energy per photon ($\hbar\omega$) times the photon flux, the spectral brightness can be rewritten in terms of photon flux, rather than energy, as

$$B_{\Delta\omega/\omega} = \frac{(\hbar\omega)^3}{4\pi^3\hbar^3 c^2}\frac{1}{e^{\hbar\omega/\kappa T}-1} \tag{8.135b}$$

now in photons per second per steradian per unit area · per unit $\Delta\omega/\omega$. Normalizing to κT and substituting standard values for $\hbar$ and c (from Appendix A), the *spectral brightness of blackbody radiation* can be expressed as

$$B_{\Delta\omega/\omega} = 3.146 \times 10^{20}\left(\frac{\kappa T}{eV}\right)^3 \frac{(\hbar\omega/\kappa T)^3}{\left(e^{\hbar\omega/\kappa T}-1\right)}\frac{\text{photons/s}}{\text{mm}^2 \cdot \text{sr} \cdot (\Delta\omega/\omega)} \tag{8.136a}$$

This can also be expressed in units previously used for the brightness of synchrotron radiation (Chapter 5, Section 5.4.6) by noting that 1 sr = 10^6 mrad^2, and that for the special case $\Delta\omega/\omega = 0.1\%$BW,

$$B_{\Delta\omega/\omega} = 3.146 \times 10^{11}\left(\frac{\kappa T}{eV}\right)^3 \frac{(\hbar\omega/\kappa T)^3}{\left(e^{\hbar\omega/\kappa T}-1\right)}\frac{\text{photons/s}}{\text{mm}^2 \cdot \text{mrad}^2 \cdot (0.1\%\text{BW})} \tag{8.136b}$$

The spectral brightness has a functional dependence of the form $x^3/(e^x-1)$, as plotted in Figure 8.20, where $x = \hbar\omega/\kappa T$. This Planckian function has a maximum

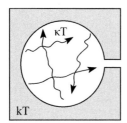

Figure 8.21 A cavity, or *hohlraum*, in which the contained radiation field is in radiative equilibrium with the surrounding wall.

value of 1.421 at $x = 2.822$, so that the peak spectral brightness occurs at a photon energy

$$\hbar\omega|_{\mathrm{pk}} = 2.822\kappa T \tag{8.137}$$

with peak spectral brightness, per steradian and per unit relative spectral bandwidth

$$B_{\Delta\omega/\omega}(2.822\kappa T) = 4.472 \times 1020 \left(\frac{\kappa T}{eV}\right)^3 \frac{\mathrm{photons/s}}{\mathrm{mm}^2 \cdot \mathrm{sr} \cdot (\Delta\omega/\omega)} \tag{8.138a}$$

In units of per square milliradian and per 0.1% bandwidth, the peak spectral brightness is

$$B_{\Delta\omega/\omega}(2.822\kappa T) = 4.472 \times 10^{11} \left(\frac{\kappa T}{eV}\right)^3 \frac{\mathrm{photons/s}}{\mathrm{mm}^2 \cdot \mathrm{mrad}^2 \cdot (0.1\% \, \mathrm{BW})} \tag{8.138b}$$

Thus for example, at a radiation temperature of $\kappa T = 100$ eV, the peak spectral brightness at 282 eV is 4.47×10^{26} (photons/s)/mm$^2 \cdot$ sr, or 4.47×10^{17} (photons/s)/ mm$^2 \cdot$ mrad$^2 \cdot$ (0.1%BW), falling to half these values to either side of the peak, at photon energies of 116 eV and 541 eV. We see that a blackbody with κT of order 100 eV is a copious radiator of EUV and soft x-ray radiation, indeed with a high spectral brightness. For laboratory plasmas we will find that this provides a useful, if idealistic, reference value for the consideration of near-thermal radiation.

An important issue for laboratory plasmas is that the lifetime of the hot dense plasma is in general very close to the duration of the heating pulse, which is typically measured in nanoseconds for such high temperatures. Thus in order to obtain substantial time averaged emission it becomes important to have a very high repetition rate, for instance, kilohertz lasers in which each individual pulse is focused to an intensity of order 10^{12} W/cm^2 or higher. These numbers will become more meaningful as this section continues.

It is often of interest to know the radiant energy flux (power per unit area) passing a given surface, as, for instance, through the hole in the cavity in Figure 8.21. Since the radiation is isotropic within the blackbody, we understand that the *net* energy flow across any given surface is zero. However, we may consider the single-sided integral of energy flux crossing a specified area in one direction, which is non-zero.

From Eq. (8.135a) we can compute the spectral intensity of radiation crossing in one direction as

$$I_{\Delta\omega/\omega} = \int_{2\pi} B_{\Delta\omega/\omega} \cos\theta \, d\Omega \tag{8.139}$$

where θ is measured from the surface normal, $d\Omega = \sin\theta \, d\theta \, d\phi = 2\pi \sin\theta$, for $0 \leq \theta \leq \pi/2$, and where $I_{\Delta\omega/\omega}$ has units of energy per unit area per unit of relative spectral bandwidth $\Delta\omega/\omega$. Since the spectral brightness is isotropic (no θ-dependence), and $\int_0^{2\pi} 2\pi \sin\theta \cos\theta d\theta = \pi$, one has

$$I_{\Delta\omega/\omega} = \frac{1}{4\pi^2\hbar^3c^2} \frac{(\hbar\omega)^4}{e^{\hbar\omega/\kappa T} - 1}$$

Integrating this over all (normalized) frequencies $d\omega/\omega$, one has the (single-sided) intensity in units of energy per unit area, passing a given interface from one side to the other,

$$I = \int_0^\infty \frac{1}{4\pi^2\hbar^3c^2} \frac{(\hbar\omega)^3}{e^{\hbar\omega/\kappa T} - 1} \hbar \, d\omega \tag{8.140a}$$

$$I = \frac{(\kappa T)^4}{4\pi^2\hbar^3c^2} \int_0^\infty \frac{x^3 \, dx}{e^x - 1} \tag{8.140b}$$

where the integral can be found in standard integral tables[53] as equal to $\pi^4/15$, so that the blackbody intensity at any interface is

$$I = \frac{\pi^2}{60c^2\hbar^3} (\kappa T)^4 \tag{8.141a}$$

or

$$\boxed{I = \sigma T^4} \tag{8.141b}$$

where

$$\sigma = \frac{\pi^2\kappa^4}{60c^2\hbar^3} \tag{8.142}$$

is the Stefan–Boltzmann constant, written in terms of the Boltzmann constant κ.

Equation (8.141b) is known as the Stefan–Boltzmann radiation law; it says that the radiant energy flux per unit area crossing any surface or interface of a blackbody is proportional to the fourth power of the absolute temperature. This applies to the energy received by a nearby surface, and to the energy radiated back if the two are in radiative equilibrium.

Figure 8.21 illustrates a small cavity, or *hohlraum*, for which the radiation field is Planckian of temperature κT, in equilibrium with the surrounding walls, also at temperature κT. For a very small hole, small enough so as not to upset the overall radiative equilibrium, the radiated intensity per unit area is given by Eq. (8.141b). This relationship is very useful for estimating an upper bound to the radiated intensity from a laser heated surface, the sun, a fireplace, or any hot object to the extent that it approximates radiative equilibrium. In many cases the emission spectrum is quite complex and we define an *equivalent blackbody temperature* by setting the radiated intensity equal to σT^4.

Equation (8.141b) can be written in terms of κT, expressed in electron volts, which is more convenient for application involving EUV and soft x-ray radiation, i.e.,

$$I = \widehat{\sigma}(\kappa T)^4 \tag{8.143a}$$

in terms of a modified Stefan–Boltzmann constant

$$\widehat{\sigma} = \frac{\pi^2}{60\hbar^3 c^2} = 1.027 \times 10^5 \frac{\text{W}}{\text{cm}^2 \cdot \text{eV}^4} \tag{8.143b}$$

In these units we readily see that a blackbody with temperature $\kappa T = 100$ eV radiates at an intensity $I \cong 1.0 \times 10^{13}$ W/cm^2, and according to Eq. (8.137) peaks at $\hbar\omega = 282$ eV. We might ask under what practical circumstance such emission characteristics might be achieved with a laboratory laser produced plasma, with finite fractional absorption of the incident laser radiation, and fractional conversion of absorbed light to re-radiated near-thermal emission. Experience with high intensity, nominally nanosecond duration laser pulses tells that the plasmas produced reach a steady state in a matter of picoseconds, during which the hot dense plasma is formed and energy partition between the charged particles, radiation field, and surrounding material is achieved, albeit briefly. Depending on various parameters, including intensity, laser wavelength, material (Z), and achieved temperature, as discussed earlier in Sections 8.4.9 and 8.4.10, this can lead to the creation of a rather efficient hot dense plasma radiator of soft x-rays and extreme ultraviolet radiation. For example, with a laser of 0.53 μm wavelength and incident intensity of 10^{14} W/cm^2 in a half nanosecond duration pulse incident on a medium to high-Z solid target, one can expect to produce a near thermal plasma with typically 80% absorption[33] of the incident laser energy, and 10% of the absorbed energy re-radiated at short wavelengths, for a total radiation conversion efficiency[49] of about 50% (see Section 8.7.6 regarding laser wavelength trends). Thus the re-radiated intensity would be 5×10^{13} W/cm^2, which for a blackbody according to Eq. (8.143a) would correspond to a temperature $T \cong 150$ eV, with a peak at 420 eV. Even with an incident laser intensity of 10^{12} W/cm^2, with similar assumptions, the equivalent blackbody temperature is reduced by only a factor of about 3.2 (fourth root), so that $\kappa T \cong 50$ eV, with peak emission at a photon energy of about 140 eV, sufficiently energetic to create the desired ionization states among the plasma ions and to radiate strongly in the EUV region of the spectrum. We discuss the departures for the idealized blackbody emission in the following sections.

8.7.2   Line Emission and Ionization Bottlenecks

Hot dense plasmas are essentially fully ionized, that is, every atom has at least one electron removed. In fact the temperatures are sufficiently high that most atoms have many fewer bound electrons than protons in the nucleus. The ionization state (number of electrons removed) depends primarily on the binding energies of the various electrons and on the electron plasma temperature. Typically the outer electrons in a multi-electron atom are held by only a few electron volts, while the core K, L, and M shell electrons (principal quantum numbers $n = 1$, 2, and 3) are closer to the nucleus and held more tightly, with binding energies of hundreds of thousands of electron volts.[54] Table 6.2 gives the binding energies calculated by Scofield[55] for selected ions. The elements listed vertically extend from a carbon nucleus ($Z = 6$) to a xenon nucleus ($Z = 54$). Shown in the body of the table is the energy in electron volts required to remove an additional electron from an ion having 1 ("hydrogen-like"), 2 ("helium-like"), . . . , 10 ("neon-like"), etc., remaining electrons.

For example, aluminum in its neutral state has 13 electrons. Table 8.2 indicates that with one electron already gone and 12 remaining ("magnesium-like" in electron configuration) the calculated energy to remove the twelfth electron is 18.8 eV. That would leave 11 electrons. The energy required to remove another electron (the eleventh) is 28.4 eV. Having removed the eleventh, there are ten remaining electrons, forming a closed shell in the neon-like ($1s^2 2s^2 2p^6$) configuration. The symmetry of the closed shell makes removal of an additional electron more difficult, and the ionization potential (binding energy) jumps significantly to 120 eV for neonlike aluminum. This significantly increased threshold for further electron removal can be considered an ionization bottleneck for a plasma of a given temperature.

For instance, if irradiation conditions (intensity, etc.) were adjusted to produce an electron temperature of 15–20 eV one could expect that relatively direct electron–ion collisions would easily remove the outermost electrons of most aluminum ions, including the eleventh electron, which has a binding energy equal to 28.4 eV. However, owing to the closed neon-like shell, removal of the tenth electron, which has a binding energy of 120 eV, would be substantially more difficult. Thus there is an *ionization bottleneck* that effectively closes, or nearly closes, the ionization process if the electron temperature and closed shell ionization threshold are well matched. As the temperature represents a distribution of energies, there are higher energy electrons well out on the tail of the distribution that can cause further ionization, but the energy distribution falls off exponentially beyond κT, so that the process quickly decreases in efficiency.

Data will be shown in Section 8.7.4 for the line emission from a laser-produced titanium ($Z = 22$) plasma at high laser intensity, with $\kappa T_e \cong 1$ keV, where the ionization thresholds (Table 8.2) permit relatively easy removal of all but the last two electrons. For example, with only three electrons remaining, the energy required to remove an additional electron is 1.4 keV, which occurs rather efficiently in a 1 keV plasma. With two remaining electrons, the ion is then in a closed shell helium-like configuration and the threshold for further ionization jumps to 6.2 keV, creating a well-defined step (bottleneck) in this sequential ionization process. As a result the plasma at this temperature consists largely of helium-like titanium ions with some lithium-like and beryllium-like ions, but little else. As we will see in Section 8.7.4, this leads to very strong line emission from helium-like titanium ions. The principal $n = 2$ to $n = 1$ transitions for helium-like (and hydrogen-like) ions, again calculated by Scofield,[55] are given in Table 8.3. For

Table 8.2 Ionization energies for selected ionic species.[55] Each column is labeled with the number of electrons bound to the ion before ionization and, in parenthesis, the symbol of the neutral atom with the same number of electrons (courtesy of J. Scofield, Lawrence Livermore National Laboratory).

Element	Ionization energy (eV)									
	1 (H)	2 (He)	3 (Li)	4 (Be)	10 (Ne)	11 (Na)	12 (Mg)	27 (Co)	28 (Ni)	29 (Cu)
6 C	490.0	392.1	64.49	47.89						
7 N	667.1	552.1	97.89	77.48						
8 O	871.4	739.3	138.11	113.90						
9 F	1103.1	953.9	185.18	157.15						
10 Ne	1362.2	1195.8	239.09	207.26	21.564					
11 Na	1648.7	1465.1	299.86	264.21	47.286	5.139				
12 Mg	1962.7	1761.8	367.5	328.0	80.143	15.035	7.646			
13 Al	2304.2	2086.0	442.0	398.7	119.99	28.447	18.828			
14 Si	2673.2	2437.7	523.4	476.3	166.42	45.12	33.64			
15 P	3070	2816.9	611.7	560.8	220.31	65.02	51.50			
16 S	3494	3224	707.0	652.1	281.00	88.05	72.59			
17 Cl	3946	3658	809.2	750.5	348.5	114.20	96.84			
18 Ar	4426	4121	918.4	855.8	422.8	143.46	124.24			
19 K	4934	4611	1034.6	968.0	503.9	175.82	154.75			
20 Ca	5470	5129	1157.7	1087.3	591.9	211.28	188.38			
21 Sc	6034	5675	1288.0	1213.6	686.6	249.84	225.13			
22 Ti	6626	6249	1425.3	1346.9	788.2	291.50	264.98			
23 V	7246	6851	1569.7	1487.3	896.6	336.3	307.9			
24 Cr	7895	7482	1721.1	1634.8	1011.8	384.2	354.0			
25 Mn	8572	8141	1879.9	1789.5	1133.8	435.2	403.2			
26 Fe	9278	8828	2045.8	1951.3	1262.7	489.3	455.6			
27 Co	10012	9544	2218.9	2120.4	1398.3	546.6	511.0	7.86		
28 Ni	10775	10289	2399.3	2296.7	1540.8	607.0	569.7	18.17	7.63	
29 Cu	11568	11063	2587.0	2480.2	1690.2	670.6	631.4	36.83	20.29	7.73
30 Zn	12389	11865	2782.0	2671.1	1846.4	737.3	696.4	59.57	39.72	17.96
31 Ga	13239	12696	2984.4	2869.4	2009.4	807.3	764.5	86.0	63.4	30.7
32 Ge	14119	13557	3194	3075	2179.3	880.4	835.8	115.9	90.5	45.72
33 As	15029	14448	3412	3288	2356.0	956.8	910.3	149.2	121.2	62.3
34 Se	15968	15367	3637	3509	2539.6	1036.3	988.1	185.5	155.4	81.7
35 Br	16937	16317	3869	3737	2730.1	1119.1	1069.1	225.4	192.8	103.0
36 Kr	17936	17296	4109	3973	2927.4	1205.2	1153.3	268.2	233.4	125.9
37 Rb	18965	18306	4357	4216	3132	1294.5	1240.8	314.2	277.1	150.7
38 Sr	20025	19345	4612	4467	3343	1387.2	1331.5	363.3	324.1	177.3
39 Y	21115	20415	4876	4726	3561	1483.1	1425.6	413.6	374.0	205.9
40 Zr	22237	21516	5147	4993	3786	1582.4	1523.0	471	427.4	236.2
41 Nb	23389	22648	5426	5268	4017	1684.9	1623.7	530	483.8	268.5
42 Mo	24572	23810	5713	5550	4256	1790.9	1727.8	592	541.7	302.6
43 Tc	25787	25004	6008	5841	4502	1900.3	1835.2	656	605.8	338.5
44 Ru	27033	26230	6312	6140	4754	2013.0	1946.1	724	671.4	376.3
45 Rh	28312	27487	6623	6447	5014	2129.2	2060.3	795	740.1	416.0
46 Pd	29623	28776	6943	6762	5280	2248.9	2178.0	869	811.8	457.5
47 Ag	30966	30097	7271	7086	5553	2372.0	2299.2	946	886.6	500.9
48 Cd	32341	31451	7608	7418	5834	2498.6	2423.9	1026	964.5	546.2
49 In	33750	32837	7953	7758	6121	2628.8	2552.1	1109	1045.4	593.3
50 Sn	35192	34257	8307	8107	6415	2762.5	2683.9	1196	1129.1	642.3
51 Sb	36668	35710	8670	8465	6717	2899.8	2819.2	1285	1215.3	693.2
52 Te	38177	37196	9041	8832	7025	3041	2958.1	1377	1306.3	746.1
53 I	39721	38716	9421	9207	7340	3185	3101	1472	1399.3	800.8
54 Xe	41300	40271	9810	9591	7663	3334	3247	1571	1495.4	857.4

the helium-like titanium ion the table shows principal emission lines at 4.727 keV and 4.750 keV for the $2p^1P_1$ and $2p^3P_1$ to $2p^1S_0$, $1s2p \to 1s^2$ transitions.

8.7.3 Sub-Kilovolt Line and Continuum Emissions

Hot dense laboratory plasmas are copious emitters of extreme ultraviolet and soft x-ray radiation, have a generally complex internal density and temperature structure, and are not in equilibrium with their surroundings, although they may reach some quasi-steady state for a brief period during their expansion. These laboratory plasmas generally exhibit an emission spectrum different from the ideal blackbody considered in Section 8.7.1. Recall that the highest electron temperature occurs in the relatively low density ($n_e \leq n_c$) absorption region. The absorbed energy is transported by random charged particle motion and radiation to a cooler, higher density ($n_e > n_c$) region. Radiation from this relatively high density region tends to dominate the emission process, generating the bulk of low to medium photon energies. As the propagation path is relatively long, through a dense plasma region, the line spectra due to the various ions tend to be smoothed to a modulated continuum. Where the modulation is relatively modest, blackbody radiation characteristics can be helpful in understanding and scaling problems. In some cases it is useful to introduce an equivalent blackbody temperature T_{eq} that would generate the same radiated intensity when integrated over a broad spectral region. We will consider such radiation in this section.

In the following section we will consider higher photon energy radiation from the hotter low density region in which the propagation path out of the plasma is shorter, involves less absorption and scattering, and results in a highly modulated spectrum with pronounced line structure characteristic of the various ionization states present. For a discussion of radiation transport in optically thick (much absorption and scattering) and optically thin (minimal absorption and scattering) regions of plasma see the book by Griem.[9]

An example of near-thermal emission from a moderate intensity high-Z laser-produced plasma is shown in Figure 8.22. The experiment, conducted by Zigler and colleagues,[56] utilized a 70 J, 3.5 ns FWHM Nd laser pulse at 1.06 μm wavelength, focused to an intensity of about 10^{14} W/cm^2 on a lanthanum ($Z = 57$) target. Dispersion of the emitted spectrum was achieved through use of a potassium ammonium phosphate (KAP) crystal and recorded on x-ray film. The observed modulation of the emission spectrum is ascribed to $4f \to 3d$ transitions in lanthanum ions of various ionization states, extending from zinc-like (30 remaining electrons, +27 charge state) to titanium-like (22 remaining electrons, +35 charge state). The various peaks consist of numerous closely packed lines in what the authors call an unresolved transition array. After correcting for the non-linear film response,[57, 58] the modulation is about 20%, peak to valley. The ionization energy for lanthanum in this range of charge states (+35 to +27) is typically several hundred electron volts. At an irradiation intensity of 10^{14} W/cm^2 the maximum equivalent blackbody temperature is about 170 eV, with a peak photon energy near 500 eV, or a wavelength of 25 Å. Thus the observed lines in Figure 8.22 are likely somewhat on the high photon energy side of the emission peak.

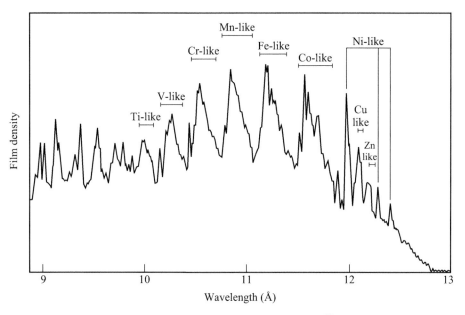

Figure 8.22 Emission spectrum from a laser-heated lanthanum target,[56] showing 4f to 3d line structure due to ions in various ionization states from zinc-like (+27 charge state) to titanium-like (+35 charge state). Laser energy is 70 J, 3.5 ns FWHM, 1.06 μm wavelength at about 10^{14} W/cm². Courtesy of A. Zigler, Hebrew University, Jerusalem.

Emission from a moderate-Z laser-produced plasma, also at moderate laser intensity, is shown in Figure 8.23. In this case a chromium target ($Z = 24$) is irradiated with 1.06 μm wavelength light, at 2×10^{14} W/cm², in a 150 psec duration pulse. Well-defined lines of neon-like C_r^{+14}, and weaker C_r^{+15}, are seen. The high photon energy limit of the observed spectrum is set by the L-absorption edge of neutral iron filter (707 eV, Appendix D). Dispersion here is provided by a lead behenate crystal ($2d \cong 120$ Å), with data recorded on x-ray film. According to Table 8.2, the ionization energy for C_r^{+13} (11 electrons) is 384 eV, while that for neon-like C_r^{+14} (10 electrons) is 1012 eV. With an irradiation intensity of 2×10^{14} W/cm² we can expect an equivalent black-body temperature $\kappa T_{eq} \cong 200$ eV, with a spectral emission peak approaching 600 eV. This is then consistent with a neon-like ionization bottleneck in which most C_r^{+13} ions are ionized further to C_r^{+14}, explaining the observation of strong C_r^{+14} lines and weak C_r^{+15} lines. According to Kelly,[54] the principal 3s to 2p and 3d to 2p emission lines|| should lie at photon energies of 586 eV, 594 eV, 660 eV, and 670 eV, respectively, very close to experimental values seen in Figure 8.23. Note that the C_r^{+15} 3s to 2p lines are of somewhat higher energy than those for C_r^{+14}, as there is one fewer electron and thus somewhat reduced screening of the nuclear charge, leading to tighter binding of the remaining electrons and somewhat larger transition energies. In general the spectral evidence here appears to be quite consistent with an electron temperature, or

|| More completely, $1s^2\ 2s\ 2p^5\ 3s$ to $1s^2\ 2s\ 2p^6$, etc.

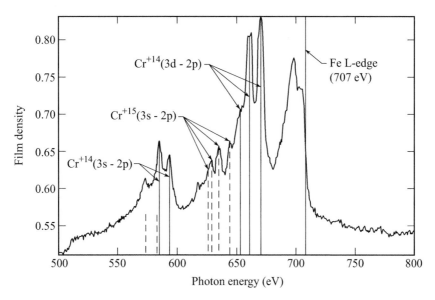

Figure 8.23 The emission spectrum recorded for a chromium target irradiated by a 1.06 μm Nd laser pulse of 150-ps duration and 2×10^{14} W/cm^2 incident intensity. Prominent lines of neon-like C_r^{+14} are evident, as well as weaker C_r^{+15} (fluorine-like) lines. Courtesy of R. Kauffman and L. Koppel, unpublished, Lawrence Livermore National Laboratory.

equivalent blackbody temperature, near 200 eV in the strong emission region of the plasma.

In many applications fine details of the emission spectra are not essential, except for diagnostic purposes. Rather, what is essential is an accurate understanding of energy transport. Toward this end it is often useful to measure the radiated energy in rather broad energy *bins* (intervals). Slivinsky, Kornblum, Tirsell, and their colleagues[59] have done this for the laser fusion program at Lawrence Livermore National Laboratory using a series of glancing incidence mirrors, K- and L-edge transmission filters, and absolutely calibrated x-ray diodes, in instruments they refer to as "Dante." These instruments have been used extensively, and provide the basic data for many studies of the conversion efficiency of laser light to soft x-rays. An interpretive summary of experiments at 1.06 μm, 0.53 μm, 0.35 μm, and 0.26 μm wavelengths is given by Mead *et al.*[36]

Time resolved studies of thermal emission from laser-produced plasmas have been conducted using a 15 ps resolution streak camera[60] and a series of glancing incidence mirrors with matching transmission filters,[61, 62] as illustrated in Figure 8.24. The combination of low pass glancing incidence mirrors and high pass transmission filters was described in Chapter 3, Figure 3.11, as a notch filter with relative spectral pass band $E/\Delta E \cong 3$ to 5. The streak camera has a slit photocathode sensitive to soft x-rays and x-rays. The emerging photoelectrons inside the tube are refocused as a slit image on a rear face phosphor plate. By optically triggering a properly timed ramp voltage the passing (time dependent) slit-shaped electron beam is swept vertically across the output phosphor screen, producing a time resolved *streak* of the slit, which is optically

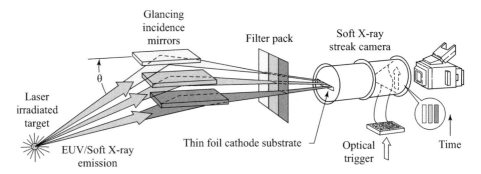

Figure 8.24 Schematic drawing of a time resolved three-channel soft x-ray spectrometer, employing a 15 ps resolution soft x-ray streak camera and three side-by-side "notch filters" nominally centered at 200 eV, 400eV, and 600 eV, each consisting of a glancing incidence soft x-ray mirror and a matching K or L-edge absorption filter.[61, 62] Courtesy of G. Stradling, LANL; R. Kauffman and H. Medecki and LLNL.

recorded. By placing information along the slit, such as three side-by-side notch filtered soft x-ray signals, one can obtain a time history of emission in the three selected photon energy bands. G. Stradling, R. Kauffman, H. Medecki, and colleagues[61, 62] conducted such studies with nominal 200 eV, 400 eV, and 600 eV channels of, respectively, a 5° carbon mirror with a carbon K-edge filter, a 5° nickel mirror with a vanadium L-edge filter, and a 3° nickel mirror with an iron L-edge filter. Figure 8.25 shows time resolved soft x-ray data in the three channels for a laser irradiated gold disk target.

Spectral integration of the three channels, also shown in Figure 8.25, has a temporal peak of radiated soft x-ray power equal to 1.5×10^{10} W, assuming isotropic radiation into 4π sr. With an emission area equal to the laser focal spot area, this gives a temporally peaked soft x-ray emission intensity of 3×10^{14} W/cm^2. The equivalent blackbody temperature required to radiate at this intensity, according to Eqs. (8.143a, b), is 230 eV, with a spectral peak at 660 eV. Similar temperatures are quoted in the article by Sigel.[63] Note that the time history is similar to that of the irradiating laser pulse, 680 ps FWHM, except for a somewhat faster rise, presumably due to heating of the initially cold target and a long temporal decay as the target cools and energy flows to the low photon energy (200 eV) channel. Thus the soft x-ray emission has a somewhat longer pulse duration of 790 ps FWHM. Note that a recording instrument with a slower response would indicate a lower peak power (same radiated energy, longer time) and thus a reduced equivalent radiation temperature. In Figure 8.25 the equivalent temperature drops to about 160 eV at times ± 0.5 ns to either side of the temporal peak. Further time resolved experiments, involving energy transport in layered targets, have been reported in the thesis by Stradling[61] in which the mirror–filter channels were replaced with narrow band multilayer mirrors (see Chapter 10) coated to match selected emission lines from various layers of the target as thermal energy arrives.

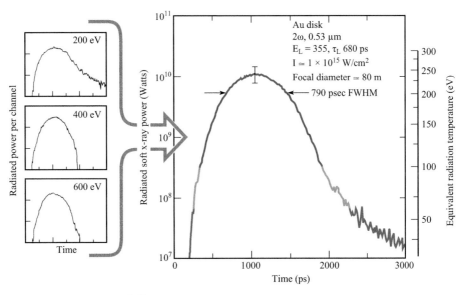

Figure 8.25 Time histories[61, 62] of soft x-ray emission in the 200, 400, and 600 eV channels for a laser irradiated gold disk target at an incident intensity of about 10^{15} W/cm^2, in a 680 ps pulse of frequency-doubled (2ω) Nd laser light at 0.53 μm wavelength. Spectral integration of the three channels shows a total soft x-ray power that peaks at about 1.5×10^{10} W, assuming isotropic radiation into 4π sr. The equivalent radiation temperature has a temperature peak at about 230 eV. Courtesy of R. Kauffman, unpublished, Lawrence Livermore National Laboratory.

8.7.4   Multi-Kilovolt Line Emission

Emission lines at higher photon energies, in the multi-kilovolt region, are generated predominantly in the sub-critical plasma where electron temperatures are high and densities are low. Here the line emissions are quite prominent, especially when dealing with K-shell spectra, as there are few ionization states (He-like and H-like), the background continuum is relatively low, and, because of the low density and high temperature, opacity effects are minimal.[64, 65] This low density high temperature region is sometimes called the plasma corona.

Figure 8.26 shows the well resolved 2–3 keV silicon K lines obtained from a 0.35 μm, 2 ns, 3×10^{14} W/cm^2 irradiation of a glass (SiO$_2$) disk, obtained by Kauffman[49] using a crystal for dispersion and x-ray recording film. The data show several prominent lines of H-like and He-like silicon merging into the continuum at longer wavelengths. Following early spectroscopic notation, α corresponds to an $n = 2$ to $n = 1$ transition, β to $n = 3$ to $n = 1$, γ to $n = 4$ to $n = 1$, etc., for each ionization state present.

Data from Matthews and colleagues[66] having higher photon energy lines from He-like and Li-like titanium are shown in Figure 8.27. The 2p1s to 1s^2 He-like lines at 4.7 keV clearly dominate the spectra, while Li-like lines are present but less intense, and the H-like 2p to 1s (H-like K$_\alpha$) is barely visible. From Table 8.2 we see that the ionization

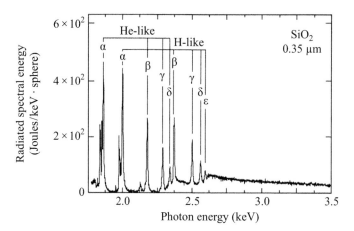

Figure 8.26 High resolution spectrum[49] of He-like and H-like silicon emission lines from a glass (SiO$_2$) disk irradiated by a 0.35 μm wavelength, 2 ns laser pulse at 3×10^{14} W/cm^2. Courtesy of R. Kauffman, Lawrence Livermore National Laboratory.

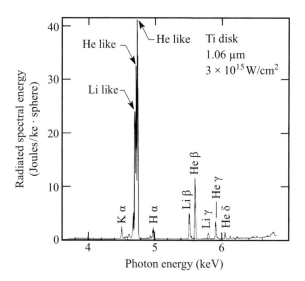

Figure 8.27 X-ray emission lines[66] from a laser-heated titanium disk, showing prominent He-like 1s2p to 1s^2 lines, less intense Li-like lines, and a barely observable H-like line. Irradiation is at 1.06 μm wavelength at an intensity of 3×10^{15} W/cm^2, in a 600 ps pulse. With 3 kJ of incident laser light, nearly 2 J is radiated in the lines at 4.75 keV. Courtesy of D. Matthews, Lawrence Livermore National Laboratory.

energy of Li-like titanium is 1.4 keV, while that of He-like titanium jumps to 6.2 keV. This is a clear example of an ionization bottleneck. The hot plasma has stripped off the outer electrons until there are only two electrons remaining (He-like). The fact that the Li-like lines are relatively weak indicates that there are fewer ions present with three electrons; thus the electron temperature is near 1 keV. The fact that the hydrogen-like

Table 8.3 Transition energies[55] for transitions from the $n = 2$ state to the $n = 1$ ground state of H- and He-like ions. Courtesy of J.Scofield, Lawrence Livermore National Laboratory.

	Transition energy (eV)			
	Hydrogen-like		Helium-like	
Element	$2p_{1/2}$	$2p_{3/2}$	$2p\ ^3P_1$	$2p\ ^1P_1$
5 B	255.17	255.20	202.78	205.4
6 C	367.5	367.5	304.3	307.8
7 N	500.3	500.4	426.3	430.7
8 O	653.5	653.7	568.7	574.0
9 F	827.3	827.6	731.5	737.8
10 Ne	1021.5	1022.0	914.9	922.1
11 Na	1236.3	1237.0	1118.8	1126.9
12 Mg	1471.7	1472.7	1343.2	1352.3
13 Al	1727.7	1729.0	1588.3	1598.4
14 Si	2004.3	2006.1	1853.9	1865.1
15 P	2301.7	2304.0	2140.3	2152.6
16 S	2619.7	2622.7	2447.3	2460.8
17 Cl	2958.5	2962.4	2775.1	2789.8
18 Ar	3318	3323	3124	3140
19 K	3699	3705	3493	3511
20 Ca	4100	4108	3883	3903
21 Sc	4523	4532	4295	4316
22 Ti	4966	4977	4727	4750
23 V	5431	5444	5180	5205
24 Cr	5917	5932	5655	5682

lines are relatively weak as well suggests that the 6.2 keV binding energy of the helium-like electrons is too high for the electron energies present in this plasma, and that the electron temperature is thus far below 6 keV. In this experiment[66] the titanium disk was irradiated at 3×10^{15} W/cm^2, with 3 kJ of 1.06 μm light in a 600 ps pulse delivered to a 450 μm diameter spot. Dispersion of the collected x-rays was achieved with a PET crystal, with data recorded on film. The dominant emission lines are at 4.727 and 4.750 keV, the $1s2p \rightarrow 1s^2$ lines of helium like titanium (see Table 8.3). In this experiment approximately 2 J was radiated in the helium-like lines, for a conversion efficiency (into 4π sr) of order 0.1% for just this doublet, which makes it an excellent source to be used as an x-ray back-lighter to visualize laser fusion implosion processes.[67]

These studies have been extended to shorter wavelengths (0.53 μm and 0.35 μm) and shorter pulse duration (60–120 ps), and a wider range of target materials, by Yaakobi et al.[68] and by Phillion and Hailey,[69] seeking to further refine the parameters for optimization of a short wavelength probe pulse, or *flash backlighter*, for high density laser fusion implosion studies.[67] In addition, because of their short wavelength, these emissions can be used to infer valuable plasma temperature and density information through the appearance of lines, line widths, line intensity ratios, and merger into the continuum.[64, 68−72]

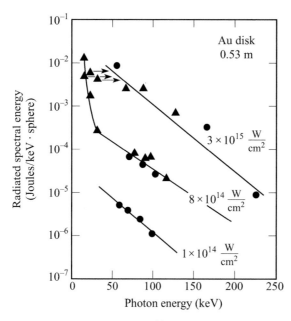

Figure 8.28 Hard x-ray emission[35] in the form of a suprathermal tail, from gold disk targets irradiated at 0.53 μm (Nd 2ω) wavelength, nominally 3.5 kJ energy in a 1 ns pulse. The hard x-ray signal is seen to rise dramatically with intensity above 10^{14} W/cm^2. Hot electron temperatures of 20–35 keV are inferred at the higher intensities. Circular data points were obtained with a K-edge filter; triangular data points with a filter–fluorescer combination. The solid line is a fit to the data. Courtesy of P. Drake, Lawrence Livermore National Laboratory, presently University of Michigan.

8.7.5   Suprathermal X-Rays

In laser irradiation experiments at the highest intensities it is common to observe a suprathermal tail of x-ray emission extending beyond 100 keV, and in some cases to several hundred keV. These high energy tails can be very strong at a laser wavelength of 1.06 μm, and even more so at 10.6 μm (CO$_2$ laser). Figure 8.28 shows typical data[35] obtained with 0.53 μm light on a gold disk target at intensities from 10^{14} W/cm^2 to 3×10^{15} W/cm^2. The data are collected with a series of K-absorption edge filters with calibrated diode detection, and with a set of filter–fluorescers with photomultiplier recording.[73] The very hard x-ray emission is sensitive to wavelength, generally described in terms of the parameter $I\lambda^2$, as suggested in Section 8.4.11. Temperatures of the suprathermal tail can be in the range of 10–40 keV at high values of $I\lambda^2$.

Figure 8.29 shows an example of data for gold disks irradiated at a nominal intensity of 10^{14} W/cm^2 to 3×10^{15} W/cm^2, for wavelengths of 1.06 μm, 0.53 μm, and 0.35 μm,[74] with nominal focal spot diameters of 150 μm. The measured photon flux in the 45–50 keV channel is nearly 100 times less for 0.35 μm wavelength irradiation than for 1.06 μm, at this intensity. Correlation with optical data indicates that the suprathermal tail is associated with stimulated Raman scattering,[35] as was described in Section 6.4.10.

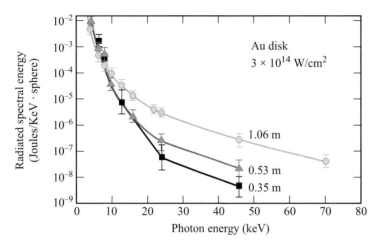

Figure 8.29 Suprathermal x-ray emission[74] from gold disk targets at 3×10^{14} W/cm², for wavelengths of 1.06 μm, 0.53 μm, and 0.35 μm. Incident energies ranged from 20 J to 30 J. Nominal pulse duration and focal spot diameters were 700 ps and 150 μm, respectively. The units assume isotropic radiation into a sphere. Courtesy of E.M. Campbell, B. Pruett, R. Turner, F. Ze, and W. Mead, Lawrence Livermore National Laboratory.

8.7.6   Laser Wavelength Trends

To conclude this section we summarize the general trends of laser light absorption and conversion to thermal and suprathermal x-rays as functions of laser wavelength and intensity on targets. The trends are illustrated for nominal nanosecond duration pulses, where the expansion time is sufficient to permit significant collisional absorption, particularly at the low range of intensities. The trends are shown for a planar gold disk target. As discussed in the previous sections, plasma properties and processes can vary widely with the electron density and its gradient, the electron temperature and its spatial variation, and the ion charge state (and thus target material) – all of which are affected by the irradiation wavelength (through n_c), intensity, and pulse duration (through the scale length l_{exp}). Thus these trends must be taken as illustrative only, but are nonetheless valuable for guiding the choice of operating parameters for specific applications. The curves are generally derived from specific references with extensive diagnostic capabilities and from sufficient data to cover large intensity and wavelength variations; the trends represented by them reflect a broad consensus developed within the international community based on measurements at many facilities. Fortunately the conversion of intense 1.064 μm laser light to its second and third harmonics, at 532 nm and 347 nm, can be done with high conversion efficiency.[75]

   Figure 8.30 shows general trends of laser light absorption as a function of intensity on a gold disk target, for nominally 1 ns duration pulses of Nd laser light at 1.06 μm and its harmonics ($n = 2, 3$) at 0.53 μm and 0.35 μm. Collisional absorption clearly favors shorter wavelength radiation where n_c, proportional to $1/\lambda^2$, is significantly higher, and thus collisional absorption, proportional to $n^2 Z$, is far more effective. At high intensities radiation pressure can depress sub-critical electron densities, thus reducing

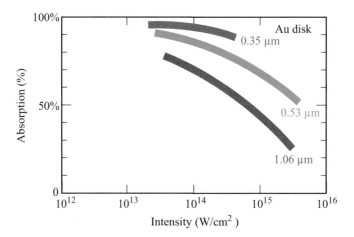

Figure 8.30 Curves showing the general trend of laser light absorption as a function of incident light intensity and wavelength, for nanosecond duration light pulses. In this example the target is a gold disk. For lower-Z material the absorption is generally less [see Eqs. (6.120)–(6.122)]. These curves are derived from the data of E.M. Campbell, R. Turner, F. Ze, C. Max, and colleagues,[33, 74−77] Lawrence Livermore National Laboratory.

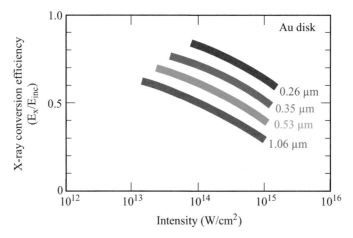

Figure 8.31 Curves showing the trend of x-ray conversion energy within the broad 0.1 keV to 1.5 keV spectral window, divided by the incident laser energy, assuming a Lambertian ($\cos \theta$) angular distribution in the emission hemisphere. Efficiency is shown as a function of intensity for Nd laser light at 1.06 μm (red) and its harmonics at 0.53 μm (green), 0.35 μm (blue), and 0.26 μm (ultraviolet) for nanosecond duration pulses. The target is a gold disk. The curves are derived largely from the data of R. Kauffman, V.W. Slivinsky, H. Kornblum, G.Tirsell, P. Lee, R. Turner, and colleagues,[36, 49, 78, 79] all of Lawrence Livermore National Laboratory.

absorption, while non-linear mechanisms increase energy losses due to scattering and deposit absorbed energy in non-thermal particle and photon distributions.

Conversion of laser light to near-thermal x-rays is addressed in Figure 8.31, as a function of laser intensity and wavelength. Conversion efficiency is defined as

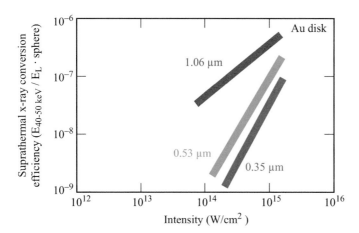

Figure 8.32 Curves showing the general trend of suprathermal x-ray generation vs. laser intensity and wavelength. The specific data here show radiated energy in a 40–50 keV spectral window, as a function of incident laser intensity for three wavelengths: 1.06 μm (red), 0.53 μm (green), and 0.35 μm (blue). The curves are derived from the data of V.W. Slivinsky, H. Kornblum, E.M. Campbell, B. Pruett, R.Turner, F. Ze, and W. Mead,[73, 74] all of Lawrence Livermore National Laboratory.

radiated energy within the broad window extending from 0.1 keV to 1.5 keV divided by the incident laser energy. The data assume a Lambertian ($\cos \theta$) angular distribution of thermal x-rays from the flat disk target. Again the example is for a gold disk irradiated with a nominally 1 ns duration pulse. The curves are of course affected by the finite absorption values from Figure 8.30. For low intensity illumination, especially at short wavelength, the conversion efficiency to thermal radiation is high. For the shorter wavelengths the absorption is at high densities where collisional thermalization is very efficient. At longer wavelengths and higher intensities non-thermal processes result in decreased absorption (increased stimulated scattering processes) and in increased radiation at multi-keV photon energies. Except for the lowest intensities shown, there is a great advantage to the use of harmonic conversion to short wavelength.

At high irradiation intensity, non-linear processes tend to dominate laser–plasma interaction physics as stimulated scattering processes drive high amplitude, high phase velocity plasma waves, trapping and accelerating some electrons to very high energy. One signature of these processes is the appearance of high energy suprathermal x-rays, such as illustrated in Figure 8.32, where the growth of 40–50 keV x-rays is shown as a function of intensity and wavelength, again for a gold disk target and a nominally nanosecond irradiation pulse. Again the message is clear: suppression of non-thermal processes is best accomplished through use of modest irradiation intensity and short wavelength.

As indicated in the literature, there are many ways to affect the specific numerical values given in the illustrations presented in this section (target material, geometry, laser bandwidth, focal spot uniformity, etc.), but the general trends as functions of wavelength and intensity are essentially universal.[80] At high laser intensities use of shorter wavelengths is mandated and widely used to limit the generation of high-energy electrons. This is permissible in large part because of the very high efficiencies with which Nd laser light at 1.06 μm wavelength can be converted to its harmonics – for example, 80% conversion to 3ω for both small and large scale lasers.[37]

On the other hand, some take advantage of even higher laser intensities to generate yet higher electron energies in the pursuit of very compact, high acceleration fields which might someday bring a dramatic reduction in the scale of free electron lasers and very high energy particle physics accelerators. Leemans and colleagues[81] have reported the use of 40 fs FWHM duration Ti:sapphire laser pulses at 810 nm wavelength and 10 Hz repetition rate, focused to intensities of 3×10^{18} W/cm^2 in pre-formed 30 μm FWHM diameter by 33 mm long hydrogen plasmas, to accelerate electrons to 1 GeV in a 4×10^{18} e/cm^3 wake field plasma. The plasmas were contained within somewhat larger diameter, 33 mm long capillary waveguides. The wake field plasma produces high amplitude plasma waves, as discussed in the previous section, here at larger amplitude due to the higher laser intensity and electric field. Acceleration to higher electron energies is envisioned with staged accelerations.

8.8   An Application: EUV Emitting Plasma for Computer Chip Lithography

A candidate technology for printing future computer chips involves radiation at a wavelength of 13.5 nm (91.8 eV), just below the L-absorption edge of silicon, and thus well matched to the use of Mo/Si multilayer interference coatings (Multilayer Mirrors, Chapter 10). This wavelength is much shorter than the presently used 193 nm, thus permitting the printing of finer, more densely packed features, enabling an increased density of memory storage and faster computations, thus continuing the general trend known as Moore's Law.[82, 83] Efforts to move to shorter wavelengths than the present 193 nm have been frustrated by significantly greater absorption in the ultraviolet. This has led to a decision in the lithography community to move to reflective optics, at considerably shorter wavelength, to transfer reference electronic patterns to a recording wafer where they are copied many times.[84–86] A full description of EUV lithography and its critical issues is given in the text by H. Kinoshita.[84] The Mo/Si multilayer mirror provides the highest known reflectivity in the sub-200 nm wavelength region, achieving a near normal incidence reflectivity as high as 70% at 13.5 nm (see Chapter 10, Multilayer Mirrors). As the optical transfer systems employed in computer chip manufacturing typically involve nine or ten mirrors, the transfer efficiency from the source to the wafer is proportional to R^N, where R is the reflectivity and N is the number of mirrors. Thus there is great advantage to obtaining the highest possible reflectance, and for this reason Mo/Si mirrors at 13.5 nm dominate the choice of wavelength for what is known as EUV lithography. Even with this high reflectivity, the transfer efficiency of the mirror system is only 3% or 4%, thus placing a significant burden on the EUV source which must deliver significantly higher power than might otherwise be required. In order to have a commercially competitive throughput of printed wafers it is estimated that the

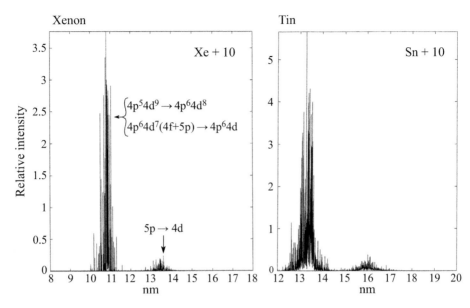

Figure 8.33 EUV emission by xenon and tin ions of +10 charge state (courtesy of G. O'Sullivan, University College Dublin).

EUV source must deliver 300–1000 W of average power[87, 88] at 13.5 nm, within the nominal 2% spectral bandwidth of a multi-mirror Mo/Si optical transfer system. Factors affecting the overall transfer efficiency and power requirements are described by V. Banine and colleagues.[89] As delivered, this radiation should be "clean," that is absent of radiation at other wavelengths and absent of plasma debris. Leading candidates for generating the 13.5 nm radiation are electrical discharges[90, 91] and laser produced plasmas[92, 93] having electron temperatures in the vicinity of 30 eV.** For the sake of brevity we give as an example here only the laser produced plasma, with details in the following paragraphs.

When heated to an appropriate electron temperature both xenon (Xe) and tin (Sn) ions radiate strongly in the desired EUV spectral range. Calculated emission spectra by G. O'Sullivan and colleagues[94] are shown in Figure 8.33. Strong emission is seen for Xe^{+10} and Sn^{+10} ions. Radiation from nearby ionization stages, such as Sn^{+7} to Sn^{+12}, also contribute to the 2% spectral acceptance band of Mo/Si. The emission occurs in what is known as "unresolved transition arrays," or UTAs, consisting of many nearby lines, primarily 4f–4d and 4d–4p.[95, 96] Creation of the various ions is dependent on the electron temperature, as illustrated in Figure 8.34, where one observes that the Sn^{+9}, Sn^{+10} and Sn^{+11} ions, which dominate emission in the desired spectral band,[95] are present in high fractions for temperatures of 30 eV to 40 eV. While the choice of inert xenon would be good for minimizing contamination of the optics, its emission within the strong reflectivity spectral region of Mo/Si is limited. Tin, on the other hand, has

---

** As a first estimate of the required temperature we note that a blackbody has a peak emission at 2.8 κT, thus a 30 eV plasma radiates strongly near 90 eV.

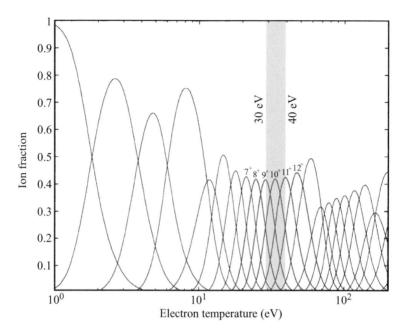

Figure 8.34 Sn ion fractions[95] for various charge states, as a function of electron temperature. The calculations assume an electron density of 1×10^{21} e/cm^3, the critical electron density for Nd laser light at 1.06 μm wavelength. Note that for electron temperatures in the range of 30 eV to 40 eV there is a strong presence of Sn^{+9}, Sn^{+10} and Sn^{+11}ions, which dominate emission in the 13.5 nm acceptance band of Mo/Si multilayer mirror systems (courtesy of J. White and G. O'Sullivan, University College Dublin).

its very similar emission characteristics spectrally shifted to an optimal position with respect to Mo/Si. Unfortunately, as the source material of choice, it provides a debris mitigation challenge to protect the optics, chamber, etc., from deposition of Sn ions, and also a challenge to recover and recirculate the source material. Among the solutions to these issues are baffles and deflecting magnetic fields. Other target elements, from Mo to Pd, radiate in as UTAs in the 6–8 nm wavelength range, when irradiated at somewhat higher laser intensities, and thus may be useful for future scaling of EUV lithography.[95]

The next question is: what laser systems are available that can provide the high focused intensity needed to produce 30–40 eV plasmas with tin target materials, and sufficiently high average power to meet wafer throughput requirements? Candidates of greatest interest include Nd and CO_2 laser systems, at 1.06 μm and 10.6 μm wavelengths, respectively. Both can operate with sufficient energy per pulse, and with pulse durations in the picosecond to nanosecond range, to achieve high peak power. The CO_2 laser has the disadvantage of a longer wavelength and thus somewhat limited focusability, but has the advantage of high repetition rates and high average power. Futhermore, for CO_2 the lower electron density, and thus lower ion density, in the interaction region reduces opacity effects (re-absorption) for the EUV radiation. Figure 8.35 shows the results of numerical calculations of electron density and electron temperature[97–99] as a function of radius for a 35 μm diameter, tin-doped[98] spherical target, illuminated at

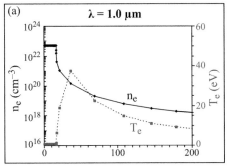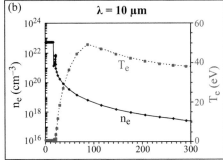

Figure 8.35 Calculated electron density and electron temperature profiles[97, 98] for (a) Nd (~ 1 μm) and (b) CO_2 (~ 10 μm) irradiation of composite spherical Sn targets of 35 μm diameter by a 10 ns duration pulse of intensity 1×10^{11} W/cm² (courtesy of M. Richardson, U. Central Florida).

1×10^{11} W/cm² with a 10 ns pulse of (a) 1.0 μm wavelength (Nd) and (b) 10 μm wavelength (CO_2) laser light. Both generate peak electron temperatures in the 30–40 eV spectral region. These temperatures can be fine tuned by varying the focused laser intensity so as to generate an optimal yield of the desired EUV in-band radiation. The electron temperatures in each case reach their peak values just below the respective critical electron density, 1×10^{21} e/cm³ for Nd and 1×10^{19} for CO_2. The gradient of electron density is important for optimal absorption of the laser light, and can be controlled by use of a prepulse of modest energy. The prepulse has another important role as it can also bring the plasma to an electron temperature just below the value needed for optimal ionization of the Sn ions. In this manner the energy of the main heating laser pulse can be more efficiently converted to the desired radiation. With a proper prepulse, less investment of the heating pulse energy is needed to reach the desired ionization state, and thus more energy is available to efficiently raise Sn ions to the excitation level which upon relaxation would then emit the desired EUV photons at a high CE. For example, the total energy[††] required to take a neutral Sn atom to a Sn^{+10} state is about 860 eV with an additional 93 eV to reach the excitation level required for emission of the desired EUV photon, a total of about 953 eV is needed, delivered collisionally by successive electron impacts. With much of the required energy investment provided by a short duration prepulse of modest energy but sufficient focused intensity, the resultant conversion efficiency of heating pulse energy to EUV plasma emission can be optimized. Additionally, the prepulse can initiate a plasma expansion by which small, Sn or Sn-composite materials, known as mass-limited targets whose size is chosen to minimize optics contamination, can expand to a size matched to the focal size of the heating pulse.[99–103] This is a balance between available laser energy per pulse, temperature-tuned laser intensity, and the energy required per atom to reach the optimum ionization state. This is particularly important for CO_2 laser systems[104, 105] because of the relatively long 10.6 μm wavelength, which limits the achievable focal spot size, typically to

---

[††] The energies required to take neutral Sn to Sn^{+1}, Sn^{+1} to Sn^{+2}, etc., to Sn^{+10} are[94, 101, 102] 7.1 eV, 14.4 eV, 29.4eV, 40.1 eV, 76.5 eV, 96.0 eV, 116.5 eV, 137.8 eV, 160.0 eV, and 183.0 eV, for a total of 861 eV.

about 80 μm FWHM diameter. Considering a CO_2 pulse energy of about 60 mJ, optimum target sizes are of order 25–30 μm diameter. For example, a solid Sn spherical target has an atomic density of $1 \times 10^{22}/cm^3$. With 860 eV needed to raise each atom to a Sn^{+10} ionization state, plus 92 eV for excitation to an appropriate EUV emitting state, a total energy of about 950 eV per ion is required. For a CO_2 pulse energy of 60 mJ, the optimum target diameter is thus about 26 μm, corresponding to about 3×10^{14} Sn atoms. As this is smaller than an 80 μm diameter CO_2 laser spot size, use of a prepulse to create and expand the Sn plasma is essential. With an electron temperature of near 40 eV and a Sn atomic mass of 118 AMU, the plasma expansion velocity [Eq. (6.69)] is about 2.3 μm/ns, thus expanding radially about 23 μm for a 10 ns prepulse, to a total diameter $(2 \times 23 + 26)$ of about 72 μm, well matched to an 80 μm FWHM diameter focal spot. If the prepulse energy is delivered at a sufficiently high intensity, about 8×10^{10} W/cm² in a 10 ps Nd pulse, it can prepare the plasma with an electron temperature of about 30 eV, so that upon arrival of the main CO_2 heating pulse minimal addition ionization is required, thus enhancing the efficient conversion of CO_2 laser light to 13.5 nm EUV radiation.

The next issue is how to deliver mass limited Sn targets of 20–30 μm diameter. The preferred approach is based on jets of liquid droplets formed by forcing liquid at high pressure through a nozzle and into a vacuum chamber. At some distance from the nozzle the jet breaks up into liquid droplets. The process is enhanced and controlled by application of a high frequency vibration to the nozzle. The technique was first utilized to enhance laser energy conversion efficiency to short wavelength radiation for soft x-ray microscopy[106] and EUV lithography[107] by Hertz and his student group at KTH university in Stockholm. The group also introduced the use of a prepulse[108] to enhance the conversion efficiency by a factor of eight with droplet targets, and was the first to demonstrate the use of Sn droplets for EUV lithography.[109] That work has been followed by others[97, 109] and is now the technique of choice for commercial developments of EUV lithography working towards a high volume manufacturing (HVM) capability.[110, 111] A stream of Sn droplets, as used in the EUV lithography exposure tool being developed by the company Gigaphoton in Japan, is shown[104] in Figure 8.36. These are 28 μm diameter Sn liquid droplets. The spacing is close enough that more than one would be exposed to an incident laser pulse, so a technique is used to select some by electrostatic charging and then deflection,[104] providing a 500 μm separation. The current Gigaphoton EUV system is dual-wavelength, utilizing a 10.6 μm wavelength, 12 kW average power CO_2 laser which delivers 60 mJ, 20 ns FWHM heating pulses to individual Sn droplets at repetition rates up to 100 kHz. A synchronized 1.064 μm wavelength Nd laser system irradiates each droplet with a 10 ps duration, 10 mJ prepulse 10 ns before arrival of the heating pulse. The CO_2 heating pulse is focused onto the expanded Sn droplet with a focal spot diameter of 80 μm, and a peak intensity of nominally 6×10^{10} W/cm², sufficient to achieve the desired electron temperature. A conversion efficiency of 2.0 % has been achieved at 100 kHz,[110] from the CO_2 heating pulses to clean, 2% in-band, EUV pulses at 13.5 nm. This corresponds to 43 W of EUV power at what is known as the intermediate focus (IF) of the multi-mirror EUV optical system.[87] Higher average power CO_2 systems are being prepared which, along with

Figure 8.36 A sequence of liquid Sn droplets of 28 μm diameter as used in the Gigaphoton[104, 105] dual-wavelength EUV lithography system. The laser system consists of a 10 ps Nd laser prepulse at 1.06 μm wavelength, followed by a 10 ns duration, CO_2 main heating pulse at 10.6 μm wavelength. As the liquid droplets are closely spaced, only some are selected and deflected into the laser–target interaction region where they are irradiated at rates up to 100 kHz. Courtesy of M. Nakano, Gigaphoton.

improvements to the conversion efficiency, are expected to deliver 250 W of clean, in-band, EUV power.[105] Cymer/ASML has made similar projections.[112, 113]. Recently ASML reported extended operation at a power in excess of 90 W with one of its EUV exposure tools at TSMC in Taiwan, resulting in the exposure of more than 1000 wafers in a single day.[114] The availability and timing of EUV lithography as a viable technology for high volume manufacturing of computer chips is critically dependent on further improvements in source power.[83, 88] It is anticipated that the first use of EUV lithography for high volume manufacturing will be in the 2017–2019 time frame, beginning at the so-called "7 nm node." At this node the lithographic half-pitch is expected to vary from 13 nm to 25 nm depending on complexity of the patterns to be printed. Microprocessor gate lengths are expected to be 14 nm at this node. Progress towards these goals can be followed online at websites such as those of Sematech, ASML, Intel, Samsung, Global Foundries, and TSMC. Further discussion of EUV lithography is given in Chapter 11, Section 11.8.

References

1. W.L. Kruer, *The Physics of Laser Plasma Interactions* (Addison-Wesley, Redwood City, CA, 1988).
2. A. Rubenchik and S. Witkowski, Editors, *Physics of Laser-Plasma*, Handbook of Plasma Physics, Vol. 3 (North-Holland, Amsterdam, 1991).
3. N. Marcuvitz, "Notes on Plasma Dynamics," New York University, unpublished (1970).
4. D. Nicholson, *Introduction to Plasma Theory* (Wiley, New York, 1983).
5. Yu.L. Klimontovich, *The Statistical Theory of Non-equilibrium Processes in a Plasma* (MIT Press, Cambridge, MA, 1967).

6. P.C. Clemmow and J.P. Dougherty, *Electrodynamics of Particles and Plasmas* (Addison-Wesley, Reading, MA, 1979).

7. H. Motz, *The Physics of Laser Fusion* (Academic Press, New York, 1979).

8. G. Bekefi, *Radiation Processes in Plasmas* (Wiley, New York, 1966).

9. H. Griem, *Principles of Plasma Spectroscopy* (Cambridge University Press, Cambridge, UK, 1997).

10. C.K. Birdsall and A.B. Langdon, *Plasma Physics via Computer Simulation* (McGraw-Hill, New York, 1985).

11. T. Stix, *The Theory of Plasma Waves* (McGraw-Hill, New York, 1962).

12. S. Ichimaru, *Statistical Plasma Physics* (Addison-Wesley, Reading, MA, 1992).

13. N.G. Basov, Yu. A. Zakharenkov, N.N. Zorev *et al.*, *Heating and Compression of Thermonuclear Targets by Laser Beam* (Cambridge University Press, Cambridge, UK, 1986).

14. T. Dolan, *Fusion Research* (Pergamon, New York, 1982).

15. P.A. Charles and F. Seward, *Exploring the X-Ray Universe* (Cambridge University Press, Cambridge, UK, 1995).

16. S. Chapman and T.G. Cowling, *The Mathematical Theory of Non-uniform Gases* (Cambridge University Press, Cambridge, UK, 1964).

17. R.B. Bird, *Transport Phenomena* (Wiley, New York, 1960).

18. L.D. Landau,"On the Vibrations of the Electron Plasma," *J. Physics (USSR)* 10, 25 (1946).

19. E.C. Jordon, *Electromagnetic Waves and Radiating Systems* (Prentice-Hall, Englewood Cliffs, NJ, 1950).

20. J.A. Stratton, *Electromagnetic Theory* (McGraw-Hill, New York, 1941).

21. J.D. Jackson, *Classical Electrodynamics* (Wiley, 1999), Third Edition, p. 325.

22. I.P. Shkarofsky, T.W. Johnston, and M.P Bachynski, *Particle Kinetics of Plasmas* (Addison-Wesley, Reading, MA, 1966).

23. J. Dawson and C. Oberman,"High-Frequency Conductivity and the Emission and Absorption Coefficients of a Fully Ionized Plasma," *Phys. Fluids* **5**, 517 (1962); J.M. Dawson, "On the Production of Plasma by Giant Pulse Lasers," *Phys. Fluids* **7**, 981 (1964).

24. T.W. Johnston and J.M Dawson,"Correct Values for High Frequency Power Absorption by Inverse Bremsstrahlung in Plasmas," *Phys. Fluids* **16**, 722 (1973).

25. V.L. Ginzburg, *Propagation of Electromagnetic Waves in Plasmas* (Pergamon Press, New York, 1970); P. Mulser,"Resonance Absorption and Ponderomotive Action," Ref. 2.

26. D.W. Forslund, J.M. Kindel, K. Lee, E.L. Lindman and R.L. Morse,"Theory and Simulation of Resonant Absorption in a Hot Plasma," *Phys. Fluids* **11**, 679 (1975).

27. C.W. Allen, *Astrophysical Quantities* (Athlone Press, London, 1997), Third Edition.

28. E.N. Parker, *Spontaneous Current Sheets in Magnetic Fields with Applications to Stellar X-Rays* (Oxford University Press, Oxford, 1994).

29. A.L. Peratt, *Physics of the Plasma Universe* (Springer-Verlag, Berlin, 1992).

30. R.L. Bowers and T. Deeming, *Astrophysics* (Jones and Bartlett, Portola Valley, CA, 1984).

31. N. Marcuvitz, "Notes on Plasma Turbulence," New York University, unpublished (1969).

32. C.S. Liu, M.N. Rosenbluth, and R.B. White,"Raman and Brillouin Scattering of Electromagnetic Waves in Inhomogeneous Plasmas," *Phys. Fluids* **17**, 1211 (1974).

33. H.A. Baldis, E.M. Campbell, and W.L. Kruer, "Laser-Plasma Interactions," Chapter 9 in *Physics of Laser-Plasma* (North Holland, Amsterdam, 1991), A. Rubenchik and S. Witkowski, Editors.

34. R.E. Turner, K. Estabrook, R.P. Williams *et al.*, "Observation of Forward Raman Scattering in Laser-Produced Plasmas," *Phys. Rev. Lett.* **67**, 1725 (1986).

35. R.P. Drake, R.E. Turner, B.F. Lasinski *et al.*, "Efficient Raman Sidescatter and Hot-Electron Production in Laser-Plasma Interaction Experiments," *Phys. Rev. Lett.* **53**, 1739 (1984); R.P. Drake, R.E. Turner, B.F. Lasinski *et al.*, "X-Ray Emission Caused by Raman Scattering in Long-Scale-Length Plasmas," *Phys. Rev. A* **40**, 3219 (1989).

36. W.F. Mead, E.K. Stover, R.L. Kauffman, H.N. Kornblum and B.F. Lasinski, "Modeling, Measurements, and Analysis of X-Ray Emission from 0.26 μm Laser-Irradiated Gold Disks," *Phys. Rev. A* **38**, 5275 (1988); R.E. Turner *et al.*, "Evidence for Collisional Damping in High Energy Raman Scattering Experiments at 0.26 Microns," *Phys. Rev. Lett.* **54**, 189 (1985).

37. R.S. Craxton, R.L. McCrory and J.M. Soures, "Progress in Laser Fusion," *Sci. Amer.* **255**, 68 (1986); B.M. Van Wonterghem, J.R. Murray, J.H. Cambell *et al.*, "Performance of a Prototype for a Large-Aperture Multipass Nd: Glass Laser for Inertial Confinement Fusion," *Appl. Opt.* **36**, 4932 (1997).

38. R.P. Drake, "Three-Wave Parametric Instabilities in Long-Scale-Length, Somewhat Planar, Laser Produced Plasmas," *Laser Part. Beams* **10**, 599 (1992); R.P. Drake, R.G. Watt and K. Estabrook,"Onset and Saturation of the Spectral Intensity of Stimulated Scattering in Inhomogeneous Laser-Produced Plasmas," *Phys. Rev. Lett.* **77**, 79 (1996).

39. D.W. Forslund, J.M. Kindel and E.L. Lindman,"Plasma Simulation Studies of Stimulated Scattering Processes in Laser-Irradiated Plasmas," *Phys. Fluids* **18**, 1017 (1975).

40. K. Estabrook, W.L. Kruer and B. Lasinski,"Heating by Raman Backscatter and Forward Scatter," *Phys. Rev. Lett.* **45**,1399 (1980).

41. G.B. Zimmerman, LLNL Report UCRL-75881, 1974 (unpublished); also G.B. Zimmerman and W.L. Kruer, "Numerical Simulation of Laser-Initiated Fusion," *Comments Plasma Phys. Controlled Fusion* **2**, 51 (1975).

42. G.H. Dahlbacka, W.C. Mead, C.E. Max and J.J. Thomson, "Calculations of Self-Generated Magnetic Fields in Parylene Disk Experiments," Laser Program Annual Report 1975, UCRL 50021–75, A.J. Glass, Editor, Lawrence Livermore National Laboratory (4975), p. 271.

43. J. Stamper,"Laser-Generated Jets and Megagauss Magnetic Fields," *Science* **281**, 1469 (1998).

44. X. Liu and D. Umstadter, "Competition between Ponderomotive and Thermal Pressures in Short-Scale-Length Laser Plasmas," *Phys. Rev. Lett.* **69**, 1935 (1992).

45. D.T. Attwood, D. Sweeny, J. Auerbach and P.H.Y. Lee, "Interferometric Confirmation of Radiation-Pressure Effects in Laser-Plasma Interactions," *Phys. Rev. Lett.* **40**,184 (1978); D.W. Sweeny, D.T. Attwood, and L.W. Coleman, "Interferometric Probing of Laser-Produced Plasmas", *Appl. Opt.* **15**, 1126 (1976).

46. D.T. Attwood, "Diagnostics for the Laser Fusion Program – Plasma Physics on the Scale of Microns and Picoseconds," *IEEE J. Quant. Electr.* **QE-14**, 909 (1978).

47. L.B. Da Silva, T.W. Barbee, R. Cauble *et al.*, "Electron Density Measurements of High Density Plasmas Using Soft X-Ray Laser Interferometry," *Phys. Rev. Lett.* **74**, 3991 (1995); A.S. Wan, T.W. Barbee, R. Cauble *et al.*, "Electron Density Measurement of a Colliding Plasma Using Soft X-Ray Laser Interferometry," *Phys. Rev. E* **55**, 6293 (1997).

48. J. Filevich, J.J. Rocca, E. Jankowska *et al.*, "Two-Dimensional Effects in Laser-Created Plasmas Measured with Soft-X-Ray Laser Interferometry," *Phys. Rev. E* **67**, 056409 (May 2003).

49. R. Kauffman, "X-Ray Radiation from Laser Plasma," Chapter 3 in *Physics of Laser Plasma* (North Holland, Amsterdam, 1991), A. Rubenchik and S. Witkowski, Editors.

50. P.A. Tipler and R.A. Llewelyn, *Modern Physics* (Freeman, New York, 2012), Sixth Edition, p. 127.

51. Ya.B. Zel'dovich and Yu.P. Rasier, *Physics of Shock Waves and High Temperature Hydrodynamic Phenomena* (Academic Press, New York, 1966).

52. C. Kittel and H. Kroemer, *Thermal Physics* (Freeman, New York, 1980).

53. I.S. Gradshteyn and I.M. Ryzhik, *Tables of Integrals, Series, and Products* (Academic Press, New York, 1994), Fifth Edition, p. 370, No. 3.411–1, and p. xxxi.

54. R.L. Kelly, "Atomic and Ionic Spectrum Lines Below 2000 Å: Hydrogen Through Krypton," *J. Phys. Chem. Ref. Data* **16**, Suppl. 1 (1987); also R.L. Kelly and L.J. Palumbo, "Atomic and Ionic Emission Lines Below 2000 Angstroms: Hydrogen Through Krypton," Naval Research Laboratory Report 7599, NRL, Washington, DC (1973).

55. J.H. Scofield, "Energy Levels for Hydrogen-, Helium-, and Neon-Like Ions," in *X-Ray Booklet*, Lawrence Berkeley Laboratory PUB-490 rev. (April 1986), D. Vaughan, Editor: also LLNL Report UCID-16848 (1975). Extended to Co, Ni, and Cu-like ions by J. Scofield (1999, private communication) in part using data from C.E. Moore, NBS Pub. NSRDS-NBS 34 (1970), and J. Sugar and A. Musgrove, *J. Chem. Phys. Ref. Data* **24**, 1803 (1995), and others referenced therein.

56. A. Zigler, M. Givon, E. Yarkoni *et al.*, "Use of Unresolved Transition Arrays for Plasma Diagnostics," *Phys. Rev. A* **35**, 280 (1987).

57. C.M. Dozier, D.B. Brown, L.S. Birks, P.B. Lyons and R.F. Benjamin, *J. Appl. Phys.* **47**, 3732 (1976).

58. B.L. Henke, F.G. Fujiwara, M.A. Tester, C.H. Dittmore and M.A. Palmer, *J. Opt. Soc. Amer. B* **1**, 828 (1984).

59. K.G. Tirsell, H.N. Kornblum and V.W. Slivinsky, "Time Resolved, Sub-keV X-Ray Measurements Using Filtered X-Ray Diodes," Report UCRL-81478, Lawrence Livermore National Laboratory; also P.H.Y. Lee and K.G. Tirsell, "X-Ray Conversion Efficiency," Laser Fusion Annual Report 1980, Report UCRL-50021–80, Lawrence Livermore National Laboratory, p. 7–10; also R.A. Heinle and K.G. Tirsell, "Filtered-Mirror Sub-keV X-Ray Measurement System," Laser Program Annual Report 1979, Report UCRL-50021–79, L.W. Coleman, Editor, Lawrence Livermore National Laboratory, p. 5–5.

60. C.F. McConaghy and L.W. Coleman, "Picosecond X-Ray Streak Camera," *Appl. Phys. Lett.* **25**, 268 (1974).

61. G.L. Stradling, D.T. Attwood, and R.L. Kauffman, "A Soft X-Ray Streak Camera," *IEEE J. Quant. Electr.* **QE-19**, 604 (1983); also G.L. Stradling, "Time Resolved Soft X-Ray Studies of Energy Transport in Layered and Planar Laser-Driven Targets," PhD thesis, Department of Applied Science, University of California, Davis (1982).

62. R.L. Kauffman, G.L. Stradling, D.T. Attwood and H. Medecki, "Quantitive Intensity Measurement Using a Soft X-Ray Streak Camera," *IEEE J. Quant. Electr.* **QE-19**, 616 (1983).

63. R. Sigel, "Laser-Generated Intense Thermal Radiation," Chapter 4 in Ref. 2.

64. H.R. Green, *Principles of Plasma Spectroscopy* (Cambridge University Press, 1997).

65. D. Milhalas and B.W. Mihalas, *Foundations of Radiation Hydrodynamics* (Oxford University Press, 1984); D. Mihalas, *Stellar Atmospheres* (Freeman, San Francisco, 1978), Second Edition.

66. D.L. Matthews, E.M. Campbell, N.H. Ceglio *et al.*, "Characterization of Laser-Produced Plasma X-Ray Sources for Use in X-Ray Radiography," *J. Appl. Phys.* **54**, 4260 (1983).

67. D.T. Attwood, N.H. Ceglio, E.M. Campbell *et al.*, "Compression Measurement in Laser Driven Implosion Experiments," p. 423 in *Laser Interaction and Related Plasma Phenomena*, Vol. 5 (Plenum, New York, 1981), H. Schwarz, H. Hora, M. Lubin, and B. Yaakobi, Editors.

68. B. Yaakobi, P. Bourke, Y. Conturie *et al.*, "High X-Ray Conversion Efficiency with Target Irradiation by a Frequency-Tripled Nd: Glass Laser," *Opt. Commun.* **38**, 196 (1981).

69. D.W. Phillion and C.J. Hailey, "Brightness and Duration of X-Ray Line Sources Irradiated with Intense 0.53-μm Laser Light at 60 and 120 ps Pulse Width," *Phys. Rev. A* **34**, 4886 (1986).

70. B. Yaakobi, D. Steel, E. Thoros, A. Hauser and B. Perry, "Direct Measurement of Compression of Laser-Imploded Targets Using X-Ray Spectroscopy," *Phys. Rev. Lett.* **39**, 1526 (1977); also B. Yaakobi, D.M. Villeneuve, M.C. Richardson *et al.*, "X-Ray Spectroscopy Measurements of Laser-Compressed, Argon Filled Shells," *Opt. Commun.* **43**, 343 (1982); B. Yaakobi, F.J. Marshall, D.K. Bradley *et al.*, "Signatures of Target Performance and Mixing in Titanium Doped, Laser-Driven Target Implosions," *Plasma Phys.* **4**, 3021 (1997).

71. C.F. Hooper, D.P. Kilcrease, R.C. Mancini *et al.*, "Time-Resolved Spectroscopic Measurements of High Density in Ar-Filled Microballoon Implosions," *Phys. Rev. Lett.* **63**, 267 (1989).

72. T.D. Shepard, C.A. Back, D.H. Kalantar *et al.*, "Te Measurements in Open- and Closed-Geometry Long-Scale-Length Laser Plasmas via Isoelectronic X-Ray Spectral Line Ratios," *Rev. Sci. Instrum.* **66**, 749 (1995).

73. V.W. Slivinsky, H.N. Kornblum and H.D. Shay, "Determination of Suprathermal Electron Distributions in Laser-Produced Plasmas," *J. Appl. Phys.* **46**, 1973 (1975).

74. E.M. Campbell, "Dependence of Laser-Plasma Interaction Physics on Laser Wavelength and Plasma Scalelength," p. 579 in *Radiation in Plasmas*, Vol. II (World Science, Singapore, 1983), B. McNamara, Editor; E.M. Campbell, B. Pruett, R.E. Turner, F. Ze, and W.C. Mead, "Suprathermal Electrons from Disks," p. 6–36 in *1981 Laser Program Annual Report*, E.V. George, Editor, Lawrence Livermore National Laboratory Report UCRL-50021–81.

75. R.S. Craxton, "High Efficiency Frequency Tripling Schemes for High Power Nd: Glass Lasers," *IEEE J. Quant. Electr.* **QE-17**, 1771 (1981).

76. C.E. Max, F. Ze, E.M. Campbell *et al.*, "Agrus and Shiva Experiments: Absorption and Stimulated Brillouin Scatter," p. 6–30 in *1981 Laser Program Annual Report*, E.V. George, Editor, Lawrence Livermore National Laboratory Report UCRL-50021–81.

77. F. Ze, E.M. Campbell, V.C. Rupert and R.E. Turner, "Target-Interaction Experiments at 0.53 μm and 0.35 μm: Absorption," pp. 7–8 in *1980 Laser Program Annual Report*, L.W. Coleman and W.F. Krupke, Editors, Lawrence Livermore National Laboratory Report UCRL-50021–80.

78. R.L. Kauffman, M.D. Cable, H.N. Kornblum and J.A. Smith, "X-Ray Conversion Efficiency," pp. 4–8 in *1985 Laser Program Annual Report*, M.L. Rufer and P.W. Murphy, Editors, Lawrence Livermore National Laboratory Report UCRL-50021–85.

79. R.E. Turner, W.C. Mead, C.E. Max *et al.*, "X-Ray Conversion Efficiency at 1ω, 2ω, and 3ω," p. 6–34 in *1981 Laser Program Annual Report*, E.V. George, Editor, Lawrence Livermore National Laboratory Report UCRL-50021–81.

80. P.D. Goldstone, S.R. Goldman, W.C. Mead *et al.*, "Dynamics of High-Z Plasmas Produced by a Short-Wavelength Laser," *Phys. Rev. Lett.* **59**, 56 (1987).

81. W.P. Leemans, B. Nagler, A.J. Gonsalves *et al.*, "GeV Electron Beams from a Centimetre-Scale Accelerator," *Nature Physics* **2**, 696 (October 2006); W. Leemans and E. Esarey, "Laser-Driven Plasma-Wave Electron Accelerators," Physics Today, 44 (March 2009); E. Esarey, C.B. Schroeder and W.P. Leemans, "Physics of Laser-Driven Plasma-Based Electron Accelerators," *Rev. Mod. Phys.* **81**, 1229 (July–September 2009); S. Steinke, J. van

Tilborg, C. Benedetti *et al.*, "Multistage Coupling of Independent Laser-Plasma Accelerators", *Nature* **530**, 190 (February 11, 2016).

82. G.E. Moore, "Lithography and the Future of Moore's Law," in *Electron-Beam, X-Ray, EUV, and Ion-Beam Submicrometer Lithographies for Manufacturing V* (SPIE, Bellinghom, WA, 1995), J. Warlaumont, Editor, *Proc. SPIE* **2437**, 2 (1995); G.E. Moore, "Cramming More Components onto Integrated Circuits," *Electr. Mag.* **114** (April 19, 1965); G.E. Moore, in *Proc. IEEE Int. Electr. Dev. Meeting* (1975).

83. *International Technology Roadmap for Semiconductors*; http://public.itrs.net; Sematech's *International Symposium on Extreme Ultraviolet Lithography*, Maastrict, Netherlands, October 2015.

84. H. Kinoshita, *Extreme Ultraviolet Lithography: Principles and Basic Technologies* (Lambert Academic Publishing, Berlin, June 2016); H. Kinoshita, R. Kaneko, K. Takei, N. Takeuchi, and S. Ishihara: Ext. Abstr. (1986, 47th Autumn Meet., Japan Society of Applied Physics, 28-ZF-15 [in Japanese]; H. Kinoshita, K. Kurihara, Y. Ishii, and Y. Torii, "Soft X-Ray Reduction Lithography Using Multilayer Mirrors," *J. Vac. Sci. Technol. B* 7, 1648 (1989); H. Kinoshita, T. Watanabe and T. Harada, "Development of Element Technologies for EUVL," *Adv. Optical Technol.* **4**(4), 319–331 (July 2015).

85. V. Bakshi, *EUV Lithography* (Wiley and SPIE, 2009); Second Edition (SPIE, 2016).

86. B. Wu and A. Kumar, "Extreme Ultraviolet Lithography: A Review," *J. Vac. Sci. Technol. B*, **25**, 1743 (November/December 2007).

87. The industry definition of acceptable EUV power is that it be centered at 13.5 nm, within a 2% spectral bandwidth, clean of both plasma debris and extraneous radiation outside this bandwidth (IR, visible, UV), and that the power be measured at what is known as the intermediate focus (IF), the entrance to the multi-mirror EUV projection optics that transfer the radiation to the mask and from there to the wafer.

88. Y. Borodovsky, Director, Advanced Lithography, Intel, "EUV Lithography at Insertion and Beyond", International Workshop on EUV Lithography (Maui, HI, 2012).

89. V.Y. Banine, K.N. Koshelev and G.H.P.M. Swinkels, "Physical Processes in EUV Sources for Microlithography," *J. Phys. D: Appl. Phys.* **44**, 253001 (June 2011); V. Banine, J. Benschop, M. Leenders and R. Moors, "The Relationship Between an EUV Source and the Performance of an EUV Lithographic System," SPIE **3997**, 126 (2000).

90. J. Pankert *et al.*, "EUV Sources for the Alpha-Tools," *SPIE* **6151** (March2006)

91. M. Corthout, Y. Teramoto and M. Yoshioka, "EXTREME Technologies: First in Tin Beta SoCoMo Ready for Wafer Exposure", Sematech International EUV Lithography Symposium (Kobe, October 2010).

92. D. Brandt *et al.*, "LPP Source System Development for High Volume Manufacturing," *SPIE* **7636**, 763611 (2010); I.V. Fomenkov *et al.* "Laser Produced Plasma Light Source for EUVL," SPIE **7636**, 763639 (2010); I. Fomenkov, "EUVL Exposure Tools for HVM: Status and Outlook", EUVL Workshop, Berkeley (June 2016).

93. H. Mizoguchi *et al.*, First Generation Laser-Produced Plasma Source System for HVM EUV Lithography," *SPIE* **7636**, 763608 (2010).

94. G. O'Sullivan and P.K. Carroll, "4d-4f Emission Resonances in Laser-Produced Plasmas," *J. Opt. Soc. Amer.* **71**, 227 (1981); G. O'Sullivan, A. Cummings, P. Dunne *et al.*, "Atomic Physics of Highly Charged Ions and the Case for Sn as a Source Material", Chapter 5 in *EUV Sources for Lithography* (SPIE, Bellingham, WA, 2005); G. O'Sullivan, B. Li, R. D'Arcy *et al.*, "Spectroscopy of Highly Ionized Ions and its Relevance to EUV and Soft X-Ray Source Development for Lithography," *J. Phys.* B **48**, 144025 (2015).

95. J. White, P. Hayden, P. Dunne *et al.*,"Simplified Modeling of 13.5 nm Unresolved Transition Array Emission of a Sn Plasma and Comparison with Experiment," *J. Appl. Phys.* **98**, 113301 (December 2005); J. White, "Opening the Extreme Ultraviolet Lithography Source Bottleneck: Developing a 13.5-nm Laser-Produced Plasma Source for the Semiconductor Industry", PhD Thesis, University College Dublin (February 2006); R. Lokasani, E. Long, O. Maguire *et al.*, "XUV Spectra of 2nd Transition Row Elements: Identification of 3d-4p and 3d-4f Transition Arrays," *J. Phys. B: At. Mol. Opt. Phys.* **48**, 245009 (November 13, 2015).

96. J. White, P. Dunne, P. Hayden, F. O'Reilly and G. O'Sullivan,"Optomizing 13.5 nm Laser-Produced Tin Plasma Emission as a Function of Laser Wavelength," *Appl. Phys. Lett.* **90**, 181502 (2007).

97. F. Jin, M. Richardson, M. Kado, A.F. Vassiliev and D. Salzmann, *SPIE* **2015**, 151(1994); presented July 2013 at SPIE San Diego.

98. M.C. Richardson, C.-S. Koay, K. Takenoshita and C. Keyser, "High Conversion Efficiency Mass-Limited Sn-Based Laser Plasma Source for EUV Lithography," *J. Vac. Sci. Technol. B* **22**, 785 (2004); M. Richardson, C.-S. Koay, K. Takenoshita *et al.*, "Laser Plasma EUV Sources based on Droplet Target Technology", Chapter 26 in *EUV Sources for Lithography* (SPIE, Bellingham, WA, 2005); C.-W. Koay, "Radiation Studies of the Tin-Doped Microscopic Droplet Laser Plasma light Source Specific to EUV Lithography", PhD thesis, University of Central Florida, May 2006.

99. M. Al-Rabban, M. Richardson, H. Scott *et al.*, "Modeling LPP Sources", Chapter 10 in *EUV Sources for Lithography* (SPIE, Bellingham, WA, 2005).

100. K. Takenoshita,"Debris Study and Mitigation on Tin-Doped Droplet Laser Plasma Source for EUV Lithography", PhD thesis, University of Central Florida, August 2006.

101. S.A. George, "Spectroscopic Studies of Laser Plasmas for EUV Sources," PhD thesis, University of Central Florida, December 2007; S.A. George, W.T. Silfvast, K. Takenoshita *et al.*, "Comparative Extreme Ultraviolet Emission Measurements for Lithium and Tin Laser Plasmas," *Optics Lett.* **32**, 997 (April 15, 2007).

102. I. Yu. Tolstikhina, S.S. Churilov, A.N. Ryabtsev and K.N. Koshelev, "Atomic Tin Data," Chapter 4 in *EUV Sources for Lithography* (SPIE, Bellingham, WA, 2005).

103. J.D. Gillaspy, "Atomic Xenon Data", Chapter 3 in *EUV Sources for Lithography*, *ibid*. For recent tin data see the NIST reference tables: www.physics.nist.gov/PhysRefData/ASD/ionEnergy.html

104. M. Nakano, T. Yabu, H. Someya *et al.*, "Sn Droplet Target Development for Laser Produced Plasma EUV Light Source," *SPIE* **6921**, 692130 (2008)

105. H. Mizoguchi, H. Nakarai, T. Abe *et al.*, "Performance of One Hundred Watt HVM LPP-EUV Source," *SPIE* **9422**, 942211 (2015); H. Mizoguchi *et al.*, "One Hundred Watt Class EUV Source Development for HVM Lithography", 2014 EUV Source Workshop (University College Dublin, Ireland, November 2014).

106. L. Rymell and H.M. Hertz, "Droplet Target for Low-Debris Laser-Plasma Soft X-Ray Generation," *Optics Commun.* **103**, 105 (November 1, 1993).

107. H.M. Hertz, L. Rymell, M. Berglund and L. Malmqvist, "Debris-Free Soft X-ray Generation Using a Liquid Droplet Laser-Plasma Source," *SPIE* **2523**, 88 (1995); B.A.M. Hansson and H.M. Hertz, "Liquid-Jet Laser-Plasma Extreme Ultraviolet Sources: From Droplets to Filaments," *J. Phys. D: Appl. Phys.* **37**, 3233 (November 2004).

108. M. Berglund, L. Rymell and H.M. Hertz, "Ultraviolet Prepulse for Enhanced X-Ray Emission and Brightness from Droplet-Target Laser Plasmas," *Appl. Phys. Lett.* **69**, 1683 (September 16, 1996).

109. P.A.C. Jansson, B.A.M. Hansson, O. Hemberg *et al.*, "Liquid-Tin-Jet Laser-Plasma Extreme Ultraviolet Generation," *Appl. Phys. Lett.* **84**, 2256 (March 29, 2004).

110. K. Nishihara, A. Sunahara, A. Sasaki *et al.*, "Advanced Laser-Produced EUV Light Source for HVM with Conversion Efficiency of 5–7% and B-Field Mitigation of Ions," *SPIE* **6921**, 692125 (April 2008).

111. Cymer: https://www.cymer.com/plasma-chamber-detail

112. A. Pirati, R. Peters, D.A. Smith *et al.*, "Performance Overview and Outlook of EUV Lithography Systems," *SPIE* **9422**, 942260 (2015).

113. A.A. Schafgans, D.J. Brown, I.V. Fomenkov *et al.*, "Performance Optimization of MOPA Prepulse LPP Light Source," *SPIE* **9422**, 942210 (2015).

114. A. Yen, TSMC; ASML press releases February 24, 2015 and April 22, 2015.

Homework Problems

Homework problems for each chapter will be found at the website:
www.cambridge.org/xrayeuv

9   Extreme Ultraviolet and Soft X-Ray Lasers

$$\frac{I}{I_0} = e^{GL} = e^{z/Lg} \qquad (9.2)$$

$$G \equiv n_u \sigma_{\text{stim}} F \qquad (9.4)$$

$$\sigma_{\text{stim}} = \frac{\pi \lambda r_e}{(\Delta\lambda/\lambda)} \left(\frac{g_l}{g_u}\right) f_{lu} \qquad (9.18)$$

$$\left.\frac{(\Delta\lambda)}{\lambda}\right|_{\text{FWHM}} = \frac{2\sqrt{2\ln 2}}{c}\sqrt{\frac{\kappa T_i}{M}} \qquad (9.19)$$

$$\frac{P}{A} = \frac{16\pi^2 c^2 \hbar (\Delta\lambda/\lambda) GL}{\lambda^4} \qquad (9.22)$$

Lasing at short wavelengths in the EUV and soft x-ray regions of the spectrum is achieved in hot dense plasmas. Temperatures of several hundred electron volts to above 1 keV are required to collisionally excite atoms (ions) to the required energy levels. As these are well above the binding energies of outer electrons, the atoms are necessarily ionized to a high degree. Upper state lifetimes are typically measured in picoseconds, so that energy delivery (pumping) must be fast. As a result high power infrared, visible, and ultraviolet lasers are generally employed to create and heat the plasma, although in some cases fast electrical discharges are employed. Population inversion is generally accomplished through selective depopulation, rather than selective population. High gain lasing requires a high density of excited state ions, thus mandating a high density plasma. Preferred electron configurations are hydrogen-like (single electron, nuclear charge $+Ze$), neon-like (10 electrons), and nickel-like (28 electrons) ions, which tend to have a large fraction of the plasma ions in a desired ionization state. The short lifetime of hot dense plasmas limits the effectiveness of cavity end mirrors, so that in general these are high gain single pass lasers, albeit with some exceptions. Lacking multipass mode control, short wavelength lasers typically are far from diffraction limited. Temporal coherence lengths, set largely by ion Doppler line broadening, are typically 10^4 waves. The pumping power necessary to produce short wavelength lasers scales as $1/\lambda^4$. Recent high gain experiments demonstrate a capability for saturated lasing throughout much of this spectral region.

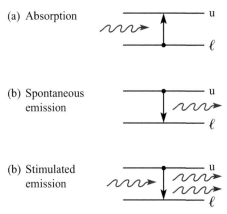

(a) Absorption

(b) Spontaneous
    emission

(b) Stimulated
    emission

Figure 9.1 The processes of absorption, spontaneous emission, and stimulated emission.

9.1   Basic Processes

Lasing involves the stimulated emission and amplification of resonant electromagnetic radiation by quantized atomic systems in identical excited states.*[1-6] In such interactions the passing radiation stimulates the excited atoms to undergo transitions from their upper state to a lower state, resulting in the emission of radiation at the same frequency and phase coherent with the stimulating radiation. For lasing to occur there must be a population inversion in which there are more atoms in the upper excited state than in the lower state.

We begin the discussion of lasing with a review of absorption, spontaneous emission, and stimulated emission of radiation involving quantized atomic states, as illustrated in Figure 9.1. For the absorption process the atom is initially in the lower energy state, labeled l in Figure 9.1(a). Incident radiation of precise energy $\hbar\omega = E_u - E_l$ causes the bound electron to oscillate, acquiring the necessary energy to make a transition to the upper energy state u. For states whose difference in energy is defined to a specificity $\Delta E/E = \Delta\omega/\omega = \Delta\lambda/\lambda$, the transition involves a large number of oscillations between the two states, with the atom eventually residing in the upper level. The number of oscillations, of order $E/\Delta E$, is typically 10^6 or more. For the spontaneous emission process, described previously in Chapter 1 and shown here in Figure 9.1(b), the atom is initially in an excited state. Perhaps due to background field fluctuations, the electron is perturbed and begins to oscillate between the upper and lower energy states, emitting radiation at frequency ω in a wavetrain of duration (in cycles) of order $E/\Delta E$, eventually residing in the lower energy state. The third process, shown in Figure 9.1(c), is that of *stimulated emission*, which occurs when incident radiation of resonant frequency ω encounters an atom already in the upper excited state. Here again the electron is caused to oscillate at the frequency ω by the incident radiation, undergoing many oscillations and thus resulting in the emission of radiation with the atom eventually residing in the

---

* The acronym "laser" is derived[3] from "light amplification through stimulated emission of radiation."

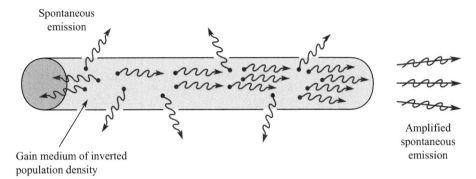

Figure 9.2 The lasing process begins with amplified spontaneous emission (ASE) in directions for which there is a long axial path length. The illustration shows amplification only toward the right, but it would actually occur in both directions. Radiation to the side leaves the gain medium after too short a propagation path to experience significant gain.

lower energy level. In this case the emitted radiation is not only at the same frequency, but is *phase coherent* with the stimulating radiation, and of the same polarization.

Lasing occurs when many atoms are initially in the same upper states and a cascading occurs in which some initial radiation, perhaps building from spontaneous emission (noise) or from an incident wave, causes the sequential stimulation of many phase coherent emissions, leading to substantial wave amplitude or energy amplification.[7–9] The spatial and temporal coherence properties of the resultant radiation will depend on the control of this initiating process. Known as *mode control*, this involves phase space, bandwidth, and polarization limitations imposed by geometry and cavity optics, a topic we take up later in this section.

The initiation of lasing is illustrated in Figure 9.2, which shows the random emission from a collection of atoms, initially all in the same excited state. Owing to variations in background field perturbations and the initial absence of a stimulating field, the atoms begin to spontaneously emit radiation at random times and in various directions. Early in the process, although at the same frequency ω, the various emissions are phase incoherent, as the initiation of the emission processes in different atoms is random and uncorrelated. As time progresses the situation changes. For emission in the lateral directions there is little chance for substantial amplification, due to the short path lengths; thus these emissions continue to be spontaneous and incoherent. However, for radiation emitted along the longer axial path there is a much increased probability of interacting with excited atoms, leading to stimulated phase coherent emission in a cascading, ever more intense propagating wave. This is the process of amplified spontaneous emission (ASE), which, with sufficient path length and density of excited atoms, evolves to lasing action. As we will see in the following section, it leads to exponential growth in the long path axial direction. In some of the earliest EUV and soft x-ray laser demonstrations,[10–13] the observation of exponential intensity growth with length, in a well-defined axial direction, was a primary diagnostic, giving clear evidence that lasing had occurred. This was contrasted with lateral emissions that grew only linearly

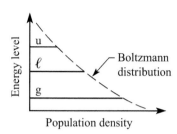
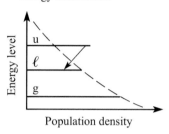

(a) Equilibrium energy distribution

(b) Non-equilibrium inverted energy distribution

Figure 9.3 Equilibrium and non-equilibrium energy distributions. Lasing requires an inverted population density (more atoms in the upper state than in the lower state), as in (b).

with path length, with little chance for amplification before exiting the active (inverted population density) region. Reviews of EUV/soft x-ray lasing are given by Daido,[14] Tallents,[15] Elton,[16] and by Suckewer and Jaeglé.[17]

The process of lasing described above is dependent on a population inversion: the presence of more atoms in the upper excited state u than in the lower state l. Without such an inversion there is likely to be more absorption of radiation than stimulated emission, leading to a net decrease of wave intensity with propagation distance. Figure 9.3 shows population density versus energy level for two cases, (a) an equilibrium situation in which the upper state u is less populated than the lower state l, and (b) a non-equilibrium situation with an inverted population distribution between the upper and lower states. The inverted case, with more atoms in the upper state than in the lower state, can lead to lasing, while the equilibrium distribution leads to net absorption. The question then is, how can an inverted population distribution be obtained?

One method is to flood the atoms with radiation of sufficiently high photon energy to excite them to higher energy levels, permitting the atoms to evolve back to the ground state through transitions to various intermediate excited states. Because of differences in decay times for the various states, some of which may be *metastable* or relatively long lived due to less favorable quantum transition rates, temporary population inversions may occur. Thus for a brief period conditions for lasing may be present. In the presence of a sufficiently high density of excited states and a sufficiently long length of such lasing material, lasing will occur. Many visible light lasers operate with nanosecond, picosecond, and shorter pulse durations. Some operate in a pulse repetition mode, or in a continuous wave (cw) mode, where the cycle of pumping, excitation, and de-excitation through lasing is constantly repeated.

Other forms of pumping the inversion, more typical of short wavelength lasers, involve collisional excitation or recombination into a higher excited state. Figure 9.4 shows several energy levels of an atom to which energy has been provided by a pump that raises the atom to a long-lived excited state u, or to a higher excited state (not shown) from which it cascades down to the state u, and eventually lases to the state labeled l. The pump could consist, for example, of a collision with an energetic electron in a plasma. The lower level is unoccupied, perhaps because of its relatively rapid

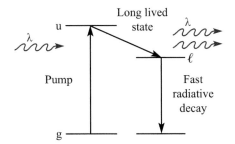

Figure 9.4 Three-level lasing between an upper state u and a lower state l. A pumping mechanism from the ground state g is also shown. Population inversion is obtained with many such atoms if the transition from u to l is relatively slow, while that from l to the ground state is relatively fast. This avoids subsequent reabsorption of the stimulated emission, or so-called radiation trapping.

radiative decay. Thus for some period of time, determined by quantum transition rates, the atom resides in the upper state, available to participate in stimulated emission. The key to producing the inverted population density is the availability of upper and lower states with sufficiently dissimilar lifetimes.

Figure 9.5 shows a somewhat more realistic energy level diagram for a particularly simple one-electron (hydrogen-like) ion. This diagram will help us understand the energetics of EUV and soft x-ray lasing. In this spectral region the photon energies of interest for laboratory lasers extend from perhaps 50 eV to 500 eV. The requisite pump energies are necessarily higher, in that the atom must be lifted from the ground state to an energy at least equal to that of the upper excited state. With such energetic pump processes the atoms will surely be ionized to some high degree, as was described in Chapter 8, Section 8.7.2. In fact all successful EUV and short wavelength lasers involve highly ionized plasmas,[10–21] either laser-produced plasmas or discharge plasmas. The simplified energy level diagram in Figure 9.5 relates to a hydrogen-like (single electron) ion of nuclear charge Z. In this case the highly ionized atom is stripped of all electrons through energetic collisions in a hot plasma. As the plasma cools, a nearby free electron recombines with the ion, into a high level excited state. The electron then drops down through the bound states, emitting characteristic radiation. In Figure 9.5 the ion is shown undergoing a stimulated $n = 3$ to $n = 2$ (Balmer α) transition, emitting a characteristic photon of energy [see Chapter 1, Eq. (1.8)]

$$\hbar\omega = (13.606 \text{ eV})Z^2 \left(\frac{1}{n_f^2} - \frac{1}{n_i^2} \right) \tag{9.1}$$

where for $n_i = 3$, $n_f = 2$, and a carbon nucleus of charge $Z = 6$ we have $\hbar\omega = 68.03$ eV (18.22 nm wavelength). Indeed, studies of this transition in hydrogen-like carbon played an important role in the early development of short wavelength lasers.[22–26] Note that the ionization energy required to remove the last electron from the carbon atom is 490 eV [$n_i = 1$, $n_f = \infty$ in Eq. (9.1)], so that preparation of such an ion for recombination laser studies requires an electron temperature of 100 eV to 200 eV.

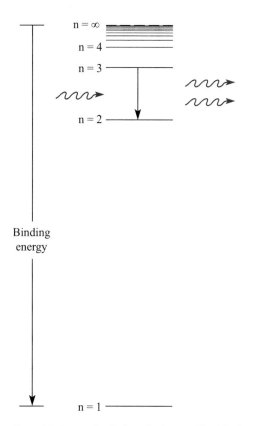

Figure 9.5 Energy levels for a hydrogen-like (single electron) ion of nuclear charge Z. Stimulated emission is shown for an $n = 3$ to $n = 2$, Balmer α transition. Energy levels scale according to Z^2. See Eq. (9.1) and Table 9.1 for values.

With regard to favorable lifetimes and oscillator strengths, summarized here in Table 9.1 for a hydrogen-like ion of nuclear charge Z, the 3d $\rightarrow$ 2p transition has an oscillator strength f_{32} of 0.696, with thus a high probability for transition, and a lifetime of 12 ps.[27] The subsequent 2p $\rightarrow$ 1s transition to the ground state has an oscillator strength of 0.416 and a lifetime of only 1.2 ps, allowing for a fast depopulation of the lower state, as desired. Note that energy levels in Eq. (9.1) scale as Z^2, a general trend for higher photon energy lasers. The ionization energies scale in the same manner, so that requisite temperatures scale as $\hbar\omega$ and Z^2 as well.

Earlier in this section reference was made to the use of cavity optics for the control of coherence, spectral bandwidth, and polarization, techniques frequently employed at visible wavelengths.[7, 9] Figure 9.6 illustrates (a) cavity optics for a visible light laser, and (b) a high gain, single pass EUV/soft x-ray laser. The visible light laser is configured for best spatial and temporal coherence. High reflectivity front and back mirrors are used to return the initial amplified spontaneous emission (ASE) many times through the gain medium (gas or solid with population inverted atoms – the "active" region). This improves energy extraction, providing a higher output power and permitting many

Table 9.1 Transitions in single electron, hydrogen-like carbon ions ($Z = 6$ nuclear charge). Photon energies, according to Eq. (9.1), scale as Z^2. Oscillator strengths are independent of Z. Radiative lifetimes,[27] which are the reciprocals of the Einstein coefficients A_{ul} given in Eq. (9.17), scale as Z^4, i.e., $\tau_Z = \tau_H / Z^4$.

Transition u–l	Photon energy $\hbar\omega$ (eV)	Wavelength λ (nm)	Oscillator strength[a] f_{lu}	Lifetime $\tau = 1/A_{ul}$ (ps)
2p –1s	367.0	3.378	0.4162	1.2
3p–1s	435.0	2.850	0.0791	
4p –1s	458.7	2.703	0.0290	
3p –2s	68.03	18.22	0.435	4.1
3s –2p	68.03	18.22	0.0136	
3d –2p	68.03	18.22	0.696	12.0
4p –2s	91.84	13.50	0.103	
4s –2p	91.84	13.50	0.0030	
4d –2p	91.84	13.50	0.122	
4p –3s	23.81	52.07	0.485	

[a] The emission oscillator strength f_{ul} (which has a negative value) and absorption oscillator strength f_{lu} are related by $f_{ul} = -(g_l/g_u)f_{lu}$.

passes for phase-space control, spectral bandwidth selection, and choice of polarization. (Phase space control refers to the product of beam diameter and divergence at the lasing wavelength. The relationship between phase-space product and full spatial coherence, $d \cdot \theta = \lambda/2\pi$ for Gaussian beams, is discussed in Chapter 4.)

Typically the rear mirror in Figure 9.6(a) would have a reflectivity of 99.9% for a visible light laser, while the partially transmitting front mirror might have a reflectivity of 90%, allowing some laser radiation to pass once for each round trip within the cavity. A transverse mode selector, typically a pinhole, blocks all ASE in the laser startup period except that which will eventually satisfy the stringent phase-space constraint for full spatial coherence – in cavity parlance a transverse electromagnetic mode, TEM$_{00}$. The accepted ASE, which passes through the pinhole aperture, is reflected back through the gain medium, amplified, and returned again by the focusing rear mirror. Through many round trip passes the phase-space-selected radiation is selectively and exponentially amplified, easily dominating random ASE in undesired directions. At the same time a longitudinal mode selector – typically two axially separated resonant thin films, forming a high spectral selectivity Fabry–Perot bandpass – is used to narrow the laser line width to as little as a single axial mode of the cavity, typically narrower than the natural line width of stimulated emission. Narrowing the spectral bandwidth $\Delta\lambda$ increases the temporal (longitudinal) coherence length $l_{\mathrm{coh}} = \lambda^2/2\Delta\lambda$, also discussed in Chapter 4.

In some cases the narrow spectrum mode selector is replaced by an axial mode-locking *saturable dye* absorber. This passes only large intra-cavity intensity spikes, which tend to include contributions from all possible axial modes. Possessing the largest possible lasing bandwidth, this tends to produce the shortest time duration pulses. This

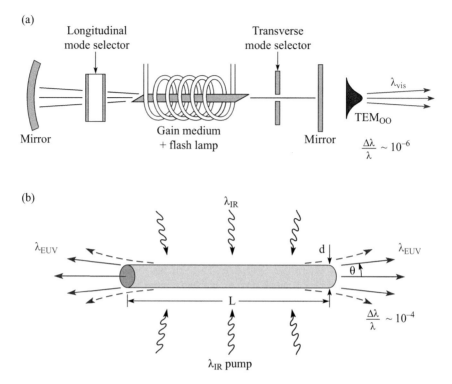

Figure 9.6 (a) Diagram of a visible light laser employing cavity defining feedback mirrors, an etalon for spectral narrowing through the enhancement of selected longitudinal cavity modes, a pinhole aperture for transverse mode control (spatial coherence), the gain medium, and a flashlamp pump to lift atoms into desired energy states. (b) Diagram of an EUV or soft x-ray laser based on high gain amplified spontaneous emission from an elongated laser-produced plasma (pump wavelength λ_{IR}) containing an inverted population density of ions in an upper excited state. The divergence of emitted radiation is set largely by the aspect ratio, d/L, but is also affected by refraction in the steep transverse density gradient plasma, as suggested by the dashed lines.

and other gain controlling techniques are commonly used to extract available laser energy in short pulses, providing high peak power output.

The gain medium shown in Figure 9.6(a) is a solid state rod, perhaps ruby, YAG, or glass, cut at Brewster's angle[†] (see Chapter 3, Section 3.6) for polarization control, and pumped by a flash lamp of incoherent light to achieve the initial population inversion. Both gaseous and solid state lasing media are common at visible wavelengths.[7, 9]

By comparison the EUV/soft x-ray laser, at least in its typical configuration as shown in Figure 9.6(b), is much simpler. In order to simultaneously achieve the requisite

---

[†] For lasing rods cut at Brewster's angle (see Chapter 3, Section 3.6), properly polarized light experiences no reflective loss at the air–solid interface, giving a slight intensity advantage over other ASE polarizations in the round trip gain competition, which rapidly leads to single polarization dominance in the multipass exponential lasing process.

high temperature and high density, energy is deposited in a short time, typically sub-picosecond to several nanoseconds in duration, generally by a high power infrared or visible laser or, for longer EUV wavelength lasers, by a short pulse, high current electrical discharge. Since light travels at a speed of 300 μm/ps, a 1 ns duration laser pulse would permit use of a single mirror at 15 cm or two mirrors each 7.5 cm from the center for full cavity operation. Generally mirror damage due to pump–laser scattering and plasma debris limit the use of cavity end mirrors at EUV and soft x-ray wavelengths, although some work has been done.[28, 81, 82] For shorter pulse duration the cavity length would be even shorter than the required gain length, typically measured in centimeters, thus excluding the possibility of multipass operation, unless regenerated plasma techniques are developed. As a consequence, present EUV/soft x-ray lasers are largely single pass, high gain devices. Without cavity mirrors, the ability to control spatial coherence is limited to geometrical considerations, such as the ratio of output aperture diameter to axial lasing length. These considerations are further complicated by refractive effects that tend to increase the divergence of laser radiation due to sharp lateral density gradients encountered in these rapidly expanding hot dense plasmas.[11, 14, 30] The subject of refraction in plasmas is discussed in Chapter 8, Section 8.6. Efforts to control the effects of refraction on short wavelength lasers, including prepulses, multiple pulses, double targets and special pump focusing techniques, are currently of great interest.[29, 31–33] In part this interest is due to the fact that the phase space product of EUV/soft x-ray lasers (the product of beam diameter and divergence) is generally much larger than the wavelength, indicating that the radiation consists of many transverse modes, and indicating minimal spatial coherence, particularly at the shorter wavelengths.[34]

Further discussion of spatial coherence, and improvements through techniques such as pinhole spatial filters, staged amplifiers, and seeding, is presented at the end of this chapter. Laser line widths, typically dominated by Doppler broadening[16] due to motion of the relatively hot lasing ions, are generally of order $\Delta\lambda/\lambda \sim 10^{-4}$, leading to temporal coherence lengths approaching a millimeter, values quite useful for many applications.

9.2   Gain

Obtaining exponential gain from stimulated emission of radiation requires a substantial population density inversion. Generally we inquire as to what difference in upper state and lower state ion densities, n_u and n_l, will lead to a substantial gain–length product GL, such that an initial emission intensity I_0 grows according to

$$\frac{I}{I_0} = e^{GL} \qquad (9.2a)$$

where I_0 is an initial intensity (power per unit area) that grows to a value I after propagating a distance L (or z) in a lasing medium of gain per unit length G.[‡] Note that G

‡ For short-wavelength lasers a more appropriate model is that of uniformly distributed spontaneous emission, amplified by stimulated emission as it propagates to the exit surface of the gain medium. In this case the

has units of inverse length, and that the "gain length", defined as one e-folding of laser intensity, is given

$$L_g \equiv 1/G \tag{9.2b}$$

The gain is often expressed in terms of atomic cross-sections for stimulated emission and absorption, σ_{stim} and σ_{abs}:

$$G = n_u \sigma_{\text{stim}} - n_l \sigma_{\text{abs}} \tag{9.3}$$

which can be written in terms of a density inversion factor F, as

$$G \equiv n_u \sigma_{\text{stim}} F \tag{9.4}$$

where the density inversion factor is given by

$$F \equiv 1 - \frac{n_l \sigma_{\text{abs}}}{n_u \sigma_{\text{stim}}} = 1 - \frac{n_l g_u}{n_u g_l} \tag{9.5}$$

and where the otherwise symmetrical cross-sections for stimulated emission and absorption may differ by statistical weights g_u and g_l, which are due to degeneracies (same energy, different quantum numbers) in the upper and lower lasing states.

Expressions for the cross-section for stimulated emission and gain can be obtained in terms of the Einstein A and B coefficients,[36-38] which appear when one considers rate equations for transitions among quantum states. For example, if we consider the rate of transitions, per unit volume, between the upper and lower states u and l, then *in radiative equilibrium* (as many transitions up as down)

$$n_u A_{ul} + n_u B_{ul} U_{\Delta\omega}(\hbar\omega) = n_l B_{lu} U_{\Delta\omega}(\hbar\omega) \tag{9.6}$$

where again n_u and n_l are the densities of atoms (ions) in the upper and lower states, A_{ul} is the spontaneous decay rate (number per second) from u to l, B_{ul} is the stimulated transition rate from u to l in the presence of radiation of spectral energy density[¶] $U_{\Delta\omega}(\hbar\omega)$, and B_{lu} is the absorption coefficient for transition from the lower state l to the upper state u. Again B_{ul} and B_{lu} would be equivalent, due to the symmetric nature of the two processes, except for the degeneracies, so that in general $g_l B_{lu} = g_u B_{ul}$. In equilibrium, $U_{\Delta\omega}$ is the Planckian distribution

$$U_{\Delta\omega}(\hbar\omega) = \frac{\hbar\omega^3}{\pi^2 c^3 (e^{\hbar\omega/kT} - 1)} \tag{9.7}$$

where the density of states follows the Boltzmann energy distribution[34]

$$\frac{n_l}{n_u} = \frac{g_l}{g_u} e^{(E_u - E_l)/\kappa T} = \frac{g_l}{g_u} e^{\hbar\omega/\kappa T} \tag{9.8}$$

---

integrated output intensity I, for an active medium gain G and length L, is $I = J_s (e^{GL} - 1)^{3/2} / G(GLe^{GL})^{1/2}$ where J_s is the spontaneously emitted power per unit volume within the lasing line. This is often referred to as the Linford formula.[35] It is widely used to assign gain values G in experimental studies of laser intensity versus length, such as we shall encounter here in Sections 9.3 and 9.4.

[¶] See Chapter 8, Eq. (8.134a). $U_{\Delta\omega}(\hbar\omega)$ has units of energy per unit volume, per unit frequency interval $\Delta\omega$, at frequency ω.

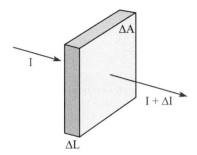

Figure 9.7 Geometry for considering laser amplification in terms of an incremental increase in intensity, ΔI.

A consistent solution of Eqs. (9.6)–(9.8) requires that the Einstein coefficients be related[36] by[§]

$$\frac{n_l}{n_u} = \frac{g_l}{g_u}e^{(E_u - E_l)/\kappa T} = \frac{g_l}{g_u}e^{\hbar\omega/\kappa T} \tag{9.9}$$

an expression we will make use of shortly.

For lasing the idea is to get out of equilibrium, creating a temporary population inversion where $n_u > n_l$, with a sufficient spectral energy density of photons at $\hbar\omega = E_u - E_l$ that the stimulated process dominates, resulting in the preferential phase-coherent emission of radiation in a narrow spectral band. For the situation where stimulated emission far exceeds spontaneous emission, the rate equation for the increase in energy per unit time, within a narrow spectral bandwidth $\Delta\omega$, can be written [in distinction to Eq. (9.6) for equilibrium] as

$$\frac{\Delta E}{\Delta t} = \Delta I \cdot \Delta A = [n_u B_{ul} U_{\Delta\omega}(\hbar\omega) - n_l B_{lu} U_{\Delta\omega}(\hbar\omega)]\, \hbar\omega \cdot \Delta A \cdot \Delta L \tag{9.10}$$

where, as shown in Figure 9.7, ΔI is the incremental increase in intensity within the resonant bandwidth of the transition from u to l, ΔA is the element of cross-sectional area, and ΔL is the thickness of the volume element in the direction of propagation, and where we have neglected the spontaneous emission term as being relatively small. The incremental increase in intensity due to stimulated emission, contributed by the population inversion in the element of length ΔL, is then

$$\Delta I = n_u F B_{ul} U_{\Delta\omega}(\hbar\omega)\hbar\omega \cdot \Delta L \tag{9.11}$$

where we have used the density inversion factor defined earlier in Eq. (9.5). Writing the radiation spectral energy density in terms of the local intensity per unit bandwidth (see Chapter 8, Section 8.7.1), we have

$$U_{\Delta\omega}(\hbar\omega) = \frac{I(\hbar\omega)}{(\Delta\omega)c} \tag{9.12}$$

[§] $A_{ul}/B_{ul} = 8\pi h\nu^3/c^3$ when expressed in terms of ν rather than ω. See Ref. 38, pp. 688, 712, and 63.

so that the incremental increase in intensity due to stimulated emission can be written as

$$\frac{\Delta I}{I} = \frac{\hbar \omega n_u F B_{ul} \cdot \Delta L}{(\Delta \omega) c} \tag{9.13}$$

Recalling the relation between A and B coefficients [Eq. (9.9)], this can be rewritten as

$$\frac{\Delta I}{I} = \frac{\pi^2 c^2 n_u F A_{ul} \cdot \Delta L}{(\Delta \omega) \omega^2} \tag{9.14a}$$

Noting that $(\Delta \omega) \omega^2 = (\Delta \omega / \omega) \omega^3 = (\Delta \lambda / \lambda)(2\pi)^3 c^3 / \lambda^3$, the increase in intensity as a function of wavelength is

$$\frac{\Delta I}{I} = \frac{\lambda^3 n_u F A_{ul} \cdot \Delta L}{8\pi c (\Delta \lambda / \lambda)} \tag{9.14b}$$

Integrating this from $\Delta L = 0$ to L, over which the intensity increases from I_0 to I, one obtains the expression previously written as Eq. (9.2a):

$$\frac{I}{I_0} = e^{GL}$$

where now the gain G is given explicitly as

$$G = \frac{\lambda^3 n_u F A_{ul}}{8\pi c (\Delta \lambda / \lambda)} \tag{9.15}$$

Note that G has units of inverse length, and that the "gain length," defined as one e-folding of laser intensity, is given by $L_g \equiv 1/G$. Recalling our earlier definition of the stimulated emission cross-section σ_{stim} in terms of the gain [Eq. (9.4)] as

$$G = n_u \sigma_{\text{stim}} F$$

we now identify the cross-section as[14]

$$\sigma_{\text{stim}} = \frac{\lambda^3 A_{ul}}{8\pi c (\Delta \lambda / \lambda)} \tag{9.16}$$

where $\Delta \lambda / \lambda$ is the full spectral bandwidth.

Quantum mechanically the Einstein A coefficient can be expressed in terms of the oscillator strength[||] f_{lu} as[9, 27]

$$A_{ul} = \frac{e^2 \omega^2}{2\pi \epsilon_0 m c^3} \left(\frac{g_l}{g_u} \right) f_{lu} \tag{9.17}$$

The cross-section for stimulated emission is then

$$\boxed{\sigma_{\text{stim}} = \frac{\pi \lambda r_e}{\Delta \lambda / \lambda} \left(\frac{g_l}{g_u} \right) f_{lu}} \tag{9.18}$$

[||] In this chapter we return to the standard use of f for the oscillator strength, with subscripts u and l to show the upper and lower states. In Chapters 2 and 3 f was used to represent the atomic scattering factor. Also, we use g here, again with u and l subscripts, to denote the degeneracy of atomic states.

where we have introduced the classical electron radius $r_e = e^2/4\pi\epsilon_0 mc^2 = 2.82 \times 10^{-13}$ cm, as described in Chapter 2, Eqs. (2.44) and (2.46).

For EUV and soft x-ray lasers the observed linewidth is dominated by Doppler broadening in the hot plasma. For a Maxwellian velocity distribution the resultant full width at half maximum (FWHM) spectral bandwidth is given by [see Chapter 8, Eq. (8.86), written for ions]

$$\left.\frac{(\Delta\lambda)}{\lambda}\right|_{\text{FWHM}} = \frac{2\sqrt{2\ln 2}\,v_i}{c} = \frac{2\sqrt{2\ln 2}}{c}\sqrt{\frac{\kappa T_i}{M}} \tag{9.19a}$$

where v_i is the rms ion thermal velocity, κT_i is the ion temperature, and M is the ion mass. Expressing κT_i in electron volts, indicated by the brackets, and the ion mass as $2m_p Z$, where m_p is the proton mass and Z is the number of protons, the Doppler broadened linewidth is

$$\left.\frac{(\Delta\lambda)}{\lambda}\right|_{\text{FWHM}} = 7.688 \times 10^{-5}\sqrt{\frac{\kappa T_i\,[eV]}{2Z}} \tag{9.19b}$$

Returning to the calculation of gain, from Eq. (9.4),

$$G = \pi r_e f_{lu}\frac{\lambda n_u F}{(\Delta\lambda/\lambda)}\frac{g_l}{g_u} \tag{9.20}$$

Expressions for gain as an explicit function of wavelength, for given line shapes, are derived by Silfvast in Ref. 6, Chapter 4.

For the previously cited example, with a 3d $\rightarrow$ 2p Balmer α transition at $\hbar\omega = 68.03$ eV (18.22 nm wavelength) in hydrogen-like carbon (see Figure 9.5 and Table 9.1), we can estimate the cross-section for stimulated emission from Eq. (9.18). For a carbon plasma initially heated to an electron temperature of several hundred electron volts to ensure full ionization, then quickly cooled to an ion temperature[12] of $\kappa T_i = 10$ eV, with an ion mass of $M = 2Zm_p = 12$ atomic mass units, the FWHM relative spectral bandwidth is $\Delta\lambda/\lambda = 7.0 \times 10^{-5}$. For the 3d to 2p transition,[27] (see Table 9.1), $f_{23} = 0.696$ and $g_l/g_u = 3/5$, so that from Eq. (9.18), $\sigma_{\text{stim}} \simeq 9.6 \times 10^{-15}$ cm^2. The associated gain, given by Eq. (9.4), is $G \simeq (9.6 \times 10^{-15}$ cm$^2)n_u F$. For exponential amplification one requires, by Eq. (9.2b), a length L such that $GL > 1$, or an inversion density–length product

$$n_u FL > 1/(9.6 \times 10^{-15}\text{ cm}^2)$$

Assuming a lasing medium of length $L = 0.3$ cm and a density inversion factor approaching unity, this requires an initial excited state ($n = 3$) ion density of greater than $3.5 \times 10^{14}/$cm^3. Calculations of ion density and excited state distributions[26] for C^{+5} (one electron) in plasmas in the assumed temperature ranges typically indicate a fraction of order 10^{-3} in the $n = 3$ excited state, thus requiring a total C^{+5} ion density of order 4×10^{17} ions/cm^3, and thus an electron density (five times greater for charge neutrality) of $n_e \simeq 2 \times 10^{18}$ e/cm^3. This electron density is below the critical value for

CO_2 and Nd lasers, 10^{19} and 10^{21} e/cm^3 respectively (see Chapter 8, Table 8.1), and thus is reasonably approached with plasma formation by either system. In the next section we discuss early lasing experiments conducted with plasmas produced by both Nd and CO_2 lasers.

Having some understanding now of the temperatures and densities required to achieve lasing, it is interesting to inquire as to what power and intensity this requires of the driver, and how these scale with lasing wavelength. To estimate the required power that must be delivered to the lasing medium (plasma) in order to maintain the inverted population density, we can write

$$P = \frac{\hbar\omega n_u F V}{\tau} \tag{9.21}$$

where $n_u F$ is the inverted population density, V is the plasma volume, and $\hbar\omega$ is the photon energy that would be emitted by spontaneous emission in a transition of lifetime τ. In fact this is a lower limit on the required power, as the pumping is far from 100% efficient, involving several ionization stages, many energy states, and a general investment in thermal energy. Observing that $\tau = 1/A_{ul}$, and using Eq. (9.15) to replace $n_u F A_{ul} = 8\pi c(\Delta\lambda/\lambda)G/\lambda^3$, the required power per unit volume of plasma is

$$\frac{P}{V} = \frac{16\pi^2 c^2 \hbar (\Delta\lambda/\lambda)G}{\lambda^4}$$

so that in terms of the gain–length product, with $V = AL$, the required power per unit area (i.e., intensity) is

$$\frac{P}{A} = \frac{16\pi^2 c^2 \hbar (\Delta\lambda/\lambda)GL}{\lambda^4} \tag{9.22}$$

Thus to maintain a population inversion density with a given gain–length product and a linewidth $(\Delta\lambda/\lambda)$ determined by the ion temperature κT_i as given in Eqs. (9.19a, b), the requisite laser intensity scales as $1/\lambda^4$. Actually, the linewidth $\Delta\lambda/\lambda \propto \sqrt{\kappa T_i}$, so that if $T_i \propto T_e$ and $\kappa T_e \propto \hbar\omega \propto 1/\lambda$, one has the proportionality $\Delta\lambda/\lambda \propto 1/\sqrt{\lambda}$. Where this is so, the required laser intensity scales[39, 40] as $1/\lambda^{4.5}$. This very rapid scaling of required pump intensity with lasing wavelength provides a significant challenge for the achievement of laser action at soft x-ray wavelengths.

9.3   Recombination Lasing with Hydrogen-Like Carbon Ions

The early history of EUV/soft x-ray lasing is associated with the pursuit of population inversion in hydrogen-like carbon plasmas, formed by the irradiation of carbon fibers at the focus of high power, short duration Nd lasers. The work was motivated in part by the observation of intense EUV lines and the possibility for population inversion as pointed out by Jaeglé, Carillon, and their colleagues,[41] as well as early theoretical predictions by Gudzenko and Shelepin[42] and others.

The first experimental inferences of population inversion, albeit small, were reported by Irons and Peacock,[22] followed by scalable threshold observations by Pert, Ramsden, and their colleagues[23, 25] and by Key, Lewis, and Lamb.[24] These pioneering experiments were all based on recombination in rapidly cooling, fully ionized carbon plasmas, described by the energy level diagram in Figure 9.5, in geometries similar to that of Figure 9.6(b). The general idea is to produce fully stripped carbon ions in a hot dense laser-produced plasma. By Eq. (9.1), removal of the last electron requires an energy of 490 eV, thus requiring an initial electron temperature of 100–200 eV, a value achievable[24, 25] with laser intensities of order 10^{13} W/cm^2 to 10^{14} W/cm^2, depending on the laser wavelength (see Chapter 8, Section 8.7.1). With rapid cooling by expansion and radiation, recombination takes place with low energy electrons populating upper excited levels, forming hydrogen-like (single electron) carbon ions. In this collisional recombination a third particle (an additional electron) is required to satisfy conservation of energy. As a consequence the rate of recombination is proportional to $n_e^2 n_i$, and thus occurs most efficiently at high electron density. Recombination is dependent on a low electron temperature; thus fast cooling is critical to this lasing technique. As the single bound electrons cascade down to lower excited states, a population inversion is created between the $n = 3$ and $n = 2$ levels due to the faster decay rates from $n = 2$ to $n = 1$, as indicated in Table 9.1. The 3d → 2p transition, with a 0.696 oscillation strength and a 12 ps lifetime, is then a good candidate for amplified spontaneous emission, as discussed in the previous section.

Suckewer and his colleagues introduced a novel[43] and successful[12, 13] approach to the idea of recombination lasing by suggesting the use of an axial magnetic field to constrain the lateral expansion of the laser-produced carbon plasma, thus tending to maintain high densities for a longer time while introducing a more favorable geometry for lasing. The basic geometry is illustrated in Figure 9.8(a). As electron densities under 10^{19} e/cm^2 are consistent with required ion densities, as discussed in the preceding section, use was made of a CO$_2$ laser, which delivered approximately 300 J in a 70 ns FWHM pulse, resulting in an incident intensity of 5×10^{12} W/cm^2 on the solid carbon disk target. At this intensity the plasma should reach a temperature approaching 100 eV at peak irradiation intensity, permitting at least a fair fraction of the desired C^{+5} density.** Furthermore, with a CO$_2$ laser wavelength of 10.6 μm, this intensity corresponds to a rather high value of $I\lambda^2$ (see Chapter 8, Section 8.4.11). This leads to the generation of a non-Maxwellian energy distribution in which the hot-electron tail may further assist in the ionization process. With a solenoidal magnetic field of 90 kG, the electrons tend to circle the axial field lines with a Larmor radius†† of order 1 μm as they expand (axially) away from the irradiated disk region, typically 1 mm in diameter. Aspect ratios (L/r) of order 10–100 were obtained with the magnetically confined plasma expansion.

** Note that many authors follow the spectroscopic notation in which C VI is equivalent to C^{+5} (neutral carbon is C I).

†† The Larmor radius, $rL = mv/eB$, is 3.7 μm for a 100 eV electron in a 90 kG, or 9 T, magnetic field.

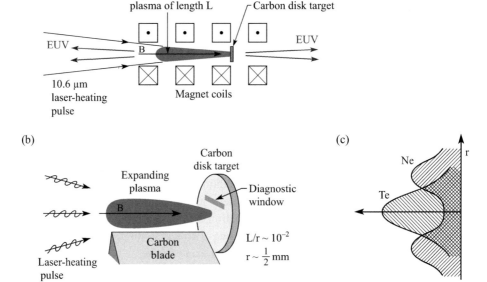

Figure 9.8 (a) A CO_2-laser-produced plasma utilized in early EUV Lasing experiments.[12] The plasma is constrained to a primary axial expansion by a solenoidal magnetic field. (b) The plasma, shown in red, is seen expanding away from a carbon disk target. A perpendicular carbon blade is shown, whose role is to enhance cooling, plasma density, and uniformity in the expanding plasma. A diagnostic window (slot) is shown in the carbon disk, which permits collection and spectral analysis of radiation in the axial direction through use of a spectrometer. A second spectrometer is used to study emissions in the transverse direction. (c) The presence of a strong axial magnetic field, through pressure balance, tends to produce an annular plasma with a density depression on-axis. Courtesy of S. Suckewer, Princeton University.

Shown in Figure 9.8(a) is the incident CO_2 laser beam irradiating a solid carbon disk target, the solenoidal magnetic field, and the expanding plasma. In these experiments one to four carbon blades [see Figure 9.8(b)], mounted perpendicular to the solid disk in off axis positions, were utilized to enhance plasma density, cooling, and uniformity in the axial direction.[13] The presence of a strong axial magnetic field tends to produce a plasma with electron temperature peaked on axis (set by the laser irradiation profile), but with density peaked off axis, as illustrated in Figure 9.8(c), tending to produce favorable lasing conditions in an annular geometry.

Spectrally resolved emissions were observed both axially and in a transverse direction. Sample spectra are shown in Figure 9.9 for a single CO_2 laser pulse. In the axial direction there is a dominant emission line identified as that of the $3 \rightarrow 2$ transition in C^{+5}. In the transverse direction this line is barely discernible. This particular experiment[13] employed a 300 J CO_2 laser pulse of nominal 80 ns duration, and four symmetrically located carbon blades. Measurements in this series showed that maximum gain occurred off axis in a 200 μm thick annular shell at a radius of 1.3–1.5 mm, as observed through the diagnostic viewing slot. The maximum gain in these experiments[13]

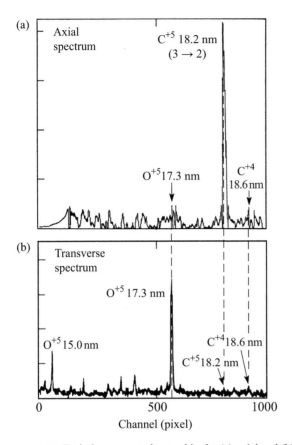

Figure 9.9 Emission spectra observed in the (a) axial and (b) transverse directions from a CO_2 laser-produced plasma formed by irradiating a carbon disk target located in a strong axial magnetic field.[13] The laser energy was 300 J in an 80 ns pulse. Courtesy of S. Suckewer, Princeton University.

corresponds to a gain–length product $GL \simeq 8$. The gain was determined by the simultaneous observation of lasing and non-lasing emission lines in C^{+5}: the $n = 3$ to $n = 2$ line at 18.22 nm, the $n = 4$ to $n = 2$ line at 13.50 nm, and the spontaneous emission $n = 2$ to $n = 1$ line at 3.378 nm used to monitor reproducibility of plasma conditions. Lasing energies of 3 mJ per pulse were recorded at a repetition rate of 0.05 Hz, within a divergence angle of 5–10 mrad.

In a continuation of these early recombination lasing experiments with hydrogen-like carbon, the Princeton group conducted a further series of experiments[44] utilizing a 25 J, 3 ns duration Nd laser pulse to irradiate carbon targets with a 100 μm by several millimeter cylindrical focus, designed to generate lasing in a direction parallel to the target surface. A stainless steel (C + Fe) blade was again used, parallel and near (0.8 mm) to the line focus, to assist in plasma cooling.[43, 44] A magnetic field parallel to the line focus was employed, but played a less essential role due to the natural plasma line shape

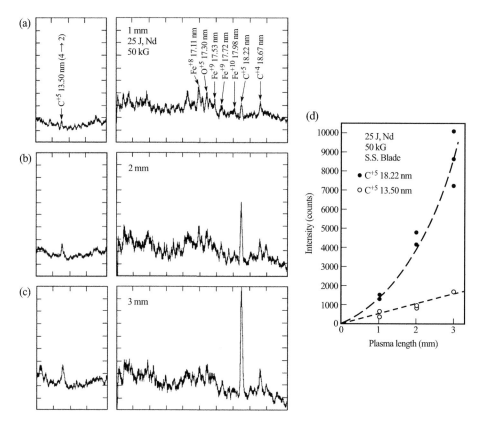

Figure 9.10 (a)–(c) Emission spectra from a hydrogen-like carbon plasma formed by a Nd: glass laser.[44] The line focus length is varied from 1 mm to 3 mm across the surface of a carbon target. Emission is observed in the elongated plasma direction. With longer plasma length the emission line at 18.22 nm grows exponentially out of the noise to become a prominent spectral feature. A companion line, less likely to lase, at 13.50 nm, is shown for comparison. (d) Relative emission intensities of C^{+5} lines at 18.22 nm (solid circles) and 13.50 nm (open circles), showing an exponential growth with plasma length for the $n = 3$ to $n = 2$ line at 18.22 nm. Nd laser energy was 25 J (15 J on target), with a magnetic field of 50 kG. The dashed curve is a theoretical fit to a gain of 8.1/cm. Growth of the 13.50 nm line ($n = 4$ to 2) is linear with length. Courtesy of S. Suckewer, Princeton University.

associated with the line focus. The laser intensity was nominally $(0.8–1) \times 10^{13}$ W/cm^2. Spectrally resolved emission lines observed in the long plasma direction are shown in Figure 9.10, for plasma lengths of 1–3 mm. For the shortest plasma length (1 mm) a number of weak emission lines are observed just above the continuum, including lines of iron (Fe) and oxygen associated with the nearby stainless steel blade. As the plasma length is increased to 2 mm and 3 mm, by elongating the line focus, the C^{+5} $n = 3$ to 2 transition at 18.22 nm grows rapidly out of the noise. A companion line at 13.50 nm, corresponding to an $n = 4$ to 2 transition in C^{+5} (see Table 9.1), shows weaker growth with plasma length. This 13.50 nm line has a shorter wavelength and smaller oscillator strength (see Table 9.1), and is thus expected by Eq. (9.20) to have a smaller gain. In

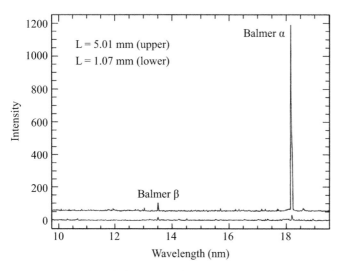

Figure 9.11 On-axis emission spectra from Nd laser irradiated carbon fibers of 7 µm diameter and different lengths.[46] The data show substantial growth for the $n = 3$ to $n = 2$ Balmer α line of C^{+5} at 18.22 nm. Laser intensity is nominally 6×10^{15} W/cm^2 in a 2 ps duration pulse. The gain length product was $GL = 6.5$ for the 5.0 mm long plasma. Courtesy of J. Zhang and M.H. Key, Rutherford-Appleton Laboratory, now at Shanghai Jiao Tong University and Lawrence Livermore National Laboratory, respectively.

Figure 9.10(d) the intensity increase of these two lines is shown as a function of plasma length. The lasing line at 18.22 nm is observed to exponentiate with a gain of $G \simeq 8.1/$cm, while the reference line at 13.50 nm grows linearly with length. The observation of exponential intensity growth with plasma length was used as proof of lasing in these early studies.[45]

Further research at the Rutherford-Appleton Laboratory[46] has led to beautifully resolved 18.22 nm lines in higher intensity, 3×10^{15} W/cm^2, 2 ps duration Nd laser irradiations of 7 µm diameter, 5 mm long carbon fibers, achieving gains up to 12.5/cm. An example of their data is shown in Figure 9.11.

9.4   Collisionally Pumped Neon-Like and Nickel-Like Lasers

Collisionally pumped lasers, involving closed shell, highly ionized ions, offer an alternative path toward high gain at short wavelengths. The technique makes use of ionization bottlenecks, associated with closed electron shells as discussed in Chapter 8, Section 8.7.2, to ensure a high density of ions in a particular ionization state. It employs cylindrical illumination of thin, elongated foils, which are laser-heated to high density and high electron temperature, as illustrated in Figure 9.12. The irradiation intensity is chosen to produce an electron temperature that is well matched to the ionization potential of the desired closed shell ion. Excited states are mostly produced by collisions with plasma electrons, but also by cascading down from higher-still Ne-like (10 electrons)

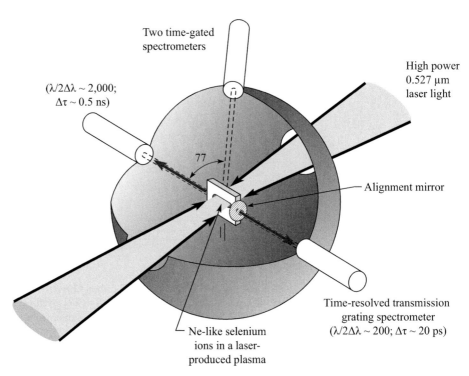

Figure 9.12 Double-sided irradiation of a thin film selenium target, which produced high gain lasing at wavelengths of 20.65 nm and 20.98 nm.[10] High gain amplified spontaneous emission (ASE) grows exponentially in the elongated plasma column, resulting in intense narrow band lasing in two opposite directions. Irradiation is by high power visible light (nominally 2.4 TW, 0.527 μm wavelength, 450 ps FWHM, 7×10^{13} W/cm², 200 μm × 1.1 cm elongated focal spot). Two time-gated spectrometers record axial and off-axis emission spectra. A time resolved transmission grating streak spectrometer records axial emission spectra in the opposite direction. Courtesy of D. Matthews, Lawrence Livermore National Laboratory.

excited states, and by recombination of overly ionized F-like ions. Population inversion is by selective depopulation of the lower lasing state, rather than by selective population of the upper state. These techniques were pioneered at Lawrence Livermore National Laboratory by Matthews,[10, 47] Rosen,[11, 48] Hagelstein, MacGowan,[49–52] and their colleagues,[53–57] based in part on the early theoretical work of Vinogradov and Shlyaptsev[58] and others.

The earliest demonstration[10, 11] of high gain at short wavelength utilizing collisionally pumped closed shell ions employed neon-like selenium ($Z = 34$, 10 electrons, net charge +24). The ionization energy required to remove an electron from an 11-electron Na-like ion is 1036 eV, while that for a 10-electron Ne-like ion is 2540 eV due to the closed shell.[‡‡] With an electron temperature of about 1 keV one can then expect a large fraction of the ions to be in the Ne-like state, with fewer in the Na-like and lower ionization states. Simulations indicate,[48] for instance, that about 20% of all ions are Ne-like,

---

[‡‡] See Chapter 8, Table 8.2, in Section 8.7.2.

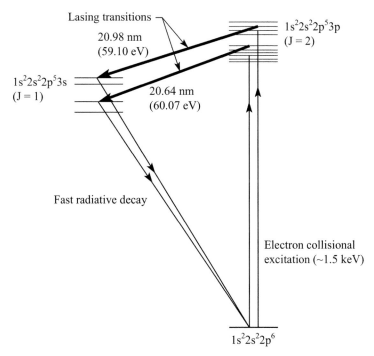

Figure 9.13 Simplified energy diagrams showing two lasing lines, at 20.64 nm and 20.98 nm, in neon-like (10 electron) selenium ions.[11] The ground state has a $1s^2\, 2s^2\, 2p^6$ configuration. The Ne-like ions are produced in a hot dense plasma. The ions are pumped into excited states by direct electron collisional excitation, and by downward cascading from overly ionized F-like ions. Population inversion results from selective depopulation. The 3p to 2p transition is a long-lived dipole forbidden transition, while the 3s to 2p is dipole allowed with a fast radiative decay of about $1/3$ ps. This creates a population inversion between the 3p and 3s states. Courtesy of M. Rosen, Lawrence Livermore National Laboratory.

a similar number are F-like (nine electrons), and the rest are dispersed over a broad range of ionization states. These observations are confirmed by experimental emission spectra.[11] For the Ne-like ions, a population inversion is produced between the $1s^2\, 2s^2$ $2p^5\, 3p$ and $1s^2\, 2s^2\, 2p^5\, 3s$ states, as shown in the energy level diagram of Figure 9.13. Both states are filled by collisions from below, and by recombination and cascading from higher level states. Population inversion results as the $2p^5\, 3p$ to $2p^6$ (ground state) transition is dipole forbidden (see Chapter 1, Section 1.3), while the $2p^5\, 3s$ to $2p^6$ transition is radiatively allowed with a high oscillation strength and a short lifetime, of order $1/3$ ps. Lasing then occurs on the 3p ($J = 2$) to 3s ($J = 1$) transitions. Electron collisional excitation from the ground state to these and higher excited states requires at least 1.5 keV,[59] also requiring a high electron temperature.

Experimental data[10] obtained with thin film selenium foils of three lengths are shown in Figure 9.14. The targets were irradiated with frequency doubled Nd laser light (0.527 μm wavelength), line focused to a nominal intensity of 7×10^{13} W/cm^2, in a pulse of 450 ps duration (FWHM). The foil targets consisted of a 750 Å thick layer of

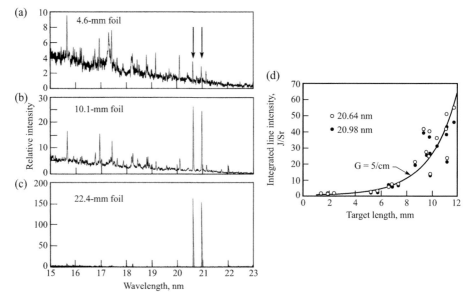

Figure 9.14 (a)–(c) Axial emission spectra are observed for selenium plasmas[10] of lengths 4.6 mm, 10.1 mm, and 22.4 mm. The Ne-like selenium lasing lines, at 20.64 nm and 20.98 nm wavelengths, are observed to grow dramatically in intensity with increasing plasma length. Data are obtained with the time-gated spectrometer shown in Figure 9.12. Observation of the same spectral range from an off-axis position shows these same lines to be very weak, lacking a sufficient propagation distance for growth by stimulated emission. (d) Exponential growth of integrated line intensity with target length provided the primary evidence for lasing on the 3p to 3s transitions at 20.64 nm and 20.98 nm wavelength in Ne-like selenium ions. Courtesy of D. Matthews and colleagues, Lawrence Livermore National Laboratory.

selenium vapor deposited on a nominally 1500 Å thick Formvar ($C_{11}H_{18}O_5$) substrate. Using a cylindrical lens, a line focus of 200 μm × 1.1 cm was obtained. With two beam illumination from opposite sides of the Novette laser, plasmas up to 2.2 cm in length were formed. Owing to the thin nature of the target material, the entire irradiated area was vaporized, creating a single elongated plasma. The combination of experimental data and computer simulations indicates[57] that at the time of lasing the electron density was about $(3–5) \times 10^{20}$ e/cm^3 with a density scale length in excess of 100 μm, electron temperature approximately 900 eV, and ion temperature approximately 400 eV.

Figure 9.14 shows time-gated axially observed emission spectra for plasma lengths of 4.6 mm, 10.1 mm, and 22.4 mm. For the 4.6 mm long plasma the 2p^5 3p to 2p^5 3s lines at 20.64 nm and 20.98 nm,[52, 53] are evident but comparable in intensity to many other emission lines, just above the background continuum. Observed at an angle away from the axis, with a companion instrument, these lines are barely discernible above the background continuum.[10] As seen in Figure 9.14(b), with a plasma length of 10.1 mm the two lasing lines begin to dominate the spectrum. For the 22.4 mm plasma length, Figure 9.14(c), the lasing lines at 20.64 nm and 20.98 nm completely dominate the observed axial emission spectra. Integrated line intensities versus target length are shown in Figure 9.14(d) for (laterally displaced) double-sided target irradiations. The exponential growth of intensity for both 20.64 nm and 20.98 nm lines is a clear diagnostic of amplification by stimulated emission. This exponential growth is observed only in the axial direction, where a sufficient gain–length (*GL*) product exists. Fits to the experimental data in Figure 9.14 indicate a gain of approximately 5/cm for the two-sided irradiations.

Further confirmation of lasing, obtained in later experiments, involved measurement of the emission line spectral shape ($\Delta\lambda$), and the observation of gain narrowing of the 20.64 nm line in time resolved high spectral resolution studies by Koch *et al.*[56, 57] Neon-like lasing was extended to relatively modest facilities by Lee, McLean, and Elton,[60] who used a 400 J, 2 ns Nd laser and solid targets to demonstrate lasing in Cu and Ge, albeit at the somewhat longer wavelengths of 27.93 nm and 23.22 nm, respectively. Collisionally pumped lasing in Ne-like electron configurations was extended to shorter wavelengths[15–17, 47, 49] using Ne-like Y at 15.50 nm and 15.71 nm, and Ne-like Mo at 13.10 nm and 13.27 nm.[¶¶]

Continuation to still shorter wavelengths using this same isoelectronic sequence requires ever higher electron temperatures and thus higher visible laser intensities, which is problematic because of power requirements for these large area targets. Recent experiments utilizing double pulse *transient excitation* offer a new and more efficient route to collisionally pumped lasing. In this technique two time-separated laser pulses are used to heat a plasma. The first is a relatively modest intensity nanosecond duration pulse that pre-forms a plasma to the desired ionization stage. The plasma is then allowed to expand for 1–2 ns, creating a larger plasma with more gentle density gradients. A second, more intense pulse, typically one picosecond in duration, is then used to rapidly heat the pre-formed plasma, collisionally exciting the existing neon-like atoms to higher excited states. Differences in radiative decay rates among the various excited states provide the desired population inversion. The more gentle density gradients reduce refraction in the plasma, increase the potential lasing volume, and permit more efficient extraction of energy from the picosecond heating pulse.

Nickels and his colleagues[61] have used this transient inversion technique with a relatively modest laser facility to achieve lasing in neon-like titanium at 32.6 nm. In their experiments a total of only about 10 J was used to obtain clear lasing results, pointing a path toward future table top capabilities. Using a 30 µm by 5 mm line focus on a solid titanium target, 7 J of 1.053 nm laser light in a 1.5 ns FWHM pulse is delivered to the target, at an intensity of about 10^{12} W/cm^2. This is followed by a 1.5 ns delayed pulse of 4 J in 0.7 ps, at a nominal intensity of 10^{15} W/cm^2. The 0.7 ps transient pulse permits a favorable inversion condition with respect to the longer lifetimes of the excited upper states. With an irradiated target length of 5 mm, a gain of 19/cm was achieved, for a gain–length product of $GL = 9.5$. Continuation of these transient population inversion experiments is described toward the end of this section in experiments involving collisionally pumped nickel-like lasers.

¶¶ A comparison of experimentally observed and calculated Ne-like lasing wavelengths is given by Nilsen and Scofield.[53]

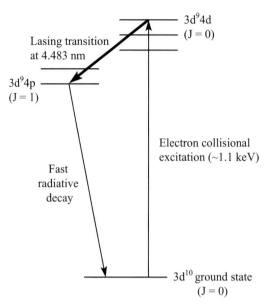

$3d^9 4d$

(J = 0)

Lasing transition
at 4.483 nm

$3d^9 4p$
(J = 1)

Electron collisional
excitation (~1.1 keV)

Fast
radiative
decay

$3d^{10}$ ground state
(J = 0)

Figure 9.15 Simplified energy level diagram[52] for Ni-like Ta^{+45}, showing the 4d to 4p lasing transition at 4.483 nm (276 eV). Following B. MacGowan *et al.*, LLNL.

A second route in the pursuit of ever shorter wavelength lasers, also utilizing collisional pumping of closed electron shells, is that involving nickel-like ions.[50–52, 54] In this case the closed-shell ground state is $1s^2\ 2s^2\ 2p^6\ 3s^2\ 3p^6\ 3d^{10}$, a 28-electron ion, with lasing between the $3d^9\ 4d$ and $3d^9\ 4p$ excited states. Key advantages of the Ni-like schemes are the lower ionization potentials of $n = 4$ levels, vs. $n = 3$ levels for Ne-like, and the lower excitation energies from the $3d^{10}$ ground state, about 1.1 keV vs. about 1.5 keV for Ne-like (from $2p^6$). Figure 9.15 shows a simplified energy level diagram[52, 54] for Ni-like tantalum ($Z = 73$, 28 electrons, net charge +45). Pumping is largely through direct electron collisional excitation and through cascading down from upper levels. Population inversion is again achieved through selective depopulation of the lower $3d^9\ 4p$ level in a fast radiative decay to the ground state.

Experimental data[51, 52] for axial emissions of Ni-like Ta^{+45} are shown in Figure 9.16. With a small increase in plasma length from 1.7 cm to 2.5 cm, the 4.483 nm line is seen to emerge dramatically from the noise. The intense line of the 2.5 cm case has a gain–length product $GL \cong 8$. Axially observed lasing lines at 5.023 nm, 4.483 nm, 4.318 nm, and 3.556 nm, observed with Ni-like ions of ytterbium, tantalum, tungsten (wolfram), and gold, are shown in Figure 9.17. The lines of Ta and W are selected as they straddle the K-absorption edge of carbon at 4.36 nm, an important feature for many scientific and technological applications, as discussed for instance by London, Rosen, and Trebes.[55] For the Ta^{+45} laser, the output energy is estimated[62] to be about 30 μJ in a 250 ps pulse, radiating into a horizontal divergence angle of about 12 mrad (FWHM).

Work on Ni-like lasers, in pursuit of saturated lasing at shorter wavelengths, continues. In a collaboration between the Institute for Laser Engineering in Osaka and

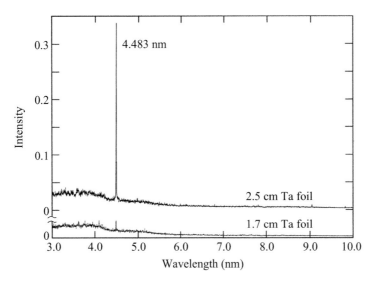

Figure 9.16 Measured axial emission spectrum[52] for 1.7 cm and 2.5 cm long tantalum foils, showing the emergence of the 4d to 4p lasing line in Ni-like Ta, at 4.483 nm wavelength (Courtesy of B. MacGowan and colleagues, Lawrence Livermore National Laboratory).

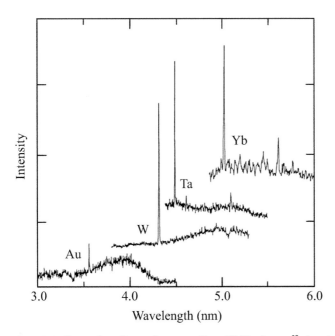

Figure 9.17 Examples of on-axis spectra from Ni-like lasers,[52] showing strong 4d to 4p lines in Yb at 5.023 nm, in Ta at 4.483 nm, in W at 4.318 nm, and in Au at 3.556 nm wavelength. (Courtesy of B. MacGowan and colleagues, Lawrence Livermore National Laboratory.)

the National Laboratory for High Power Lasers and Physics in Shanghai, Kato, Wang, Daido, and their colleagues[29, 63, 64] have employed multiple pulse irradiation techniques to control refraction, and multilayer mirrors to provide feedback, in experiments involving Ni-like lasing in Nd ($Z = 60$) at 7.905 nm wavelength. Their technique employs four pulses of 1.053 μm, each of nominal 100 ps duration and 7×10^{13} W/cm^2 intensity. The first pulse is used to pre-form the plasma, allowing free expansion to set a relatively long density scale length so as to minimize refractive turning during subsequent irradiations. Cylindrical focusing is used on side-by-side slab targets, illuminated from opposite sides and aligned for double plasma path length. The two targets are separated axially by 3 cm center to center, are curved to compensate for refraction, and are laterally displaced. The double targets are irradiated with this sequence of four pulses, in each case with one target irradiation delayed by 100 ps to enhance gain in one direction ($\Delta \tau = \Delta l / c$). A Ru–B$_4$C multilayer mirror,[§§] with 7% reflectivity at 7.9 nm wavelength, is placed 6 cm from the center of the closest target. This separation matches the 400 ps interval between pulses (2×6 cm$/c$). With feedback from the mirrors, lasing is enhanced on the third and fourth pulses. The combination of multipass lasing in refraction-compensated, quasi-traveling-wave illumination (delayed by one target) of solid Nd targets generates a sequence of three 130 ps duration pulses, of 40 μJ energy each,[64] at 7.905 nm wavelength in a nickel-like neodymium plasma.

Numerical simulations of plasma conditions at the peak of the third pulse, obtained using a one-dimensional (radial) computer code, are reproduced here in Figure 9.18. The figure shows radial profiles of electron density, electron temperature, ion temperature, percentage of Ni-like Nd^{+32} ions, and predicted gain. A maximum gain of 4.8/cm is predicted during the third pulse, for a double length target of 4.6 cm. Maximum gain is predicted to occur at an expansion radius 81 μm from the initial surface, in a radial region about 40 μm wide, radiating into a divergence angle of about 3 mrad. At maximum gain, the computer code predicts an electron temperature of 820 eV, an electron density of 2.8×10^{20} e/cm^3, an ion temperature of about 340 eV, and a Ni-like Nd^{+32} ion fraction of about 35%.

Saturation of Ni-like lasers at a variety of wavelengths, as short as 7.355 nm in Sm, has been achieved in experiments at the Rutherford-Appleton Laboratory using 75 ps, 1.05 μm laser heating pulses, cylindrically focused to $(1–4) \times 10^{13}$ W/cm^2 irradiation intensity. Zhang, MacPhee, Lin, and their colleagues[32, 65] have employed double pulses (2.2 ns separation) to irradiate side-by-side slab targets, axially aligned and oppositely illuminated, up to 3.6 cm total length. With the resultant opposed density gradients, refraction in the sequential plasmas is partially compensated, with resultant divergence angles of 1–2 mrad. These saturation results are particularly interesting in that they are obtained with modest laser intensities (2×10^{12} W/cm^2) that are accessible at many smaller laser facilities, thus showing a path towards wider access to short wavelength lasing.

Figure 9.19 shows the axial emission spectrum of a single 18 mm long curved silver target.[65] The spectrum is dominated by the 4d $\rightarrow$ 4p ($J = 0 \rightarrow 1$) lasing line of

---

[§§] See Chapter 10.

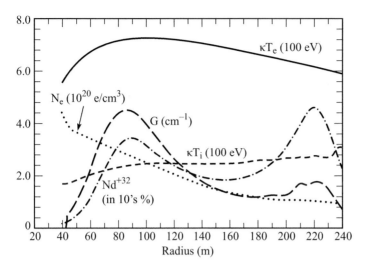

Figure 9.18 Radial distributions of electron temperature, electron density, ion temperature, percentage of Nd ions in the Ni-like ionization stage, and predicted gain, at the peak of the third pulse in a multipulse irradiation of coaxial Nd slab targets at nominal 7×10^{13} W/cm^2, 100 ps, cylindrically focused Nd laser pulses.[29] The simulations are based on experiments conducted with the GEKKO XII laser at the Institute of Laser Engineering in Osaka. At peak gain (expansion radius 81 μm) the predicted values are $\kappa T_e = 820$ eV, $n_e = 2.8 \times 10^{20}$ e/cm^3, $\kappa T_i \cong$ 220 eV, 35% of Nd ions in the Ni-like ionization state, and gain 4.8/cm. Courtesy of Y. Kato, H. Daido, ILE, Osaka University, and S. Wang and colleagues, NLHPLP, Shanghai.

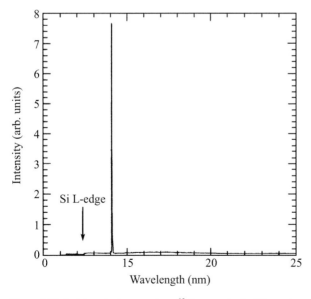

Figure 9.19 Axial emission spectrum[65] from a single 18 mm long curved silver target, showing the dominant Ni-like Ag 4d to 4p lasing line at 13.89 nm. Courtesy of J. Zhang, Rutherford-Appleton Laboratory, presently at Shanghai Jiao Tong University.

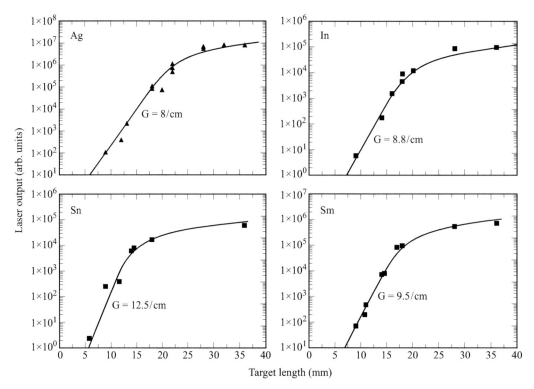

Figure 9.20 Laser output versus target length,[66] showing saturation for nickel-like silver (13.89 nm), indium (12.59 nm), tin (11.98 nm), and samarium (7.355 nm). Gains are shown in inverse centimeters. For these experiments refraction compensating double targets were irradiated by a pair of 75 ps, 2×10^{13} W/cm^2 Nd pulses (1.053 μm) separated by 2.2 ns. Courtesy of J.Y. Lin and colleagues, Rutherford-Appleton Laboratory, with permission of Elsevier Science.

nickel-like Ag at 13.89 nm.||| The small signal gain in these experiments was about 7.2/cm. Saturation, as shown in Figure 9.20, was achieved with a gain–length product of $GL \cong 1.6$. Based on best estimates of source size (43 μm ×57 μm FWHM) and divergence (1.5 mrad × 3.5 mrad FWHM), a pulse peak spectral brightness of 1.1 × 10^{25} photons/s · mm^2 · mrad is estimated. This is based on a laser output energy of 90 μJ in a single 43 ps pulse, corresponding to an output intensity of 69 GW/cm^2. Saturation curves[66] for nickel-like lasing in Ag, In, Sn, and Sm, at wavelengths of 13.89 nm, 12.59 nm, 11.98 nm, and 7.355 nm, respectively, are shown in Figure 9.20. The Sm laser at 7.355 nm has a smaller angular divergence than silver (1.3 mrad versus 2.8 mrad) and a significantly higher output intensity, 2.5×10^{11} W/cm^2.

The efficiency with which saturation is achieved in these Ni-like lasers has recently been improved by an order of magnitude using picosecond transient inversion techniques combined with a true traveling wave illumination. In these experiments, also

---

||| The wavelengths quoted here follow computational values of Scofield and MacGowan,[54] and those of Y. Li *et al.*,[54] and are within the uncertainties of experimental measurements.

at the Rutherford-Appelton Laboratory, MacPhee, Lewis, Pert, and colleagues[67] use a 280 ps FWHM, nominal 10^{13} W/cm^2 prepulse followed 550 ps later by a 3 ps FWHM, nominal 10^{15} W/cm^2 heating pulse (both 1.053 µm wavelength) to drive a transient population inversion. By using off-axis illumination the transient heating pulse travels along the gain medium at the same velocity as the exponentially growing short wavelength laser pulse. Using tin coated strips on solid glass targets, very high gains of 31/cm were obtained on the Ni-like Sn line at 11.98 nm wavelength. Full saturation was obtained[67] with a target length of only 1 cm, yielding an output energy of about 60 µJ in a nominal 3 ps duration pulse, for a peak power of order 20 MW. The measured beam divergence was approximately 6 mrad. The total energy input for this experiment was approximately 30 J, most of it in the prepulse. The achievement of saturated lasing with modest pump energy, at these relatively short wavelengths, portends well for the development of table top soft x-ray lasers, especially as further improvements in irradiation efficiency are realized.

9.5   Compact EUV Lasers

While relatively large visible and infrared laser drivers are used to form laser produced plasmas for the generation of soft x-ray and EUV laser radiation, discharge tube plasmas provide an alternative route, at least for EUV lasing that may not require as high a density or temperature. Rocca and his colleagues[68–74] at Colorado State University have developed an electrical-discharge-driven EUV laser operating at 46.86 nm*** wavelength on a 3p to 3s transition ($J = 0$ to 1) in Ne-like argon.[69] The discharge, in a 500–700 mTorr argon gas, is driven by a 70 ns, 37 kA peak current from a 3 nF capacitor. The high current pulse drives a $\mathbf{J} \times \mathbf{B}$ radial compression[74] of the plasma, as shown in Figure 9.21. The discharge occurs through the argon plasma in a 16 cm long, 4 mm diameter capillary tube. Electron densities of $(5–8) \times 10^{18}$ e/cm^3 are obtained, with electron temperatures of 65–90 eV, across a plasma column of order 300 µm diameter. The elongated plasma column thus has an aspect ratio of up to 1000: 1. Examples of axial emission spectra were reported for plasma lengths of 3, 6, and 12 cm. For the 3 cm plasma length, many lines were observed, including several identified as Al-like (13 electrons) and Mg-like argon (12 electrons). The Ne-like Ar^{+8} 3p $\rightarrow$ 3s line is not particularly intense. With a 6 cm plasma length, amplified spontaneous emission intensified the 3p $\rightarrow$ 3s line to prominence. With a 12 cm plasma length the Ne-like Ar^{+8} 3p $\rightarrow$ 3s line at 46.86 nm had grown to an intensity of 100 times that of all other lines.[69] The early data matched an exponential gain of about 0.6/cm, giving a gain–length product $GL \simeq 7$. Later studies of discharge generated lasing in Ne-like Ar measured the near- and far-field patterns of the laser radiation,[71] observing an exit beam size of 150–300 µm diameter (FWHM) and a divergence angle (2θ) of 2–5 mrad (FWHM), depending on discharge gas pressure, with the smaller numbers corresponding to the highest initial argon pressure (750 mtorr). The smallest combination corresponds to a

---

*** As a result of spectral resolution limits the wavelength is generally described as 46.9 nm in the literature.

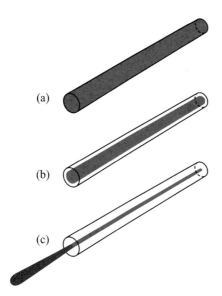

(a)

(b)

(c)

Figure 9.21 A fast high current electrical discharge compresses and heats a plasma column, shown in red, for use in EUV laser generation experiments.[68] Current density increases in the sequence from (a) to (b) and (c). The optimum electron density and temperature for lasing are obtained near the end of the compression. After J. Rocca *et al.*, Colorado State University, Fort Collins.

phase space of approximately 14 times diffraction limited at this wavelength.[†††] Strong refraction by the sharp lateral density gradients are believed to reduce the divergence, and thus phase space, of the growing laser beam. These phase space observations suggested a breakthrough in the spatial coherence of EUV lasers and led to measurements using Young's double-pinhole technique (see Chapter 4, section 4.5). Two-pinhole interference measurements[72] made by Liu *et al.*, shown here in Figure 9.22, illustrate the growth of spatial coherence as the capillary length is increased from (a) 18 cm to (b) 27 cm and finally to (c) 36 cm. One observes the modulation of the interference patterns growing from barely observable (a) for the 18 cm long discharge, to substantial (b) for 27 cm, and to a very high degree of modulation in (c), indicating nearly full spatial coherence for a discharge plasma length of 36 cm. The measurements were made with a pair of 10 μm diameter pinholes separated by 200 μm. Other measurements were made with separations of 300 μm and 700 μm. From the measured interference patterns, analysis of the degree of spatial coherence, $\mu(\Delta x)$, indicates a transverse coherence length of 300 μm, comparable to the diameter of the lasing beam. For these experiments the argon gas pressure was 440 mTorr and the 40 ns current pulse reached ~25 kA. The laser pulse energy[73, 74] at 46.86 nm was 0.88 mJ for a 34.5 cm plasma column length, corresponding to an average laser power of 3.5 mW at a repetition rate of 4 Hz. The laser pulse was measured to be approximately 1.5 ns FWHM in duration and the

---

[†††] See Chapter 4, Eq. (4.7). For a radiation source of Gaussian spatial distribution and Gaussian far-field angular distribution, the diffraction limited product of source size (*d*) and divergence angle (2θ) is given in terms of FWHM values as $(d \cdot 2\theta)_{\text{FWHM}} = 0.4413\lambda$.

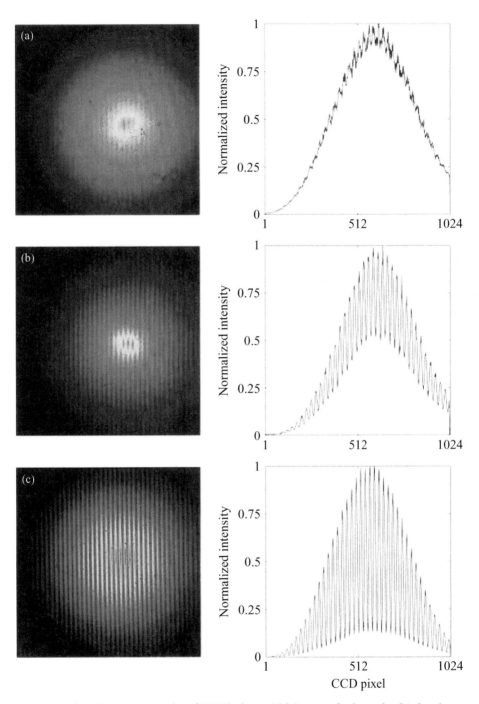

Figure 9.22 The coherence properties of EUV lasing at 46.86 nm, on the 3p → 3s, $J = 0$ to 1 transition in neon-like argon is investigated. Young's double-pinhole interference patterns are shown[72] for discharge plasma lengths of (a) 18 cm, (b) 27 cm, and (c) 36 cm. The latter indicates a high degree of spatial coherence, nearly diffraction limited, with a transverse coherence length comparable to the laser beam diameter of about 300 μm. Courtesy of Y.W. Liu, UC Berkeley, and J. Rocca, Colorado State University.

spectral bandwidth was estimated to be $\lambda/\Delta\lambda \leq 1 \times 10^{-4}$. The corresponding longitudinal (temporal) coherence length was 230 μm. The pinhole mask was located 40 cm from the exit plane of the laser, and a 1024×1024 pixel array, EUV sensitive CCD detector was located 300 cm beyond the mask. This truly tabletop laser (0.4 m × 1 m) opened a new chapter in the development of highly coherent short wavelength lasers, a topic discussed further in the next section.

In an effort to achieve shorter wavelength lasing with compact lasers, Rocca and his colleagues[75-77] have pursued use of the Ni-like lasing scheme, as discussed earlier, but now utilizing various techniques to make maximum use of available pump energy and to dramatically reduce the size of the requisite facility. The general approach includes use of solid targets of various materials, illuminated by two pulses, a normal incidence prepulse of modest intensity designed to create a shallow gradient electron density profile in the underdense interaction region. This is then followed several hundred picoseconds later by a glancing incidence high intensity heating pulse. The prepulse has a doubly important role, first to create a gently sloped electron density profile in the sub-critical interaction region which significantly enhances efficient absorption of the main laser heating pulse, and second to creat a long gain distance for the growing EUV laser beam in this region of optimal electron density and temperature.

Following the prepulse, the energy of the near-IR main heating pulse is delivered with cylindrical focusing optics to a highly elongated region along the target surface, creating a very long, high aspect ratio gain region. Because of refraction, the choice of glancing incidence angle, typically 20° to 40°, sets the highest electron density to be reached,[78] an important parameter for a collisionally pumped laser. Because the EUV/SXR laser pulse has a somewhat faster phase velocity than the 800 nm IR pump, and because excited state lifetimes in the Ni-like ions are relatively short, an additional efficiency is obtained by the use of traveling wave illumination. At glancing incidence the IR wavefront advances at an angle-dependent phase velocity[76] $v = c/\cos\theta$, equaling $1.1c$ at $\theta = 26°$, and thus can be used to match the phase velocity of the faster EUV/SXR laser pulse. This is much like the crests of an ocean wave which run at a higher phase velocity along a beachfront when incident at an angle. In these experiments the prepulse and heating pulse are all derived from the same 800 nm wavelength Ti:sapphire laser, but with different pulse durations managed with differing pulse stretching and compression techniques. Nominal intensities and durations are $2–3 \times 10^{12}$ W/cm^2 and 100 ps FWHM for the 800 nm, normal incidence prepulse, followed after a 100–500 ps delay by a $3–6 \times 10^{14}$ W/cm^2, 800 nm, glancing incidence heating pulse of 1–8 ps FWHM duration. The choice of specific laser and plasma parameters depends on the particular target material and thus the Ni-like ions to be generated and the EUV lasing wavelength of interest. These then determine the electron temperature required to reach the desired Ni-like ionization state. With cylindrical focusing optics, the goal is to achieve a very long aspect ratio lasing region, typically 30 μm wide by 4–6 mm long, with a relatively uniform electron temperature in the vicinity of 400–850 eV and an electron density in the range of $2–6 \times 10^{20}$ e/cm^3, all depending on the energy required to reach and excite the Ni-like ion state of interest. Sophisticated numerical modeling codes, which describe coupled plasma physics with steep spatial and temporal gradients, ionization dynamics, and extensive spectroscopic capabilities, are used to plan these experiments.[95]

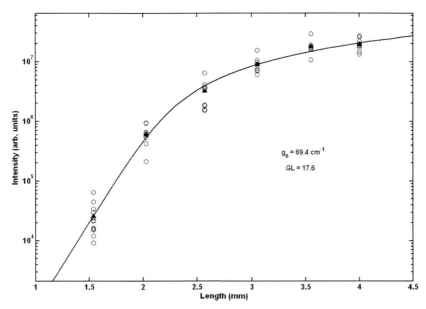

Figure 9.23 The gain curve[75] for a tabletop Ni-like Cd laser at 13.2 nm, showing saturation within a propagation distance of 4 mm. Illumination of the solid Cd target consisted of a normal incidence double-prepulse of 365 mJ total energy and 120 ps duration, followed after a 200 ps delay by a 23° glancing incidence 1 J, 8 ps FWHM heating pulse, all at 800 nm wavelength. The small signal gain was 69/cm, and the gain length product was $GL = 18$. The output energy of the laser is just under a half microjoule per 5 ps duration pulse. Courtesy of J. Rocca, Colorado State University.

Figure 9.23 shows an example of a gain curve for Ni-like Cd lasing on the 4d–4p transition at 13.2 nm wavelength, with a small signal gain of 69/cm and a gain-length product GL = 18, reaching saturation at a plasma length of 4 mm, with an output energy of nearly a half microjoule per pulse, in a pulse of about 5 ps FWHM duration, and at a repition rate of 5Hz. Experimental lasing data have been obtained for a series of iso-electronic Ni-like ions extending from Ru ($Z = 44$), to Pd ($Z = 46$), Ag ($Z = 47$), Cd ($Z = 48$), Sn ($Z = 50$), Sb ($Z = 51$), Te ($Z = 52$), La ($Z = 57$), Ce ($Z = 58$), Pr ($Z = 59$), Nd ($Z = 60$), and Sm ($Z = 62$), as seen in Figure 9.24. Many of these lasers have reached saturation with plasma lengths of only 4–6 mm. While the detailed numerical simulations are necessary, it is also worthwhile to have a simple estimate of the dominant energetics. For example, inspection of Table 8.2 shows that removal of the twentyninth electron to form a 28 electron Ni-like ion requires 501 eV for Ag, and 546 eV for Cd. From Section 8.7.1, the quasi-equilibrium black body model predicts that the intensity required to reach a particular temperature scales as $I = \hat{\sigma}(\kappa T)^4$ [Eq. (8.143)]. From the example given in that section, where an intensity of 1.0×10^{13} W/cm^2 is required to achieve a temperature of 100 eV, one estimates by the fourth power scaling that an intensity of 2.6×10^{14} W/cm^2 is required to produce an electron temperature of 400 eV. As the average electron energy in such a plasma is $1.5\kappa T_e$, or 600 eV, this temperature is well matched to the energy required to remove the last electron to form Ni-like Ag (501 eV) and Ni-like Cd (546 eV). Higher electron temperature, and thus higher

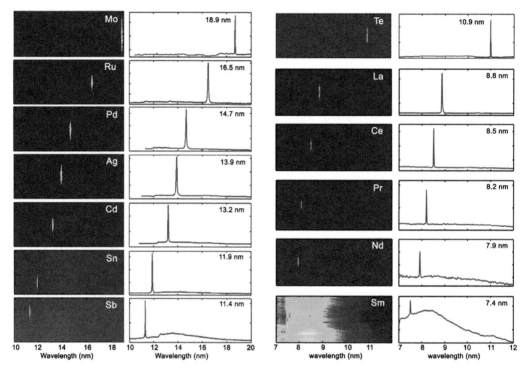

Figure 9.24 On-axis emission spectra from tabletop Ni-like lasers, ranging in wavelength from 16.5 nm to 7.4 nm. The column on the left[76] represents lasing in Ni-like Mo, Ru, Pd, Ag, Cd, Sn, and Sb. Illumination in all cases consisted of a normal incidence 120 ps duration prepulse, followed 100–400 ps later by a glancing incidence 1 J/8 ps FWHM heating pulse focused with cylindrical optics to an approximately 30 μm wide by 4 mm long region at an incident intensity of about 1×10^{14} W/cm^2, all at 800 nm wavelength and 5 Hz repition rate. The column on the right[77] represents lasing in Te, La, Ce, Pr, Nd, and Sm. Illumination at 800 nm was similar to those on the left, but with a 4 J, 210 ps, 6×10^{12} W/cm^2 prepulse followed by a 4 J, 1–3 ps FWHM heating pulse focused with a traveling wave illumination at an incident intensity of 6×10^{14} W/cm^2 across a 30 μm wide by 6.4 mm long region. The illumination parameters varied somewhat to match required conditions for the various targets. Courtesy of J. Rocca, Colorado State University.

illumination intensity, is required for lasing at shorter wavelengths with higher-Z Ni-like atoms. The average power of these tabletop EUV/SXR lasers is enhanced by operation at 100 Hz, as described in the literature.[79, 80]

9.6   Spatially Coherent EUV and Soft X-Ray Lasers

The phase space of EUV and soft x-ray lasers tends to be large due to the hot-dense nature of the plasma which causes expansion in all directions. Furthermore, it is difficult to maintain cavity optics for many pass amplification, as would be required for transverse mode control. This is due to the short gain lifetime which requires that cavity end-mirrors be kept close, where debris from the hot expanding plasma damages the mirrors and limits their lifetime.[81–83] As a result laser radiation in this region tends to be characterized by a large phase space product, $d \cdot \theta \gg \lambda/2\pi$. Where this is true, substantial spatial filtering is required for applications involving spatial coherence. Spectral filtering, on the other hand, is generally not required, as these lasers naturally radiate with a very narrow linewidth[84] typically with $\Delta\lambda/\lambda$ of order 10^{-4}. The coherent power available after spatial filtering can be written as [Eq. (5.76)]:

$$P_{\text{coh}} = \frac{(\lambda/2\pi)^2}{(d_x\theta_x)(d_y\theta_y)} P_{\text{laser}} \qquad (9.23)$$

where P_{laser} is the radiated laser power occupying an elliptical phase-space $(d_x\theta_x)(d_y\theta_y)$. Recalling the nickel-like tantalum laser[50] of Figures 9.15 and 9.16, radiating 100 kW in a single 250 ps pulse at a wavelength of 4.483 nm, we can estimate the coherent power that would be available after spatial filtering. With a source diameter d estimated at 100 μm and a divergence half angle θ of 6 mrad, the phase space product $d \cdot \theta$ is approximately 700 times larger than $\lambda/2\pi$. The single pulse coherent power available after spatial filtering would be about $100 \, \text{kW}/(700)^2$, or 200 mW. With a pulse duration of 250 ps, this corresponds to an energy of 40 pJ, or about 10^6 spatially coherent photons within a temporal (longitudinal) coherence length of about 25 μm.

An option to post-lasing pinhole spatial filtering might be to arrange effective mode control within the laser itself, as in the conventional visible light laser seen in Figure 9.6. Spatial filtering within an oscillator would be the most energy efficient way to proceed, as atoms (ions) would be stimulated to radiate only within the desired phase space. This, however, would require multipass lasing between cavity mirrors, which in general is difficult with hot dense plasmas as the lasing medium as they tend to spew energetic plasma particles coating or destroying the cavity optics. For a half-nanosecond duration plasma, this permits a propagation path length of only 15 cm (at the speed of light); thus for just one round trip, the cavity end mirrors[81, 82] would be just 7.5 cm apart and an intracavity pinhole would at most be only a few centimeters from a very intense pump pulse. Such a configuration would tend to vaporize the pinhole and create an additionally complicated refractive medium within the cavity. Another approach is to use a sequence of collinear laser amplifiers in a chain that incorporates a pinhole spatial filter following the first stage.[83–86] Thus the second and sequential amplification stages are driven by single mode spatially coherent radiation, rather than growing from noise as in the first stage, and have the possibility of growing to high coherent power levels. Such staging requires precise timing and alignment as well as careful attention to the control of transverse gradients in the amplification stages, which would lead to increased angular divergence and thus reduced spatial coherence. Nishikino et al.[87] at the Kansei Advanced Photon research Center have used a two-stage oscillator–amplifier configuration to demonstrate a high degree of spatial coherence on the 4d–4p lasing transition in Ni-like Ag. They used a Nd:glass laser at 1.053 μm and two solid silver targets separated by 200 mm, irradiating each with both a prepulse and a heating pulse. The first target received a 1 J/300 ps prepulse followed 600 ps later by a 10 J/4 ps heating pulse.

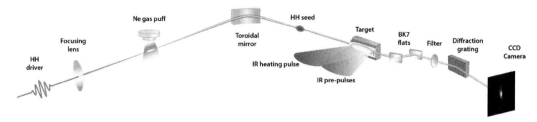

Figure 9.25 An arrangement whereby two IR prepulses and an IR heating pulse create plasma conditions optimum for EUV/Soft x-ray lasing, much as in the previous section, but here an additional 50 fsec high harmonic pulse, generated at high intensity in a Ne or Ar pulsed gas jet (the HHG process) enters the plasma nearly tangential to the target surface, timed to coherently seed the laser just as the amplification process begins. The spatial coherence of the high harmonic imparts that coherence to the EUV/SXR laser, while the laser amplifies only that portion of the high harmonic that lies within its narrow spectral linewidth. Furthermore, this technique imparts a phase coherence beyond simple phase-space constraints, as is discussed in Chapter 4, Section 4.6. Figure courtesy of J. Rocca, Colorado State University and Ph. Zeitun, Laboratory of Optical Applications, Palaiseau.

The second target received a 1.5 J/300 ps prepulse followed 600 ps later by a 12 J/12 ps heating pulse. Illumination of the second target was delayed by approximately 670 ps with respect to the first. Travelling wave illumination irradiated target areas of 20 μm by 6.5 mm for all pulses. The target separation distance of 200 mm was selected so that divergence of the oscillator radiation would lead to an overfilling of the active lasing region of the amplifier stage, resulting in effective spatial filtering of seed pulse to the second stage. The measured 50 μm FWHM beam diameter and 200 μrad FWHM beam divergence of the amplified laser radiation at 13.9 nm corresponds to a phase space volume [Eq. (4.7)] of about 1.5 times the diffraction limit, indicating a high degree of spatial coherence. This was confirmed by Young's double-slit interference experiments which demonstrated spatial coherence across the full beam. The measured output power was 25 nJ, corresponding to 2×10^9 ph/pulse, in an 8 ps pulse with a spectral bandwidth $\Delta\lambda/\lambda \leq 10^{-4}$. The gain-length product was $GL = 5.1$, with a total gain of 170 for the 6.5 mm lasing distance in the amplifier stage. Further ideas are discussed in the literature.[88]

A very different route toward spatially coherent EUV/SXR lasing, originally explored by Dittmire et al.,[89] has recently been demonstrated by the groups of Zeitun[90, 92] in Palaiseau and Rocca[91, 93–96] in Fort Collins, Colorado, in which they combine the best features of narrow linewidth EUV/SXR lasers with the spatial coherence of a high harmonic seed pulse (see Chapter 7, Laser High Harmonic Generation). Figure 9.25 shows the basic arrangement in which normal incidence IR prepulses creates a modest gradient electron density profile in the sub-critical region of a solid target, followed by a higher intensity IR heating pulse at glancing incidence, as was described in the previous section. But here an additional high harmonic pulse enters the already prepared plasma, nearly tangential to the target surface, where it will act as an above threshold injected seed to phase-coherently drive the lasing process. Having a much higher frequency than the IR pulses, the high harmonic is less affected by plasma refraction and thus can enter the plasma nearly tangential ($\sim 10°$) to the target surface. The high harmonic pulse is generated by focusing a nominally 50 fs, 800 nm Ti:sapphire laser pulse at high intensity into a neon or argon pulsed gas jet, as has been described in Chapter 7.

A specific example of the high harmonic seeding of a SXR/EUV laser is described by Wang and colleagues[93, 95] in which they seed a Ni-like silver laser, which radiates in a narrow ($\lambda / \Delta \lambda \leq 10^{-4}$) spectral line at 13.9 nm, with the spectrally broader fiftyninth harmonic of a 815 nm, 50 fs Ti:sapphire pulse generated at high intensity in a neon gas jet. All pulses are derived from the same Ti:sapphire laser: two 815 nm normal incidence prepulses, an 815 nm near grazing incidence heating pulse, and the fiftyninth harmonic laser seeding pulse. Each, however, is temporally shaped as needed for the three different functions. For the Ni-like Ag experiments two relatively low intensity prepulses were used to create a smooth, low gradient, sub-critical electron density plasma with an electron temperature of about 90 eV, sufficient to bring the Ag ions to a degree of ionization approaching that of the Ni-like state. The normal incidence prepulses were 10 mJ and 350 mJ, 5 ns apart and 120 ps FWHM duration each, followed 200–300 ps later by a nominally 0.9J/6.7 ps heating pulse at 23° incidence angle. All were focused with cylindrical optics to a 30 µm wide by 4.1 mm long region of the solid Ag target. The fiftyninth harmonic, 50 fs seed pulse entered the plasma at a small angle from the surface, timed for maximum gain in the lasing process. Gains of more than 400 were observed, with the most energetic pulses reaching 75 nJ. The pulse duration was broadened during the lasing process to a bandwidth limited value of about 1 ps. Understanding the evolution of the seed pulse to output pulse spectral banwidth and concomitant pulse duration is critically dependent on the use of computer codes that can follow the non-equilibrium dynamics in an evolving plasma of sharp spatial and temporal profiles.[95] The authors used Young's double-slit technique to confirm that the output laser pulse had a high degree of spatial coherence. In this same publication[93] the authors reported similar results for Ni-like Mo lasing at 18.9 nm. These experiments demonstrate that modest scale EUV/SXR lasers are capable of generating short, intense pulses that are spatially, temporally, and phase coherent at these short wavelengths.

The group of Depresseux, Sebban, Zeitoun and their colleaques at Ecole Polytechnique have extended these table-top coherently pumped SXR/EUV atomic laser techniques to the femtosecond regime using an electron density gating technique to extinguish the gain and thus limit the pulse duration.[97] The process involves four laser pulses, all derived from an 813 nm/5 mJ/1 Hz/30 fsec Ti:sapphire laser. A 30 fs duration 'ignitor' pulse and a delayed 600 ps IR 'heater' pulses are used to efficiently create a several mm long, preformed electron density channel in a krypton jet plasma. A frequency doubled 407 nm wavelength Mach–Zehnder interferometer is used to monitor and help optimize the electron density profile within the plasma column. The high radial wall, low axial electron density channel is then used to guide an intense, 5×10^{18} W/cm^2/1.36 J/30 fs IR pump beam through the plasma, creating and collisionally exciting Kr^{+8} Ni-like ions for lasing on a 4d–4p transition at 32.8 nm wavelength. Lasing is induced by a spectrally overlapping, 350 fs, $n = 25$ high harmonic (HHG) seed pulse created in a 40 mbar argon gas cell. Arrival timing of the seed pulse at the excited Kr^{+8} plasma column is controlled by an optical delay line. The amplified HHG pulse, spectrally narrowed to 32.8 nm by the gain width of the nickel-like transition, achieved a peak gain of 1200. By controlling the temporal overlap of the longer HHG seed and the 30 fs IR heating pulse it was possible to bring the seed pulse through the plasma at a time when the rapidly rising electron density was collisionally over-ionizing the krypton ions. It is in this manner that the laser amplification process was rapidly gated off, or extinguished, on a 30 fs temporal scale. At peak gain the emitted 32.8 nm laser pulse was measured to have an energy of 2 μJ per pulse (3×10^{11} photons/pulse) and a duration of 450 fs FWHM, breaking new ground for spatially and spectrally coherent table-top lasers.[97]

References

1. J.P. Gordon, H.J. Zeiger and C.H Townes,"The Maser – New Type of Microwave Amplifier, Frequency Standard and Spectrometer," *Phys. Rev.* **99**, 1264 (1955).
2. N.G. Basov and A.M. Prokhorov, "3-Level Gas Oscillator," *Zh. Eksp. Teor. Fiz. (Moscow)* **27**, 431, (1954); also **28**, 249 (1955).
3. A.L. Schawlow and C.H. Townes, "Infrared and Optical Masers," *Phys. Rev.* **112**, 1940 (1958); also see C.H. Townes, *How the Laser Happened* (Oxford University Press, 1999).
4. A.M. Prokhorov, "Molecular Amplifier and Generator for Submillimeter Waves," *Zh. Eksp. Teor. Fiz. (Moscow)* **34**, 1658 (1958); also "Quantum Electronics," *Science* **149**, 828 (August 20, 1965).
5. T. Maiman, "Stimulated Optical Radiation in Ruby," *Nature* **187**, 493 (August 6, 1960).
6. A. Javan, W.R. Bennett, and D.R. Herriott, "Population Inversion and Continuous Optical Maser Oscillation in a Gas Discharge Containing a He–Ne Mixture," *Phys. Rev. Lett.* **6**, 106 (1961).
7. A.D. Siegman, *Lasers* (Univ. Sci. Books, Mill Valley, CA, 1986). In particular see pp. 28–35 regarding the relationship between the field description of stimulated transitions and the resultant coherence and directionality properties of the radiation.
8. M. Sargent, M.O. Scully and W.E. Lamb, *Laser Physics* (Addison-Wesley, Reading, MA, 1974). In particular, see pp. 38–41 regarding an ensemble of stimulated oscillators and the resultant directionality of the emitted photons.
9. W.T. Silfvast, *Laser Fundamentals* (Cambridge University Press, Cambridge, 2004), Second Edition.
10. D.L. Matthews, P.L. Hagelstein, M.D. Resen *et al.*, "Demonstration of a Soft X-Ray Amplifier," *Phys. Rev. Lett.* **54**, 110 (1985).
11. M.D. Rosen, P.L. Hagelstein, D.L Matthews *et al.*, "Exploding-Foil Technique for Achieving a Soft X-Ray Laser," *Phys. Rev. Lett.* **54**, 106 (1985).
12. S. Suckewer, C.H. Skinner, H. Milchberg, C. Keane and D. Voorhees,"Amplification of Stimulated Soft X-Ray Emission in a Confined Plasma Column," *Phys. Rev. Lett.* **55**, 1753 (1985).
13. S. Suckewer, D.H. Skinner, D. Kim *et al.*, "Divergence Measurements of Soft X-Ray Laser Beam," *Phys. Rev. Lett.* **57**, 1004 (1986).
14. H. Daido,"Review of Soft X-Ray Laser Researches and Developments," *Rep. Prog. Phys.* **65**, 1513–1576 (September 18, 2002).
15. G.J. Tallents, "The Physics of Soft X-Ray Lasers Pumped by Electron Collisions in Laser Plasmas", *J. Phys. D: Appl. Phys.* **36**, R259–R276 (July 16, 2003).
16. R.C. Elton, *X-Ray Lasers* (Academic Press, San Diego, 1990).

17. S. Suckewer and P. Jaeglé, "X-Ray Laser: Past, Present, and Future", *Laser Phys. Lett.* **6**, 411 (June 2009); P. Jaeglé, *Coherent Sources of XUV Radiation* (Springer, Heidelberg, 2006).

18. J. Rocca, C. Menomi and M. Marconi, Editors, *X-Ray Lasers 2014* (Springer, Heidelberg, 2016).

19. S. Sebban, S. Gauthier, J. Ros and Ph. Zeitoun, Editors, *X-Ray Lasers 2012* (Springer, Heidelberg, 2014).

20. J. Lee, C.-H. Nam and K.A. Janulewicz, Editors, *X-Ray Lasers 2010* (Springer, Heidelberg, 2011).

21. C. Lewis and D. Riley, Editors, *X-Ray Lasers 2008* (Springer, Heidelberg, 2009).

22. F.E. Irons and N.J. Peacock, "Experimental Evidence for Population Inversion in C^{+5} in an Expanding Laser-Produced Plasma," *J. Phys. B.: Atom. Molec. Phys. (London)* **7**, 1109 (1974).

23. R.J. Dewhurst, D. Jacoby, G.J. Pert and S.A. Ramsden, "Observation of a Population Inversion in a Possible Extreme Ultraviolet Lasing System," *Phys. Rev. Lett.* **37**, 1265 (1976).

24. M.H. Key, C.L.S. Lewis and M.J. Lamb, "Transient Population Inversion at 18.2 nm in a Laser Produced C VI Plasma," *Optics Comm.* **28**, 331 (1979).

25. D. Jacoby, G.J. Pert, S.A. Ramsden, L.D. Shorrock and G.J. Talents, "Observation of Gain in a Possible Extreme Ultraviolet Lasing System," *Optics Comm.* **37**, 193 (1981).

26. G.J. Pert, "Model Calculations of XUV Gain in Rapidly Expanding Cylindrical Plasmas," *J. Phys. B.: Atom. Molec. Phys. (London)* **9**, 3301 (1976); part II, **12**, 2076 (1979).

27. A. Corney, *Atomic and Laser Spectroscopy* (Oxford University Press, Oxford, 1977), pp. 106–115.

28. N.M. Ceglio, D.G. Stearns, D.P. Gaines, A.M. Hawryluk and J.E. Trebes, "Multipass Amplification of Soft X-Rays in a Laser Cavity," *Optics Lett.* **13**, 108 (1988).

29. Y. Kato, S. Wang, H. Daido *et al.*, "Generation of Intense X-Ray Laser Radiation at 8 nm in Ni-like Nd Ions," *SPIE* **3156**, 2 (1997).

30. M.D. Rosen and P.L. Hagelstein, "X-Ray Lasing: Theory," p. 2 in *Energy and Technology Review* (Lawrence Livermore National Laboratory, November 1985), UCRL-52000–85–11.

31. G.J. Pert, S.B. Healy, J.A. Plowes and P.A. Simms, "Effects of Density Structures and Pulse Temporal Shaping," p. 260 in *X-Ray Lasers 1996* (Inst. Phys., Bristol, 1996), S. Svanberg and C.-G. Wahlström, Editors; S.B. Healy, G.F. Cairns, C.L.S. Lewis, G.J. Pert and J.A. Plowes, "A Computational Investigation of the Ne-like Germanium Collisionally Pumped Laser Considering the Effect of Prepulses," *IEEE J. Select. Top. Quant. Electr.* **1**, 949 (1995); J.A. Plowes, G.J. Pert, S.B. Healy and A. Toft, "Beam Modelling for X-Ray Lasers," *Opt. and Quant. Elect. (London)* **28**, 219 (1996).

32. J. Zhang, A.G. MacPhee, J. Lin *et al.*, "Recent Progress in Nickel-Like X-Ray Lasers at RAL," p. 53 in Ref. 20.

33. Y. Li, P. Lu, J. Maruhn *et al.*, "Study of Prepulse-Induced Ne- and Ni-Like X-Ray Lasers," p. 21 in Ref. 20.

34. T.-N. Lee, S.-H. Kim and H.-J. Shin, "Comparison of X-Ray Lasers and Third Generation Synchrotron Radiation Sources," p. 250 in Ref. 20.

35. G.J. Linford, E.R. Peressini, W.R. Sooy and M.L. Spaeth, "Very Long Lasers," *Appl. Opt.* **13**, 379 (1974).

36. A. Einstein, "Zur Quantentheorie der Strahlung," *Phys. Z.* **18**, 121 (1917); English translation, "On the Quantum Theory of Radiation," *The Old Quantum Theory* (Pergamon, Oxford, 1967), D. Ter Haar, Editor, p. 167.

37. R. Loudon, *The Quantum Theory of Light* (Oxford University Press, Oxford, 1983), Sections 1.5, 2.4, and 2.12.

38. R. L. Liboff, *Introductory Quantum Mechanics* (Addison-Wesley, Reading, MA, 2002), Fourth Edition, Chapter 13, Sections 7 and 9.

39. A.V. Vinogradov and I.I. Silverman,"On the Possibility of Lasers in the UV and X-Ray Ranges", *Sov. Phys. JETP* **36**, 1115 (1973).

40. M.H. Key, "Laboratory Production of X-Ray Lasers," *Nature* **316**, 314 (July 25, 1985).

41. P. Jaeglé, A. Carillon, P. Dhez *et al.*, "Experimental Evidence for the Possible Existence of a Stimulated Emission in the Extreme UV Range," *Phys. Lett. (Netherlands)* **36A**, 167 (1971); B. Rus, A. Carillon, P. Dhez *et al.*, "Efficient, High Brightness Soft X-ray Laser at 21.2 nm," Phys. Rev. A **55**, 3858 (1997).

42. L.I. Gudzenko and L.A. Shelepin,"Radiation Enhancement in a Recombining Plasma," *Dokl. Akad. Nauk. SSSR* **160**, 1296 (1965). [*Sov. Phys. Dokl.* **10**, 147 (1965).]

43. S. Suckewer and H. Fishman,"Conditions for Soft X-Ray Lasing Action in a Confined Plasma Column," *J. Appl. Phys*. **51**, 1922 (1980).

44. D. Kim, C.H. Skinner, G. Umesh and S. Suckewer, "Gain Measurements at 18.22 nm in C VI Generated by a Nd: Glass Laser," *Optics Lett*. **14**, 665 (1989).

45. V.A. Bhagavatula and B. Yaakobi, "Direct Observation of Population Inversion Between Al^{+11} Levels in a Laser-Produced Plasma," *Optics Comm*. **24**, 331 (1978).

46. J. Zhang, M.H. Key, P.A. Norreys *et al.*, "Experiments of High Gain C vi X-Ray Lasing in Rapidly Recombining Plasmas," *X-Ray Lasers 1994* (Amer. Inst. Phys., New York, 1994), D.C. Eder and D.L. Matthews, Editors, p. 80.

47. D.L. Matthews, M. Eckart, D. Eder *et al.*, "Review of Livermore's Soft X-Ray Laser Program," p. C6–1 in *X-Ray Lasers* (J. de Physique, Tome 47, Editions de Physique, Les Udis, France, 1986), P. Jaelge and A. Sureau, Editors.

48. M.D. Rosen J.E. Trebes, B.J. MacGowan *et al.*, "Dynamics of Collisional-Excitation X-Ray Lasers," *Phys. Rev. Lett*. **59**, 2283(1987).

49. B.J. MacGowan, M.D. Rosen, M.J. Eckart *et al.*, "Observation of Soft X-Ray Amplification in Neon-like Molybdenum," *J. Appl. Phys*. **61**, 5243 (1987).

50. B.J. MacGowan, S. Maxon, P.L. Hagelstein *et al.*, "Demonstration of Soft X-Ray Amplification in Nickel-like Ions," *Phys. Rev. Lett*. **59**, 2157 (1987).

51. B.J. MacGowan, S. Maxon, L.B. DaSilva *et al.*, "Demonstration of X-Ray Amplifiers Near the Carbon K Edge," *Phys. Rev. Lett*. **65**, 420 (1990).

52. B.J. MacGowan, L.B. DaSilva, D.J. Fields *et al.*, "Short Wavelength X-Ray Laser Research at the Lawrence Livermore National Laboratory," *Phys. Fluids B* **4**, 2326 (1992).

53. J. Nilsen and J. Scofield, "Wavelengths of Neon-like 3p → 3s X-Ray Laser Transitions," *Phys. Scripta* **49**, 558 (1994).

54. J.H. Scofield and B.J. MacGowan, "Energies of Nickel-like 4d to 4p Laser Lines," *Phys. Scripta*, **46**, 361 (1992); and Y. Li, J. Nilsen, J. Dunn *et al.*, "Wavelengths of the Ni-Like 4d 1S0–4p 1P1 X-Ray Laser Line," *Phys. Rev. A* **58**, R2668 (October 1998).

55. R.A. London, M.D. Rosen and J.E. Trebes, "Wavelength Choice for Soft X-Ray Laser Holography of Biological Samples," *Appl. Opt*. **28**, 3397 (1989).

56. J.A. Koch, B.J. MacGowan, L.B. DaSilva *et al.*, "Observation of Gain-Narrowing and Saturation Behavior in Se X-Ray Laser Line Profiles," *Phys. Rev. Lett*. **68**, 3291 (1992).

57. J.A. Koch, B.J. MacGowan, L.B. DaSilva *et al.*, "Experimental and Theoretical Investigation of Neon-like Selenium X-Ray Laser Spectral Linewidths and Their Variation with Amplification," *Phys. Rev. A* **50**, 1877 (1994).

58. A.V. Vinogradov and V.N. Shlyaptsev, "Amplification of Ultraviolet Radiation in a Laser Plasma," *Kvant. Elektr. (Moscow)* **10**, 2325 (1983); *Sov. J. Quant. Electr.* **13**, 1511 (1983).

59. R.L. Kelly, *Atomic and Ionic Spectrum Lines Below 2000 Angstroms: Hydrogen through Krypton, Part II (Mn–Kr)* (Amer. Inst. Phys., New York, 1987), published as *J. Phys. Chem. Ref. Data* **16**, Suppl. 1 (1987). See p. 1332.

60. T.N. Lee, E.A. McLean, and R.C. Elton, "Soft X-Ray Lasing in Neon-Like Germanium and Copper Plasmas," *Phys. Rev. Lett.* **59**, 1185 (1987).

61. P.V. Nickels, V.N. Shlyaptsev, M. Kalachnikov *et al.*, "Short Pulse X-Ray Laser at 32.6 nm Based on Transient Gain in Ne-Like Titanium," *Phys. Rev. Lett.* **78**, 2748 (1997).

62. B.J. MacGowan, L.B. DaSilva, D.J. Fields *et al.*, "Short Wavelength Nickel-Like X-Ray Laser Development," p. 221 in Ref. 16.

63. S. Wang, Z. Lin, Y. Gu *et al.*, "Intense Nickel-Like Neodymium X-Ray Laser at 7.9 nm with a Double-Curved Slab Target," *Jpn. J. Appl. Phys. Lett.* **37**, L1 234 (1998).

64. S. Sebban, N. Sakaya, H. Daido *et al.*, "Optimization of a Ni-Like Silver Collisional Soft X-Ray Laser Using Prepulse Technique and Double Target Geometry," in *X-Ray Lasers 1998* (Inst. Phys., Bristol, England, 1999), K. Kato, H. Takuma, and H. Daido, Editors.

65. J. Zhang, A.G. MacPhee, J. Nilsen *et al.*, "Demonstration of Saturation in a Ni-Like Ag X-Ray Laser at 14 nm," *Phys. Rev. Lett.* **78**, 3856 (1997).

66. J.Y. Lin, G.J. Tallents, J. Zhang *et al.*, "Gain Saturation of Ni-Like X-Ray Lasers," *Optics Comm.* **158**, 55 (1998); R. Smith *et al.*, *Phys. Rev. A* **59**, R47 (1999).

67. R.E. King, G.J. Pert, S.P. McCabe *et al.*, "Saturated X-Ray Lasers at 196 and 73 Å Pumped by a Picosecond Traveling-Wave Excitation," *Phys. Rev. A* **64**, 053810 (October 2001).

68. J.J. Rocca, O.D. Cortázar, B. Szapiro, K. Floyd and F.G. Tomasel, "Fast-Discharge Excitation of Hot Capillary Plasmas for Soft X-Ray Amplifers," *Phys. Rev. E* **47**, 1299 (1993); J.J. Rocca, V. Shlyaptsev, F.G. Tomasel *et al.*, "Demonstration of a Discharge Pumped Table-Top Soft X-Ray Laser," *Phys. Rev. Lett.* **73**, 2192 (October 17, 1994).

69. J.J. Rocca, F.G. Tomasel, M.C. Marconi *et al.*, "Discharge-Pumped Soft-X-Ray Laser in Neon-Like Argon", *Phys. Plasmas* **2**, 2547 (June 1995); J.J. Rocca, M.C. Marconi, J.L.A. Chilla *et al.*, "Discharge-Driven 46.9 nm Amplifier with Gain–Length Approaching Saturation," *IEEE J. Sel. Top. Quant. Electr.* **1**, 945 (1995).

70. J.J. Rocca, D.P. Clark, J.L.A. Chilla and V.N. Shlyaptsev,"Energy Extraction and Achievement of the Saturation Limit in a Discharge-Pumped Table-Top Soft X-Ray Amplifier," *Phys. Rev. Lett.* **77**, 1476 (1996).

71. C.H. Moreno, M.C. Marconi, V.N. Shlyaptsev *et al.*, "Two-Dimensional Near-Field and Far-Field Imaging of a Ne-like Ar Capillary Discharge Table-Top Soft X-Ray Laser," *Phys. Rev. A* **58**, 1509 (1998).

72. Y. Liu, M. Seminario, F.G. Tomasel *et al.*, "Achievement of Essentially Full Spatial Coherence in a High-Average-Power Soft-X-Ray Laser", *Phys. Rev. A* **63**, 033802 (March 2001); Yanwei Liu, "Coherence Properties of EUV/Soft X-Ray Sources", PhD Thesis, Applied Science and Technology Graduate Group, December 2003.

73. B.R. Benware, C.D. Macchieto, C.H. Moreno and J.J. Rocca,"Demonstration of a High Average Power Table-Top Soft X-Ray Laser," *Phys. Rev. Lett.* **81**, 5804 (1998).

74. C.D. Macchietto and B.R. Benware, "Generation of MilliJoule-Level Soft-X-Ray Laser Pulses at a 4-Hz Repetition rate in a Highly Saturated Tabletop capillary Discharge Amplifier," *Optics Lett.* **24**, 1115 (August 15, 1999).

75. J.J. Rocca, Y. Wang, M.A. Larotonda *et al.*, "Saturated 13.2 nm High-Repetition-Rate Laser in Nickel-like Cadmium," *Optics Lett.* **30**, 2581 (October 1, 2005).

76. Y. Wang, M.A. Larotonda, B.M. Luther *et al.*, "Demonstration of High-repetition-Rate Table-top Soft-X-Ray Lasers with Saturated Output at Wavelengths down to 13.9 nm and gain down to 10.9 nm," *Phys. Rev. A* **72**, 053807 (November 8, 2005); A. Aquila, D. Bleiner, J. Balmer and S. Bajt, "Polarization Measurements of Plasma Excited X-ray lasers," *SPIE* **8140**, 81400Z (2011).

77. D. Alessi, Y. Wang, B.M. Luther *et al.*, "Efficient Excitation of Gain-Saturated Sub-9-nm-Wavelength Tabletop Soft-X-Ray Lasers and Lasing Down to 7.36 nm," *Phys. Rev. X*, **1**, 021023 (December 27, 2011).

78. R. London, "Beam Optics of Exploding Foil Plasma X-Ray lasers," *Phys. Fluids* **31**, 184 (January 1988).

79. B.A. Reagan, K.A. Wernsing, A.H. Curtis *et al.*, "Demonstration of a 100 Hz Repetition Rate Gain-Saturated Diode-Pumped Table-Top Soft X-Ray Laser," *Optics Lett.* **37**, 3624 (September 1, 2012).

80. B.A. Reagan, M. Berrill, K.A. Wernsing, C. Baumgarten, M. Woolston and J.J. Rocca, "High-Average-Power, 100-Hz-repetition-rate, Tabletop Soft X-Ray Lasers at Sub-15-nm Wavelengths", *Phys. Rev. A* 89, 053820 (15 May 2014).

81. N. M. Ceglio, D.G. Stearns, D.P. Gaines, A.M. Hawryluk and J.E. Trebes, "Multipass Amplification of Soft X-Rays in a Laser Cavity," *Optics Lett.* **13**, 108 (1988); N.M. Ceglio, D.P. Gaines, J.E. Trebes, R.A. London and D.G. Stearns, "Time Resolved Measurement of Double Pass Amplification of Soft X-Rays," *Appl. Opt.* **7**, 5022 (1988); N.M. Ceglio, D.P. Gaines, D.G. Stearns and A.M. Hawryluk, "Double Pass Amplification of Laser Radiation at 131 Å," *Optics Comm.* **69**, 285 (1989); T. Haga, M. Tinone, M. Shimada, T. Ohkubo and A. Ozawa, "Soft X-ray Multilayer Beam Splitters," *J. Synchrotron Rad.* **5**, 690 (1998).

82. B.J. MacGowan, S. Mrowka, T.W. Barbee *et al.*, "Investigation of Damage to Multilayer Optics in X-Ray Laser Cavities: W/C, WRe/C, WC/C, Stainless Steel/C, and $Cr_3 C_2$/C Mirrors," *J. X-Ray Sci. Techn.* **3**, 231 (1992).

83. D. Attwood, "Comparative Features of Partially Coherent X-Ray Sources," in *Proceedings of the First Symposium on the Applications of Laboratory X-Ray Lasers, Asilomar, February 1985*, N.M. Ceglio, Editor; published by Lawrence Livermore National Laboratory as CONF-850293-Abstracts.

84. M.D. Rosen, J.E. Trebes and D.L. Matthews, "A Strategy for Achieving Spatially Coherent Output from Laboratory X-Ray Lasers," *Comments Plasma Phys. Fusion* **10**, 245 (1987).

85. R.A. London, M. Strauss and M.D. Rosen, "Model Analysis of X-Ray Laser Coherence," *Phys. Rev. Lett.* **65**, 563 (July 30, 1990); see also R.A. London, P. Amendt, M. Strauss *et al.*, "Coherent X-Ray Lasers for Applications," p. 363 in *X-Ray Lasers 1990* (Inst. Phys., Bristol, England, 1990), G.J. Tallents, Editor.

86. J.A. Koch, B.J. MacGowan, L.B. DaSilva *et al.*, "Selenium X-Ray Laser Line Profile Measurements," in *X-Ray Lasers 1992* (Inst. Phys., Bristol, England, 1992), E.E. Fill, Editor; also J.A. Koch *et al.*, *Phys. Rev. Lett.* **68**, 3291 (1992).

87. M. Nishikino, M. Tanaka, K. Nagashima *et al.*, "Demonstration of a Soft-X-Ray Laser at 13.9 nm with Full Spatial Coherence," *Phys. Rev. A* **68**, 061802(R) (December 30, 2003).

88. G.J. Pert, S.B. Healy, J.A. Plowes and P.S. Simms, "Effects of Density Structures and Temporal Shaping of the Pump Pulse in X-Ray Lasing," in *X-Ray Lasers 1996*, S. Svanberg and C.-G. Wahlström, Editors, p. 260.; and F. Albert, A. Carillon, P. Jaeglé *et al.*, "New Approach for Measurement of the X-Ray Laser Transverse Coherence," ibid, p. 427.

89. T. Dittmire, M.H.R Hutchinson, M.H. Key *et al.*, "Amplification of XUV Harmonic Radiation in a Gallium Amplifier," *Phys. Rev. A* **51**, R4337 (June 1995).

90. Ph. Zeitun, G. Faivre, S. Sebban *et al.*, "A High-Intensity Highly Coherent Soft X-Ray Femtosecond Laser Seeded by a High Harmonic Beam," *Nature* **431**, 426 (September 23, 2004).

91. Y. Wong, E. Granados, M.A. Larotonda *et al.*, "High-Brightness Injection-seeded Soft-X-Ray-Laser Amplifier Using a Solid Target," *Phys. Rev. Lett.* **97**, 123901 (September 22, 2006).

92. I.R. Al'miev, O. Larroche, D. Benredjem *et al.*, "Dynamical Description of Transient X-Ray Lasers Seeded with High-Order Harmonic Radiation through Maxwell-Bloch Numerical Simulations," *Phys. Rev. Lett.* **99**, 123902 (September 21, 2007).

93. Y. Wang, E. Granados, F. Pedaci *et al.*, "Phase-Coherent, Injection-Seeded, Table-Top Soft-X-Ray Lasers at 18.9 nm and 13.9 nm," *Nature Photon.* **2**, 94 (February 2008).

94. E. Pedaci, Y. Wang, M. Berrill *et al.*, "Highly Coherent Injection-Seeded 13.2 nm Tabletop Soft X-Ray Laser," *Optics Lett.* **33**, 491 (March 1, 2008).

95. M. Berrill, D. Alessi, Y. Wang *et al.*, "Improved Beam Characteristics of Solid-State Soft X-Ray Laser Amplifiers by Injection Seeding with High Harmonics," *Optics Lett.* **35**, 2317 (July 15, 2010).

96. Y. Wang, S. Wang, E. Oliva *et al.*, "Gain Dynamics in a Soft X-Ray Laser Amplifier Perturbed by a Strong Injected X-ray Field," *Nature Photon.* **8**, 381 (May 2014).

97. A. Depresseux, E. Oliva, J. Gautier *et al.*, "Table-top Femtosecond Soft X-ray Laser by Collisional Ionization Gating," *Nature Photonics* **9**, 817 (December 2015).

Homework Problems

Homework problems for each chapter will be found at the website:
www.cambridge.org/xrayeuv

10 X-Ray and Extreme Ultraviolet Optics

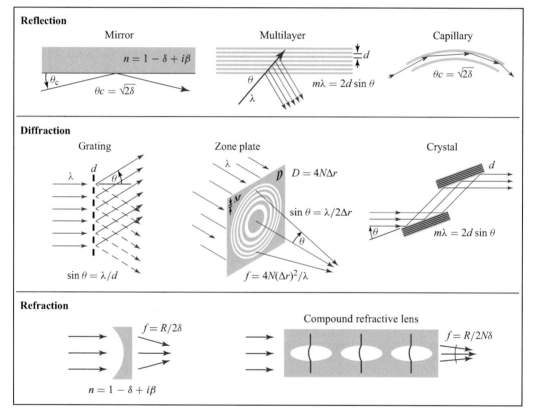

In this chapter common forms of reflective, diffractive, and refractive optics for use at extreme ultraviolet, soft x-ray and hard x-ray wavelengths are described. These include flat and curved mirrors, multilayer coated mirrors, transmission and reflection gratings, zone plates, natural crystals for dispersion and monochromatization, and compound refractive lenses.

10.1   Introduction

X-ray and EUV optics can shape or manipulate radiation spatially or temporally. The ability of x-ray and EUV optics to reflect, refract, and diffract radiation is based on the same physical principles of light interaction with matter described in Chapters 1–3. Materials used to fabricate x-ray and EUV optics have photon energy specific properties, captured in the complex atomic scattering factors, and succinctly by the refractive index as was described in Eqs. (3.12) and (3.13). Combinations of different materials, in defined geometries, can lead to optical elements whose purposes include radiation transport, focusing, imaging, energy (wavelength) dispersion, polarization control, spatial and spectral filtering for coherence control, each for use in a wide range of scientific experiments and technical applications.

Because the selection of materials and their properties play a key role in the performance of optical elements, we briefly review the refractive index, which is written as $n = 1 - \delta + i\beta$, where typically $\delta, \beta \ll 1$. The small deviation from unity for the refractive index of x-rays and EUV differs significantly from visible wavelengths. The real part, δ, is much less than unity, and the wave vector bends away from the surface normal as it enters the material from vacuum, rather than towards the normal. This makes possible total external reflection for x-rays and EUV, as opposed to total internal reflection for visible light. The real part, δ, also determines the phase shift of a wave within a material. Furthermore, the imaginary part, β, indicates absorption which is significant in this spectral range, particularly at EUV and soft x-ray wavelengths, and to a lesser degree for very hard x-rays. Indeed, for sufficiently hard x-rays all materials are transparent. Sample absorption lengths are given in Table 1.1 for various materials and for photon energies of 100 eV, 1 keV, and 10 keV. The absorption lengths range from tens of nanometers in the EUV to millimeters and even centimeters for hard x-rays.

The main properties of reflection, diffraction, and refraction are briefly reviewed here. Reflection describes the property whereby incident radiation reflects off a surface at an angle equal to the angle of incidence. Single surface normal incidence reflectivity [Eq. (3.50)], $R_\perp \simeq (\delta^2 + \beta^2)/4$, is very small throughout this spectral region. In order to experience a high reflectivity, the angle of incidence must be less than the critical angle for total external reflection [Eq. (3.41)], renumbered here as

$$\theta_c = \sqrt{2\delta} \tag{10.1}$$

This angle is quite small, typically measured in degrees or fractions of a degree, depending on wavelength and material. Small angles of incidence such as this are described as being of grazing or glancing incidence. Diffraction is the redirection of light, for example by a crystal or other well-defined structure, such as a zone plate, grating, or a simple pinhole aperture. Natural crystals do this very efficiently for hard x-rays. For diffractive structures, the geometric pattern and materials composition will redirect the beam in a well-defined manner determined by wavefront interference and propagation. Refraction is the turning of light through a medium, or sharply at an interface between materials. The turning of light from an angle ϕ to ϕ' at an interface is governed by Snell's law [Eq. (3.38)], $\sin \phi' = \sin \phi / n$. Because the index of refraction of all materials is very close to unity in this spectral range, refractive turning effects are quite small. Nevertheless, designs for optics such as the compound refractive lens allow

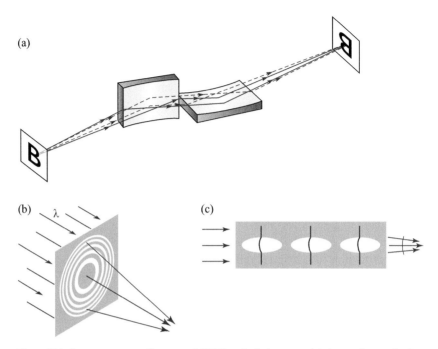

Figure 10.1 Common types of x-ray and EUV optical elements. (a) A set of curved mirrors, known as a Kirkpatrick–Baez pair, which sequentially reflect radiation at glancing incidence to form a real image as shown, or to focus radiation to a small spot.[1, 2] (b) A Fresnel zone plate lens which by diffraction focuses radiation to a small spot, or forms an image. (c) A compound refractive lens with sequential refractive elements can be used to focus hard x-rays.[3]

the utilization of refraction in the hard x-ray region where absorption can be relatively small.

In combination with emerging fabrication technology, optical elements continue to evolve significantly to meet the demands and opportunities provided by modern sources of x-ray, soft x-ray, and EUV radiation. Representative optics that demonstrate the use of reflection, diffraction and refraction are illustrated in Figure 10.1. In this chapter we describe optical elements of each type, the principles upon which they operate, and related applications. These include reflective optics such as curved mirrors, multilayer coated mirrors, and capillary optics; diffractive optics such as gratings, zone plates, and crystals; and compound refractive lenses.

10.2   Reflective X-Ray Optics

X-ray mirrors follow the law of reflection, that the angle of reflection equals the angle of incidence. An x-ray mirror is composed of a highly reflective polished surface, with or without a thin film coating, which can be shaped to accommodate focusing capabilities in either one or two dimensions. The optical shape of the mirror is known as the "figure" and the characteristics of the surface quality and roughness as the "finish." The quality of the figure and the finish determine the mirror's performance. For reflective optical systems, the challenge has been to achieve the desired figure for focusing with a suitable bending, cutting, polishing, or coating technique, while maintaining ångström (Å) level finish across a broad range of spatial frequencies.

Mirrors are commonly used at synchrotron, free electron laser, and HHG facilities for beam transport or focusing. Because total external reflection of x-rays occurs only at small angles, mirrors can be very long, in some cases longer than 1 m. Coating the mirror with a thin film of a material, such as gold, increases the critical angle, and results in a decreased mirror length. The thin film coating can also provide spectral filtration as a low pass filter. This can be used for applications such as defining a beamline's spectral range or for rejection of higher order harmonic wavelengths, as with undulator radiation. In addition, the mirror can be coated with a multilayer thin film for spectral bandpass control, as will be discussed in Section 10.3. Curved x-ray mirrors used for focusing hard x-rays regularly achieve a sub-micron spot size, in some cases as small as 10 nm.

The shape of the mirror determines its focusing properties. A plane mirror reflects radiation without focusing while curved mirrors can focus the radiation. For example, a mirror with an ellipsoidal shape is ideal for focusing radiation from one point to another. Kirkpatrick–Baez mirror pairs[1] are two curved mirrors placed at right angles to one another where the first mirror focuses the incoming radiation to a line, and the second mirror focuses the line radiation to a point. Shaping of the mirrors can be performed by mechanical bending,[4] grinding, and polishing, and more recently by chemical reaction techniques using the controlled delivery of ultrafine reactive particles.[5] During use, thermal variations caused by radiation induced heating can deform mirrors, thus requiring countermeasures such as water cooling through internal ducts. Figure errors affect the focusing properties in a manner similar to that of common optical aberrations.

The finish of the mirror is set by the surface roughness. This is affected by the manufacturing technique and subsequent ability to polish the mirror to extreme smoothness. Roughness of the finish generates unwanted scattering, removing photons from the desired direction and sending them in other directions, thus affecting both throughput and contrast.

Using geometrical optics, we first consider the reflection of x-rays from a smooth surface. We recall from Chapter 3 that an important characteristic of the law of reflection is its achromatic nature, that the angle of reflection equals the angle of incidence for all wavelengths. This means that the focal length of a mirror is independent of incoming wavelengths. This property has many implications. For example, a short pulse from an x-ray free electron laser (FEL) or from laser high harmonic generation (HHG), which has a large spectral bandwidth, can be focused with achromatic mirrors to a single spot since the focal length of the mirror is the same for the entire spectral bandwidth. Reflection from a planar surface was described in Chapter 3, Section 3.5, as a function of angle, photon energy and polarization. In addition to full formulas applicable to all these parameters, specialized formulas for normal and glancing incidence were obtained, as quoted in the previous section. The material used to coat a mirror will be a major factor

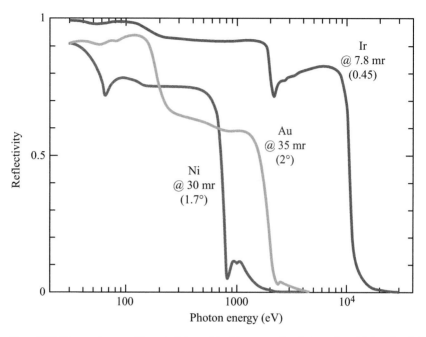

Figure 10.2 X-ray mirrors with materials and incidence angles that produce high reflectivity and relatively sharp energy cutoffs[6, 7] at 700 eV, 2 keV, and 10 keV. These can be used both to redirect radiation and also to act as low pass filters, absorbing higher energy radiation.

determining its reflective characteristics. Figure 10.2 shows reflectivity as a function of photon energy for mirrors of nickel, gold, and iridium, which for the chosen angles of incidence have relatively sharp cutoffs at 700 eV, 2 keV and 10 keV.[6, 7] These curves illustrate the high reflectivity achievable at glancing incidence, and their capability for low pass spectral filtering. The latter can be useful in combination with an absorption edge filter to produce a "notch filter" as was illustrated in Figure 3.11. The low pass energy cutoff is also useful in many applications, such as for blocking higher photon energy bending magnet radiation in a soft x-ray microscope, blocking unwanted higher harmonics of undulator radiation when operating at K values greater than 1, selecting a particular high harmonic, or defining a spectral acceptance window as in an x-ray telescope.

Regarding curved surfaces, an ellipsoidal mirror has the ideal shape for focusing from one point to another point as illustrated in Figure 10.3. Radiation emitted at various angles from one focus will all come to a focus at the other foci. Although this mirror possesses an ideal shape, fabricating single ellipsoid mirrors is non-trivial. Polishing optics to a fine finish while maintaining the figure works well with flats and spheres, but is more challenging with other shapes. This has led to an alternative solution for focusing radiation using what is known as Kirkpatrick–Baez (KB) mirror pairs.[1] KB optics consist of a set of two mirrors aligned perpendicular to each other as illustrated in Figure 10.1(a), each having an elliptical shape and focusing radiation in one direction. KB mirror pairs are commonly used in demagnifying x-ray nanoprobes with the source placed at one of the foci and the sample under study placed far away at the second focus. In some

Figure 10.3 An elliptical optic has the property of focusing all radiation from one focus to the other. Glancing incidence angles less than or equal to the critical angle are required. An example would be an x-ray nanoprobe with the source at one point and the sample at the other. Note that the angular spread (NA) is larger on the focusing side, permitting demagnification to a smaller spot size. In some cases an ellipse with a multilayer interference coating is used, graded to meet the Bragg condition at every point on the surface. This diagram is not to scale; typical eccentricities are several thousand.

cases the surface is uncoated for broad spectral (achromatic) coverage. In other cases the optic is coated with a graded multilayer coating which meets the Bragg condition at every point on the surface within a narrow spectral bandwidth. Use of a multilayer coating provides larger angles of incidence and reflection, thus increasing the effective numerical aperture (NA), reducing the focal spot size, and reflecting only over a prescribed spectral window, typically 3%–5% wide. The disadvantage is that different optics and coatings are needed for each spectral band of interest. Axisymmetric conic section pairs,[8, 9] such as the Wolter type 1 paraboloid/hyperboloid combination, are used in orbiting x-ray telescopes like the Chandra X-ray Observatory.[10, 11] Figure 10.4(a) shows Chandra's four Wolter type 1 axisymmetric nested pairs[12] (eight mirrors total), each having an iridium coating.[13] With these optics Chandra covers deep space with a spectral range from 0.1 to 10 keV. A composite x-ray/UV/visible image of the Messier 51 spiral galaxy system, known as the "Whirlpool Galaxy," is shown in (b). Chandra is a major tool for studying distant galaxies, including the dynamics of super massive black holes (SMBHs) within those galaxies.[14]

Several methods exist for the fabrication of high precision mirrors, including grinding and polishing, bending a polished flat, and use of mandrels. One of the highest performing figuring techniques is elastic emission machining (EEM), developed at Osaka University.[5] In EEM, the surface of a powder, such as silica, is used to chemically react with the mirror surface. Contact and separation between the silica powder and the mirror surface causes atom-level amounts of the mirror surface to be removed. This numerically controlled chemical processing technique involves a jet of high purity water delivering ultra small silica particles to the optical substrate. The technique is capable of preparing flat and aspheric surfaces to sub-nanometer peak-to-valley accuracy across spatial scales of 300 μm or more, with less than 2 Å rms roughness. As it is a relatively slow process, it is utilized as the final figuring and surface smoothing step following a faster but coarser figuring step known as chemical vaporization machining (CVM). In this way, the polishing of x-ray mirrors is analogous to smoothing a wooden surface with finer and finer grit sandpaper until achieving a very smooth surface. Interferometric monitoring of the surface, comparing it to a design reference, is also critical.

Yamauchi, Mimura, Matsuyama and their colleagues at Osaka University have developed a wide range of KB systems[15–21] based on use of EEM and CVM[5] as discussed above.[22] These include a two mirror elliptic KB system employing a broadband coating; a four bounce KB system in which the first "pre-focus" set provides a smaller source

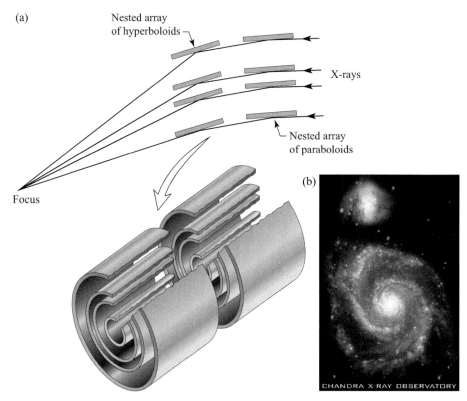

Figure 10.4 (a) Four nested pairs of paraboloid/hyperboloid, Wolter type 1, x-ray optics for the study of deep space with the orbiting Chandra X-Ray Observatory[10–13]. Photon energies cover the range from 0.1 to 10 keV. Optics diagram courtesy of D. Schwartz[12] of the Harvard-Smithsonian Center for Astrophysics. (b) A composite image of the spiral galaxy Messier 51, known as the "Whirlpool Galaxy," which is about 26 million light years from earth.[14] The composite image combines x-ray data from Chandra (purple), UV light from the Galaxy Evolution Explorer (GALEX; blue), visible light from the Hubble Telescope (green), and IR light from the Spitzer Space Telescope (red). Courtesy of the Harvard-Smithsonian Center for Astrophysics, and NASA.

and larger NA for the second set; use of graded multilayer coatings to increase the NA and improve the focus of elliptic optics; and the use of adaptive optics to compensate for low spatial frequency figure errors. Shown in Figure 10.5 is a two mirror elliptical KB system for use as a 15 keV hard x-ray fluorescence nanoprobe on an undulator beamline at SPring-8.[17, 18] The mirror substrates are each 100 mm long Si (111) single (CZ) crystals, each having a 50 nm platinum coating. The incidence angle on the first mirror is 3.80 mrad (0.218°), providing a sharp cutoff just above 20 keV. The figure of the first mirror is an ellipse of 23.9 m semi-major axis and 13.1 mm semi-minor axis, thus an eccentricity of ~1800. The second mirror has a somewhat shorter semi-minor axis and an eccentricity of ~2500. The working distance is 10 mm. The beamline includes a double crystal Si (111) monochromator set for a bandpass of 1.4×10^{-4} at 15 keV. The achieved beam size with a 10 µm diameter upstream aperture is nearly diffraction limited at 36 nm horizontal by 48 nm vertical, with a photon flux of 6×10^9 ph/s. With

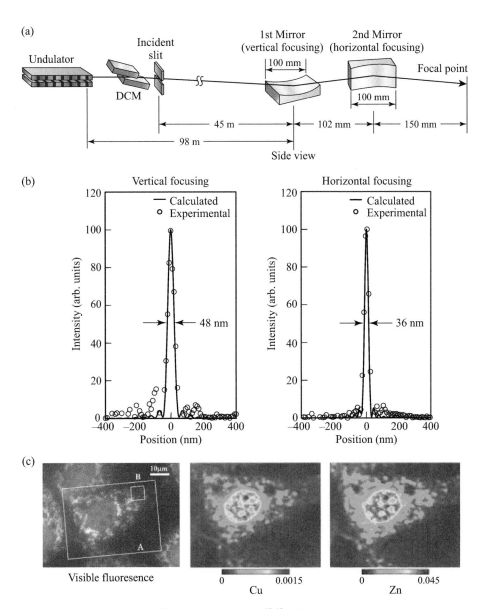

Figure 10.5 (a) The nanofocus fluorescence beamline[17, 18] at SPring-8 employs two Pt coated elliptical mirrors capable of focusing 15 keV undulator to near the diffraction limit of 30 nm (h) by 50 nm (v) with a 10 μm diameter upstream aperture. The photon flux within this essentially diffraction limited focal spot was 6×10^9 ph/s with the double crystal Si (111) monochromator set for a spectral bandpass of 1.4×10^{-4}. Higher flux in a larger focal spot is obtainable with larger upstream apertures. (b) Measured focal spot widths were 36 nm (h) by 48 nm (v) with 10 μm upstream slits. (c) Elemental distribution maps[18] of Cu and Zn, are seen largely within the nucleus of a single NIH/3T3 cell. The data were collected on the scale of sub-cellular organelles. The spatial scale bar is 10 μm. The color scales are in femtograms (fg) as shown for each. Elemental maps were also reported for P, S, Cl, Ca, and Fe. A visible light fluorescence microscope image is shown for reference and complementarity. Courtesy of S. Matsuyama and K. Yamauchi, Osaka University.

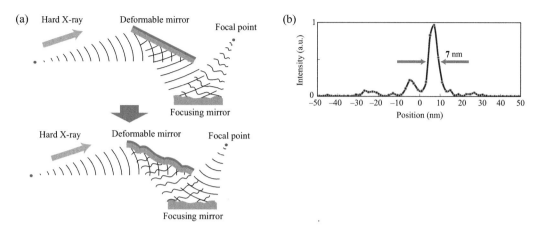

Figure 10.6 (a) The technique of active wavefront correction[15] using an upstream deformable mirror and a feedback system based on measurements of the resultant focal spot. (b) The measured one-dimensional focal width achieved at 20 keV (0.62 Å) with the combination of an upstream deformable mirror and a downstream, multilayer coated elliptical optic. Further experiments by the Osaka group achieved similar results in a two-dimensional spot.[16] Courtesy of H. Mimura, University of Tokyo, and K. Yamauchi, Osaka University.

a larger upstream aperture, and thus larger focal spot, higher photon flux is available, for example 3×10^{11} ph/s with a 130 nm × 230 nm focal spot, or 4×10^{12} ph/s with a 570 nm × 980 nm focal spot.

In a separate experiment, Mimura, Yamauchi, and their colleagues at Osaka University and SPring-8, have achieved a 7 nm one-dimensional (1D) focusing of 20 keV undulator radiation using an upstream deformable mirror and a multilayer coated elliptical mirror.[15] The upstream deformable mirror uses a 150 mm long silicon substrate with 18 pairs of ceramic piezoelectric actuators fitted to the rear surface, allowing for in situ wavefront compensation for surface irregularities with periods as short as 16 mm. The deformable mirror was fabricated for figure and finish with CVM, EEM, and microstitching interferometry, as above, and then coated with platinum for 4 mrad illumination at 20 keV. The downstream focusing optic was coated with a d-space graded platinum/carbon multilayer coating having 20 layer pairs for high reflectivity at 20 keV and an incidence angle of 7 mrad. The larger Bragg angle of the multilayer coatings provided an increased numerical aperture of $NA = 4.6 \times 10^{-3}$, so that with a wavelength of 6.2 Å the diffraction limited (1D) Rayleigh resolution is $0.50\lambda/NA = 6.7$ nm. Figure 10.6 illustrates the technique of active wavefront compensation, plus a measured 7 nm (1D) focal width.[15] Further results, with two-dimensional focusing of 20 keV undulator radiation, have achieved a two-dimensional focal spot of 7 nm (h) by 8 nm (v).[16] Using a four-mirror KB system at SACLA, the Osaka group was able to achieve a very high FEL focal intensity with 9.9 keV photons, a 7 fs pulse radiation, and a 30 nm by 55 nm focal spot.[21] The uncoated upstream KB optics pair produced an intermediate focal spot of increased divergence, which then illuminated a downstream, platinum coated final focusing elliptical KB pair. With this tight focus and short pulse duration the focused FEL intensity reached 10^{20} W/cm^2, opening new opportunities for non-linear x-ray studies.[21]

10.3   Multilayer Mirrors

Multilayer interference coatings, often referred to as multilayer mirrors, or simply as "multilayers," are formed by depositing alternating layers of two materials of differing refractive index that form long-term stable interfaces. Often the two materials are of alternating high and low atomic number (Z) in order to maximize the difference in electron density and thus refractive index. The coatings are largely amorphous within individual layers, and reflection conforms to Bragg's law, renumbered here as

$$m\lambda = 2d \sin \theta \qquad (10.2)$$

for a periodicity d equal to the thickness of one bilayer pair, as discussed in Chapter 3, Section 3.9, Eq. (3.91). Multilayer coatings have the great advantage of being adaptable to curved surfaces, enabling their use as reflective optics in EUV and soft x-ray microscopes, telescopes, and other applications. At EUV wavelengths the coatings permit the achievement of high normal incidence reflectivity, within a modest spectral bandwidth. A normal incidence reflectivity of approximately 70% has been achieved[23] in the EUV with Mo/Si just below the L_3-absorption edge of silicon at 99 eV. The spectral bandpass for these mirrors is of order $1/N$, where N is the number of layer pairs, generally between 30 and 50 for high reflectivity. For normal incidence reflection, individual layers are each about $\lambda/4$ thick thus limiting possibilities for soft x-rays and hard x-rays where each layer is only a few atoms thick. Lack of interface perfection on this spatial scale, due principally to nano-roughness and interdiffusion, greatly reduces collective interference and thus diminishes achievable normal-incidence reflectivity at these shorter wavelengths. Off-normal incidence, however, offers further opportunities with glancing incidence reflection at angles several times larger than the critical angle, θ_c, and with control over spectral bandwidth, typically a few percent. Figure 10.7(a) shows a TEM side view of a high quality Mo/Si multilayer mirror. The reflectivity of a similar, Mo/B$_4$C/Si multilayer mirror achieved a near normal incidence reflectivity of 70% at a wavelength of 13.5 nm, corresponding to a photon energy of 91.8 eV, just below the L_3-edge of Si.[23] Additional coating techniques for the EUV suppress corrosion and nanocrystallization.[24]

An example of multilayer coated normal incidence optics used to obtain high resolution astronomical images is illustrated in Figure 10.8(a). Fine details of solar activity are seen by observing EUV emission across the full solar disk with the Lockheed/NASA Atmospheric Imaging Assembly (AIA) telescope, part of the geo-synchronous Solar Dynamics Observatory (SDO), which was launched February 11, 2010.[25] The telescope provides ten spectral windows: seven EUV pass bands, two UV, and one visible band for alignment and stabilization. Of the seven EUV channels, six emphasize coverage of various ionization states of iron (Fe^{+7} to Fe^{+23}) to obtain fine temperature sensitivity. These pass bands are centered at wavelengths of 9.39, 13.10, 17.11, 19.35, 21.13, and

(a)

(b)

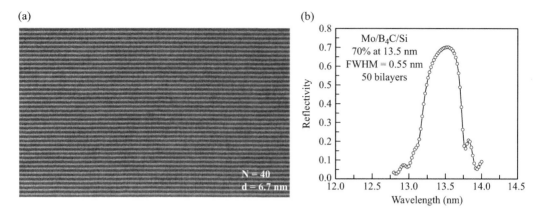

Figure 10.7 (a) A high quality Mo/Si multilayer mirror with 40 layer pairs of 6.70 nm d-spacing designed for operation at a wavelength of 13.5 nm. (b) Near normal incidence reflectivity of 70% was obtained at a wavelength of 13.5 nm with a similar mirror that also included B_4C diffusion barriers between all layers.[23] Courtesy of S. Bajt of Lawrence Livermore National Laboratory, presently at DESY Laboratory, Hamburg.

33.54 nm, employing multilayer coatings of Mo/Y, Mo/Si, and SiC/Si, combined with various absorption filters. Images are recorded with a 4096×4096 CCD array, covering a 41 arcmin (12 mrad) field of view, approximately 1.28 times the sun's diameter), with a resolution of 0.6 arcsec (2.9 µrad). The image in Figure 10.8(a), obtained at a wavelength of 30.38 nm, the He^{+1} doublet, shows fine coronal loop detail at the solar limb. Exposure times are seconds, with continuous coverage of months, thus providing a powerful tool for studies of magnetic field embedded plasma dynamics across the full solar disk.[25]

In the fabrication and testing of multilayer mirrors, a common method for measuring d-spacing, and also inferring interface sharpness, is to perform an angle dependent reflectivity scan with short wavelength radiation, typically Cu K_α at 1.54 Å. An example of such data is shown in Figure 10.9, for a tungsten–carbon multilayer mirror of 36.1 Å period, where several Bragg peaks are evident.[26] The amplitudes of the peaks can be understood in terms of a spatial Fourier transform of the multilayer density distribution. For example, if in Figure 10.9(a), the density $n_e(z)$ were sinusoidal, we would expect to see only a single Bragg peak ($m = 1$) in an angular scan. However, if $n_e(z)$ were a symmetric step function we would expect to see a series of Bragg peaks corresponding to $m = 1, 3, 5, \ldots$, with peak intensities declining as $1/(m\pi)^2$. If the interfaces were not so sharp we would expect the higher order peaks to vanish and intermediate orders to diminish in amplitude. If the density function were asymmetric, due to unequal thicknesses of the high- and low-Z materials, we would expect even orders ($m = 2, 4, \ldots$) to appear. Thus a large number of higher-order Bragg peaks indicates a sharp interface, while the presence of even orders indicates an asymmetry within the bilayer. In Figure 10.9(b) experimentally observed Bragg peaks to $m = 5$ are quite clear. However, the intensities decline more rapidly than $1/(m\pi)^2$, indicating that the density profile

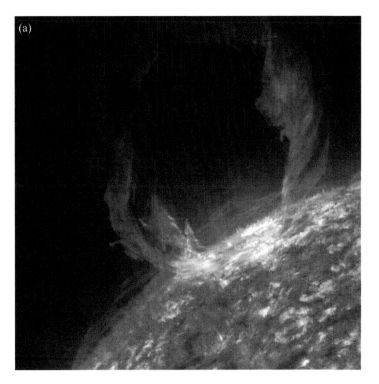

(a)

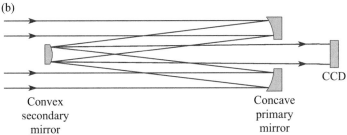

(b)

CCD

Convex
secondary
mirror

Concave
primary
mirror

Figure 10.8 (a) An EUV image of the sun showing a coronal loop of magnetized plasma near the solar limb.[25] Courtesy of J.R. Lemen, Lockheed Martin, and NASA. (b) A Cassegrain telescope employing multilayer coated normal incidence optics. Multilayer coatings were provided by R. Soufli, LLNL and D.L. Windt, Reflective X-ray Optics.

is less pronounced than a square wave, which is not surprising when one considers interface roughness, interpenetration during the coating process, and the small number of atomic monolayers contributing to each layer pair (about seven per layer in this case).

The appearance of even orders in Figure 10.9(b) indicates that the multilayer coating is somewhat asymmetric. In fact this coating was designed to have the tungsten layers somewhat thinner than the carbon layers, which has the effect of decreasing absorption and thus enhancing diffraction to the first order.[27, 28] In designing multilayer coatings

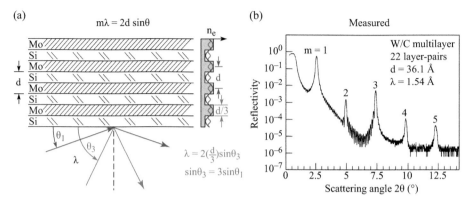

Figure 10.9 (a) A simple diagram for understanding the relationship between observed reflection orders of a multilayer and the spatial frequency components of the density profile within the multilayer. The angular scan of reflectivity provides a Fourier analysis of the density structure. (b) Measured angular scan (2θ) of reflectivity for a tungsten–carbon multilayer[26] at $\lambda = 1.54$ Å (Cu K$_\alpha$). The mirror consists of 22 W/C bilayer pairs of 36.1 Å period. In addition to total external reflection at angles below $1°$, one observes a strong first order ($m = 1$) Bragg peak at about $2.5°$, and less intense peaks corresponding to $m = 2, 3, 4$, and 5. Part (b) is courtesy of J.H. Underwood, Lawrence Berkeley National Laboratory.

an important parameter is the ratio of high-Z material thickness to bilayer period, represented here by the Greek letter Γ, i.e.,

$$\Gamma = \frac{\Delta t_H}{\Delta t_H + \Delta t_L} = \frac{\Delta t_H}{d} \tag{10.3}$$

where Δt_H is the thickness of the high-Z material and Δt_L is the thickness of the low-Z material. In general it is best if the low-Z material acts simply as a "spacer," with as little absorption as possible. In fact, the optical constants of the low-Z material, δ_L and β_L, should be as small as possible to provide the greatest refractive index contrast at the interfaces. In that limit, or an approximation to that limit, the high-Z layers provide both the scattering and absorption. The tradeoff, then, is to obtain sufficiently strong scattering, through refractive index contrast at the interfaces, while minimizing the absorption by reducing the thickness of the high-Z layer. Vinogradov and Zeldovich[27] have studied this and find, for normal incidence, an optimized value Γ_{opt} given by

$$\tan(\pi \, \Gamma_{opt}) = \pi \left[\Gamma_{opt} + \frac{\beta_L}{\beta_H - \beta_L} \right] \tag{10.4}$$

where β_L and β_H are the absorptive components of refractive index for the low-Z(L) and high-Z(H) materials. For the W/C multilayer in Figure 10.9, at 1.54 Å wavelength, using values of β derived from the tables of Henke *et al.*[6, 7], Eq. (10.4) indicates that the optimum value is $\Gamma_{opt} \cong 0.1$. For the specific case considered here, with $d = 36.1$ Å, the ideal thickness of the tungsten layers would be about 3.6 Å, or about 1.5 monolayers. This is not very realistic with currently achievable interface definition. Factors such as interface roughness, compound formation (tungsten carbide, molybdenum

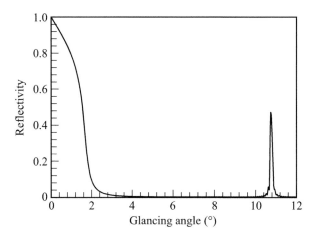

Figure 10.10 Computed reflectivity[30, 31] versus glancing incidence angle for a tungsten–carbon multilayer mirror of 100 bilayers, with periodicity $d = 2.25$ nm, $\Gamma = 1/3$, at a wavelength of 0.834 nm (1.49 keV). The angle θ is measured from the surface. Total external reflectance is shown below 2°, and a strong first order ($m = 1$) Bragg peak is seen at just under 11°. Courtesy of J. Underwood, Lawrence Berkeley National Laboratory.

silicide, etc.), cross-interface interpenetration, crystallite formation, etc., limit the minimum layer thickness to at least several monolayers. For such practical reasons, typically achieved values of Γ are more generally in the region of 0.3 to 0.5 for small d-spacing coatings. For further discussion, see the text by Spiller.[29]

A soft x-ray multilayer reflectivity curve is shown in Figure 10.10. It shows calculated reflectivity as a function of glancing incidence angle θ, for a W/C multilayer mirror at 0.834 nm wavelength.[30] The multilayer coating in this calculation has 100 layer pairs, with $d = 2.25$ nm and $\Gamma = 1/3$. High reflectivity due to total external reflection is indicated for angles less than 2°, and a strong first order Bragg peak at an incidence angle just under 11°. The Bragg peak is predicted to have a reflectivity of about 70%. The larger angle for multilayer reflection offers great advantages for both soft and hard x-ray optical applications. The increased numerical aperture (NA) permits improved focusing and imaging resolution, as in laboratory nanoprobes[26, 18] and astronomical telescopes.[9, 25] The angular bandpass in this soft x-ray example is about 1.5%, indicating that the spectral bandpass, for a given incidence angle, would also be about 1.5%. This proves useful in various applications where modest but well-defined spectral selectivity is desired, such as for plasma diagnostics, EUV/soft x-ray astronomy, and EUV/soft x-ray laser line isolation. Measured reflectivity values are typically somewhat less than calculations due to interface roughness, interpenetration of the two materials during fabrication, and compound formation.

Normal incidence reflection requires nominally $\lambda/4$ thicknesses for each layer. Thus for soft x-rays, with wavelengths of just a few nanometers, each layer would be a fraction of a nanometer thick, corresponding to only a few atomic monolayers. To be more

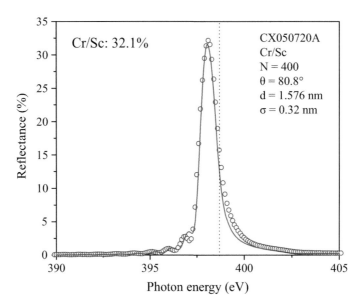

Figure 10.11 Reflectivity curve for a near-normal incidence soft x-ray $Cr/B_4C/Sc$ multilayer mirror having 400 layer pairs with a d-spacing of 1.576 nm.[32] The peak reflectivity is 32% at 398 eV (3.12 nm). Courtesy of E.M. Gullikson, Lawrence Berkeley National Laboratory.

specific, a normal incidence multilayer mirror designed to operate just below the nitrogen K-edge at 410 eV (3.02 nm wavelength in vacuum) would require individual layers of about 0.75 nm, corresponding to only a few atomic monolayers. Clearly this is difficult to achieve in non-crystalline material combinations. In experiments utilizing B_4C interface barriers between primary materials, Gullikson[32] has achieved near normal incidence reflectivities of 32% with Cr/Sc at 398 eV, 17% with Cr/Ti at 453 eV, and 9.1% with Cr/V just below 512 eV. The near-normal incidence (9.2°) reflectivity curve for a $Cr/B_4C/Sc$ multilayer mirror with a peak reflectivity of 32% at 398 eV is shown in Figure 10.11. It has 400 layer pairs, each with a *d*-spacing of 1.576 nm.

In the limit where material interfaces are well defined, it is possible to calculate reflectivities using the techniques developed in Chapter 3, Section 3.5, where reflection and refraction at well-defined interfaces were described in terms of increments in refractive index. In fact, finite transition zones can be addressed using these same techniques. Following Underwood,[31] Figure 10.12 illustrates the method for calculating multilayer reflection. The calculation is performed for incident radiation at an angle θ from the surface, applying the boundary conditions that tangential components of E and H must be continuous at each interface, as employed in Chapter 3 for a single interface.

Each material is characterized by a refractive index, described in Eqs. (3.9) and (3.12) as

$$n = 1 - \delta + i\beta = 1 - \frac{n_a r_e \lambda^2}{2\pi} \left(f_1^0 - i f_2^0 \right) \qquad (10.5)$$

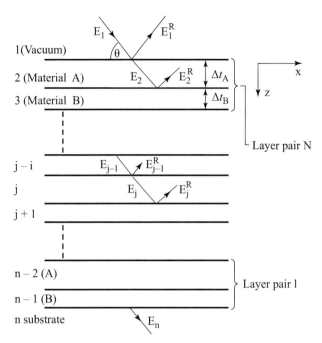

Figure 10.12 Computational model[31] for calculating multilayer reflectivity, shown here as N bilayer pairs of two materials, A and B, of thickness Δt_A and Δt_B, where the multilayer period is $d = \Delta t_A + \Delta t_B$. Courtesy of J. Underwood, Lawrence Berkeley National Laboratory.

where n_a is the density in atoms per unit volume, r_e is the classical electron radius, λ is the wavelength, and f_1^0, f_2^0 are the real and imaginary components of the complex atomic scattering function (see Appendix C).[5, 6, 33, 34] Calculations of reflected and refracted waves are then performed at each interface and summed. The calculations are done separately for perpendicular (s) and parallel (p) polarization. An example of calculated reflectance was shown in Figure 10.10. To account for the differences between experimental results and idealized (perfect interfaces) calculations, researchers sometimes invoke a Debye–Waller factor[28] in the reflection coefficients that is meant to represent in some fashion the diffuse nature of the interface. Originally introduced to take account of thermal motion of atoms in a crystalline lattice, it is used in multilayer characterization as a catch-all factor representing the spatial scale of interface broadening and roughness, generally in the form of a Gaussian distribution of rms value σ, to either side of the ideal interface.

Multilayer interference coatings for EUV and x-ray wavelengths have been successfully fabricated using evaporation,[29] sputtering,[35–38] epitaxial growth,[39] and laser-plasma deposition techniques.[40] For applications involving curved surfaces, especially where uniformity or d-spacing control is important, sputtering is particularly attractive and widely used. Figure 10.13(a) shows the basic sputtering technique and geometry. Blank mirror substrates are placed face down over openings in a circular table that rotates above sputtering targets for the two desired materials, labeled here as "high Z"

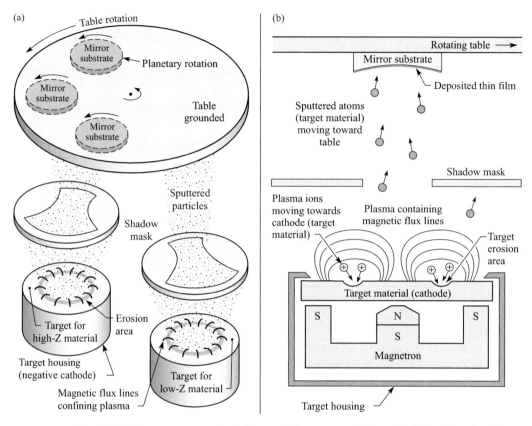

Figure 10.13 Magnetron sputtering is illustrated for two materials, one high-Z and one low-Z. Positive argon ions from a magnetically confined plasma are attracted to the negative cathode, where they sputter away (knock free) neutral atoms of the desired material. These neutral atoms move freely, some reaching the mirror substrates passing above. Shadow masks and planetary rotations are used to improve uniformity and control desired d-space variation. (b) Ion and neutral atom motions are depicted for the magnetron sputtering process. Drawing courtesy of J. Bowers and J. Underwood, Lawrence Berkeley National Laboratory.

and "low Z." In this manner a bilayer pair is deposited with each revolution of the table. In addition to table rotation, the substrates are also rotated, at higher angular speed, in localized planetary rotations so as to improve coating uniformity. Shadow masks[36–38] are often employed to further improve uniformity or control specified d-space variation, as shown schematically in Figure 10.13(a). Further sputtering heads are used to provide interface barriers between the low- and high-Z materials.

The magnetron sputtering process described here involves a plasma discharge maintained just above the desired target materials. The discharge is maintained between the grounded rotation table (anode) and the negative target (cathode) housings, with most of the voltage drop very close to the target. A magnetron structure, shown in Figure 10.13(b), provides magnetic flux lines that tend to hold the plasma in the vicinity of the target material. Typically the plasma is formed in a low pressure argon gas. Argon ions are attracted to the negative target housing, colliding and knocking free desired atoms for the coating process. The uncharged atoms then freely move away in all directions, perhaps restricted by a shadow mask, many of them reaching the substrate passing above and adhering as a sputtered thin film. The table rotation speed is set to provide a single layer of the desired thickness on each pass. Stable control of the plasma properties through gas pressure and well-regulated voltages make this a highly reproducible process. Nonetheless, the use of witness plates and Cu K_α measurements of d-spacing is normal practice. In some cases, in situ x-ray monitoring is also employed. Often it is desirable to employ a third material as a *capping layer*, at the end of the process, to control oxidation upon eventual exposure to air.[41, 42] Additional materials serving as diffusion barriers are sometimes used to limit interpenetration of atoms of one type into the preceding layer during deposition, for example two or three monolayers of B_4C between layers of Mo and Si, as used for the mirror described in Figure 10.7(b) that achieved 70% reflectivity.

Multilayer mirrors provide a powerful tool for controlling and measuring the polarization properties of radiation at these short wavelengths, based on reflection properties near 45° incidence angle.[43-46] As seen in Chapter 3, Section 3.6, there is an angle at which p-polarized radiation (electric field vector **E** lying in the plane of incidence) is not reflected at an interface, or is minimally reflected in the case of a lossy material ($\beta \neq 0$). At this same angle the s-polarization can have a relatively large reflection. For a single interface this angle is known as Brewster's angle, and for short wavelength radiation it is very close to $\pi/4$ radians. As seen in Chapter 3, Eq. (3.60), Brewster's angle for EUV/soft x-ray radiation incident from vacuum onto a material of refractive index $n = 1 - \delta + i\beta$ is given by

$$\phi_B \simeq \frac{\pi}{4} - \frac{\delta}{2} \tag{10.6}$$

where ϕ_B is measured from the surface normal. For multilayer mirrors the reflection process is somewhat different because the radiation interacts sequentially with two alternating materials. As a consequence the radiation experiences small refractive turnings toward and away from the surface normal, in response to small positive and negative changes in refractive index. At each interface, however, the result is similar in that for p-polarized radiation a minimal reflection occurs at each sequential interface, preserving the essence and advantage of Brewster's angle. Thus with multilayer mirrors designed for use at Brewster's angle the difference in reflectivities can be very large, approaching unity for s-polarized radiation and orders of magnitude less for p-polarized radiation. As a result, Brewster angle multilayer mirrors are often used to control and analyze polarization content of EUV, and to some extent soft x-ray, radiation.[46] They are used, for example, to study the magnetic properties of materials like Fe, Co and Ni by Faraday rotation of transmitted soft x-rays probing near their respective L-edges.[47]

Multilayer coatings have been extended to very high photon energies, that is very hard x-rays, even above 100 keV.[48-51] Finally, we note that during high thermal loading, for example with undulator or wiggler radiation, multilayer mirrors can be heated to high temperatures, potentially leading to degradation of otherwise stable interfaces. Toward

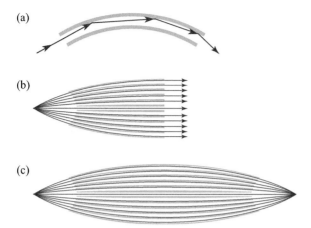

(a)

(b)

(c)

Figure 10.14 (a) A single hollow glass capillary uses total external reflection, from its internal surface, to turn x-rays through a large angle. Polycapillary lenses as shown in (b) can be used to collect and collimate x-rays. By combining two such polycapillary optics, as in (c), one can collect, transport, and reimage radiation from one plane to another. Courtesy of S. Dabagov, INFN Frascati.

this end Ziegler,[52] Takenaka,[53] and others have studied multilayer mirrors at high x-ray intensity. Recent comments regarding advances in multilayer coatings are given by Soufli[54] and by Kuznetsov, Bijkerk and colleagues.[54]

10.4   Capillary Optics

X-ray capillary optics are small hollow glass capillaries, either single or arrayed, that are curved and tapered through heating and pulling techniques. This permits x-ray transport by total external reflection along the internal surfaces at grazing incidence.[55–58] The principle of utilizing single glass capillary optics to focus the light to a point was first recorded in the 1950s and developed since then. Figure 10.14(a) illustrates how a single hollow glass capillary uses total external reflection from the interior surface, to turn x-rays through a large angle. Figure 10.14(b) shows a bundle of such capillaries, heated and drawn into a particular shape to form a polycapillary lens which collects and collimates x-rays. Figure 10.14(c) shows how a combination of these polycapillary optics can be used to collect, transport, and reimage or focus x-rays.

   Single-bounce elliptically shaped capillary optics operate using the same reflection properties discussed in the previous section on mirrors, where the source is located at one focus and the x-ray focal point at the other. Single-bounce capillary optics are commonly used at photon energies of several keV to tens of keV. The energy dependence of the critical angle allows the capillary optic to act as a lowpass spectral filter. Shown in Figure 10.15 is an elliptically shaped single capillary, shaped perhaps by drawing a solid glass fiber in the region of a glass bubble, stretching it to the desired shape, then cutting

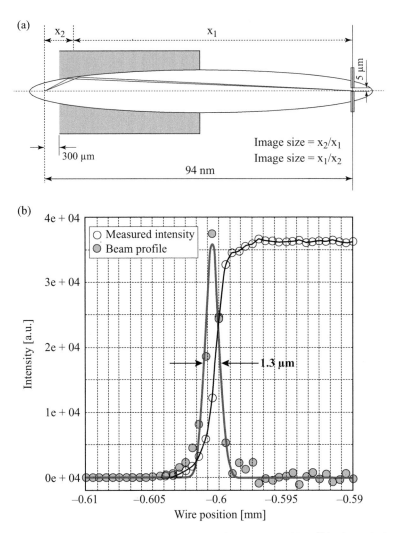

Figure 10.15 The diagram of a single, elliptically shaped capillary[56] is shown in (a). Incident radiation passing through a 5 μm diameter pinhole, located at one foci, ~~is~~ converges at the other, 300 μm beyond the cut edge of the optic. The pinhole size is set so that only a small section of the elliptical surface is illuminated, thus limiting aberrations. The knife edge measured vertical focal spot profile of 1.3 μm is shown in (b).[56] Courtesy of A. Erko, Helmholtz-Zentrum Berlin.

the ends as shown.[56] Similar to the case of x-ray mirrors, the performance of these optics is determined by the quality of the inner reflection surface, its smoothness, shape errors, and roughness. Such an optic has been used at BESSY II in Berlin to focus broad spectrum bending magnet radiation.[56] Only a small portion of the elliptical surface is used in order to limit the effect of surface irregularities and thus obtain a small spot with minimal divergence. The x-rays were focused to a 1.3 μm × 2.1 μm focal spot. A knife edge measured vertical focal spot profile is shown in Figure 10.15(b). Higher flux can

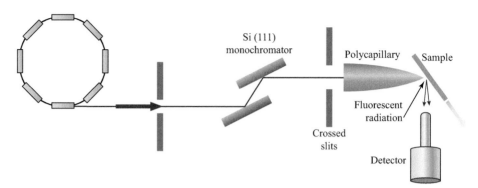

Figure 10.16 A polycapillary optic is used to collect radiation from a larger source area and deliver it to a smaller area of the sample for fluorescence or other characterization techniques.[55] The optics are achromatic so that the degree of monochromatization is set only as is required for the detection technique to be used. Following D. Bilderback, Cornell University.

be obtained by using an upstream zone plate to focus radiation through the pinhole. The combination of this two step focusing with off-axis illumination can provide a further reduced focal spot and the reduction of unwanted straight through photons, as reported by the Snigerevs, Erkos, Bjeomikhovs, and their colleagues.[59]

A recent application of a single-bounce capillary optic is its use as a hollow-cone condenser for a full field soft x-ray zone plate microscope,[60] which is described in a following section. The glass capillary accepts only a narrow angle of radiation defined by the critical angle and the length of the glass capillary. In this manner, a hollow cone shaped illumination can be created, as required for such a microscope.

In contrast to a single-bounce capillary optic, a polycapillary consists of a bundle of thousands of tightly bound small glass capillaries in which x-rays undergo multiple bounces within each capillary. The bundle of tightly bound capillaries can be heated and drawn into a desired shape for a specific function. This type of optic permits a tradeoff between resolution and efficiency, for example producing a lower resolution focal spot than a single capillary optic, but providing a greater collection area and thus a higher photon flux. These polycapillary optics are commonly used with both compact x-ray tubes and synchrotron radiation sources for collecting, transporting, and re-imaging the radiation to a smaller spot, typically tens of micrometers to millimeters in diameter.

A common application of polycapillary optics involves x-ray fluorescence mapping[55] of elements within complex materials, where radiation from an x-ray tube or synchrotron is collected, transported, and brought to a small area of the sample, as shown for example in Figure 10.16. In this manner the sample receives a spatially constrained, concentrated x-ray flux as compared to an unfocused source, permitting fluorescence mapping to be performed with higher spatial resolution, greater efficiency, and reduced measurement time. Furthermore, a polycapillary can be used to collect the emitted fluorescence and deliver it to a detector system. By shaping the bundle, radiation can be collected in one geometry and delivered in another, for example collecting with a circular shape and delivering in a narrow rectangular shape matched to the detector.

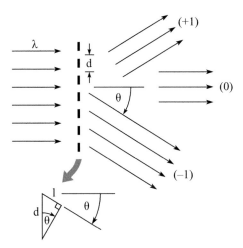

Figure 10.17 Diffraction from a transmission grating showing only the 0th and ±1st orders. Higher orders are omitted for clarity. Constructive interference of the diffracted radiation occurs at angles where the path length increases by λ for each additional period d of the grating, such that $\sin\theta = \lambda/d$ in first order.

10.5 Transmission Gratings

Transmission gratings are widely used as spectroscopic tools at x-ray and EUV wavelengths. Astronomers use them near the focus of x-ray telescopes to isolate various spectral features for elemental analysis and for estimating temperatures. They are widely used in plasma diagnostics, combined with streak cameras to measure time-dependent temperature, and in the EUV to exclude unwanted out-of-band radiation. A simple diagram of diffraction from a transmission grating is seen in Figure 10.17, showing only the 0th and ±1st orders. Higher orders are omitted for clarity. Constructive interference of the diffracted radiation occurs at angles where the path length increases by λ for each additional period d of the grating, as indicated in the illustration.

$$\sin\theta = \frac{\lambda}{d} \tag{10.7}$$

This occurs for both positive and negative angles, giving rise to the ±1st orders of the grating, in addition to the 0th order in the forward direction. Higher orders will be generated at angles θ_m, corresponding to increased path lengths $m\lambda$, such that

$$\sin\theta_m = \frac{m\lambda}{d} \tag{10.8}$$

where $m = 0, \pm1, \pm2, \pm3, \ldots$. For radiation incident on the grating at an angle θ_i, measured from the normal, one readily shows that the condition for constructive interference is

$$\boxed{\sin\theta - \sin\theta_i = \frac{m\lambda}{d}} \tag{10.9}$$

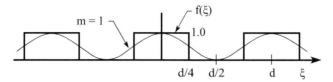

Figure 10.18 Representation of a unit absorption transmission grating in terms of its Fourier components. Each component m of the expansion represents an equivalent thin grating, where the coefficient c_m is related to the amount of energy diffracted to a given order m, and where the period d/m is related to the angle of diffraction for that order. Only the first term ($m = 1$) in the expansion is shown. The first coefficients are $c_0 = \frac{1}{2}$, $c_1 = 1/\pi$, $c_2 = 0$, $c_3 = -1/3\pi$, etc., as derived in the text.

where again $m = 0, \pm 1, \pm 2, \pm 3, \ldots$. Equation (10.9) is known as the *grating equation*,[61] and Eq. (10.7) is clearly a special case for normal incidence, $\sin \theta_i = \pi/2$.

The fraction of incident energy diffracted into the various orders depends on the nature of the periodic structure, i.e., the sharpness of profile, the bar to space ratio (line width as a fraction of grating period), and the complex refractive index, which affects absorption and phase shift in the grating. For a transmission grating having opaque lines of width equal to half the grating period, as illustrated in Figure 10.18, one can represent the transmission function in a Fourier series expansion, taking even (cosine) terms only for the coordinate choice taken:

$$f(\xi) = \sum_{m=-\infty}^{\infty} c_m \cos\left(\frac{2\pi m\xi}{d}\right) \tag{10.10}$$

with coefficients

$$c_m = \frac{1}{d} \int_{-d/2}^{d/2} f(\xi) e^{-2\pi i m\xi/d} \, d\xi$$

where $f(\xi) = 1$ in the interval $|\xi| \le d/4$, and $= 0$ in the interval $d/4 < \xi \le d/2$. Substituting for $f(\xi)$, noting that $e^{-i\theta} = \cos\theta - i \sin\theta$ (Appendix D) and that the sine term does not contribute in this even interval, the integral for the coefficient becomes

$$c_m = \frac{2}{d} \int_0^{d/4} \cos\left(\frac{2\pi m\xi}{d}\right) d\xi$$

$$c_m = \frac{\sin(m\pi/2)}{m\pi} \tag{10.11}$$

By L'Hospital's rule, $c_0 = \frac{1}{2}$. The even order coefficients are all zero, due to the symmetry of the problem and the choice of coordinate origin. The odd order coefficients are $c_m = 1/\pi, -1/3\pi, 1/5\pi, \ldots$, for $m = \pm 1, \pm 3, \pm 5, \ldots$, respectively.

We can now represent the single rectangular grating of unit absorption by a superposition of thin cosine gratings of increasing spatial frequency $k_m = 2\pi m/d$ and transmission c_m. Each such grating corresponds to one term in the expansion, leading to radiation of the various diffractive orders m, at angles θ_m described earlier in Eq. (10.8), and associated electric fields $E_m = c_m E_0$, where E_0 is the incident electric field at the grating. From Chapter 3, Eqs. (3.19)–(3.20), it follows that the intensities of the various diffractive orders are given by

$$I_m = \sqrt{\epsilon_0/\mu_0}|E_m|^2 = |c_m|^2 I_0 \qquad (10.12)$$

so that the efficiencies $\eta_m = I_m/I_0$ for diffraction to the various orders are proportional to $|c_m|^2$, and thus from Eq. (10.11)

$$\eta_m = \begin{cases} 0.25 & m = 0 \\ 1/m^2\pi^2 & m \ \ \text{odd} \\ 0 & m \ \ \text{even} \end{cases} \qquad (10.13)$$

For an opaque transmission grating of equally wide lines and spaces, 25% of the incident energy is in the 0th order, approximately 10% is diffracted to each of the ±1st orders, and so forth, while the grating itself absorbs 50% of the incident energy.

For phase gratings the opaque lines are replaced by partially transmitting materials to reduce absorptive losses. For materials and wavelengths for which β/δ is minimal, and for thicknesses that permit a relative propagation phase shift approaching π, this can lead to a significant enhancement of diffraction efficiency, a factor of four in the limit of a π-phase change with $\beta/\delta \to 0$.

The coefficients in Eq. (10.13) correspond to a symmetric grating of equal line and space widths, permitting a representation involving only even cosine functions. For an asymmetric grating involving unequal line and space widths, odd sine functions would also appear. An example would be a grating with line widths equal to $1/3$ the grating period and open spaces of width equal to $2/3$ of the grating period. In such cases the asymmetry leads to non-zero even order sine terms, i.e., finite values of $|c_m|^2$ for $m = \pm2, \pm4$, etc. This is analogous to the discussion of even multilayer diffraction orders for asymmetric coatings of $\Gamma \neq 0.5$, as discussed in Section 10.3 and seen in Figure 10.9.

An extensive literature exists on the subject of diffraction from transmission and reflection gratings. In particular see Born and Wolf[61] for an extensive discussion, Hecht[62] for a tutorial on blazed reflection gratings, and both Morrison[63] and Michette[64] for descriptions of transmission and reflection gratings, blazed gratings, and zone plates, at short wavelengths. The subject of reflection gratings is considered further in Section 10.7.

10.6   The Fresnel Zone Plate Lens

At short wavelengths, particularly in the soft and hard x-ray regions extending from 4 nm (300 eV) to 0.2 Å (60 keV), Fresnel zone plate lenses are of great interest because of their ability to form images at very high spatial resolution,[63–71] approaching the diffraction limit. A zone plate can be represented as a circular diffraction grating with zones of equal area but decreasing width towards its outer diameter. With equal areas all zones have equal diffracting power. Like the transmission gratings considered in Section 10.5, the zone plate diffracts incoming radiation to many orders, plus (+) and

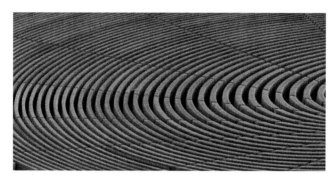

Figure 10.19 A silicon Fresnel zone plate lens for x-ray microscopy.[109] The outer zone width is $\Delta r = 100$ nm, the thickness, $h = 10$ μm, and the aspect ratio is 100:1. The number of zones is $N = 150$.

minus (−), with constructive interference determining angular deflections that depend on the local period. With proper zone radii and separations the zone plate can focus a positive order, such as $m = 1$, to a spot with many of the properties of a visible light refractive lens, but in this case by diffraction rather than refraction. Like gratings, Fresnel zone plate lenses can also operate in higher order, such as $m = 3$, offering a higher numerical aperture with increased spatial resolution, albeit with a loss of efficiency.

The lens can also provide point-to-point imaging, again by diffraction and constructive interference, and furthermore can image a full three-dimensional object by point-to-point local interference. In this sense we refer to the structure as a Fresnel zone plate lens. An example is seen in Figure 10.19. X-ray lenses are presently capable of forming two-dimensional images at a spatial resolution of ten nanometers, and three-dimensional tomographic images at 30–40 nm resolution, depending on object depth. These diffractive lenses are especially valuable when used to project full images to a CCD detector in a full field x-ray microscope, and to provide a small focal spot for probing material in a scanning transmission x-ray microscope. More recently they have been used with additional functionality, such as a spiral zone plate for adding angular momentum to the probing radiation, various phase contrast capabilities, or structured illumination for improved spatial resolution, as will be shown later in Figure 10.26.

10.6.1   Zone Plate Parameters

Figure 10.20 illustrates the general technique for point to point imaging with a Fresnel zone plate lens. In its simplest form the zone plate consists of alternating opaque and transparent zones, essentially a circular grating, with radial zones located such that the increased path lengths through sequential transparent zones differ by one wavelength each and thus add in phase at the image point. In this manner, on a point by point basis, the image of a full two-dimensional object can be formed by interference in the image plane, with localized spatial coherence. The smallest possible spot size that can be formed at P is obtained with spatially coherent illumination of the zone plate, a

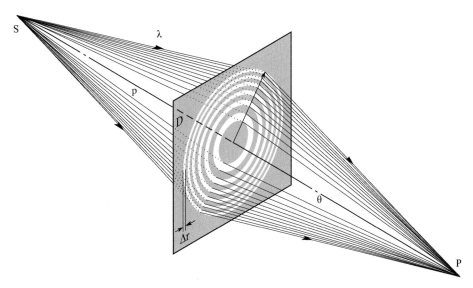

Figure 10.20 A Fresnel zone plate diffractive lens forms an x-ray image at point P, of a source at point S. The lens has a diameter D and outer zone width Δr. The object and image distances are p and q, respectively.

subject of interest for the formation of scanning spot microscopes, which are described in the following paragraphs. Also of interest is use of partially coherent radiation and its advantage in image formation and resolution.

The focusing properties of a Fresnel zone plate lens can be understood by considering the first order diffraction from a circular grating with the zonal periods adjusted so that at increasing radius from the optic axis the periods become shorter, and thus by Eq. (10.7) the diffraction angle becomes larger, thus permitting a real first-order focus, as illustrated in Figure 10.21. The path length from each zone to the focus is increased by a half wavelength. Alternate zones are chosen to be opaque to assure only positive interference of radiation from the various zones at the focal point. If one draws a right triangle with the focal length f as one side, the radius of any zone r_n as a second side, and the hypotenuse of length $f + n\lambda/2$, then by the Pythagorean theorem the zonal radii are given by

$$f^2 + r_n^2 = \left(f + \frac{n\lambda}{2} \right)^2 \tag{10.14}$$

which upon expansion and consolidation of like terms becomes

$$r_n^2 = n\lambda f + \frac{n^2\lambda^2}{4} \tag{10.15}$$

The term $n^2\lambda^2/4$, which represents spherical aberration, can be ignored for $f \gg n\lambda/2$, which we will see shortly corresponds to a lens of small numerical aperture NA $= \sin\theta = \lambda/(2\Delta r) \ll 1$, as often used at x-ray wavelengths. Where this is not the case, for

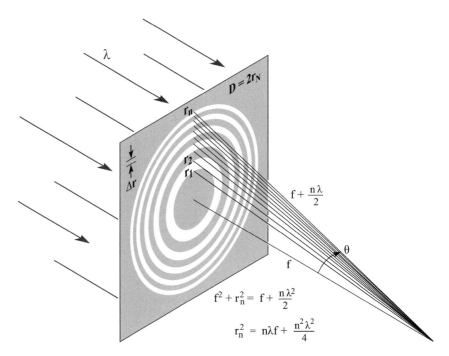

$$D = 2r_N$$

$$f + \frac{n\lambda}{2}$$

$$f$$

$$\theta$$

$$f^2 + r_n^2 = f + \frac{n\lambda^2}{2}$$

$$r_n^2 = n\lambda f + \frac{n^2\lambda^2}{4}$$

Figure 10.21 A Fresnel zone plate lens with plane wave illumination, showing only the convergent + 1 order of diffraction. Sequential zones of radius r_n are specified such that the incremental path length to the focal point is $n\lambda/2$. Alternate zones are opaque in this transmission zone plate. With a total number of zones N the zone plate lens is fully specified. Lens characteristics such as the focal length (f), diameter (D), and numerical aperture (NA) are described in the text in terms of λ, N, and Δr, the outer zone width.

example with a larger NA optic at an EUV wavelength, the additional term is retained. For the low NA case Eq. (10.14) simplifies to

$$r_n \simeq \sqrt{n\lambda f} \qquad (10.16)$$

showing that a real first-order focus is achieved when successive zones increase in radius by $\sqrt{n}$, providing the desired prescription by which the radial grating period must decrease in order to provide a common focus. The earliest known record regarding the demonstration of focusing light with alternately opaque Fresnel zones is that of Lord Rayleigh in 1871.[65]

We can now obtain expressions for the lens diameter D, focal length f, numerical aperture NA = $\sin\theta$, spatial resolution, and depth of focus. We choose to do this in terms of the wavelength λ, the total number N of zones, and the outer zone width Δr. We do this from an experimental point of view. In designing an experiment the wavelength is often a first priority, driven by the elemental composition of the material or sample under study and their characteristic absorption edges. In microscopy the next priority is the spatial resolution required to see features of interest. For zone plate lenses the spatial resolution limit is set by the outer zone width Δr, and illumination characteristics, as we will see shortly. As our third choice we take N, the total number of zones. As we will see in the following paragraphs, zone plate lenses are highly chromatic, that is, the focal length of the lens varies strongly with wavelength. Thus for precise imaging it is necessary to restrict the illumination spectral bandwidth, $\Delta\lambda/\lambda$. We will see shortly that there is an inverse relationship between $\Delta\lambda/\lambda$ and N, the total number of zones. Thus the total number of zones N will be restricted by the relative spectral bandwidth of the illuminating radiation. With this motivation we proceed in the following paragraphs to develop relationships for f, D, NA, resolution, and depth of focus in terms of λ, Δr, and N.

We begin by defining the *outer zone width* for $n \to N$,

$$\Delta r \equiv r_N - r_{N-1} \tag{10.17}$$

where N is the total number of zones, i.e., the sum of both opaque and transparent zones (twice the number of radial periods). The outer zone width Δr provides a very convenient parameter, and is useful in expressions for other lens parameters. We now write the square of Eq. (10.16) twice, once for r_N and once for r_{N-1}, and subtract as follows:*

$$r_N^2 - r_{N-1}^2 = N\lambda f - (N-1)\lambda f = \lambda f$$

Using the definition of Δr given in Eq. (10.17), one also has for the left side of the above equation

$$r_N^2 - (r_N - \Delta r)^2 = 2r_N\,\Delta r - (\Delta r)^2 \simeq 2r_N\,\Delta r$$

since $\Delta r \ll r_N$ for large N. Combining the above two equations, one obtains

$$2r_N\,\Delta r \simeq \lambda f$$

or

$$D\,\Delta r \simeq \lambda f \tag{10.18}$$

From Eq. (10.16) we note that $\lambda f \simeq r_N^2/N$, so that with Eq. (10.18) one has $D\,\Delta r \simeq \frac{r_N^2}{N} = \frac{D^2}{4N}$, thus

$$\boxed{D \simeq 4N\,\Delta r} \tag{10.19}$$

The focal length can then be obtained from Eq. (10.18) as

$$f \simeq \frac{D\,\Delta r}{\lambda}$$

or in combination with Eq. (10.19)

$$\boxed{f \simeq \frac{4N(\Delta r)^2}{\lambda}} \tag{10.20}$$

---

* Note that the area of successive zones, $\pi(r_n^2 - r_{n-1}^2) = \pi\lambda f$, is a constant, at least within the long focal length, small NA approximation, going from Eq. (10.15) to Eq. (10.16). Thus the areas of all zones are equal and contribute equally to intensity at the focus.

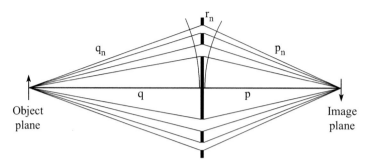

Figure 10.22 Point by point imaging with a Fresnel zone plate lens is illustrated. Successive propagation paths are increased by $\lambda/2$.

This is a very important relationship for the design of zone plate microscope lenses in that it shows that the focal length scales directly with the number of zones, with the square of the outer zone width (which largely sets the resolution), and inversely with the wavelength, thus introducing a strong chromatic effect.

The numerical aperture (NA) of a lens in vacuum is defined as

$$NA = \sin\theta$$

where θ is the half angle measured from the focal point on the optic axis back to the lens periphery, as illustrated here in Figure 10.21. Thus the numerical aperture of a zone plate lens is given by $NA = r_N/f = D/2f$, or from Eq. (10.18)

$$\boxed{NA \simeq \frac{\lambda}{2\Delta r}} \tag{10.21}$$

which is a particularly simple form that will be convenient when considering spatial resolution. A related quantity is the lens *F-number*, which we will abbreviate as *F#*, defined as

$$F\# \equiv \frac{f}{D}$$

or again using Eq. (10.18)

$$F\# \simeq \frac{\Delta r}{\lambda} \tag{10.22}$$

We will return to these parameters in the following section on spatial resolution, depth of focus, and chromatic aberration.

In the preceding paragraphs we have considered the focusing conditions for a zone plate lens with plane wave illumination, as illustrated in Figure 10.21. Next we consider point to point imaging of an object at a finite distance q from the zone plate, to an image plane at a distance p, as illustrated in Figure 10.22.

Again the successive zones, alternately transmissive and opaque, are constructed so as to add $\lambda/2$ to successive path lengths, so that

$$q_n + p_n = q + p + \frac{n\lambda}{2}$$

where for modest numerical aperture lenses

$$q_n = \left(q^2 + r_n^2\right)^{1/2} \simeq q + \frac{r_n^2}{2q}$$

$$p_n = \left(p^2 + r_n^2\right)^{1/2} \simeq p + \frac{r_n^2}{2p}$$

so that

$$q + \frac{r_n^2}{2q} + p + \frac{r_n^2}{2p} \simeq q + p + \frac{n\lambda}{2}$$

$$\frac{r_n^2}{2q} + \frac{r_n^2}{2p} \simeq \frac{n\lambda}{2}$$

$$\boxed{\frac{1}{q} + \frac{1}{p} \simeq \frac{1}{f}} \tag{10.23}$$

where from Eq. (10.16), $f = r_n^2/n\lambda$. Equation (10.23) relates the image and object distances to the focal length as for an ordinary visible light refractive lens. Similarly, one can show that the transverse magnification is

$$\boxed{M = \frac{p}{q}} \tag{10.24}$$

We now have a basic understanding of how a Fresnel zone plate can be used both to focus radiation and to form a real image of an extended object using first order diffraction.

Recall, however, that a transmission grating generates many orders, thus complicating the use of a zone plate lens and leaving only a fraction of the available photons for the primary purposes of a given experiment. The procedure, illustrated in Figure 10.21, of adding a path length of $n\lambda/2$ for constructive interference of sequential zones in first order can be extended to the higher orders ($m = 2, 3, \dots$) by adding path lengths $mn\lambda/2$. The first three positive and negative orders are illustrated in Figure 10.23.

Calculating the diffraction efficiency of the various orders is accomplished by following the same procedures used in preceding paragraphs for the first order ($m = 1$). The zone radii corresponding to phase advances for higher-order diffracted waves are given by

$$r_n^2 \simeq mn\lambda f_m \tag{10.25}$$

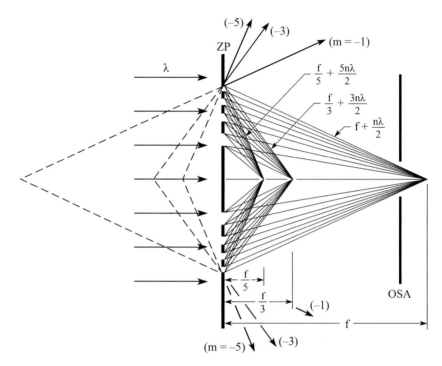

Figure 10.23 Zone plate diffractive focusing is illustrated for the first three positive orders. An order sorting aperture (OSA), of the type that would be used to block all but the first order, is also shown. Negative orders ($m = -1, -3, -5$) are shown as solid lines diverging from the optical axis, and projected backward to virtual foci (behind the lens) by dashed lines.

for zones $n = 0, 1, 2, \ldots$ and diffractive orders $m = 0, \pm 1, \pm 2, \ldots$ The corresponding focal lengths are given by

$$f_m = \frac{f}{m} \qquad (10.26)$$

where we note that the negative orders give rise to virtual foci of negative focal length. The diffraction efficiencies of the various orders can be analyzed much like the transmission grating efficiencies of the previous section [see Eqs. (10.10) and (10.11)]. For the case of a transmission zone plate of unit absorption in the opaque zones, one can represent the transmission function in a Fourier series expansion in terms of r^2, as suggested by Eq. (10.25). The sketch in Figure 10.24 is useful for visualizing the Fourier decomposition and identifying the periodicity in r^2. Following Goodman,[72] we expand the transmission function in terms of γr^2, taking only odd (cosine) terms for the chosen coordinates, so that

$$f(\gamma r^2) = \sum_{m=-\infty}^{\infty} c_m \cos(m \gamma r^2) \qquad (10.27)$$

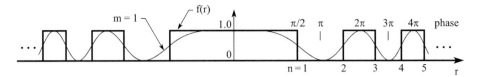

Figure 10.24 Representation of a Fresnel zone plate as a transmission grating in terms of the radius squared. Only the first term ($m = 1$) is shown.

where from Figure 10.24 we see[†] that $\gamma = \pi / \lambda f$. This can be written as

$$f(u) = \sum_{m=-\infty}^{\infty} c_m \cos(mu)$$

where $u = \gamma r^2 = \pi r^2 / \lambda f$, and where the Fourier coefficients are given by

$$c_m = \frac{1}{2\pi} \int_{-\pi}^{\pi} f(u) \cos(mu)\, du \qquad (10.28)$$

For the alternately opaque and transmissive zones of interest here the transmission function $f(u) = 1$ for $0 \le u \le \pi/2$, and $f(u) = 0$ for $\pi/2 < u \le \pi$ (see Figure 10.24), so that

$$c_m = \frac{1}{\pi} \int_0^{\pi/2} \cos mu\, du$$

or

$$c_m = \frac{\sin(m\pi/2)}{m\pi} \qquad (10.29)$$

where $m = 0, \pm 1, \pm 2, \pm 3, \ldots$. This is identical to the result obtained earlier for the linear transmission grating. As we observed in that case, the diffraction efficiencies to the various orders are given by [Eq. (10.12)]

$$I_m = |c_m|^2 I_0$$

so that for a Fresnel zone plate of alternately opaque and transmissive zones the diffraction efficiencies are given by

$$\eta_m = \begin{cases} 0.25 & m = 0 \\ 1/m^2\pi^2 & m \text{ odd} \\ 0 & m \text{ even} \end{cases} \qquad (10.30)$$

where half the incident energy is absorbed by the opaque zones. The efficiency to the first-order focus is thus about 10%, another 10% goes to the divergent $m = -1$ order, approximately 1% goes to the divergent third order ($m = 3$, virtual focus), etc., while 50% of the incident radiation is absorbed and 25% is transmitted in the forward direction

---

[†] A radial phase shift of 2π corresponds to a difference $\Delta n = 2$ in the zone plate (one opaque, one transmissive). From Eq. (10.25), for $m = 1$, this gives an argument in the expansion parameter $\gamma(r_n^2 - r_{n-2}^2) = 2\lambda f \gamma = 2\pi$, or $\gamma = \pi/\lambda f$.

Note that in Figure 10.24, the phase shift between $n = 2$ and $n = 4$ corresponds to $7\pi/2 - 3\pi/2 = 2\pi$.

($m = 0$). As in the case of the transmission grating considered in Section 10.5, the even orders do not contribute in the symmetric case where successive zone areas are equal.

The decreasing efficiency with increasing order m has an interesting explanation. Within a given transmissive zone n, the even orders of m cancel at the focus, so that only odd orders ($m = \pm 1, \pm 3, \pm 5, \dots$) need be considered. For the odd orders of $|m| > 1$, the diffraction efficiency will be decreased by factors of $1/m^2$ relative to $m = 1$ because of canceled contributions within each transparent zone. For instance, the third-order focus will receive field contributions from three sub-regions within each zone, two of which will cancel, leaving only a $1/3$ contribution to the electric field at the third-order focus ($1/m$). As the intensity is proportional to E^2, the intensity will be diminished by $1/9$, i.e., by $1/m^2$. Likewise, for the fifth-order focus, four of the five sub-zone contributions will cancel (in pairs), leaving only a $1/5$ contribution to the field, or a $1/25$ contribution to the intensity.

It is possible to increase the efficiency of zone plates in several ways. By replacing opaque zones with phase reversal zones, where the goal is to achieve a $\lambda/2$ phase shift with minimal absorption, as first suggested by Lord Rayleigh.[65, 67] In this manner the electric fields at focus can be increased by up to a factor of two, and thus the intensities (and efficiencies) by up to a factor of four. The required thickness Δt for obtaining a phase shift of π for a given material and wavelength is given by [see Chapter 3, Eq. (3.29)]

$$\Delta t = \frac{\lambda}{2\delta} \tag{10.31}$$

Of course, for x-rays and EUV radiation, where the refractive index is $n = 1 - \delta + i\beta$, and where β/δ is non-negligible, this factor of four cannot be realized. Nonetheless significant improvements are possible. Kirz[71] has calculated the optimum zone plate thickness as a function of the parameter β/δ. He finds, for example, that the optimum thickness decreases to about 0.9 of that given in Eq. (10.31) for $\beta/\delta = 0.2$, and about 0.8 at $\beta/\delta = 0.5$. Enhanced diffraction efficiencies of zone plate lenses to first order, based on partial phase contributions in the material zones, have been reported in the literature.[73-76] The relative electric field after propagating through a finite thickness Δt is given by [Chapter 3, Eq. (3.17)]

$$\frac{E}{E_0} = e^{-\pi \beta/\delta} \tag{10.32}$$

From this the efficiencies to various orders can be calculated.[63] Note that values of δ and β can be calculated as a function of photon energy knowing f_1^0 and f_2^0, which are reproduced in Appendix C for several common materials.

The foregoing analysis is based on diffraction by binary zone plates, that is, structures having uniform thicknesses normalized to unity or zero. Further possibilities for improved efficiency employ approximations to theoretically optimized phase profiles, generally referred to as blazed zone plates, or blazed gratings, as discussed in the literature[77, 78] and sketched here in Figure 10.25. Efficiencies can be calculated for

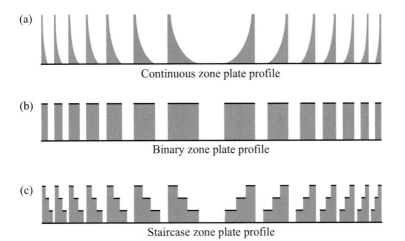

(a)

Continuous zone plate profile

(b)

Binary zone plate profile

(c)

Staircase zone plate profile

Figure 10.25 (a) An idealized phase zone plate with shaped zone profiles that result in 100% efficiency to first order ($m = 1$) in the approximation that $\beta \to 0$. (b) A binary zone plate of the type considered in previous paragraphs. In the lossless approximation, $\beta \to 0$, this structure can reach 40% efficiency to first order. With hard x-rays, where for some materials $\beta/\delta \ll 1$, values approaching 40% are possible. (c) A four step approximation to the optimized continuous profile. In the lossless limit this four step approximation achieves an efficiency of 80%, less of course with finite β/δ. These approximate considerations apply as well to linear transmission gratings.

these structures using Fourier transform techniques as described earlier in this section, and for linear transmission gratings in the previous section. Maximum efficiencies for blazed and phase zone plates can be significantly higher than for simple binary absorption structures. In these calculations the blazed profile, with complex refractive index, is considered in place of the binary absorption zone plate. Squares of the resulting Fourier series coefficients then describe efficiencies to the various orders. An ideal transmission zone plate would have a spatially continuous quadratic profile, as shown in Figure 10.25(a). This profile phase shifts the incoming x-ray wavefront piecewise by 2π across each period, for example generating a near-spherical convergent wavefront in a focusing geometry. In the lossless limit ($\beta \to 0$) this ideal quasi-continuous phase profile, if fabricated, would diffract all the incoming light to a single diffraction order, with a lens efficiency of 100%. A degree of finite absorption (β/δ) would of course lower the efficiency accordingly. Owing to fabrication limitations, what is often produced for the x-ray regime is a binary zone plate, which is a first approximation to the continuous phase profile, as sketched here in Figure 10.25(b). The binary zone plate with fully attenuating (large β) zones, as described earlier, would achieve a maximum efficiency of 10%, or if fabricated and used in a pure phase material ($\beta \to 0$) would achieve an efficiency of 40%. In this lossless limit the full period contributes to the diffracted order, rather than just the open half-period, thus doubling the electric field and quadrupling the detected intensity, for a four-fold increase of efficiency. With real materials having a complex refractive index at x-ray and EUV wavelengths, binary zone plate efficiencies lie between that of the fully attenuating and phase only conditions. For soft x-rays there

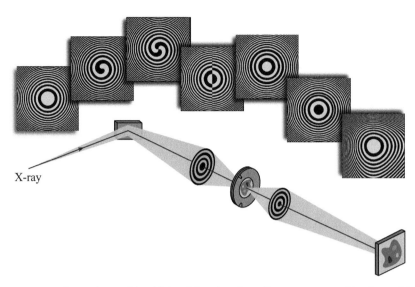

Figure 10.26 Zone plates with additional functionality offer new opportunities for probing and imaging. Combined function zone plates superimpose basic annular zone plate patterns for focusing or imaging with an additional pattern which addresses a second function of interest. Examples shown here include a normal zone plate, two zone plates that add varying amounts of orbital angular momentum, differential phase contrast, two versions of Zernike phase contrast, and one for extended depth of field in tomographic imaging.[84–87]

can be useful gain due to the phase effect. For example, a 250 nm thick nickel zone plate can achieve a first order diffraction efficiency just over 20% at a photon energy of 500 eV.[79] For hard x-rays, where β/δ can be quite small, it is possible to achieve significant improvements with binary zone plates where, for example, a 10 μm thick gold zone plate could provide an efficiency approaching 40% at 50 keV.

For further gains in efficiency, fabrication of zone plates must incorporate a better fit to the approximation of a continuous phase profile zone plate.[77, 78] For example, a staircase profile which approximates the desired profile in a finite number of steps, as illustrated in Figure 10.25(c). These steps can be fabricated by using multiple step, well-aligned, electron beam pattern writings and sequential lithographic processing, or by using mechanical stacking of separately fabricated zone plates, again with accurate alignment methods. A four-level staircase zone plate profile such as illustrated in Figure 10.25(c) could potentially reach an efficiency of 80% efficiency in the ideal case of minimal absorption and near perfect steps. Several groups[80–83] have made and tested multistep structures for use with soft x-rays of finite β/δ.

Because a zone plate can be thought of generally as a distribution of phase profiles, localized displacement of the zones can be understood as phase shifts within the generated wavefront local to that section. Prescribed phase deviations will then propagate and interfere within the wavefront. Thus phase structures can be added to the zone plate pattern to create an optic that generates focal plane or projected properties of interest. Several examples of dual functionality zone plates are shown in Figure 10.26.

In addition to simple phase addition methods, x-ray holographic elements can also be designed through iterative algorithms where the desired output properties are modelled. Wavefront propagation between the desired output beam, the holographic element, and the input beam is performed iteratively until the algorithm converges to a solution. Constraints on optical element dimensions such as smallest feature sizes, aperture size, and aperture shape are included in the algorithm to ensure accuracy. Applications of such optics include the shaping of x-ray beams for imaging applications, creation of orbital angular momentum, other tailored properties, or encoded illumination techniques.[84–87]

10.6.2   Diffraction of Radiation by a Zone Plate Lens

To understand the limiting spatial resolution of a lens, set by the wavelength of radiation and the numerical aperture (NA), it is necessary to have knowledge of the focal plane intensity distribution due to the incident radiation. We begin with a point source an infinite distance away, which can be approximated by plane wave illumination. One can then consider two such point sources, bring them close to each other, and set some criterion for the separation distance that renders the two just resolvable. This brings us to the subject of *diffraction*, previously introduced in Chapter 4, Section 4.7, where there are features comparable in size to the irradiating wavelength that cause radiation to propagate in directions different from those given by geometrical considerations. Of particular interest here is the diffraction of short-wavelength radiation by zone plates, which are much used in this region of the spectrum. Of course, diffraction is a basic electromagnetic phenomenon, well known at longer wavelengths. For instance, when a common refractive lens is used to focus visible laser light, the limited lens aperture causes some angular spread of the radiation, resulting in a finite width of the focal spot and some nearby ringing due to interfering fields in the focal plane. In this section we will discuss analogous diffractive limitations that occur with x-ray and EUV zone plate lenses. The closely related diffraction by small pinholes was described in detail in Chapter 4, Section 4.7. The introductory comments, very brief history, and references to more extensive reviews in that section are also useful here.

As pointed out in Section 4.7, there is a long history of mathematical development on the theory of diffraction that successfully predicts physical observations. Most notable is the scalar theory of diffraction developed by Kirchhoff in 1882, extending the earlier work of Huygens (1690) and Fresnel (1818), and leading to what is now known as the *Fresnel–Kirchhoff diffraction formula*.[88–93] For small-angle scattering in the near-forward direction, $\theta \cong \lambda/d \ll 1$, where d is a characteristic dimension, this can be written as[89]

$$E(x, y) = \frac{-i}{\lambda} \iint \frac{E(\xi, \eta)e^{ikR}}{R} \, d\xi \, d\eta \tag{10.33}$$

where $k = 2\pi/\lambda$, $E(x, y)$ is the electric field observed at a distant point P(x, y), $E(\xi, \eta)$ is the incident field as a function of position in the aperture plane $(z = 0)$, and R is the distance from each point (ξ, η) in the source plane to a point (x, y) in the observer plane. Basically this states that the field detected in a distant plane is obtained by summing the contributions from every point in the aperture plane, accounting for its propagation distance R and phase e^{ikR}, as if each point were a secondary source of radiation. The finite aperture introduces unbalanced contributions near boundaries, leading to interference effects specific to the geometry and clearly dependent on the wavelength.

When a Fresnel zone plate is placed within the aperture a large number of diffracted orders are generated. As we saw earlier the various orders come to focus on axis at focal distances given by Eq. (10.25)

$$r_n^2 = mn f_m$$

with diffraction efficiencies given by Eq. (10.30). It is possible then to evaluate the Fresnel–Kirchhoff integral for this case by replacing the zone plate with an infinite series of thin lenses, one for each order.[89] The stepwise radial phase advance (or retardation), $mn\lambda/2$, of the wavefront associated with each zone and each order, as seen in Figure 10.27(b), is then given by

$$\Delta\phi_{m,n}(\rho) = k\,\Delta l = k\left(\frac{mn\lambda}{2}\right)$$

which by Eq. (10.25) can be written in terms of the radius ρ_n as

$$\Delta\phi_{m,n} = \frac{k\rho_n^2}{2f_m} \tag{10.34}$$

where $f_m = f/m$, and where f is the first-order ($m = +1$) focal length. For a zone plate of many zones, for instance[94] $N > 100$, the first-order wavefront can be approximated by a continuous radial phase advance

$$\phi(\rho) = \frac{k\rho^2}{2f} \tag{10.35}$$

where $f = r_1^2/\lambda$, or equivalently in terms of the outer zone width Δr and number of zones N, $f = 4N(\Delta r)^2/\lambda$. Thus the electric field in the aperture, as it appears in Eq. (10.33), can be written as

$$E(\xi, \eta) = \sum_m \frac{E_0}{|m\pi|} e^{-ik\rho^2/2f_m} \tag{10.36}$$

where the coefficients $|1/m\pi|$ correspond to the diffractive efficiencies of the non-zero orders, as given in Eq. (10.30), and again $f_m = f/m$.

The Fresnel–Kirchhoff diffraction formula can then be approximated for each of the various orders as

$$E_m(x, y) = \frac{-iE_0}{m\pi\lambda} \iint \frac{e^{-ik(\xi^2+\eta^2)/2f_m}\, e^{ikR}}{R}\, d\xi\, d\eta \tag{10.37}$$

where as seen earlier in Eq. (4.26) for a first-order focus at $z = f$, with $f \gg x, y$ and $f \gg \xi, \eta$

$$R \simeq f + \frac{x^2}{2f} + \frac{\xi^2}{2f} - \frac{\xi x}{f} + \frac{y^2}{2f} + \frac{\eta^2}{2f} - \frac{\eta y}{f} \tag{10.38}$$

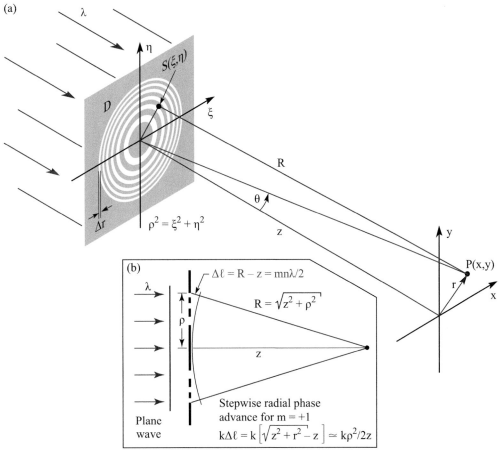

Figure 10.27 (a) Geometry for the description of diffraction of radiation from a Fresnel zone plate. The geometry and nomenclature is similar to that of Figure 4.18. The zones are described as a function of the radius ρ in the ξ, η-plane; however, for consistency with the zone plate formulae developed in Section 10.6, we take the outermost open zone to have a width Δr, and take D as the zone plate diameter. The inset (b) illustrates the wavefront curvature (phase advance) of the $m = +1$ diffracted order, which is brought to focus a distance $z = f$ from the zone plate.

for the geometry shown in Figure 10.27. With this expansion of R the integral for the first-order field becomes

$$E_1(x, y) = \frac{-iE_0}{\pi \lambda f} e^{ikf} e^{ik(x^2+y^2)/2f} \iint e^{-ik(\xi x+\eta y)/f} d\xi \, d\eta$$

where the $-ik(\xi^2 + \eta^2)/2f$ term due to the zone plate phase advance in the first order has exactly canceled the $+ik(\xi^2 + \eta^2)2z$ term from the expansion of R at $z = f$. The conversion to polar coordinates for this axisymmetric geometry is the same as in

Section 4.7, [see Fig. 4.18 and Eqs. (4.26)–(4.29)]. The resultant diffraction formula for the first-order field becomes

$$E_1(r) = \frac{-2iE_0}{\lambda f} e^{ikf} e^{ikr^2/2f} \int_0^a J_0(k\rho\theta)\,\rho\,d\rho$$

where $\theta = r/f$, $a = D/2$ is the outer radius of the zone plate, and where J_0 is the Bessel function of the first kind, order zero. The integral is of a standard form, as we saw in Eq. (4.31). The first-order field in the focal plane at $z = f$ is then

$$E_1(\theta) = \frac{-2ia^2 E_0}{\lambda f} e^{ikf} e^{ik\theta^2 f/2} \frac{J_1(ka\theta)}{ka\theta} \tag{10.39}$$

where $J_1(v)$ is the Bessel function of the first kind, order one, and where the above can be written in terms of the focal plane radius by substituting $r = f\theta$ for fixed focal length f.

The corresponding focal plane intensity, for the first order, $m = +1$, is obtained by writing the first-order intensity as $I_1(\theta) = \sqrt{\epsilon_0/\mu_0}|E_1(\theta)|^2$, with the result that the focal plane intensity of the zone plate is given by

$$\boxed{\frac{I_1(\theta)}{I_0} = N^2 \left| \frac{2J_1(ka\theta)}{ka\theta} \right|^2} \tag{10.40}$$

where $I_0 = \sqrt{\epsilon_0/\mu_0}|E_0|^2$, is the illumination intensity of the aperture, N is the total number of zones in the zone plate, $a = D/2$, and by Eq. (10.16), written for $n \to N$, we have $D^2 = 4N\lambda f$. The function $|2J_1(v)/v|^2$, the so-called Airy function that was seen earlier in Figure 4.19, is unity on axis, has its first null at $ka\theta = 3.832$, and oscillates beyond that with ever decreasing amplitude. Thus according to Eq. (10.40), the first-order focus of the zone plate lens has an on-axis intensity N^2 greater than the incident intensity. The radial focal plane intensity pattern (Airy pattern) is shown in Figure 10.28.

Note that the first null occurs for $kD\theta/2 = 3.832$ so that with focal plane radius $r = f\theta$ in Figure 10.27, the radius of the first null occurs at

$$\boxed{r_{\text{rull}} = \frac{0.610\lambda}{\text{NA}}} \tag{10.41}$$

where in the small angle approximation ($\rho, r \ll z$) we have taken $\text{NA} = \sin\theta \cong \theta$. This is a very well-known result that plays a significant role in determining the resolution of an ideal lens, limited only by the wavelength λ and the numerical aperture.[‡] In this ideal case the lens performance is described as *diffraction limited*. In the following sections we consider the Rayleigh criterion for determining resolution in this limit, as well as practical limitations due to various effects. We also describe how such

---

[‡] Of somewhat practical interest we note for a lens of $\text{NA} = \lambda/2\Delta r$, the Airy null diameter of the focal spot is $2r_{\text{null}} = 2.440\Delta r$ and the Airy full width at half maximum diameter is $1.029\Delta r$ FWHM.

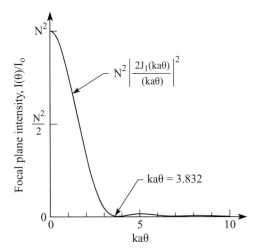

Figure 10.28 The Airy pattern describing focal plane intensity of a zone plate lens at its first-order focus. I_0 is the illumination intensity at the zone plate, N is the total number of zones, $k = 2\pi/\lambda$, $a = D/2$ is the lens radius, $\theta = r/f$ is measured from the center of the lens away from the optic axis, r is the radius in the focal plane, and f is the focal length.

lenses are fabricated. Various microscopes in which they are used are described in Chapter 11.

The analyses above are all based on *thin zone plate theory* in that they do not take account of finite wavelength effects in thick structures with narrow zones, as would occur for high aspect ratio, high spatial resolution structures. In practice the zones are very thick compared to the wavelength, and experience finite transmittance due to finite values of β/δ. Maser, Schneider, and Schmahl at Göttingen have described the effects of diffraction through thick structures mathematically using a coupled-wave analysis in which incident radiation couples to many modes within a waveguide-like structure.[95–98]

The Moscow X-Ray Optics Group of Popov, Kopylov, and Vinogradov[99] has taken a different approach. They note that at these very short wavelengths the refractive index is very close to unity, so that in calculating diffracted fields it is possible to factor out the rapidly varying z-dependence, e^{ikz} in the Fresnel–Kirchhoff equation, leaving a complex field amplitude with a relatively slow spatial dependence. The resultant field satisfies a parabolic wave equation similar to that encountered in quantum mechanics for a complex potential. For a refractive index near unity, the transverse propagation of energy evolves slowly, permitting numerically efficient solutions for arbitrary diffractive structures. With this approach it is not only possible to consider optimized zone structures, but also to efficiently calculate off-axis imaging properties, including aberrations, in thick structures.

Finally, we note that although pinhole and zone plate diffraction formulae are quite similar in form, both involving the Airy pattern $|2J_1(v)/v|^2$, the effects are quite different. For the pinhole the Airy pattern is expanding radially in the far field to characteristic dimensions much larger than the original pinhole size, while in the case of the zone plate lens the Airy pattern is formed in the focal plane with characteristic lateral dimensions (focal spot size) much smaller than the zone plate diameter.

10.6.3   Spatial Resolution of a Zone Plate Lens

One measure of the resolution of a lens is the minimum discernible separation of two mutually incoherent point sources. This in turn depends on the so-called *point spread function* of the lens, that is, the image plane intensity distribution due to a distant point source. For an ideal lens, including an ideal zone plate of many zones, the point spread function is an Airy pattern [Eq. (10.40), Figure 10.28], whose lateral extent (spread) depends on both wavelength and lens numerical aperture. The famous Rayleigh resolution, which we discuss in the following paragraph, involves two such point sources, each producing independent Airy patterns. The Rayleigh resolution criterion corresponds to the two being brought sufficiently close that the first null of one pattern is just aligned with the peak intensity of the other, resulting in a small but discernible dip in the summed intensity distribution. More generally, one finds that the resolution depends on illumination and can, of course, be limited by lens imperfections, mounting errors, noise, and other factors. For a Fresnel zone plate lens two additional limitations to resolution are the focal length's dependence on wavelength, a blur due to excessive spectral width (chromatic aberration), and zone placement errors, which we discuss in the paragraphs that follow. Other common aberrations,[69] such as astigmatism and coma, can also limit resolution and image fidelity, as discussed for zone plate lenses by Vladimirsky,[78] Morrison,[63] and Michette.[64]

We begin the study of spatial resolution by considering two point sources of quasi-monochromatic radiation, each producing an Airy intensity pattern in the image plane of the lens, as illustrated in Figure 10.29(a). In each case the radius to the first null is given by Eq. (10.41). Two physical situations that would approximate this mathematically ideal limit are those of (1) a sub-resolution point of emission in the object plane of a microscope, perhaps created by a backlighted pinhole of very small dimension, and (2) a distant star observed at an angular extent such that it is well below resolvable limits. In each case an image plane intensity distribution approximating that in Figure 10.28 would be obtained in near-perfect experimental conditions.

One could then imagine two such point sources of equal intensity in a noise-free back-ground. If the two were sufficiently separated, by a large distance Δl, as shown in Figure 10.29(a), the image would contain two distinct Airy patterns and we would conclude that the two sources of emission were well resolved. Next, one could imagine the two sources of emission being closer to each other, so that the Airy patterns overlapped. In the consideration here the two are assumed mutually incoherent, so that the fields do not form a time averaged interference pattern; rather, the intensities add. As the two Airy patterns are brought closer to each other, the intensity between the two peaks rises so that it becomes more difficult to discern the two. By Rayleigh's criterion the two are said to be just resolved when the first null of one overlaps the peak of the other, so that $\Delta l = r_{\text{null}}$, as illustrated in Figure 10.29(b). In this case the central Rayleigh dip

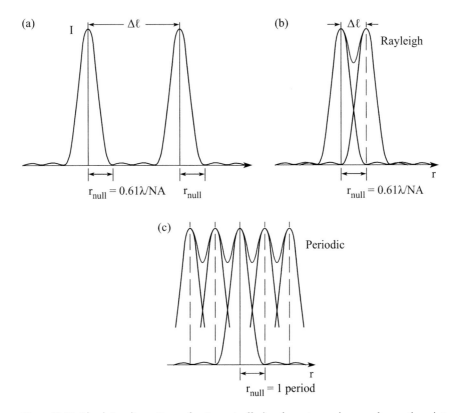

Figure 10.29 Airy intensity patterns due to mutually incoherent, quasi-monochromatic point sources at the image plane of a lens of numerical aperture NA. In (a) the two are well separated and easily resolved. In (b) the two are just resolved by the Rayleigh criterion. In (c) several form a periodic intensity pattern.

in intensity is 26.5%. The corresponding *Rayleigh resolution* limit, for Δl just equal to r_{null} in Figure 10.29(b), is

$$\Delta r_{Rayl.} = \frac{0.610\lambda}{NA} \tag{10.42}$$

From Eq. (10.21), the numerical aperture of a zone plate lens is $NA = \lambda/2\Delta r$, so that

$$\Delta r_{Rayl.} = 1.22\Delta r \tag{10.43}$$

where Δr is the outermost zone width of the zone plate lens. If the objects are brought any closer, the intensity dip rapidly diminishes, and the two are considered unresolved. Although other criteria exist, the Rayleigh criterion is a well-known benchmark, and one that is difficult to achieve at x-ray and EUV wavelengths. Note that a distant point source such as this corresponds to coherent illumination by a uniform spherical wave across the aperture, or to a good approximation by a plane wave if sufficiently distant.

There is an advantage, however, that accrues with non-plane wave illumination, that is, partially coherent illumination. We discuss this in the following paragraphs. As periodic patterns are also commonly used in tests of resolution, we also show in Figure 10.29(c) the intensity pattern that would result for a series of mutually incoherent point sources laid side by side with the respective Airy peaks and nulls overlapped in the Rayleigh manner. Here we see that the period of the resulting intensity pattern is just equal to what we have taken as the Rayleigh criterion separation, i.e.,

$$1 \text{ period} \simeq \Delta r_{\text{Rayl.}} = \frac{0.610\lambda}{\text{NA}} \tag{10.44}$$

where the intensity modulation is not greatly different than that for the just-resolved points.

Equation (10.44) is interesting for the insight it provides with regard to resolution tests involving periodic patterns. For instance, gratings of equally wide bars and spaces are frequently used for such tests, essentially to measure how the peak to valley intensity modulation varies as a function of decreasing period. Although one must be careful here to properly account for partial coherence of the illumination, Eq. (10.44) clearly states that the Rayleigh criterion corresponds to a full period of the intensity pattern, so that the identification of individual line or space widths, corresponding to a half period, should be numerically equal to half the Rayleigh resolution, and should be understood as such. As an example, Figure 10.30 shows an x-ray zone plate image of a sliced multilayer test pattern with half-periods extending from 20.6 nm to 7.8 nm.[100] The images were obtained in third order ($m = 3$) at 1.77 nm wavelength (700 eV) with a Au zone plate of outer zone width (Δr) equal to 20 nm. Illumination is partially coherent utilizing an elliptical hollow core capillary with a ratio of numerical apertures[¶] $\sigma \cong 0.16$. As seen in Figure 10.30(b) the pattern with an 11 nm half-period has a modulation of about 20% compared to the fully resolved longer periods. Further discussion of spatial resolution and its measurement is deferred, as these are affected by the illumination (partial coherence) pattern of the optical system as discussed in Chapter 11.

Another common test of resolution is the knife-edge test. This is a particularly attractive test because of its ease of implementation, but requires a sharp absorbing edge on the scale of the focal spot to be measured, which can be quite challenging. Nonetheless it is widely used and can be quite valuable. In Figure 10.31 an effort is made to correlate the knife-edge test with the Rayleigh criterion for the case of a point source illumination. A focal plane Airy distribution is numerically scanned across a mathematically perfect knife-edge. The resultant intensity profile is shown to the right. Upon comparison, it is observed that the Rayleigh criterion corresponds to a 10%–90% intensity variation. The 25%–75% intensity variation overstates spatial resolution, relative to the Rayleigh criterion by a factor of two. Nonetheless, the 25%–75% criterion is often used because it is readily identified even with noisy data. As it corresponds to a half-Raleigh period it is approximately equivalent to using the half-period of a grating test pattern.

¶ The ratio of numerical apertures, σ, and its role in image formation are discussed in Chapter 11, Section 11.2.

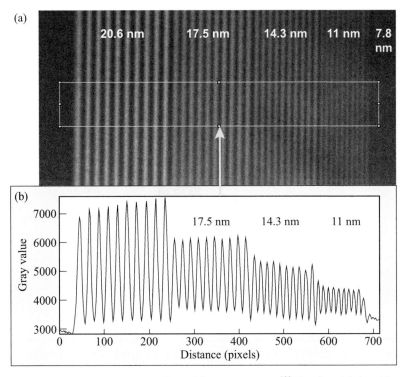

Figure 10.30 (a) An x-ray image of a multilayer test pattern[100] showing side-by-side patterns of 20.6 nm, 17.5 nm, 14.3 nm, 11.0 nm, and 7.8 nm half period. The soft x-ray image was obtained in third order ($m = 3$) with a $\Delta r = 20$ nm Au zone plate, at 700 eV (1.77 nm wavelength) using the full field x-ray microscope at the HZB/BESSY II synchrotron radiation facility in Berlin. The 11 nm period pattern shows a modulation of 20%. The 7.8 nm period pattern was not resolved. Courtesy of S. Rehbein and G. Schneider, Helmholtz-Zentrum Berlin.

10.6.4   Depth of Focus and Spectral Bandwidth

The depth of focus (DOF) of a lens or imaging system is the permitted displacement, away from the focal or image plane, for which the intensity on axis is diminished by some permissibly small amount, or image resolution is only slightly degraded. For the focal plane of a perfect circular lens, with plane wave illumination, simulations show[61] that the on-axis intensity of the Airy pattern decreases by only 20% when the observation plane is displaced from the ideal focal plane (smallest focal spot, thus highest axial intensity, $z = f$ in Figure 10.27) by an amount

$$\boxed{\Delta z = \pm \frac{1}{2} \frac{\lambda}{(\text{NA})^2}}$$ (10.45)

or

$$\boxed{\Delta z = \pm 2(\Delta r)^2/\lambda}$$ (10.46)

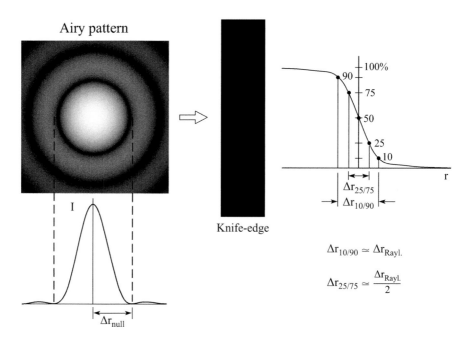

Figure 10.31 Simulation of a knife-edge test with an Airy image plane intensity distribution. The laterally integrated intensity is shown as a function of position as a calculated Airy pattern is scanned across a mathematically sharp knife edge. A radial knife edge translation producing an intensity variation from 10% to 90% is approximately equal to the Rayleigh resolution. A translation producing a 25%–75% intensity variation is approximately equivalent to half the Rayleigh resolution.

where for the zone plate lens Δr is the outer zone width, and NA $\cong \lambda / 2\Delta r$. We observe that the depth of focus is inversely proportional to the square of the numerical aperture. Thus for lenses of large NA, which potentially achieve the smallest focal spot according to Eq. (10.41) or best spatial resolution according to Eq. (10.42), the depth of focus is very short, leaving little latitude for error. In microscopes and other imaging systems, this can also limit the region of highest resolution to a depth less than the thickness of the object being studied. As an example, a zone plate lens with an outer zone width of 20 nm, has a depth of focus at $\lambda = 2.4$ nm of $\Delta z = \pm 0.33$ μm. Thus the focal region in which the axial intensity is not diminished by more than 20% is restricted to 0.66 μm. Thus in an imaging microscope, as will be discussed in Chapter 11, an extended three-dimensional object would have only a section 0.66 μm thick in focus; material outside that region would be seen with diminished resolution. This is currently an issue for soft x-ray tomographic imaging of single biological cells, which either operate with a larger outer zone width, which improves DOF by $(\Delta r)^2$, or perhaps use a shorter wavelength, or study a smaller object. An illustration of depth of focus is presented in Figure 10.32, including isometric views of the intensity distribution at best focus, the Airy pattern, and in places two and four depths of focus away.[101]

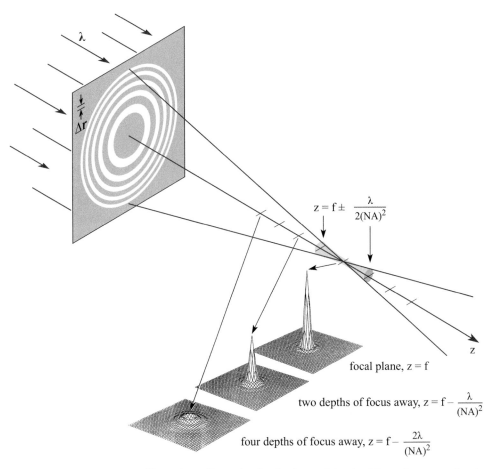

Figure 10.32 An illustration of intensity distribution and depth of focus for a zone plate lens with monochromatic plane wave illumination. At the focal plane the radiation is described by an Airy pattern, as given in Eq. (10.40). Within a nearby region of displacement $\Delta z = \pm\lambda/2(\text{NA})^2$, referred to as the depth of focus, the on-axis intensity decreases by only 20%. Further departures from the focal plane, such as the planes at twice and four times this amount, show significantly decreasing peak intensity. The simulated zone plate has $\Delta r = 30$ nm, $N = 300$, and $\lambda = 2.4$ nm. Numerical simulations courtesy of N. Iskander,[101] UC Berkeley and Lawrence Berkeley National Laboratory.

We observed earlier in Eq. (10.20) that the zone plate focal length has a strong wavelength dependence

$$f \simeq \frac{4N(\Delta r)^2}{\lambda}$$

Thus we can imagine that a source of spectral bandwidth $\Delta\lambda$ might have a sufficient focal length variation to exceed the depth of focus due to diffraction. We consider now the bandwidth necessary to avoid this situation. The derivative of f with respect to wavelength is

$$df \simeq \frac{-4N(\Delta r)^2}{\lambda^2} \, d\lambda$$

Combining these two equations one obtains a relation between spectral width and focal distance increment,

$$\Delta f = -f \cdot \frac{\Delta\lambda}{\lambda}$$

That is, the focal length is now spread by an amount Δf due to the finite spectral bandwidth $\Delta\lambda$. This focal shift is symmetric, $\pm\Delta f/2$ about the monochromatic focal plane at f, with longer wavelengths coming to focus closer to the lens and shorter wavelengths farther away. Equating this chromatic focal shift to the diffractive depth of focus given in Eq. (10.46), one has

$$\pm\frac{2(\Delta r)^2}{\lambda} = \pm\frac{f}{2} \cdot \frac{\Delta\lambda}{\lambda}$$

or, using Eq. (10.20),

$$\frac{2(\Delta r)^2}{\lambda} = \frac{2N(\Delta r)^2}{\lambda} \cdot \frac{\Delta\lambda}{\lambda}$$

yielding the condition of maximum spectral bandwidth,

$$\boxed{\frac{\Delta\lambda}{\lambda} \leq \frac{1}{N}} \qquad (10.47)$$

where $\Delta\lambda$ is the spectral bandwidth and N is the total number of zones (opaque and transmissive) in the zone plate.

Thus to avoid chromatic aberration, so that the focal plane intensity is not significantly diminished within the depth of focus, we require that the relative spectral bandwidth $\Delta\lambda/\lambda$ be less than or equal to one over the number of zones, i.e., less than $1/N$. For a typical zone plate lens with several hundred zones this requires a relative spectral bandwidth proportionately less than 1%. To obtain near-diffraction-limited resolution, limited only by wavelength and numerical aperture as given in Eq. (10.42), a polychromatic source of broader spectral bandwidth must be monochromaticized to the extent indicated by Eq. (10.47).

10.6.5   Extending Spatial Resolution Using Partially Coherent Illumination

In Section 10.6.3 we considered the ability to discern two mutually incoherent point sources, each producing an independent Airy intensity pattern in the image plane of the lens. The criterion for just resolving these two points is that their separation should be such that the first intensity nulls of each overlap the other's central intensity peak. Here we consider the effect of illumination on the ability to resolve a periodic structure in the object plane of a lens, as illustrated in Figures 10.33(a) and 10.34(a), such that the diffracted radiation at an angle $\theta \cong \lambda/d$ is just captured by the lens and thus contributes

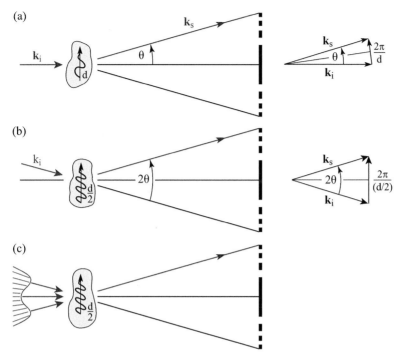

Figure 10.33 The effect of illumination on achievable resolution is suggested through the use of scattering diagrams. In (a) normal incidence plane wave illumination, scattered through an angle θ and collected by the lens of NA $= \sin \theta$, permits one to "see" features of characteristic dimension (period) d. In (b) the illumination is oblique, entering the sample at an angle θ, allowing radiation scattered through an angle 2θ to be collected, and thus permitting one to "see" smaller features of characteristic dimension $d/2$, effectively gaining a factor of 2 in resolution. In practice the gains in resolution are more modest. For the gain of 2 a narrow angular cone of illumination would be required, thus sacrificing available photon flux and compromising the contrast of lower spatial frequency features. In (c) tailoring the angular distribution of the illuminating radiation is suggested as a method to provide a broad spatial frequency response within the imaging system. A qualitative analysis of possible improvements in resolution, as described by Hopkins, is presented in Ref. 61.

to image formation in the image plane (not shown). This situation can be improved, however, by bringing the incident radiation in at an oblique angle, as suggested in Figures 10.33(b) and 10.34(c).

Figures 10.33 and 10.34 describe illumination advantages from the perspective of general scattering diagrams, as was discussed in Chapter 2, and diffraction from simple gratings, as discussed in Section 10.5, respectively. In each case it is clear that in the limit of extreme illumination and collection, it is possible to improve resolution by a factor approaching 2 beyond that corresponding to coherent illumination. In practice a factor of ~1.5 is common. Thus, for example, one might imagine a zone plate microscope[§] employing a condenser that provides an annulus of incident radiation of

---

[§] Or a visible light microscope with a high NA refractive objective lens.

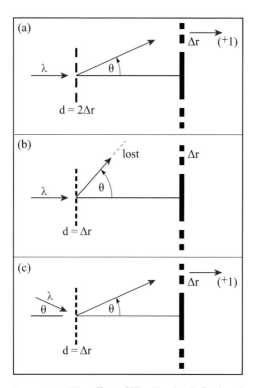

Figure 10.34 The effect of illumination is further demonstrated for the specific case of a zone plate lens (outer zone width Δr, local period $2\Delta r$) collecting radiation diffracted from a transmission grating. In (a) the grating period is $d = 2\Delta r$, allowing normal incidence radiation at wavelength λ to be captured by the lens. In (b) the grating period is reduced by a factor of 2, causing diffraction at twice the angle and resulting in loss of the radiation, i.e., the radiation is not captured by the lens and thus does not contribute to an image of the grating. In (c) the illumination is oblique, at an incidence angle θ, so that diffraction from the shorter period grating is again captured by the lens. Thus, as in Figure 10.33, we see that controlling the angular distribution of incident radiation can provide improved resolution ($d = \Delta r$ vs. $d = 2\Delta r$) up to a factor 2 in the limit. Details depend on the illumination (condenser) and collection (lens) numerical apertures, the detailed angular distributions therein, the system throughput (photon flux), and the desired contrast as a function of feature size in the object. This topic is discussed further in Chapter 11.

large numerical aperture, combined with a zone plate objective lens with its innermost zones blocked, so that only a narrow annulus of radiation is collected, corresponding to diffraction of the highest spatial frequencies in the object or sample. This would permit one, in principle, to achieve a spatial resolution exceeding the coherent case by a factor approaching 2. Thus in Figure 10.34(c), with obliquely incident radiation, it is possible to "see" a grating of period $d = \Delta r$, rather than $d = 2\Delta r$ as in case (a) for normal incidence illumination, where Δr is the outer zone width of the lens.

The control of angular illumination as a method of enhancing the spatial resolution of microscopes and other imaging or printing (lithographic) optics is discussed in the literature.[61] The application to x-ray microscopes, including both the ratio of condenser to objective lens numerical aperture and the effect of objective lens central obstruction, is considered by Jochum and Meyer-Ilse.[102] Further discussion of high spatial resolution imaging, to 10 nm, is presented in Chapter 11.

10.6.6   Zone Plate Fabrication

Advances in nanofabrication capabilities, electron beam lithography and thin-film deposition have allowed diffractive and reflective x-ray optics to be improved over time. Zone plates are primarily fabricated using a combination of electron-beam lithography, etching, and electroplating, an outgrowth of techniques developed for the semiconductor industry which for decades has been involved in the manufacture of integrated circuit (IC) electronic devices based on complex material structures with nanometer features.[103–108] Fabrication involves writing a desired pattern in some recording medium, with an electron beam,[107–110] typically 100 keV electron energy, with a beam size of a few nm in diameter. As seen in earlier sections, zone plates work on the basis of interference of radiation from across the structure, which therefore requires an electron-beam placement accuracy of at least $\Delta r/3$. In addition, these structures must be sufficiently thick to provide suitable efficiency, primarily through the absorption of soft x-rays, or additionally with a strong phase shifting component for harder x-rays. For soft x-rays zone plate efficiencies are typically around 10% to first order, but vary depending on the choice of material, the duty cycle, substrate attenuation, thickness, and grading of the zones. Aspect ratios of only 3:1 or 4:1 are common for the highest-resolution zone plates, which limits the diffraction efficiency, but recent progress[109] utilizing vertical directionality controlled metal assisted chemical etching (V-MACE) has produced structures with aspect ratios greater than 100:1. This opens new opportunities, particularly for hard x-rays where significant absorption and phase advance are now possible, depending on x-ray energy. The process steps are shown in Figure 10.35. An example of a high aspect zone plate fabricated with this technique is shown in Figure 10.36.

Several additional methods have been used to obtain narrower but taller structures, including physical zone plate stacking, zone doubling through atomic layer deposition, and multiple patterning, a technique used widely throughout the semiconductor industry. With smallest zone widths of 10 nm, volume effects of the zone plate must be considered, and tapering of zones is required to achieve high efficiency. Current expectations are that zone plate imaging can be extended to less than 10 nm, but its utility in scientific applications at that resolution will depend on object contrast, radiation dose, depth-of-field, and available working distance.

10.6.7   Multilayer Laue Lenses

A Multilayer Laue Lens (MLL) is a linear zone plate that focuses x-rays in a one-dimensional (1D) line focus. The MLL was conceived as a method to overcome aspect

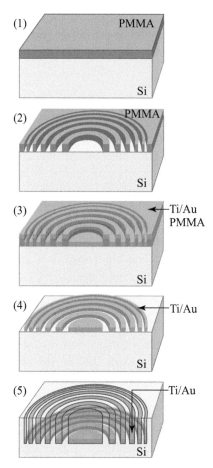

Figure 10.35 The V-MACE process of fabricating a high aspect x-ray zone plate is shown.[109] Step (1) A high resolution resist is spun onto the wafer. (2) The desired pattern is written with electron beam lithography and developed. (3) A thin layer of metal catalyst, here Ti/Au, is evaporated onto the sample. (4) A liftoff is performed leaving the metal layer on top of the silicon wafer. (5) A solution of cooled etchant, consisting of HF, H_2O_2, and H_2O, is placed onto the sample. An etchant front proceeds vertically downward, removing silicon beneath the metal mask. The metal mask sinks with the etching front creating structured trenches of silicon and leaving a relatively thick silicon mold. The zone plate pattern can then be metalized using techniques such as electroplating or atomic layer deposition.

ratio limitations due to fabrication limits. MLLs are fabricated using multilayer coating techniques with varied d-spacings that correspond to the zones of a zone plate for constructive interference at the focus. These are coated, sectioned, and polished to an equivalent outermost zone width as small as 4 nm, with several thousand zones and a coating thickness greater than 10 μm, thus providing an extremely high aspect ratio when used in a transmission geometry. Diagrams of various MLL 1D and 2D geometries as described by Yan et al.,[111] are illustrated in Figure 10.37. The geometries show successively sophisticated MLLs presently under development with ever higher aspect ratios

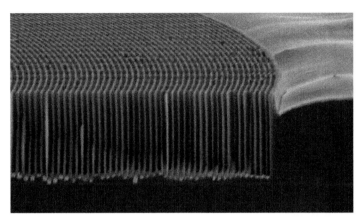

Figure 10.36 A high aspect ratio silicon zone plate for high efficiency with hard x-rays.[109] The outer zone width is 100 nm and the height is 6.6 μm, for an aspect ratio of 66:1. The number of zones $N = 150$.

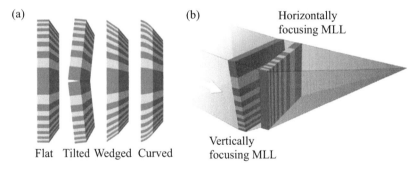

Figure 10.37 (a) One-dimensional Multilayer Laue Lenses (MLL) with flat, tilted, wedged, and curved zones. (b) Crossed MLLs for two-dimensional focusing.[111] Courtesy of H. Yan, National Synchrotron Light Source, Brookhaven National Laboratory.

and smaller outermost zone widths. Higher efficiencies and better resolution require the use of wedged, and eventually curved, multilayer coatings with d-spacings which satisfy the Bragg condition at every point on the surface. A second pair of such lenses used in the orthogonal direction can provide two-dimensional focusing as with the Kirkpatrick–Baez optics discussed in Section 10.1. Kang et al.[112] have used side-by-side 1D multilayer Laue lenses in a tilted geometry to achieve a 16 nm line focus with a 5 nm outer zone width, achieving a 31% efficiency at 20 keV (0.64 Å). Huang et al.[113] have extended this with a wedged MLL having graded d-spacings in the propagation direction, focusing 14.6 keV radiation at the APS near Chicago. In that experiment ptychographic wavefront reconstruction is consistent with a 1D line focus of 26 nm and an efficiency of 27%. Morgan, Bajt, and colleagues[114] have reported fabrication and testing of a 1D wedged MLL with varied d-spacings in the propagation direction matching the Bragg condition across the lens aperture. In experiments at PETRA III in Hamburg, their ptychographic reconstruction of the wavefront is consistent with an 8 nm line focus at 22 keV (0.56 Å). An extensive review of multilayer Laue lenses is given by Yan et al.,[111] including diagrams of various MLL 1D and 2D geometries which are reproduced here in Figure 10.37.

Topics not discussed here but which may be of interest to the reader are the fabrication of so-called "jelly roll," or sputtered–sliced zone plates[115] and Bragg–Fresnel zone plates.[116] The sputtered zone plates are made using multilayer fabrication techniques, in this case coating a rotating wire with alternating materials. The goal is to achieve very narrow outer zones with high aspect ratio. Bragg–Fresnel zone plates combine diffraction of hard x-rays by a crystal with an imprinted zone plate pattern.

10.7   Reflection Gratings

Reflection gratings combine both reflective and diffractive effects. Reflective gratings are commonly used at grazing incidence in soft x-ray and EUV monochromator and spectrometer applications such as isolating particular laser high harmonics, narrowing the spectral width of undulator and free electron laser radiation, measuring time resolved laser line width narrowing, and as a diagnostic for spectroscopic analysis of plasma emissions. Reflective gratings are often produced by mechanical ruling methods where a series of grooves are physically cut at specified angles. In some instances, lithographic patterning and anisotropic etching are also used. These gratings provide dispersion of different wavelengths according to the *grating equation,*[117–120]

$$m\lambda = d(\sin\theta_m - \sin\theta_i)(m = 0, \pm 1, \pm 2, \pm 3) \tag{10.48}$$

where θ_i is the angle of incidence measured from the surface normal, θ_m is the angle of the *m*th reflected order measured from the normal, λ is the wavelength, d is the local groove spacing, and m is the diffraction order. The equation can be derived by tracking the various path lengths as was done earlier for a transmission grating. The geometry is illustrated in Figure 10.38.

A particularly useful monochromator/spectrometer design involves the use of a varied line space (VLS) plane grating[121–126] operating with convergent radiation in a manner that greatly simplifies wavelength scanning, significantly reduces aberrations, maintains a fixed focus and fixed exit direction, is very compact, and greatly reduces cost. For wavelength tuning the plane grating is rotated about an axis parallel to the central groove. This simple rotation is the only movement. Through grating design, it is possible to greatly reduce aberrations and increase wavelength (photon energy) resolution for both monochromators and spectrometers. Spectral resolutions to several times 10^4 have been obtained.[122, 125] An example of a VLS grating monochromator was seen in the soft x-ray coherence beamline of Figure 5.31. Rather than equidistant grooves of constant period along the ruled surface, the grating can be ruled in a manner such that the period varies with position depending on properties of the incoming and outgoing radiation, such as wavelength, the degree of convergence, output spectral width, and other factors. With a ruling machine there is great freedom with respect to the groove

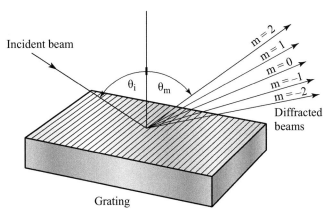

Incident beam

Diffracted beams

Grating

Figure 10.38 A reflection grating for x-rays and EUV radiation.[120] The angles of incidence and reflection are described by Eq. (10.48). Courtesy of J. Underwood, Lawrence Berkeley National Laboratory.

density and variations thereof. To form the grating an optically flat fused silica substrate is typically coated with a thin layer of Au, Pt, Ir, and Ni. Furthermore, the grooves are cut at a "blazed" angle that greatly enhances the concentration of energy into a specified diffraction order, providing high grating efficiency for a band of wavelengths. In this geometry the groove profiles are shaped with approximately flat faced facets, tilted so that the angle of incidence equals the angle of reflection with respect to the surface normal, thus greatly increasing the efficiency of a diffracted order in that same direction. The grooves can be blazed by the shape of the cutting tool. Efficiencies of 30%–40% are common in the EUV, somewhat lower for soft x-rays.[124]

Varied line space gratings have been created with both ruling machines[127] and holographic interference techniques.[128] Holographically written patterns offer the advantage of reduced high order content and reduced stray light, but with near sinusoidal patterning they have a reduced capability for blazing. A comparison of mechanically and holographically ruled gratings is given in Reference 129.

10.8 Crystal Optics

Crystal optics provide a major tool for experiments with hard x-rays having photon energies in the 6–60 keV range. Natural crystals of high purity and near perfect atomic arrangement are cut, polished, and in some cases thinned, to create x-ray optical components. A major use of these crystals is as spectral dispersion elements for use in hard x-ray monochromators or spectrometers. A crystal monochromator is a device that is used to select the wavelength needed for a particular experiment, while a spectrometer is a device used to disperse different wavelengths to different angles, for example for studies of absorption or emission spectra. Common x-ray techniques such as x-ray scattering and imaging require some level of monochromaticity in the beam. Because x-ray sources often generate a larger bandwidth than required for many scientific studies and measurement techniques, an x-ray monochromator is often used to spectrally narrow the beam to meet experimental requirements.

An x-ray monochromator consists of a crystal oriented to reflect from a particular set of planes (111, 200, etc.) depending on wavelength and desired degree of monochromaticity.[130–137] Spectral energy resolution, angular divergence, intensity, polarization, and in some cases time structure, can be controlled and manipulated using an x-ray crystal. Crystal monochromators follow Bragg's Law of diffraction [see Chapter 3, Section 3.9, Eq. (3.91)], renumbered here as

$$m\lambda = 2d \sin\theta \qquad (10.49)$$

where θ is measured from the crystal planes and d is the distance between those planes.

Depending on the crystal type, the d-spacings of its planes, orientation of the crystal planes, and angle of incidence selected, the diffracted photon energy can be selected. Smaller d-spacings allow for shorter wavelengths and higher photon energy selection. The diagram in Figure 10.39 shows a two-dimensional view of various atomic planes of NaCl, illustrating combinations of angles, wavelengths, and d-spacings that satisfy the Bragg equation.[138]

Because it is possible to obtain large, near perfect crystals of high purity, and because the x-rays form an interference pattern at the Bragg condition within the crystal that minimizes their interaction with core level (absorbing) electrons, it is possible, in fact common, to achieve a reflectivity of greater than 90%. This inter-plane interference effect was described in Chapter 3, Section 3.9, and illustrated in Figure 3.18. Figure 10.40 shows calculated reflectivity curves[133] for a Si(111) crystal at various photon energies. Note that for 20 keV photons the acceptance width is 7.8 μrad FWHM, increasing to 30 μrad FWHM for 10 keV. For comparison, the angular half-width of the central radiation cone for 10 keV undulator radiation at Argonne's APS is 11 μrad, so that the full central cone of undulator radiation can pass with very high efficiency at 10 keV. Undulator central cone angles are similar at all third generation hard x-ray facilities, somewhat smaller for higher values of γ. The relative spectral resolution achievable with an x-ray crystal monochromator depends on the number of reflection planes and thus the absorption length. For a Bragg reflection from Si(111) with $d = 3.1355$ Å, a crystal just 10 μm thick has approximately 30 000 planes so that with minimal absorption, as occurs with inter-plane interference and a photon energy greater than 10 keV, spectral resolving powers greater than 10^4 are routinely obtained. The spectral resolving power of a crystal is described by $\Delta E/E = \Delta\lambda/\lambda = \Delta\theta/\tan\theta$, obtained by taking a derivative of the Bragg equation for fixed d. Values of order 10^{-4} to 10^{-5} are routinely obtained with low Miller indices,[‖] for example (111) or (220), and significantly narrower with higher indicies.[131] Intrinsic imperfections in the crystal such as dislocations, point defects, stacking faults, or other crystal defects can affect the performance of the monochromator.

[‖] Miller indices are the *hkl* integers which describe the various crystallographic planes.

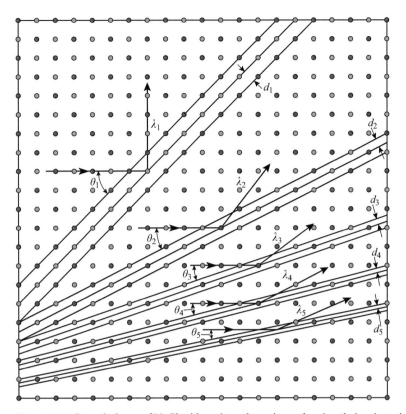

Figure 10.39 Crystal planes of NaCl with various d-spacings, showing their orientation and how they would reflect x-rays of various wavelengths that satisfy the Bragg condition. Following F. Richtmyer, E. Kennard, and T. Lauritsen.[138] The highlighted planes are described by Miller indices (220) for d_1, (240) for d_2, (260) for d_3, etc.[135]

Fixed exit angle Si(111) double crystal monochromators (DCMs) are very common components of hard x-ray beamlines at synchrotron facilities. This configuration is widely used due to its high throughput, typically greater than 90%, and its ability to maintain a constant x-ray beam direction during photon energy tuning. In the DCM setup, the two crystals are mounted parallel to each other and utilized in series.[130, 131, 137] A double crystal monochromator as used in the APS hard x-ray coherent imaging beamline is seen in Figure 5.32.

10.9   Compound Refractive Lenses

In the x-ray regime, as we saw in Chapter 3, the refractive index is given by $n = 1 - \delta + i\beta$, where δ and $\beta \ll 1$. This is in contrast to visible light where refractive indices are typically ~ 1.5 in glass, with significant refractive effects and minimal absorption. The x-ray region is thus characterized in general as having minimal refractive effects (other than total external reflection) and significant absorption. The refractive effect of a single

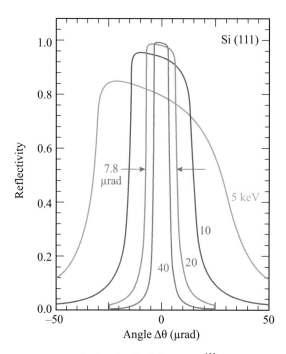

Figure 10.40 Calculated reflectivity curves[133] for a Si(111) crystal, $d = 3.1355$ Å, with incident x-ray energies of 5, 10, 20, and 40 keV. Note that for the higher photon energies the reflectivity approaches 100%. At 20 keV, the Darwin width is 7.8 μrad FWHM, increasing to 30 μrad FWHM at 10 keV. Courtesy of B. Batterman and D. Bilderback, Cornell University.

x-ray lens is therefore very small. Furthermore, with the real part of the refractive index being less than 1, a concave x-ray refractive lens is required to produce convergent radiation, as seen in the frontpiece figure for this chapter. For hard x-rays, where δ and β are both very small, and $\beta/\delta \ll 1$, it is possible to have a small refractive effect with minimal absorption. For example, at 20 keV, silicon[6, 7] has $\delta \cong 2.3 \times 10^{-6}$ and $\beta \cong 7.1 \times 10^{-8}$.

The focal length of a refractive lens is given by the lens maker's formula[139]

$$\frac{1}{f} = (n-1)\left(\frac{1}{R_1} - \frac{1}{R_2}\right) \tag{10.50}$$

where f is the focal length, n the refractive index, R_1 is the radius of curvature of the first surface, and R_2 that of the second surface. For a bi-concave lens R_1 is negative and R_2 is positive. For hard x-rays with $n \cong 1 - \delta$, the focal length of the bi-concave lens with equal radii is given by

$$f = R/2\delta \tag{10.51}$$

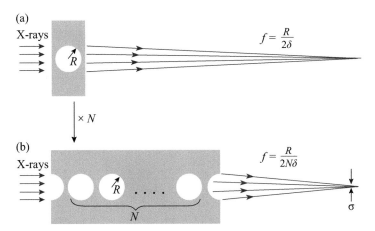

Figure 10.41 A single x-ray refractive lens with focal length $f = R/2\delta$, and a compound refractive lens[140] of N lens elements having a focal length $f = R/2N\delta$. Courtesy of A. Snigerev, I. Snigereva (ESRF) and B. Lengeler (RWTH, Aachen).

so that for a Si lens at 20 keV, with a radius of curvature of 100 μm, the focal length is 22 m. In an important first paper Snigerev *et al.*[140] proposed a linear sequence of N circular holes, which they called a *compound refractive lens (CRL)*, as shown in Figure 10.41, which has a cumulative focal length

$$f = R/2N\delta \tag{10.52}$$

In that first paper they demonstrated the use of 30 cylindrical lenses drilled with a radius of 300 μm in aluminum, which with $\delta = 2.8 \times 10^{-6}$ at 14 keV, yielded a focal length of 1.8 m, which they confirmed in their measurements. We see then that compound refractive lenses consist of a series of refractive lenses which bend the light, each ever so slightly. Thus a sequence of lenses results in a much reduced focal length such that CRLs are now commonly used in hard x-ray beamlines. A tradeoff in compound refractive lens design involves decreased focal length versus acceptable transmission.

A large number of lenses decreases the focal length, however, transmission is proportionally decreased. A low Z, and thereby less absorbing material, such as Be, B, C, Al, or Si, depending on photon energy, is more desirable for these lenses. Later papers[141–145] have pursued the successful use of beryllium, aluminum, and other low-Z materials. Beryllium offers the advantage of much reduced absorption, for example[6, 7] $\beta/\delta \sim 5 \times 10^{-5}$ at 20 keV. Note that CRL lenses are somewhat chromatic as delta (δ) is a weak function of photon energy.

A compound refractive lens can be fabricated several ways, including drilling a series of holes into a low-Z material, resulting in cylindrical surfaces, as was done for the first lens. However, parabolic surfaces offer the advantage of eliminating spherical aberrations and thus can provide a tighter focus using the same, or larger aperture. For parabolic shapes other fabrication techniques are required, such as electron beam or visible/UV light photolithography.[145] The image seen in Figure 10.42 shows an array

(a)   (b)

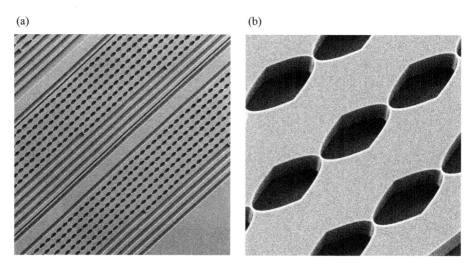

Figure 10.42 (a) An SEM image of several side-by-side parabolic compound refractive lenses formed in silicon. Individual lenses (b) have an aperture of 50 μm, depth of 50 μm, and length of 100 μm. The lens to lens separation is 2 μm. Side-by-side sets of CRLs provide an ability to maintain a constant focal length of 100 mm for discretely different photon energies extending from 10 keV to 50 keV in 5 keV steps. The CRLs have been tested for use in the range of 10 keV to 80 keV. A best focal spot of 150 nm FWHM was achieved at 50 keV.[145] Courtesy of I. Snigireva, ESRF.

of side-by-side parabolic CRLs formed in silicon, known as an Integrated Lens System (ILS). With differing numbers of individual lenses per CRL, the ILS array provides a means for maintaining a constant focal length for differing photon energies, as would be needed for spectral probing of sample properties. For the set of parabolic lenses in Figure 10.42 there are ten side-by-side CRLs, covering the photon energy range from 10 keV to 50 keV in 5 keV steps, each CRL having a 100 mm focal length. The number of individual lens elements, however, varies from 12 to 392, with many more lens elements required for the higher photon energies where δ is much smaller. Each individual lens element has an aperture of 50 μm and a length of 100 μm. The separation distance between individual elements, in the direction of the optical axis, is 2 μm. The focal spot size of these CRLs has been measured from 10 keV to 80 keV using knife-edge techniques at ESRF. The results show a spot size of 150 nm FWHM at 50 keV, and generally smaller than 200 nm FWHM across the spectral range.[145]

Compound refractive lenses have been used with synchrotron radiation for beamline alignment and focusing hard x-rays to micron scale dimensions. More recently they have been used with x-ray free electron lasers.[146] A recent scientific application of a CRL in full field x-ray microscopy employed seven one-dimensional Be lenses, each having a 200 μm apex radius of curvature, and an effective aperture of 860 μm, achieving a spatial resolution of 150 nm at 17 keV (0.73 Å) in dark field geometry.[147] In its first test the CRL hard x-ray microscope was used to study grains along dislocations in aluminum during annealing and recrystallization.[147]

References

1. P. Kirkpatrick and A.V. Baez, "Formation of Optical Images by X-rays," *J. Opt. Soc. Amer.* **38**, 766 (1948).

2. J.H. Underwood, "Imaging Properties and Aberrations of Spherical Optics and Nonspherical Optics", p. 145 in *Vacuum Ultraviolet Spectroscopy I* (Academic Press, 1998), J.A. Samson and D.L. Ederer, Editors.

3. A. Snigerev, V. Kohn, I. Snigereva and B. Lengeler, "A Compound Refractive Lens for Focusing High-Energy X-rays," *Nature* **384**, 49 (November 7, 1996).

4. J.H. Underwood and D. Turner, "Bent Glass Optics," *SPIE* **106**, 125 (1977).

5. K. Yamauchi, H. Mimura, K. Inagaki and Y. Mori, "Figuring with Subnanometer-Level Accuracy by Numerically Controlled Elastic Emission Machining," *Rev. Sci. Instrum.* **73**(11), 4028 (November 2002); Y. Mori, K. Yamauchi, K. Yamamura and Y. Sano, "Development of Plasma Chemical Vaporization Machining," *Rev. Sci. Instrum.* **71**(12), 4627 (December 2000); K. Yamauchi *et al.*, "Nearly Diffaction-limited Line Focusing of a Hard-X-ray Beam with an Elliptically Figured Mirror," *J. Synchr. Rad.* **9**, 313 (2002).

6. B.L. Henke, E.M. Gullikson and J.C. Davis, "X-Ray Interactions: Photoabsorption, Scattering, Transmission and Reflection at $E = 50$–$30,000$ eV, $Z = 1$–92," *Atomic Data and Nucl. Data Tables* **54**, 181 (1993).

7. E.M. Gullikson, https://www.cxro.LBL.gov/optical_constants

8. J.H. Underwood, "X-Ray Optics," *Amer. Scientist* **66**(4), 476 (1978); J.H. Underwood, "X-ray Optics," *Encyclopedia of Applied Physics* **23**, 525 (Wiley, 1998); J.H. Underwood and T.W. Barbee, "Soft X-Ray Imaging with a Normal Incidence Mirror," *Nature* **294**, 429 (1981).

9. B. Aschenbach, "X-ray Telescopes," *Rep. Prog. Phys.* **48**, 579 (1985).

10. M. Weisskopf, B. Brinkman, C. Canizares *et al.*, "An Overview of the Performance and Scientific results from the Chandra X-Ray Observatory," *Pubs. Astronom. Soc. Pacific* **114**, 1 (January 2002); M. Santos-Lleo, N. Schartel, H. Tananbaum, W. Tucker and M.C. Weisskopf, "The First Decade of Science with Chandra and XMM-Newton," *Nature* **462**, 997 (December 24, 2009).

11. NASA, Chandra orbiting x-ray observatory; https://www.nasa.gov/mission_pages/chandra/main/index.html

12. D. Schwarz, "The Chandra X-Ray Observatory," *Rev. Sci. Instrum.* **85**, 061101 (2014).

13. D.E. Graessle, R. Soufli, A.J. Nelson *et al.*, "Iridium Optical Constants from Synchrotron Reflectance Measurements over 0.05–to–12 keV X-ray Energies," *SPIE* **5538**, 72 (2004).

14. NASA and the Harvard-Smithsonian Center for Astrophysics, Messier M-51, the "Spiral Galaxy": http://chandra.harvard.edu/photo/2014/m51; http://www.nasa.gov/mission_pages/chandra/multimedia/spiral-galaxy-m51.html; Q.D. Wang *et al.*, "Dissecting X-Ray-Emitting Gas Around the Center of Our Galaxy," *Science* **342**, 981 (August 30, 2013); J.D. Schnittman, "The Curious Behavior of the Milky Way's Central Black Hole," *Science* **341**, 964 (August 30, 2013); "NASA's Chandra Finds Supermassive Black Hole Burping Nearby" (January 5, 2016), https://www.nasa.gov/mission_pages/chandra/nasa-s-chandra-finds-supermassive-black-hole-burping-nearby.html (E. Schlegel, C. Jones, M. Maracheck and L. Vega, American Astronomical Society, Kissimmee, Fl, submitted for publication).

15. H. Mimura, S. Hanada, T. Kimura *et al.*, "Breaking the 10 nm Barrier in Hard-X-ray Focusing," *Nature Phys.* **6**, 122 (February 2010).

16. K. Yamauchi, H. Mimura, T. Kimura *et al.*, "Single-Nanometer Focusing of Hard X-rays by Kirkpatrick–Baez Mirrors," *J. Phys.: Cond. Matter* **23**, 394206 (September 15, 2011).

17. S. Matsuyama, H. Mimura, H. Yumoto *et al.*, "Development of Scanning X-ray Fluorescence Microscope with Spatial Resolution of 30 nm Using Kirkpatrick-Baez Mirror Optics," *Rev. Sci. Instrumen.* **77**, 103102 (2006); S. Matsuyama *et al.*, "Development of Mirror Manipulator for Hard-X-ray Nanofocusing at the Sub-50 nm Level," *Rev. Sci. Instrum.* **77**, 093107 (2006).

18. S. Matsuyama, M. Shimura, H. Mimura *et al.*, "Trace Element mapping of a Single Cell Using a Hard X-ray Nanobeam Focused by a Kirkpatrick–Baez Mirror System," *X-Ray Spectrom.* **38**, 89 (November 26, 2009).

19. T. Goto, H. Nakamori, T. Kimura *et al.*, "Hard X-ray Nanofocusing Using Adaptive Focusing Optics Based on Piezoelectric Deformable Mirrors," *Rev. Sci. Instrum.* **86**, 043102 (2015).

20. H. Yumoto, H. Mimura, T. Koyama *et al.*, "Focusing of X-ray Free-Electron Laser Pulses with Reflective Optics," *Nature Photonics* **7**, 43 (January 2013); T. Pardini, D. Cocco and S.P. Hau-Riege, "Effect of Slope Errors on the Performance of Mirrors for X-ray Free Electron Laser Applications," *Optics Express*, **23**(25), 31889 (December 14, 2015).

21. H. Mimura, H. Yumoto, S. Matsuyama *et al.*, "Generation of 1021 W/cm^2 Hard X-ray Laser Pulses with Two-Stage Reflective Focusing System," *Nature Commun.* **5**, 3539 (April 30, 2014).

22. K. Yamauchi *et al.*, Microstitching Interferometry for X-ray Reflective Optics," *Rev. Sci. Instrum.* **74**(5), 2894 (May 2003).

23. S. Bajt, D.G. Stearns and P.A. Kearney, "Investigation of the Amorphous-to-Crystalline Transition in Mo/Si Multilayers," *J. Appl. Phys.* **90**(2), 1018 (July 15, 2001); S. Bajt, J.B. Alameda, T.W. Barbee *et al.*, "Improved Reflectance and Stability of Mo-Si Multilayers," *Opt. Eng.* **41**(8), 1797 (August 2002).

24. R. Soufli, M. Fernández-Perea, S.L. Baker *et al.*, "Spontaneously Intermixed Al-Mg Barriers Enable Corrosion Resistant Mg/SiC Multilayer Coatings," *Appl. Phys. Lett.* **101**(4), 043111 (July 23, 2012); Q. Zhong, Z. Zhang, R. Qi *et al.*, "Enhancement of the Reflectivity of Al/Zr Multilayers by a Novel Structure," *Optics Express* **21**(12), 14399 (June 17, 2013).

25. J.R. Lemen *et al.*, "The Atmospheric Imaging Assembly (AIA) on the Solar Dynamics Observatory (SDO)," *Solar Phys.* **275**, 17 (2012).

26. J.H. Underwood, A.C. Thompson, Y. Wu and R.D. Giaque, "X-Ray Microprobe Using Multilayer Mirrors," *Nucl. Instrum. Meth. A* **266**, (1988); Y. Wu, "Phase Transition and Equation of State of CsI Under High Pressure and the Development of a Focusing System for X-Rays," PhD Thesis, Physics Department, University of California, Berkeley 1990; D. Clery, "New Synchrotrons Light Up Microstructure of Earth," *Science* **277**, 1220 (August 29, 1997).

27. A.V. Vinogradov and B. Ya. Zeldovich, "X-Ray and Far UV Multilayer Mirrors: Principles and Possibilities," *Appl. Opt.* **16**, 89 (1977).

28. E. Spiller, "Reflective Multilayer Coatings for the Far UV Region," *Appl. Opt.* **15**, 2333 (1976).

29. E. Spiller, *Soft X-Ray Optics* (SPIE, Bellingham, WA, 1994); "Reflecting Optics: Multilayers", p.271 in *Vacuum Ultraviolet Spectroscopy I* (Academic Press, 1998), J.A. Samson and D.L. Ederer, Editors.

30. J.H. Underwood, "Multilayer Mirrors for X-Rays and Extreme UV," *Optics News* **12**, 20 (OSA, Washington, DC, March 1986); J.H. Underwood and E.M. Gullikson, "Beamline for

Measurement and Characterization of Multilayer Optics for EUV Lithography," *SPIE* **3331**, 52 (1998).

31. J.H. Underwood and T.W. Barbee, "Layered Synthetic Microstructures as Bragg Diffractors for X-Rays and Extreme Ultraviolet: Theory and Predicted Performance," *Appl. Opt.* **20**, 3027 (1981).

32. E.M. Gullikson, F. Salmassi, A.L. Aquila and F. Dollar, "Progress in Short Period Multilayer Coatings for Water Window Applications" (unpublished, June 20,2008), http://escholarship .org/uc/item/8hv7q0hj; Q. Huang *et al.*, "High Reflectance Cr/V Multilayer with B4C Barrier Layer for Water Window Wavelength Region," *Optics Lett.* **41**, 701 (February 15, 2016).

33. R. Soufli and E.M. Gullikson, "Reflectance Measurements on Clean Surfaces for the Determination of Optical Constants of Silicon in the Extreme Ultraviolet-Soft X-Ray Region," *Appl. Optics.* **36**, 5499 (1997); R. Soufli and E.M. Gullikson, "Absolute Photoabsorption Measurements of Molybdenum in the Range 60 to 930 eV for Optical Constant Determination," *Appl. Opt.* **37**, 1713 (1998).

34. R. Soufli, "Optical Constants of Materials in the EUV/Soft X-Ray Region for Multilayer Mirror Applications," PhD thesis, Electrical Engineering and Computer Science, University of California, Berkeley (1997).

35. T.W. Barbee and D.L. Keith, "Synthesis of Metastable Materials by Sputter Deposition Techniques," p. 93 in *Synthesis and Properties of Metastable Phases*, E.S. Machlin and T.J. Rowland, Editors (Metallurgical Society, Amer. Inst. Mech. Eng., Warrendale, PA, 1980); J.H. Underwood, T.W. Barbee and D.C. Keith, "Layered Synthethic Microstructures: Properties and Applications in X-Ray Astronomy," *SPIE* **184**, 123 (1979).

36. J.A. Folta, S. Bajt, T.W Barbee, Jr. *et al.*, "Advances in Multilayer Reflective Coatings for Extreme Ultraviolet Lithography," *SPIE* **3676** (1999); D.G. Stearns, R.S. Rosen and S.P. Vernon, "Multilayer Mirror Technology for Soft X-Ray Projection Lithography," *Appl. Optics* **32**, 6952 (1993); P.A. Kearney, C.E. Moore, S.I. Tan, S.P. Vernon and R.A. Levesque, "Mask Blanks for Extreme Ultraviolet Lithography: Ion Beam Sputter Deposition of Low Defect Density Mo/Si Multilayers," *J. Vac. Sci. Technol. B* **15**, 2452 (1997).

37. D.L. Windt and W.K. Waskiewicz, "Multilayer Facilities Required for Extreme-Ultraviolet Lithography," *J. Vac. Sci. Technol. B* **12**, 3826 (1994).

38. J.B. Kortright, E.M. Gullikson and P.E. Denham, "Masked Deposition Techniques for Achieving Multilayer Period Variations Required for Short-Wavelength (68 Å) Soft X-Ray Imaging Optics," *Appl. Opt.* **32**, 6961 (1993).

39. C.M. Falco and J.M. Slaughter, "Characterization of Metallic Multilayers for X-Ray Optics," *J. Magn. and Magn. Materials* **126**, 3 (1993); J.A.R. Ruffner, J.M. Slaughter, J. Eickmann and C.M. Falco, "Epitaxial-Growth and Surface-Structure of (0001)Be on (111)Si," *Appl. Phys. Lett.* **64**, 31 (1994).

40. S.V. Gaponov, S.A. Gusev, B.M. Luskin, N.N. Salashchenko and E.S. Gluskin, "Long-Wave X-Ray Radiation Mirrors," *Opt. Commun.* **38**, 7 (1981).

41. J.H. Underwood, E.M. Gullikson, and K. Nguyen "Tarnishing of Mo/Si Multilayer X-Ray Mirrors," *Appl. Opt.* **32**, 6985 (1993).

42. S. Bajt, H.N. Chapman, N. Nguyen *et al.*, "Design and Performance of Capping Layers for Extreme Ultraviolet Multilayer Mirrors," *Appl. Optics* **42**, 5750 (2003); S. Bajt, N.V. Edwards and T.E. Madey, "Properties of Ultrathin Films Appropriate for Optics Coating Layers in Extreme Ultraviolet Lithography (EUVL)," *Surface Sci. Rept.* **63**, 73 (2007).

43. P. Dhez, "Polarizers and Polarimeters in the X-UV Range," *Nucl. Instrum. Meth. A* **261**, 66 (1987).

44. J.B. Kortright and J.H. Underwood, "Multilayer Optical Elements for Generation and Analysis of Circularly Polarized X-Rays," *Nucl. Instrum. Meth. A* **291**, 272 (1990).

45. M. Yamamoto, K. Mayama, H. Kimura, Y. Gato and M. Yanagihara, "Thin Film Ellipsometry at a Photon Energy of 97 eV," *J. Electr. Spectrosc. Rel. Phenom.* **80**, 465 (1996).

46. A.L. Aquila, "Development of Extreme Ultraviolet and Soft X-ray Multilayer Optics for Scientific Studies with Femtosecond and Attosecond Sources", PhD Thesis, Applied Science and Technology Program, University of California, Berkeley, 2009.

47. J.B. Kortright, M. Rice, and R. Carr, "Soft X-Ray Faraday Rotation at Fe L2, 3 Edges," *Phys. Rev. B* **51**, 10240 (1995); M.J. Freiser, "A Survey of Magnetooptic Effects," *IEEE Trans. Magn.* **MAG-4**, 152 (1967); J.C. Suits, "Magnetooptical Properties," Chapter 9 in *Handbook of Magnetic Materials* (Reinhold, New York, 1968), P.A. Albertos and F.E. Luborsky, Editors.

48. D.L. Windt, S. Donguy, C.J. Hailey *et al.*, "W/SiC X-ray Multilayers Optimized for use Above 100 keV," *Appl. Optics* **42**(3), 2415 (May 1, 2003).

49. F.A. Harrison *et al.*, "The Nuclear Spectroscopic Telescope Array (NuStar) High-Energy X-ray Mission," *Astrophys. J.* **770**, 103 (June 20, 2013).

50. M. Fernández-Perea, M.-A. Descalle, R. Soufli *et al.*, "Physics of Reflective Optics for the Soft Gamma-Ray Photon Energy Range," *Phys. Rev. Lett.* **111**, 027404 (July 12, 2013).

51. N. F. Brejnholt, R. Soufli, M.-A. Descalle *et al.*, "Demonstration of Multilayer Reflective Optics at Photon Energies Above 0.6 MeV," *Optics Expr.* **22**(13), 15364 (June 30, 2014).

52. E. Ziegler, "Multilayers for High Heat Load Synchrotron Applications," *Opt. Engr.* **34**, 445 (1995).

53. H. Takenaka, T. Kawamura and Y. Ishii, "Heat Resistance of Mo/Si, $MoSi_2$/Si, and Mo_5Si_3/Si Multilayer Soft X-Ray Mirrors," *J. Appl. Phys.* **78**, 5227 (1995); H. Takenaka, H. Ito, T. Haga and T. Kawamura, "Design and Fabrication of High Heat-Resistant Mo/Si Multilayer Soft X-Ray Mirrors with Interleaved Barrier Layers," *J. Synchr. Rad.* **5**, 708 (1998).

54. R. Soufli, "Breakthroughs in Photonics 2013: X-ray Optics", *IEEE* **6**(2), 0700606 (April 2014); D.S. Kuznetsov, A.E. Yakshin, J.M. Sturm *et al.*, "High-Reflectance La/B-Based Multilayer Mirror for 6.X nm Wavelength," *Optics Lett.* **40** (16), 3778 (August 15, 2015).

55. D.H. Bilderback, "Review of Capillary X-ray Optics from the 2nd International Capillary Optics Meeting," *X-Ray Spectrum.* **32**, 195 (2003).

56. A. Erko and I. Zizak, "Hard X-ray Microspectroscopy at Berliner Elektronenspeicherring für Synchrotronstrahlung II," *Spectrochimica Acta B* **64**, 833 (2009); A.I. Erko, M. Idir, T. Christ and A.G. Michette, *Modern Developments in X-Ray and Neutron Optics* (Springer, Berlin, 2008).

57. S.B. Dabagov, "Channeling of Neutral Particles in Micro- and Nanocapillaries," *Uspekhi* **46**, 1053–1075 (October 2003); S.B. Dabagov, M.A. Kumakov and S.V. Nikitina, "On the Interference of X-rays in Multiple Reflection Optics," *Phys. Lett.* **A203**, 279 (1995).

58. C.A. MacDonald, *Introduction to X-ray Physics, Optics and Applications* (Princeton University Press, 2016); C.A. MacDonald and W.M. Gibson, "Applications and Advances in Polycapillary Optics," *X-Ray Spectrum.* **32**, 258 (2003).

59. A. Snigerev, A. Bjeoumikhov, A. Erko *et al.*, "Two-Step Hard X-ray Focusing Combined Fresnel Zone Plate and Single-Bounce Ellipsoidal Capillary," *J. Synchr. Rad.* **14**, 326 (2007).

60. G. Schneider, P. Guttmann, S. Rehbein, S. Werner and R. Follath, "Cryo X-ray Microscope with Flat Sample Geometry for Correlative Fluorescence and Nanoscale Tomographic Imaging," *J. Struct. Bio.* **177**, 212 (2012).

61. M. Born and E. Wolf, *Principles of Optics* (Cambridge University Press, New York, 1999), Seventh Edition. Diffraction is discussed in Chapter 8, Section 8.6.

62. E. Hecht, *Optics* (Addison-Wesley, Reading, MA, 2002), Fourth Edition. Zone plates are discussed in Section 10.3.5, p. 495; Section 6.3, p. 253, considers geometrical aberrations.

63. G.R. Morrison, "Diffractive X-Ray Optics," Chapter 8 in *X-Ray Science and Technology* (Inst. Phys., Bristol, 1993). Grating efficiencies are discussed in Section 8.2.2, p. 313.

64. A.G. Michette, *Optical Systems for Soft X-Rays* (Plenum, London, 1986).

65. Lord Rayleigh, "Wave Theory," p. 429 in *Encylopaedia Brittanica*, Ninth Edition, Vol. 24, (1988); Rayleigh's first entry in his notebook, describing the first successful demonstration, is dated 11 April 1871.

66. J.L. Soret, "Concerning Diffraction by Circular Gratings," *Ann. Phys. Chem.* **156**, 99 (1875).

67. R.W. Woods, *Physical Optics* (Macmillian, New York, 1911; Opt. Soc. Amer., Washington, DC, 1988). See comments regarding phase zone plates on pp. 37–39.

68. A.V. Baez, "A Study in Diffraction Microscopy with Special Reference to X-Rays," *J. Opt. Soc. Amer.* **42**, 756 (1952).

69. S J. Ojeda-Castañeda and C. Gómez-Reino, Editors, *Selected Papers on Zone Plates* (SPIE, Bellingham, WA, 1996).

70. G. Schmahl and D. Rudolph, "High Power Zone Plates as Image Forming Systems for Soft X-Rays" (in German), *Optik* **29**, 577 (1969); B. Nieman, D. Rudolph and G. Schmahl, "Soft X-Ray Imaging Zone Plates with Large Zone Numbers for Microscopic and Spectroscopic Applications," *Opt. Commun.* **12**, 160 (1974); G. Schmahl, D. Rudolph, P. Guttmann and O. Christ, "Zone Plates for X-ray Microscopy", p. 63 in *X-Ray Microscopy* (Springer-Verlag, Berlin, 1984), G. Schmahl and D. Rudolph, Editors; G. Schmahl, D. Rudolph, P. Guttmann and O. Christ, "Status of the Zone Plate Microscope," *SPIE* **316**, 100 (1982).

71. J. Kirz, "Phase Zone Plates for X-Rays and the Extreme UV," *J. Opt. Soc. Amer.* **64**, 301 (1974).

72. J.W. Goodman, *Introduction to Fourier Optics* (McGraw-Hill, New York, 1996), Second Edition, p. 124.

73. E. Anderson and D. Kern, "Nanofabrication of Zone Plates for X-Ray Microscopy," p. 75 in *X-Ray Microscopy III* (Springer-Verlag, Berlin, 1992).

74. W. Yun, B. Lai, Z. Cai *et al.*, "Nanometer Focusing of Hard X-rays by Phase Zone Plates," *Rev. Sci. Instrum.* **70**(5), 2238 (May 1999); R.P. Winarski *et al.*, "A Hard X-ray Beamline for Nanoscale Microscopy," *J. Synchr. Rad.* **19**, 1056 (2012).

75. C. David, S. Gorelick, S. Rutishauser *et al.*, "Nanofocusing of Hard X-ray Free Electron Laser Pulses Using Diamond Based Fresnel Zone Plates," *Scientific Reports* **1**, 57 (August 8, 2011).

76. S. Gorelick, J. Vila-Comamala, V.A. Guzenko *et al.*, "High-Efficiency Fresnel Zone Plates for Hard X-rays by 100 keV e-Beam Lithography and Electroplating," *J. Synchr. Rad.* **18**, 442 (May 2011).

77. R. Tatchyn, "Optimum Zone Plate Theory and Design," p. 40 in *X-Ray Microscopy* (Springer-Verlag, Berlin, 1984).

78. Y. Vladimirsky, "Zone Plates", p. 289 in *Vacuum Ultraviolet Spectroscopy I* (Academic Press, 1998), J.A. Samson and D.L. Ederer, Editors.

79. T. Schliebe and G. Schneider, "Zone Plates in Nickel and Germanium for High Resolution X-ray Microscopy", p. IV-3 in *X-Ray Microscopy and Spectromicroscopy* (Springer-Verlag, Berlin, 1998), J. Thieme, G. Schmahl, D. Rudolph and E. Umbach, Editors.

80. E. DiFabrizio, F. Romanato, M. Gentill *et al.*, "High-Efficiency Multilevel Zone Plates for keV X-rays," *Nature* **401**, 895 (October 28, 1999).

81. Y. Fang, M. Feser, A. Lyon *et al.*, "Nanofabrication of High Aspect Ratio 24 nm X-ray Zone Plates for X-ray Imaging Applications," *J. Vac. Sci. Technol. B* **25**(6), 2004 (November/December 2007).

82. S. Werner, S. Rebein, P. Guttmann and G. Schneider, "Three-Dimensional Structured On-Chip Stacked Zone Plates for Nanoscale X-ray Imaging with High Efficiency," *Nano Research* **7**(4), 528 (Tsinghua University Press, April 2014).

83. O. v. Hofsten, M. Bertilson, J. Reinspach *et al.*, "Sub-25-nm Laboratory X-ray Microscopy Using a Compound Fresnel Zone Plate," *Optics Lett.* **34**(17) (September 1, 2009); F. Uhlén, D. Nilsson, A. Holmberg *et al.*, "Damage Investigation on Tungsten and Diamond Diffractive Optics at a Hard X-ray Free-Electron Laser," *Optics Expr.* **21** (7), 8051 (April 8, 2013).

84. A.E. Sakdinawat, "Contrast and Resolution Enhancement Techniques for Soft X-ray Microscopy," PhD Thesis, Joint Bioengineering Program, University of California, Berkeley, and University of California, San Francisco, 2008.

85. C. Chang, A. Sakdinawat, P. Fischer, E. Anderson and D. Attwood, "Single-Element Objective Lens for Soft X-ray Differential Interference Contrast Microscopy," *Optics Lett.* **31**, 1564 (2006).

86. A. Sakdinawat and Y. Liu, "Soft-X-ray Microscopy Using Spiral Zone Plates," *Optics Letters* **32**, 2635–2637 (2007).

87. A. Sakdinawat and Y. Liu, "Phase Contrast Soft X-ray Microscopy Using Zernike Zone Plates," *Optics Express* **16**, 1559–1564 (2008).

88. M. Born and E. Wolf, *Optics*, see Reference 61.

89. J.W. Goodman, *Fourier Optics*, see Reference 72.

90. J.D. Jackson, *Classical Electrodynamics* (Wiley, 1999), Third Edition, Chapter 10, Sections 10.5 to 10.9.

91. E. Hecht, *Optics* (Addison-Wesley, 2002), Fourth Edition, Chapter 10, p. 443, and Section 10.3.5, p. 495.

92. G. Fowles, *Introduction to Modern Optics* (Dover, New York, 1975), Second Edition. Available in paperback. A compact book with good diagrams and good explanation. See Chapter 5 regarding diffraction.

93. A. Sommerfield, *Optics* (Academic Press, New York, 1964), of Lectures on Theoretical Physics, Vol. IV. See Chapter V on the theory of diffraction, particularly Sections 37 and 38 regarding the rigorous solutions to the problem of diffraction by a straightedge.

94. A.G. Michette, *Optical Systems for Soft X-Rays* (Plenum, London, 1986), p. 175.

95. J. Maser and G. Schmahl, "Coupled Wave Description of the Diffraction by Zone Plates with High Aspect Ratios," *Opt. Commun.* **89**, 355 (1992).

96. G. Schneider, "Zone Plates with High Efficiency in High Orders of Diffraction Described by Dynamical Theory," *Appl. Phys. Lett.* **71**, 2242 (1997).

97. G. Schneider and J. Maser, "Zone Plates as Imaging Optics in High Diffraction Orders Described by Coupled Wave Theory," p. IV-71 in *X-Ray Microscopy and Spectromicroscopy* (Springer-Verlag, Berlin, 1998), J. Thieme, G. Schmahl, D. Rudolph and E. Umbach, Editors.

98. H. Yan, J. Maser, A. Macrander *et al.*, "Takagi-Taupin Description of Dynamical Diffraction from Diffractive Optics with Large Numerical Aperture," *Phys. Rev. B* **76**, 115438 (2007).

99. Y.V. Kopylov, A.V. Popov and A.V. Vinogradov, "Application of the Parabolic Wave Equation to X-Ray Diffraction Optics," *Opt. Commun.* **118**, 619 (1995).

100. S. Rehbein, P. Guttmann, S. Werner and G. Schneider, "Characterization of the Resolving power and Contrast Transfer Function of a Transmission X-ray Microscope with Partially Coherent Illumination", *Opt. Express* **20**(6), 5830 (March 12, 2012); G. Schneider, P. Guttmann, S. Rehbein, S. Werner and R. Follath, "Cryo X-ray Microscope with Flat Sample Geometry for Correlative Fluorescence and Nanoscale Tomographic Imaging," *J. Struct. Biol.* **177**, 212 (2012).

101. N. Iskander (unpublished), using the computer code ZCALC written by E. Anderson.

102. L. Jochum and W. Meyer-Ilse, "Partially Coherent Image Formation with X-Ray Microscopes," *Appl. Optics* **34**, 4944 (1995).

103. D. Shaver, D. Flanders, N. Ceglio and H. Smith, "X-Ray Zone Plates Fabricated Using Electron-Beam and X-Ray Lithography," *J. Vac. Sci. Techn.* **16**, 1626 (1979).

104. N. Ceglio, "The Impact of Microfabrication Technology on X-Ray Optics," p. 210 in *Low Energy X-Ray Diagnostics* (Amer. Inst. Phys, New York, 1981), D. Attwood and B. Henke, Editors.

105. M.A. McCord and M.J. Rooks, "Electron Beam Lithography," Chapter 2, pp. 139–249, in *Handbook of Microlithography, Micromachining, and Microfabrication, Vol. 1: Microlithography* (SPIE, Bellingham, WA, 1997), P. Rai-Choudhury, Editor; G. Owen and J.R. Sheats, "Electron Beam Lithography Systems," pp. 367–401 in *Micro-Lithography: Science and Technology* (Marcel Dekker, New York, 1998), J.R. Sheats and B.W. Smith, Editors.

106. D. Kern, P. Coane, R. Acosta *et al.*, "Electron Beam Fabrication and Characterization of Fresnel Zone Plates for Soft X-Ray Microscopy," *SPIE* **447**, 204 (1984).

107. E.H. Anderson, "Specialized Electron Beam Nanolithography for EUV and X-ray Diffractive Optics," *IEEE J. Quant. Electr.* **42**(1) (January 2006).

108. J. Reinspach, M. Lindbloom, M. Bertilson, O. Von Hofsten and H.M. Hertz, "13 nm High Efficiency Nickel-Germanium Soft X-ray Zone Plates," *J. Vac. Sci. Techn. B* **29**(1), 011012 (January/February 2011).

109. C. Chang and A. Sakdinawat, "Ultra-High Aspect Ratio High-Resolution Nanofabrication for Hard X-ray Diffractive Optics," *Nature Commun.* **5**, 4243 (June 27, 2014).

110. S. Gorelick, J. Vila-Comamala, V.A. Guzenko *et al.*, "High-Efficiency Fresnel Zone Plates for Hard X-rays by 100 keV E-beam Lithograppy and Electroplating," *J. Synchrotron Rad.* **18**, 442 (2011); J. Vila-Comamala, S. Gorelick, E. Färm *et al.*, "Ultra-High Resolution Zone-Doubled Diffractive X-ray Optics for the Multi-keV Regime," *Optics Expr.* **19**(1), 175 (January 3, 2011); J. Vila-Comamala, K. Jefimovs, J. Raabe *et al.*, "Advanced Thin Film Technology for Ultrahigh Resolution X-ray Microscopy," *Ultramicroscopy* **109**, 1360 (2009); K. Jefimovs, J. Vila-Comamala, T. Pilvi *et al.*, "Zone-Doubling Technique to Produce Ultrahigh-Resolution X-ray optics," *Phys. Rev. Lett.* **99**, 264801 (December 31, 2007).

111. H. Yan, R. Conley, N. Bouet and Y.S. Chu, "Hard X-ray Nanofocusing by Multilayer Laue Lenses," *J. Phys. D.: Appl. Phys.* **47**, 263001 (2014).

112. H.C. Kang, H. Yan, R.P. Winarski *et al.*, "Focusing of Hard X-rays to 16 Nanometers with a Multilayer Laue Lens," *Appl. Phys. Lett.* **92**, 221114 (2008)

113. X. Huang, H. Yan, E. Nazaretski *et al.*, "11 nm Hard X-ray Focus from a Large-Aperture Multilayer Laue Lens," *Sci. Repts.* **3**, 3562 (December 20, 2013).

114. A.J. Morgan, M. Prasciolu, A. Andrejczuk *et al.*, "High Numerical Aperture Multilayer Laue Lenses," *Sci. Repts.* **5**, 09892 (June 1, 2015).

115. N. Kamijo, S. Tamura, Y. Suzuki *et al.*, "Fabrication of Hard X-ray Sputtered-Sliced Fresnel Zone Plate", p. IV-65 in *X-Ray Microscopy and Spectromicroscopy* (Springer-Verlag, Berlin, 1998), J. Thieme, G. Schmahl, D. Rudolph and E. Umbach, Editors.

116. A.I. Erko, M. Idir, T. Krist and A.G. Michette, *Modern Developments in X-Ray and Neutron Optics* (Springer, Berlin, 2008).

117. M. Born and E. Wolf, *Principles of Optics*, Section 8.6, p. 449 (Cambridge University Press, 1999), Seventh Edition.

118. G.W. Stroke, "Diffraction Gratings," pp. 473 in *Handbuch der Physik, XXIX, Optical Instruments*, S. Flügge, Editor.

119. E. Hecht, *Optics* (Addison-Wesley, 2002), Fourth Edition, p. 479.

120. J.H. Underwood, "X-Ray Optics," pp. 525–540 in *Encyclopedia of Applied Physics*, Volume 23 (Wiley, 1998).

121. M.C. Hettrick and S. Bowyer, "Variable Line-Space Gratings: New Designs for Use in Grazing Incidence Spectrometers," *Appl. Optics* **22**(24), 3921 (December 15, 1983).

122. M. Hettrick, J. Underwood, P. Batson and M. Eckart, "Resolving Power of 35,000 in the Extreme Ultraviolet Employing a Grazing Incidence Spectrometer," *Appl. Opt.* **27**, 200 (January 15, 1988).

123. M.C. Hettrick, "In-Focus Monochromator: Theory and Experiment of a New Grazing Incidence Mounting," *Appl. Optics* **29**(31), 4531 (November 1, 1990).

124. M.C. Hettrick and J.H. Underwood, "Varied-Space Grazing Incidence Gratings in High Resolution Scanning Spectrometers", p. 237 in *Short Wavelength Coherent Radiation: Generation and Applications* (Amer. Instit. Phys. Conf. Proc. 147, 1986), D.T. Attwood and J. Bokor, Editors.

125. J.H. Underwood and J.A. Koch, "High-Resolution Tunable Spectrograph for X-ray Laser Linewidth Measurements with a Plane Varied-Line Spacing Grating," *Appl. Optics* **36**(21), 4913 (July 20, 1997).

126. J.H. Underwood, "Spectographs and Monochromators Using Varied Line Spacing Gratings" pp. 55–72, in *Vacuum UV Spectroscopy II* (Academic Press, 1998), J.A. Samson and D.L. Ederer, Editors.

127. T. Harada and T. Kita, "Mechanically Ruled Aberration Corrected Concave Gratings," *Appl. Optics* **19**, 3987 (1980).

128. K. Amemiya, Y. Kitajima, T. Ohta and K. Ito, "Design of a Holographically Recorded Plane Grating with a Varied Line Spacing for a Soft X-ray Grazing Incidence Monochromator," *J. Synchr. Rad.* **3**, 287 (1996).

129. J.H. Underwood, E.M. Gullikson, M. Koike and S. Mrowka, "Experimental Comparison of Mechanically Ruled and Holographically Recorded Varied-Line Spacing Gratings," *SPIE* **3150**, 40 (1997).

130. J. Als-Neilson and D. McMorrow, *Elements of Modern X-Ray Physics* (Wiley, 2011), Second Edition, Chapter 6.

131. Y. Shvyd'ko, *X-Ray Optics* (Springer, Berlin, 2004).

132. R. W. James, *The Optical Properties of the Diffraction of X-Rays* (Bell, 1950).

133. B.W. Batterman and D.H. Bilderback, pp. 105–151, "X-Ray Monochromators and Mirrors," in *Handbook on Synchrotron Radiation*, Volume 3 (Elsevier Science, 1991), G. Brown and D. E. Moncton, Editors.

134. T. Matsushita and H. Hashizume, "X-Ray Monochromators," pp. 261–314 in *Handbook of Synchrotron Radiation*, Volume 1b (North Holland, Amsterdam, 1983), E.E. Koch, Editor.

135. B.D. Cullity and S.R. Stock, *Elements of X-Ray Diffraction* (Pearson, 2001), Third Edition; R. Gronsky, MSE/UC Berkeley, private communication.

136. B.E. Warren, *X-Ray Diffraction* (Dover, 1990); D.E. Sands, *Introduction to Crystallography* (Dover, 2014).

137. D.M. Mills, "X-Ray Optics for Third-Generation Synchrotron Radiation Sources," in *Third-Generation Hard X-Ray Synchrotron Radiation Sources: Source Properties, Optics, and Experimental Techniques* (Wiley, 2002), D.M. Mills, Editor.
138. F.K. Richtmyer, E.H. Kennard and T. Lauritsen, *Introduction to Modern Physics* (McGraw-Hill, NY, 1955), Fifth Edition, Chapter 8.
139. E. Hecht, *Optics*, Reference 62, p. 158.
140. A Snigerev, V. Kohn, I. Snigireva and B. Lengeler, "A Compound Refractive Lens for Focusing High-Energy X-Rays," *Nature* **384**, 49 (November 7, 1996).
141. A Snigerev, V. Kohn, I. Snigireva, A Souvorov and B. Lengeler, "Focusing High-Energy X-Rays by Compound Refractive Lenses," *Appl. Optics* **37**(4), 653 (February 1, 1998).
142. B. Lengeler, J. Tümmler, I. Snigireva, A Souvorov and C. Raven, "Transmission and Gain of Singly and Doubly Focusing Refractive X-Ray Lenses," *J. Appl. Phys.* **84**(11), 5855 (December 1, 1998).
143. B. Lengeler, C.G. Schroer, B. Benner *et al.*, "Parabolic Refractive X-Ray Lenses: A Breakthrough in X-Ray Optics," *Nucl. Instrum. Meth. A*, **467–468**, 944 (2001).
144. B. Lengeler, C.G. Schroer, M. Kuhlmann *et al.*, "Refractive X-Ray Lenses," *J. Phys. D: Appl. Phys.* **38**, A218 (2005).
145. A. Snigerev, I. Snigereva, M. Grigoriev *et al.*, "High Energy X-ray Nanofocusing by Silicon Planar Lenses," *J. Phys.: Conf. Series* **186**, 012072 (2009).
146. M. Chollet, R. Alonso-Mori, M. Cammarata *et al.*, "The X-ray Pump-Probe Instrument at the Linac Coherent Light Source," *J. Synchrotron Rad.* **22**, 503 (2015).
147. H. Simons, A. King, W. Ludwig *et al.*, "Dark-Field X-ray Microscopy for Multiscale Structural Characterization," *Nature Commun.* **6**, 6098 (January 14, 2015).

Homework Problems

Homework problems for each chapter will be found at the website: www.cambridge.org/xrayeuv

11 X-Ray and EUV Imaging

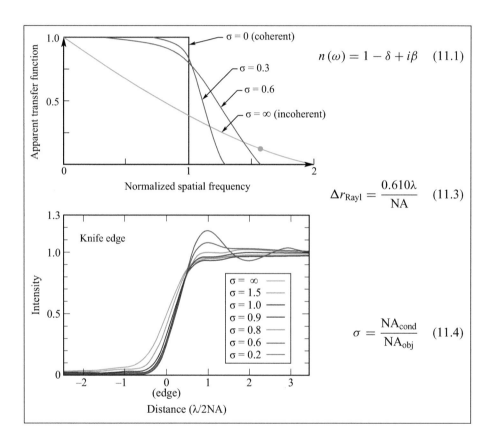

$$n(\omega) = 1 - \delta + i\beta \quad (11.1)$$

$$\Delta r_{\mathrm{Rayl}} = \frac{0.610\lambda}{\mathrm{NA}} \quad (11.3)$$

$$\sigma = \frac{\mathrm{NA_{cond}}}{\mathrm{NA_{obj}}} \quad (11.4)$$

11.1   Introduction

Among the first features of x-rays appreciated in the days and months following their discovery by Röntgen[1, 2], was their value for imaging otherwise hidden objects, most famously the first image of Frau Röntgen's hand, then quickly followed in the weeks and months ahead in laboratories around the world.[3, 4] This simple shadow casting remains one of the most powerful uses of x-rays, seen on a daily basis in hospitals and medical clinics, dental offices, and airport security stations. Today x-rays are used to determine structural properties of samples at a wide range of length scales, from the atomic level in crystallography to the macroscale level in medical imaging and industrial non-destructive testing. Whether on a macroscale or an atomic scale, image formation must have certain qualities to be useful. The primary role of an image is to replicate the location, amplitude, and spatial features of content within the original object. To do this x-rays need to be sufficiently energetic to *penetrate the material* or objects of interest, such as the set of metallic weights within Röntgen's wooden box, the bones within our bodies, or some nanoscale structures of interest. Additionally there needs to be sufficient *contrast* between objects of interest and their surroundings, and sufficient *resolution* to discern features of interest. This generally requires harder x-rays for macroscopic or high-Z materials, and softer x-rays for nanoscale structures. These features are considered in the following paragraphs. Whether x-ray optics are required or not, in a given situation, will depend on the complementary roles of radiation collection and quality image formation. In scientific studies the use of x-ray imaging is often accompanied by the use of complementary techniques to form a more complete understanding of dominant features or processes. These would include use of other wavelengths, such as infrared or visible light, and other techniques such as scattering, fluorescence, or scanning tip probes.

The manner by which x-rays and EUV radiation interact with matter is crucial to image formation. As we have seen in earlier chapters, x-rays interact with individual atoms by both absorption and scattering, leading to the emission of photoelectrons, Auger electrons, fluorescence emission (Chapter 1, Figures 1.2 and 1.4), and redirected (scattered) photons of unshifted (Chapter 2) and shifted energy (Chapter 8). The collective interaction of x-rays with material containing many atoms is described by the refractive index,

$$n(\omega) = 1 - \delta + i\beta \qquad (11.1)$$

which has both refractive (δ) and absorptive (β) components, as described earlier in Chapter 3, Eq. (3.12), and reproduced here for convenience as Eq. (11.1). The real part, δ, describes changes in the wave's phase velocity [Eqs. (3.8)–(3.17)] due to imposed oscillations of free and bound electrons, which lead to wavefront phase variations as the x-rays propagate through the material. In a relatively uniform material this can be observed as interference effects in the forward direction, as in interferometry. In transversely non-uniform materials this leads to refraction and diffraction, depending on transverse scale lengths. These can all affect observables in an image, as we will see in the following paragraphs. The spatial mapping of these interactions, or wavefront perturbations, determines the content of an image, what we see in the image, and what we can learn from that. The easiest form of image contrast to understand is caused by two-dimensional variations of absorption within the object due to positional differences of density and elemental content. At each point the intensity after transmission is given by Eq. (1.3a), renumbered here as Eq. (11.2)

$$\frac{I}{I_0} = e^{-\rho\mu x} \qquad (11.2)$$

where ρ is the mass density, x is the propagation distance, $\mu = \mu(E, Z)$ is the photon energy and element dependent absorption coefficient, $E = \hbar\omega$ is the photon energy, and Z is the atomic number. Values of μ as a function of photon energy are given in Appendix C for several elements. Values for all naturally occurring elements from $Z = 1$ to 92 are given in the literature.[5] Recall from Chapter 3, Section 3.2, that the absorption coefficient μ, the imaginary part of the refractive index β, the absorption cross-section σ_{abs}, and the absorptive portion of the atomic scattering factor $f_2^0(\hbar\omega)$ are all linearly related. Owing to bond-related atomic energy level shifts in molecules, typically affecting p-shell electrons, absorption spectroscopy can display molecular features within the broad elemental signature. This provides a valuable spectroscopic tool, for example in the environmental and cultural heritage sciences. Using techniques of spectromicroscopy, which combine nanoscale x-ray imaging with detailed spectroscopic scans, it is possible to identify not only which elements are present in a sample (O, N, Fe, . . .) but also chemical bonding information, for instance, in terms of oxidation states such as Fe^{II} or Fe^{III}, which can then help to identify specific molecules which might be present through comparison with reference spectral signatures. Examples are given in Sections 11.4 and 11.5. Because transition rates in magnetic materials, such as Fe, Co, and other transition metals, have spin related angular momentum requirements, it is also possible to observe nanoscale magnetic orientations in those materials using soft x-ray absorption microscopy.[6]

Phase shifts due to localized variations in refractive index can contribute strongly to x-ray image formation. For example, refraction or diffraction from local gradients or sharp edges will generate an angular spread of the local wavefront, even with very minimal coherence of just a wave or two. This leads to localized interference effects, causing local intensity variations which have the effect of enhancing visibility in these areas of the image.

In addition to absorption and phase effects, *fluorescence* also offers a powerful tool for localizing specific elements within an image. See Chapter 1, Figure 1.2(c). One example of this is the use of a hard x-ray nanoprobe, as was illustrated in Chapter 10, Figure 10.5. In this technique a focused beam of hard x-rays is scanned across a sample, knocking out K-shell electrons from the various atoms present, which then relax to lower energy level orbitals, typically in a 2p–1s transition, emitting photons of characteristic energy for those elements present. Spectrometry of the detected photon energies and their intensities is then electronically combined with knowledge of scanning positions to form elemental maps of the atoms present. As there is no background x-ray noise (no bremsstrahlung) these measurements can be done to parts per billion (ppb) and femtogram accuracies. The technique is widely used in materials characterization studies. *Spectromicroscopy* is the combination of high spatial resolution imaging with high spectral resolution at each point, for example using near edge x-ray absorption fine structure (NEXAFS), which provides additional information on elemental oxidation states and localized molecular structures.

X-ray *photoemission*, as illustrated in Chapter 1, Figures 1.2(b, d), offers another means of determining elemental and chemical distributions based on x-ray illumination. Because the range of electrons is generally very short in all materials (see Chapter 1,

Figure 1.5) the technique is widely used in the surface sciences to determine atomic and molecular profiles within an escape depth of order 1 nm. Atomic structure near the surface can be deduced by observing photoemission angular variations due to the interference of electron wave functions radiated by neighboring atoms, sometimes referred to as electron holography (see Chapter 1, reference 12). As an example of photoemission spectromicroscopy, measurements of Si 2p energy levels were used to reveal the distributions of Si, SiO_2, and $TiSi_2$ at the surface of a specially prepared computer chip.[7] Angle resolved photoemission spectroscopy, with tunable incident x-ray energies is another major tool, for example to study solid state effects such as superconductivity in materials where correlated electron motions are believed to significantly affect transport properties.[8] Recently this has been extended to "NanoARPES," which combines a nanoscale illumination spot with angle resolved photoemission techniques.[9]

Image formation based on x-ray *diffraction* from atomic planes, often referred to as x-ray crystallography,[10–14] has been a primary scientific tool for understanding atomic scale structures for the biological and materials sciences since shortly after the work of von Laue, Ewald, Darwin, and the Braggs from 1912 through the early 1930s.[14, 15] Indeed, it was x-ray crystallographic data that led to understanding the structure and function of DNA[16] For the crystallographic technique well-collimated x-rays are incident upon a sample that is rotated in front of the beam. Multi-order diffraction patterns, each corresponding to satisfaction of the Bragg equation for the apparent spacings, are recorded at each orientation, as was illustrated earlier in Chapter 10, Figure 10.39. The collection of diffraction spot angular locations and intensities are then numerically analyzed to form a computed three-dimensional structure.[10] Today the pharmaceutical industry uses protein crystallography as a major tool for drug discovery, employing undulator radiation at synchrotron facilities worldwide. The technique of femtosecond nanocrystallography has recently emerged as a method suitable for use with crystallized material too small for characterization at synchrotron facilities. This technique, known as "diffract and destroy"[17] employs a stream of identical nanocrystals, irradiated sequentially by intense femtosecond x-ray FEL pulses, destroying each in the process but capturing pre-destruction diffracted x-rays for subsequent analysis.[18] An early example of the technique is illustrated in Chapter 6, Figure 6.17.

Scattering of x-rays is fundamental to image formation. Small structures scatter to large angles, and large structures to small angles, which we can understand in terms of Fourier transforms of density structures. As a diagnostic technique, scattering offers the great advantage that it can "see" smaller features as it is not bound to the angular collection confines of a lens's numerical aperture. In some situations, because of geometry or repetitive atomic planes, individual scatterings evolve to a collective "diffraction". But for non-crystalline objects with varied scale size substructures, generally amorphous in nature, the scatterings reveal only distributions of size, and to some extent elemental and chemical content. Image formation for such arbitrary material requires collection of the x-rays, knowledge of the angular distributions, and knowledge of the phase associated with each angle as the integrated phase gives the distance from the scattering feature to the image plane position. The primary method for image formation is use of an optic, such as a zone plate lens, a reflective optical system such as a Kirkpatrick–Baez mirror pair, or perhaps a capillary optic, all of which were described in Chapter 10. In each of these cases the optic forms an image by bringing together in the image plane all rays of equal path length (accumulated phase), within the resolution, depth of field, and efficiency of the optic. Use of *coherent scattering* offers additional possibilities. For example, using spatially filtered undulator radiation, spatially coherent laser high harmonic radiation, or free electron laser radiation (Chapters 5, 6, and 7, respectively), it is possible to collect the x-ray scatterings and numerically reconstruct an image without the use of an optic. With a CCD electronic array detector, the angular content is known from locations on the CCD. But the phase at each point is not known. An emerging technique, known as *Coherent Diffraction Imaging* (CDI),[19, 20] attempts to numerically determine the phase at each point by sequentially and iteratively calculating Fourier transforms between the detector and object planes, back and forth, applying known constraints in each plane (data at the CCD, lateral extent in the object plane), beginning the process with random assumptions of phase at the CCD, and continuing the iterations for thousands of cycles. The technique is relatively new for x-rays as it requires spatially coherent radiation, as well as large computational resources. It offers the potential advantage of avoiding optical inefficiencies, numerical aperture and depth of field limitations, and aberrations associated with an imaging optic or optical system. In addition to requiring a high degree of spatial coherence, CDI also requires well-defined object plane constraints, and numerical techniques that provide convergence to an accurate, asymptotic solution. The technique and early results are discussed further in Section. 11.6.

A related technique, also involving coherent illumination and scattering/diffraction is known as *ptychography*.[21] This technique is related to scanning transmission x-ray microscopy (see Section 11.4, Figure 11.14) in that the sample is raster scanned in front of a spatially coherent x-ray beam. In this case, however, the beam is not necessarily nanoscale, and the resultant spatial resolution is not set by the beam size. Rather the scattered radiation is collected with a two-dimensional detector, much as in diffraction, and reconstructed from distributed angular content. With a beam size larger than the scanning stage steps, there is much redundancy in the collected data, permitting image reconstruction to a significantly improved spatial resolution, as has recently been demonstrated.[22] As with the CDI technique above, ptychography requires a high degree of coherence and considerable computational resources. It also requires a lengthy data collection time, due both to the reduced photon flux after spatial filtering, and requirements associated with partially redundant scanning of sample areas. This technique is also described further in Section 11.6.

11.2   Spatial Resolution and Contrast

Two important qualities of an image are its spatial frequency content, which is related to *spatial resolution*, and *contrast* as a function of spatial frequency. Spatial resolution, which provides a measure of how small a feature we can accurately characterize, has been discussed to some degree in Chapter 10, Section 10.6.3, with regard to zone plate microscopes with varying illumination. Here we further consider the question "what

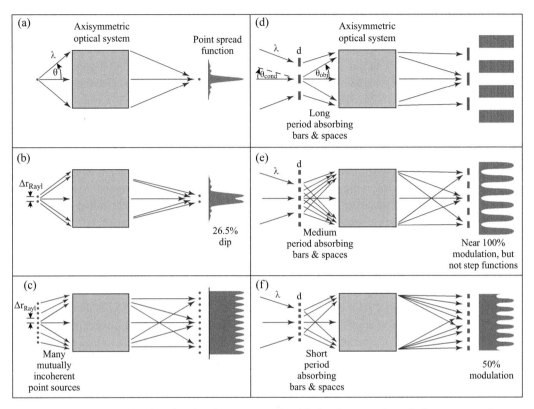

Figure 11.1 Sketches showing the response of an ideal, axisymmetric optical system to various sources of radiation emitted from the object plane and projected to a conjugate image plane. In (a), (b), and (c) the emission is from mutually incoherent point sources. In (d), (e), and (f) there are patterns of 1:1 absorbing bars and spaces of varying spatial period, illuminated by convergent radiation. The responses show image plane intensity modulations that vary with the spatial frequency of the object plane patterns, with high modulation for long spatial periods but declining modulations for spatial periods approaching the response function of the imaging system.

are the finest features that can be resolved when closely spaced?". Contrast relates to how well we are able to discern features within an image, to what degree they are visible in the presence of local background and nearby features? To that end we discuss modulation transfer* properties of an imaging system as a function of spatial frequency. Figure 11.1 sketches the response of an optical system, perhaps a single lens, to various sources of radiation at its object plane. In panel (a) there is a single point source, in (b) there are two mutually incoherent point sources, and in (c) there is a set of side-by-side, mutually incoherent point sources. In each case the radiation is captured by the optical system which then projects it back to form an image in the conjugate plane.

---

* Modulation is defined as $M = [(I_{max} - I_{min})/2]/I_{ave}$, where $I_{ave} = (I_{max} + I_{min})/2$. Thus $M = (I_{max} - I_{min})/(I_{max} + I_{min})$. Alternative names are contrast, C, and visibility, V. (See O. Hecht, Chapter 10, Reference 62, Eqs. (11.89) and (12.1).)

The transmitted and scattered radiation is captured by the optical system and respective responses are projected back to the conjugate image plane. The image plane intensity pattern in (a) is referred to as the "point response function" of the optical system. It is an Airy pattern as was calculated in Section 10.6.2 and illustrated in Figure 10.28. The response in (b), for two just resolved, mutually incoherent point sources, was seen earlier in Figure 10.29. This is the classic case described by Rayleigh in which the null of one intensity pattern overlays the peak of the other, resulting in a 26.5% central intensity dip. Note that the Rayleigh case applies to two mutually incoherent point sources so that the electric fields are uncorrelated and it is the intensities that add. Shown in panel (c) is a series of mutually incoherent point sources, each separated by the Rayleigh resolution Δr_{Rayl}, thus reinforcing our earlier observation that the Rayleigh resolution corresponds to a period. Panels (d), (e), and (f) show a set of side by side absorbing rectangular lines and spaces of equidistant (1:1) widths, illuminated by convergent radiation. In (d) the pattern of lines and spaces is of sufficiently low spatial frequency that the optical system produces an image plane intensity pattern of rectangular shapes with near 100% modulation. In (e) the pattern is of an intermediate spatial frequency which the optical system largely replicates, but with somewhat reduced modulation and rounded corners. In panel (f) the spatial frequency of the lines and spaces is higher, approaching the optical system's response function such that the resultant image modulation is reduced to 50%. For patterns of yet shorter spatial frequency the image plane modulation would fall rapidly to zero. In this limit the diffracted radiation is simply not being captured by the aperture of the optical system. With convergent illumination, seen in panels (d) through (f), the system's ability to accurately image structures is improved.

As we saw in Chapter 10, Section 10.6.3, for an axisymmetric optical system, when two Airy patterns of equal intensity are overlaid with the peak of each overlaying the first null of the other, there is a central dip of 26.5% ; see Figures 10.29 and Figure 11.1(b). Assuming the intensity pattern is of a minimal noise level, and that there is a low and relatively uniform background, this provides a good criterion, known widely as the Rayleigh criterion, for "resolving" the two objects within the image. Closer objects would have less of a central dip. Inequality of the two point source intensities, or minimal photon count with finite signal-to-noise ratio (S/N), or a non-uniform background would compromise the resolving capability. From the analysis of Chapter 10 we learned that the Rayleigh criterion, given in Eq. (10.42) and renumbered here as Eq. (11.3), can be expressed as

$$\Delta r_{\text{Rayl}} = \frac{0.610\lambda}{\text{NA}} \tag{11.3}$$

where λ is the wavelength and $\text{NA} = \sin\theta$ is the numerical aperture of the objective lens. Note that the Rayleigh criteria corresponds to point source radiators, which are mutually incoherent so that it is their intensities that add, not their fields, resulting in a peak-to-peak period in the image plane, as can be seen most clearly in Figures 11.1(b) and (c). Although it is common when characterizing an optical system with 1:1 line/space ratio test patterns to quote the half-period, the spatial resolution is properly quoted as

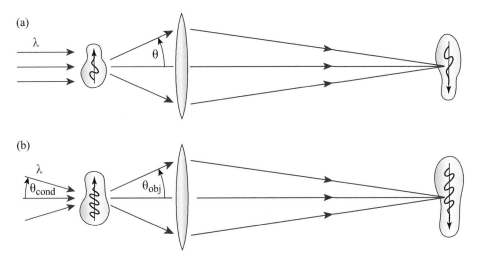

Figure 11.2 (a) Spatially coherent radiation illuminates and scatters from an object. The optical system, represented here by a single lens, then collects the radiation and projects it to form an image in the conjugate plane. With partially coherent illumination, shown in (b) as convergent radiation from a source of finite size (not shown), it is possible for portions of the incident radiation to scatter through larger angles and still be captured by the optical system, thus contributing to the image. Larger scattering angles are caused by smaller objects, thus with partially coherent illumination it is possible to see smaller objects.

a period. Often it is convenient to characterize focal spot resolution using a scanning knife-edge technique. A rough comparison of the knife edge focal spot test with the Rayleigh double-Airy pattern was obtained earlier, as seen in Figure 10.31, where it was observed that a 10%–90% knife edge intensity criterion is approximately equivalent to the Rayleigh resolution (full period), whereas a 25%–75% value is approximately equivalent to half-Rayleigh.

While the Rayleigh criterion is a very valuable tool, applicable in many situations, a description accounting for illumination and spatial frequency content can provide further insights. Figure 11.2 introduces the subject of illumination and how it can affect achievable spatial resolution. Spatially coherent illumination is illustrated in Figure 11.2(a) and partially coherent illumination in Figure 11.2(b).

The ability to print fine, high contrast features can be significantly affected by the degree of coherence within the optical system. If there exists a localized degree of spatial coherence, diffraction from adjacent features in the object plane will interfere in the image plane, significantly modifying the recorded patterns. With electric fields adding, even within a small region of coherence, ringing will be observed near sharp features and as a result contrast can vary dramatically for spatial structures that contribute to constructive or destructive interference. The parameter used to characterize the degree of partial coherence[23] in an optical system is

$$\sigma = \frac{\mathrm{NA_{cond}}}{\mathrm{NA_{obj}}} \qquad (11.4)$$

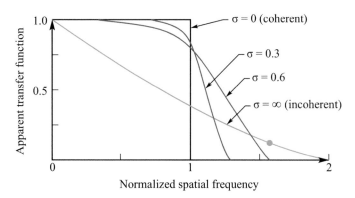

Figure 11.3 The effective modulation transfer function of a perfect imaging system, as a function of normalized spatial frequency for varying degrees of partial coherence. The partial coherence parameter $\sigma = 0$ corresponds to coherent illumination. Fractional values of σ correspond to partial coherence, and values ≥ 1 are effectively incoherent.[23] Courtesy of A. Neureuther, UC Berkeley. A dot has been added to the $\sigma = \infty$ curve corresponding to a Rayleigh modulation of 0.13. This occurs at a normalized spatial frequency of approximately 1.6.

where $NA_{cond} = \sin \theta_{cond}$ is the illumination numerical aperture and $NA_{obj} = \sin \theta_{obj}$ is the collection numerical aperture of the optical system. Both angles are shown in Figure 11.2(b). The coherent limit corresponds to $\sigma = 0$, as occurs for instance with a uniform plane wave illumination of $NA_{cond} = 0$, as seen in part (a) of the figure. The incoherent limit corresponds to an illumination cone equal to or larger than the collection NA of the imaging optic. For $\sigma \geq 1$ the illumination is effectively incoherent. The advantages of partially incoherent illumination were discussed in Chapter 10, Section 10.6.5, Figures 10.33 and 10.34, and are illustrated here in Figure 11.2(b). Basically radiation enters with some angular spread, thus permitting scattering of some portion of the radiation through larger angles, yet still collected for image formation. In principle, with angles of incidence as large as the collection angle, scattering through twice the angle of the coherent case can be realized, corresponding to resolved spatial features with half the size, albeit with vanishing contrast as the limit is approached.

In order to form an accurate image, the optical system must be able to transfer all relevant spatial frequencies with fidelity to the image plane, up to its spatial frequency (resolution) limit. However, it is difficult to compare the general transfer properties of an optical system for both coherent and incoherent illumination due to possible interference effects. In the incoherent limit, diffraction from various features is imaged independently, and the process is accurately described by a linear *modulation transfer function* (MTF). In the coherent limit, electric fields interfere and the transfer function is non-linear, depending on detailed aspects of the object to be imaged. Nonetheless, there is value in presenting an *apparent transfer function*,[23] as in Figure 11.3, that attempts to describe the resultant contrast achievable with patterns of various spatial frequencies, as a function of the partial coherence factor σ, understanding that low σ curves are object dependent. In the coherent limit ($\sigma = 0$), the transfer function is flat out to

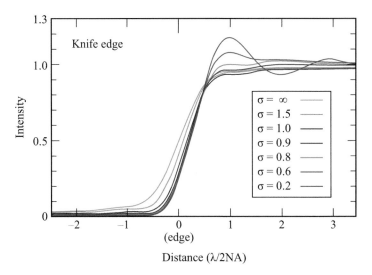

Figure 11.4 One-dimensional calculations of intensity versus distance, in normalized units of $\lambda/(2\,\mathrm{NA})$, across the image of a sharp knife edge as a function of the partial coherence parameter, σ. Courtesy of M. O'Toole and A. Neureuther,[24] University of California, Berkeley.

a very sharp cutoff at a normalized spatial frequency of one; incident and diffracted wavefronts are both well defined, and the wave is either fully captured by the lens or not at all. In the incoherent limit ($\sigma \geq 1$), there is incident radiation at a wide variety of incidence angles, which when diffracted are captured to varying degrees, out to twice the spatial frequency of the coherent case. Transfer of modulation out to higher spatial frequencies in the incoherent case is due to the large angle of incidence, θ_{cond}, which permits radiation to be collected after scattering through twice the angle available in the coherent limit. Diagrams illustrating these points with diffraction gratings were discussed in Chapter 10, Section 10.6.5 with regard to resolution in x-ray microscopes. Modulation transfer functions are shown in Figure 11.3 for several values of the partial coherence parameter, σ. For increasingly larger angles of incidence σ grows towards unity, extending the transfer capability to larger spatial frequencies, albeit with ever diminishing efficiency. This extends to a normalized spatial frequency factor of two, but at zero transferred intensity in that limit. This corresponds to the incoherent limit where $\mathrm{NA}_{\mathrm{cond}} = \mathrm{NA}_{\mathrm{obj}}$, $\sigma = 1$, shown effectively as $\sigma \to \infty$ in Figure 11.3.

Figure 11.4 shows a numerical simulation of the image across a sharp knife edge for varying degrees of partial coherence, as calculated by O'Toole and Neureuther.[24] Notice the substantial interference effects (ringing) for cases with a high degree of spatial coherence (low σ), and the rather smoother, but somewhat broader profiles for $\sigma \geq 1$. Optimal values of σ are object dependent. For this knife edge $\sigma \cong 0.9$ appears to provide sharpest spatial resolution.

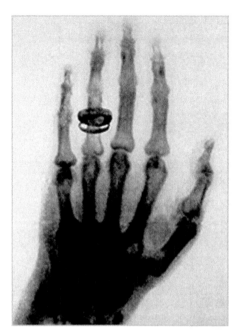

Figure 11.5 An x-ray image recorded by Röntgen on January 23, 1896 following his first public lecture in which he described his new discoveries at the Würzburg Physical-Medical Society. The hand is that of A. von Kölliker. Within weeks high-resolution images were recorded around the world, including some for medical diagnoses.[3, 4]

11.3   Projection Imaging

The first x-ray images were recorded by Wilhem Conrad Röntgen in his University of Würzburg laboratory on December 22, 1895, only a few weeks after his initial observation on November 8, 1895 of the penetrating properties of what he called "x-rays".[1, 2] The very first image, somewhat blurred, was of Frau Röntgen. The second, that same day, also of Frau Röntgen, was of higher quality. The source of x-rays was electron impact at the anode of a simple electrical discharge tube operated at sufficiently high voltage. The tube was surrounded by thick black paper. X-rays were initially observed with a fluorescent screen. Photographic plates were then used to record images of objects in close proximity, such as reference metal weights within an enclosing wooden box, a metal enclosed magnetic compass, and images of various side-by-side objects that showed transmittance to be a decreasing function of atomic weight and material thickness. Observations were also made of fluorescence from various materials. The new rays were not affected by magnetic fields, and it appeared at that time that they could not be refracted or reflected.[1, 2] Following his first public lecture on January 23, 1896 he recorded the x-ray image of a hand,[3] shown here in Figure 11.5. Within weeks very high quality x-ray images were recorded in laboratories around the world, and soon after in medical clinics.[3, 4] This near proximity technique with contrast based on simple absorption [Chapter 1, Section 1.2, Eq. (1.3a)] remains the most widely used type of x-ray image recording today, with common applications to medical imaging, dental imaging, and non-destructive imaging in widely varying industrial applications. The radiation source is commonly an x-ray tube producing a mixture of bremsstrahlung and line radiation from a stationary anode, or a rotating anode or a microfocus metal jet anode for higher x-ray flux. Fluorescent scintillators and film are still widely used, but electronic area detectors such as charge-coupled-devices (CCDs) and complementary-metal–oxide semiconductor (CMOS) detectors are ever more widely used, sometimes in combination with a scintillator for hard x-rays.

11.3.1 Tomographic Imaging

Today it is possible to provide high resolution x-ray images based on slices of three-dimensional (3D) data sets obtained using computed tomography (CT) techniques. The CT technique, invented by Godrey Hounsfield in 1972,[25, 26] involves taking a large number of two-dimensional (2D) x-ray images at various observation angles, typically at angular intervals of $1°$ or less, covering the full $180°$ (or nearly so). The observed parameter is the x-ray absorption coefficient, μ. The 2D images (projections), one at each observation angle, are Fourier transformed using the *Fourier slice theorem*, to a spatial frequency distribution along the respective observation angles. The modified Fourier transform used is known as a *Radon transform*. One method to reconstruct the image is use of the *filtered back projection method*. In this method, the data are first "filtered" by weighting the data in the Fourier domain. Because the Fourier slice theorem produces a transform of the object along a single line for each projection, the resulting data in the Fourier plane are more densely sampled in the lower frequencies than in the higher frequencies. To compensate, a weighting or "filtering" of the data is performed, which involves multiplying each slice or line of the Fourier transform by an amount proportional to the area needed to properly fill Fourier space. This can be visualized as a wedge-shaped weighting function, as shown by the dashed green lines in Figure 11.6(b). After the data are filtered (weighted), a "back projection" is performed by inverse Fourier transforming each modified slice back to real space. The results are then summed in real space to obtain a full 3D data set describing the object. A pixel-by-pixel interpolation among nearby back projections is used to smooth the 3D data set. From this 3D real space data set, 2D slice images can be visualized through any plane in real space. For small-scale objects the sample is normally rotated in front of the x-ray beam, while for large-scale objects, such as clinical imaging of a human patient, a gantry rotates the source-detector path. At hard x-ray synchrotron facilities, with short wavelengths and small angular divergence, diffractive and refractive blurring is minimal, permitting the acquisition of x-ray images with spatial resolutions of order one to several microns without the need for optics. Nanoscale soft x-ray tomography, where diffractive effects are stronger, does require optics, as discussed in Section 11.5.2.

Although filtered back projection is the most widely used method for reconstruction, it is sometimes insufficient for use in imperfect imaging conditions, such as

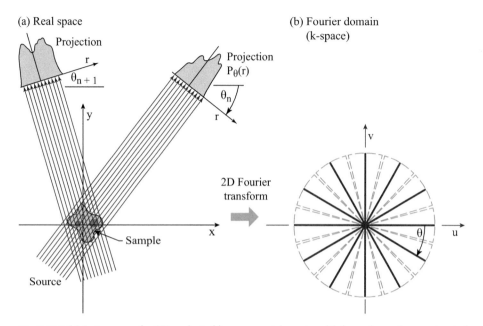

Figure 11.6 (a) In tomography 2D projected images are taken at multiple angles and reconstructed using Fourier transform techniques to form a 3D, real space data set from which arbitrary 2D slice images can be obtained. Ignoring diffraction, line integrals of attenuation through the object represent slices through the origin in Fourier space, shown as purple lines in (b). Using the *Fourier slice theorem* each of these angles is identified with the corresponding projection at that same angle in real space. Image reconstruction using filtered back projection or other iterative reconstruction methods produce a 3D image in real space. The ability to accurately reconstruct a 3D data set is dependent on how well one can sample and obtain sufficient information to fill Fourier space. To improve the image reconstruction process, before back projecting to real space, Fourier space is filled by broadening ("filtering") each of the narrow (purple) lines. This filling of Fourier space is done by weighting each narrow line with a pie shaped wedge, as shown by the dashed green lines in (b). The filtered k-space data for each wedge is then Fourier transformed back to real space. Image reconstruction involves summing in real space these Fourier transformed back projections.

images containing a significant amount of noise. This results in low signal to noise data which affects the accuracy of the reconstruction process. In such cases, iterative algorithms such as those based upon maximum likelihood of algebraic reconstruction are utilized.[27–29]

CT images can be used in combination with another imaging modality to provide "multi-modality" imaging where several types of information can be gathered and correlated. The first multi-modality imaging method combined CT and single photon emission computed tomography (SPECT), CT used to provide structural information, and SPECT used to provide functional information. Later, other types of multi-modality techniques were explored, including with positron emission tomography and magnetic resonance imaging, CT/PET and CT/MRI, leading to a greatly improved diagnostic and treatment tool for a wide range of medical cases. At synchrotron radiation facilities,

(a)

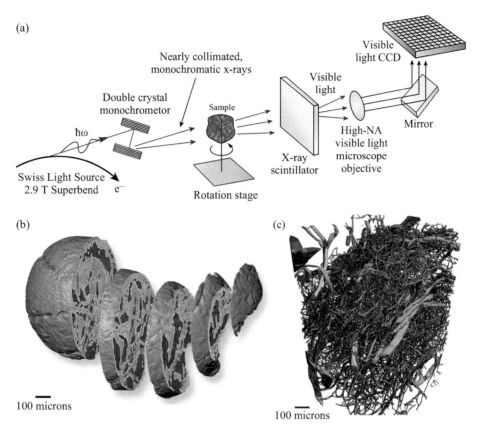

(b)   (c)

100 microns

100 microns

Figure 11.7 (a) The Tomcat beamline at the Swiss Light Source which uses x-rays from a superbend (2.9 keV, 11 keV E_{crit}) and a double crystal monochromator, shown here illuminating a rotating sample and an x-ray scintillator, followed by optical transfer to a visible light CCD. (b) Three-dimensional tomographic renderings of a 0.8 mm diameter, approximately 500 million year old fossilized embryo from southern china, showing fine internal detail, including spinal structures and an axial tube that may be part of a developing digestive tract. The tomographic data set was obtained with 17.5 keV x-rays and a spatial resolution of approximately 1.5 μm.[31] (c) Slice image from a 13.5 keV tomographic data set showing detailed microvascular architecture of blood vessels within a mouse brain.[32] The data set consisted or 1001 2D projections at 80 ms per view. The spatial resolution was about 1.5 μm. Courtesy of M. Stampanoni, PSI and ETH Zürich, and for (b) with P. Donoghue, University of Bristol, and for (c) with T. Krucker, Novartus.

for scientific visualization, multimodality techniques such as CT/fluorescence tomography have been developed.[30]

A capability for absorption based projection tomography of modest sized objects, developed at the Swiss Light Source, is outlined in Figure 11.7.[31, 32] The Tomcat beamline there uses hard x-rays from a 2.9 Tesla super-bend magnet, monochromatized by a double crystal monochromator, to illuminate rotating samples. X-rays of 8–45 keV are available. After passing through the rotating sample projected x-rays illuminate a two-dimensional x-ray-to-visible light scintillator. The visible light pattern from the rear of the scintillator is optically transferred to a visible light CCD with 0.74 μm square pixels. Typical 3D data collection times are several minutes with 20 keV x-rays. Shown in Figure 11.7(b) are slices from a CT reconstruction of an approximately 500 million year old fossilized embryo from Southern China.[31] The embryo is of great interest as it was formed in the pre-Cambrian period, just at the start of an explosion of different life forms on Earth. The objects themselves are very fragile. Their inner structures cannot be determined by sectioning as they simply crumble. X-ray tomography provides a non-destructive path to understanding its full 3D inner structure. The fossilized embryo in Figure 11.7(b) is about 0.8 mm in diameter and was probed at 17.5 keV, resulting in 2D slice images resolved to a spatial resolution of about 1.5 μm. The images show fine detail of developing anatomical features, including spinal and other evolving internal structures. Shown on a different scale in Figure 11.7(c) is a 2D slice image showing microvascular architecture for blood flow in a mouse brain.[32] It was obtained by taking 1001 2D images at 13.5 keV, also at the SLS Tomcat beamline.

Further examples of absorption based tomography are presented in Section 11.5 on full-field imaging with zone plate lenses. Operating primarily with soft x-rays these microscopes take a broad angular set of 2D images, each resolved to several tens of nanometers, to form full 3D data sets from which reconstructed 2D slice images are formed. Tomographic data sets with striking 2D slice images of single biological cells are regularly published by active groups in Berlin, Berkeley, and Barcelona. Depending on object size and wavelength, depth of focus limits can affect achievable spatial resolution in nanoscale tomographic imaging, although some mitigation is possible with stacked imaging.[33] In an early application of absorption tomography to the study of semiconductor integrated circuits (ICs) at Taiwan's NSRRC, 3D data sets and high resolution 2D slice images were obtained of interconnecting tungsten plugs which provide electrical connections between layers of multilevel integrated circuits.[34] Tomographic principles can be utilized in many other imaging geometries and modalities, including diffraction imaging, reflection, magnetic resonance imaging, and others. Examples of tomographic diffraction imaging will be discussed in Section 11.6.

11.3.2   Phase Enhanced Imaging

A further extension of early techniques introduces phase enhanced imaging at x-ray and EUV wavelengths. The first implementation utilized a compact, polychromatic x-ray source, and a wide separation between object and recording detector to permit diffraction from sharp edges to diverge somewhat, creating new opportunities for phase enhanced imaging.[35, 36] The technique is now known as propagation based imaging (PBI). Often contrast mechanisms in propagation based imaging include contributions from both absorption and phase contrast. With a divergent source, magnification is also possible. Important characteristics of the source include its energy spectra, the focal spot size of the electrons at the anode, and the emitted photon flux. Together these parameters also describe the source brightness which is important in imaging and focusing experiments, for example as discussed in Chapter 5, Section 5.4.6. The general technique of propagation based imaging is illustrated in Figure 11.8(a). An example of the

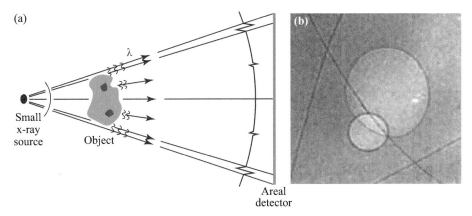

Figure 11.8 (a) A diagram illustrating the technique of propagation based imaging (PBI) in which a small x-ray source illuminates an object with divergent radiation. The projected two-dimensional image is affected by absorption and phase variations along each ray path, and by interference effects due to diffraction and refraction within the object and at its boundaries. The phase and diffractive effects can significantly enhance image contrast, particularly at sharp edges. The angles shown are greatly exaggerated for clarity. Typical angles are of order 1 mrad or smaller. (b) An image of air bubbles within a polymer glue ("Tarzan's Grip") obtained with a spatially filtered compact x-ray source radiating broadly from 20 to 60 keV.[35] The image is essentially due to pure phase effects. The dark ring at the periphery of each bubble image is due to overlapping, polychromatic, interference effects. The image width in (b) is 2 mm. Following S. Wilkens, CSIRO, Victoria.

technique, obtained with broad band polychromatic radiation (20–60 keV) and spatially filtered radiation, is shown in Figure 11.8(b) for air bubbles in a polymer glue.[35] The source to object distance was 0.2 m and the object to image distance was 1.2 m, giving an image plane magnification of six. The fine focus, compact source diameter of 20 μm at a distance of 0.2 m from object, gave a coherence patch, according to Chapter 4, Eq. (4.7a), of a fraction of a micron at any point on the source, thus providing requisite spatial filtering. The image in Figure 11.8(b), magnified by six, shows a phase-darkened edge of the bubble, broadened further to about 20 μm by a 1.2 m propagation distance (~3 μm back projected to the object plan). While increasing the source to object distance improves spatial coherence in small patches around edges and small features of interest, this comes at the cost of reduced photon flux in that same area. The exposure time for the image in (b) was 8 minutes.

These phase effects are further enhanced for hard x-rays by what is known as diffraction enhanced imaging (DEI) in which an x-ray crystal is placed between the object and the detector plane to provide fine tuning of angular acceptance. In Figure 11.8(a) this would appear as a Bragg crystal located before the detector plane, reflecting x-rays to a 2D detector moved to an appropriate location and tilted orientation. The purpose of this "analyzer crystal" is to transmit only a narrow angular band of x-rays, chosen to transmit only the diffracted radiation, thus significantly enhancing image contrast. Typical angular acceptance is set by the photon energy and crystal planes, but is typically of order 10 μrad, thus more appropriate to the use of synchrotron radiation. By rotating the

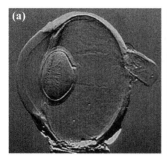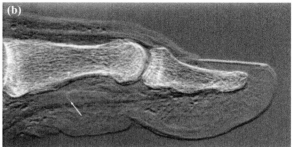

Figure 11.9 (a) A diffraction enhanced image (DEI) of a porcine eye obtained with 40 keV x-rays and a Si(333) analyzer crystal using bending magnet radiation at NSLS. The image shows fine detail of the eye's cornea, iris, retina, optical nerve, and vortex veins, most of which were not discernible with conventional absorption radiography.[38] (b) A DEI image of a human thumb obtained with tungsten K_α x-rays at 59.32 and 57.98 keV from a 1 kW compact x-ray source using Si(333) crystals for both monochromator and analyzer. The image shows fine detail of the soft tissue, the tendon connected to the tip of the bone, the fingernail, and a clear skin edge.[39] Images courtesy of D. Chapman, University of Saskatchewan, Z. Zhong, NSLS, and for (a) with M. Kelly, Cleveland Clinic and for (b) with C. Partham, University of North Carolina, Chapel Hill.

crystal through its rocking curve one can select for reflection radiation an undiffracted image, or images based on diffraction to one direction or the other at the low and high energy sides of the rocking curve. The radiation is necessarily monochromatic, within the crystal acceptance, and to some degree spatially coherent, sufficient to coherently illuminate the diffractive edges and other features of interest.[37–41] Figure 11.9 shows two results obtained using diffraction enhanced imaging. In Figure 11.9(a) one sees a DEI image of a porcine eye obtained with 40 keV x-rays and a Si(333) analyzer crystal at Brookhaven National Laboratory's NSLS. The image shows fine detail of the eye's cornea, iris, retina, optical nerve, and vortex veins, most of which were not discernible with conventional absorption radiography.[38] Shown in Figure 11.9(b) is a DEI image of a human thumb obtained with 59.32 and 57.98 keV W-K_α x-rays from a 1 kW compact source, and a Si(333) analyzer crystal.[39] Note the fine soft tissue detail, the tendon connected to the tip of the bone, the finger nail, and the clear skin edge. An x-ray image plate with 100 μm square pixels was used. The exposure time was 11 hours, but would be reduced to 20 minutes with a 60 kW compact source.

11.3.3   Grating Based X-Ray Imaging

A further method for hard x-ray phase sensitive imaging is based on use of an interferometric grating pair, the first grating acting as a beam-splitter, the second as an analyzer. The technique offers advantages for use with compact sources which are characterized by both broad angular emissions and wide photon energy bandwidth.[40–46] With this technique the sample is placed just in front of the first grating, as shown in Figure 11.10, and a detector is placed immediately behind the second grating. The technique is sensitive to the full properties represented by the refractive index, $n = 1 - \delta + i\beta$, providing

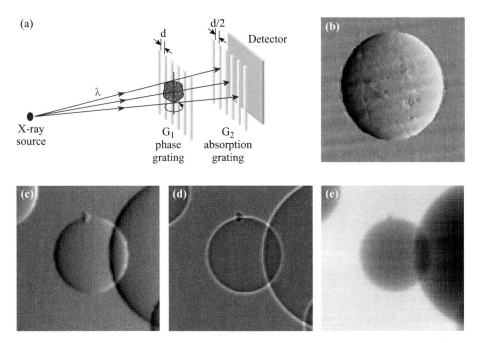

Figure 11.10 (a) A diagram illustrating grating based imaging (GBI). A compact, broad-band source is used. A near-π phase grating of period d is used as a beam splitter, generating radiation of equal amplitude in two directions with no zero order. Its first two orders largely overlap due to the ratio of $\lambda/d \sim 10^{-7}$, forming an interference pattern of period $d/2$. The second, G_2, is an absorption grating of period $d/2$, which serves as an analyzer. The rotatable object is placed just in front of the first grating. The areal detector is placed just behind the second grating. Following T. Zhou, KTH, Stockholm.[45] (b) The phase image of a 1.2 mm diameter plastic sphere obtained with spatially coherent 1 Å x-rays, monochromatized with a Si(111) DCM and GBI at SPring-8, and recorded with a phosphor screen optically coupled to a visible light CCD. Small bubbles are seen within the sphere.[40] Panels (c)–(e), images of a 100 μm diameter polystyrene sphere obtained by GBI with 17 ± 0.5 keV wiggler radiation at the Swiss Light Source, using Zr and Si absorption filters and a YAG scintillator, optically coupled to a visible light detector.[42] A partial image of a larger sphere appears to the right in each. The images display phase gradient in (c) with clear edge enhancement, quantitative phase in (d), and absorption in (e). Image (b) courtesy of A. Momose, University of Tokyo, presently Tohoku University; and images (c), (d), and (e) courtesy of T. Weitkamp, Paul Scherrer Institute.

quantitative measures of both absorption and phase shift along path integrals through the sample. As the wavefront propagates through the sample it is modified by absorption, phase shifts, and turned locally by gradients of refractive index. These variations are to be detected by the interferometer. For opaque gratings, diffraction of all orders recombine to form a sequence of bright interference patterns, self-images of the grating, in what are known as Talbot planes,[40, 41] at downstream intervals of $z_T = d^2/\lambda$. For phase gratings self-image interference patterns are formed with a period $d/2$ at distances $d^2/2\lambda$. The second grating is best located at one of the Talbot planes. The technique requires minimal spatial coherence, across a lateral distance approximately

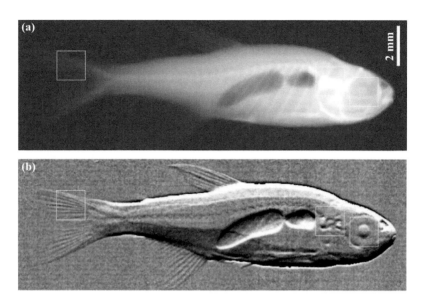

Figure 11.11 Absorption (a) and phase (b) x-ray images at 22.4 keV of a small fish obtained using a three-grating version of Talbot GBI interferometric imaging and a compact source of extended (8 mm) width.[46] Both were obtained from the same data set. The panels are each 2 cm wide. The phase image shows far greater detail, as seen for instance in the boxed areas of the tail and eye. Courtesy of F. Pfeiffer, PSI, now at the Technical University of Munich.

equal to a period of the grating, and only minimal temporal coherence, permitting relatively broad spectral content, far more than would be acceptable with a crystal technique. Spatial resolution is approximately equal to the period of the first grating, which is beyond the capability of present two-dimensional detectors. This is a significant challenge which has been solved by placing a half-period absorption grating in a Talbot plane, with the areal detector just behind it. It is in this manner that the second grating, G_2, acts as an analyzer. By scanning the second grating, in fractional grating steps, it transforms wavefront amplitude and phase information to recorded intensity as a function of analyzer position. By taking sequential images, or phase and absorption data sets, as the object is rotated, a 3D data set and subsequent 3D reconstructions of the object can be obtained. Pioneering efforts by Momose,[40] Weitkamp,[42] and their colleagues are described in the literature, and examples of their early work shown in Figures 11.10(b) and (c)–(e) respectively. In both papers synchrotron radiation is used but the focus is on future use of compact x-ray sources.

A variant of two grating interferometric imaging is the use of a third grating, G_0, placed just in front of a larger area compact source. The purpose of G_0 is to create individual beamlets each having a high degree of spatial coherence, but being mutually incoherent, and all contributing equally to the interference pattern at the analyzer grating. The geometry and techniques is essentially the same as seen in Figure 11.10 except with an absorption grating, G_0, placed just in front of a 1 kW x-ray source of extended width, 8 mm (h) by 0.4 mm (v). Pioneered by F. Pfeifer and colleagues,[46] an example of their early work is shown in Figure 11.11. Figures 11.11(a) and (b) show comparative absorption and phase contrast images of a small (~2 cm long) fish recorded with a relatively broad (~10%) x-ray spectrum, centered at 22.4 keV, from a conventional laboratory source. A YAG fluorescence screen was optically coupled to a visible light CCD to record the images. The spatial resolution was estimated to be about 30 μm. The total exposure time, averaging over 27 raw images to reduce noise, was 27×40 s. Fine details of the fin and the eye are seen in the phase image. The combination of source size and power provides a figure of merit for grading based imaging. Recent papers have explored the use of a liquid metal jet, with a focused electron beam, as a compact, 40 W, small area x-ray source, and have used it to compare propagation and grating based imaging with regard to radiation dose, image quality, and exposure time.[44, 45, 47]

11.3.4   Fluorescence Imaging

Fluorescence imaging methods are used for spatially resolved elemental analyses in broadly varying scientific and engineering fields, including the identication of impurities in materials, geological samples, trace elements in biological samples, authentification of historical documents, and studies of coloring compounds used in historic paintings, pottery glazes, and other items of cultural heritage. In a major form of implementation the process involves sample irradiation by a scanning, focused, hard x-ray beam and energy resolved detection of characteristic photon emissions, as was outlined in Chapter 1, Figure 1.2(b, c). The process is most efficient for K-shell, hard x-ray emissions from high-Z atoms, as seen in Figure 1.3, but is also possible to some degree for lower photon energy emissions. A major advantage of the technique is that the characteristic fluorescent emissions have clearly identified photon energies, making elemental identification relatively simple. Furthermore the hard x-ray emissions are capable of transmission through relatively thick host materials, often operating in air. For photon-in, photon-out experiments, such as a scanning nanoprobe with a high efficiency quantum detector, there is essentially no background radiation. For such experiments photon counting statistics are the major source of noise and it is commonly possible to achieve part-per-billion (ppb) and femtogram (fg) elemental detection sensitivities.[48] Often fluorescence identification is combined with other techniques such as x-ray absorption near edge structure (XANES) spectroscopy for chemical speciation at selected sample positions, and x-ray diffraction for structural information. A very flexible beamline for such studies at the European Synchrotron Radiation Facility (ESRF) in Grenoble is illustrated in Figure 11.12(a).[49] The beamline covers the spectral range from 2 to 10 keV, utilizes hard x-ray undulator radiation, a high spectral resolution double crystal monochromator (DCM), a scanning transmission x-ray microscopy (STXM) capability with various focusing optics, and detection of fluorescence, XANES, and diffraction from micron scale samples, μ-XRF, μ-XANES, and μ-XRD, respectively. Both zone plate and KB optics are available for the STXM; zone plates for high spectral resolution and KB mirrors for high efficiency and broad spectral coverage with a focal spot of 0.35×0.71 μm. Typical fluorescence operation with KB optics provides an irradiation of 7×10^{10} photons/s, with high efficiency capillary plus Bragg crystal detection in a nominal

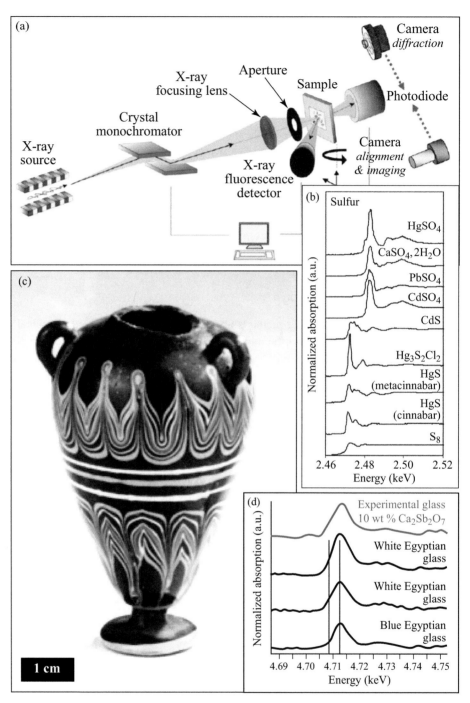

Figure 11.12 (a) The x-ray microscopy beamline ID21 at ESRF. The beamline utilizes both 32 mm and 42 mm period undulators to cover the 2–10 keV spectral region, a double crystal monochromator with choice of Bragg crystals, the choice of a zone plate lens or KB mirrors for focusing, a scanning sample stage, and detectors for fluorescence, XANES, and diffraction. (b) An example of reference sulfur (S) XANES spectroscopic signatures as used to identify unknown compounds, (c) a photograph of an Egyptian glass vase c. 1500 BC, and (d) XANES data used to identify calcium antimonite, $Ca_2Sb_2O_7$, as an important compound contributing to the distinctive "Egyptian blue" color.[50, 49] Courtesy of M. Cotte and J. Susini, ESRF.

30 eV spectral bandwidth at 6 keV.[49] Shown in Figure 11.12(b) is a series of μ-XANES spectra taken with reference sample compounds containing sulfur (S), as used for example by M. Cotte, J. Susini, and their colleagues at ESRF to identify unknown components through a library of element specific spectroscopic signatures. An example from one of their studies is shown in Figure 11.12(c), an Egyptian vase, *c.* 1500 BC, among the earliest known to use decorative glass. As shown in Figure 11.12(d), the particularly distinctive "Egyptian blue" glaze was identified as containing nanoscale calcium antimonite ($Ca_2Sb_2O_7$) by comparing the XANES signature of a small blue chip with various reference compounds.[50]

Other cultural heritage studies involving x-ray fluorescence include the analysis of historic artwork, the paint materials used, and studies of decorative tiles used centuries ago in cultural sites around the world. An example of interest is the reuse of canvas by the Dutch artist Vincent van Gogh. An early version of his painting *Head of a Woman* was found using x-ray fluorescence by a team of Dutch, French, and German scientists, and art restoration experts, buried beneath his *Patch of Grass (1887).*[51] Figure 11.13(a) shows a reproduction of van Gogh's *Patch of Grass*, with a square region marked by fine red lines where the painting was studied for characteristic fluorescence emission and XANES. Various paint compounds, and the elements they contain, are known to produce various colors. Mercury (Hg) and trace elements of antimony (Sb) were of particular interest as known compounds, available at that time, that were used to produce flesh-like tones. The painting was transported from the Kröller-Müller Museum in The Netherlands for studies using monochromatic 38.5 keV x-rays at DESY's DORIS III synchrotron in Hamburg. Figure panel (b) shows detected fluorescent emissions from 3 to 34 keV, taken at locations within the red box, and (c) shows XANES spectroscopic scans taken near the K-edge of antimony (Sb). A microscopic sample was transported to ESRF's ID21 for high resolution Sb L_3 and L_1-edge XANES studies, as shown in panels (d) and (e). These comparisons with reference compounds were used by Dik *et al.* to conclude that "Naples Yellow," $Pb(SbO_3)_2.Pb_3(Sb_3O4)_2$, was one of those used by van Gogh to achieve flesh-like tones in facial areas.[51] Panel (f) shows their reconstruction of the artist's hidden *Head of a Woman* based on the elemental and chemical identifications made by fluorescence emissions and XANES spectroscopy.[51] Fluorescence studies using laboratory x-ray sources were used to complement portions of these studies with *in-museum* measurements.

A further study using x-ray fluorescence was conducted at the Synchrotron Light Research Institute, Siam Photon Laboratory, Thailand, where mosaic glasses have been widely used over the centuries to decorate historical temples and palaces. Klysubun *et al.*[52, 53] used several complementary techniques, including electron beam spectrometry and laser ablation mass spectrometry for determining molecular concentrations present, and synchrotron based XANES spectroscopy to identify specific oxidation states in elements of interest. For example, in studies of decorative mosaics at the Temple of the Emerald Buddha in Bangkok they identified use of a lead glass containing particular elements that affect the striking colors observed, for example, the oxidation state Co^{2+}, which imparts a blue color, Pb content that can change the blue to turquoise and green depending on PbO concentration, and Fe^{3+}, which can impart a yellow color.

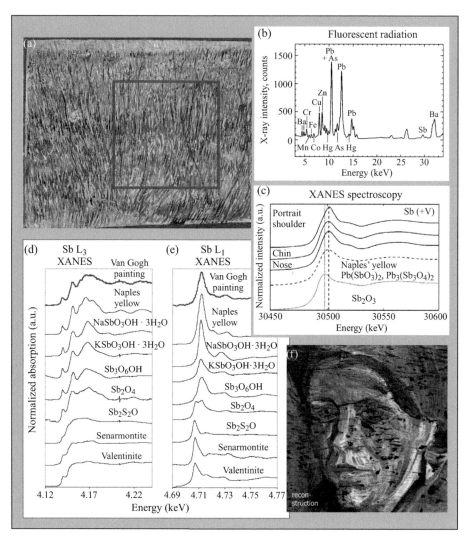

Figure 11.13 (a) A reproduction of Vincent van Gogh's *Patch of Grass (1887)*, with a fine red lined box outlining an area studied for an earlier, covered over and now hidden version of an early *Head of a Woman* painting.[51, 49] (b) Samples of characteristic fluorescence emissions showing many detected elements, as indicated. (c) XANES data near the K-edge of antimony (Sb) corresponding to various points in the underlying painting. Panels (d) and (e) show high resolution Sb L_3 and L_1-edge XANES data of various compounds. These were used to identify "Naples Yellow" as the compound used by van Gogh to help effect flesh like tones in facial areas [top two traces in (d)]. (f) A tritonal color non-destructive reconstruction of an early, over painted version of van Gogh's *Head of a Woman* based on the identification of white, yellow, and red pigments and their locations.[51] Courtesy of J. Dik, Delft University of Technology, M. Cotte, ESRF, and the Kröller-Müller Museum, Otterlo, The Netherlands.

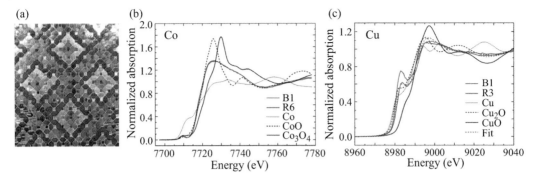

Figure 11.14 (a) A nineteenth-century inlaid mosaic sample taken from the Royal Chapel within the Temple of the Emerald Buddha, Bangkok, Thailand.[52] (b, c) Sample XANES spectroscopic data at the Co and Cu K-edges, respectively, for the antique glass (B1), a reference glass (R6 and R3, respectively), and standard references as indicated.[53] Courtesy of W. Klysubun, Synchrotron Light Research Institute, Nakhon Ratchasima, Thailand.

They took extensive XANES data at the K-edges of Co, Fe, Mn, and Cu, comparing the spectroscopic signatures with that from various reference samples. Their measurements were made using bending magnet radiation, a double crystal monochromator (DCM) using a Ge(220) crystal and a 13 element germanium fluorescence detector.[52] A mosaic studied, and sample XANES data, are shown in Figure 11.14. Phase imaging and fluorescence x-ray techniques are now widely used to study works of archival heritage.[54, 55] For studies which do not require high spatial resolution, compact laboratory X-ray fluorescence spectrometers can provide significant scientific results. An example is identification of the young Egyptian King Tutankhamun's iron dagger as being meteoritic in origon, which is important because iron objects were rare and prized[55] during the Bronze Age in which he lived.

11.4   Scanning X-Ray Microscopy

In this section we consider the application of zone plate lenses to high spatial resolution x-ray microscopy, first with scanning techniques in this section, then with full-field techniques in the following section. With short wavelengths, most commonly in the soft x-ray region from 0.4 nm to 4 nm, and high quality lenses of relatively high numerical aperture (NA $= \lambda/2\Delta r$, Eq. (10.21)), these microscopes are capable of achieving spatial resolutions of order 10 nm.[56-60] The corresponding photon energies of 0.3–3 keV span the primary (K-shell) and secondary (L-shell) resonances of half the elements in the periodic chart, providing natural contrast mechanisms for elemental (Z) and even chemical bond mapping.[61-65] Operation at higher photon energies, to 10 keV and higher, is also possible.[58, 66] In their most common usage, these microscopes are used for the formation of images based on differential absorption and phase shift of transmitted radiation from relatively thick samples (to 10 μm). Detection of photoemitted electrons of well-defined energy (incident photon energy minus binding energy) within the electron

(a) Scanning Transmission X-ray Microscope

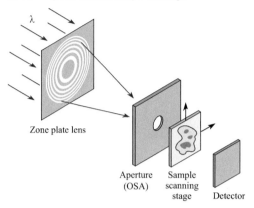

(b)

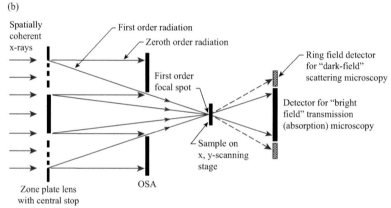

Figure 11.15 (a) A scanning transmission x-ray microscope (STXM) in which the sample is raster scanned two-dimensionally past a focused spot. A point detector records an electronic signal proportional to the transmitted x-ray intensity. An image is then constructed electronically, pixel by pixel, revealing a 2D map of sample absorption. (b) The STXM utilizes a thick central zone plate stop and an order sorting aperture (OSA) to block all but the first diffraction ($m = +1$) order from reaching the sample. Note that while the STXM is most commonly used to measure absorption with a point detector in "bright field" mode, it can also be used to detect scattering at larger angles with a surrounding ring or quadrant detector, in what is known as "darkfield" mode.[75, 76] Diagrams follow J. Kirz, H. Rarback, C. Jacobsen, and colleagues,[71–74] then of Stony Brook University and Brookhaven National Laboratory.

range of an accessible surface has also been demonstrated.[67–69] At the higher range of incident photon energies the detection of characteristic fluorescent emission is also possible, and is commonly used in a wide range of applications.[70]

A scanning soft x-ray microscope (STXM) in its most common embodiment is illustrated in Figure 11.15(a). In this microscope spatially coherent radiation illuminates a zone plate lens, which forms a first-order focal spot at the sample plane. The lens basically images a distant x-ray source, spatially filtered to appear as a near-point source, at the sample. If properly illuminated and mounted, the zone plate provides an Airy pattern focal spot intensity distribution, with resolution set by the outer zone width according to Eq. (10.42). Spatially coherent illumination is most commonly obtained by spatially filtering undulator radiation, as was described previously in Chapters 4 and 5, Figures 4.8 and 5.30. The radiation transmitted through the sample is detected by a fast x-ray detector (a photodiode) as the sample is raster scanned past the focal spot, revealing absorption as a function of position. An image of arbitrary size is then constructed by correlating sample position with electronic signal. The zeroth and other orders are prevented from reaching the sample by the combination of a thick zone plate central stop and a thick order sorting aperture (OSA),[71] as shown in Figure 11.15(b). Radiation dose to the sample is minimized by incurring the zone plate lens losses (inefficiency) upstream, before radiation reaches the sample. This is especially important for the study of radiation sensitive materials such as polymers and environmental samples. For biological samples, which are also very sensitive to radiation dose, radiation damage is largely limited through the use of quick freeze cryogenic sample techniques, which maintain the integrity of internal structures at scale sizes relevant to present resolving capabilities. Development of the scanning transmission x-ray microscope (STXM) was pioneered by Kirz, Rarback, Jacobsen, and their colleagues[71-74], then of Stony Brook University and the National Synchrotron Light Source (NSLS). Significant contributions to stage control and image resolution were later made by T. Tyliszczak at the Advanced Light Source (ALS).[63]

The scanning microscope is very flexible in that it can be used in several modes of operation. In the transmission mode, as described above, it can be used to record sample absorption versus position, repeated at various wavelengths for elemental and chemical analyses at spectral resolution set by the upstream monochromator. In this mode of operation the technique is referred to as spectromicroscopy.[61-63] It can also be used in a fluorescence or luminescence mode[64, 65, 70] in which the incident radiation excites or indirectly causes emission of radiation that reveals the chemical nature of the sample, again as a function of scanned position. A third mode of operation is that of detecting photoelectron emission as a function of position. Combining the latter with photoelectron spectroscopy, at each scanned position, can provide a powerful tool for the study of surface composition and chemistry.[67-69] Large sample areas are possible with appropriate scanning. Exposure times can be relatively long with the scanning microscope, despite the use of an undulator, due to the significant loss of flux incurred through spatial filtering. Typical exposure times are several minutes for a 400×400 pixel array. Future "diffraction limited" storage ring s (see Chapter 5, Figure 5.24 caption) will provide smaller electron beam size and divergence, reducing losses due to spatial filtering, and thus reducing image acquisition times.

Note that coherence has several roles associated with these scanning techniques. The first is providing the ability to focus to the smallest possible focal spot. But additionally there is the possibility of providing an encoded illumination through use of a holographic diffractive optical element (see text near Figure 10.26) such as to provide various forms of phase contrast, add orbital angular momentum, increase depth of field, or enhance spatial resolution.[77] Furthermore, the ability to record coherent scattering

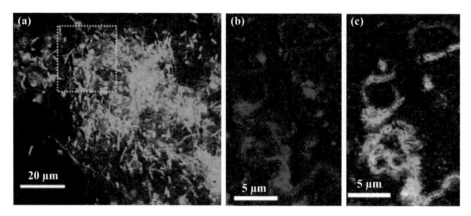

Figure 11.16 An environmental sample showing biofilm distributions of (a) Ni(II) in blue overlaid on O 1s data in green generally representing biomaterial, (b) Fe(III) in red and Fe(II) in blue, and (c) Mn(II) in red, Mn(III) in green and Mn(IV) in blue.[62] The data are based on water samples from the South Saskatchewan River, Saskatoon, Canada. Scanning soft x-ray spectromicroscopy was performed at the ALS in Berkeley. Courtesy of A. Hitchcock, McMaster University.

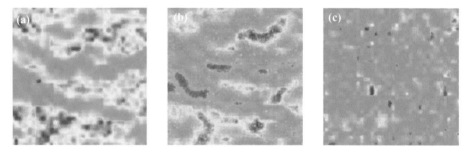

Figure 11.17 Characteristic fluorescent emission of (a) Mn L_α, (b) Cu L_α, and (c) Mg L_α, part of a study of potential electrochemical replacements for platinum electrodes in energy storage devices.[65] The data were obtained with the TwinMic scanning x-ray microscope at the Elettra-Sincrotrone Trieste. The TwinMic microscope is unique in that it has capabilities for high spatial-spectral resolution spectromicroscopy and fluorescence detection with a wide variety of sample environments.[78] Courtesy of M. Kiskinova, Sincrotrone Trieste.

from the sample on a larger detector, a form of diffractive imaging known as "ptychography," opens new opportunities as discussed further in Section 11.6.

11.4.1   Environmental Spectromicroscopy

A particular advantage of the STXM, as can be seen in Figure 11.15, is that the inefficiency of the focusing lens is upstream, resulting in minimal dose to the sample. This is particularly advantageous in studies of polymers and environmental samples which are radiation sensitive but where cryofixation, as used for biological samples, is not desired. Figure 11.16 shows data obtained by Hitchcock and colleagues as part of a broad study of bioremediation technigues, in this case of metal uptake by biomaterials in the Saskatoon river.[62]

11.4.2   Complementary STXM Techniques

The TwinMic STXM at Sincrotrone Trieste is unique in that it combines abilities for high spatial and spectral resolution spectromicroscopy with nanoscale resolution fluorescence, a combination which provides additional insights in widely varying fields of application. Figure 11.17 shows the use of scanning microscopy utilizing the detection of characteristic L_α, 2p $\rightarrow$ 1s, radiation to study spatial distributions of potential replacement materials for platinum electrodes in energy storage devices.[64, 65] Other applications include agricultural, biological, and materials studies.[78]

11.5   Full-Field X-Ray Microscopy

The full-field x-ray microscope is illustrated in Figure 11.18. In this microscope the zone plate lens forms a full two-dimensional image on an array detector. The image is formed by localized interference on a point by point basis. Its operational use has much in common with the visible light (refractive) microscope found in every biology laboratory. The achievable spatial resolution is set largely by the lens outer zone width Δr, but can be improved somewhat through optimized partial coherence, or use of higher order illumination, as was discussed in Section 10.6.3 and illustrated in Figure 10.30. Because of wavelength dispersion in the diffractive process, known optically as chromatic aberration, the illuminating radiation must be of relatively narrow spectral bandwidth, with $\Delta\lambda/\lambda$ less or equal to $1/N$, where N is the number of zones, as described previously in Eq. (10.47). As shown in Figure 11.18(a), the incident x-rays pass through the sample, where they are partially absorbed with a spatial variation dependent on the atoms present, their spatial distribution and spectral absorbing features, and the incident photon energy. After transmission and absorption by the sample, the emerging radiation is diffracted by the lens to form a first-order ($m = +1$) image at the CCD. Larger areas are possible by stitching together an array of individual CCD images.[79] Magnifications are typically several thousand, set by the CCD pixel size, typically 25 μm, and the desired pixel size projected back to the sample plane, typically 10 nm. Image contrast is natural, set by elemental and chemical absorption features within the sample, without the need for contrast enhancing additives. The zeroth order, and other zone plate orders, are blocked by "hollow cone" illumination in combination with a thick central zone, an OSA, and detection geometry, as illustrated in Figure 11.18(b). Phase contrast imaging[80] has been demonstrated in soft x-ray microscopy using a back focal plane annular phase plate,[81] whose position is indicated in Figure 11.18(b), and more recently by use of a dual function Zernike zone plate which achieves the same result.[77] Development of the full-field soft x-ray microscope was pioneered by Schmahl, Rudolph, Niemann, and their colleagues[82–86] at George-August University in Göttingen, first using bending magnet radiation at LURE in France, and later at the BESSY synchrotron facility in Berlin.

(a) Full-Field X-ray Microscope

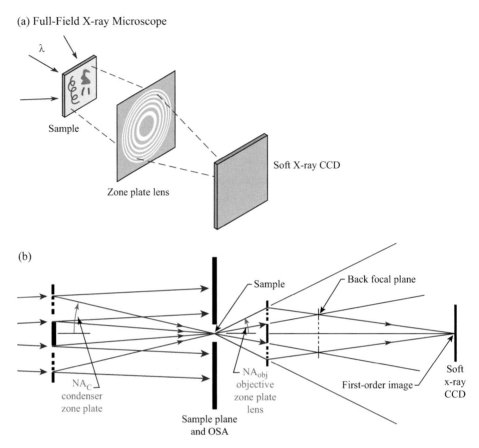

Figure 11.18 (a) A full-field x-ray microscope in which a complete two-dimensional image is formed at the CCD, recording the spatial distribution of absorption within the sample. (b) The full-field x-ray microscope utilizes a condenser optic, shown here as a zone plate with a thick central stop, combined with an order sorting aperture in the sample plane, to illuminate the sample with first-order radiation. An objective zone plate lens with high numerical aperture collects transmitted and diffracted radiation from the sample, forming a high resolution, high magnification ($M \sim 400$–2000) image at the CCD. The back focal plane of the micro zone plate is available for use of an annular phase plate.[83–85] Following G. Schmahl, D. Rudolph, B. Niemann, and colleagues, George-August University, Göttingen.

11.5.1   Two-Dimensional Nanoscale Imaging

The primary advantage of the soft x-ray microscope is its simplicity and its ability to form x-ray images at the highest spatial resolution. Because it requires only minimal spatial coherence, a coherence patch equal in size to the spatial resolution, it can form images with bending magnet radiation, do so with exposure times of just a few seconds, and with compact x-ray sources as well. The use of undulator radiation provides higher photon flux within the acceptable $1/N$ spectral width, permitting spectromicroscopic studies which combine high spatial and spectral resolutions, as well as

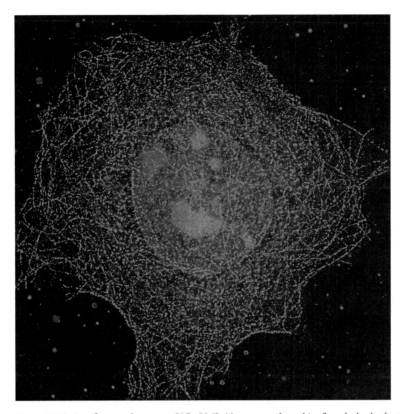

Figure 11.19 A soft x-ray image at 517 eV (2.40 nm wavelength) of a whole, hydrated, cryofixed mouse epithelial cell (EPH4) obtained with the XM-1 full-field microscope and bending magnet radiation at the Advanced Light Source in Berkeley.[88] The image is color coded according to the degree of absorption. The microtubule network of structural proteins, made evident by high absorption of the silver enhanced gold tags, is color coded blue. The cell nucleus and nucleoli, characterized by moderate absorption, are coded orange. The less absorbing, aqueous regions of the cell are color coded black. The silver enhanced gold is part of a molecular double label that permits cross correlation with visible light fluorescence in a confocal microscope. Courtesy of C. Larabell, UCSF and LBNL, and W. Meyer-Ilse, Lawrence Berkeley National Laboratory.

other functionalities. Owing to relatively modest zone plate efficiencies, typically 10%–20% for soft x-rays, and up to 40% with hard x-rays, the radiation dose to the sample is larger than desired for radiation sensitive materials such as polymers and biological samples. For biological samples this is largely obviated through the use of cryogenic sample holders, which maintain structural integrity at high radiation dose for features as small as several nanometers.[87–91, 60]

Figure 11.19 shows a whole, hydrated, cryofixed, 8–10 μm thick mouse epithelial cell imaged with 517 eV (2.40 nm wavelength) with bending magnet radiation and the XM-1 microscope[88] at the ALS in Berkeley. Sample preparation employed a double labeling technique for visible light fluorescence in a confocal microscope and with enhanced absorption in the soft x-ray microscope. Labeling with gold particles of 1.4 nm diameter, enhanced with silver to 50 nm, was used to form easily observed

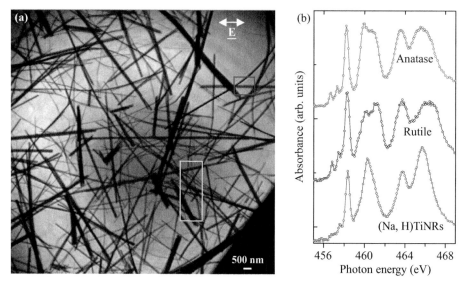

Figure 11.20 (a) A spectromicroscopic soft x-ray image at 460 eV of a set of randomly oriented sodium titanate nanoribbons. The image was obtained with the full-field x-ray transmission microscope at BESSY II in Berlin, using undulator radiation, a high spectral resolution spherical grating monochromator, and a high efficiency capillary condenser, which together provide high photon flux within a narrow spectral band.[92] For these images the simultaneous spatial and temporal resolutions were 25 nm (half-pitch) and $E/\Delta E = 2 \times 10^4$. Shown in (b) is a comparison of NEXAFS spectra, in the vicinity of the titanium L-edges, for sodium titanate nanoribbons and reference samples of rutile and anatase, common forms of titanium dioxide. The notation (Na,H)TiONRs refers to Na or H ionic bonding to oxygen ions at the ends of a molecule containing several TiO_2 molecules, such as $Na_2Ti_3O_7$ or its hydrogenated replacement $H_2Ti_3O_7$. NR refers to nanoribbons.[93] The red and yellow sections in (a) were used to show that there are orientation/polarization dependent spectral features associated with electron motion in nanocrystalline structures. Courtesy of C. Bittencourt, U. Antwerp, P. Guttmann and G. Schneider, BESSY II/HZB.

x-ray absorption markers and thus permit visualization of the microtubule network of structural proteins as evidenced by the "strings of blue pearls" in Figure 11.19. The outer zone width of the objective lens was 35 nm and that of the condenser 55 nm, for a partially coherent illumination characterized by $\sigma = 0.64$. The measured knife edge spatial resolution was 36 nm (Rayleigh-like, 10%–90%; 21 nm, 25%–75%). Typical exposure time is 1 s.

An example of high spatial and high spectral resolution soft x-ray spectromicroscopy for material science is shown in Figure 11.20(a) where an image of randomly oriented crystalline sodium titanate nanoribbons is shown for a photon energy of 460 eV. The elongated nanoribbons were typically 70–180 nm wide. Near-edge x-ray absorption spectroscopy (NEXAFS)[92] near the L-edges of titanium, nominally 454–460 eV but affected by local bonding, is shown for the nanoribbons in Figure 11.20(b) and compared with well studied anatase and rutile forms of crystalline TiO_2 reference powders. Observations included variations of spectra as a function of size and orientation. Various forms of titanate nanostructures are commercially important as the white

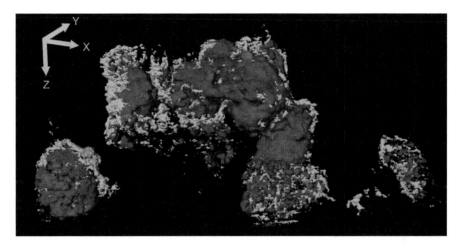

Figure 11.21 A three-dimensional tomographic slice image of Ni and NiO concentration distributions obtained near the 8333 keV Ni K-edge in a study of battery discharge dynamics. The study uniquely combines hard x-ray, three-dimensional imaging with high spectral resolution x-ray absorption near edge structure (XANES) spectroscopy.[94] Concentrations of metallic nickel are shown in green, and of NiO in red. The measurements were made with the SLAC SSRL synchrotron facility using their Xradia, now Zeiss Microscopy, full-field x-ray microscope. Courtesy of F. Mierer, University of Utrecht, and P. Pianetta, SSRL/SLAC/Stanford.

material in paints and ceramics, use as photocatalysts, and recently as potential anode material in ion battery research.

11.5.2 Nanoscale 3D Tomographic Imaging

High resolution, three-dimensional (3D) imaging is possible using tomographic techniques, as discussed in Section 11.3.1, in which successive 2D images are taken as the sample is rotated about a central axis, then reconstructed numerically to build a full 3D data set from which any 2D "slice image," or surface observation, can be viewed. Typically a two-dimensional image is recorded for each 1° of rotation, each taking about 1 s. The full-field x-ray microscope at SLAC's SSRL synchrotron facility has been used to combine high spatial resolution, hard x-ray, full-field tomographic microscopy, with high spectral resolution x-ray absorption near edge structure (XANES) spectroscopy. In a demonstration experiment electrochemical battery discharge dynamics in the vicinity of a largely NiO, Li-ion composite electrode was studied.[94] The spatial resolution for the studies was 30–40 nm, with a field of view of 30 μm and a depth of field (DOF) of order 50 μm, suitable for nanoscale imaging with relatively large particles. The relative spectral resolution for the experiment was $\Delta E/E \sim 10^{-4}$. Spectroscopic data were taken every 0.5 eV, across a range from 8250 to 8600 eV, encompassing the Ni K-edge at 8333 eV, and tomographically from −67° to +67° in 1° steps. The total acquisition time for the full XANES 3D x-ray tomographic microscopy data set for this demonstration experiment was 17.5 h. The data collection time could be substantially reduced by taking an optimized spectral data set with just three points in a Ni/NiO system. Compared with the experiment illustrated in Figure 11.21, which took 700 very

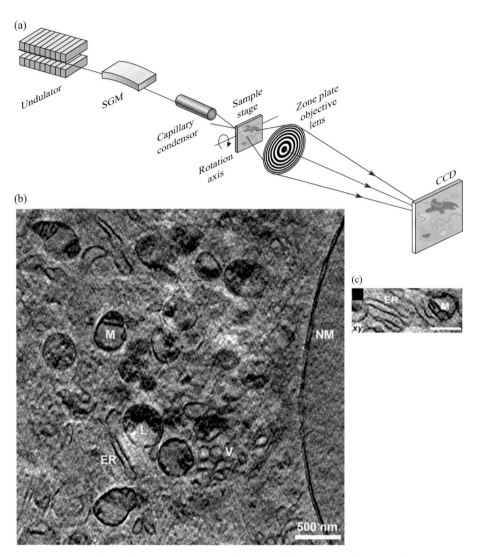

Figure 11.22 (a) A full-field x-ray microscope with a capability for three-dimensional, soft x-ray tomography. A series of two-dimensional images taken at various angles are combined into a self-consistent 3D tomographic data set, from which 2D "slice images" can be reconstructed at any point and at any angular orientation.[95] (b) A sample 2D slice image taken from a full 3D tomographic data set obtained with an approximately 5 μm wide adenocarcinoma mouse cell using the full-field transmission x-ray microscope at BESSY II in Berlin. The sample was whole, fully hydrated, cryofixed and mounted on a flat sample holder. Two-dimensional (2D) images were taken at 1° angular intervals over ±60°. The image shown in (b) was obtained at 510 eV with a 25 nm outer zone width objective lens and partially coherent radiation at $\sigma = 0.43$. Much cell ultrastructure is seen, including mitochondria (M), lysosomes (L), vesicles (V), the nucleus, the double layer nuclear membrane (NM), and nuclear pores (NP), and nuclear membrane channels (NMC). The double layer nuclear membrane is observed to have a 36 nm peak-to-peak (full period) separation. (c) In a close up view the endoplasmic reticulum (ER) is clearly seen. Tomographic slices and movies are available at www.cambridge.org/xrayeuv . Courtesy of G. Schneider, BESSY II/HZB, and J. McNally, National Cancer Institute, NIH, now of BESSY II/HZB.

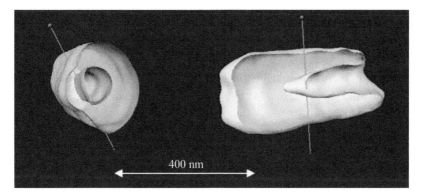

400 nm

Figure 11.23 Two two-dimensional slice images of a nominally 50–80 nm diameter void which developed within an approximately 250 nm × 500 nm interconnecting tungsten plug in an operating computer chip due to excessive electron current density. The images were derived from a 10.5 keV 3D tomographic data set at Taiwan's NSRRC.[34] The slice images were resolved to approximately 60 nm. Courtesy of G.-C. Yin and M.-T. Tang, NSRRC, Taiwan.

fine spectral steps, this would reduce the total data time by a factor of order 200, to approximately 1 h.

Tomography of single biological cells is possible by combining full-field microscopy with cryofixation of the sample in order to mitigate against radiation damage.[87–91] Several groups have been actively pursuing biological applications using tomographic microscopy.[90, 95, 96] A diagram illustrating the technique is shown in Figure 11.22(a). The sample can be rotated in a low-Z capillary tube, or on a 2D stage; advantages accrue for both. An example of a 2D image taken from a full 3D data set is shown in Figure 11.22(b). The image was obtained using 510 eV (2.43 nm wavelength) undulator radiation at the BESSY II synchrotron facility in Berlin.[96] For these images the microscope utilized a $\Delta r = 25$ nm objective zone plate lens and a capillary condenser (see Chapter 10, Section 10.4), for a partial coherence factor of $\sigma = 0.43$. The sample under study is an approximately 5 μm wide adenocarcinoma mouse cell. Much cell ultrastructure is seen, including mitochondria (M), lyosomes (L), vesicles (V), the nucleus, the double layer nuclear membrane (NM), nuclear pores (NP), and nuclear membrane pores (NMP). In a close-up view, Figure 11.22(c), the endoplasmic reticulum (ER) is clearly seen. In other views, not shown here, nucleoli (Nu) and nuclear membrane channels (NMC) are seen.[95] Spatial resolution for this image is about 18 nm (half-period). A tomographic reconstruction of the data showing 3D rotation and sequential 2D slices can be seen online at www.cambridge.org/xrayeuv.

In an early application of nanoscale absorption tomography to the study of semiconductor integrated circuits (ICs) Yin, Tang, Song, and colleagues at Taiwan's NSRRC obtained 3D images of the interconnecting tungsten plugs which provide electrical connections between layers of multilevel integrated circuits.[34] For their measurements they used thick (890 nm), 50 nm outer zone width, gold zone plates, taking 141 2D images, from −70° to + 70°, at a photon energy of 10.5 keV, just above the W-L_3 absorption edge. They were able to identify and measure the development of large internal voids formed by the impact of high electron current densities driving W atoms out of the region. This is an issue of great interest to the semiconductor industry as it evolves towards chips with higher current densities and ever smaller feature sizes. The NSRRC team observed the formation of large, potentially catastrophic, internal voids during operational times of several hours in nominally 250 nm diameter by 500 nm long tungsten plugs.[34] Two-dimensional slice images of a 50–80 nm diameter void inside a tungsten plug are shown in Figure 11.23.

11.6   Diffractive Imaging Techniques

X-ray diffractive imaging began with crystallographic techniques, first with naturally occurring materials and later with crystallized biological samples. The recent availability of spatially and temporally coherent x-rays and EUV radiation offers new possibilities for the imaging of amorphous, nanoscale objects using coherent scattering combined with numerical reconstruction techniques. In the following sections we discuss two techniques which are currently receiving much attention, coherent diffractive imaging[19, 20, 97–107] and ptychographic imaging.[21, 22, 108–110, 113, 114] Criterion for the evaluation of spatial resolution with coherent and partially coherent imaging systems is addressed in Reference 112.

11.6.1   Coherent Diffractive Imaging

For the technique of coherent diffractive imaging (CDI)[21, 22] one records the scattering of spatially coherent x-rays as a function of angle in the far-field with a distant, 2D detector, such as a CCD. To reconstruct an image of the scattering object one needs to know both the intensity and phase as a function of position. That is, the intensity and phase must be known everywhere in the distant plane in order to calculate the electric fields there. If known, the fields can be Fourier transformed back to the object plane allowing a numerical image reconstruction. The power of CDI is in the iterative phase retrieval techniques used to determine phase at each point. These techniques are based on projection algorithms where the common solution is found by projecting between two feasible sets as illustrated in Figure 11.24. The phase is determined by numerical iterations, initially assuming random phase values at each pixel of the CCD. Fourier transforms of fields are then calculated sequentially and iteratively between the detector and object planes, going back and forth, while applying known constraints in each plane; intensity data at the CCD and lateral extent of the sample in the object plane.[99, 100] Iterations are continued for many cycles until convergence to stable values is achieved, at which point image reconstruction is viable. In general, the phasing algorithms do not converge to a unique solution. Rather, the algorithms are run multiple times, with differing initial random phases, and solutions are averaged, or in some cases tighter constraints are imposed. Several types of projection algorithms, including Gerchberg and Saxton's error reduction method,[98] Fienup's Hybrid Input Ouput (HIO) algorithm,[99] Elser's Difference Map algorithms,[100] Bautschke, Combettes and Luke's ASR, HPR, and RASR

(a)

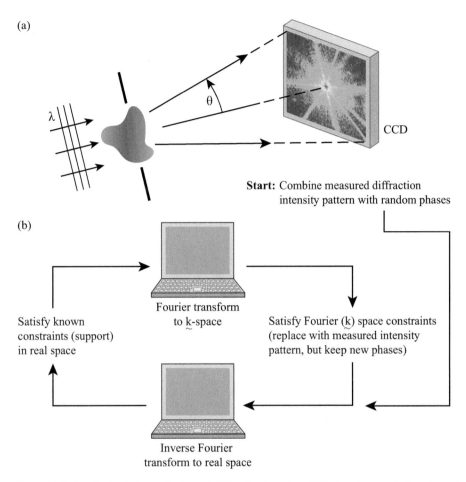

Start: Combine measured diffraction intensity pattern with random phases

(b)

Satisfy known constraints (support) in real space

Fourier transform to k-space

Satisfy Fourier (k) space constraints (replace with measured intensity pattern, but keep new phases)

Inverse Fourier transform to real space

Figure 11.24 (a) The technique of coherent diffractive imaging (CDI) involves spatially coherent radiation illuminating and scattering from a sample, then detection by a distant, 2D detector. An aperture may be used to better define the sample-illumination intensity region. (b) Beginning with the measured x-ray flux at each CCD pixel, and assigning random initial phases to each pixel, the calculated fields are then Fourier transformed back to the sample plane (real space) where they are compared to the known lateral extent of the object and its illumination and corrected accordingly. The region in which the sample has finite amplitude, perhaps defined by its absorption plus a restricting mask, is referred to as its "support." The adjusted fields are then Fourier transformed back to the detector plane where they are compared to measured intensities and corrected as necessary. The process is continued until convergence of phases is achieved. From the pixel-by-pixel convergent phase and amplitude of the electric fields an image is numerically reconstructed.[98–102] This process is repeated many times to assure convergence to a consistent solution (image) with differing initial random phases.

algorithms[101], and Marchesini's Shrinkwrap.[102] The CDI technique offers the potential advantage of avoiding optical inefficiencies, numerical aperture and depth of field limitations, as well as aberrations associated with an imaging optic or optical system. In addition to requiring a high degree of spatial coherence,[103] CDI also requires

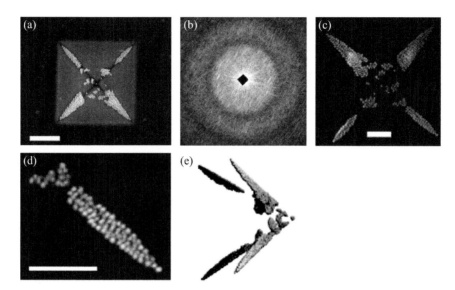

Figure 11.25 An early demonstration of x-ray coherent diffraction imaging (CDI).[102] (a) Top-down SEM view of a 2.5 μm wide, 1.8 μm high, pyramid shaped indentation in a 100 nm thick silicon nitride membrane, with 50 nm diameter gold spheres clustered along the inner edges. (b) X-ray diffraction pattern obtained with 588 eV (2.11 nm), spatially coherent, undulator radiation at the ALS. (c) Numerically reconstructed image. (d) Enlarged region of (c) showing individual gold spheres. (e) Iso-surface rendering of the reconstructed 3D image. Scale bar in (a) is 1 μm. Scale bar in (c) and (d) is 500 nm. Courtesy of S. Marchesini, A. Barty and H.N. Chapman, LLNL, now at LBNL, DESY, and DESY/U. Hamburg, respectively.

well-defined object plane constraints,[102] and numerical techniques that provide convergence to a consistent and accurate solution (taking account of experimental noise).[102] For x-ray imaging the technique can be used with spatially filtered undulator radiation, coherent laser high harmonic radiation, and coherent free electron laser radiation (Chapters 5, 6, and 7, respectively). While the technique is sometimes described as "lensless," a high quality lens is required to provide a spatially coherent wavefront at the sample with a well-defined intensity profile.

An early demonstration of 3D coherent diffractive imaging at high spatial resolution was achieved by Marchesini, Barty, Chapman and colleagues, who imaged clusters of 50 nm diameter gold spheres clustered along the inner edges of a pyramid shaped silicon nitride membrane.[102] The 2.5 × 2.5 × 1.8 μm high pyramid was formed by anisotropic etching in silicon followed by deposition of a 100 nm thick coating of Si_3N_4. Clusters of 50 nm Au balls were then dragged into place within liquid droplets which subsequently evaporated. Figure 11.25(a) shows an SEM image of the pyramid structure and clustered 50 nm Au balls. Spatially filtered 750 eV (1.65 nm) undulator radiation was used at the ALS to illuminate the sample with an estimated flux of 8×10^9 photons/μm²·s within a spectral bandwidth of $\lambda/\Delta\lambda \cong 600$. Diffraction patterns were recorded on an x-ray sensitive, 1300 × 1340, 20 μm square pixel array CCD, at a distance of 142 mm beyond the sample. Diffraction patterns were recorded at 1° intervals by rotating the sample from $-57°$ to $+66°$, and in $0.5°$ intervals within the central $19°$. To enable the detection of weak scattering from small features, observed at larger angles, data were taken over a large dynamic range by stepping the central stop to larger sizes and increasing the associated collection times. To accomplish this, data was taken in ten or more exposures at each observation angle, in steps of 0.1, 1, 10, and 60 s, for a cumulative exposure time of 73 s per observation angle. This permitted recording signals ranging from 1.9 photons/pixel at the largest angles (smallest feature sizes), to 1.1×10^5 photons/pixel at the center of the diffraction pattern. The acquisition time for a full 3D data set was approximately 3 h. Figure 11.25(b) shows a typical recorded diffraction pattern. Full 3D images were reconstructed from the 3D data using a hybrid input-output (HIO) algorithm,[99, 102] and numerous 3D fast Fourier transforms (FFTs), as discussed above, much memory storage, and use of the group's "shrink wrap" algorithm.[102] The shrink wrap algorithm improves convergence to a consistent far-field diffraction pattern/real space reconstructed image by reducing the samples lateral extent ("the support"') following each iteration. Typically 1000 k-space to real space iterations are required for convergence of the reconstructed image. Confirmation of convergence to a consistent solution is typically achieved by repeating the process hundreds of times with new assignments of random initial phases.[99, 102] Sample reconstructions are shown in Figures 11.25(c)–(e). In panel (d) an expanded view shows a cluster of clearly resolved 50 nm Au spheres. Panel (e) shows an iso-surface rendering from the 3D reconstructed image.

A major goal of CDI is the three-dimensional imaging of isolated biological structures such as proteins, viruses, organelles, and perhaps a small living cell. Typical sizes of 30–300 nm are envisioned, perhaps to 1 µm, thus requiring spatial resolutions well below 10 nm and exposure times measured in femtoseconds for "diffract and destroy" studies.[17] Photon energies approaching 10 keV offer the advantage of smaller diffraction angles. Two recent studies[104, 105] have made significant strides in this direction, all involving sequentially delivered biological samples irradiated by femtosecond duration, x-ray FEL pulses (see Chapter 6, Sections 6.4–6.9). In the paper by van der Schot, Chapman, Hajdu, and colleagues,[104] 517 eV (2.40 nm), nominally 70 fs FEL pulses were delivered at 120 Hz to a 3 µm × 7µm focal spot, each pulse containing approximately 1.1×10^{11} photons/(µm)^2 within a nominally 0.2% relative spectral bandwidth, at SLAC's LCLS. Live, micron-sized, cyanobacteria (*Cyanobium gracile*) were delivered in an aerosolized wet helium jet to the FEL focal area, also at 120 Hz. Diffracted photons were detected by two pairs of side-by-side CCDs, each panel being 512 × 1024 pixels, with a space between panels to pass the zero-order incident beam. The well capacity of each pixel was approximately 2800 photons at 517 eV. Single diffraction patterns were obtained for each cell as it traversed the x-ray interaction region. The reconstructions consisted of 5000 iterations, which were repeated 400 times with different initial phase assumptions. Figure 11.26 shows a series of recorded diffraction patterns, reconstructed 2D images (electron density distributions), and calculated Nomarski-like images as they would appear at 2.40 nm wavelength. The total number of detected photons contributing to each diffraction pattern varies between 0.5 and 5 million. The scale bar is 1 mm for each reconstructed image. Diffraction patterns

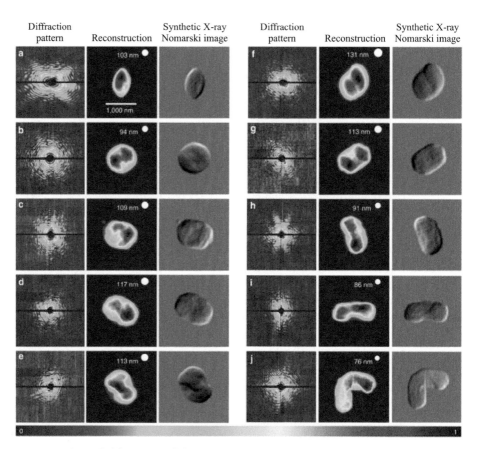

| Diffraction pattern | Reconstruction | Synthetic X-ray Nomarski image | Diffraction pattern | Reconstruction | Synthetic X-ray Nomarski image |

Figure 11.26 Recorded femtosecond duration x-ray diffraction patterns are shown for nominally 1 μm wide cyanobacteria cells, the numerically reconstructed CDI x-ray images for each of those patterns, and Nomarski equivalent differential interference contrast (DIC) images calculated from the same x-ray data sets.[104] The order of presentation in (a) through (j) is by object size. The diffraction pattern for each cell was obtained with a single 517 eV, 70 fs x-ray FEL pulse at SLAC's LCLS (see Chapter 6, Section 6.4). A 1 μm scale bar is shown in panel (a). Estimated spatial resolution is indicated for each reconstruction, e.g., 103 nm in panel (a). Courtesy of G. van der Schot, T. Eckberg, and J. Hajdu, Uppsala University.

were recorded to angles associated with 4 nm spatial period. Reconstructed image resolutions were conservatively estimated based on a $1/e$ decrease of the calculated phase retrieval transfer function. Reconstructed spatial resolution values (full period), ranging from 76 to 131 nm, are given for each image, and are illustrated as a white dot. The observed power spectral density was reported to decay in proportion to $1/k^{3.3}$, where $k \equiv 1/d$ is the spatial frequency. The authors estimate that extension of these results to nanometer spatial resolution will require an increased photon flux density to order 10^{12} to 10^{13} photons/(μm)^2, an increase of 10–100, preferably with 3–10 keV photons. Larger objects will require larger total photon flux and a narrow spectral bandwidth, as might be achieved with self-seeding as demonstrated at LCLS (see Chapter 6, Section 6.6). Improved focal spot size profile control, as well as sample tracking and directional injection, could help to optimize available photons. Extension to single particle, 3D reconstructions would require acquisition of simultaneous diffraction patterns at multiple angles.

Ekeberg, Chapman, Hajdu, and colleagues have extended the above study in a proof-of-concept experiment in which sequential, randomly oriented, similar but not necessarily identical giant mimivirus particles were examined using these same CDI techniques.[105] For this study 1.2 keV (1 nm), nominally 70 fs pulses were utilized at LCLS. A total of 198 diffraction patterns were obtained. Comparison of the 3D image reconstructions showed that the 198 mimivirus particles were identical to within the estimated 125 nm (full period) resolution. The authors suggest further studies of reproducible biological objects in the 100–200 nm range which would include HIV, influenza, and herpes viruses.[105] Very tight constraints on spatial coherence are in general required to assure reconstructed convergence to a proper solution. These constraints, and the possibility for allowing some degree of partial coherence, are discussed by Williams, Nugent and colleagues.[103] CDI is also being applied in scientific and technical applications using coherent EUV high harmonic generation (HHG, Chapter 7). As examples, new reflection CDI applications include morphological characterization of biological cells,[106] and defect detection in EUV masks for computer chip lithography (see Section 11.7).[107]

11.6.2   Ptychography

A related technique, also involving coherent illumination and scattering/diffraction is known as ptychography.[21] Unlike CDI, which targets imaging of small, isolated structures at the molecular level, ptychography offers the possibility of nanoscale imaging of much larger structures. It is related to scanning transmission x-ray microscopy (see Section 11.4, Figure 11.15) in that the sample is illuminated by a spatially coherent x-ray beam. In this case, however, the focused beam is larger and does not limit the spatial resolution. Diffracted radiation is recorded on a two-dimensional CCD, rather than simply recording transmitted x-rays as with a STXM. By scanning the sample in a grid of small, overlapping steps and collecting diffraction patterns from each, redundant information is obtained which with reconstruction algorithms allows formation of the image. With angular rotation of the sample and sufficient 2D images one can obtain 3D slice images by tomographic ptychography. The technique was first explored using electron microscopy and is now used as an x-ray imaging technique. Ptychography provides a potential route towards higher resolution imaging of nanoscale objects as the numerical aperture is set by the angle subtended by the detector (and the wavelength). Thus the resultant spatial resolution is not limited by available outer zone width, the associated lens numerical aperture, lens aberrations, or depth of focus, and is not determined by the illuminating beam size as with STXM. With small scanning stage steps there is much redundancy in the collected data, permitting image reconstruction with an improved spatial resolution, as has recently been demonstrated.[22] As with the CDI technique above, ptychography requires a high degree of spatial coherence. Ptychography also requires a lengthy data collection time, due to the reduced photon flux following undulator spatial filtering, the need for multiple redundant scanning of sample areas,

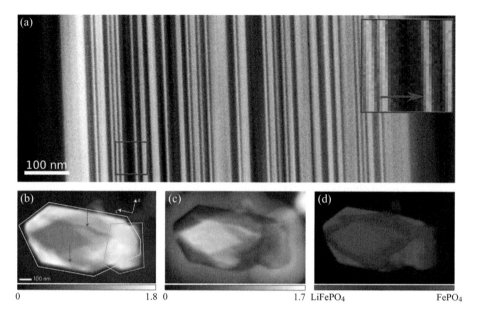

Figure 11.27 (a) A ptychographically reconstructed image of a two-dimensional test pattern showing 5 nm half-period lines and spaces and an approximately 6 nm wide isolated line within the red square. It was obtained with spatially coherent 710 eV (1.74 nm) undulator radiation at the ALS.[22] The experiments utilized a STXM with a 60 nm outer zone width focusing lens, and a 1340×1300 array CCD having 20 μm square pixels, and located 8 cm downstream of the sample. Diffraction patterns were taken in 40 nm steps on a rectangular grid. The total data acquisition time was approximately 30 s. Ptychographic absorption and phase maps are shown in (b) and (c), respectively, for a $LiFePO_4$ crystalline nanoplate (details in the text). (d) Chemical mapping of $LiFePO_4$ and $FePO_4$ obtained by taking spectroscopic data sets to either side of the Fe L_3-edge, at photon energies extending from 700 eV to 716 eV.[22] Courtesy of D. Shapiro, LBNL.

the need for high dynamic range detection to record and reconstruct weak high spatial frequency features. The technique also requires a high signal to noise ratio, well above detector noise, for successful numerical reconstructions. Tomographic imaging requirements further exacerbate data acquisition times and radiation dose to the sample.

Early experiments demonstrating the imaging and resolution capabilities of x-ray ptychography for structured and amorphous materials, one related to energy storage applications the other to sub-cellular biological imaging, are shown in Figures 11.27 and 11.28, respectively. In Figure 11.27(a) one sees the reconstructed image of a micronscale 2D test pattern having line and space test pattern of various periods. To the left side of the red-boxed area is a well resolved periodic pattern of 5 nm half-period. The data for the image in (a) were obtained at 710 eV (1.75 nm) using spatially filtered undulator radiation at the ALS 11.0.2 STXM.[22] The microscope employed a 60 nm outer zone width and an incident coherent photon flux of 4×10^7 ph/s. Diffraction patterns were recorded on a rectangular grid in steps of 40 nm using a 1340×1300 CCD with 20 μm square pixels. Double exposures of 30 ms and 150 ms duration were taken at

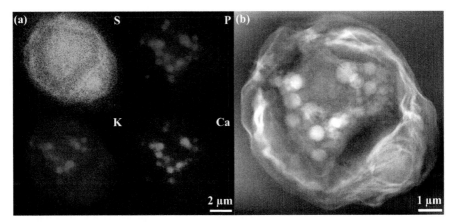

Figure 11.28 Simultaneous x-ray fluorescence and ptychographic imaging of a single *chalmydomanas alga* cell, obtained with partially coherent 5.2 keV photons at the APS.[108] (a) Fluorescence data show elemental distributions of P, S, K, and Ca within the cell to a resolution of about 70 nm set by the illumination focal spot size. The four elements have K absorption edges at 2.145, 2.472, 3.608, and 4.038 keV, respectively, all below the 5.2 keV energy of incident photons. (b) A two-dimensional phase image of the cell. The image was obtained using continuous data acquisition ptychography[109] with more than 40 000 diffraction patterns. The total data acquisition time was approximately 75 min. Courtesy of J. Deng and C. Jacobsen, Northwestern University.

each point to obtain an extended dynamic range. The data acquisition time for the image shown in (a) was approximately 30 s. The nominally 2.6 cm wide CCD was located 8 cm downstream of the sample, corresponding to a detector NA of 0.17, and a detector half-angle limited spatial resolution of 6.3 nm (full period) at 1.75 nm wavelength. Shown in Figures 11.27 (b) and (c) are the ptychographically reconstructed optical density (absorption, 710 eV) and phase maps (709.2 eV) of a partially delithiated $LiFePO_4$ crystalline nanoplate, of interest for the study of energy storage devices. The images are 7.6 μm × 12 μm wide. The data sets for (b) and (c) were obtained with spatially filtered bending magnet radiation illuminating a different STXM at the ALS. A 60 nm outer zone width illumination lens was also used, but in this case with a reduced spatially coherent photon flux of 5×10^5 ph/s. The data acquisition times were approximately 13 h, equivalent to about 10 min with the above quoted undulator coherent photon flux. Shown in Figure 11.27(d) is a colored composition map showing relative concentrations from $LiFePO_4$ to $FePO_4$, obtained by taking spectroscopically resolved data sets for photon energies extending from 700 eV to 716 eV, covering spectral regions to either side of the Fe L_3-edge at 706.8 eV.

An example of ptychograhic biological imaging is shown in Figure 11.28 for an approximately 7 μm diameter, single cryo fixed *chalmydomanas alga* cell.[108] The image was obtained at a single viewing angle using spatially filtered coherent undulator radiation at APS beamline 21-ID-D. A double crystal monochromator provided 1/5000 spectral resolution at 5.2 keV (2.8 Å), while a 50 μm wide slit provided spatial coherence in the horizontal direction and the 20 μm FWHM electron beam size provided spatial coherence in the vertical direction.[‡] The multi-keV photon energy permits deep sample penetration, clearer phase effects, and increased depth of field. A 160 μm diameter zone plate with a 70 nm outer zone width focused 3×10^8 ph/s to a nominal spot size of 72 nm FWHM. More than 40 000 diffraction patterns were taken at this single angle using a continuous data acquisition technique.[109] The total data collection time was approximately 75 min and the delivered radiation dose was estimated to be 5×10^7 Gray. The reconstructed phase image in Figure 11.28(b) was trimmed to 8.5 μm × 9 μm. The two-dimensional spatial resolution is estimated to be about 26 nm half-period. The presence of overlapping substructure within the relatively thick cell affects the ability to delineate fine features, as observed in Figure 11.28(b). Employing tomographic ptychography in future studies might alleviate this limitation through the availability of thin slice images. Taking data for many angles will, however, require substantially increased data collection times and radiation dose, of order several hundred, and more if improved spatial resolution is required. Advances in accelerator technology may provide some relief as far as data collection time, particularly with newly developed diffraction limited storage rings (DLSRs), such as recently demonstrated at MAX IV in Sweden[¶]. At MAX IV this has led to a reduced electron beam size by a factor greater than 10 in the horizontal and a factor of 2 in the vertical, and an emission solid angle reduced by an additional factor of 4, for an overall increase of brightness of order 100. Improvements in spectral efficiency, by collecting a larger fraction of the undulator spectral bandwidth than is presently accepted with double crystal monochromators, could provide an additional gain of 30. Storage ring and monochromator advances such as these have the potential to provide a path towards significantly reducing exposure times for ptychographic tomography,[110] and indeed for all other imaging experiments. However, mitigation of the increased radiation dose will remain a significant challenge.

While there is some hope that dose fractionalization[111] might provide some relief, viable reconstructions require high signal to noise within each 2D pixel, likely limiting use of this technique. The issues of coherent photon flux, data collection time, spatial resolution, and radiation dose for tomographic ptychography x-ray imaging are discussed further in papers by Dierolf, Thibault, Pfeiffer, and colleagues at the Technical University of Munich and the Swiss Light Source[113], and by Holler, Färm and colleagues at the Swiss Light Source and the University of Helsinki, respectively.[114] For example, in the paper by Hotter *et al.* achievement of radiation damage limited, 16 nm (10%–90%) isotropic 3D resolution, is reported for an artificially prepared 6 μm diameter porous glass cylindrical structure. Pores of the glass structure were coated with Ta_2O_5 for contrast. The data were obtained with 720 viewing angles at a wavelength of 2 Å, and a delivered radiation dose of 1.2×10^{10} Gray. The authors believe that the dose can be reduced by a factor of six while maintaining achievable spatial resolution.

[‡] See Chapter 5, Section 5.6, and Figure 5.33 for discussion of spatial filtering at a similar APS undulator beamline.

[¶] A brief discussion of DLSRs can be found in Chapter 5, in the caption of Figure 5.24, and references therein.

11.7   Optical System for EUV Lithography

In this section we describe the optical system installed in the first commercial EUV lithography step-and-scan system for printing computer chip patterns. Known as an EUV wafer scanner, it is manufactured by the Dutch company ASML,[115] with multi-layer coated, EUV reflective optics by Zeiss.[116–118] The scanner and optics are shown in Figure 11.29. The optics are multilayer coated for high reflectivity at 13.5 nm. The ASML model NXE 3300 utilizes a pulsed EUV emitting plasma source created by CO_2 laser pulses sequentially irradiating liquid tin (Sn) droplets, as described previously in Chapter 8, Section 8.8. Radiation from the plasma is collected by a partially enclos-ing, large solid angle (5 Sr), multilayer coated first mirror.[119] This mirror plus three others (four in total) constitute the illuminating optical system[116, 117] for the reflective mask. In addition to collecting the radiation, the illumination optics shape it into an arc shaped 1.5–2 mm $\times$ 26 mm illumination pattern at the mask. The three-dimensional (3D) multilayer mask pattern that is to be copied is fabricated on a six inch square by ¼ inch thick substrate made of a low thermal expansion material (LTEM). In operation the mask is scanned above the illumination ring. The projected 4:1 reduced image is then printed at the wafer using the step-and-scan as illustrated in Figure 11.29(c). A system of six, near-normal incidence, multilayer coated, aspheric mirrors collects the wavefront from the mask and transfers it to a 4:1 demagnified, 26 mm $\times$ 33 mm field on the synchronously scanned wafer.[116, 118] This lithographic system is called a step-and-scan system as it sequentially scans and records the pattern, then steps to a fresh area of the wafer and repeats the copying process. Projection optics in the ASML model NXE 3300 have a 0.33 NA, a wavefront error of 0.30 nm rms ($\lambda/45$), and partially coherent illumination characterized by $\sigma \cong 0.9$ for standard operation. Variable partial coherent illumination, with $\sigma = 0.2$–0.9, will be available depending on the nature of the pattern to be printed. Mo/Si multilayer coatings, with interface barriers and a capping layer, as was described in Chapter 10, Section 10.3, are used to provide a nominally 2% wide spectral bandpass at 13.5 nm, with reflectivity approaching 70%. The full optical train (all mirrors, full aperture) achieves an average reflectivity of 64% per mirror. Special mirror coatings have been developed that additionally suppress out-of-band UV radia-tion.[120] The optics have a high spatial frequency surface roughness of 0.1 nm rms, and mid-spatial frequency§ values between 0.06 nm and 0.08 nm rms, corresponding to a 4% flare specification[121, 122] for printing 2 µm wide lines. Flare is the unwanted, small angle scattered radiation that reduces contrast in the printed pattern. The wafer size is 300 mm in diameter, permitting the exposure of nearly one hundred 26 $\times$ 33 mm pat-terns. It is covered with an EUV recording material, a molecular photoresist, typically with sensitivity in the range of 10–60 mJ/cm^2, selected in a tradeoff among spatial res-olution, sensitivity, and line-edge-roughness of the printed patterns. With the currently

---

§ For EUV wavelengths, mid-spatial frequencies extend from $2\pi/1$ mm to $2\pi/1$ µm, largely affecting image contrast. Lower spatial frequencies are considered figure errors, due to aberrations. Higher spatial frequen-cies (surface roughness, from $2\pi/1$ µm to $2\pi/10$ nm, scatter light out of the optical system and thus largely affect reflectivity.[121, 122]

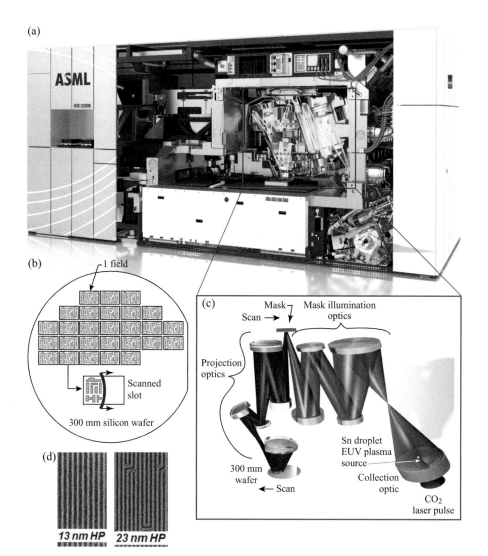

Figure 11.29 (a) The ASML NXE 3300 EUV lithography wafer scanner for the production of next generation computer chips.[115] The system uses a CO_2 pulsed laser and a jet of liquid tin droplets to generate 13.5 nm radiation which is collected, shaped, and directed at an EUV reflective mask upon which is the pattern to be copied for a particular chip layer. The reflected wavefront is then optically transferred and imaged to a recording wafer. The collection/mask illumination system consists of four mirrors. The mask to wafer projection optics, which transfer an image of the mask pattern at 4:1 demagnification to a recording wafer, consists of six near-normal incidence, multilayer coated mirrors.[116–118] (b) An illustration of the step-and-scan method of printing by which a scanned image is formed at the wafer, then stepped to a clean area and repeating the process there. (c) A diagram showing the Zeiss EUV reflective optical system as used in the NXE 3300 scanner.[116–118] (d) A single exposure, printed test pattern showing parallel 13 nm half-period lines and spaces, and a somewhat more complex printed test pattern with 23 nm half-period lines and spaces. Courtesy of V. Banine, ASML, and W. Kaiser, Zeiss. A description of a generic tin droplet-plasma source for EUV lithography is given in Chapter 8, Section 8.8. An overview of EUV lithography is presented in the book by H. Kinoshita.[124]

available 100 W EUV source power (in-band, no debris, IF), and a resist sensitivity of
15 mJ/cm^2, the NXE 3300 is capable of printing 80 wafers/h. Sample single exposure
test patterns are shown in Figure 11.29(d, e). Further advances in source power will
permit the use of less sensitive but higher spatial resolution, better line-edge-roughness
resists, and/or increased wafer throughput for higher throughput manufacturing of computer chips at the "7 nm node" in the year 2017 and beyond. At this node it is anticipated
that chips will be fabricated with half-pitch dimensions of 13–25 nm depending on pattern complexity, for example logic, flash or DRAM. High performance microprocessor
(MPU) chips are expected to have a 14 nm wide physical gate width.[123]

References

1. W.C. Röntgen, "On a New Kind of Rays," Sitzungsberichte der Würzburger Physik-Medic
   Gesellschaft (December 28, 1895); translations appeared in *Nature* **53** (1369), 274 (January
   23, 1896); and in *Science* **III** (59), 227 (February 14, 1896).
2. W.C. Rontgen, "A New Form of Radiation," *Science* **III** (72)726 (May 15, 1896).
3. O. Glasser, *Wilhelm Conrad Rontgen and the Early History of the Roentgen Rays* (Thomas
   Books, Baltimore, Md and Springfield, IL, 1934).
4. R.F. Mould, *A Century of X-rays and Radioactivity in Medicine* (IOP, Bristol, 1993).
5. B.L. Henke, E.M. Gullikson, and J.C. Davis, "X-Ray Interactions: Photoabsorption, Scattering, Transmission, and Reflection at E = 50–30,000 eV, Z = 1−92," *Atomic Data and Nucl.
   Data Tables* 54, 181 (1993). Updated values are maintained by E.M. Gullikson at www.cxro
   .LBL.gov/optical_constants .
6. J. Stöhr and H.C. Siegmann, *Magnetism: From Fundamentals to Nanoscale Dynamics*
   (Springer, Heidelberg, 2006).
7. S. Singh, "X-Ray Photoemission Spectromicroscopy and its Application to the Study of
   Patterned Titanium Silicide," PhD thesis, Physics Department, University of Wisconsin–
   Madison (1996); S. Singh, H. Solak, N. Krasnoperov *et al.*, "An X-Ray Spectromicroscopic
   Study of the Local Structure of Patterned Titanium Silicide," *Appl. Phys. Lett.* **71**, 55 (1997).
8. S. Hüfner, *Photoelectron Spectroscopy: Principles and Applications* (Springer, Berlin,
   1996).
9. E. Rotenberg and A. Bostwick, "MicroARPES and nanoARPES at Diffraction-Limited
   Light Sources: Opportunities and Performance Gains," *J. Synchr. Rad.* **21**, 1048 (September
   2014).
10. J. Als-Neilson and D. McMorrow, *Elements of Modern X-Ray Physics* (Wiley, 2011),
    Second Edition, Chapter 6.
11. Y. Shvyd'ko, *X-Ray Optics* (Springer, Berlin, 2004).
12. R. W. James, *The Optical Properties of the Diffraction of X-Rays* (Bell, 1950).
13. B.D. Cullity and S.R. Stock, *Elements of X-Ray Diffraction* (Pearson, 2001), Third Edition.
14. B.E. Warren, *X-Ray Diffraction* (Dover, 1990); D.E. Sands, *Introduction to Crystallography*
    (Dover, 2014).
15. W.H. Bragg and W.L. Bragg, "The Reflection of X-rays by Crystals," *Proc. Royal Society
    (London)* **88** (605), 428 (July 1, 1913); W.L. Bragg, "The Specular Reflection of X-rays,"
    Nature **90** (2250), 410 (December 12, 1912).
16. J.D. Watson and F.H.C. Crick, "Molecular Structure of Nucleic Acids," *Nature* **171**, 737
    (April 25, 1953); M.H.F. Wilkins, A.R. Stokes and H.R. Wilson, "Molecular Structure
    of Deoxypentose Nucleil Acids," ibid, 738; R.E. Franklin and R.G. Gosling, "Molecular

Configuration in Sodium Thymonucleate," ibid, 740; J.D. Watson, *Double Helix* (Norton Critical Edition), G.S. Stent, Editor; M. Ridley, *Francis Crick, Discoverer of the Genetic Code* (HarperCollins, Atlas Books, Eminent Lives, 2006).

17. R. Neutze, R. Wouts, D. van der Spoel, E. Weckert and J. Hajdu, "Potential for Biomolecular Imaging with Femtosecond X-ray Pulses," *Nature* **406**, 752 (August 17, 2000); H. N. Chapman *et al.*, "Femtosecond Time-Delay X-Ray Holography," *Nature* **448**, 676 (2007).

18. H.N. Chapman *et al.*, "Femtosecond X-Ray Protein Nanocrystallography," *Nature* **470**, 73 (2011); K. Ayyer *et al.*, "Molecular Diffractive Imaging Using Imperfect Crystals," *Nature* **530**, 202 (February 11, 2016).

19. I.K. Robinson, I.A. Vartanyants, G.J. Williams and M.A. Pitney, "Reconstruction of the Shapes of Gold Nanocrystals Using Coherent X-ray Diffraction," *Phys. Rev. Lett.* **87** (19), 195505 (November 5, 2001).

20. H. N. Chapman *et al.*, "Femtosecond Diffractive Imaging with a Soft-X-Ray Free-Electron Laser," *Nature Phys*. **2**, 839 (2006); J.C.H. Spence, U. Weierstall and H.N. Chapman, "X-Ray Lasers for Structural and Dynamic Biology," *Rep. Progr. Phys.* **75**, 102601 (2012).

21. P. Thiebault, M. Dierolf, A. Menzel *et al.*, "High-Resolution Scanning X-ray Diffraction Microscopy", *Science* **321**, 379 (July 18, 2008); J.M. Rodenburg, A.C. Hurst, A.G. Cullis *et al.*, "Hard-X-Ray Lensless Imaging of Extended Objects," *Phys. Rev. Lett.* **98** (3), 034801 (January 19, 2007).

22. D.A. Shapiro, Y.-S. Yu, T. Tyliszczak *et al.*, "Chemical Composition Mapping with Nanometre Resolution by Soft X-ray Microscopy," *Nature Photonics* **8**, 765 (October 2014).

23. A.R. Neureuter, University of California, Berkeley. Also see M. Selin, "3D X-ray Microscopy: Image formation, Tomography and Instrumentation," Ph.D Thesis, KTH Royal Institute of Technology, Stockholm, Figure 3.2.

24. M.M. O'Toole and A.R. Neureuter, "The Influence of Partial Coherence on Projection Printing," *SPIE* **174**, 22 (1979); J.W. Goodman, *Introduction to Fourier Optics* (McGraw-Hill, New York, 1996), Second Edition, p. 159.

25. G.N. Hounsfield, "A Method of and Apparatus for Examination of a Body by Radiation such as X-ray or Gamma Radiation", Patent Specification1283915, The Patent Office (UK, 1972); G.N. Hounsfield, "Computerized Transverse Axial Scanning (Tomography) I. Description of System," *British J. Radiology* **46**, 1016 (1973); G.N. Hounsfield, "Computed Medical Imaging", Nobel Lecture (December 8, 1979).

26. A.C. Kak and M. Slaney, *Principles of Computerized Tomographic Imaging* (IEEE Press, New York, 1988); J. Hsieh, *Computed Tomography: Principles, Design, Artifacts and Recent Advances* (SPIE Press, 2003).

27. A.P. Dempster, N.M. Laird and D.B. Rubin, "Maximum Likelihood from Incomplete Data via the EM Algorithm," *J. Royal Statist. Soc. B* **39**(1), 1–38 (1977).

28. R. Gordon, R. Bender and G.T. Herman, "Algebraic Reconstruction Techniques (ART) for Three-Dimensional Electron Microscopy and X-ray Photography," *J. Theoret. Biology* **29**(3), 471 (December 1970).

29. F. Natterer, *The Mathematics of Computerized Tomography* (B.G. Teubner, Stuttgart, 1986).

30. M. D. de Jonge and S. Vogt, "Hard X-ray Fluorescence Tomography: An Emerging Tool for Structural Visualization," *Current Opinion Struct. Biology* **20**, 606 (2010).

31. P.C.J. Donoghue, S. Bengston, X.-P. Dong *et al.*, "Synchrotron X-ray Tomographic Microscopy of Fossil Embryos," *Nature* **442**, 680 (August 10, 2006).

32. S. Heinzer, G. Kuhn, T. Krucker *et al.*, "Novel Three-Dimensional Analysis Tool for Vascular Trees Indicates Complete Micro-Networks, not Single Capillaries, as the

Angiogenic Endpoint in Mice Overexpressing Human $VEGF_{165}$ in the Brain," *Neuroimage* **39**, 1549 (2008); C. Hintermuller, J.S. Coats, A. Obenhaus *et al.*, "3D Quantification of Brain Microvessels Exposed to Heavy Particle Radiation," *9th Int'l Conf. X-ray Microsc.* (2009), doi: 10.1088/1742–6596/186/1/012087; C. Hintermuller, F. Marone, A. Isenegger and M. Stampanoni, "Image Processing Pipeline for Synchrotron-Radiation-Based Tomographic Microscopy," *J. Synchr. Rad.* **17**, 550 (2010).

33. M. Selin, E. Fogelqvist, S. Werner and H.M. Hertz, "Tomographic Reconstruction in Soft X-ray Microscopy using Focus-Stack Back-Projection," *Optics Lett.* **40**(10), 2201 (May 15, 2015).

34. G.-C. Yin, M.-T. Tang, Y.-F. Song *et al.*, "Energy-Tunable Transmission X-ray Microscope for Differential Contrast Imaging with Near 60 nm Resolution Tomography," *Appl. Phys. Lett.* **88**, 241115 (2006); D. Attwood, "Nanotechnology Comes of Age," *Nature* **442**, 642 (August 10, 2006).

35. S.W. Wilkins, T.E. Gureyev, D. Gao, A. Pogany and A.W. Stevenson, "Phase-Contrast Imaging Using Polychromatic Hard X-rays," *Nature* **384**, 335 (November 28, 1996); A. Pogany, D. Gao and S.W. Wilkens, "Contrast and Resolution in Imaging with a Microfocus X-ray Source," *Rev. Sci. Instrum.* **68**(7), 2774 (July 1997).

36. P. Cloetens, R. Barrett, J. Baruchei, J.-P. Guigay and M. Schlenker, "Phase Objects in Synchrotron Radiation Hard X-ray Imaging," *J. Phys. D: Appl. Phys.* **29**, 133 (1996).

37. D. Chapman, W. Thomlinson, R.E. Johnson *et al.*, "Diffraction Enhanced X-ray Imaging," *Phys. Med. Bio.* **42**, 2015 (1997); Z. Zhong, W. Thomlinson, D. Chapman and D. Sayers, "Implementation of Diffraction-Enhanced Imaging Experiments: at the NSLS and APS," *NIMA* **450**, 556 (2000).

38. M.E. Kelly, D.J. Coupal, R.C. Beavis *et al.*, "Diffraction-Enhanced Imaging of a Porcine Eye," *Can. J. Opthalmology* **42**(5), 731 (2007).

39. C. Partham, Z. Zhong, D.M. Connor, L.D. Chapman and E.D. Pisano, "Design and Implementation of a Compact Low-Dose Diffraction Enhanced Medical Imaging System," *Acad. Radiol.* **16**, 911 (2009).

40. A. Momose, S. Kawamoto, I. Koyama *et al.*, "Demonstration of X-ray Talbot Interferometry," *Jpn. J. Appl. Phys.* **42**(2, 7B), L866 (July 15, 2003); "Phase Tomography by X-ray Talbot Interferometry for Biological Imaging," *Jpn. J. Appl. Phys.* **45**(6A), 5254 (2006); W. Yashiro, Y. Takeda, A. Takeuchi, Y. Suzuki and A. Momose, "Hard X-ray Phase-Difference Microscopy using a Fresnel Zone Plate and a Transmission Grating," *Phys. Rev. Lett.* **103**, 180801 (October 30, 2009).

41. Lord Rayleigh, "On Copying Diffraction-Gratings," *Phil. Magazine* XXV (Series 5)**11**, 196 (1881).

42. T. Weitkamp, A. Diaz, C. David *et al.*, "X-ray Phase Imaging with a Grating Interferometer," *Optics Express* **13** (16), 6296 (August 8, 2005).

43. J.M. Kim, I.H. Cho, S.Y. Lee *et al.*, "Observation of the Talbot Effect Using Broadband Hard X-ray Beam," *Optics Express* **18**(24), 24975 (November 22, 2010).

44. T. Thüring, T. Zhou, U. Lundström *et al.*, "X-ray Grating Interferometry with a Liquid-Metal-Jet Source," *Appl. Phys. Lett.* **103**, 091105 (2013).

45. T. Zhou, U. Lundström, T. Thüring *et al.*, "Comparison of Two X-ray Phase Contrast Imaging Methods with a Microfocus Source," *Optics Express* **21**(25), 030183 (December 16, 2013); W. Vågberg, D.H. Larsson, M. Li, A. Arner and H.M. Hertz, "X-ray Phase-Contrast Tomography for High-Spatial-Resolution Zebrafish Muscle Imaging," *Science Rept.* **5**, 16625 (November 13, 2015).

46. F. Pfeiffer, T. Weitkamp, O. Bunk and C. David, "Phase Retrieval and Differential Phase-Contrast Imaging with Low-Brilliance X-ray Sources," *Nature Physics* **21**, 258 (April 2006).

47. www.excillum.com

48. A.C. Thompson, J.H. Underwood, Y. Wu *et al.*, "Elemental Measurements with an X-ray Microprobe of Biological and Geological Samples with Femtogram Sensitivity," *NIMA* **266**, 318 (1988).

49. M. Cotte, J. Susini, J. Dik and K. Janssens, "Synchrotron-Based X-ray Absorption Spectroscopy for Art Conservation: Looking Back and Looking Forward," *Accts. Chem. Res.* **43**(6), 705 (June 2010).

50. S. Lahlil, I. Biron, M. Cotte, J. Susini and N. Menguy, "Synthesis of Calcium Antimonate Nano-crystals by the 18th Dynasty Egyptian Glassmakers," *Appl. Phys. A: Mat. Sci. Process.* **98**, 1 (2010).

51. J. Dik, K. Janssens, G. van der Snickt *et al.*, "Visualization of a lost Painting by Vincent van Gogh using Synchrotron Radiation Based X-ray Fluorescence Elemental mapping," *Anal. Chem.* **80**, 6436 (August 15, 2008).

52. W. Klysubun, C.A. Hauzenberger, B. Ravel *et al.*, "Understanding the Blue Color in Antique Mosaic Mirrored Glass from the Temple of the Emerald Budda, Thailand," *X-ray Spectrom.* **44**, 116 (2015).

53. B. Ravel, G.L. Carr, C.A. Hauzenberger and W. Klysubun, "X-ray and Optical Spectroscopic Study of the Coloration of Red Glass used in 19th Century Decorative Mosaics at the Temple of the Emerald Budda," *J. Cult. Heritage* **16**, 315 (2015).

54. V. Mocella, E. Brun, C. Ferrero and D. Delattre, "Revealing Letters in Rolled Herculaneum Papyri by X-ray Phase Contrast Imaging," *Nature Commun.* **6**, 5895 (January 20, 2015).

55. D. Commeli, M. D'Orazio, L. Folco *et al.*, "The Meteoric Origin of Tutankhamon's Iron Dagger Blade," *Meteoritics and Planetary Science* **1**, 9 (June 2016).

56. S. Rehbein, P. Guttmann, S. Werner and G. Schneider, "Characterization of the Resolving power and Contrast Transfer Function of a Transmission X-ray Microscope with Partially Coherent Illumination," *Opt. Express* **20**(6), 5830 (March 12, 2012).

57. W. Chao, B.D. Harteneck, J.A. Liddle, E.H. Anderson and D.T. Attwood, "Soft X-ray Microscopy at a Spatial Resolution Better than 15 nm," *Nature* **435**, 1210 (June 30, 2005).

58. J. Vila-Comamala, K. Jefimovs, J. Raabe *et al.*, "Advanced Thin Film Technology for Ultra-high Resolution X-ray Microscopy," *Ultramicroscopy* **109**, 1360 (2009).

59. W. Chao, P. Fischer, T. Tyliszczak *et al.*, "Real Space Soft X-ray Imaging at 10 nm Spatial Resolution," *Optics Express* **20**(9), 9777 (April 23, 2012).

60. H.M. Hertz, O. von Hofsten, M. Bertilson *et al.*, "Laboratory Cryo Soft X-ray Microscopy," *J. Struct. Bio.* **177**, 267 (2012).

61. H. Ade, X. Zhang, S. Cameron *et al.*, "Chemical Contrast in X-ray Microscopy and Spatially resolved XANES Spectroscopy of Organic Specimens," *Science* **258**, 972 (November 6, 1992); C. Jacobsen, G. Flynn, S. Wirick and C. Zimba, "Soft X-ray Spectroscopy from Image Sequences with Sub-100 nm Spatial Resolution," *J. Microsc.* **197**, 173 (2000); M. Lerotic and C. Jacobsen, "Cluster Analysis of Soft X-ray Spectromicroscopy Data," *Ultramicrosc.* **100**, 35 (2004); H. Ade and H. Stoll, "Near-Edge X-ray Absorption Fine-Structure Microscopy of Organic and Magnetic Materials," *Nature Materials* **8**, 281 (April 2009).

62. A.P. Hitchcock, J.J. Dynes, J.R. Lawrence *et al.*, "Soft X-ray Spectromicroscopy of Nickel Sorption in a Natural River Biofilm," *Geobiology* **7**, 432 (2009); M. Obst, J. Wang and A.P. Hitchcock, "Soft X-ray Spectro-Tomography Study of Cyanobacterial Biomineral Nucleation," *Geobiology* **7**, 577 (2009).

63. J.J. Dynes, T. Tyliszczak, T. Araki *et al.*, "Speciation and Quantitative Mapping of Metal Species in Microbial Biofilms using Scanning Transmission X-ray Microscopy," *Environ. Sci. Technol.* **40**(5), 1556 (2006); J.D. Denlinger, E. Rotenberg, T. Warwick *et al.*, "First Results from the SpectroMicroscopy Beamline at the Advanced Light Source," *Rev. Sci. Instrum.* **66** (2), 1342 (February 1995).

64. B. Kaulich, P. Thibault, A. Gianoncelli and M. Kiskinova, "Transmission and Emission X-ray Microscopy: Operation Modes, Contrast Mechanisms and Applications," *J. Phys.: Condens. Matter* **23**, 083002 (2011).

65. B. Bozini, A. Gianoncelli, L. Gregoratti and M. Kiskinova, "Recent Advances of Synchrotron-Based Scanning X-ray Microscopy in Addressing Properties of Morphologically Complex Materials: Electrosyntheses and Aging of Co/PPy Electrocatalyst", *Int'l Sympos. Atomic Level Charact. Mater. Devices, ALC'15*, Matsue, Japan (October 2015).

66. C. Chang and A. Sakdinawat, "Ultra-High Aspect Ratio High-Resolution Nanofabrication for Hard X-ray Diffractive Optics," *Nature Commun.* **5**, 4243 (June 27, 2014).

67. H. Ade, "Development of a Scanning Photoemission Microscope," PhD thesis, Physics Department, SUNY, Stony Brook (1990); H. Ade, C.-H. Ko, E. Johnson and E. Anderson, "Improved Images with the Scanning Photoemission Microscope at the National Synchrotron Light Source," *Surface and Interface Anal.* **19**, 17 (1992).

68. C. Capasso, W. Ng, A.K. Ray-Chaudhuri *et al.*, "Scanning Photoemission Spectromicroscopy on MAXIMUM Reaches 0.1 Micron Resolution," *Surface Sci.* **287/88**, 1046 (1993); A.K. Ray-Chaudhuri, "Development of a Scanning Photoemission Microscope Based on Multilayer Optics and its Initial Application to GaAs (110) Surface Studies," PhD thesis, Electrical and Computer Engineering Department, University of Wisconsin–Madison (1993).

69. E. Rotenberg and A. Bostwick, "MicroARPES and nanoARPES at Diffraction-Limited Light Sources: Opportunities and Performance Gains," *J. Synchr. Rad.* **21**, 1048 (September 2014); M.A. Olmstead, R.I.G. Uhrberg, R.D. Bringans and R.Z. Bachrach, "Photoemission Study of Bonding at the CaF_2-on-Si(111) Interface," *Phys. Rev. B* **35** (14), 7526 (May 15, 1987); E. Rotenberg, J.D. Denlinger and M.A. Olmstead, "Altered Photoemission Satellites at CaF_2- and SrF_2-on-Si(111) Interfaces," *Phys. Rev. B* **53** (3), 1584 (January 15, 1996).

70. M. Cotte, J. Susini, J. Dik and K. Janssens, "Synchrotron-Based X-ray Absorption Spectroscopy for Art Conservation: Looking Back and Looking Forward," *Accts. Chem. Res.* **43**(6), 705 (June 2010).

71. H. Rarback, D. Shu, S.C. Feng *et al.*, "Scanning X-Ray Microscope with 75-nm Resolution," *Rev. Sci. Instrum.* **59**, 52 (1988).

72. J. Kirz, C. Jacobsen, S. Lindaas *et al.*, "Soft X-Ray Microscopy at the National Synchrotron Light Source," p. 563 in *Synchrotron Radiation in the Biosciences* (Oxford University Press, 1994), B. Chance, J. Deisenhofer, T. Sasaki *et al.*, Editors.

73. J. Kirz, C. Jacobsen and M. Howells, "Soft X-Ray Microscopes and Their Biological Applications," *Q. Rev. Biophys.* **28**, 1 (1995).

74. C. Jacobsen, S. Williams, E. Anderson *et al.*, "Diffraction-Limited Imaging in a Scanning Transmission X-Ray Microscope," *Opt. Commun.* **86**, 351 (1991).

75. G.R. Morrison and M.T. Browne, "Dark-Field Imaging with the Scanning Transmission X-ray Microscope," *Rev. Sci. Instrum.* **63**, 611 (1992).

76. H.N. Chapman, C. Jacobsen and S. Williams, "A Characterization of Dark-Field Imaging of Colloidal Gold Labels in a Scanning Transmission Microscope," *Ultramicroscopy* **62**,

191 (1996); M.P.K. Feser, "Scanning Transmission X-ray Microscopy with a Segmented Detector" PhD Thesis, Physics Department, SUNY Stony Brook (2002).

77. A.E. Sakdinawat, Contrast and Resolution Enhancement Techniques for Soft X-ray Microscopy", PhD Thesis, Joint Bioengineering Program, University of California, Berkeley, and University of California San Francisco (2008).

78. www.elettra.trieste.it/elettra-beamlines/twinmic.html

79. B.W. Loo, W. Meyer-Ilse and S.S. Rothman, "Automatic Image Acquisition, Calibration and Montage Assembly for Biological X-ray Microscopy", *J. Microsc.* **197**(2), 185 (February 2000).

80. F. Zernike, "How I Discovered Phase Contrast," Nobel Lecture, Stockholm, December 1953, in *Science* **121**, 345 (1955); also "Phase Contrast, A New Method for the Microscopic Observation of Transparent Objects, Part I," *Physica* **9**, 686 (1942).

81. G. Schmahl, P. Guttmann, G. Schneider *et al.*, "Phase Contrast Studies of Hydrated Specimens with the X-Ray Microscope at BESSY," p. 196 in *X-Ray Microscopy IV* (Bogorodskii Press, Chernogolovka, Russia, 1994), V.V. Aristov and A.I. Erko, Editors.

82. G. Schmahl and D. Rudolph, "High Power Zone Plates as Image Forming Systems for Soft X-Rays" (in German), *Optik* **29**, 577 (1969).

83. B. Nieman, D. Rudolph and G. Schmahl, "Soft X-Ray Imaging Zone Plates with Large Zone Numbers for Microscopic and Spectroscopic Applications," *Opt. Commun.* **12**, 160 (1974); B. Niemann, D. Rudolph and G. Schmahl, "X-Ray Microscopy with Synchrotron Radiation," *Appl. Opt.* **15**, 1883 (1976); G. Schmahl, D. Rudolph, P. Guttmann and O. Christ, "Zone Plates for X-ray Microscopy," p. 63 in *X-Ray Microscopy* (Springer-Verlag, Berlin, 1984), G. Schmahl and D. Rudolph, Editors.

84. G. Schmahl, D. Rudolph, P. Guttmann and O. Christ, "Status of the Zone Plate Microscope", *SPIE* **316**, 100 (1982); G. Schmahl, D. Rudolph, B. Niemann and O. Christ, "Zone Plate X-Ray Microscopy," *Q. Rev. Biophys.* **13**, 297 (1980).

85. D. Rudolph, B. Niemann, G. Schmahl and O. Christ, "The Göttingen X-Ray Microscope and X-Ray Microscopy Experiments at the BESSY Storage Ring," p. 192 in *X-Ray Microscopy* (Springer-Verlag, Berlin, 1984), G. Schmahl and D. Rudolph, Editors.

86. G. Schmahl, D. Rudolph, B. Niemann *et al.*, "Natural Imaging of Biological Specimens with X-Ray Microscopes," p. 538 in *Synchrotron Radiation in the Biosciences* (Oxford University Press, 1994), B. Chance, J. Deisenhofer, T. Sasaki *et al.*, Editors.

87. G. Schneider, B. Niemann, P. Guttmann, D. Rudolph and G. Schmahl, "Cryo X-Ray Microscopy," *Synchr. Rad. News* **8**, 19 (1995).

88. W. Meyer-Ilse, D. Hammamoto, A. Nair *et al.*, "High Resolution Protein Localization using Soft X-ray Microscopy," *J. Microsc.* **201**, 395 (March 2001).

89. G. Schneider, "Cryo X-ray Microscopy with High Spatial Resolution in Amplitude and Phase," *Ultramicroscopy* **75**, 85 (1998); G. Schneider *et al.*, *Surf. Sci. Lett.* **9**, 177 (2002).

90. C.A. Larabell and M.A. Le Gros, "X-ray Tomography Generates 3-D Reconstructions of the Yeast, *Saccharomyces cerevisiae*, at 60-nm Resolution," *Molec. Biol. Cell.* **15**, 957 (March 2004); M.A. Le Gros, G. McDermott and C. Larabell, "X-ray Tomography of Whole Cells," *Curr. Opt. Struct. Biol.* **15**, 593 (2005); D.Y. Parkinson, G. McDermott, L.D. Elkin, M.A. Le Gros and C.A. Larabell, "Quantitative 3-D Imaging of Eukaryotic Cells using Soft X-ray Tomograph," *J. Struct. Biol.* **162**, 380 (2008).

91. J.L. Carrascosa, F.J. Chicón, E. Pereiro *et al.*, "Cryo-X-ray Tomography of Vaccina Virus Membranes and Inner Compartments," *J. Struct. Biol.* **168**, 234 (2009).

92. P. Guttmann, C. Bittencourt, S. Rehbein *et al.*, "Nanoscale Spectroscopy with Polarized X-rays by NEXAFS-TXM," *Nature Photon.* **6**, 25 (January 2012); P. Guttmann and C. Bittencourt, "Overview of Nanoscale NEXAFS Performed with Soft X-ray Microscopes," *Beilstein J. Nanotechn.* **6**, 595 (2015).

93. C. Bittencourt, M. Rutar, P. Umek *et al.*, "Molecular Nitrogen in N-Doped TiO_2 Nanoribbons," *RSC Adv.* **5**, 23350 (2015); M. Rutar *et al.*, *Beilstein J. Nanotechn.* **5**, 831 (2015); I. Arusenko *et al.*, *Acta Crystallogr. Sect. B: Struct. Sci.* **67**, 218 (2011).

94. F. Meirer, J. Cabana, Y. Liu *et al.*, "Three-Dimensional Imaging of Chemical Phase Transformations at the Nanoscale with Full-Field Transmission X-ray Microscopy," *J. Synchr. Rad.* **18**, 773 (2011).

95. G. Schneider, P. Guttmann, S. Heim *et al.*, "Three-Dimensional Cellular Ultrastructure Resolved by X-ray Microscopy", *Nature Meth.* **7**(12), 985 (December 2010); G. Schneider, P. Guttmann, S. Rehbein, S. Werner and R. Follath, "Cryo X-ray Microscope with Flat Sample Geometry for Correlative Fluorescence and Nanoscale Tomographic Imaging," *J. Struct. Biol.* **177**, 212 (2012); C. Hagan *et al.*, *Cell* **163**, 1692 (December 17, 2015).

96. E. Pereiro and F.J. Chichón, "Cryo Soft X-ray Tomography of the Cell," *eLS Wiley* (November 2014).

97. J. Miao, P. Charalambous, J. Kirz and D. Sayre, Extending the Methology of X-ray Crystallography to Allow Imaging of Micrometre-Sized Non-Crystalline Specimens," *Nature* **400**, 342 (1999).

98. R.W. Gerchberg and W.O. Saxton, "A Practical Algorithm for the Determination of Phase from Image and Diffraction Plane Pictures," *Optik* **35**, 237 (1972).

99. J.R. Fienup, "Reconstruction of an Object from the Modulus of its Fourier Transform," *Optics Lett.* **3**, 27 (July 1, 1978); J.R. Fienup, "Phase Retrieval Algorithms: A Comparison," *Applied Optics* **21**, 2758 (August 1, 1982).

100. V. Elser, "Phase Retrieval by Integrated Projections," *J. Opt. Soc. Amer. A* **20**, 40 (2003)

101. H. Bautschke, P.L. Combettes and D.R. Luke, *J. Opt. Soc. Amer. A* **19**, 1344 (2002).

102. S. Marchesini, H.N. Chapman, S.P. Hau-Riege *et al.*, "Coherent X-ray Diffractive Imaging: Applications and Limitations," *Optics Express* **11**, 2344 (September 22, 2003); S. Marchesini *et al.*, "X-ray Image Reconstruction from a Diffraction Pattern Alone," *Phys. Rev. B* **68**, 140101 (October 1, 2003); H.N. Chapman *et al.*, "High-Resolution ab initio Three-Dimensional X-ray Diffraction Microscopy," *J. Opt. Soc. Am. A* **23**, 1179 (May 2006).

103. G.J. Williams, H.M. Quincy, A.G. Peele and K.A. Nugent,"Coherent Diffractive Imaging and Partial Coherence," *Phys. Rev. B* **75**, 104102 (2007).

104. G. van der Schot *et al.*, "Imaging Single Cells in a Beam of Live Cyanobacteria with an X-ray Laser," *Nature Commun.* **6**, 5704 (February 11, 2015); L. Strüder *et al.*, "Large-Format, High-Speed, X-ray pnCCDs Combined with Electron and Ion Imaging Spectrometers in a Multipurpose Chamber for Experiments at 4th Generation Light Sources," *NIM A* **614**, 483 (March 8, 2010).

105. T. Ekeberg *et al.*,"Three-Dimensional Reconstruction of the Giant Mimivirus Particle with an X-ray Free-Electron Laser," *Phys. Rev. Lett.* **114**, 098102 (March 6, 2015).

106. M.W. Zürch, "High-Resolution Extreme Ultraviolet Microscopy: Imaging of Artificial and Biological Specimens with Laser-Driven Ultrafast XUV Sources", PhD. Thesis, Friedrich Schiller University of Jena (2014); Springer Theses: Recognizing Outstanding PhD Research (Springer, 2015).

107. T. Harada, M. Nakasuji, M. Tada *et al.*, "Critical Dimension Measurement of an Extreme-Ultraviolet Mask Utilizing Coherent Extreme-Ultraviolet Scatterometry Microscope at NewSUBARU," *Jpn. J. Appl. Phys.* **50**, 06GB03–1 (2011).

108. J. Deng, D.J. Vine, S. Chen *et al.*, "Simultaneous Cryo X-ray Ptychographic and Fluorescence Microscopy of Green Algae," *PNAS* **112**(8), 2314 (February 24, 2015).

109. J. Deng, Y.S.G. Nashed, S. Chen *et al.*, "Continuous Motion Scan Ptychography: Characterization for Increased Speed in Coherent X-ray Imaging," *Optics Express* **23**(5), 5438 (2015).

110. M.D. de Jonge, C.G. Ryan and C.J. Jacobsen, "X-ray Nanoprobes and Diffraction-Limited Storage Rings: Opportunities and Challenges of Fluorescence Tomography of Biological Specimens," *J. Synchr. Rad.* **21**, 1031 (2014).

111. P.F. McEwen, K.H. Downing and R.M. Glaeser, "The Relevance of Dose-Fractionation in Tomography of Radiation Sensitive Specimens," *Ultramicroscopy* **60**, 357 (1995).

112. R. Horstmeyer, R. Heintzmann, G. Popescu, L. Waller and C. Yang, "Standardizing the Resolution Claims for Coherent Microscopy," *Nature Photonics* **10**, 68 (February 2016).

113. M. Dierolf, A. Menzel, P. Thibault *et al.*, "Ptychographic X-ray Computed Tomography at the Nanoscale," *Nature* **467**, 436 (September 23, 2010).

114. M. Holler, A. Diaz, M. Guizar-Sicairos *et al.*, "X-ray Ptychographic Computed Tomography at 16 nm Isotropic 3D Resolution," *Scientific Reports* **4**, 3857 (January 24, 2014).

115. M. van den Brink, Plenary Presentation, EUVL Symposium, Maastricht, Netherlands (October 2015); www.ASML.com

116. W. Kaiser, Keynote Address, EUVL Symposium, Maastricht, Netherlands (October 2015); www.zeiss.com/semiconductor-manufacturing-technology/en_de/products-solutions/semiconductor-manufacturing-optics/lithography-at-13-5-nanometers-euv-.html

117. M. Lowisch, P. Kuerz, H.-J. Mann, O. Natt and B. Thuering, "Optics for EUV Production," *SPIE* **7636**, 736–2 (April 23, 2010).

118. M. Lowisch, P. Kuerz, O. Conradi *et al.*, "Optics for ASML's NXE:3300B Platform," *SPIE* **8679**, 8679–52 (February 24, 2013).

119. N.R. Böwering, A.I. Ershov, W.F. Marx *et al.*, "EUV Source Collector," *SPIE* **6151**, 61513R (2006).

120. Q. Huang, D.M. Paardekooper, E. Zoethout *et al.*, "UV Spectral Filtering by Surface Structured Multilayer Mirrors," *Optics Lett.* **39**(5), 1185 (March 1, 2014).

121. R. Soufli, S.L. Baker, E.M. Gullikson *et al.*, "Review of Substrate Materials, Surface Metrologies and Polishing Techniques for Current and Future-Generation EUV/X-ray Optics," *SPIE* **8501**, 850102 (2012).

122. U. Dinger, G. Seitz, S. Schulte *et al.*, "Fabrication and Metrology of Diffraction Limited Soft X-ray Optics for EUV Metrology," *SPIE* **5193**, 18 (2004).

123. International Technology Roadmap for Semiconductors (ITRS) Roadmap, Executive Summary (2013); http://www.itrs.net/Links/2013ITRS/Summary2013.htm

124. H. Kinoshita, *Extreme Ultraviolet Lithography: Principles and Basic Technologies* (Lambert Academic Publishing, Berlin, June 2016).

Homework Problems

Homework problems for each chapter will be found at the website:
www.cambridge.org/xrayeuv

Appendix A
Units and Physical Constants

A.I   The International System of Units (SI)

Table A.1 SI base units.[1,2]

Quantity	Name of unit	Unit symbol
Length	meter	m
Mass	kilogram	kg
Time	second	s
Electric current	ampere	A
Thermodynamic temperature	kelvin	K
Amount of substance	mole	mol
Luminous intensity	candela	cd

Table A.2 SI prefixes.

Factor	Prefix	Symbol	Factor	Prefix	Symbol
10^{24}	yotta	Y	10^{-1}	deci	d
10^{21}	zetta	Z	10^{-2}	centi	c
10^{18}	exa	E	10^{-3}	milli	m
10^{15}	peta	P	10^{-6}	micro	μ
10^{12}	tera	T	10^{-9}	nano	n
10^{9}	giga	G	10^{-12}	pico	p
10^{6}	mega	M	10^{-15}	femto	f
10^{3}	kilo	k	10^{-18}	atto	a
10^{2}	hecto	h	10^{-21}	zepto	z
10^{1}	deka	da	10^{-24}	yocto	y

Table A.3 Examples of derived units.

Quantity (symbol)	Name of unit	Unit symbol	Equivalent
Plane angle (θ)	radian	rad	m/m = 1
Solid angle (Ω)	steradian	sr	$m^2/m^2 = 1$
Velocity (v)			m/s
Acceleration (a)			m/s^2
Frequency (f)	hertz	Hz	s^{-1}
Force (F)	newton	N	$kg \cdot m/s^2$
Pressure (P)	pascal	Pa	N/m^2
Energy (ε)	joule	J	$N \cdot m, kg \cdot m^2/s^2$
Momentum (p)			$N \cdot s, kg \cdot m/s$
Power (P)	watt	W	J/s
Electric charge (q)	coulomb	C	$A \cdot s$
Electric potential (V)	volt	V	J/C, W/A
Resistance (R)	ohm	Ω	V/A
Capacitance (C)	farad	F	C/V
Magnetic flux (ϕ)	weber	Wb	$V \cdot s$
Inductance (L)	henry	H	Wb/A
Electric field strength (E)			V/m, N/C
Electric displacement (D)			C/m^2
Magnetic flux density (B)	tesla	T	$Wb/m^2, N/(A \cdot m)$
Magnetic field strength (H)			A/m
Temperature (T)	degree Celsius	°C	K −273.15
Intensity (I)			$J \cdot s^{-1} \cdot m^{-2}$
Brightness (B)			$J \cdot s^{-1} \cdot m^{-2} \cdot rad^{-2}$

Table A.4 Conversion factors.

Length	Angstrom	$1 \text{ Å} = 10^{-10}$ m
Length	astronomical unit	$1 au = 1.496 \times 10^{11}$ m
Length	parsec (1 arcsec × 1 au)	$1 pc = 3.086 \times 10^{16}$ m
Length	light year	$1 ly = 9.461 \times 10^{15}$ m
Area	barn	$1 \text{ barn} = 10^{-28} m^2$
Volume	liter	$1 L = 10^{-3} m^3 = 1000 cm^3$
Plane angle	degree	$1° = (\pi/180) rad \cong 17.45$ mrad
	arcminute	$1' = 1/60° = (\pi/10\,800) rad \cong 290.9$ μrad
	arcsecond	$1'' = 1/60' = (\pi/648\,000) rad \cong 4.848$ μrad
Solid angle	steradian	$1 sr = 1 rad^2 = 10^6 mrad^2$
Mass	atomic mass unit	$m_\mu = 1.660\,538\,21 \times 10^{-27}$ kg
		1 dalton = 1 amu
Time	minute	1 min = 60 s
	hour	1 h = 60 min = 3600 s
	day	1 d = 24 h = 86 400 s
Pressure	standard atmosphere	1 atm = 101 325 Pa
		1 atm = 760 mmHg = 760 torr
	bar	$1 bar = 10^5$ Pa
Acceleration	gravitational accel. at earth's surface	$g = 9.806\,65 m/s^2$
Energy	electron volt	$1 eV = 1.602\,176\,565 \times 10^{-19}$ J
	calorie	1 cal = 4.1868 J
Absorbed energy	gray	1 Gy = 100 rad = 1 J/kg
Magnetic flux density	gauss	$1 G = 10^{-4} T = 10^{-4} Wb/m^2$
Wavelength in vacuum	photon energy	$\lambda \cdot \hbar\omega = hc = 1239.841\,93 eV \cdot nm$
Molar definition	atomic mass unit	$m_u \cdot N_A = 1 g = 10^{-3}$ kg (exactly)

A.2   Physical Constants[3, 4]

Table A.5

Quantity	Symbol	Value	Units
Speed of light in vacuum	c	299 792 458 (exactly)	$m \cdot s^{-1}$
Permeability of vacuum	μ_0	$4\pi \times 10^{-7}$ (exactly)	$N \cdot A^{-2}$
Permittivity of vacuum	ϵ_0	$1/(\mu_0 c^2) = 8.854\,187\,817\ldots$	$10^{-12}\,F \cdot m^{-1}$
Impedance of vacuum ($\mu_0 c$)	Z_0	$376.730\,313\,461\,\Omega$	
Planck's constant	h	4.135 667 516	$10^{-15}\,eV \cdot s$
Planck's constant/2π	$\hbar$	6.582 119 28	$10^{-16}\,eV \cdot s$
Electron charge	e	1.602 176 565	$10^{-19}\,C$
Electron mass	m	9.109 382 91	$10^{-31}\,kg$
Electron rest energy	mc^2	0.510 998 928	MeV
Proton mass	m_p	1.672 621 844	$10^{-27}\,kg$
Neutron mass	m_n	1.674 926 816	$10^{-27}\,kg$
Atomic mass unit $\left[m\left(^{12}C\right)/12 \right]$	m_u	1.660 538 921	$10^{-27}\,kg$
Rydberg constant $\left(me^4/32\pi^2\epsilon_0^2\hbar^2 \right)$	$R_\infty hc$	13.605 692 53	eV
Bohr radius ($4\pi\epsilon_0\hbar^2/me^2$)	α_0	0.529 177 2109	$10^{-10}\,m$
Classical electron radius $\left(e^2/4\pi\epsilon_0 mc^2 \right)$	r_e	2.817 940 3267	$10^{-15}\,m$
Thomson cross-section $\left(8\pi r_e^2/3 \right)$	σ_e	0.665 245 8734	$10^{-28}\,m^2$
Fine-structure constant $\left(e^2/4\pi\epsilon_0\hbar\hbar c \right)$	α	7.297 352 5698	10^{-3}
Compton wavelength ($\hbar/mc$)	λ_C	386.159 268 00	$10^{-15}\,m$
Bohr magneton ($e\hbar/2m$)	μ_B	5.788 381 8066	$10^{-5}\,eV \cdot T^{-1}$
Nuclear magneton ($e\hbar/2m_p$)	μ_N	3.152 451 2605	$10^{-8}\,eV \cdot T^{-1}$
Avogadro's number	N_A	6.022 141 29	$10^{23}\,mol^{-1}$
Boltzmann constant (R/N_A)	κ	8.617 3324	$10^{-5}\,eV \cdot K^{-1}$
Stefan–Boltzmann constant $[(\pi^2/60)\kappa^4/\hbar^3 c^2]$	σ	5.670 373	$10^{-8}\,W \cdot m^{-2} \cdot K^{-4}$
Universal (molar) gas constant	R	8.314 510(70)	$J \cdot mol^{-1} \cdot K^{-1}$
Molar volume (ideal gas) (RT/P, 273.15 K, 101 325 Pa)	V_m	22 413 968	$cm^3 \cdot mol^{-1}$
Loschmidt's number (N_A/Vm)	n_L	2.686 763(23)	$10^{25}\,m^{-3}$
Photon energy–wavelength product	hc	1239.841 93	$eV \cdot nm$

References

1. R.A. Nelson, "Guide for Metric Practice," *Phys. Today BG* **15** (August 2002).
2. Inst. Electr. Electron. Eng., "American National Standard for Metric Practice," ANSI/IEEE Std. 268–1992 (IEEE, New York, 1992).
3. P.J. Mohr, B.N. Taylor and D.B. Newell, "The Fundamental Physical Constants," *Phys. Today*, p. 52 (July 2007). These constants are next expected to be updated in 2018.
4. http://physics.nist.gov/constants

Appendix B
Electron Binding Energies, Principal K- and L-Shell Emission Lines, and Auger Electron Energies

Table B.1 Electron binding energies in electron volts for the elements in their natural forms.[a]

Element	K1s	L_12s	$L_22p_{1/2}$	$L_32p_{3/2}$	M_13s	$M_23p_{1/2}$	$M_33p_{3/2}$	$M_43d_{3/2}$	$M_53d_{5/2}$	N_14s	$N_24p_{1/2}$	$N_34p_{3/2}$
1 H	13.6											
2 He	24.6[b]											
3 Li	54.7[b]											
4 Be	111.5[b]											
5 B	188[b]											
6 C	284.2[b]											
7 N	409.9[b]	37.3[b]										
8 O	543.1[b]	41.6[b]										
9 F	696.7[b]											
10 Ne	870.2[b]	48.5[b]	21.7[b]	21.6[b]								
11 Na	1070.8[c]	63.5[c]	30.4[c]	30.5[b]								
12 Mg	1303.0[c]	88.6[b]	49.6[c]	49.2[c]								
13 Al	1559.6	117.8[b]	72.9[b]	72.5[b]								
14 Si	1838.9	149.7[b]	99.8[b]	99.2[b]								
15 P	2145.5	189[b]	136[b]	135[b]								
16 S	2472	230.9[b]	163.6[b]	162.5[b]								
17 Cl	2822.4	270.2[b]	202[b]	200[b]								
18 Ar	3205.9[b]	326.3[b]	250.6[b]	248.4[b]	29.3[b]	15.9[b]	15.7[b]					
19 K	3608.4[b]	378.6[b]	297.3[b]	294.6[b]	34.8[b]	18.3[b]	18.3[b]					
20 Ca	4038.5[b]	438.4[c]	349.7[c]	346.2[c]	44.3[c]	25.4[c]	25.4[c]					
21 Sc	4492.8	498.0[b]	403.6[b]	398.7[b]	51.1[b]	28.3[b]	28.3[b]					
22 Ti	4966.4	560.9[c]	461.2[c]	453.8[c]	58.7[c]	32.6[c]	32.6[c]					
23 V	5465.1	626.7[c]	519.8[c]	512.1[c]	66.3[c]	37.2[c]	37.2[c]					
24 Cr	5989.2	695.7[c]	583.8[c]	574.1[c]	74.1[c]	42.2[c]	42.2[c]					
25 Mn	6539.0	769.1[c]	649.9[c]	638.7[c]	82.3[c]	47.2[c]	47.2[c]					
26 Fe	7112.0	844.6[c]	719.9[c]	706.8[c]	91.3[c]	52.7[c]	52.7[c]					
27 Co	7708.9	925.1[c]	793.3[c]	778.1[c]	101.0[c]	58.9[c]	58.9[c]					
28 Ni	8332.8	1008.6[c]	870.0[c]	852.7[c]	110.8[c]	68.0[c]	66.2[c]					
29 Cu	8978.9	1096.7[c]	952.3[c]	932.5[c]	122.5[c]	77.3[c]	75.1[c]					
30 Zn	9658.6	1196.2[b]	1044.9[b]	1021.8[b]	139.8[b]	91.4[b]	88.6[b]	10.2[b]	10.1[b]			
31 Ga	10367.1	1299.0[b]	1143.2[c]	1116.4[c]	159.5[c]	103.5[c]	103.5[c]	18.7[c]	18.7[c]			
32 Ge	11103.1	1414.6[b]	1248.1[b]	1217.0[b]	180.1[b]	124.9[b]	120.8[b]	29.0[b]	29.0[b]			
33 As	11866.7	1527.0[b]	1359.1[b]	1323.6[b]	204.7[b]	146.2[b]	141.2[b]	41.7[b]	41.7[b]			
34 Se	12657.8	1652.0[b]	1474.3[b]	1433.9[b]	229.6[b]	166.5[b]	160.7[b]	55.5[b]	54.6[b]			
35 Br	13473.7	1782.0[b]	1596.0[b]	1549.9[b]	257[b]	189[b]	182[b]	70[b]	69[b]			

Table B.1 (*cont.*)

Element	K1s	$L_1 2s$	$L_2 2p_{1/2}$	$L_3 2p_{3/2}$	$M_1 3s$	$M_2 3p_{1/2}$	$M_3 3p_{3/2}$	$M_4 3d_{3/2}$	$M_5 3d_{5/2}$	$N_1 4s$	$N_2 4p_{1/2}$	$N_3 4p_{3/2}$
36 Kr	14325.6	1921.0	1730.9^b	1678.4^b	292.8^b	222.2^b	214.4	95.0^b	93.8^b	27.5^b	14.1^b	14.1^b
37 Rb	15199.7	2065.1	1863.9	1804.4	326.7^b	248.7^b	239.1^b	113.0^b	112^b	30.5^b	16.3^b	15.3^b
38 Sr	16104.6	2216.3	2006.8	1939.6	358.7^c	280.3^c	270.0^c	136.0^c	134.2^c	38.9^c	20.3^c	20.3^c
39 Y	17038.4	2372.5	2155.5	2080.0	392.0^b	310.6^b	298.8^b	157.7^c	155.8^c	43.8^b	24.4^b	23.1^b
40 Zr	17997.6	2531.6	2306.7	2222.3	430.3^c	343.5^c	329.8^c	181.1^c	178.8^c	50.6^c	28.5^c	27.7^c
41 Nb	18985.6	2697.7	2464.7	2370.5	466.6^c	376.1^c	360.6^c	205.0^c	202.3^c	56.4^c	32.6^c	30.8^c
42 Mo	19999.5	2865.5	2625.1	2520.2	506.3^c	411.6^c	394.0^c	231.1^c	227.9^c	63.2^c	37.6^c	35.5^c
43 Tc	21044.0	3042.5	2793.2	2676.9	544^b	445^b	425^b	257^b	253^b	68^b	39^c	39^b
44 Ru	22117.2	3224.0	2966.9	2837.9	586.2^c	483.5^c	461.4^c	284.2^c	280.0^c	75.0^c	46.5^c	43.2^c
45 Rh	23219.9	3411.9	3146.1	3003.8	628.1^c	521.3^c	496.5^c	311.9^c	307.2^c	81.4^b	50.5^c	47.3^c
46 Pd	24350.3	3604.3	3330.3	3173.3	671.6^c	559.9^c	532.3^c	340.5^c	335.2^c	87.6^b	55.7^c	50.9^c
47 Ag	25514.0	3805.8	3523.7	3351.1	719.0^c	603.8^c	573.0^c	374.0^c	368.0^c	97.0^c	63.7^c	58.3^c
48 Cd	26711.2	4018.0	3727.0	3537.5	772.0^c	652.6^c	618.4^c	411.9^c	405.2^c	109.8^c	63.9^c	63.9^c
49 In	27939.9	4237.5	3938.0	3730.1	827.2^c	703.2^c	665.3^c	451.4^c	443.9^c	122.7^c	73.5^c	73.5^c
50 Sn	29200.1	4464.7	4156.1	3928.8	884.7^c	756.5^c	714.6^c	493.2^c	484.9^c	137.1^c	83.6^c	83.6^c
51 Sb	30491.2	4698.3	4380.4	4132.2	946^c	812.7^c	766.4^c	537.5^c	528.2^c	153.2^c	95.6^c	95.6^c
52 Te	31813.8	4939.2	4612.0	4341.4	1006^c	870.8^c	820.8^c	583.4^c	573.0^c	169.4^c	103.3^c	103.3^c
53 I	33169.4	5188.1	4852.1	4557.1	1072^c	931^b	875^b	631^b	620^b	186^b	123^b	123^b
54 Xe	34561.4	5452.8	5103.7	4782.2	1148.7^b	1002.1^b	940.6^b	689.0^b	676.4^b	213.2^b	146.7	145.5^b
55 Cs	35984.6	5714.3	5359.4	5011.9	1211^b	1071^b	1003^b	740.5^b	726.6^b	232.3^b	172.4^b	161.3^b
56 Ba	37440.6	5988.8	5623.6	5247.0	1293^b	1137^b	1063^b	795.7^b	780.5^b	253.5^c	192.1	178.6^c
57 La	38924.6	6266.3	5890.6	5482.7	1362^b	1209^b	1128^b	853^b	836^b	247.7^b	205.8	196.0^b
58 Ce	40443.0	6548.8	6164.2	5723.4	1436^b	1274^b	1187^b	902.4^b	883.8^b	291.0^b	223.2	206.5^b
59 Pr	41990.6	6834.8	6440.4	5964.3	1511.0	1337.4	1242.2	948.3^b	928.8^b	304.5	236.3	217.6
60 Nd	43568.9	7126.0	6721.5	6207.9	1575.3	1402.8	1297.4	1003.3^b	980.4^b	319.2^b	243.3	224.6
61 Pm	45184.0	7427.9	7012.8	6459.3	–	1471.4	1356.9	1051.5	1026.9	–	242	242
62 Sm	46834.2	7736.8	7311.8	6716.2	1722.8	1540.7	1419.8	1110.9^b	1083.4^b	347.2^b	265.6	247.4
63 Eu	48519.0	8052.0	7617.1	6976.9	1800.0	1613.9	1480.6	1158.6^b	1127.5^b	360	284	257
64 Gd	50239.1	8375.6	7930.3	7242.8	1880.8	1688.3	1544.0	1221.9^b	1189.6^b	378.6^b	286	270.9
65 Tb	51995.7	8708.0	8251.6	7514.0	1967.5	1767.7	1611.3	1276.9^b	1241.1^b	396.0^b	322.4^b	284.1^b
66 Dy	53788.5	9045.8	8580.6	7790.1	2046.8	1841.8	1675.6	1332.5	1292.6^b	414.2^b	333.5^b	293.2^b
67 Ho	55617.7	9394.2	8917.8	8071.1	2128.3	1922.8	1741.2	1391.5	1351.4	432.4^b	343.5	308.2^b
68 Er	57485.5	9751.3	9264.3	8357.9	2206.5	2005.8	1811.8	1453.3	1409.3	449.8^b	366.2	320.2^b
69 Tm	59398.6	10115.7	9616.9	8648.0	2306.8	2089.8	1884.5	1514.6	1467.7	470.9^b	385.9^b	332.6^b
70 Yb	61332.3	10486.4	9978.2	8943.6	2398.1	2173.0	1949.8	1576.3	1527.8	480.5^b	388.7^b	339.7^b
71 Lu	63313.8	10870.4	10348.6	9244.1	2491.2	2263.5	2023.6	1639.4	1588.5	506.8^b	412.4^b	359.2^b
72 Hf	65350.8	11270.7	10739.4	9560.7	2600.9	2365.4	2107.6	1716.4	1661.7	538^b	438.2^c	380.7^c
73 Ta	67416.4	11681.5	11136.1	9881.1	2708.0	2468.7	2194.0	1793.2	1735.1	563.4^c	463.4^c	400.9^c
74 W	69525.0	12099.8	11544.0	10206.8	2819.6	2574.9	2281.0	1871.6	1809.2	594.1^c	490.4^c	423.6^c
75 Re	71676.4	12526.7	11958.7	10535.3	2931.7	2681.6	2367.3	1948.9	1882.9	625.4	518.7^c	446.8^c
76 Os	73870.8	12968.0	12385.0	10870.9	3048.5	2792.2	2457.2	2030.8	1960.1	658.2^c	549.1^c	470.7^c
77 Ir	76111.0	13418.5	12824.1	11215.2	3173.7	2908.7	2550.7	2116.1	2040.4	691.1^c	577.8^c	495.8^c
78 Pt	78394.8	13879.9	13272.6	11563.7	3296.0	3026.5	2645.4	2201.9	2121.6	725.4^c	609.1^c	519.4^c
79 Au	80724.9	14352.8	13733.6	11918.7	3424.9	3147.8	2743.0	2291.1	2205.7	762.1^c	642.7^c	546.3^c
80 Hg	83102.3	14839.3	14208.7	12283.9	3561.6	3278.5	2847.1	2384.9	2294.9	802.2^c	680.2^c	576.6^c
81 Tl	85530.4	15346.7	14697.9	12657.5	3704.1	3415.7	2956.6	2485.1	2389.3	846.2^c	720.5^c	609.5^c
82 Pb	88004.5	15860.8	15200.0	13035.2	3850.7	3554.2	3066.4	2585.6	2484.0	891.8^c	761.9^c	643.5^c
83 Bi	90525.9	16387.5	15711.1	13418.6	3999.1	3696.3	3176.9	2687.6	2579.6	939^c	805.2^c	678.8^c
84 Po	93105.0	16939.3	16244.3	13813.8	4149.4	3854.1	3301.9	2798.0	2683.0	995^b	851^b	705^b
85 At	95729.9	17493	16784.7	14213.5	4317	4008	3426	2908.7	2786.7	1042^b	886^b	740^b

(*cont.*)

Table B.1 (*cont.*)

Element	K1s	L₁2s	L₂2p₁/₂	L₃2p₃/₂	M₁3s	M₂3p₁/₂	M₃3p₃/₂	M₄3d₃/₂	M₅3d₅/₂	N₁4s	N₂4p₁/₂	N₃4p₃/₂
86 Rn	98404	18049	17337.1	14619.4	4482	4159	3538	3021.5	2892.4	1097[b]	929[b]	768[b]
87 Fr	101137	18639	17906.5	15031.2	4652	4327	3663	3136.2	2999.9	1153[b]	980[b]	810[b]
88 Ra	103921.9	19236.7	18484.3	15444.4	4822.0	4489.5	3791.8	3248.4	3104.9	1208[b]	1057.6[b]	879.1[b]
89 Ac	106755.3	19840.	19083.2	15871.0	5002	4656	3909	3370.2	3219.0	1269[b]	1080[b]	890[b]
90 Th	109650.9	20472.1	19693.2	16300.3	5182.3	4830.4	4046.1	3490.8	3332.0	1330[b]	1168[b]	966.4[c]
91 Pa	112601.4	21104.6	20313.7	16733.1	5366.9	5000.9	4173.8	3611.2	3441.8	1387[b]	1224[b]	1007[b]
92 U	115606.1	21757.4	20947.6	17166.3	5548.0	5182.2	4303.4	3727.6	3551.7	1439[b]	1271[b]	1043.0[c]

Element	N₄4d₃/₂	N₅4d₅/₂	N₆4f₅/₂	N₇4f₇/₂	O₁5s	O₂5p₁/₂	O₃5p₃/₂	O₄5d₃/₂	O₅5d₅/₂	P₁6s	P₂6p₁/₂	P₃6p₃/₂
48 Cd	11.7[c]	10.7[c]										
49 In	17.7[c]	16.9[c]										
50 Sn	24.9[c]	23.9[c]										
51 Sb	33.3[c]	32.1[c]										
52 Te	41.9[c]	40.4[c]										
53 I	50[b]	50[b]										
54 Xe	69.5[b]	67.5[b]	–	–	23.3[b]	13.4[b]	12.1[b]					
55 Cs	79.8[b]	77.5[b]	–	–	22.7	14.2[b]	12.1[b]					
56 Ba	92.6[c]	89.9[c]	–	–	30.3[c]	17.0[c]	14.8[c]					
57 La	105.3[b]	102.5[b]	–	–	34.3[b]	19.3[b]	16.8[b]					
58 Ce	109[b]	–	–	–	37.8	19.8[b]	17.0[b]					
59 Pr	115.1[b]	115.1[b]	–	–	37.4	22.3	22.3					
60 Nd	120.5[b]	120.5[b]	–	–	37.5	21.1	21.1					
61 Pm	120	120	–	–	–	–	–					
62 Sm	129	129	–	–	37.4	21.3	21.3					
63 Eu	133	127.7[b]	–	–	31.8	22.0	22.0					
64 Gd	140.5	142.6[b]	–	–	43.5[b]	20	20					
65 Tb	150.5[b]	150.5[b]	–	–	45.6[b]	28.7[b]	22.6[b]					
66 Dy	153.6	153.6	–	–	49.9	29.5	23.1					
67 Ho	160[b]	160[b]	–	–	49.3[b]	30.8[b]	24.1[b]					
68 Er	167.6[b]	167.6[b]	–	–	50.6[b]	31.4[b]	24.7[b]					
69 Tm	175.5[b]	175.5[b]	–	–	54.7[b]	31.8[b]	25.0[b]					
70 Yb	191.2[b]	182.4[b]	–	–	52.0[b]	30.3[b]	24.1[b]					
71 Lu	206.1[b]	196.3[c]	8.9[b]	7.5[b]	57.3[b]	33.6[b]	26.7[b]					
72 Hf	220.0[c]	211.5[c]	15.9[c]	14.2[c]	64.2[c]	38[b]	29.9[b]					
73 Ta	237.9[c]	226.4[c]	23.5[c]	21.6[c]	69.7[c]	42.2[b]	32.7[b]					
74 W	255.9[c]	243.5[c]	33.6[c]	31.4[c]	75.6[c]	45.3[b]	36.8[b]					
75 Re	273.9[c]	260.5[c]	42.9[b]	40.5[c]	83[c]	45.6[b]	34.6[b]					
76 Os	293.1[c]	278.5[c]	53.4[c]	50.7[c]	84[c]	58[b]	44.5[c]					
77 Ir	311.9[c]	296.3[c]	63.8[c]	60.8[c]	95.2[b]	63.0[b]	48.0[c]					
78 Pt	331.6[c]	314.6[c]	74.5[c]	71.2[c]	101[c]	65.3[b]	51.7[c]					
79 Au	353.2[c]	335.1[c]	87.6[c]	83.9[c]	107.2[b]	74.2[c]	57.2[c]					
80 Hg	378.2[c]	358.8[c]	104.0[c]	99.9[c]	127[c]	83.1[c]	64.5[c]	9.6[c]	7.8[c]			
81 Tl	405.7[c]	385.0[c]	122.2[c]	117.8[c]	136.[b]	94.6[c]	73.5[c]	14.7[c]	12.5[c]			
82 Pb	434.3[c]	412.2[c]	141.7[c]	136.9[c]	147[b]	106.4[c]	83.3[c]	20.7[c]	18.1[c]			
83 Bi	464.0[c]	440.1[c]	162.3[c]	157.0[c]	159.3[b]	119.0[c]	92.6[c]	26.9[c]	23.8[c]			
84 Po	500[b]	473[b]	184[b]	184[b]	177[b]	132[b]	104[b]	31[b]	31[b]			
85 At	533[b]	507[b]	210[b]	210[b]	195[b]	148[b]	115[b]	40[b]	40[b]			

Table B.1 (*cont.*)

Element	$N_4d_{3/2}$	$N_5d_{5/2}$	$N_64f_{5/2}$	$N_74f_{7/2}$	O_15s	$O_25p_{1/2}$	$O_35p_{3/2}$	$O_45d_{3/2}$	$O_55d_{5/2}$	P_16s	$P_26p_{1/2}$	$P_36p_{3/2}$
86 Rn	567[b]	541[b]	238[b]	238[b]	214[b]	164[b]	127[b]	48[b]	48[b]	26		
87 Fr	603[b]	577[b]	268[b]	268[b]	234[b]	182[b]	140[b]	58[b]	58[b]	34	15	15
88 Ra	635.9[b]	602.7[b]	299[b]	299[b]	254[b]	200[b]	153[b]	68[b]	68[b]	44	19	19
89 Ac	675[b]	639[b]	319[b]	319[b]	272[b]	215[b]	167[b]	80[b]	80[b]	–	–	–
90 Th	712.1[c]	675.2[c]	342.4[c]	333.1[c]	290[b]	229[b]	182[b]	92.5[c]	85.4[c]	41.4[c]	24.5[c]	16.6[c]
91 Pa	743[b]	708[b]	371[b]	360[b]	310[b]	232[b]	232[b]	94[b]	94[b]	–	–	–
92 U	778.3[b]	736.2[c]	388.2[b]	377.4[c]	321[b]	257[b]	192[b]	102.8[c]	94.2[c]	43.9[c]	26.8[c]	16.8[c]

[a] Electron binding energies for the elements in their natural forms, as compiled by G.P. Williams, Brookhaven National Laboratory, in Ref. 1, Chapter 1. The energies are given in electron volts relative to the vacuum level for the rare gases and for H_2, N_2, O_2, F_2, and Cl_2; relative to the Fermi level for the metals; and relative to the top of the valence bands for semiconductors. Values are based largely on those given by J.A. Bearden and A.F. Barr, "Reevaluation of X-Ray Atomic Energy Levels," *Rev. Mod. Phys.* **39**, 125 (1967); corrected in 1998 by E. Gullikson (LBNL, unpublished). For further updates consult the website www.cxro.LBL.gov/optical_constants.

[b] From M. Cardona and L. Lay, Editors, *Photoemission in Solids I: General Principles* (Springer-Verlag, Berlin, 1978).

[c] From J.C. Fuggle and N. Mårtensson, "Core-Level Binding Energies in Metals," *J. Electron. Spectrosc. Relat. Phenom.* **21**, 275 (1980).

Table B.2 Photon energies, in electron volts, of principal K- and L-shell emission lines.[a]

Element	$K\alpha_1$	$K\alpha_2$	$K\beta_1$	$L\alpha_1$	$L\alpha_2$	$L\beta_1$	$L\beta_2$	$L\gamma_1$
3 Li	54.3							
4 Be	108.5							
5 B	183.3							
6 C	277							
7 N	392.4							
8 O	524.9							
9 F	676.8							
10 Ne	848.6	848.6						
11 Na	1040.98	1040.98	1071.1					
12 Mg	1253.60	1253.60	1302.2					
13 Al	1486.70	1486.27	1557.45					
14 Si	1739.98	1739.38	1835.94					
15 P	2013.7	2012.7	2139.1					
16 S	2307.84	2306.64	2464.04					
17 Cl	2622.39	2620.78	2815.6					
18 Ar	2957.70	2955.63	3190.5					
19 K	3313.8	3311.1	3589.6					
20 Ca	3691.68	3688.09	4012.7	341.3	341.3	344.9		
21 Sc	4090.6	4086.1	4460.5	395.4	395.4	399.6		
22 Ti	4510.84	4504.86	4931.81	452.2	452.2	458.4		
23 V	4952.20	4944.64	5427.29	511.3	511.3	519.2		
24 Cr	5414.72	5405.509	5946.71	572.8	572.8	582.8		
25 Mn	5898.75	5887.65	6490.45	637.4	637.4	648.8		
26 Fe	6403.84	6390.84	7057.98	705.0	705.0	718.5		
27 Co	6930.32	6915.30	7649.43	776.2	776.2	791.4		
28 Ni	7478.15	7460.89	8264.66	851.5	851.5	868.8		
29 Cu	8047.78	8027.83	8905.29	929.7	929.7	949.8		
30 Zn	8638.86	8615.78	9572.0	1011.7	1011.7	1034.7		

(*cont.*)

Table B.2 (*cont.*)

Element	$K\alpha_1$	$K\alpha_2$	$K\beta_1$	$L\alpha_1$	$L\alpha_2$	$L\beta_1$	$L\beta_2$	$L\gamma_1$
31 Ga	9251.74	9224.82	10 264.2	1097.92	1097.92	1124.8		
32 Ge	9886.42	9855.32	10 982.1	1188.00	1188.00	1218.5		
33 As	10 543.72	10 507.99	11 726.2	1282.0	1282.0	1317.0		
34 Se	11 222.4	11 181.4	12 495.9	1379.10	1379.10	1419.23		
35 Br	11 924.2	11 877.6	13 291.4	1480.43	1480.43	1 525.90		
36 Kr	12 649	12 598	14 112	1586.0	1586.0	1636.6		
37 Rb	13 395.3	13 335.8	14 961.3	1694.13	1692.56	1752.17		
38 Sr	14 165	14 097.9	15 835.7	1806.56	1804.74	1871.72		
39 Y	14 958.4	14 882.9	16 737.8	1922.56	1920.47	1995.84		
40 Zr	15 775.1	15 690.9	17 667.8	2042.36	2039.9	2124.4	2219.4	2302.7
41 Nb	16 615.1	16 521.0	18 622.5	2165.89	2163.0	2257.4	2367.0	2461.8
42 Mo	17 479.34	17 374.3	19 608.3	2293.16	2289.85	2394.81	2518.3	2623.5
43 Tc	18 367.1	18 250.8	20 619	2424.0	–	2536.8	–	–
44 Ru	19 279.2	19 150.4	21 656.8	2558.55	2554.31	2683.23	2836.0	2964.5
45 Rh	20 216.1	20 073.7	22 723.6	2696.74	2692.05	2834.41	3001.3	3143.8
46 Pd	21 177.1	21 020.1	23 818.7	2838.61	2833.29	2990.22	3171.79	3328.7
47 Ag	22 162.92	21 990.3	24 942.4	2984.31	2978.21	3150.94	3347.81	3519.59
48 Cd	23 173.6	22 984.1	26 095.5	3133.73	3126.91	3316.57	3528.12	3716.86
49 In	24 209.7	24 002.0	27 275.9	3286.94	3279.29	3487.21	3713.81	3920.81
50 Sn	25 271.3	25 044.0	28 486.0	3443.98	3435.42	3662.80	3904.86	4131.12
51 Sb	26 359.1	26 110.8	29 725.6	3604.72	3595.32	3843.57	4100.78	4347.79
52 Te	27 472.3	27 201.7	30 995.7	3769.33	3758.8	4029.58	4301.7	4570.9
53 I	28 612.0	28 317.2	32 294.7	3937.65	3926.04	4220.72	4507.5	4800.9
54 Xe	29 779	29 458	33 624	4109.9	–	–	–	–
55 Cs	30 972.8	30 625.1	34 986.9	4286.5	4272.2	4619.8	4935.9	5280.4
56 Ba	32 193.6	31 817.1	36 378.2	4466.26	4450.90	4827.53	5156.5	5531.1
57 La	33 441.8	33 034.1	37 801.0	4650.97	4634.23	5042.1	5383.5	5788.5
58 Ce	34 719.7	34 278.9	39 257.3	4840.2	4823.0	5262.2	5613.4	6052
59 Pr	36 026.3	35 550.2	40 748.2	5033.7	5013.5	5488.9	5850	6322.1
60 Nd	37 361.0	36 847.4	42 271.3	5230.4	5207.7	5721.6	6089.4	6602.1
61 Pm	38 724.7	38 171.2	43 826	5432.5	5407.8	5961	6339	6892
62 Sm	40 118.1	39 522.4	45 413	5636.1	5609.0	6205.1	6586	7178
63 Eu	41 542.2	40 901.9	47 037.9	5845.7	5816.6	6456.4	6843.2	7480.3
64 Gd	42 996.2	42 308.9	48 697	6057.2	6025.0	6713.2	7102.8	7785.8
65 Tb	44 481.6	43 744.1	50 382	6272.8	6238.0	6978	7366.7	8102
66 Dy	45 998.4	45 207.8	52 119	6495.2	6457.7	7247.7	7635.7	8418.8
67 Ho	47 546.7	46 699.7	53 877	6719.8	6679.5	7525.3	7911	8747
68 Er	49 127.7	48 221.1	55 681	6948.7	6905.0	7810.9	8189.0	9089
69 Tm	50 741.6	49 772.6	57 517	7179.9	7133.1	8101	8468	9426
70 Yb	52 388.9	51 354.0	59 370	7415.6	7367.3	8401.8	8758.8	9780.1
71 Lu	54 069.8	52 965.0	61 283	7655.5	7604.9	8709.0	9048.9	10 143.4
72 Hf	55 790.2	54 611.4	63 234	7899.0	7844.6	9022.7	9347.3	10 515.8
73 Ta	57 532	56 277	65 223	8146.1	8087.9	9343.1	9651.8	10 895.2
74 W	59 318.24	57 981.7	67 244	8397.6	8335.2	9672.35	9961.5	11 285.9
75 Re	61 140.3	59 717.9	69 310	8652.5	8586.2	10 010.0	10 275.2	11 685.4

Table B.2 (*cont.*)

Element	$K\alpha_1$	$K\alpha_2$	$K\beta_1$	$L\alpha_1$	$L\alpha_2$	$L\beta_1$	$L\beta_2$	$L\gamma_1$
76 Os	63 000.5	61 486.7	71 413	8911.7	8841.0	10 355.3	10 598.5	12 095.3
77 Ir	64 895.6	63 286.7	73 560.8	9175.1	9 099.5	10 708.3	10 920.3	12 512.6
78 Pt	66 832	65 112	75 748	9442.3	9361.8	11 070.7	11 250.5	12 942.0
79 Au	68 803.7	66 989.5	77 984	9713.3	9628.0	11 442.3	11 584.7	13 381.7
80 Hg	70 819	68 895	80 253	9988.8	9897.6	11 822.6	11 924.1	13 830.1
81 Tl	72 871.5	70 831.9	82 576	10 268.5	10 172.8	12 213.3	12 271.5	14 291.5
82 Pb	74 969.4	72 804.2	84 936	10 551.5	10 449.5	12 613.7	12 622.6	14 764.4
83 Bi	77 107.9	74 814.8	87 343	10 838.8	10 730.91	13 023.5	12 979.9	15 247.7
84 Po	79 290	76 862	89 800	11 130.8	11 015.8	13 447	13 340.4	15 744
85 At	81 520	78 950	92 300	11 426.8	11 304.8	13 876	–	16 251
86 Rn	83 780	81 070	94 870	11 727.0	11 597.9	14 316	–	16 770
87 Fr	86 100	83 230	97 470	12 031.3	11 895.0	14 770	14 450	17 303
88 Ra	88 470	85 430	100 130	12 339.7	12 196.2	15 235.8	14 841.4	17 849
89 Ac	90 884	87 670	102 850	12 652.0	12 500.8	15 713	–	18 408
90 Th	93 350	89 953	105 609	12 968.7	12 809.6	16 202.2	15 623.7	18 982.5
91 Pa	95 868	92 287	108 427	13 290.7	13 122.2	16 702	16 024	19 568
92 U	98 439	94 665	111 300	13 614.7	13 438.8	17 220.0	16 428.3	20 167.1
93 Np	–	–	–	13 944.1	13 759.7	17 750.2	16 840.0	20 784.8
94 Pu	–	–	–	14 278.6	14 084.2	18 293.7	17 255.3	21 417.3
95 Am	–	–	–	14 617.2	14 411.9	18 852.0	17 676.5	22 065.2

[a] Photon energies in electron volts of some characteristic emission lines of the elements of atomic number $3 \leq Z \leq 95$, as compiled by J. Kortright, "Characteristic X-Ray Energies," in *X-Ray Data Booklet* (Lawrence Berkeley National Laboratory Pub-490 Rev. 2, 1999). Values are largely based on those given by J.A. Bearden, "X-Ray Wavelengths," *Rev. Mod. Phys.* **39**, 78 (1967), which should be consulted for a more complete listing. Updates may also be noted at the website http://www-cxro.lbl.gov/data-booklet/

Table B.3 Auger energy curves, in electron volts, for elements of atomic number $3 \leq Z \leq 92$. Only dominant energies are given, and only for principal Auger peaks. The literature should be consulted for detailed tabulations, and for shifted values in various common compounds.[1-3] (Courtesy of Physical Electronics, Inc.[1])

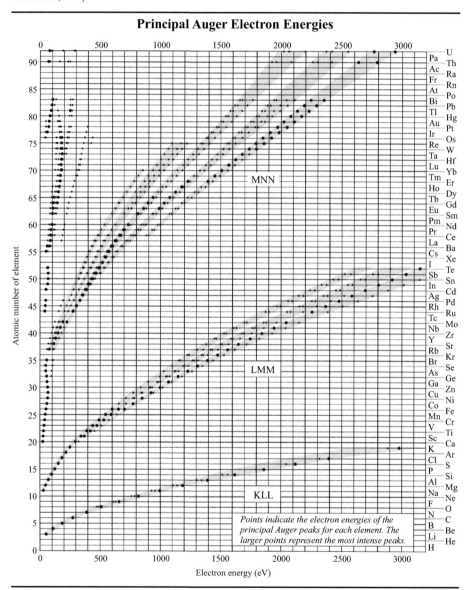

References

1. K.D. Childs, B.A. Carlson, L.A. Vanier *et al.*, *Handbook of Auger Electron Spectroscopy* (Physical Electronics, Eden Prairie, MN, 1995), C.L. Hedberg, Editor.
2. J.F. Moulder, W.F. Stickle, P.E. Sobol and K.D. Bomben, *Handbook of X-Ray Photoelectron Spectroscopy* (Physical Electronics, Eden Prairie, MN, 1995).
3. D. Briggs, *Handbook of X-Ray and Ultraviolet Photoelectron Spectroscopy* (Heyden, London, 1977).

Appendix C
Atomic Scattering Factors, Atomic Absorption Coefficients, and Subshell Photoionization Cross-Sections

Table C.1 Atomic scattering factors f_1^0 and f_2^0 in the approximation $\Delta k \cdot \Delta r_s \to 0$, for several common elements in their natural form. As described in Chapter 2, Section 2.7, this approximation is satisfied for forward scattering or for long wavelengths (greater than the Bohr radius), and denoted here by the superscript zero. Also given are values for the absorption coefficient μ in cm^2/g, as described in Chapter 3, Section 3.2. Values are from Henke, Gullikson, and Davis.[1] Their procedure is to obtain μ as a function of energy by making a fit to the best available experimental data, for each element as it is commonly found in nature. Values of $f_2^0(\omega)$ are obtained using a relationship equivalent to Eq. (3.26) as given here in Chapter 3, Section 3.2. Values for $f_1^0(\omega)$ are calculated using Kramers–Kronig relations, equivalent to that given as Eq. (3.85a) in Chapter 3, Section 3.8.

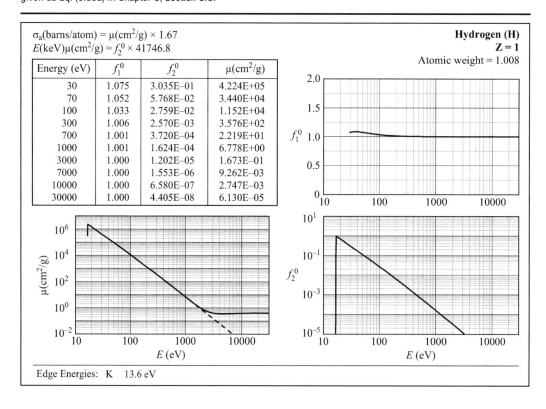

$\sigma_a(\text{barns/atom}) = \mu(\text{cm}^2/\text{g}) \times 1.67$

$E(\text{keV})\mu(\text{cm}^2/\text{g}) = f_2^0 \times 41746.8$

Hydrogen (H)
Z = 1
Atomic weight = 1.008

Energy (eV)	f_1^0	f_2^0	$\mu(\text{cm}^2/\text{g})$
30	1.075	3.035E–01	4.224E+05
70	1.052	5.768E–02	3.440E+04
100	1.033	2.759E–02	1.152E+04
300	1.006	2.570E–03	3.576E+02
700	1.001	3.720E–04	2.219E+01
1000	1.001	1.624E–04	6.778E+00
3000	1.000	1.202E–05	1.673E–01
7000	1.000	1.553E–06	9.262E–03
10000	1.000	6.580E–07	2.747E–03
30000	1.000	4.405E–08	6.130E–05

Edge Energies: K 13.6 eV

Table C.1 (*cont.*)

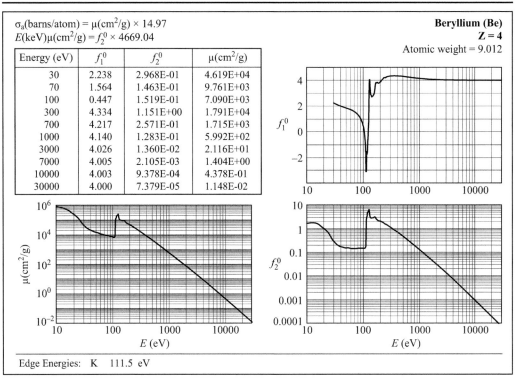

$\sigma_a(\text{barns/atom}) = \mu(\text{cm}^2/\text{g}) \times 14.97$
$E(\text{keV})\mu(\text{cm}^2/\text{g}) = f_2^0 \times 4669.04$

Beryllium (Be)
Z = 4
Atomic weight = 9.012

Energy (eV)	f_1^0	f_2^0	$\mu(\text{cm}^2/\text{g})$
30	2.238	2.968E-01	4.619E+04
70	1.564	1.463E-01	9.761E+03
100	0.447	1.519E-01	7.090E+03
300	4.334	1.151E+00	1.791E+04
700	4.217	2.571E-01	1.715E+03
1000	4.140	1.283E-01	5.992E+02
3000	4.026	1.360E-02	2.116E+01
7000	4.005	2.105E-03	1.404E+00
10000	4.003	9.378E-04	4.378E-01
30000	4.000	7.379E-05	1.148E-02

Edge Energies: K   111.5 eV

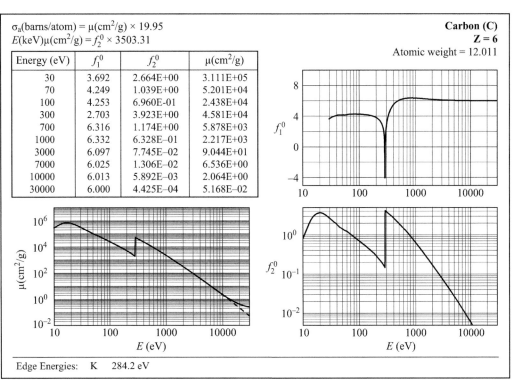

$\sigma_a(\text{barns/atom}) = \mu(\text{cm}^2/\text{g}) \times 19.95$
$E(\text{keV})\mu(\text{cm}^2/\text{g}) = f_2^0 \times 3503.31$

Carbon (C)
Z = 6
Atomic weight = 12.011

Energy (eV)	f_1^0	f_2^0	$\mu(\text{cm}^2/\text{g})$
30	3.692	2.664E+00	3.111E+05
70	4.249	1.039E+00	5.201E+04
100	4.253	6.960E-01	2.438E+04
300	2.703	3.923E+00	4.581E+04
700	6.316	1.174E+00	5.878E+03
1000	6.332	6.328E-01	2.217E+03
3000	6.097	7.745E-02	9.044E+01
7000	6.025	1.306E-02	6.536E+00
10000	6.013	5.892E-03	2.064E+00
30000	6.000	4.425E-04	5.168E-02

Edge Energies: K   284.2 eV

(*cont.*)

Table C.1 (*cont.*)

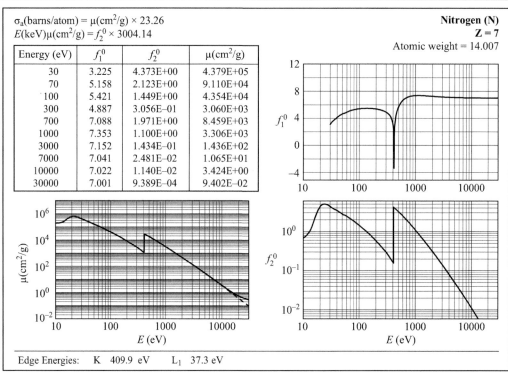

$\sigma_a(\text{barns/atom}) = \mu(\text{cm}^2/\text{g}) \times 23.26$
$E(\text{keV})\mu(\text{cm}^2/\text{g}) = f_2^0 \times 3004.14$

Nitrogen (N)
Z = 7
Atomic weight = 14.007

Energy (eV)	f_1^0	f_2^0	$\mu(\text{cm}^2/\text{g})$
30	3.225	4.373E+00	4.379E+05
70	5.158	2.123E+00	9.110E+04
100	5.421	1.449E+00	4.354E+04
300	4.887	3.056E–01	3.060E+03
700	7.088	1.971E+00	8.459E+03
1000	7.353	1.100E+00	3.306E+03
3000	7.152	1.434E–01	1.436E+02
7000	7.041	2.481E–02	1.065E+01
10000	7.022	1.140E–02	3.424E+00
30000	7.001	9.389E–04	9.402E–02

Edge Energies: K   409.9 eV   L$_1$ 37.3 eV

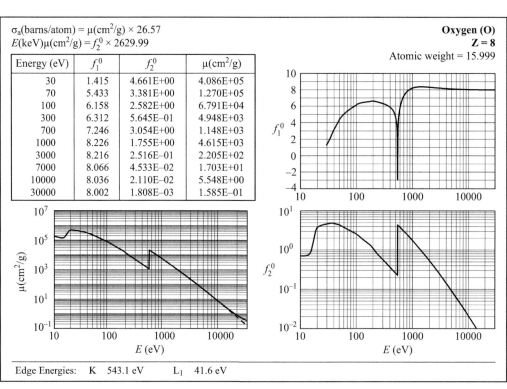

$\sigma_a(\text{barns/atom}) = \mu(\text{cm}^2/\text{g}) \times 26.57$
$E(\text{keV})\mu(\text{cm}^2/\text{g}) = f_2^0 \times 2629.99$

Oxygen (O)
Z = 8
Atomic weight = 15.999

Energy (eV)	f_1^0	f_2^0	$\mu(\text{cm}^2/\text{g})$
30	1.415	4.661E+00	4.086E+05
70	5.433	3.381E+00	1.270E+05
100	6.158	2.582E+00	6.791E+04
300	6.312	5.645E–01	4.948E+03
700	7.246	3.054E+00	1.148E+03
1000	8.226	1.755E+00	4.615E+03
3000	8.216	2.516E–01	2.205E+02
7000	8.066	4.533E–02	1.703E+01
10000	8.036	2.110E–02	5.548E+00
30000	8.002	1.808E–03	1.585E–01

Edge Energies: K   543.1 eV   L$_1$ 41.6 eV

Table C.1 (*cont.*)

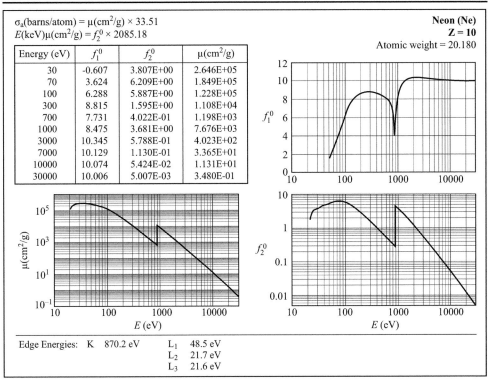

$\sigma_a(\text{barns/atom}) = \mu(\text{cm}^2/\text{g}) \times 33.51$
$E(\text{keV})\mu(\text{cm}^2/\text{g}) = f_2^0 \times 2085.18$

Neon (Ne)
Z = 10
Atomic weight = 20.180

Energy (eV)	f_1^0	f_2^0	$\mu(\text{cm}^2/\text{g})$
30	-0.607	3.807E+00	2.646E+05
70	3.624	6.209E+00	1.849E+05
100	6.288	5.887E+00	1.228E+05
300	8.815	1.595E+00	1.108E+04
700	7.731	4.022E-01	1.198E+03
1000	8.475	3.681E+00	7.676E+03
3000	10.345	5.788E-01	4.023E+02
7000	10.129	1.130E-01	3.365E+01
10000	10.074	5.424E-02	1.131E+01
30000	10.006	5.007E-03	3.480E-01

Edge Energies: K 870.2 eV L_1 48.5 eV
                              L_2 21.7 eV
                              L_3 21.6 eV

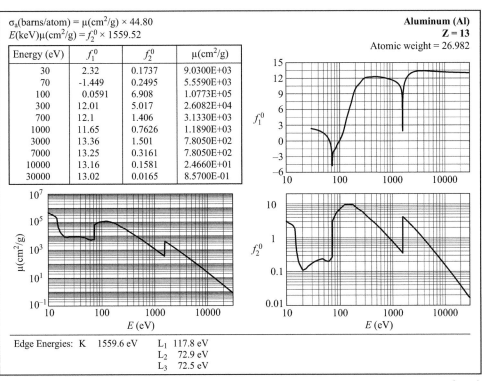

$\sigma_a(\text{barns/atom}) = \mu(\text{cm}^2/\text{g}) \times 44.80$
$E(\text{keV})\mu(\text{cm}^2/\text{g}) = f_2^0 \times 1559.52$

Aluminum (Al)
Z = 13
Atomic weight = 26.982

Energy (eV)	f_1^0	f_2^0	$\mu(\text{cm}^2/\text{g})$
30	2.32	0.1737	9.0300E+03
70	-1.449	0.2495	5.5590E+03
100	0.0591	6.908	1.0773E+05
300	12.01	5.017	2.6082E+04
700	12.1	1.406	3.1330E+03
1000	11.65	0.7626	1.1890E+03
3000	13.36	1.501	7.8050E+02
7000	13.25	0.3161	7.8050E+02
10000	13.16	0.1581	2.4660E+01
30000	13.02	0.0165	8.5700E-01

Edge Energies: K 1559.6 eV L_1 117.8 eV
                              L_2 72.9 eV
                              L_3 72.5 eV

(*cont.*)

Table C.1 (*cont.*)

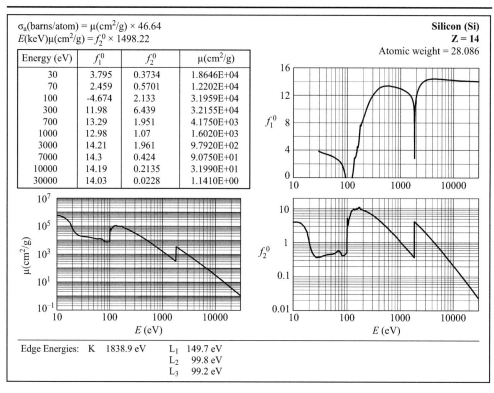

σ_a(barns/atom) = μ(cm²/g) × 46.64
E(keV)μ(cm²/g) = f_2^0 × 1498.22

Silicon (Si)
Z = 14
Atomic weight = 28.086

Energy (eV)	f_1^0	f_2^0	μ(cm²/g)
30	3.795	0.3734	1.8646E+04
70	2.459	0.5701	1.2202E+04
100	-4.674	2.133	3.1959E+04
300	11.98	6.439	3.2155E+04
700	13.29	1.951	4.1750E+03
1000	12.98	1.07	1.6020E+03
3000	14.21	1.961	9.7920E+02
7000	14.3	0.424	9.0750E+01
10000	14.19	0.2135	3.1990E+01
30000	14.03	0.0228	1.1410E+00

Edge Energies: K   1838.9 eV L_1 149.7 eV
                                   L_2 99.8 eV
                                   L_3 99.2 eV

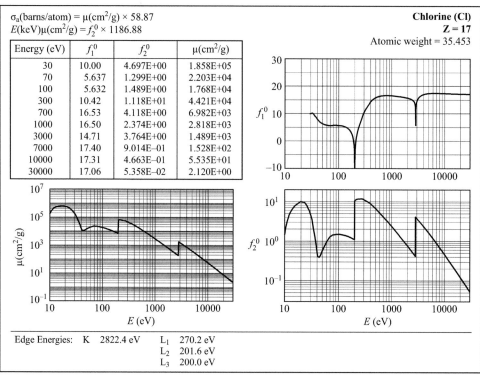

σ_a(barns/atom) = μ(cm²/g) × 58.87
E(keV)μ(cm²/g) = f_2^0 × 1186.88

Chlorine (Cl)
Z = 17
Atomic weight = 35.453

Energy (eV)	f_1^0	f_2^0	μ(cm²/g)
30	10.00	4.697E+00	1.858E+05
70	5.637	1.299E+00	2.203E+04
100	5.632	1.489E+00	1.768E+04
300	10.42	1.118E+01	4.421E+04
700	16.53	4.118E+00	6.982E+03
1000	16.50	2.374E+00	2.818E+03
3000	14.71	3.764E+00	1.489E+03
7000	17.40	9.014E-01	1.528E+02
10000	17.31	4.663E-01	5.535E+01
30000	17.06	5.358E-02	2.120E+00

Edge Energies: K   2822.4 eV L_1 270.2 eV
                                   L_2 201.6 eV
                                   L_3 200.0 eV

Table C.1 (*cont.*)

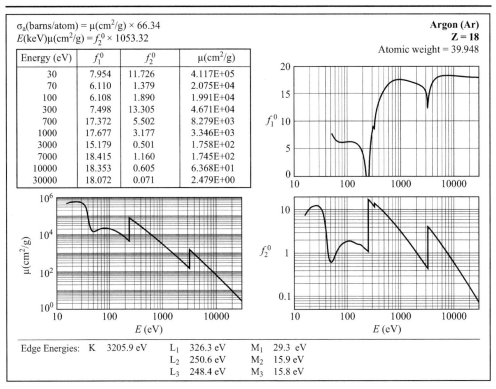

σ_a(barns/atom) = μ(cm^2/g) × 66.34
E(keV)μ(cm^2/g) = f_2^0 × 1053.32

Argon (Ar)
Z = 18
Atomic weight = 39.948

Energy (eV)	f_1^0	f_2^0	μ(cm^2/g)
30	7.954	11.726	4.117E+05
70	6.110	1.379	2.075E+04
100	6.108	1.890	1.991E+04
300	7.498	13.305	4.671E+04
700	17.372	5.502	8.279E+03
1000	17.677	3.177	3.346E+03
3000	15.179	0.501	1.758E+02
7000	18.415	1.160	1.745E+02
10000	18.353	0.605	6.368E+01
30000	18.072	0.071	2.479E+00

Edge Energies: K 3205.9 eV   L$_1$ 326.3 eV   M$_1$ 29.3 eV
                             L$_2$ 250.6 eV   M$_2$ 15.9 eV
                             L$_3$ 248.4 eV   M$_3$ 15.8 eV

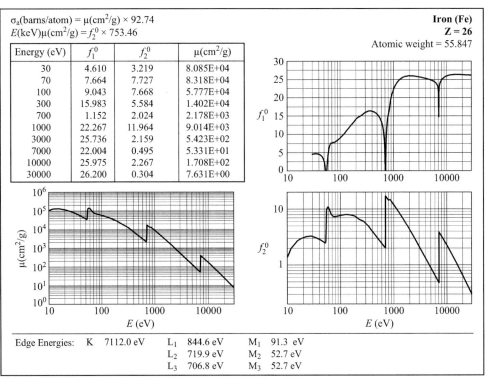

σ_a(barns/atom) = μ(cm^2/g) × 92.74
E(keV)μ(cm^2/g) = f_2^0 × 753.46

Iron (Fe)
Z = 26
Atomic weight = 55.847

Energy (eV)	f_1^0	f_2^0	μ(cm^2/g)
30	4.610	3.219	8.085E+04
70	7.664	7.727	8.318E+04
100	9.043	7.668	5.777E+04
300	15.983	5.584	1.402E+04
700	1.152	2.024	2.178E+03
1000	22.267	11.964	9.014E+03
3000	25.736	2.159	5.423E+02
7000	22.004	0.495	5.331E+01
10000	25.975	2.267	1.708E+02
30000	26.200	0.304	7.631E+00

Edge Energies: K 7112.0 eV   L$_1$ 844.6 eV   M$_1$ 91.3 eV
                             L$_2$ 719.9 eV   M$_2$ 52.7 eV
                             L$_3$ 706.8 eV   M$_3$ 52.7 eV

Table C.1 (*cont.*)

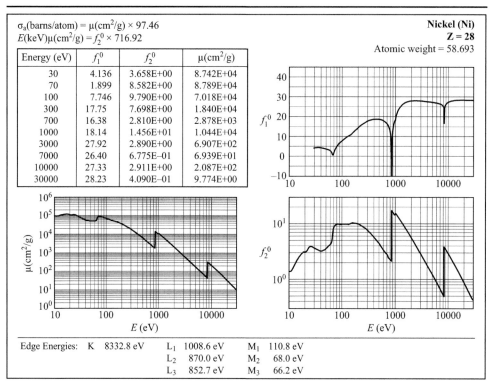

$\sigma_a(\text{barns/atom}) = \mu(\text{cm}^2/\text{g}) \times 97.46$
$E(\text{keV})\mu(\text{cm}^2/\text{g}) = f_2^0 \times 716.92$

Nickel (Ni)
Z = 28
Atomic weight = 58.693

Energy (eV)	f_1^0	f_2^0	$\mu(\text{cm}^2/\text{g})$
30	4.136	3.658E+00	8.742E+04
70	1.899	8.582E+00	8.789E+04
100	7.746	9.790E+00	7.018E+04
300	17.75	7.698E+00	1.840E+04
700	16.38	2.810E+00	2.878E+03
1000	18.14	1.456E+01	1.044E+04
3000	27.92	2.890E+00	6.907E+02
7000	26.40	6.775E−01	6.939E+01
10000	27.33	2.911E+00	2.087E+02
30000	28.23	4.090E−01	9.774E+00

Edge Energies: K   8332.8 eV L_1 1008.6 eV M_1 110.8 eV
                                     L_2 870.0 eV M_2 68.0 eV
                                     L_3 852.7 eV M_3 66.2 eV

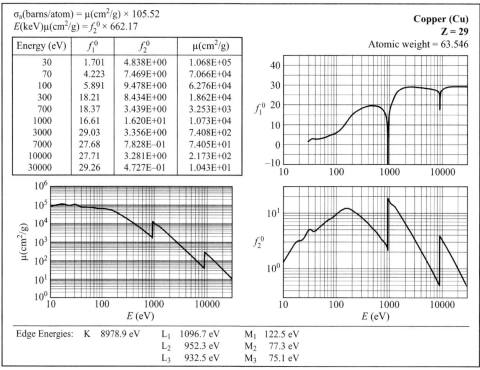

$\sigma_a(\text{barns/atom}) = \mu(\text{cm}^2/\text{g}) \times 105.52$
$E(\text{keV})\mu(\text{cm}^2/\text{g}) = f_2^0 \times 662.17$

Copper (Cu)
Z = 29
Atomic weight = 63.546

Energy (eV)	f_1^0	f_2^0	$\mu(\text{cm}^2/\text{g})$
30	1.701	4.838E+00	1.068E+05
70	4.223	7.469E+00	7.066E+04
100	5.891	9.478E+00	6.276E+04
300	18.21	8.434E+00	1.862E+04
700	18.37	3.439E+00	3.253E+03
1000	16.61	1.620E+01	1.073E+04
3000	29.03	3.356E+00	7.408E+02
7000	27.68	7.828E−01	7.405E+01
10000	27.71	3.281E+00	2.173E+02
30000	29.26	4.727E−01	1.043E+01

Edge Energies: K   8978.9 eV L_1 1096.7 eV M_1 122.5 eV
                                     L_2 952.3 eV M_2 77.3 eV
                                     L_3 932.5 eV M_3 75.1 eV

Table C.1 (*cont.*)

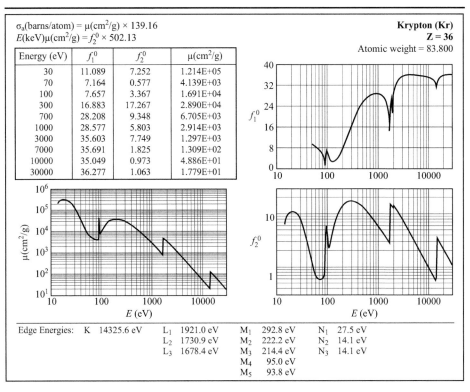

σ_a(barns/atom) = μ(cm^2/g) × 139.16
E(keV)μ(cm^2/g) = f_2^0 × 502.13

Krypton (Kr)
Z = 36
Atomic weight = 83.800

Energy (eV)	f_1^0	f_2^0	μ(cm^2/g)
30	11.089	7.252	1.214E+05
70	7.164	0.577	4.139E+03
100	7.657	3.367	1.691E+04
300	16.883	17.267	2.890E+04
700	28.208	9.348	6.705E+03
1000	28.577	5.803	2.914E+03
3000	35.603	7.749	1.297E+03
7000	35.691	1.825	1.309E+02
10000	35.049	0.973	4.886E+01
30000	36.277	1.063	1.779E+01

Edge Energies: K 14325.6 eV | L$_1$ 1921.0 eV | M$_1$ 292.8 eV | N$_1$ 27.5 eV
| | L$_2$ 1730.9 cV | M$_2$ 222.2 eV | N$_2$ 14.1 eV
| | L$_3$ 1678.4 eV | M$_3$ 214.4 eV | N$_3$ 14.1 eV
| | | M$_4$ 95.0 eV |
| | | M$_5$ 93.8 eV |

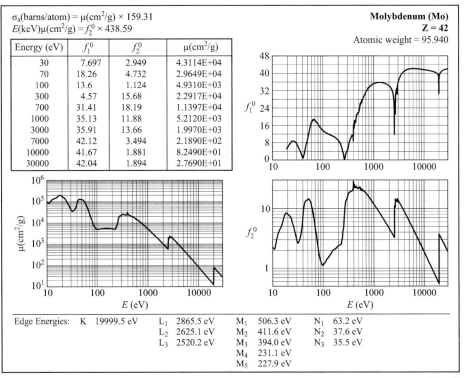

σ_a(barns/atom) = μ(cm^2/g) × 159.31
E(keV)μ(cm^2/g) = f_2^0 × 438.59

Molybdenum (Mo)
Z = 42
Atomic weight = 95.940

Energy (eV)	f_1^0	f_2^0	μ(cm^2/g)
30	7.697	2.949	4.3114E+04
70	18.26	4.732	2.9649E+04
100	13.6	1.124	4.9310E+03
300	4.57	15.68	2.2917E+04
700	31.41	18.19	1.1397E+04
1000	35.13	11.88	5.2120E+03
3000	35.91	13.66	1.9970E+03
7000	42.12	3.494	2.1890E+02
10000	41.67	1.881	8.2490E+01
30000	42.04	1.894	2.7690E+01

Edge Energies: K 19999.5 eV | L$_1$ 2865.5 eV | M$_1$ 506.3 eV | N$_1$ 63.2 eV
| | L$_2$ 2625.1 eV | M$_2$ 411.6 eV | N$_2$ 37.6 eV
| | L$_3$ 2520.2 eV | M$_3$ 394.0 eV | N$_3$ 35.5 eV
| | | M$_4$ 231.1 eV |
| | | M$_5$ 227.9 eV |

(*cont.*)

Table C.1 (*cont.*)

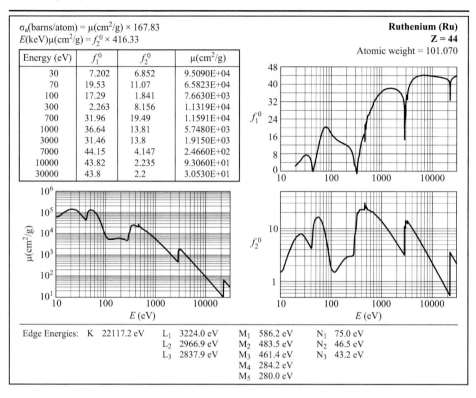

$\sigma_a(\text{barns/atom}) = \mu(\text{cm}^2/\text{g}) \times 167.83$
$E(\text{keV})\mu(\text{cm}^2/\text{g}) = f_2^0 \times 416.33$

Ruthenium (Ru)
Z = 44
Atomic weight = 101.070

Energy (eV)	f_1^0	f_2^0	$\mu(\text{cm}^2/\text{g})$
30	7.202	6.852	9.5090E+04
70	19.53	11.07	6.5823E+04
100	17.29	1.841	7.6630E+03
300	2.263	8.156	1.1319E+04
700	31.96	19.49	1.1591E+04
1000	36.64	13.81	5.7480E+03
3000	31.46	13.8	1.9150E+03
7000	44.15	4.147	2.4660E+02
10000	43.82	2.235	9.3060E+01
30000	43.8	2.2	3.0530E+01

Edge Energies: K 22117.2 eV L₁ 3224.0 eV M₁ 586.2 eV N₁ 75.0 eV
L₂ 2966.9 eV M₂ 483.5 eV N₂ 46.5 eV
L₃ 2837.9 eV M₃ 461.4 eV N₃ 43.2 eV
M₄ 284.2 eV
M₅ 280.0 eV

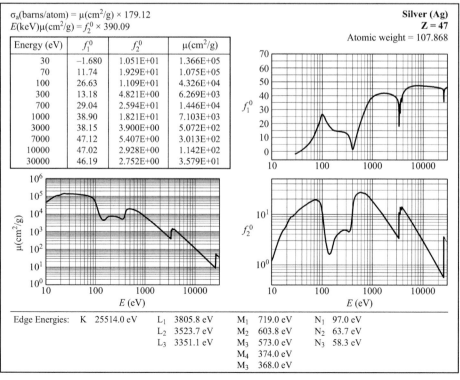

$\sigma_a(\text{barns/atom}) = \mu(\text{cm}^2/\text{g}) \times 179.12$
$E(\text{keV})\mu(\text{cm}^2/\text{g}) = f_2^0 \times 390.09$

Silver (Ag)
Z = 47
Atomic weight = 107.868

Energy (eV)	f_1^0	f_2^0	$\mu(\text{cm}^2/\text{g})$
30	−1.680	1.051E+01	1.366E+05
70	11.74	1.929E+01	1.075E+05
100	26.63	1.109E+01	4.326E+04
300	13.18	4.821E+00	6.269E+03
700	29.04	2.594E+01	1.446E+04
1000	38.90	1.821E+01	7.103E+03
3000	38.15	3.900E+00	5.072E+02
7000	47.12	5.407E+00	3.013E+02
10000	47.02	2.928E+00	1.142E+02
30000	46.19	2.752E+00	3.579E+01

Edge Energies: K 25514.0 eV L₁ 3805.8 eV M₁ 719.0 eV N₁ 97.0 eV
L₂ 3523.7 eV M₂ 603.8 eV N₂ 63.7 eV
L₃ 3351.1 eV M₃ 573.0 eV N₃ 58.3 eV
M₄ 374.0 eV
M₃ 368.0 eV

Table C.1 (*cont.*)

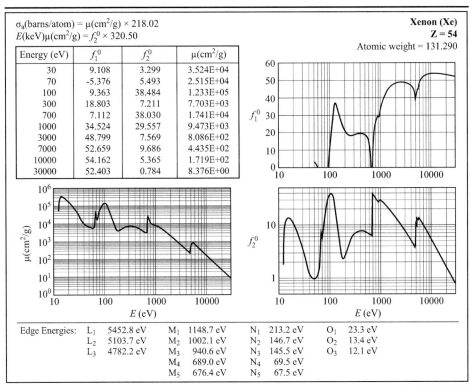

σ_a(barns/atom) = μ(cm²/g) × 218.02
E(keV)μ(cm²/g) = f_2^0 × 320.50

Xenon (Xe)
Z = 54
Atomic weight = 131.290

Energy (eV)	f_1^0	f_2^0	μ(cm²/g)
30	9.108	3.299	3.524E+04
70	-5.376	5.493	2.515E+04
100	9.363	38.484	1.233E+05
300	18.803	7.211	7.703E+03
700	7.112	38.030	1.741E+04
1000	34.524	29.557	9.473E+03
3000	48.799	7.569	8.086E+02
7000	52.659	9.686	4.435E+02
10000	54.162	5.365	1.719E+02
30000	52.403	0.784	8.376E+00

Edge Energies:						
L_1	5452.8 eV	M_1	1148.7 eV	N_1	213.2 eV	O_1	23.3 eV
L_2	5103.7 eV	M_2	1002.1 eV	N_2	146.7 eV	O_2	13.4 eV
L_3	4782.2 eV	M_3	940.6 eV	N_3	145.5 eV	O_3	12.1 eV
		M_4	689.0 eV	N_4	69.5 eV		
		M_5	676.4 eV	N_5	67.5 eV		

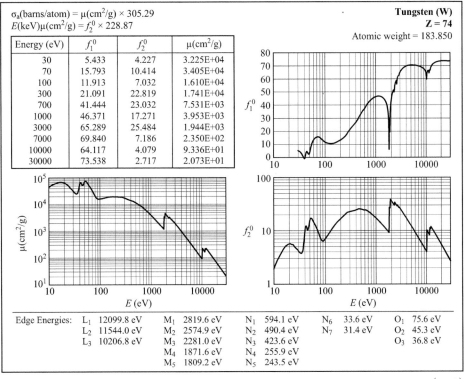

σ_a(barns/atom) = μ(cm²/g) × 305.29
E(keV)μ(cm²/g) = f_2^0 × 228.87

Tungsten (W)
Z = 74
Atomic weight = 183.850

Energy (eV)	f_1^0	f_2^0	μ(cm²/g)
30	5.433	4.227	3.225E+04
70	15.793	10.414	3.405E+04
100	11.913	7.032	1.610E+04
300	21.091	22.819	1.741E+04
700	41.444	23.032	7.531E+03
1000	46.371	17.271	3.953E+03
3000	65.289	25.484	1.944E+03
7000	69.840	7.186	2.350E+02
10000	64.117	4.079	9.336E+01
30000	73.538	2.717	2.073E+01

Edge Energies:								
L_1	12099.8 eV	M_1	2819.6 eV	N_1	594.1 eV	N_6	33.6 eV	O_1	75.6 eV
L_2	11544.0 eV	M_2	2574.9 eV	N_2	490.4 eV	N_7	31.4 eV	O_2	45.3 eV
L_3	10206.8 eV	M_3	2281.0 eV	N_3	423.6 eV			O_3	36.8 eV
		M_4	1871.6 eV	N_4	255.9 eV				
		M_5	1809.2 eV	N_5	243.5 eV				

(*cont.*)

Table C.1 (*cont.*)

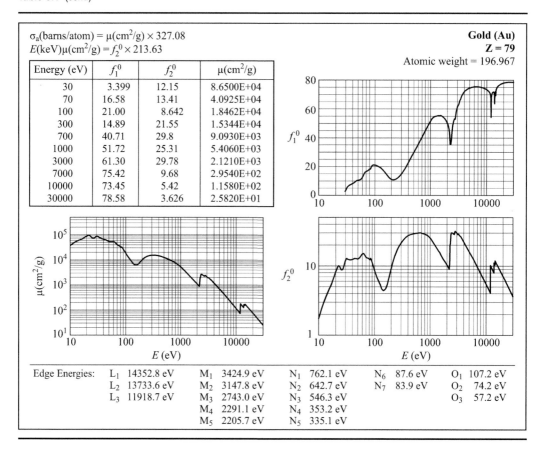

$$\sigma_a(\text{barns/atom}) = \mu(\text{cm}^2/\text{g}) \times 327.08$$
$$E(\text{keV})\mu(\text{cm}^2/\text{g}) = f_2^0 \times 213.63$$

Gold (Au)
Z = 79
Atomic weight = 196.967

Energy (eV)	f_1^0	f_2^0	$\mu(\text{cm}^2/\text{g})$
30	3.399	12.15	8.6500E+04
70	16.58	13.41	4.0925E+04
100	21.00	8.642	1.8462E+04
300	14.89	21.55	1.5344E+04
700	40.71	29.8	9.0930E+03
1000	51.72	25.31	5.4060E+03
3000	61.30	29.78	2.1210E+03
7000	75.42	9.68	2.9540E+02
10000	73.45	5.42	1.1580E+02
30000	78.58	3.626	2.5820E+01

Edge Energies:	L_1 14352.8 eV	M_1 3424.9 eV	N_1 762.1 eV	N_6 87.6 eV	O_1 107.2 eV
	L_2 13733.6 eV	M_2 3147.8 eV	N_2 642.7 eV	N_7 83.9 eV	O_2 74.2 eV
	L_3 11918.7 eV	M_3 2743.0 eV	N_3 546.3 eV		O_3 57.2 eV
		M_4 2291.1 eV	N_4 353.2 eV		
		M_5 2205.7 eV	N_5 335.1 eV		

Table C.2 Atomic subshell photoemission cross-sections, calculated for isolated atoms by Yeh and Lindau.[2]

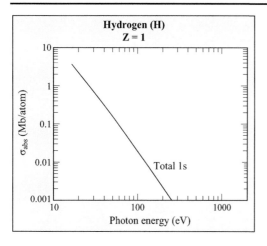

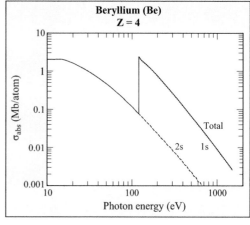

Table C.2 (*cont.*)

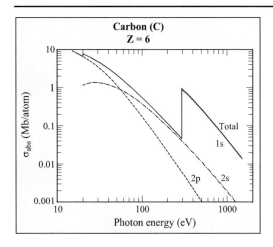

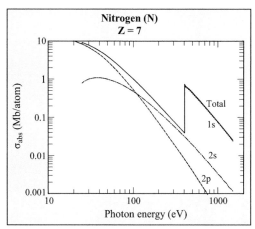

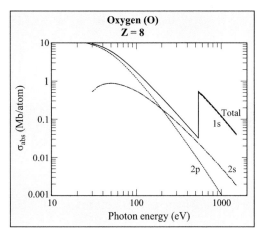

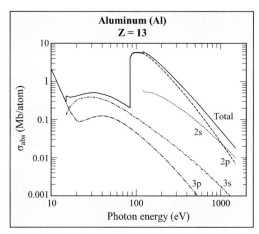

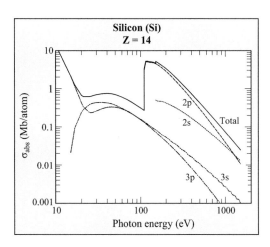

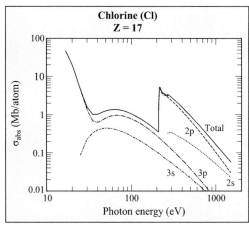

(*cont.*)

Table C.2 (*cont.*)

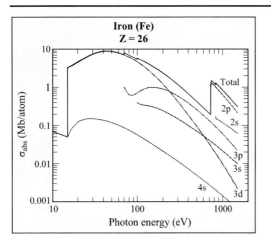

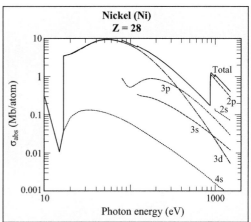

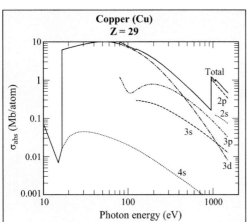

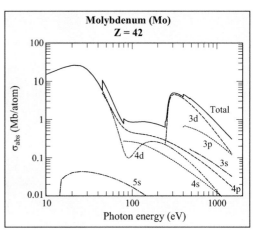

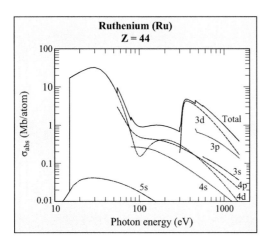

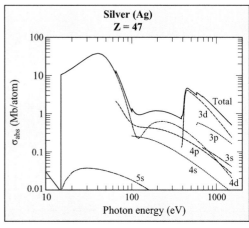

Table C.2 (*cont.*)

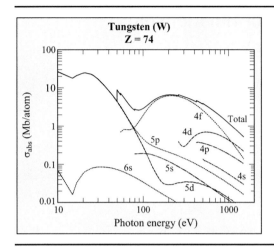

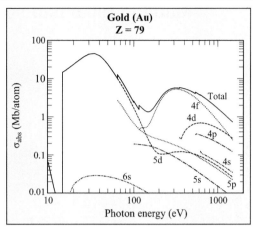

References

1. B.L. Henke, E.M. Gullikson and J.C. Davis, "X-Ray Interactions: Photoabsorption, Scattering, Transmission, and Reflection at E = 50–30,000 eV, Z = 1–92," *Atomic Data and Nucl. Data Tables* **54**, 181 (1993). Current updates are maintained by E.M. Gullikson at http://www.cxro .LBL.gov/optical_constants

2. J.-J. Yeh and I. Lindau, "Atomic Subshell Photoionization Cross Sections and Asymmetry Parameters: $1 \leq Z \leq 103$," *Atomic Data and Nucl. Data Tables* **32**, 1–155 (1985); J.-J. Yeh, *Atomic Calculation of Photoionization Cross-Sections and Asymmetry Parameters* (Gordon and Breach, Langhorne, PA, 1993); I. Lindau, "Photoemission Cross Sections," Chapter 1, p. 3, in *Synchrotron Radiation Research: Advances in Surface and Interface Science*, Vol. 2 (Plenum, New York, 1992), R.Z. Bachrach, Editor; J.-J. Yeh, "Metal/Silicon Interfaces and their Oxidation Behavior – A Photoemission Spectroscopy Analysis," PhD Thesis in Applied Physics, Stanford University (1987).

Appendix D
Mathematical and Vector Relationships

D.1   Vector and Tensor Formulas

$$\mathbf{A} \cdot (\mathbf{B} \times \mathbf{C}) = \mathbf{B} \cdot (\mathbf{C} \times \mathbf{A}) = \mathbf{C} \cdot (\mathbf{A} \times \mathbf{B}) \tag{D.1}$$

$$\mathbf{A} \times (\mathbf{B} \times \mathbf{C}) = (\mathbf{A} \cdot \mathbf{C})\mathbf{B} - (\mathbf{A} \cdot \mathbf{B})\mathbf{C} \tag{D.2}$$

$$\nabla \cdot (f\mathbf{A}) = f\nabla \cdot \mathbf{A} + \mathbf{A} \cdot \nabla f \tag{D.3}$$

$$\nabla \times (f\mathbf{A}) = f\nabla \times \mathbf{A} + \nabla f \times \mathbf{A} \tag{D.4}$$

$$\nabla \cdot (\mathbf{A} \times \mathbf{B}) = \mathbf{B} \cdot (\nabla \times \mathbf{A}) - \mathbf{A} \cdot (\nabla \times \mathbf{B}) \tag{D.5}$$

$$\nabla \times (\mathbf{A} \times \mathbf{B}) = \mathbf{A}(\nabla \cdot \mathbf{B}) - \mathbf{B}(\nabla \cdot \mathbf{A}) + (\mathbf{B} \cdot \nabla)\mathbf{A} - (\mathbf{A} \cdot \nabla)\mathbf{B} \tag{D.6}$$

$$\nabla \times (\nabla \times \mathbf{A}) = \nabla(\nabla \cdot \mathbf{A}) - \nabla^2\mathbf{A} \tag{D.7}$$

$$\nabla \times \nabla f = 0 \tag{D.8}$$

$$\nabla \cdot (\nabla \times \mathbf{A}) = 0 \tag{D.9}$$

$$\nabla \cdot (\mathbf{A}\mathbf{B}) = \mathbf{B}(\nabla \cdot \mathbf{A}) + (\mathbf{A} \cdot \nabla)\mathbf{B} \tag{D.10}$$

Integral relations for a vector field $\mathbf{B}$ over a volume V, an area A with (normal) vector differential component $d\mathbf{A}$, and line contour element s of local vector component $d\mathbf{s}$. A circle indicates an integral over a closed contour or over a closed surface:

$$\iiint_V \nabla \cdot \mathbf{B} \, dV = \oiint_A \mathbf{B} \cdot d\mathbf{A} \tag{D.11}$$
$$\text{(Gauss's divergence theorem)}$$

$$\iint_A (\nabla \times \mathbf{B}) \cdot d\mathbf{A} = \oint_S \mathbf{B} \cdot d\mathbf{s} \tag{D.12}$$
$$\text{(Stokes's theorem)}$$

D.2 Series Expansions

Assuming $x \ll 1$ and m real, except where stated otherwise,

$$f(x) = f(s) + f'(s)(x-s) + \frac{f''(s)}{2!}(x-s)^2 + \cdots$$

$$+ \frac{f^n(s)(x-s)^n}{n!} \qquad (x-s) \ll 1$$

$$f(x) = f(0) + xf'(0) + \frac{x^2}{2!}f''(0) + \cdots$$

$$\sin x = x - \frac{x^3}{3!} + \frac{x^5}{5!} - \cdots$$

$$\cos x = 1 - \frac{x^2}{2!} + \frac{x^4}{4!} - \cdots$$

$$\tan x = x + \frac{x^2}{3} + \frac{2x^5}{15} + \cdots \qquad \left(x^2 < \frac{\pi^2}{4} \right)$$

$$\sqrt{1-x} = 1 - \frac{x}{2} - \frac{1}{2^2 2!}x^2 - \frac{1 \cdot 3}{2^3 3!}x^3 + \cdots$$

$$(1-x^2)^{-1/2} = 1 + \frac{x^2}{2} + \frac{1 \cdot 3}{2^2 2!}x^4 + \cdots$$

$$(1+x)^m = 1 + \frac{m}{1!}x + \frac{m(m-1)}{2!}x^2 + \frac{m(m-1)(m-2)}{3!}x^3 + \cdots$$

$$e^x = 1 + x + \frac{x^2}{2!} + \frac{x^3}{3!} + \cdots + \frac{x^n}{n!}$$

$$\frac{1}{1-x} = 1 + x + x^2 + x^3 + \cdots$$

$$e^{\sin x} = 1 + x + \frac{x^2}{2!} - \frac{3x^4}{4!} - \frac{8x^5}{5!} - \cdots \qquad x^2 < \frac{\pi^2}{4}$$

$$e^{\cos x} = e\left(1 - \frac{x^2}{2!} + \frac{4x^4}{4!} - \frac{31x^6}{6!} + \cdots \right) \qquad x^2 < \frac{\pi^2}{4}$$

$$\sinh x = x + \frac{x^3}{3!} + \frac{x^5}{5!} + \cdots$$

$$\cosh x = 1 + \frac{x^2}{2!} + \frac{x^4}{4!} + \cdots$$

$$\tanh x = x - \frac{x^3}{3} + \frac{2x^5}{15} - \cdots$$

$$J_0(x) = 1 - \frac{x^2}{2^2(1!)^2} + \frac{x^4}{2^4(2!)^2} - \frac{x^6}{2^6(3!)^2} + \cdots$$

$$J_1(x) = \frac{x}{2} - \frac{x^3}{2^3 1! 2!} + \frac{x^5}{2^5 2! 3!} - \frac{x^7}{2^7 3! 4!} + \cdots$$

For complex $z = x + iy$:

$$\sin z = z - \frac{z^3}{3!} + \cdots$$

$$\cos z = 1 - \frac{z^2}{2!} + \cdots$$

$$\sin z = z + \frac{z^3}{3!} + \frac{z^5}{5!} + \cdots$$

$$\cosh z = 1 + \frac{z^2}{2!} + \frac{z^4}{4!} + \cdots$$

$$\tan z = z - \frac{z^3}{3} + \frac{2z^5}{15} - \cdots$$

$$e^z = 1 + z + \frac{z^2}{2!} + \cdots$$

D.3   Trigonometric Relationships

$$e^{\pm i\theta} = \cos\theta \pm i\sin\theta, \qquad \theta \text{ real}$$

$$\sin(\alpha \pm \beta) = \sin\alpha\cos\beta \pm \cos\alpha\sin\beta$$

$$\cos(\alpha \pm \beta) = \cos\alpha\cos\beta \mp \sin\alpha\sin\beta$$

$$\tan(\alpha \pm \beta) = \frac{\tan\alpha \pm \tan\beta}{1 \mp \tan\alpha\tan\beta}, \qquad \tan\left(\frac{\pi}{4} + \beta\right) = \frac{1 + \tan\beta}{1 - \tan\beta}$$

$$\sin 2\alpha = 2\sin\alpha\cos\alpha$$

$$\cos 2\alpha = \begin{cases} \cos^2\alpha - \sin^2\alpha \\ 2\cos^2\alpha - 1 \\ 1 - 2\sin^2\alpha \end{cases}$$

$$\cos\frac{\alpha}{2} = \sqrt{\frac{1}{2}(1 + \cos\alpha)}$$

$$\sin\frac{\alpha}{2} = \sqrt{\frac{1}{2}(1-\cos\alpha)}$$

$$2\sin\alpha\cos\beta = \sin(\alpha+\beta) + \sin(\alpha-\beta)$$

$$2\cos\alpha\cos\beta = \cos(\alpha+\beta) + \cos(\alpha-\beta)$$

$$2\sin\alpha\sin\beta = \cos(\alpha-\beta) - \cos(\alpha+\beta)$$

$$\sin\alpha + \sin\beta = 2\sin\tfrac{1}{2}(\alpha+\beta)\cos\tfrac{1}{2}(\alpha-\beta)$$

$$\sin 3\alpha = 3\sin\alpha - 4\sin^3\alpha$$

$$\cos 3\alpha = 4\cos^3\alpha - 3\cos\alpha$$

$$\frac{a}{\sin\alpha} = \frac{b}{\sin\beta} = \frac{c}{\sin\gamma}$$

$$a^2 = b^2 + c^2 - 2bc\cos\alpha$$

$$\tan^2\theta + 1 = \sec^2\theta, \qquad 1 + \cot^2\theta = \csc^2\theta$$

$$\sinh x = \frac{e^x - e^{-x}}{2}, \qquad \sinh(-x) = -\sinh x$$

$$\cosh x = \frac{e^x + e^{-x}}{2}, \qquad \cosh(-x) = \cosh x$$

$$\tanh x = \sinh x/\cosh x$$

$$\cosh^2 x - \sinh^2 x = 1$$

In spherical coordinates the angle between two vectors is given by

$$\cos\Theta = \cos\theta_1 + \cos\theta_2 + \sin\theta_1 \sin\theta_2 \cos(\phi_2 - \phi_1)$$

where (θ_1, ϕ_1) and (θ_2, ϕ_2) are the respective polar and azimuthal angular pairs.

D.4   Definite Integrals

Assuming x real, $a > 0$:

$$\int_0^\infty e^{-ax}dx = \frac{1}{a}$$

$$\int_c^d xe^{-ax}dx = -\frac{e^{-ax}}{a^2}(ax+1)\Big|_c^d$$

$$\frac{1}{\sqrt{2\pi}\sigma} \int_{-\infty}^{\infty} e^{-x^2/2\sigma^2} dx = 1$$

$$\int_{0}^{\infty} (\cos bx)e^{-a^2x^2} dx = \frac{\sqrt{\pi}}{2a} e^{-b^2/4a^2}$$

$$\int_{0}^{\infty} xe^{-x^2} dx = \frac{1}{2}$$

$$\int_{0}^{\infty} x^2 e^{-x^2} dx = \frac{\sqrt{\pi}}{4}$$

$$\int_{0}^{\infty} e^{-ax} \left\{ \begin{matrix} \cos mx \\ \sin mx \end{matrix} \right\} dx = \frac{\left\{ \begin{matrix} a \\ m \end{matrix} \right\}}{a^2 + m^2}$$

$$\int_{0}^{\infty} \frac{\sin x}{x} dx = \frac{\pi}{2}$$

$$\int_{0}^{\infty} \frac{\cos x}{x} dx = \infty$$

$$\int_{0}^{\infty} \frac{\tan x}{x} dx = \frac{\pi}{2}$$

$$\int_{0}^{\pi} \sin^2 mx \, dx = \int_{0}^{\pi} \cos^2 mx \, dx = \frac{\pi}{2}$$

$$\int_{0}^{\infty} \frac{\sin^2 x \, dx}{x^2} = \frac{\pi}{2}$$

$$\int_{-\pi}^{\pi} \cos nx \cos mx \, dx = \begin{cases} 0 & n \neq m \\ \pi & n = m \end{cases}$$

$$\int_{-\pi}^{\pi} \sin nx \cos mx \, dx = 0$$

D.5   Functions of a Complex Variable

$$z = x + iy = r(\cos\theta + i\sin\theta) = re^{i\theta}$$
$$e^z = e^x(\cos y + i\sin y)$$
$$e^{iz} = \cos z + i\sin z$$
$$\sqrt{i} = \pm\frac{\sqrt{2}}{2}(1 + i)$$

$$\sin z = \frac{e^{iz} - e^{iz}}{2i}, \quad \sin(-z) = -\sin z$$

$$\cos z = \frac{e^{iz} + e^{-iz}}{2}, \quad \cos(-z) = \cos z$$

$$\sinh z = \frac{e^z - e^{-z}}{2} = -i \sin(iz)$$

$$\cosh z = \frac{e^z + e^{-z}}{2} = \cos(iz)$$

$$\tanh z = \frac{\sinh z}{\cosh z}$$

$$\cos^2 z + \sin^2 z = 1$$

$$\cosh^2 z - \sinh^2 z = 1$$

$$\sin 2z = 2 \sin z \cos z$$

$$\sin z = \sin x \cosh y + i \cos x \sinh y$$
$$= \sin x \cos iy + \cos x \sin iy$$

$$\cos z = \cos x \cosh y - i \sin x \sinh y$$
$$= \cos x \cos iy - \sin x \sin iy$$

$$\sin z = \sinh x \cos y + i \cosh x \sin y$$

$$\cosh z = \cosh x \cos y + i \sinh x \sin y$$

$$\cos iy = \cosh y$$

$$\sin iy = i \sinh y$$

$$\frac{d}{dz}e^z = e^z$$

$$\frac{d}{dz}\sin z = \cos z$$

$$\frac{d}{dz}\cos z = -\sin z$$

$$\frac{d}{dz}\cosh z = \sinh z$$

$$\frac{d}{dz}\sinh z = \cosh z$$

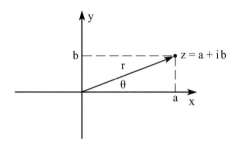

$$r = \sqrt{a^2 + b^2}$$
$$\cos\theta = \frac{a}{r} = \frac{a}{\sqrt{a^2 + b^2}}$$
$$\sin\theta = \frac{b}{r} = \frac{b}{\sqrt{a^2 + b^2}}$$

$$\sqrt{z} = \sqrt{a + ib} = r^{1/2}e^{i\theta/2} = r^{1/2}\left(\cos\frac{\theta}{2} + i\sin\frac{\theta}{2}\right)$$

$$\sqrt{a + ib} = (a^2 + b^2)^{1/4}\left[\sqrt{\tfrac{1}{2}(1 + \cos\theta)} + i\sqrt{\tfrac{1}{2}(1 - \cos\theta)}\right]$$

$$\sqrt{a + ib} = (a^2 + b^2)^{1/4}\left[\sqrt{\frac{1}{2}\left(1 + \frac{a}{\sqrt{a^2 + b^2}}\right)} + i\sqrt{\frac{1}{2}\left(1 - \frac{a}{\sqrt{a^2 + b^2}}\right)}\right]$$

$$\sqrt{a + ib} = \frac{1}{\sqrt{2}}\left[\sqrt{\sqrt{a^2 + b^2} + a} + i\sqrt{\sqrt{a^2 + b^2} - a}\right]$$

similarly

$$\sqrt{a - ib} = \frac{1}{\sqrt{2}}\left[\sqrt{\sqrt{a^2 + b^2} + a} - i\sqrt{\sqrt{a^2 + b^2} - a}\right]$$

Cauchy Integral Formula

$$\oint \frac{f(z)}{z - z_0} dz = 2\pi i f(z_0)$$

D.6   Fourier Transforms and Fourier Transform Pairs[11-13]

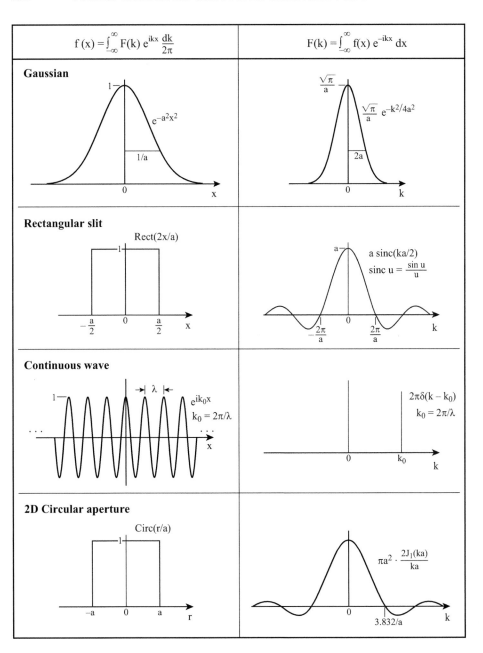

| $f(x) = \int_{-\infty}^{\infty} F(k) e^{ikx} \frac{dk}{2\pi}$ | $F(k) = \int_{-\infty}^{\infty} f(x) e^{-ikx} dx$ |

Gaussian

$e^{-a^2 x^2}$

$\frac{\sqrt{\pi}}{a} e^{-k^2/4a^2}$

Rectangular slit

$\text{Rect}(2x/a)$

$a \, \text{sinc}(ka/2)$

$\text{sinc} \, u = \frac{\sin u}{u}$

Continuous wave

$e^{ik_0 x}$

$k_0 = 2\pi/\lambda$

$2\pi\delta(k - k_0)$

$k_0 = 2\pi/\lambda$

2D Circular aperture

$\text{Circ}(r/a)$

$\pi a^2 \cdot \frac{2J_1(ka)}{ka}$

D.7   The Dirac Delta Function

The Dirac delta function $\delta(x)$ has the properties that it is zero for all x except $x = 0$, where it is infinite:

$$\delta(x) = \begin{cases} 0 & x \neq 0 \\ \infty & x = 0 \end{cases}$$

with the normalization condition

$$\int_{\text{all } x} \delta(x)dx = 1$$

With a displaced origin the function is

$$\delta(x - a) = \begin{cases} 0 & x \neq a \\ \infty & x = a \end{cases}$$

which has the *sifting property*

$$\int_{\text{all } x} f(x)\delta(x - a)dx = f(a)$$

Integrating by parts, one can show that the derivative function $\delta'(x - a)$ has the property

$$\int_{\text{all } x} f(x)\delta'(x - a)dx = -f'(a)$$

A shorthand notation for the delta function in three dimensions is, in Cartesian coordinates,

$$\delta(\mathbf{r}) \equiv \delta(x)\delta(y)\delta(z)$$

D.8   The Cauchy Principal Value Theorem

The Cauchy principal value of an integral is defined as the integral over a restricted range that avoids a contribution from an isolated singularity. Thus, for example, if a function $f(x)$ has a singularity at $x = a$, in the interval $0 < a < \infty$, the principal value of the integral would be written as

$$\mathcal{P}\int_0^\infty f(x)dx = \lim_{\delta \to 0}\left[\int_0^{a-\delta} f(x)dx + \int_{a+\delta}^\infty f(x)dx\right]$$

As a specific example, we consider the function $f(x) = 1/x$, where x represents a physical quantity with a small imaginary component ϵ. Forming real and imaginary components

$$\frac{1}{x \mp i\epsilon} = \frac{x}{x^2 + \epsilon^2} \pm i\frac{\epsilon}{x^2 + \epsilon^2}$$

the first term is to behave like $1/x$ everywhere except at $x = 0$, in the limit that ϵ goes to zero. Exactly at $x = 0$ this first term is zero, for arbitrarily small but finite ϵ. Thus this first term represents the Cauchy principal portion of the function $1/x$. The second imaginary term behaves like a Dirac delta function in the limit that ϵ goes to zero. It can be integrated using a standard trigonometric substitution. Performing this integration, one obtains the relationship

$$\lim_{\epsilon \to 0} \frac{1}{x \mp i\epsilon} = \mathcal{P}\left(\frac{1}{x}\right) \pm i\pi\,\delta(x)$$

where $\mathcal{P}(1/x)$ behaves like the function $1/x$ everywhere except at $x = 0$, where it is zero.

References

1. J. Stratten, *Electromagnetic Theory* (McGraw-Hill, New York, 1941).
2. J.D. Jackson, *Classical Electrodynamics* (Wiley, New York, 1998), Third Edition.
3. D.R. Nicholson, *Introduction to Plasma Theory* (Wiley, New York, 1983).
4. S. Solimeno, B. Crosignani and P. Di Porto, *Guiding, Diffraction and Confinement of Optical Radiation* (Academic, New York, 1986).
5. I.S. Gradshteyn and I.M. Ryzhik, *Tables of Series and Products* (Academic, New York, 1994), Fifth Edition.
6. P.M. Morse and H. Feshbach, *Methods of Theoretical Physics* (McGraw-Hill, New York, 1953).
7. H.B. Dwight, *Tables of Integrals and Other Mathematical Data* (MacMillan, New York, 1961), Fourth Edition.
8. G. Arfken, *Mathematical Methods for Physicists* (Academic, New York, 1985), Third Edition.
9. E. Kreyszig, *Advanced Engineering Mathematics* (Wiley, New York, 1993), Seventh Edition.
10. M. Abramowitz and I.A. Stegun, *Handbook of Mathematical Functions* (Dover, New York, 1972).
11. D.C. Champeney, *Fourier Transforms and their Physical Applications* (Academic Press, NY, 1973).
12. R.N. Bracewell, *The Fourier Transform and Its Applications* (McGraw-Hill, NY, 1986), Second Edition.
13. I.N. Sneddon, *Fourier Transforms* (Dover, New York, 1995).

Appendix E
Some Integrations in k, ω-Space

In Chapter 2 the electric field radiated by an accelerated charge is calculated using Fourier–Laplace transform techniques that involve a four-dimensional integration in $\mathbf{k}$, ω-space, essentially a summation of responses at all frequencies ω and all wavenumbers $k = 2\pi/\lambda$ in all directions (i.e., the wave vector $\mathbf{k}$). The radiated electric field at a position $\mathbf{r}$ and at a time t is expressed in Chapter 2, Eq. (2.22) in integral form, as

$$\mathbf{E}(\mathbf{r}, t) = \frac{ie}{\epsilon_0} \int_{\mathbf{k}} \int_{\omega} \frac{\omega \mathbf{v}_T(\omega) e^{-i(\omega t - \mathbf{k} \cdot \mathbf{r})}}{\omega^2 - k^2 c^2} \frac{d\omega \, d\mathbf{k}}{(2\pi)^4} \tag{2.22}$$

To aid us in the k-space integration we utilize the vector coordinates shown in Figure E.1, which is identical to Figure 2.3, with the addition that it shows explicitly the vector electric field $\mathbf{E}_i$ associated with the incident wave that is to be scattered by a single free electron. For a modest electric field, such that we can ignore the $\mathbf{v} \times \mathbf{B}$ term in the Lorentz force, the acceleration of the electron is given by $m\mathbf{a} = -e\mathbf{E}_i$, and in this harmonic analysis where $\mathbf{a} = d\mathbf{v}/dt = -i\omega \mathbf{v}$, the induced electron velocity is given by $\mathbf{v} = -ie\mathbf{E}/m\omega$, so that $\mathbf{E}_i$, $\mathbf{a}$, and $\mathbf{v}$ all have the same vector direction. To perform the k-space integration in Eq. (2.22) we note that the transverse component of velocity is given by $\mathbf{v}_T = -\mathbf{k}_0 \times (\mathbf{k}_0 \times \mathbf{v})$, with scalar magnitude $v_T = |\mathbf{v}_T| = |\mathbf{v}| \sin \Theta$, where Θ is measured from the direction of acceleration to the direction of observation $\mathbf{k}_0$. The k-space integration in Eq. (2.22) is thus performed with Θ treated as a fixed quantity. In essence it represents the observation direction $\mathbf{r}$ in $\mathbf{E}(\mathbf{r}, t)$, while the integration is performed over the k-space coordinates. For fixed polarization direction $\mathbf{E}_i$ and fixed observation direction $\mathbf{k}_0$, the angle Θ is constant and thus $\mathbf{v}_T$ passes through the k-space integrals. The integral expression for $\mathbf{E}(\mathbf{r}, t)$ is then

$$\mathbf{E}(\mathbf{r}, t) = \frac{ie}{\epsilon_0} \int_{\omega} \omega \mathbf{v}_T(\omega) e^{-i\omega t} \underbrace{\int_{k} \left[\frac{e^{i\mathbf{k} \cdot \mathbf{r}}}{(\omega + kc)(\omega - kc)} \frac{d\mathbf{k}}{(2\pi)^3} \right]}_{\text{a function } G(\omega; \mathbf{r})} \frac{d\omega}{2\pi}$$

where we have introduced an arbitrarily named function $G(\omega; \mathbf{r})$ to represent the requisite k-space integration in what follows. To perform the integration we use spherical

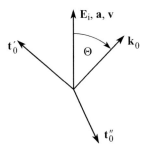

Figure E.1 Vector coordinates for scattering calculations involving a point-source electron caused to oscillate with an acceleration **a** and a velocity **v** by an incident electric field of polarization $\mathbf{E}_i$. The radiation is calculated for an observation direction $\mathbf{k}_0$. The angle Θ is measured from the direction of acceleration **a** to the direction of observed scattering $\mathbf{k}_0$. The unit vectors are defined by $k_0 \times t_0' = t_0''$. Spherical coordinates (k, θ, ϕ) are oriented around $\mathbf{k}_0$.

coordinates oriented around $\mathbf{k}_0$, as introduced* in Chapter 2:

$$d\mathbf{k} = k^2 \underbrace{\sin\theta \, d\theta \, d\phi}_{d\Omega} \, dk \tag{2.23a}$$

where

$$0 \le k \le \infty \tag{2.23b}$$

$$0 \le \theta \le \pi \tag{2.23c}$$

$$0 \le \phi \le 2\pi \tag{2.23d}$$

so that for a vector **r**, at polar angle θ to **k**, the phase term that occurs in Eq. (2.22) becomes

$$\mathbf{k} \cdot \mathbf{r} = kr\cos\theta \tag{2.23e}$$

The first integral, $G(\omega; \mathbf{r})$, can then be evaluated as

$$G(\omega; \mathbf{r}) = \int_0^\infty \int_0^{2\pi} \int_0^\pi \frac{e^{i\mathbf{k}\cdot\mathbf{r}}}{(\omega + kc)(\omega - kc)} \frac{k^2 \sin\theta \, d\theta \, d\phi \, dk}{(2\pi)^3}$$

which upon integration over ϕ becomes

$$G(\omega; \mathbf{r}) = \frac{-1}{(2\pi c)^2} \int_0^\infty \frac{1}{\left(k + \frac{\omega}{c}\right)\left(k - \frac{\omega}{c}\right)} \left[\int_0^\pi \frac{e^{ikr\cos\theta} ikr\sin\theta \, d\theta}{ikr}\right] k^2 dk$$

$$G(\omega; \mathbf{r}) = \frac{-1}{(2\pi c)^2} \int_0^\infty \frac{1}{\left(k + \frac{\omega}{c}\right)\left(k - \frac{\omega}{c}\right)} \left[\int_0^\pi \frac{e^{ikr\cos\theta} d(ikr\cos\theta)}{-ikr}\right] k^2 dk$$

---

* Recall that $d\mathbf{k}$ is not a vector, but rather shorthand notation for a volume element in differential space. For instance, in rectangular coordinates $d\mathbf{k} = dk_x \, dk_y \, dk_z$.

Then using $\int_a^b e^u du = e^u |_a^b$,

$$G(\omega; \mathbf{r}) = \frac{-1}{(2\pi c)^2} \int_0^\infty \frac{1}{\left(k + \frac{\omega}{c}\right)\left(k - \frac{\omega}{c}\right)} \underbrace{\left[\frac{e^{ikr \cos\theta} |_0^\pi}{-ikr}\right]}_{\frac{e^{-ikr} - e^{ikr}}{-ikr}} k^2 dk$$

$$G(\omega; \mathbf{r}) = \frac{-i}{(2\pi c)^2 r} \int_0^\infty \frac{(e^{-ikr} - e^{ikr})}{\left(k + \frac{\omega}{c}\right)\left(k - \frac{\omega}{c}\right)} k \, dk$$

$$G(\omega; \mathbf{r}) = \frac{i}{(2\pi c)^2 r} \left\{\left[\int_0^\infty \frac{e^{ikr} k \, dk}{\left(k + \frac{\omega}{c}\right)\left(k - \frac{\omega}{c}\right)}\right] - \left[\int_0^\infty \frac{e^{-ikr} k \, dk}{\left(k + \frac{\omega}{c}\right)\left(k - \frac{\omega}{c}\right)}\right]\right\}$$

To perform these integrations using the *Cauchy integral formula* (see Appendix D, Ref. 9)

$$\oint \frac{f(z)}{z - z_0} dz = 2\pi \, i \, f(z_0)$$

we need to extend the k-integration from $-\infty$ to $+\infty$. To achieve this we change variables in the second integral above, replacing k by $-k'$, so that

$$G(\omega; \mathbf{r}) = \frac{i}{(2\pi c)^2 r} \left\{\left[\int_0^\infty \frac{e^{ikr} k \, dk}{\left(k + \frac{\omega}{c}\right)\left(k - \frac{\omega}{c}\right)}\right] - \left[\int_0^{-\infty} \frac{e^{ik'r} k' \, dk'}{\underbrace{\left(-k' + \frac{\omega}{c}\right)\left(-k' - \frac{\omega}{c}\right)}_{\left(k' - \frac{\omega}{c}\right)\left(k' + \frac{\omega}{c}\right)}}\right]\right\}$$

We note that the integration $-\int_0^{-\infty}$ can be replaced by $+\int_{-\infty}^0$, so that the k-integration can now be written compactly as

$$G(\omega; \mathbf{r}) = \frac{i}{(2\pi c)^2 r} \left[\int_{-\infty}^{+\infty} \frac{e^{ikr} k \, dk}{\left(k + \frac{\omega}{c}\right)\left(k - \frac{\omega}{c}\right)}\right]$$

which can now be evaluated by closing the contour integral and using the Cauchy integral formula. Thus we identify $f(k) = k e^{ikr}/(k + \omega/c)$ and close the integration contour with a very large semicircle in the upper half of the complex k-plane, such that the integrand goes to zero along the added semicircular path. Note that in the complex k-plane we have $k = k_r + i k_i$, so that $e^{ikr} = e^{ik_r r} e^{-k_i r}$. For radiated fields at large distances r, the factor $e^{-k_i r}$ goes to zero for $k_i > 0$. Thus we close the contour in the upper half plane where this added semicircular path closes the contour but adds nothing to the integral.

Note that in Figure E.2 the poles at $k = \omega/c$ and $-\omega/c$ are shown shifted slightly off axis to indicate wave decay (rather than growth) as the waves propagate in their respective directions. This is justified by noting that there is always some absorption or scattering loss in real physical systems.

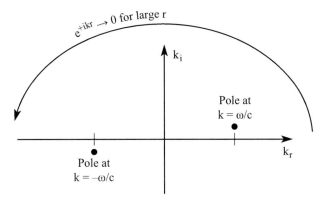

Figure E.2. Integration in the complex *k*-plane.

The integration of $G(\omega; \mathbf{r})$ then becomes

$$G(\omega; \mathbf{r}) = \frac{i}{(2\pi c)^2 r} \oint \underbrace{\frac{f(k)\,dk}{k - \frac{\omega}{c}}}_{2\pi i f\left(\frac{\omega}{c}\right)}$$

Recalling that we defined $f(k) = k e^{ikr} / \left(k + \frac{\omega}{c}\right)$, we have

$$G(\omega; \mathbf{r}) = \frac{i}{(2\pi c)^2 r} \left[2\pi i \frac{\left(\frac{\omega}{c}\right) e^{i\omega r/c}}{\frac{\omega}{c} + \frac{\omega}{c}} \right]$$

$$G(\omega; \mathbf{r}) = \frac{-e^{i\omega r/c}}{4\pi c^2 r}$$

Having completed the *k*-integrations involved in our function $G(\omega; \mathbf{r})$, we return to Eq. (2.22), where the expression for $\mathbf{E}(\mathbf{r}, t)$ now involves only a frequency integration

$$\mathbf{E}(\mathbf{r}, t) = \frac{ie}{\epsilon_0} \int_{-\infty}^{\infty} \omega \mathbf{v}_T(\omega) e^{-i\omega t} \left[\frac{-e^{i\omega r/c}}{4\pi c^2 r} \right] \frac{d\omega}{2\pi}$$

or

$$\mathbf{E}(\mathbf{r}, t) = \frac{e}{4\pi\epsilon_0 c^2 r} \int_{-\infty}^{\infty} \underbrace{(-i\omega)\,\mathbf{v}_T(\omega) e^{-i\omega(t - r/c)}}_{\frac{d}{dt}\left[\mathbf{v}_T(\omega) e^{-\omega(t - r/c)}\right]} \frac{d\omega}{2\pi} \qquad (2.24)$$

$$\mathbf{E}(\mathbf{r}, t) = \frac{e}{4\pi\epsilon_0 c^2 r} \frac{d}{dt} \int_{-\infty}^{\infty} \underbrace{\mathbf{v}_T(\omega) e^{-i\omega(t - r/c)} \frac{d\omega}{2\pi}}_{\mathbf{v}_T\left(t - \frac{r}{c}\right)}$$

where the last notation recognizes the transform of $\mathbf{v}_T(\omega)$ in the variable $t' = t - r/c$.
  Identifying the acceleration as

$$\mathbf{a}_T\left(t - \frac{r}{c}\right) = \frac{d}{dt} \mathbf{v}_T\left(t - \frac{r}{c}\right)$$

the electric field associated with the radiated wave can be written as

$$E(\mathbf{r}, t) = \frac{e\mathbf{a}_T \left(t - \frac{r}{c}\right)}{4\pi \epsilon_0 c^2 r} \qquad (2.25)$$

which is the form given in Chapter 2 as Eq. (2.25). The physical interpretation of this expression is described in the text of Chaper 2, following Eq. (2.25).

Appendix F
Lorentz Space-Time Transformations

In our studies of radiation from charged particles moving at velocities approaching that of light, a number of interesting phenomena are observed, such as the searchlight effect wherein radiation from the charged particle is constrained to a very narrow forward radiation cone. Furthermore, the calculation of detailed angular radiation patterns, in the frame of reference moving with the charged particle, and wavelength distributions are readily accomplished.[1] The results can then be transformed back to the laboratory, or observer, frame of reference. For instance, the calculation of undulator radiation reduces to use of the well-known formula for so-called *dipole radiation* from a simple oscillating electron. With this approach we need solve Maxwell's equations for only the simplest radiating system, a small amplitude oscillating electron. This approach is not only simple to follow, but gives valuable physical insights to the radiation process and the parameters that characterize it.

In order to relate calculations in one frame of reference to those in another frame of reference when the relative speed between the two approaches that of light, we must make use of the Lorentz space-time transformations, which provide relationships between spatial and temporal scales in the two frames of reference, and are consistent both with Einstein's postulates of special relativity and with all known experiments (see Ref. 2 for a discussion of the Lorentz transformations and their reduction to Galilean transformations as $v/c \to 0$).

The *Lorentz space-time transformations* between coordinates (x, y, z, t) in frame of reference S and coordinates (x', y', z', t') in frame of reference S', which moves at velocity v with respect to S in the z, z' direction (see Figure F.1), are as follows:

$$z = \gamma(z' + \beta c t') \tag{F.1a}$$

$$t = \gamma\left(t' + \frac{\beta z'}{c}\right) \tag{F.1b}$$

$$y = y' \text{ and } x = x' \tag{F.1c}$$

and

$$z' = \gamma(z - \beta c t) \tag{F.2a}$$

$$t' = \gamma\left(t - \frac{\beta z}{c}\right) \tag{F.2b}$$

$$y' = y \text{ and } x' = x \tag{F.2c}$$

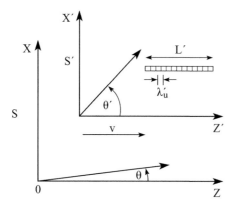

Figure F.1 Two frames of reference S and S' with relative velocity v, which approaches the speed of light in vacuum, c.

where c is the velocity of light in vacuum,

$$\beta \equiv \frac{v}{c} \qquad \text{(F.3)}$$

and

$$\gamma \equiv \frac{1}{\sqrt{1 - \beta^2}} \qquad \text{(F.4)}$$

Note that $\beta < 1$, and that for β approaching zero, γ approaches unity, and thus the Galilean transformations are obtained, i.e., $t = t'$ and $z = z' + vt'$.

We use these relationships to transform charged particle trajectories and radiation characteristics between natural frames of reference. We find it convenient to have at hand angular relationships that follow from Eqs. (F.1a–c)–(F.4), angle dependent Doppler shifts that follow therefrom, and expressions for the well-known time dilation and (Lorentz) length contraction characteristics of observations made between relativistically related coordinate systems.

F.1   Frequency and Wavenumber Relations

To develop relationships between frequency and wavelength as observed in the two coordinate systems S and S', we consider a propagating wave observed from both systems. For such a wave the amplitude varies according to a phase factor

$$e^{i\phi} = e^{i(\omega t - \mathbf{k} \cdot \mathbf{r})}$$

where the wavenumber $|\mathbf{k}| = 2\pi/\lambda$ and where $\omega = 2\pi f$. In our coordinate systems S and S', the phase can be written as

$$\phi = \omega t - k_z z - k_x x - k_y y \qquad \text{(F.5a)}$$

$$\phi' = \omega' t' - k'_z z' - k'_x x' - k'_y y' \qquad \text{(F.5b)}$$

Since the phase has a particular value at some given space-time point (it might be at the crest of wave amplitude), it must be the same in both coordinate systems, and thus we can set Eq. (F.5a) equal to Eq. (F.5b), viz.,

$$\omega t - k_z z - k_x x - k_y y = \omega' t' - k'_z z' - k'_x x' - k'_y y'$$

The desired frequency and wavenumber (wavelength) relationships can be obtained by substituting the Lorentz relationships [Eqs. (F.2a–c)] into the identical phase relationship above. Making appropriate substitutions for t', z', x', and y' in the phase relationship, we have

$$\omega t - k_z x - k_x x - k_y y = \omega' \left[\gamma \left(t - \frac{\beta z}{c} \right) \right] - k'_z [\gamma (z - \beta ct)] - k'_x x - k'_y y$$

or by rearranging terms

$$\omega t - k_z z - k_x x - k_y y = [\omega' \gamma + \gamma \beta c k'_z] t - \left[\gamma k'_z + \gamma \omega' \frac{\beta}{c} \right] z - k'_x x - k'_y y$$

As this relationship must hold for arbitrary space-time position, we can match coefficients term by term in the above equation. Doing so, we obtain

$$\omega = \gamma (\omega' + \beta c k'_z) \tag{F.6a}$$

$$k_z = \gamma \left(k'_z + \frac{\beta \omega'}{c} \right) \tag{F.6b}$$

$$k_x = k'_x \quad \text{and} \quad k_y = k'_y \tag{F.6c}$$

and the inverse transformations

$$\omega' = \gamma (\omega - \beta c k_z) \tag{F.7a}$$

$$k'_z = \gamma \left(k_z - \frac{\beta \omega}{c} \right) \tag{F.7b}$$

$$k'_x = k_x \quad \text{and} \quad k'_y = k_y \tag{F.7c}$$

which relate wave propagation characteristics of frequency and wavenumber in two frames of reference moving at relativistic speed with respect to each other. They are commonly referred to as the *energy–momentum relations* because for a photon $E = \hbar \omega$ and $\mathbf{p} = \hbar \mathbf{k}$. Note that the relationships between ω and ω' are dependent on the angle of observation in that what appears in Eqs. (F.6) and (F.7) is the axial component of the wave vector (**k**), i.e., k_z.

The relationships take on quite a simple form if we decompose the wave vector in both reference frames, and introduce the angles θ and θ', representing the propagation direction measured from the common z, z' axis. This is illustrated in Figure F.2.

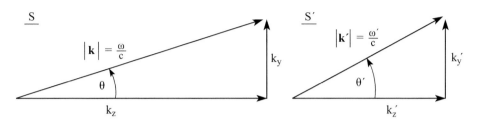

Figure F.2 The wave vector **k** for a wave propagating at an angle θ to the z-axis in the S frame of reference, and the same wave as observed in the S' frame of reference, with wave vector **k'** at an angle θ' to the (common) z-axis.

Utilizing the vector identifications evident in Figure F.2, the expression relating ω and ω' [Eq. (F.6a)] becomes

$$\omega = \gamma(\omega' + \beta c k'_z) = \gamma\left(\omega' + \beta c \frac{\omega'}{c}\cos\theta'\right)$$

$$\boxed{\omega = \omega'\gamma(1 + \beta\cos\theta')} \tag{F.8a}$$

and from Eq. (F.7)

$$\omega' = \gamma(\omega - \beta c k_z) = \gamma\left(\omega - \beta c \frac{\omega}{c}\cos\theta\right)$$

or

$$\boxed{\omega' = \omega\gamma(1 - \beta\cos\theta)} \tag{F.8b}$$

Equations (F.8a) and (F.8b) describe the relativistically correct form for the Doppler shift (Refs. 3 and 4) of frequency when there is motion between the source and receiver (observer) that approaches the velocity of light.

A familiar form of Eq. (F.8a) is that for a source $\omega' = \omega_s$ moving toward the observer directly along the z-axis, where θ' is zero and $\cos\theta'$ is unity. Identifying the frequency $\omega = \omega_R$ as that received in the observer's frame of reference, one has

$$\omega = \omega'\gamma(1 + \beta)$$

Noting that

$$\gamma = \frac{1}{\sqrt{1 - \beta^2}} = \frac{1}{\sqrt{1 - \beta}\sqrt{1 + \beta}}$$

one obtains the familiar form (Ref. 2)

$$\omega_R = \omega_S\sqrt{\frac{1 + \beta}{1 - \beta}} \quad (\text{forward, } \theta = 0)$$

or equivalently

$$f_R = f_S \sqrt{\frac{1 + \beta}{1 - \beta}}$$

as is sometimes seen.

F.2   Angular Transformations

Angular transformations between the two frames of reference (Ref. 3) can be obtained from the energy–momentum transformations, Eqs. (F.6) and (F.7). Again using the propagating wave depicted in Figure F.2, we can write

$$\cos \theta = \frac{k_z}{k} = \frac{k_z c}{\omega} = \frac{\gamma \left(k_z' + \frac{\beta \omega'}{c} \right) c}{\gamma \left(\omega' + \beta c k_z' \right)}$$

$$\cos \theta = \frac{\gamma \left[\frac{\omega' \cos \theta'}{c} + \frac{\beta \omega'}{c} \right] c}{\gamma \left[\omega' + \beta c \cdot \frac{\omega'}{c} \cos \theta' \right]}$$

or

$$\boxed{\cos \theta = \frac{\cos \theta' + \beta}{1 + \beta \cos \theta'}} \qquad \text{(F.9a)}$$

Similarly

$$\cos \theta' = \frac{k_z'}{k'} = \frac{k_z' c}{\omega'} = \frac{\gamma \left(k_z - \frac{\beta \omega}{c} \right) c}{\gamma \left(\omega - \beta c k_z \right)}$$

$$\cos \theta' = \frac{\gamma \left[\frac{\omega \cos \theta}{c} - \frac{\beta \omega}{c} \right] c}{\gamma \left[\omega - \beta c \frac{\omega \cos \theta}{c} \right]}$$

or

$$\boxed{\cos \theta' = \frac{\cos \theta - \beta}{1 - \beta \cos \theta}} \qquad \text{(F.9b)}$$

which gives us equations relating θ and θ' as a function of $\beta = v/c$. In similar fashion we can define $\sin \theta' = k_y'/k'$ to obtain

$$\sin \theta' = \frac{k_y' c}{\omega'} = \frac{k_y c}{\gamma (\omega - \beta c k_z)}$$

$$\sin \theta' = \frac{(k \sin \theta) c}{\gamma (\omega - \beta c k \cos \theta)}$$

or

$$\sin \theta' = \frac{\sin \theta}{\gamma \, (1 - \beta \cos \theta)} \qquad \text{(F.10a)}$$

Then defining $\sin \theta = k_y/k = k_y c/\omega$ and using Eq. (F.6), one finds in similar fashion that

$$\sin \theta = \frac{\sin \theta'}{\gamma \, (1 + \beta \cos \theta')} \qquad \text{(F.10b)}$$

Noting that $\tan \theta = \sin \theta / \cos \theta$ and that $\tan \theta' = \sin \theta' / \cos \theta'$, one finds that

$$\tan \theta = \frac{\sin \theta'}{\gamma \, (\cos \theta' + \beta)} \qquad \text{(F.11a)}$$

$$\tan \theta' = \frac{\sin \theta}{\gamma \, (\cos \theta - \beta)} \qquad \text{(F.11b)}$$

which could also be obtained by starting with $\tan \theta = k_y/k_z$, etc.

The $\tan \theta$ relation is convenient for illustrating the searchlight effect characteristic of synchrotron radiation from relativistic electrons for which $\gamma \gg 1$. For highly relativistic electrons experiencing significant acceleration, even if the radiation pattern is rather broad – for instance, $0 < \theta' < \pi/4$ in the electron (S') frame of reference – Eq. (F.11a) indicates that all this will be folded into a very narrow forward radiation cone, with half angle of order $1/2\gamma$.

F.3   The Lorentz Contraction of Length

The Lorentz transformations contain information regarding differences of apparent length of the same object seen from two frames of reference, moving relative to each other at a relativistic speed. This applies not only to the traditional meter stick, but, more relevant to our studies, to periodic magnet structures. Figure F.1 includes such an object with length L' in S', and substructure of periodicity λ'_u.

To understand the apparent differences in length, as seen in S and S' at a given instant of time, consider an object at rest in frame S with endpoints (z'_1, t') and (z'_2, t') in S', and (z_1, t), (z_2, t) in S. According to the Lorentz transformations, Eqs. (F.1a–c), the positions are related by

$$z_1 = \gamma(z'_1 + \beta ct')$$

and

$$z_2 = \gamma(z'_2 + \beta ct')$$

Subtracting, we have

$$z_1 - z_2 = \gamma [z_1' - z_2' + \beta ct' - \beta ct']$$

and thus

$$z_1 - z_2 = \gamma (z_1' - z_2')$$

Defining $L = z_1 - z_2$ and $L' = z_1' - z_2'$, we have

$$L = \gamma L'$$

or

$$\boxed{L' = L/\gamma} \qquad (F.12)$$

which succinctly describes the Lorentz contraction of length. An object of length L as seen in frame of reference S appears significantly shorter (L/γ) in the frame of reference S' that is moving at a relative velocity approaching that of light. Similarly, a structure of periodicity λ_u in S will be observed as one of periodicity λ_u/γ in S'. Note that $\beta = v/c$ and $\gamma = 1/\sqrt{1 - \beta^2}$, so that for β approaching unity (v approaching c), γ is very large and the contraction factor can be orders of magnitude. On the other hand, for non-relativistic velocities β approaches zero, γ approaches unity, and there is little or no contraction.

F.4   Time Dilation

We next ask, what is the relationship between time intervals as observed from a given position? The position of observation is described as z in S, and as z' in S'. Referring to the beginning of the time interval as t_1', and the end as t_2' in S', and as t_1 and t_2 respectively in S, we can write from Eqs. (F.1a–c) that

$$t_1 = \gamma \left(t_1' + \frac{\beta z'}{c} \right)$$

and

$$t_2 = \gamma \left(t_2' + \frac{\beta z'}{c} \right)$$

Subtracting, and referring to the time intervals as $\Delta t = t_2 - t_1$ and $\Delta t' = t_2' - t_1'$, one has

$$\Delta t = \gamma \, \Delta t'$$

or

$$\boxed{\Delta t' = \Delta t/\gamma} \qquad (F.13)$$

expressing algebraically what is known as *time dilation*. Equation (F.13) leads to the statement that "(relativistically) moving clocks run slow." In the famous *twin paradox*, although a time interval might be $\Delta t = 1$ second on earth, about the time of a heartbeat, the moving twin on a relativistic rocketship sees the elapsed time as much shorter, of order 1 second/γ, far too short for a heartbeat. Thus the relativistically moving twin ages more slowly.

F.5   Transforming $dP'/d\Omega'$ to $dP/d\Omega$

In the sections dealing with the transformation of radiated power per unit solid angle from S' to S, there are several places where the algebra becomes tedious and distractive. Here, we have treated several of these items separately.

(1) **The $\sin^2 \Theta'$ factor:** Dipole radiation involves a $\sin^2 \Theta'$ dependence in the S' frame. For the fundamental with oscillation in the x'-direction, this is written in θ', ϕ' as

$$\sin^2 \Theta' = 1 - \sin^2\theta'\cos^2\phi'$$

From our section on the Lorentz transformations, Eq. (F.10a), in terms of time averaged values γ^* and β^* (see Chapter 5, Sections 5.4 and 5.5),

$$\sin \theta' = \frac{\sin \theta}{\gamma^*(1 - \beta^* \cos \theta)}$$

Noting the γ-relations developed earlier [Eqs. (5.26) and (5.67)], we have

$$\beta^* \simeq 1 - \frac{1}{2\gamma^{*2}}$$

The denominator of Eq. (F.10a) takes the form for small θ of order $1/\gamma$:

$$1 - \beta^* \cos \theta \simeq 1 - \beta^* \left(1 - \frac{\theta^2}{2} + \cdots \right)$$

$$1 - \beta^* \cos \theta \simeq 1 - \left(1 - \frac{2}{2\gamma^{*2}} \right)\left(1 - \frac{\theta^2}{2} \right)$$

$$1 - \beta^* \cos \theta \simeq \frac{1}{2\gamma^{*2}} + \frac{\theta^2}{2}$$

$$1 - \beta^* \cos \theta \simeq \frac{1}{2\gamma^{*2}}(1 + \gamma^{*2}\theta^2)$$

Substitution into Eq. (F.10a) gives

$$\sin \theta' \simeq \frac{\theta}{\gamma^* \left(\frac{1}{2\gamma^{*2}} (1 + \gamma^{*2}\theta^2) \right)}$$

or

$$\sin \theta' \simeq \frac{2\gamma^*\theta}{1 + \gamma^{*2}\theta^2} \tag{F.14}$$

Then Eq. (5.36) becomes

$$\sin^2 \Theta' = 1 - \sin^2 \theta' \cos^2 \phi'$$

$$\sin^2 \Theta' \simeq 1 - \frac{(2\gamma^* \theta)^2}{(1 + \gamma^{*2}\theta^2)^2} \cos^2 \phi$$

$$\sin^2 \Theta' \simeq \frac{(1 + \gamma^{*2}\theta^2)^2 - 4\gamma^{*2}\theta^2 \cos^2 \phi}{(1 + \gamma^{*2}\theta^2)^2}$$

$$\sin^2 \Theta' \simeq \frac{1 + 2\gamma^{*2}\theta^2(1 - 2\cos^2 \phi) + \gamma^{*4}\theta^4}{(1 + \gamma^{*2}\theta^2)^2} \tag{F.15}$$

as was used in Eq. (5.44).

(2) **Transformation from $d\Omega'$ to $d\Omega$:** As a next step, we wish to Lorentz-transform the solid angle

$$d\Omega' = \sin \theta' d\theta' d\phi'$$

to a θ, ϕ-coordinate system. Since ϕ and ϕ' are in planes perpendicular to the relativistic motion between the S and S' frames, the simple relationship $\phi' = \phi$ is true. Hence,

$$d\phi' = d\phi$$

By Eq. (F.14) we have a relation between θ and θ'. Now, all that is needed is a relationship between $d\theta'$ and $d\theta$. Toward this end, we can again use Eq. (F.14). This time, we take a derivative on both sides:

$$\cos \theta' d\theta' \simeq \frac{(1 + \gamma^{*2}\theta^2)(2\gamma^* d\theta) - 2\gamma^* \theta(2\gamma^{*2}\theta d\theta)}{(1 + \gamma^{*2}\theta^2)^2}$$

$$\cos \theta' d\theta' \simeq \frac{2\gamma^*(1 - \gamma^{*2}\theta^2)}{(1 + \gamma^{*2}\theta^2)^2} d\theta$$

To eliminate $\cos \theta'$, we note that

$$\cos^2 \theta' = 1 - \sin^2 \theta' \simeq 1 \frac{(2\gamma^* \theta)^2}{(1 + \gamma^{*2}\theta^2)^2}$$

When written with a common denominator, expanded, and like terms collected, this gives

$$\cos \theta' \simeq \frac{1 - \gamma^{*2}\theta^2}{1 + \gamma^{*2}\theta^2}$$

Combining equations involving $\cos \theta'$ above, one has

$$\frac{1 - \gamma^{*2}\theta^2}{1 + \gamma^{*2}\theta^2} d\theta' \simeq \frac{2\gamma^*(1 - \gamma^{*2}\theta^2)}{(1 + \gamma^{*2}\theta^2)^2} d\theta'$$

or

$$d\theta' \simeq \frac{2\gamma^*}{1 + \gamma^{*2}\theta^2} d\theta \tag{F.16}$$

The equation for solid angle then becomes

$$d\Omega' \simeq \sin\theta' d\theta' d\phi \simeq \frac{4\gamma^{*2}}{(1 + \gamma^{*2}\theta^2)^2} \cdot \theta \, d\theta \, d\phi \tag{F.17}$$

as was used in Eq. (5.45).

(3) **The $d\Omega$ integrals:** In calculating the power radiated, it is necessary to integrate $dP/d\Omega$ [Chapter 5, Eq. (5.47)] over all solid angles. Because integration over ϕ contributes 2π to the integral [see paragraph before Eq. (5.50a)], the integration reduces to

$$\int_0^\pi \int_0^{2\pi} \frac{1 + \gamma^{*4}\theta^4}{(1 + \gamma^{*2}\theta^2)^5} \theta \, d\theta \, d\phi = \frac{2\pi}{\gamma^{*2}} \int_0^\pi \frac{(1 + \gamma^{*4}\theta^4)(\gamma^*\theta)d(\gamma^*\theta)}{(1 + \gamma^{*2}\theta^2)^5}$$

Substituting

$$u = \gamma^*\theta$$

$$x = 1 + u^2$$

$$dx = 2u \, du$$

the integrand becomes $1 + u^4 = 1 + (u^2)^2 = 1 + (x - 1)^2 = x^2 - 2x + 2$. Hence,

$$\int_0^\pi \int_0^{2\pi} \frac{1 + \gamma^{*4}\theta^4}{(1 + \gamma^{*2}\theta^2)^5} \theta \, d\theta \, d\phi = \frac{2\pi}{\gamma^{*2}} \cdot \frac{1}{2} \int_1^{(1+\pi^2\gamma^{*2})} \frac{(x^2 - 2x + 2) \, dx}{x^5}$$

$$\int_0^\pi \int_0^{2\pi} \frac{1 + \gamma^{*4}\theta^4}{(1 + \gamma^{*2}\theta^2)^5} \theta \, d\theta \, d\phi = \frac{\pi}{\gamma^{*2}} \int_1^\infty \left(\frac{1}{x^3} - \frac{2}{x^4} + \frac{2}{x^5} \right) dx$$

$$\int_0^\pi \int_0^{2\pi} \frac{1 + \gamma^{*4}\theta^4}{(1 + \gamma^{*2}\theta^2)^5} \theta \, d\theta \, d\phi = \frac{\pi}{\gamma^{*2}} \left[\frac{1}{(-2x^2)} + \frac{2}{(3x^3)} + \frac{2}{(-4x^4)} \right]_1^\infty$$

Therefore the integral taking us from Eq. (5.47) to Eq. (5.50) becomes

$$\int_0^\pi \int_0^{2\pi} \frac{1 + \gamma^{*4}\theta^4}{(1 + \gamma^{*2}\theta^2)^5} \theta \, d\theta \, d\phi = \frac{\pi}{3\gamma^{*2}} \tag{F.18}$$

References

1. J.D. Jackson, "The Impact of Special Relativity on Theoretical Physics," *Physics Today* **40**, 34 (May 1987).
2. P.A. Tipler and R.A. Llewelyn, *Modern Physics* (Freeman, New York, 2012), Sixth Edition.
3. G. Joos, *Theoretical Physics* (Hafner, New York, 1934), pp. 233–235.
4. L. Landau and E. Lifshitz, *The Classical Theory of Fields* (Addison-Wesley, Reading, MA, 1951), p. 121.

Appendix G
Some FEL Algebra

G.1   Slow and Fast Phase for Energy Transfer, θ_s and θ_f

The product cosine terms in Eq. (6.16) lead to frequency mixing, which is seen more readily in terms of sum and difference frequencies, as in Eq. (6.17). One of those two terms varies very rapidly with distance z, so rapidly that it cancels its own contribution to energy exchange when averaged over just a few cycles. The other varies very slowly, contributing in a discernible way only over many cycles. Below is the algebra that clarifies these points. From the definitions of θ_s and θ_f, seen below in Eq. (6.17):

$$\theta_s = (k_u + k)z - \omega t \tag{6.17b}$$

$$\theta_f = (k_u - k)z + \omega t \tag{6.17c}$$

$$\frac{d\theta_s}{dt} = (k_u + k)\frac{dz}{dt} - \omega$$

$$\frac{d\theta_f}{dt} = (k_u - k)\frac{dz}{dt} + \omega$$

Averaging over a full undulator period, $\frac{dz}{dt} = \bar{v}_z = c\left(1 - \frac{1+\frac{K^2}{2}}{2\gamma^2}\right)$, as seen previously in Eq. (5.25), thus

$$\frac{d\theta_s}{dt} = (k_u + k)c\left(1 - \frac{1 + \frac{K^2}{2}}{2\gamma^2}\right) - \omega$$

$$\frac{d\theta_f}{dt} = (k_u - k)c\left(1 - \frac{1 + \frac{K^2}{2}}{2\gamma^2}\right) + \omega$$

$$\frac{d\theta_s}{dt} = k_u c\left[1 + \frac{k}{k_u}\left(1 - \frac{1 + \frac{K^2}{2}}{2\gamma^2}\right) - \frac{k}{k_u}\right]$$

$$\frac{d\theta_f}{dt} = k_u c\left[1 - \frac{k}{k_u}\left(1 - \frac{1 + \frac{K^2}{2}}{2\gamma^2}\right) + \frac{k}{k_u}\right]$$

$$\frac{d\theta_s}{dt} = k_u c\left[1 - \frac{k}{k_u}\left(\frac{1 + \frac{K^2}{2}}{2\gamma^2}\right)\right]$$

$$\frac{d\theta_f}{dt} = k_u c\left[1 + \frac{k}{k_u}\left(\frac{1 + \frac{K^2}{2}}{2\gamma^2}\right)\right]$$

Introducing the normalized electron energy, $\eta \equiv \frac{\gamma - \gamma_0}{\gamma_0}$, Eq. (6.20), such that $\gamma = \gamma_0 \eta + \gamma_0 = \gamma_0(1 + \eta)$, and $\gamma^2 \cong \gamma_0^2(1 + 2\eta)$ for $\eta \ll 1$, then

$$\frac{d\theta_s}{dt} = k_u c \left[1 - \frac{k}{k_u} \left(\frac{1 + \frac{K^2}{2}}{2\gamma_0^2(1 + 2\eta)} \right) \right]$$

$$\frac{d\theta_f}{dt} = k_u c \left[1 + \frac{k}{k_u} \left(\frac{1 + \frac{K^2}{2}}{2\gamma_0^2 (1 + 2\eta)} \right) \right]$$

Noting that $\frac{k}{k_u} \left(\frac{1 + \frac{K^2}{2}}{2\gamma_0^2} \right) = \frac{1}{\lambda} \cdot \frac{\lambda_u \left(1 + \frac{K^2}{2} \right)}{2\gamma_0^2} = 1$ by the undulator equation (6.1), and $1/(1 + 2\eta) \simeq 1 - 2\eta$ for $\eta \ll 1$, the equations simplify to

$$\frac{d\theta_s}{dt} = 2k_u c \eta$$

$$\frac{d\theta_f}{dt} = 2k_u c$$

or

$$\boxed{\frac{d\theta_s}{dt} = 2\omega_u \eta} \tag{6.18}$$

$$\boxed{\frac{d\theta_f}{dt} = 2\omega_u} \tag{6.19}$$

Note that the fast variation is at twice the undulator frequency, $2\omega_u$, thus exchanging energy both ways, twice per undulator period of travel, cancelling significant energy exchange between particles and field. The slow variation, $d\theta_s/dt$, is smaller by the factor $\eta \ll 1$, exchanging energy slowly and consistently over many periods, and thus can contribute significantly to wave growth.

G.2   The One-Dimensional Wave Equation

The wave equation for an x-polarized electric field, uniform in y and z, and no transverse gradients, Eq. (2.7), is

$$\left[\frac{\partial^2}{\partial t^2} - c^2 \frac{\partial^2}{\partial z^2} \right] E_x(z, t) = -\frac{1}{\epsilon_0} \frac{\partial J_x(z, t)}{\partial t} \tag{6.24}$$

where $c^2 = 1/\mu_0 \epsilon_0$ and

$$J_x(z, t) = -e n_e(z) V_x(z, t) \tag{6.25}$$

Previously we considered the seed wave to be of constant amplitude E_0, as in Eq. (6.14), but now we wish to consider the seed wave to be slowly growing as it traverses along many undulator periods in the FEL amplifier. Towards this end we introduce the complex electric field amplitude, $\tilde{E}_x$, where the superscript tilde indicates its complex nature.

Separating the rapidly varying phase of the wave from the slowly varying, z-dependent field amplitude, we now write

$$E_x(z, t) = \tilde{E}_x(z)e^{-i(\omega t - kz)} \tag{6.26}$$

where $E_x(0) = E_0$. The time derivative in the wave equation algebrizes* to

$$\frac{\partial^2}{\partial t^2}E_x(z, t) = (i\omega)^2 E_x(z, t) = (-i\omega)^2 \tilde{E}_x(z)e^{-i(\omega t - kz)}$$

For the spatial derivatives, which involve the product of two z-dependent quantities,

$$\frac{\partial}{\partial z}E_x(z, t) = \tilde{E}_x(z) \cdot (ik)e^{-i(\omega t - kz)} + \frac{\partial \tilde{E}_x(z)}{\partial z}e^{-i(\omega t - kz)}$$

and then

$$\frac{\partial^2}{\partial z^2}E_x(z, t) = -k^2 \tilde{E}_x(z)e^{-i(\omega t - kz)} + 2ik\frac{\partial \tilde{E}_x(z)}{\partial z}e^{-i(\omega t - kz)} + \frac{\partial^2 \tilde{E}_x(z)}{\partial z^2}e^{-i(\omega t - kz)}$$

so that the one-dimensional wave equation becomes

$$\left[-\omega^2 \tilde{E}_x(z) + k^2 c^2 \tilde{E}_x(z) - 2ikc^2\frac{\partial \tilde{E}_x(z)}{\partial z} - c^2\frac{\partial^2 \tilde{E}_x(z)}{\partial z^2}\right]e^{-i(\omega t - kz)} = -\frac{1}{\varepsilon_0}\frac{\partial J_x(z, t)}{\partial t}$$

where the first two terms cancel by the dispersion relation, $\omega = kc$, leaving

$$\left[2ik\frac{\partial \tilde{E}_x(z)}{\partial z} + \frac{\partial^2 \tilde{E}_x(z)}{\partial z^2}\right]e^{-i(\omega t - kz)} = \mu_0\frac{\partial J_x(z, t)}{\partial t}$$

for the slowly growing field amplitude, $\tilde{E}(z)$. Since the field grows slowly, over many wavelengths, the gradient with respect to z, $(\partial/\partial z)^{-1} \gg \lambda = 2\pi k$, or $\partial/\partial z \ll k$, so that the second term in equation (6.26) is small. *The one-dimensional wave equation for the slowly growing field is then*

$$2ik\frac{\partial \tilde{E}_x(z)}{\partial z}e^{-i(\omega t - kz)} = \mu_0\frac{\partial J_x(z, t)}{\partial t}$$

or

$$\frac{\partial \tilde{E}_x}{\partial z} = -i\frac{\mu_o}{2k}e^{i(\omega t - kz)}\frac{\partial J_x(z, t)}{\partial t} \tag{6.27}$$

---

* Similar to the techniques employed in Chapter 2.

Appendix H
Ionization Rates of Noble Gas Atoms as a Function of Laser Intensity and Pulse Duration at 800 nm Wavelength

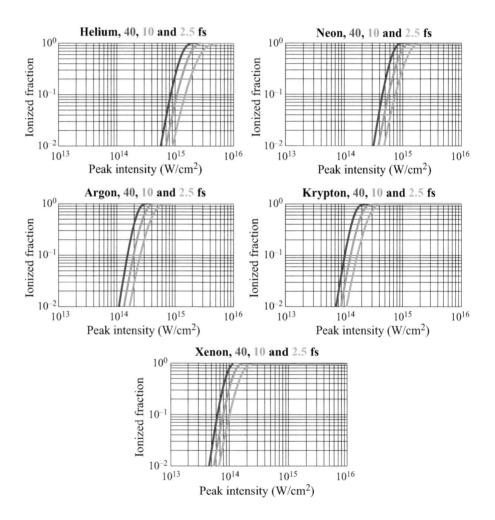

Ionization rates of noble gas atoms are shown here as a function of laser intensity and pulse duration, all at 800 nm wavelength. These curves follow the ADK model,[1] employing the slowly varying envelope approximation (SVEA). For the shortest pulse duration, the SVEA becomes increasingly invalid, and the actual electric field waveform, including the carrier envelope phase (CEP), needs to be taken into account. The

2.5 fs curves plotted here should thus be regarded as an approximate visualization, illustrating that shorter pulse durations permit higher peak intensities before the atoms are fully ionized. The curves were prepared by C. Ott, Max Planck Institute for Nuclear Physics, Heidelberg.

Reference

1. M.V. Ammosov, N.B. Delna and V.P. Kraĭnov, "Tunnel Ionization of Complex Atoms in an Alternating Electromagnetic Field," *Soviet Physics JETP* 64(6), 1191 (December 1986).

Index